MODELING IN THE NEUROSCIENCES

AF734888

Specimen Copy Not For Sale

MODELING IN THE NEUROSCIENCES

From Ionic Channels to Neural Networks

Edited by Roman R. Poznanski

Advanced Research Laboratory
Hitachi, Ltd., Japan

HARWOOD ACADEMIC PUBLISHERS
Australia • Canada • France • Germany • India • Japan • Luxembourg
Malaysia • The Netherlands • Russia • Singapore • Switzerland

10 JUN 1999

591.18 MOD

Copyright © 1999 OPA (Overseas Publishers Association) N.V. Published by license under the Harwood Academic Publishers imprint, part of The Gordon and Breach Publishing Group.

All rights reserved.

No part of this book may be reproduced or utilized in any form or by any means, electronic or mechanical, including photocopying and recording, or by any information storage or retrieval system, without permission in writing from the publisher.
Printed in India.

Amsteldijk 166
1st Floor
1079 LH Amsterdam
The Netherlands

British Library Cataloguing in Publication Data

A catalogue record for this book is available from the British Library.

ISBN 90-5702-284-2

Cover design by Jessica Cotterell.

The micrograph of gap junctionally connected amacrine cells from the Japanese dace retina on the cover appears courtesy of Dr Soh Hidaka.

CONTENTS

PREFACE

Over the last few years people have tried to interpret the 'best' methods of modeling the nervous system. A clear example has been the use of non-analytical methods such as compartmental models to gain insight into neural function. The purpose of this book is to change these interpretations and explore new avenues of theoretical work in order to proceed towards more powerfully realistic models. For example, the compartmental approach, generates a false sense of biological realism and is especially inappropriate in modeling large circuits of neurons. For this reason *Modeling in the Neurosciences* has been vindicated due to the need for alternative models to explain neural activity. Such neural activity within a large network of neurons is determined by the excitability properties of single neurons and their interactions (both synaptic and extrasynaptic). To understand how interconnected neurons process information, knowledge of the basic electrical properties of the single neuron *from ionic channels to neural networks*, the subtitle of the book, is mandatory. The ultimate goal of such a 'bottom-up' approach is to model neural networks that include detailed biophysical properties based on reduced models which capture the essence of neuronal information processing.

The neural activity of a network involving a million or more dynamically active neurons would benefit from analytical solutions at the individual neuron level. In most circumstances, reduced models can be solved analytically and so contemplating future work that involves modeling large networks of neurons, it is clear that a gradual shift from compartmental modeling to analytical models will result in breakthroughs in finding new principles and mechanisms. Furthermore, the importance of the analytical approach in theory stems from the fact that such results can establish more conclusively the intricate interaction underlying the physical process than a series of computational simulations. The development of analytical results to problems in neuroscience can also be of great value in validating the computational simulations used to explore the boundaries of complexity arising in the neurosciences. Thus it seems appropriate to produce a book devoted to the further elaboration of linear theory by use of analytical techniques and also to introduce the nonlinear theory that often requires numerical approximation. In this context, the analytical and numerical techniques are assumed to be essential tools in the construction of models in the neurosciences.

The book is not a 'cookbook' or 'handbook' on neural modeling. It brings together in a single volume elements of modeling in the neurosciences essential for realistic attempts at modeling neurons and in particular neural networks. The chapters in this book do not represent *all* modeling work being done, as modeling in the neurosciences is an ongoing process covering a wide selection of unsolved interdisciplinary problems. Nevertheless, the selected chapters in this book have been specifically chosen to provide the necessary inspiration for the experienced modeler or experimental worker who is interested in contemporary models in the neurosciences. The present book differs from other currently available texts on neural modeling that emphasize the 'computational' aspects of neuroscience. New algorithms and a new generation of powerful UNIX workstations applied to the study of neuronal information processing is not what this book is all about. The analogy that information processing in the dendrites or at the network level is equivalent to some sort of 'computation' has been stressed far too often in 'computational neuroscience'. However, this book avoids the dependence of theoretical insight on computers and software packages now available on the World Wide Web. This is because a computer simulation without an underlying theory is of heuristic value, not easily implementable or testable because a computer simulation is usually generated from a pre-existing model; conceptual errors in the model are not easily picked up in a simulation as inappropriate choice of parameters is usually seen as the most likely fault. Instead, this book replaces the notion of computation with functionalism and in doing so, develops models that can account for physiological mechanisms at the biophysical, cellular and network levels. Physiological mechanisms, such as voltage-dependent ion channels, calcium-dependent release, neurotransmitter receptor dynamics and second-messenger effects, Hebbian synapses, back-propagating sodium and calcium spikes, as well as cable properties and mechanisms governing electric current flow in neurons and neural networks are discussed in this book.

The scope of the book is suited to experimentalists with an interest in the quantitative aspects of neuroscience who want to be informed about the possibilities of mathematical and computer modeling as a tool to verify and test their data and make predictions concerning their working hypotheses. Graduate students and research workers in neural modeling who are interested in learning new methods, and techniques on problems which are closely supported by experimental facts will also profit from this book. The book is not an introductory volume as some knowledge of neural modeling, including neural anatomy, physiology and biochemistry is assumed as well as familiarity with methods of solving differential equations, and background knowledge in statistics and applied probability, and nonlinear dynamical systems theory is required. It is suitable as a companion text to Johnston and Wu's *Foundations of Cellular Neurophysiology* (MIT Press, 1995) for graduate and advanced undergraduate students taking a course in analytical neuroscience. It is also suitable as a reference book on neural modeling or as a supplementary text to Jack, Noble and Tsien's *Electric Current Flow in Excitable Cells* (Oxford University Press, 1983) for a one-semester course in quantitative neuroscience to graduate level students in medicine (neuroscience) or as a supplementary text to MacGregor's *Neural and Brain Modeling* (Academic Press, 1987) for a one-semester course in neural modeling to graduate level students in engineering and the information sciences or as a supplementary text to Tuckwell's *Introduction to Theoretical Neurobiology* (Cambridge University Press, 1988) for a one-semester course in mathematical neurobiology to graduate level students in the mathematical sciences.

It was a pleasure working with all the contributors on this project that began in March 1996 and I thank them all for their diligent effort towards the final completion of this book. The realization of this book would not have been possible without the generous financial support from Monbusho and the facilities provided by Toho University where most of the initial work was completed. More recently, the final proof reading was undertaken at HARL and I thank Kohyu Fukunishi for his support and encouragement. Lee Segel provided advice on the preparation of this book. The sections pertaining to chapters in the book of the introductory chapter were rephrased from abstracts submitted by most contributors of the book. Special thanks to Dennis Turner who provided detailed feedback, as well as to Jaap van Pelt and Jay Rosenberg for their help. The more philosophical views have matured from discussions with Henry Poznanski, and more recently with Alik Gutman. Finally, the staff of Harwood Academic Publishers should also be commended on a task well done, in bringing to fruitful completion a monograph which on many occasions was delayed by the insertion of new references and revisions at the proof reading stage in the book's production.

CONTRIBUTORS

Armantas Baginskas
Laboratory of Neurophysiology
Kaunas Medical Academy
9 Mickeviciaus Street
LT 3000 Kaunas, Lithuania
E-mail: arbagi@kma.lt

Jonathan Bell
Department of Mathematics
State University of New York at Buffalo
Buffalo, New York 14214, USA
E-mail: bell@sparcofbell.math.buffalo.edu

Roberto M. Cesar, Jr.
Instituto de Fisica de São Carlos
Universidade de São Paulo
13560–970 São Carlos, Brazil
E-mail: pinda@ultra3000.ifsc.sc.usp.br

Ying Chen
Eli Lilly & Company Ltd
Lilly Research Center
Windlesham, Surrey GU20 6PH, UK
E-mail: chen_ying@lilly.com

Regina C. Coelho
Instituto de Fisica de São Carlos
Universidade de São Paulo
13560–970 São Carlos, Brazil
E-mail: reginac@ultra3000.ifsc.sc.usp.br

Luciano da F. Costa
Instituto de Fisica de São Carlos
Universidade de São Paulo
13560–970 São Carlos, Brazil
E-mail: luciano@ultra3000.ifsc.sc.usp.br

Robert Costalat
CREARE-INSERM Neurosciences et Modelisation
Université Pierre et Marie Curie
Boite 23, 9, quai Saint-Bernard F-75252
Paris Cedex 05, France
E-mail: robert.costalat@snv.jussieu.fr

Bruno Delord
CREARE-INSERM Neurosciences et Modelisation
Université Pierre et Marie Curie
Boite 23, 9, quai Saint-Bernard F-75252
Paris Cedex 05, France
E-mail: bruno.delord@snv.jussieu.fr

Shinji Doi
Department of Electrical Engineering
Osaka University
Suita, Osaka 565, Japan
E-mail: doi@pwr.eng.osaka-u.ac.jp

Dominique M. Durand
Neural Engineering Center
Department of Biomedical Engineering
Case Western Reserve University
Cleveland, Ohio 44106–4912, USA
E-mail: dxd6@po.cwru.edu

Jonathan D. Evans
Department of Mathematics
The University of Wales
Penglais, Aberystwyth
Ceredigion SY23 3BZ, UK
E-mail: jdevans@compuserve.com

Alan M.N. Fu
Department of Electrical Engineering
University of Sydney
Sydney NSW 2006, Australia
E-mail: alanf@ee.usyd.edu.au

Lloyd L. Glenn
Department of Mental Health, Anatomy and Cell Biology
East Tennessee State University
Johnson City, Tennessee 37614–0658, USA
E-mail: glennl@etsu.east-tenn-st.edu

Aron Gutman
Laboratory of Neurophysiology
Kaunas Medical Academy
9 Mickeviciaus Street, LT 3000 Kaunas, Lithuania
E-mail: arbagi@kma.lt

Jorn Hounsgaard
Department of Medical Physiology
The Panum Institute, Bldg. 12-5-9
University of Copenhagen
Blegdamsvej 3, DK-2200, Copenhagen N, Denmark
E-mail: j.hounsgaard@mfi.ku.dk

Junko Inoue
Department of Computer Science
Koka Women's College
Nishikyogoku, Kyoto 615, Japan
E-mail: jdoi@koka.ac.jp

Akito Ishihara
Physiological Engineering Laboratory
Department of Information and Computer Sciences
Toyohashi University of Technology
Tempaku, Toyohashi 441, Japan
E-mail: ishihara@bpel.tutics.tut.ac.jp

Hiroyuki Ishii
Center for Experimental Brain Research
National Institute for Physiological Sciences
Myodaiji, Okazaki 444, Japan
E-mail: ishii@nips.ac.jp

Yoshimi Kamiyama
Department of Information Science and Technology
Aichi Prefectural University
Nagakute, Aichi 480, Japan
E-mail: Kamiyama@ist.aichi-pu.ac.jp

Jeffrey R. Knisley
Department of Mathematics
East Tennessee State University
Johnson City, Tennessee 37614–0658, USA
E-mail: knisleyj@etsuarts.etsu.edu

Jian-Chen Lin
Department of Biomedical Engineering
Case Western Reserve University
Clevelend, Ohio 44106–4912, USA
E-mail: dxd6@po.cwru.edu

Hiroyuki Mino
Department of Information Sciences
Toho University
Funabashi, Chiba 274, Japan
E-mail: mino@is.sci.toho-u.ac.jp

James M. Ogden
Division of Neuroscience and Biomedical Systems
University of Glasgow
Glasgow G12 8QQ, UK
E-mail: 905723og@udcf.gla.ac.uk

Toshihiro Ogura
Physiological Engineering Laboratory
Department of Information and Computer Sciences
Toyohashi University of Technology
Tempaku, Toyohashi 441, Japan
E-mail: ogura@bpel.tutics.tut.ac.jp

Roger D. Orpwood
Bath Institute of Medical Engineering
University of Bath
c/o The Wofson Center
Royal United Hospital
Bath BA1 3NG, UK
E-mail: r.d.orpwood@bath.ac.uk

Roman R. Poznanski
Advanced Research Laboratory
Hitachi, Ltd.
Hatoyama, Saitama 350-03, Japan
E-mail: poznan@harl.hitachi.co.jp

Jay R. Rosenberg
Division of Neuroscience and Biomedical Systems
University of Glasgow
Glasgow G12 8QQ, UK
E-mail: gpaa07@udcf.gla.ac.uk

Shunsuke Sato
Department of Biophysical Engineering
Osaka University
Toyonaka, Osaka 560, Japan
E-mail: sato@bpe.es.osaka-u.ac.jp

Charles E. Smith
Biomathematics Graduate Program
Department of Statistics
North Carolina State University
Raleigh, North Carolina 27695-8203, USA
E-mail: bmasmith@eos.ncsu.edu

Gytis Svirkis
Laboratory of Neurophysiology
Kaunas Medical Academy
9 Mickeviciaus Street
LT 3000 Kaunas, Lithuania
E-mail: arbagi@kma.lt

Natasha Svirskiene
Laboratory of Neurophysiology
Kaunas Medical Academy
9 Mickeviciaus Street, LT 3000 Kaunas, Lithuania
E-mail: arbagi@kma.lt

Julia S. Tanaka
Instituto de Fisica de São Carlos
Universidade de São Paulo
13560-970 São Carlos, Brazil
E-mail: julia@ultra3000.ifsc.sc.usp.br

Bassel Tawfik
Department of Systems and
Biomedical Engineering
University of Cairo
Cairo, Egypt
E-mail: bassel@vax1-eng.cairo.eun.eg

Dennis A. Turner
Neurosurgery and Neurobiology
Duke University Medical Center
Durham, North Carolina 27710, USA
E-mail: turne008@mc.duke.edu

Shiro Usui
Physiological Engineering Laboratory
Department of Information and
Computer Sciences
Toyohashi University of Technology
Tempaku, Toyohashi 441, Japan
E-mail: usui@bpel.tutics.tut.ac.jp

Osamu Umino
Department of Information Sciences
Toho University
Funabashi, Chiba 274, Japan
E-mail: umino@is.sci.toho-u.ac.jp

Harry B.M. Uylings
Netherlands Institute for Brain Research
Meibergdreef 33, 1105 AZ Amsterdam,
The Netherlands
E-mail: h.uylings@nih.knaw.nl

Jaap van Pelt
Netherlands Institute for Brain Research
Meibergdreef 33, 1105 AZ Amsterdam,
The Netherlands
E-mail: j.van.pelt@nih.knaw.nl

Mike West
Institute of Statistics and Decision Sciences
Duke University
Durham, North Carolina 27708–0251, USA
E-mail: mw@stat.duke.edu

Howard V. Wheal
Neuroscience Research Group
University of Southampton
Southampton S016 7PX, UK
E-mail: h.v.wheal@soton.ac.uk

Rex R. Whitehead
Department of Physics and Astronomy
University of Glasgow
Glasgow G12 8QQ, UK
E-mail: rrw@physics.gla.ac.uk

Tetsuya Yagi
Neurosystems Laboratory
Kyushu Institute of Technology
Iizuka, Fukuoka 820, Japan
E-mail: yagi@ces.kyutech.ac.jp

Hong Yan
Department of Electrical Engineering
University of Sydney
Sydney NSW 2006, Australia
E-mail: yan@ee.usyd.edu.au

CHAPTER 1
INTRODUCTION TO MODELING IN THE NEUROSCIENCES

Roman R. Poznanski

Evolution through millions of years of variation and selection has produced an immensely complex structure- the human brain, believed to contain about 100 billion neurons interconnected with a high degree of specificity totaling about 1000 trillion connections. The ability of the brain to process information which ultimately leads to behavior stems from this high degree of complexity. These functional connections between neurons as well as their morphological structure and electrical characteristics is the result of repeated cycles of variation and selection which are both genetic in origin and driven by physiological plasticity. Thus the brain is a Darwinian neural network, and a processor of information governed by Darwinian evolutionary selectionist mechanisms (Edelman, 1987; Sporns and Tonomi, 1994). The mystery of how the brain works must clearly be dependent on the interaction between the circuitry and properties of its constituent elements, the neurons. In this respect, models of single neurons are essential to the gradual development of higher brain function. These models must span from the level of individual synaptic connections to synaptic integration and dendritic function, and to eventually the exploration of high level multineuronal realistic network functioning.

This book does not deal with molecular models of permeability changes in nerve membranes or relationship between ion channels structure and function on the molecular level, nor does it deal with selectionist models, connectionist models or pyschophysical models found in cognitive neurosciences. There already exists good introductory texts in such fields (see Kandel, 1976; Levine, 1991; Smith, 1996). Neither will this book attempt to provide an explanation of behavior in terms of single neurons, i.e. the neurophysiology of the mind (see Horridge, 1977) or to provide a blueprint for building 'artificial' brains or assume that the brain 'computes'. These notions have their origins in cybernetics (see Wiener, 1961) and more recently in computational neuroscience (see Durbin et al., 1989; McKenna et al., 1992; Koch, 1998), and so a clear differentiation between computation is expected from readers, since *modeling in the neurosciences* represents a movement away from the computational doctrines towards a noncomputational neuroscience (see e.g. Globus, 1992). This movement is necessary if we want to embrace neural Darwinism (Edelman, 1987; 1993) and challenge the metaphor of the brain being entirely mechanical (Barlow, 1990). The metaphor that the brain is a computer whose sole function is to transform incoming neuroelectric signals into neuronal output is gratuitously weak as it requires a 'mind' controlling the 'thinking machine' (see Bennett, 1997 for a modern viewpoint). A more closer definition befitting this metaphor should be a Darwinian 'machine' as advocated by Calvin (1987).

Another weakness in the metaphor is that the brain does not turn on and off like some sort of computational machine. The activity of the brain is interwoven with the spontaneous activity generated when the brain is not evoked by any external incoming stimuli. Chapters 2, 3, and 4 deal with such spontaneous activity, for three different sources of variability: (i) ionic channel current fluctuations, (ii) synaptic release of neurotransmitter, and (iii) dendritic morphogenesis. By these modeling studies it is possible to identify levels of biological organization at which structures and dynamics can be described in terms of random events. It is then plausible to assume that at such a level the phenomena are not governed by a well-defined underlying organizing plan but are the result of many contributing underlying processes. These findings are relevant in issues of the organizational plan of the brain, the involvement of genes, and aspects of self-organization. By understanding the mechanisms governing cognition are firmly based upon the processes of natural selection and variability among others, the computational or metaphysical approach can be abandoned. This will be a pinnacle achievement for modeling in the neurosciences.

In the past decade great advances have been achieved in the study of ion channels (see Hille, 1992); electrophysiologists are discovering new ideas and concepts that had been impossible to determine only a decade ago. For example, patch clamp recording allows pico-amp ionic currents through a membrane channel to be directly recorded (see Sakmann and Neher, 1995). In Chapter 2 by Mino fluctuations of current caused by the gating mechanisms of ionic channels observed at all neuronal membranes with the patch clamp method are analyzed by applying methods of statistical inference for the analysis of the data. Two situations are considered: 1) stationary ionic channel current fluctuations (ICFs) observed during a constant application of chemical transmitter, and 2) non-stationary ICFs observed during repetitive stimulation. A Markov process represents the ion channel gating mechanism of a finite number of channels (see Tuckwell, 1989). Important parameters of the stochastic gating mechanism of single ionic channels are found from the statistical analysis of the ionic channel current fluctuation data. In particular, the number of ionic channels activated, current amplitude of single ionic channels, and state transition rates of the gating mechanism. Electrophysiological measurements usually create an additional source of noise that is Gaussian. In this noisy situation third order statistics are insensitive to the additive Gaussian noise component, and can provide information on the parameters of single ionic channels from the stationary ICFs data. The method for non-stationary ICFs is based on the second-order regression model including not only the ICF components but also the Gaussian noise component.

The activation of a multiple number of synaptic sites that are spatially distributed along the dendrites of a neuron contributes to the overall physiological response. The physiological efficacy of responses at each synaptic site can be defined by the following critical parameters: 1) different amounts of neurotransmitter released both between junctional sites and over time, 2) the probability of realease, 3) postsynaptic site response amplitude ('quantal' amplitude), 4) the degree of fluctuation around that characteristic amplitude ('quantal' variance) distributed both over time and between sites, 5) the efficacy of electrotonic conduction between the synaptic site of origin and the output or summation site, and 6) variable postsynaptic ligand binding with a ligand receptor mismatch. These parameters all vary as a highly dynamic process, fluctuating with the history of use at each synapse over short and long time periods, and between adjacent synaptic sites on dendritic trees. Such heterogeneity between synaptic responses can lead to summated synaptic responses which are difficult to interpret in terms of the behavior of each individual synaptic site. Statistical models for analysis of synaptic function, i.e. the physiological behavior of individual synaptic sites (consisting of a single postsynaptic density and presynaptic release region) have become increasingly sophisticated to account quantitatively for the diverse sources of variability. Chapter 3 by Turner, Chen, West and Wheal presents various statistical modeling approaches to synaptic data analysis, including binomial, Poisson and more complex (less restrictive) methods, to infer on the characteristics of synaptic release under appropriate physiological conditions. These approaches use several algorithms or statistical methods to gain insight into the data, including maximum likelihood and a novel Bayesian approach to investigate many of these sources of variability. Both constrained physiological preparations and statistical models have defined the presence of heterogeneity between synaptic sites and Hebbian mechanisms which may contribute to synaptic plasticity in the CNS.

The variation in dendritic morphology emerges during neurite outgrowth from the dynamic behavior of growth cones in the process of branching and lengthening of dendritic segments. In Chapter 4 by van Pelt and Uylings the question of how dendritic morphology emerges from the basic actions of growth cones is addressed and whether randomness in growth cone behavior is a sufficient condition for generating the typical variability in dendritic shapes, found in neurons of different types and in different species. This question is very difficult or even impossible to answer experimentally. By modeling growth cone actions, and simulating the growth process, one gets an understanding how these actions lead to the emergence of morphological characteristics and the morphological variation. Furthermore, based on these findings one can analyze quantitatively experimentally reconstructed dendritic trees and can make predictions of the time course of the branching and migration process of the growth cones during development of these dendrites. The dendritic growth cone models provide a way of producing random dendritic geometries with realistic morphological variations. These random trees can be used in studies of functional consequences of dendritic geometries and the construction of realistic morphoelectrical models of neurons. There is a great need for such tools in computational studies of neural structure–function relationships, since in recent years computer software developers have created a rise in the application of compartmental modeling originally developed in the 1960s, aiming at modeling neurons with increasing structural realism (see Bower, 1992; Bower and Beeman, 1994). An important aspect of dendritic geometry is its embedding in 3D space. Branch angles between individual segments are in this respect important parameters, but compartmental models of single neurons are lacking this structural realism, leaving a 1D caricature of the cells. The only description of compartmental models with branch angles appears fragmented in Macgregor's books (see Macgregor, 1987; 1993). The critical aspect of including branch angles is to explore the relationship of single neurons to extracellular space and other volume elements in the neuropil, including processes and glia. The realistic complication of a finite extracellular space is rarely treated in single neuron models because of the difficulties in portraying the neuron in three dimensions and surpassing the infinite media assumption so commonly used. However, critical physiological questions regarding both electrical function and chemical changes require that specific neuronal geometry embedded in a three dimensional neuropil be explicitly considered. Thus, model neurons generated randomly are an excellent supplement to realistic neurons obtained from three dimensional reconstruction allowing for future modeling based on 3D characterization of neurons and networks to be included in new and more realistic future compartmental models of neurons (see for example Chapter 18 for a unique approach in this direction).

Another point concerning the interdependence between neuronal structure and function is that the relationship may not be made out to be perfect. For example, if one function of dendrites is to maximize the surface area for synaptic contact then why are not all the dendritic surface area covered with synaptic knobs? Furthermore, there is no experimental evidence to suggest that all neurons in the brain have dendrites with generic functions. In fact, selectionist theories see the variability within population of neurons as an important key in the development of brain function (see Sporns and Tononi, 1994). It is most likely that dendritic trees of some specific neuronal types are influenced by the intrinsic membrane voltage-dependent conductances on dendritic branches and the topology of the tree, while other neuron types simply convey synaptic inputs to the cell body providing dendrites with two different modes of integration (cf. Yuste and Tank, 1996). Indeed, it may turn out that the variability hidden behind the morphologies of dendritic trees is influenced by the dynamic behavior of growth cones (i.e. migration and branching) governed by undirected random mechanisms during development, such as exploratory interactions of filopodia with their local environment, receptor mediated transmembrane signaling to cytoplasmic regulatory systems, the dynamically instable depolymerization of cytoskeletal elements, and activity-dependent processes. A hypothetical neuron possessing no dendritic tree, only a large spherical soma that accommodates all the synaptic input would make a highly unrealistic target for such random connections to emerge. Thus it may be that structure alone cannot guide us to function!

Models are indispensable, because (i) they can provide a means to formulate a hypothesis in a quantitative manner, (ii) they allow the study of the consequences of a hypothesis in many or all its aspects, (iii) they allow statistical testing of a hypothesis and its consequences, and (iv) they may lead to new experiments that would not otherwise have been considered. The use of models to help elucidate neural behavior has grown exponentially over the last decade (see Koch and Segev, 1998), but worthwhile models that can predict experimental behavior are not so common, and more often are built on the successes of earlier models that were able to satisfactorily predict experimental phenomena albeit in less detail. Some examples are the Hodgkin–Huxley (1952) model for initiation and conduction of action potentials in axons built on the membrane theories of Bernstein, and the Rall (1962) model for the distribution of synaptic potentials in the dendritic tree of a neuron built on the core-conductor theories of Weber. Thus, modeling can predict all sorts of wonderful new theories and clarify functions of neuronal activity, but must be based on a firm experimental platform, otherwise the outcome remains in the realm of neurocomputing or computational neuroscience.

Chapters 5–10 deal with reduced models of single neurons, and the role of information processing in neural function and the latest methods to measure parameters (i.e. inverse problems). Finding a simple but equivalent representation for complex dendritic trees arises from the common observation that neurons, the basic anatomical building blocks of the nervous system, come in a bewildering variety of shapes and sizes (as discussed in Chapter 3). Although the importance of appropriate analytical tools for investigating the functional consequences of neuron geometry in signal processing in the nervous system is recognized (cf. Major et al., 1993), incorporating reduced models based on analytical methods in biologically realistic neural networks is not available. One approach in understanding neuron geometry entails replacing the given dendritic tree with a multiple branched or even unbranched equivalent structure. Such models might be used as the basis for inferring the position of activated synaptic inputs on dendritic trees, given the geometry of the tree and a somatically recorded voltage transient, and to explore how neuron geometry contributes to synaptic integration. For example, what is the likely efficacy of synaptic inputs on different parts of the dendritic tree. Consequently, the spread of current in cable structures is a major theme in neuroscience and represents the largest component of the book.

In Chapter 5 by Evans an overview is given of the multiple equivalent cylinder model used to describe the passive electrical behavior of neurons. The lumping of all branches of a dendrite into an equivalent cable is assumed. Such lumping does not affect the interesting function that may occur in individual branches and can still capture the essential synaptic integration that may occur in the dendritic tree. Separate models have been specifically developed to investigate the role of dendritic branching, dendritic tapering, nonuniform electrical parameters and dendritic spines on the voltage spread within passive neurons. A mathematical analysis of such a multiple cable model is given and standard techniques are used such as Green's functions and Fourier methods to obtain exact and approximate analytical solutions to the multi-cylinder somatic shunt cable model. The approach presented is generally well suited to the study of such linear models and extends to passive cable models that involve branching and tapering of dendrites as shown in the next chapter.

A complementary approach presented in Chapter 6 by Glenn and Knisley to obtain the time-dependent membrane potential distribution in dendritic trees of arbitrary geometry with continuously tapering dendrites with arbitrary diameter, membrane time constants, lengths, and rates of tapering. The analytical solution is expressed as a Fourier cosine series, and is shown to be more compact than solutions based on error functions; numerical calculations using compartmental models are used to approximate the solution. Application of this model to spinal motoneurons shows the average electrotonic length to be between 0.4 and 0.8 length constants. The multiple equivalent cable model can be regarded as a model of intermediate complexity because it is not as simple as the tapering equivalent cable model, but less complex than the arbitrary branching models, preserving the size differences between dendritic trees. Therefore, cable modeling of dendritic neurons can provide quantitative analysis of signal spread expressed in terms of analytical solutions, however it has its limitations because of the complicated mathematical expressions needed to model neurons with branching patterns in neural network models.

The need for reduced yet realistic models is clear, but the question of whether the reduced model is realistic enough (physiologically plausible) is not obvious to the novice modeler. The question of when a membrane voltage distribution with a single cable is truly equivalent to that within a dendritic tree is answered in Chapter 7 by Ogden, Rosenberg and Whitehead using a Lanczos method commonly used in mathematical physics. The electrical mapping between dendritic tree and equivalent cable overtakes any previously developed theories on this topic, and for the first time shows how passive dendritic trees with dendrites of arbitrary geometry can be replaced with equivalent nonbranching cables. The Lanczos method uses a discrete representation of trees and their equivalent cables, and so is well suited for implementation in numerical studies of neural networks via finite-difference methods and also removes the physiological restrictions imposed in attempts to model continuous equivalent cables. This chapter presents a detailed account of the Lanczos procedure for the first time and is simply a tour de force and a triumph for analytical neuroscience. Examples of equivalent cables are given to illustrate the wide variety of structures that can be generated. It is shown that Rall's equivalent cylinder represents a very restricted subset of the general class of equivalent cable. In addition, a slight extension to the original conditions for an equivalent cylinder demonstrates the conditions under which certain empirical extensions of Rall's result, which permits changes in diameter, are equivalent.

The Rall model of a neuron is a classic model, but the major modification that allows this model to remain functional during electrophysiological experiments requires the inclusion of a somatic-shunt parameter, to compensate for the leakage of current as a result of the sharp electrode piercing the somatic membrane. Chapter 8 by Durand, Tawfik and Lin addresses the inverse problem for the somatic-shunt model using three methods: nonlinear parameter estimation algorithms, gradient optimization methods, and Fast Fourier Transform algorithm, to recover the parameters of the model in the presence of noise and contaminated data. When a signal is injected at the soma and a response is recorded, these methods are shown to be capable of rejecting biological and external interferences such as noise and artifacts in order to obtain accurate estimation of electrotonic parameters of the neuron model. The three methods are applied to the somatic-shunt cable model for neurons and the robustness of the methods is tested by estimating the errors in the estimated parameters in the presence of large amount of noise and missing data points.

Another inverse problem, dealing with the effects of gap junctions on estimates of parameters of single neurons is presented in Chapter 9 by Poznanski and Umino. A form of neural interaction that has attracted considerable attention, for example the synchronous activity in hippocampal networks or the spread of noise in retinal networks is referred to as 'electrotonic' interaction, and it involves the spread of current through gap junctions. The quantitative analysis of gap junctions in neuroscience is still in its infancy, and it will be an essential component of future brain research investigations. Their roles have been associated with various pathological diseases of the nervous system, in memory, and noise reduction in sensory centers of the brain, for example. Yet no distributed model of a dendrodendritically coupled neurons exists. It is shown using patch-clamp experimental data from hippocampal granule cells that coupling between dendrites increases the dendritic-to-somatic conductance ratio and electrotonic length. Such effects basically produce a functionally elongated dendritic arborization that is less compact with a smaller soma. These results could have important functional implications in electrotonic network models such as discussed in the next chapter.

In the outer retina, gap junctions between a large number of neurons form a syncytial layer of cells, where response behavior is determined by the network rather than individual neurons. The spatio-temporal properties of electrical signals conducted in such electrotonic networks models of neural syncytia are presented in Chapter 10 by Yagi. The analytical solutions obtained provide intuitive understanding on how visual information processing in the retina is dependent on the electrical properties of the neural syncytium. Important questions concerning visual information processing in a double layer neural syncytium such as edge detection are shown using regularization theory to be characterized by minimization of a cost function, derived from the circuit model of the interacting multiple layers of neural syncytia. Such an analysis shows that dendrodendritic electrical synapses and graded potentials can be used to process information without the need for a threshold mechanism associated with the generation of action potentials. Syncytial integration rather than synaptic integration on dendrites of neurons is more influential signal processing mechanism here, as the signals are shunted by the syncytial nature of the network (see Poznanski and Umino, 1997), and therefore electrotonic networks in the retina function using different mechanisms from chemically interconnected neural networks in the brain. One such example of dichotomy is the tonic release of neural transmitter from axon terminals in retinal cells, that is not in response to presynaptic spikes but to graded presynaptic activity.

The retina is an ideal model system for investigating neural network information processing capabilities, since progress in retinal research has revealed detailed mechanisms from molecular to network levels using physiological, morphological and pharmacological techniques. However, the detailed experimental data is not enough to understand the relationship between the microscopic characteristics of the individual cellular elements and their integrative functions. A combination of experimental work and modeling is necessary to clarify the yet unresolved complexities involved in the processing of sensory information to higher brain centers. Chapter 11 by Usui, Kamiyama, Ishii, Ogura, and Ishihara presents a model of the constituent cells in the outer retina in terms of their ionic mechanisms under space and voltage-clamp conditions to understand ionic current dynamics in network function. Since the ionic mechanism is fundamental for generating the electrical response of neurons, a computer model based on these characteristics makes it possible to analyze nonlinear dynamical response properties and to understand the role of a particular ionic channel, and thus provide key insights for elucidating the underlying mechanisms of the network. That is, mechanisms mediating various dynamic features of voltage and or light responses observed in retinal cells can be quantitatively described by changes in the current–voltage relation of the ionic currents in the cell. From this point of view, ionic current models of photoreceptors, horizontal cells and bipolar cells will undoubtedly improve our understanding of these retinal neurons. The functional role of ionic currents during a light response is presented as well as the relationship between the characteristics of neurons (i.e. ionic currents, neurotransmitters, second messengers and synaptic connectivity) at the biophysical and the cellular level.

The notion that the sole influence on a single neuron is not only dependent on synaptic generated signals in the form of spikes, but also on extrasynaptic graded signals due to the interactions encountered with other neurons through the extracellular space has recently been re-iterated (see Bullock, 1997). For example, the relative importance of nonspiking and extrasynaptic graded signals on the influence of back propagating action potentials in dendritic trees has not been fully addressed. At present researchers using compartmental models are ignorant of such finer points in the relation between structure and function. Perhaps it is important to explain why the compartmental modeling approach is less subtle than the approaches discussed in Chapters 5 and 6. Generally the assumption of membrane isopotentiality in each compartment will not hold in models with small number of compartments and even more unlikely to hold if the dendrites contain active membrane properties, as the voltage gradients will be large, and the membrane potential will not remain constant unless the compartment size approaches zero, in which case the continuous approach comes to the forefront. In nearly all models of neural networks, information transfer between neurons is solely due to synaptic transmission and at the single neuron level, most models assume that the extracellular space is isopotential. Although chemical synaptic transmission is the main source of information transfer, other mechanisms such as electrical field interactions (ephaptic transmission) in the extracellular space can also contribute to interaction between neurons, which has been demonstrated to occur experimentally, under pathological, and sometimes physiological, circumstances. To model these extracellular interactions between assemblies of closely packed neurons, a known field theory is presented in Chapter 12 by Costalat and Delord. They describe biophysical equations of electrical field interactions and apply such modeling to the hippocampus as an example where field interactions are known to occur (see e.g. Traub and Miles, 1991). It is shown that this approach is strongly correlated to the so-called 'Current Source Density Analysis' used in the study of extracellular fields. Both analytical and numerical methods are presented for investigating field effects in nerve trunks as a result of synaptic potential activation, and in populations of neurons with active membranes.

The input/output relationships of single neurons are complex and yet crucial to our understanding of the behavior of the brain. Many aspects of the basic functioning of neurons have been elucidated by neuroscientists but experimental studies that can explore the overall behavior of the cells in response to controlled inputs are difficult to arrange. For example, with the back propagation of action potentials that carry information about the cell's firing activity into the dendrites it is not known exactly how far spikes propagate into the dendritic arbors of neurons. Furthermore, functional models of single cells are also important as it is becoming clear that the behavior of networks of cells, and ultimately hierarchies of networks, are crucially dependent on the behavior of the single neuron and the interactions found between neurons. In particular, the way networks of cells are able to process patterns of information is of great interest, and inextricably linked to the way individual cells deal with pattern handling capabilities of single cells, and cortical pyramidal cells in particular. Chapter 13 by Orpwood presents numerical techniques for modeling dendrites, spines, active conductances and receptor dynamics in cortical pyramidal neurons, it also contains a simpler model which can easily be incorporated into network level modeling studies so as to gain a better understanding of how neurons in these particular cortices of the brain process information. The model is based on the actual geometry of a layer 5 cortical pyramidal cell and uses several measured properties of that particular cell in its construction. The simpler model is shown to have the same core properties that are demonstrated by the more fully developed model, so that networks of such reduced cells can be examined without using unacceptable amounts of computing time. The chapter demonstrates how the behavior of this numerical model is acceptably close to that measured in experimental work. It demonstrates the ability of the network to classify patterns and to generate responses that parallel the core processes of perception. A discussion is presented on the effects of back propagating spikes into the neuron's branches and its effect on cortical cell functioning.

It was well known from the work in the 1960s that there are multiple sites in the dendrites for action potential initiation with different thresholds, and that they back propagated (antidromically) at least to the base of the main dendrite (see e.g. Schmitt and Worden, 1979 for an overview). These dendritic spikes in large neurons have been known to occur from extracellular recordings, but the advent of the patch-clamp has enabled these spikes to be recorded within the large dendrites of neurons. If spikes are conducted throughout the dendritic tree (Schiller et al., 1997; Kamondi et al., 1998) then problems occur with the Hodgkin–Huxley equations governing the propagation of spikes along excitable fibers with uniform geometrical properties. Examples of problems that lead to anomalous behavior in such excitable fibers, include echo waves, conduction block, and antidromic propagation. Chapter 14 by Bell addresses such problems and provides a lucid mathematical exposition of the necessary conditions to support conduction in active dendritic trees based on two classic examples of inhomogeneities in neural fibers—increase in diameter in a dendritic process and its effects on propagation and conduction behavior along a fiber possessing active spines. The other aspect of excitable properties of dendrites, which is equivalent, is either a sudden or gradual change in the density of voltage-dependent conductances. Such a change in effect mimics a diameter change since propagation may be affected or discontinuous. In particular, the effects of a subthreshold gradient of voltage-dependent conductances on ordinary synaptic inputs also remains to be explored.

In cat and turtle motoneurons it is known that the input current–voltage hysteresis is caused by two mechanisms: wind-up and bistability of the distal dendritic membrane. The bistability is ascribed to dendritic slow inward current, that is slowly developing due to depolarization-dependent facilitation, so-called 'wind-up' is observed after a long-lasting depolarization or repeated depolarizations of short duration. This new phenomenon may involve calcium channels and NMDA receptors. In Chapter 15 by Baginskas, Gutman, Hounsgaard, Svirskiene and Svirskis a semi-quantitative theory for neurons with such bistable dendrites with wind-up of the slow inward current is presented and supported by experimental findings from motoneurons and other CNS neurons. It is proved that stable depolarization in such dendrites develops much more slowly than the wind-up of the membrane current. The theory accounts for the nonlinear response properties mediated by calcium channels in both spinal motoneurons and cerebellar Purkinje cells. The theory includes both channel and cable properties of the phenomena reflected by the complex regulation of calcium channels. It could have a useful application for the problem of motor control and other CNS functions. The functional implication for small neurons to use dendritic bistability is discussed with emphasis on the enormous implications the modeling could have in the construction of neural network models of neurons with bistable dendrites.

Modeling brain activity as a dynamical system can be done when gross simplifications are made to the structure of individual neurons and their connections. Chapters 16 and 17 deal with nonlinear dynamics of neuronal systems at the single neuron and network level, respectively. Nonlinear dynamics is becoming an important area of research in neurocomputing, but can still be useful to neuroscience (Haken, 1996). Neuroscientists who have insufficient background in topics like Hopf bifurcation theory, may find this part of the book more difficult to understand. For readers with insufficient knowledge of dynamical systems the next two chapters provide an introductory exposition to this area of analytical neuroscience, but may be skimmed through without too much loss in content.

In Chapter 16 by Doi, Inoue, Sato and Smith a simplified (piecewise-linearized) Bonhoeffer–van der Pol or FitzHugh–Nagumo model of a neuron is considered. This and other similar models overcome the complexity arising from the Hodgkin–Huxley theory (see Chapter 14), and take on a simplified representation of the ionic current, enabling the dynamics of the system to be studied *albeit* in a qualitative manner. In the presence of sinusoidal input, the model shows various patterns of phase-lockings and bifurcations depending on the period and amplitude of the input signal. Noise effects on the phase lockings and bifurcations in the model are investigated. The system is usually analyzed by the one-dimensional mapping of a 'phase' variable. In the presence of membrane noise or graded dendritic fluctuations of membrane potential, stochastic phase-lockings and bifurcations are analyzed in terms of the probability density evolution by the operator and its spectral characteristics. The phase-space modeling allows phenomena such as neuronal firing, oscillation, and bursting to be predicted without any knowledge of channel kinetics distribution. Such reduced models of neurons allow the complexity of a nonlinear system to be modeled without too much assistance from a computer!

Neither the 'back-propagation' learning algorithms (modern version of the *perceptrons*) nor the connectionist models (see Gluck and Rumelhart, 1990) are discussed in this book because it is felt that the biological realism in such models is lacking (see Crick, 1989). The concept of 'back-propagation' based on the steepest-descent techniques in optimization theory does not provide insight into the general principles of how the behavior of the network is created. Connectionist models, represent a gross over simplification of how biological networks function. It should be stressed that biologically realistic neural networks have still not been developed, although attempts have been made to construct such models, but the progress has been slow (see e.g. Abbott and Kepler, 1990; Abbott, 1991; Rolls and Treves, 1998). Associative memory neural networks of which the Hopfield model is a classic example (see Hopfield, 1982) are biologically unrealistic, nevertheless the model has the potential of implementing biological realism and presents a good starting point for shedding light on general principles about the workings of the brain. This is because the Hopfield model is inductive rather than deductive so its usefulness rests in theory (i.e. analytical neuroscience) rather than computational neuroscience or neurocomputing. The stable states of a Hopfield-type neural network reflect the information stored in the network and are important in studying the dynamical behavior of the network. In Chapter 17 by Fu and Yan a statistical method which takes to account both the mean and variance of noise terms as well as the randomness of the main overlap is shown to predict the stable states of an associative memory network with a large pattern rate after about 40 iterations. This theoretical analysis represents a more realistic modeling effort for predicting the stable state of a Hopfield-type neural network than previously attempted.

John Eccles once said: 'How much do we need to go into this detail on the whole dendritic tree of the neuron?... I think all this complexity is statistically resolved in the end.' There is no experimental evidence to disclaim the fact that there is a statistical washout of synaptic signals in the dendrites of some neurons. This can happen when intrinsic voltage-dependent conductances amplify synaptic input in the distal regions of a dendritic tree or are shunted by the bombardment of the incoming signals (see Cash and Yuste, 1998). The answer lies in the diverse functional roles of neurons. Given that neuroscience is a 'soft' science the notion of neurons performing computations seems ill-suited for brain function. For example, experiments on direction selective retinal ganglion cells (see Barlow and Levick, 1965), and more recently Livingstone (1998), where a veto-like operation was shown to be an important physiological mechanism. The mechanism of direction selectivity in the mammalian retina has remained a benchmark for 'computational neuroscience' (see Sejnowski et al., 1988), yet the synaptic mechanism remains unknown. If micropharmacological techniques reveal random organization of receptors on dendrites of DS ganglion cells, the foundations of 'computational neuroscience' undoubtedly will be shaken. A haphazard spatial arrangement of the excitatory and inhibitory synapses on dendrites of DS cells will rule out any notion of computation as is required with stringent and highly symmetrical morphological requirements on the positioning of synapses, such as the 'on-path' rule put forward in postsynaptic schemes of DS (see Poznanski, 1990). If there is a statistical washout of synaptic signals in dendrites, both the electrical characteristics and the topology of the dendritic tree would not have an important influence in generating the output of the neuron. A statistical washout of synaptic signals in dendrites could indicate that signals are carried to other neurons in the network via dyad synapses or dendrodendritic interactions, for example, as seen between amacrine cells in the retina (see Poznanski, 1992), then they may function in an autonomous manner independent from the spike generating soma region and the incorporation of these properties into network models would not produce more realistic and enhanced information processing capabilities in networks which rely on a simple threshold mechanism as the elementary logical operation. In this scenario in which the relevant input–output relations are limited to particular regions of the dendrites, the present neural network models would be inadequate in representing the functional aspects of neuronal architecture.

Irrespective of whether a statistical washout of signals in dendrites occurs, voltage-dependent properties of ionic channels in dendritic membranes (see Johnston et al., 1996; Midtgaard, 1996) are likely to participate in dendritic integration of synaptic inputs (see Yuste and Tank, 1996) to generate their 'rich' behavior such as decremental conduction or graded spikes (see Lorente de No, 1960). For example, it has been taken for granted that graded or subthreshold synaptic potentials are produced by channels that show weak or no voltage dependence (known as 'leakage channels'), however it may be that influences from voltage-dependent channels are involved in the modulation of subthreshold activity (see Stuart and Sakmann, 1995; Taylor et al., 1995). What is more interesting is the possibility of action potentials generated within branches of neurons (see Adams, 1992; Regehr et al., 1993) propagating towards the soma. Such models would need the 'formal' neuron models used in today's neural networks to be drastically modified in terms of logic operations (see Scott, 1973). The inclusion of all the detail on the distribution of ion channels and their gating mechanisms modeled as an adjustment of synaptic weights and the threshold mechanism associated with spike generation used in today's neural network models is not sufficient *albeit* unrealistic.

In light of the above, the closing Chapter 18 by Costa, Cesar Jnr, Coelho and Tanaka conveys the notion of combining the morphological properties of networks with the integrative properties of single neurons by presenting an integrated approach to the analysis, modeling, and summation of morphologically more realistic neural cells and structures. Is it necessary to consider dendritic structure of neurons in the context of a network so that the spatial-temporal input can be accurately determined or is the strength of the incoming signal near the axon hillock all that is required? This question and other related questions need to be answered to obtain a full synthesis of modeling realistic neural networks. For example, what information about neurons is required to construct morphologically realistic network models? One suggestion is to incorporate into network models reduced models that contain the minimum essential properties revealed by experiment. The integrate-and-fire models proposed by Abbott and Kepler (1990) is the first step towards replacing the simpler McCulloch–Pitts (1943) 'point' neurons used in present neural network models. There is no doubt that reduced models of single neurons as discussed in this book will in the next few years constitute neural network models of the future. For example the reduced models outlined in this book can distinguish the spread of synaptic potential from one branch to another branch of a tree structure with a single equivalent cable structure and together with modeling involving the inclusion of voltage-dependent conductances, it may be possible to investigate the effect of a few synapses impinging on the distal ends of the tree with such models to see whether such inputs could activate voltage-sensitive conductances and be amplified, so that eventually they may be modulated by synaptic inputs impinging along the trunk of the tree, thereby influencing the output of the neuron. A mean-field approach can then be used to predict the behavior of large networks (see Abbott, 1991).

Of importance is to model 'realism' at the network level without the need to model 'realistic' single neurons with great amounts of detail. In this book less emphasis is put on the use of general purpose simulators as the principle way to model 'structurally realistic' single neurons and networks of neurons. Firstly, the availability of software packages has led to an over enthusiastic use in neuronal modeling by non-specialists who have little or no training in numerical methods. Indeed, packaged models coupled with modern computer technology can be easily abused by the novice modeler (see e.g. Katz et al., 1998). It is sometimes argued that the user of a package does not need to understand the detailed numerical aspects. This may be true for a perfect model, but neuronal models are far from ideal and the only way the user can assess the validity of predictions and the likely reason for discrepancies is by having a complete understanding of the model and procedures for solving it. The pinnacle of computational neuroscience is unlikely to proceed further in understanding the real nervous system than for instance a program for simulating responses of neurons of arbitrary geometry with voltage-dependent ion-channels (Bush and Sejnowski, 1993; Cai et al., 1997; Hines and Carnevale, 1997), but no breakthroughs can result from computations if models are based on faulty assumptions (see earlier comment on compartmental models). Secondly, in neurobiology identity is rare, so care must be taken to avoid making conclusions for all neuronal types (although in this book models represent generic neurons to some extent due to the limited space available in a single volume). The main tenant is that biological realism of neurons based on anatomical and physiological data produces many degrees of freedom such as a huge set of parameters that must be constrained by the physiology, creating 'brute force' parameter searches (see e.g. Bhalla and Bower, 1993). Thus to provide such detailed 'realism' would seem unrealistic in the modeling of large networks of neurons even with today's super-computers. The point overlooked in modeling 'realistic' neurons, is that 'realism' can be gained from modeling abstract type or reduced models of neurons, while modeling large networks of 'realistic' neurons seems unrealistic. This is because a detailed bio-physical model of a single neuron *per se* is far too complicated to bridge the gap between structure and function at the network level.

The goal of the book is to model the brain through an understanding of the relationship between structure and function at the neuronal level. At the neuronal level modeling in the neurosciences is summarized by the following triadism: 1) models of biophysically complex single neurons; 2) reduced models of single neurons; and 3) models of spatially distributed networks. Although excellent texts are available on quantitative neuroscience (see Jack et al., 1983) it is necessary to cover new perspectives in modeling so as to gain fresh insight into neuronal phenomena in a single volume. The chapters covered in this book symbolize an effort to foster a quantitative, theoretical approach as a natural complement to experimental investigations in the neurosciences. Not only is this book proof that neuroscience can be analytical, but insight gathered from these chapters should provide the necessary spark for the development of realistic models in the neurosciences as we approach a new millennium. It is tempting to speculate that in the new millennium biologically inspired androids based on Darwinian evolutionary principles (see Reeke et al., 1990 for a review) will allow us to rediscover that the brain after all does not compute!

REFERENCES

Abbott, L.F. (1991) Realistic synaptic inputs for model neural networks. *Network* **2**, 245–258.

Abbott, L.F. and Kepler, T.B. (1990) Model neurons: from Hodgkin-Huxley to Hopfield. In, Statistical Mechanics of Neural Networks (L. Garrido, ed.) Springer-Verlag: Berlin, pp. 5–18.

Adams, P.R. (1992) The platonic neuron gets the hots. *Curr. Biol.* **2**, 625–627.

Barlow, H.B. (1990) The mechanical mind. *Annu. Rev. Neurosci.* **13**, 15–24.

Barlow, H.B. and Levick, W.R. (1965) The mechanism of directionally sensitive units in rabbit's retina. *J. Physiol. (Lond.)* **781**, 477–504.

Bennett, M.R. (1997) The Idea of Consciousness: Synapses and the Mind. Harwood Academic Publishers: Amsterdam.

Bhalla, U.S. and Bower, J.M. (1993) Exploring parameter space in detailed single neuron models. *J. Neurophysiol.* **69**, 1948–1965.

Bower, J.M. (ed.) (1992) Modeling the Nervous System. Special Issue, Vol. 15, *Trends Neurosci.*, Elsevier Science: New York.

Bower, J.M. and Beeman, D. (1994) The Book of Genesis: Exploring Realistic Neural Models with the General Neural Simulation System. TELOS: New Jersey.

Bullock, T.H. (1997) Signals and signs in the nervous system: the dynamic anatomy of electrical activity is probably information reach. *Proc. Natl. Acad. Sci. USA* **94**, 1–6.

Bush, P.C. and Sejnowski, T.J. (1993) Reduced compartmental models of neocortical pyramidal cells. *J. Neurosci. Meth.* **46**, 159–166.

Cai, Y., Walsh, E.J. and McGee, J. (1997) A simple program for simulating the responses of neurons with arbitrarily structured and active dendritic trees. *J. Neurosci. Meth.* **74**, 27–35.

Calvin, W.H. (1987) The brain as a Darwin machine. *Nature* **330**, 33–34.

Cash, S. and Yuste, R. (1998) Input summation by cultured pyramidal neurons is linear and position-independent. *J. Neurosci.* **18**, 10–15.

Crick, F. (1989) The recent excitement about neural networks. *Nature* **337**, 129–132.

Durbin, R., Miall, C. and Mitchison, G. (eds.) (1989) The Computing Neuron. Addison-Wesley: New York.

Eccles, J.C. (1993) How the Self Controls its Brain. Springer-Verlag: Berlin.

Edelman, G.M. (1987) Neural Darwinism: The Theory of Neuronal Group Selection. Basic Books: New York.

Edelman, G.M. (1993) Neural Darwinism: selection and reentrant signaling in higher brain function. *Neuron* **10**, 115–125.
Fohlmeister, J.F. and Miller, R.F. (1997) Mechanisms by which cell geometry controls repetitive impulse firing in retinal ganglion cells. *J. Neurophysiol.* **78**, 1948–1964.
Globus, G.G. (1992) Toward a noncomputational cognitive neuroscience. *J. Cognit. Neurosci.* **4**, 299–310.
Gluck, M.A. and Rumelhart, D.E. (eds.) (1990) Neuroscience and Connectionist Theory. Lawrence Erlbaum Assoc. Inc.: New Jersey.
Haken, H. (1996) Principles of Brain Functioning: A Synergetic Approach to Brain Activity, Behavior and Cognition. Springer: Berlin.
Hille, B. (1992) Ionic Channels of Excitable Membranes. 2nd edition. Sinauer: Sunderland.
Hines, M.L. and Carnevale, N.T. (1997) The NEURON simulation environment. *Neural Comp.* **9**, 1179–1209.
Hodgkin, A.L. and Huxley, A.F. (1952) A quantitative description of membrane current and its application to conduction and excitation in nerve. *J. Physiol. (Lond.)* **117**, 500–544.
Hopfield J.J. (1982) Neural networks and physical systems with emergent collective computational abilities. *Proc. Natl. Acad. Sci. USA* **79**, 2554–2558.
Horridge, G.A. (1977) Mechanistic teleology and explanation in neuroethology. *Biosci.* **27**, 725–731.
Jack, J.J.B., Noble, D. and Tsien, R.W. (1983) Electrical Current Flow in Excitable Cells. Clarendon Press: Oxford.
Johnston, D., Magee, J.C., Colbert, C.M. and Christie, B.R. (1996) Active properties of neuronal dendrites. *Annu. Rev. Neurosci.* **19**, 165–186.
Kamondi, A., Acsady, L. and Buzsaki, G. (1998) Dendritic spikes are enhanced by cooperative network activity in the intact hippocampus. *J. Neurosci.* **18**, 3919–3928.
Kandel, E.R. (1976) Cellular Basis of Behavior: An Introduction to Behavioral Neurophysiology. W.H. Freeman & Co.: San Francisco.
Katz, L.C., Welicky, M. and Dalva, M. (1998) Retraction-Relationships between local synaptic connections and orientation domains in primary visual cortex. *Neuron* **20**, 820.
Koch, C. (1998) Biophysics of Computation: Information Processing in Single Neurons. Oxford University Press: New York.
Koch, C. and Segev, I. (eds.) (1998) Methods in Neuronal Modeling: From Ions to Networks. 2nd edition. MIT Press: Cambridge, MA.
Levine, D.S. (1991) Introduction to Neural and Cognitive Modeling. Lawrence Erlbaum Assoc. Pub: New Jersey.
Livingstone, M.S. (1998) Mechanisms of direction selectivity in Macaque V1. *Neuron* **20**, 509–526.
Lorente de No, R. (1960) Decremental conduction and summation of stimuli delivered to neurons at distant synapses. In, Structure and Function of the Cerebral Cortex. (D.B. Tower and J.P. Schade, eds.) Elsevier: New York, pp. 278–281.
Macgregor, R.J. (1987) Neural and Brain Modeling. Academic Press: San Diego.
Macgregor , R.J. (1993) Theoretical Mechanics of Biological Neural Networks. Academic Press: Boston.
Major, G., Evans, J.D. and Jack, J.J.B. (1993) Solutions for transients in arbitrarily branching cables. *Biophys. J.* **65**, 423–468.
McCulloch, W.S. and Pitts, W. (1943) A logical calculus of the ideas immanent in nervous activity. *Bull. Math. Biophys.* **5**, 115–133.
McKenna, T., Davis, J. and Zornetzer, S.F. (eds.) (1992) Single Neuron Computation. Academic Press: London.
Midtgaard, J. (1996) Active membrane properties and spatiotemporal synaptic integration in dendrites of vertebrate neurones. *Acta. Physiol. Scand.* **157**, 395–401.
Poznanski, R.R. (1990) Analysis of a postsynaptic scheme based on a tapering equivalent cable model. *IMA J. Math. Appl. Med. & Biol.* **7**, 175–197.
Poznanski, R.R. (1992) Modelling the electrotonic structure of starburst amacrine cells in the rabbit retina: a functional interpretation of dendritic morphology. *Bull. Math. Biol.* **54**, 905–928.
Poznanski, R.R. and Umino, O. (1997) Syncytial integration by a network of coupled bipolar cells in the retina. *Prog. Neurobiol.* **53**, 273–291.
Rall, W. (1962) Theory of physiological properties of dendrites. *Ann. NY Acad. Sci.* **96**, 1071–1092.

Reeke, G.N. Jr., Finkel, L.H. and Edelman, G.M. (1990) Selective recognition automata. In, An Introduction to Neural and Electronic Networks. (S.F. Zornetzer et al., eds.) Academic Press: Orlando, FL.

Reichardt, W.E. and Poggio, T.A. (eds.) (1981) Theoretical Approaches in Neurobiology. The MIT Press: Camb., Mass.

Regehr, W., Kehoe, J., Ascher, P. and Armstrong, C. (1993) Synaptically triggered action potentials in dendrites. *Neuron* **11**, 145–151.

Rolls, E.T. and Treves, A. (1998) Neural Networks and Brain Function. Oxford University Press: Oxford.

Sakmann, B. and Neher, E. (eds.) (1995) Single Channel Recording. 2nd edition. Plenum Press: New York.

Schiller, J., Schiller, Y., Stuart, G. and Sakmann, B. (1997) Calcium action potentials restricted to distal apical dendrites of rat neocortical pyramidal neurons. *J. Physiol. (Lond.)* **505**, 605–616.

Schmitt, F.O and Worden, F.G. (eds.) (1979) The Neurosciences: Fourth Study Program. MIT Press: Camb., Mass.

Scott, A.C. (1973) Information processing in dendritic trees. *Math. Biosci.* **18**, 153–160.

Sejnowski, T.J., Koch, C. and Churchland, P.S. (1988) Computational neuroscience. *Science* **241**, 1299–1306.

Smith, C.U.M. (1996) Elements of Molecular Neurobiology. 2nd edition. John Wiley & Sons: New York.

Sporns, O. and Tononi, G. (eds.) (1994) Selectionism and the Brain. Academic Press: New York.

Stuart, G. and Sakmann, B. (1995) Amplification of EPSPs by axosomatic sodium channels in neocortical pyramidal neurons. *Neuron* **15**, 1065–1076.

Taylor, G.C., Coles, J.A. and Eilbeck, J.C. (1995) Conditions under which Na^+ channels can boost conduction of small graded potentials. *J. theoret. Biol.* **172**, 379–386.

Traub, R.D. and Miles, R. (1991) Neuronal Networks of the Hippocampus. CUP: Cambridge.

Tuckwell, H.C. (1989) Stochastic Processes in the Neurosciences. SAIM: Philadelphia.

Wiener, N. (1961) Cybernetics, or Control and Communication in the Animal and the Machine. John Wiley & Sons: New York.

Yuste, R. and Tank, D.W. (1996) Dendritic integration in mammalian neurons. *Neuron* **16**, 701–716.

CHAPTER 2
STATISTICAL ANALYSIS OF IONIC CHANNEL CURRENT FLUCTUATIONS

Hiroyuki Mino

1. INTRODUCTION

Since the discovery of membrane current fluctuations at the frog's neuromuscular junction, so-called 'noise', in early 1970s (Katz and Miledi, 1972), many neurophysiologists have been interested in the underlying mechanism of membrane current fluctuations (Anderson and Stevens, 1973; Anderson et al., 1978; Cull-Candy et al., 1988). This is because the early works done by Katz and Miledi (1972) suggested that there exist 'pores' distributed over the cell membrane, and these pores play a role in the flow of ionic currents from outside into inside the cell membrane. Since the pores allow flow of ionic currents, it has been called 'ionic channels', of which a schematic illustration is depicted in Figure 2.1. Likewise, these pores are randomly closed and open depending on whether the binding of chemical transmitters (e.g. acetylcholine at the neuromuscular junction) to receptors, generates a specific current waveform like a random telegraph signal. That is, the closed–open gating mechanism of ionic channels is a basis of producing membrane current fluctuations, 'noise'. To express that situation more explicitly, the term 'ionic channel current fluctuations (ICFs)' has been used rather than membrane current fluctuations. Figure 2.2 depicts a schematic illustration of single ionic channel currents, and also the superimposed process of those ionic channel currents. The superimposed process of single ionic channel currents is called the ICFs. In Figure 2.2, a case of stationary ICFs (SICFs) is shown.

From the viewpoint of stochastic modeling and statistical analysis, it is worthwhile to obtain a mathematical expression for the ionic channel gating mechanism. Although some mathematical models have been proposed as candidates for ionic channel gating mechanisms (e.g. Liebovitch and Itoh, 1991; Liebovitch and Todorov, 1996), at present the continuous-time homogeneous Markov chain is the most plausible model for that mechanism (Colquhoun and Hawkes, 1977, 1981, 1982; Neher and Stevens, 1977; Fishman, 1985; Colquhoun and Sakmann, 1985; Colquhoun and Ogden, 1988). The homogeneous Markov chain is a class of Markov processes (Tuckwell, 1989; Papoulis, 1991) and is applied to the cases where there are a finite number of discrete 'states' possessing specific values. Let us consider the case of ionic channel gating mechanism. The most basic kinetic scheme is the closed–open scheme

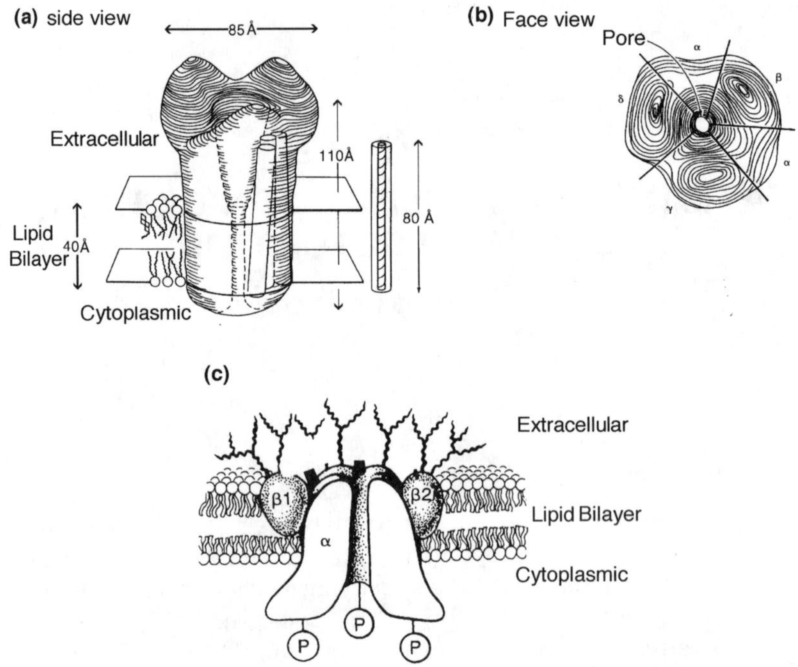

Figure 2.1 Two reconstructions of the vertebrate nicotinic AChR, based on a combination of electron microscopy and x-ray diffraction. (a) Receptor molecule in lipid bilayer showing extensive protrusion into the extracellular space. Cylinders indicate dimensions of presumed α-helical portions of the peptide chain. (b) The five subunits are tentatively identified with a pore formed between them (from Kistler et al., 1982). (c) A cross section of a hypothetical sodium channel consisting of a single transmembrane α subunit of 260 kD in association with a β_1 subunit of 36 kD and a β_2 subunit of 33 kD. The intracellular surface of the α subunit is phosphorylated by multiple protein kinases (P) (from Catterall, 1993).

shown in Figure 2.4. The 'closed state' having a current value of zero is one of the states, and the 'open state' is another state having a current value of the unit amplitude, u. As is shown in the top trace of single ionic channel currents in Figure 2.2, first it stays in the closed state taking a value of zero, in the next moment it moves to the open state taking a value of u and stays there for a while, in the next moment it comes back to the closed state taking a value of zero and stays for a while, and it continues. As long as the process in the closed–open scheme is a homogeneous Markov chain, the staying time in the states is a random variable, and is exponentially distributed with the parameters of state transition rates. In the closed–open scheme, these state transition rates are denoted by α—transitioning from the open state to the closed state, and β—transitioning from the closed state to the open state (see Figure 2.4). Thus, the unit current amplitudes and the state transition rates are important parameters for specifying the ionic channel gating mechanisms modeled by the homogeneous Markov chain.

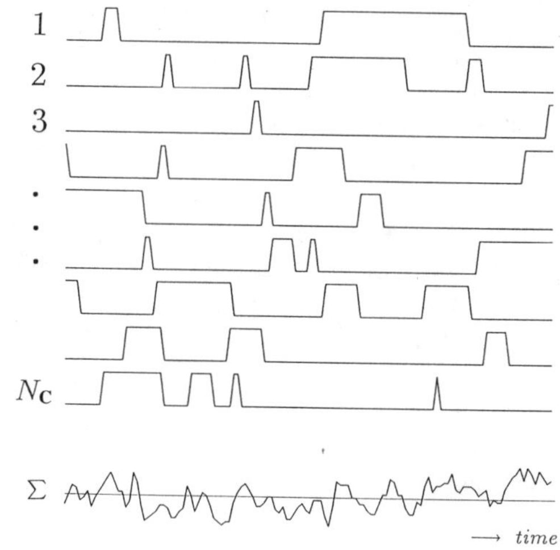

Figure 2.2 Individual single ionic channel currents and the sum of those currents. SICFs, $y(t)$, consisting of N_c channels are depicted at the bottom of this figure.

In practical situations in which the density of ionic channels is not low enough for picking up only one channel within the patch pipette by using the patch clamp method (Hamill et al., 1981) (it depends on cell preparations), it is crucial to estimate the parameters of single ionic channels from observation of the superimposed single ionic channel currents (i.e. ICFs), measured with the voltage-clamp method (Takeuchi and Takeuchi, 1959) or the whole cell clamp method (Hamill et al., 1981). In that case, statistical methods have played an important role in estimating parameters such as the state transition rates of the gating mechanism, the number of ionic channels being activated, and the unit current amplitude of single ionic channels (Mino, 1993a). This is because these parameters of interest are related to the statistics, e.g. the mean and variance of ICFs, and likewise the auto-covariance function and power spectra of ICFs. For instance, in the case of SICFs as shown in the bottom trace of Figure 2.2, the power spectrum has been positively utilized (Anderson and Stevens, 1973; Cull-Candy et al., 1988; Traynelis and Jaramillo, 1998) to estimate the kinetic parameters of single ionic channels, since the power spectrum can be expressed as the Lorentz type spectrum (DeFelice, 1981; Johnston and Wu, 1995), which is described by the state transition rates. On the other hand, for the analysis of non-stationary ICFs (NST-ICFs) as shown in Figure 2.7, the mean and variance which are functions of time because of non-stationarity have been utilized under the assumption that NST-ICFs are binomially distributed, since those statistics are related to the important parameters including the number of ionic channels being activated, and the unit current amplitude of single ionic channels (Sigworth, 1980, 1984; Ohmori, 1981; Keynes, 1994). Hence, it can be seen that statistical methods are important tools for estimating the parameters of single ionic channels from observation of SICFs or NST-ICFs.

In estimating the statistics involving the information on the parameters of single ionic channels, we usually encounter the case where a sample realization of SICFs or NST-ICFs is inevitably corrupted by unobservable additive Gaussian noise caused by amplifiers and/or the micro-electrode. That is, it is often difficult to accurately estimate the parameters of single ionic channels from the statistics of SICFs or NST-ICFs. In that noisy situation, it is worthwhile for the case of SICFs to note that third-order statistics (defined in the next section) are insensitive to the additive Gaussian noise component and provide us with the information on the parameters of single ionic channels as well as on that of the auto-covariance function and the power spectrum (Mino, 1993b). This implies that the use of third-order statistics makes it possible to estimate the parameters of single ionic channels without any compensation for the Gaussian noise component. It is also interesting for the case of NST-ICFs to note that the combination of mean and variance may give an alternative expression that is useful in analyzing the noisy sample realizations to the conventional one developed by Sigworth (1980). Actually, the combination of the mean and variance of binomially distributed NST-ICFs with the additive Gaussian noise exploits the second-order regression model, whose coefficients explicitly correspond not only to the parameters of single ionic channels, but to the variance of the additive Gaussian noise (Mino, 1993c). That is, one coefficient can absorb the additive Gaussian noise component without any interference to the parameters of single ionic channels.

In this Chapter, statistical methods of estimating the parameters of SICFs and NST-ICFs from realizations without any compensation for the additive Gaussian noise component are presented. Although this Chapter discusses current fluctuations at single ionic channels (i.e. random signal processing and its application to estimating the parameters of ionic channels) it should be made clear the microscopic behavior of ionic channels leading to macroscopic membrane currents as observed by the Hodgkin–Huxley equations (see e.g. Clay and Defelice, 1983; Armstrong and Hille, 1998) is not discussed. In section 2, the statistical method using the third-order statistics for SICFs is explored where the theoretical framework of the method, a recommended procedure, and an illustrative numerical example are presented. In section 3, the statistical method for NST-ICFs is developed where the relationship between the mean and variance of NST-ICFs is expressed as the second-order regression model, and an estimation procedure is described as well as the results of numerical simulations in some situations. In section 4 discussions are given, followed by the conclusion and future perspectives in section 5.

2. STATISTICAL METHOD OF ANALYZING SICFs IN THE PRESENCE OF GAUSSIAN NOISE

In this section, a method of estimating the parameters of single ionic channels using the third-order statistics of the noisy SICFs is proposed, and furthermore the validity of the proposed method is assessed through computer simulations.

2.1 Third-Order Cumulants of Noisy SICFs

Let a realization of the noisy SICFs $z(t)$ be represented as:

$$z(t) = y(t) + g(t), \tag{2.1}$$

where $y(t)$ and $g(t)$ respectively stand for the SICFs and the zero-mean additive Gaussian noise due to amplifiers and/or micro-electrodes in which SICFs are assumed to be a non-Gaussian stationary process, and the Gaussian noise is assumed to be a stationary process. Figure 2.2 depicts the SICFs which are the superimposed individual single ionic channel currents, and Figure 2.3 gives us a graphical interpretation of (2.1), i.e. $z(t)$ is corrupted by the Gaussian noise. Since the third-order cumulants of stationary Gaussian processes are zero ($E[g(t)g(t+\tau_1)g(t+\tau_2)]=0$, where E[ ] is the expectation operation) (Nikias and Raghuveer, 1987; Tuckwell, 1989; Mendel, 1991), the third-order cumulants of the noisy SICFs, $c_3^z(\tau_1,\tau_2)(=E[(z(t)-E[z(t)])(z(t+\tau_1)-E[z(t)])(z(t+\tau_2)-E[z(t)])])$, are represented as:

$$c_3^z(\tau_1,\tau_2)=c_3(\tau_1,\tau_2), \tag{2.2}$$

where $c_3(\tau_1,\tau_2)$ denotes the third-order cumulants of SICFs. This means that it is not necessary to subtract the third-order cumulants of the additive Gaussian noise from those of the noisy SICFs, and can be applied to estimating the parameters of single ionic channels without any compensation for the Gaussian noise component.

Here SICFs are assumed to be the continuous-time homogeneous Markov chain having N_s states, the third-order cumulants of SICFs are expressed in terms of the eigenvalues of the state transition rate matrix as (Mino, 1993b):

$$\begin{aligned} c_3(\tau_1,\tau_2)=N_c\Bigg[&\sum_{k_1=1}^{N_s-1}\sum_{k_2=1}^{N_s-1}\sum_{i_1=1}^{N_s}\sum_{i_2=1}^{N_s}\sum_{i_3=1}^{N_s} u_{i_1}u_{i_2}u_{i_3}p_{i_1}f_{i_1i_2}^{(k_1)}f_{i_2i_3}^{(k_2)}e^{\lambda_{k_1}\tau_1}e^{\lambda_{k_2}(\tau_2-\tau_1)} \\ &-\sum_{k=1}^{N_s-1}\sum_{i_1=1}^{N_s}\sum_{i_2=1}^{N_s}\sum_{i_3=1}^{N_s} u_{i_1}u_{i_2}u_{i_3}p_{i_1}\,p_{i_3}f_{i_1i_2}^{(k)}e^{\lambda_{k_2}\tau_2}\Bigg], \end{aligned} \tag{2.3}$$

where N_c denotes the number of independently superimposed homogeneous Markov chain, λ_k's $(k=1,2,\ldots,N_s-1)$ are non-zero eigenvalues of the state transition rate matrix, $p_j=f_{ij}^{(N_s)}$, and

$$f_{ij}^{(k)}=\prod_{l=1,l\neq k}^{N_s}\frac{\mathbf{Q}-\lambda_l\mathbf{I}}{\lambda_k-\lambda_l}, \tag{2.4}$$

where $\mathbf{I}$ stands for the identity matrix, and $\mathbf{Q}$ denotes state-transition-rate matrix $(N_s\times N_s)$ and specifies the transition from the ith state to the jth state in which the sum of transition rates of the same row equals zero:

$$\mathbf{Q}=\begin{bmatrix} -\sum\limits_{i,i\neq 1}k_{1i} & k_{12} & \cdots & k_{1N_s} \\ k_{21} & -\sum\limits_{i,i\neq 2}k_{2i} & \cdots & k_{2N_s} \\ \vdots & \vdots & & \vdots \\ k_{N_s1} & k_{N_s2} & \cdots & -\sum\limits_{i,i\neq N_s}k_{N_si} \end{bmatrix}.$$

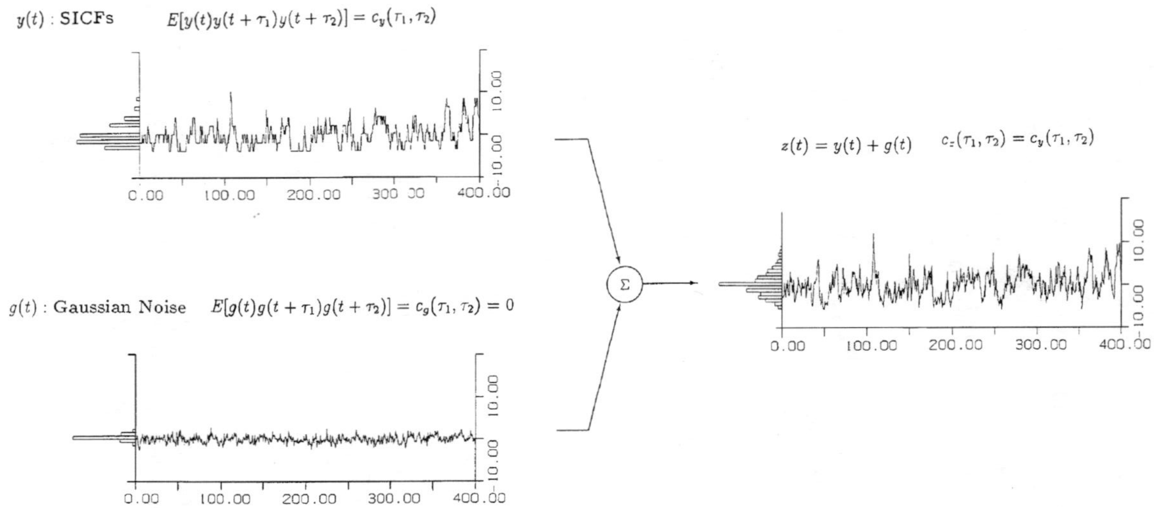

Figure 2.3 SICFs $y(t)$ and Gaussian noise $g(t)$ are depicted at the top left and the bottom left, respectively, as well as the histogram of those data. The noisy SICFs $z(t)$ $(=y(t)+g(t))$ are shown at the right side of this figure. This figure also indicates that the third-order cumulants of the noisy SICFs are equal to those of SICFs without corruptions of the Gaussian noise.

Setting $\tau_1=0$ in (2.3) yields one-variable third-order cumulants, $c_3(0,\tau_2)$ $(\equiv c_3(\tau_2))$:

$$c_3(\tau_2)=\sum_{k=1}^{N_s-1} A_k e^{-\eta_k|\tau_2|} \quad (-\infty \leqslant \tau_2 \leqslant \infty), \tag{2.5}$$

where $\eta_k=-\lambda_k$ $(k=1,2,\ldots,N_s-1)$, and

$$A_k=N_c \sum_{i_1=1}^{N_s} \sum_{i_2=1}^{N_s} \sum_{i_3=1}^{N_s} u_{i_1}u_{i_2}u_{i_3} p_{i_1} f_{i_2i_3}^{(k)} \left(\sum_{k_1=1}^{N_s-1} f_{i_1i_2}^{(k_1)} - p_{i_2} \right). \tag{2.6}$$

We note again that the one-variable third-order cumulants of SICFs, $c_3(\tau_2)$, are equal to those of the SICFs corrupted by the additive colored Gaussian noise, $c_3^z(\tau_2)$, according to (2.2).

2.2 Parameterization of SICFs in Terms of Third-Order Cumulants Equivalence

In order to parameterize SICFs, a linear system is constructed so that the third-order cumulants of the linear system output is equivalent to those of SICFs (Mino, 1993b). The impulse response function of the linear system of this sort can be expressed as N_s-1 sum of those impulse response functions of the first-order ordinary differential equations, driven by a non-Gaussian white noise having non-zero skewness, $w_k(t)$. The linear system output is obtained by:

$$y(t)=\sum_{k=1}^{N_s-1} y_k(t)=\sum_{k=1}^{N_s-1} \int_{-\infty}^{t} e^{-\eta_k(t-\tau)} w_k(\tau)\,d\tau. \tag{2.7}$$

Then, discretizing $y(t)$ with the sampling interval Δt yields the following expression for $y[n]$ $(=y(n\Delta t))$, where $n \in \mathbf{Z}$:

$$y[n]=\sum_{k=1}^{N_s-1} y_k[n]=\sum_{k=1}^{N_s-1} \frac{e_k[n]}{(1-e^{-\eta_k\Delta t}z^{-1})}=\frac{g[n]}{\prod_{k=1}^{N_s-1}(1-e^{-\eta_k\Delta t}z^{-1})}, \tag{2.8}$$

where

$$g[n]=\sum_{l=1}^{N_s-1} e_l[n] \prod_{k=1,k\neq l}^{N_s-1} (1-e^{-\eta k\Delta t}z^{-1}) \tag{2.9}$$

and $e_k[n]$ has the following properties:

$$E[e_k[n]]=0, \tag{2.10}$$

$$E[e_k[l]e_k[m]e_k[n]]=\begin{cases} A_k(1-e^{-3\eta_k\Delta t}) & (l=m=n), \\ 0 & (\text{otherwise}), \end{cases} \tag{2.11}$$

in which $e_i[n]$ is independent of $e_j[n]$ for $i \neq j$. It follows from (2.8) and (2.9) that the discretized process can be expressed as the Autoregressive process (AR(N_s-1); Box and Jenkins, 1976), driven by the sum of N_s-1 Moving Average processes (MA(N_s-2); Box and Jenkins, 1976). The discretized process may also be regarded as the Autoregressive Moving Average process (ARMA(N_s-1, N_s-2)). It may imply that the techniques of ARMA time series identifications can be applied to estimating the state transition rates of the homogeneous Markov chain, i.e. the gating mechanism of single ionic channels.

2.3 Procedure of Estimating the Parameters of SICFs

The parameters of SICFs can be estimated from AR parameters of the discretized process, as shown in the previous section. Practically, the AR parameters may be estimated by solving the set of N_s-1 modified Yule–Walker (MYW) equations (Friedlander and Porat, 1984, 1990) consisting of the estimated third-order cumulants, $\hat{c}_3[l]$ ($l \equiv \tau_2 \Delta t$, $l \geqslant 0$). However, since the parameters estimated from the set of N_s-1 equations often have large variations, the over-determined MYW (Friedlander and Sharman, 1985) that solves sets of equations greater than N_s-1 may be more useful than the conventional MYW method. The over-determined MYW is considered to improve the accuracy of the estimates because it uses long lags of third-order cumulants which contain important information on the AR parameters. In practical situations, the number of over-determined MYW equations, N_o, is recommended to be set between 5 and 15, depending on the kinetic parameters and the number of ionic channel states under investigation. We can summarize the estimation procedure as follows:

Step 1: Observe the discretized SICFs, corrupted by additive Gaussian noise, $z[n]$; $n=1,2,\ldots,N$, and calculate the third-order cumulants, $\hat{c}_3[l]$:

$$\hat{c}_3[l]=\frac{1}{N}\sum_{i=1}^{N}(z[k]-\hat{\mu}_z)^2(z[k+l]-\hat{\mu}_z), \tag{2.12}$$

where $\hat{\mu}_z$ denotes the estimated mean of $z[n]$.

Step 2: Form vector $\mathbf{r}$ ($1\times N_o$) and matrix $\mathbf{C}$ ($(N_s-1)\times N_o$):

$$\mathbf{r}=-[\hat{c}_3[N_s-1], \hat{c}_3[N_s], \ldots, \hat{c}_3[N_o+N_s-2]]^{\mathrm{T}} \tag{2.13}$$

and

$$\mathrm{C}=\begin{bmatrix} \hat{c}_3[N_s-2] & \hat{c}_3[N_s-3] & \cdots & \hat{c}_3[0] \\ \hat{c}_3[N_s-1] & \hat{c}_3[N_s-2] & \cdots & \hat{c}_3[1] \\ \vdots & \vdots & \vdots & \vdots \\ \hat{c}_3[N_o+N_s-3] & \hat{c}_3[N_o+N_s-4] & \cdots & \hat{c}_3[N_o-1] \end{bmatrix}. \tag{2.14}$$

Step 3: Obtain $\hat{\mathbf{a}}$ by solving the least-square problem:

$$\hat{\mathbf{a}} = -[\mathbf{C}^{\mathrm{T}}\mathbf{C}]^{-1}\mathbf{C}^{\mathrm{T}}\mathbf{r}, \tag{2.15}$$

where

$$\hat{\mathbf{a}} = [\hat{a}_1, \hat{a}_2, \ldots, \hat{a}_{N_s-1}]^{\mathrm{T}}. \tag{2.16}$$

Step 4: Estimate the non-zero eigenvalues of the transition rate matrix η_k's by solving the characteristic equation of AR(N_s-1) system such that:

$$1+\hat{a}_1 z^{-1}+\hat{a}_2 z^{-2}+\cdots+\hat{a}_{N_s-1} z^{-(N_s-1)} = \prod_{k=1}^{N_s-1}(1-e^{-\eta_k}z^{-1})=0. \tag{2.17}$$

Step 5: If the two-state scheme is being used as shown in Figure 2.4, then estimate α:

$$\hat{\alpha} = \hat{\eta}_1 \quad (\alpha \gg \beta) \tag{2.18}$$

or if the three-state scheme is being used as shown in Figure 2.5, then estimate k_+ and k_-, given $\hat{\alpha}$ ($\alpha \gg \beta$):

$$\hat{k}_+ = (\hat{\eta}_1+\hat{\eta}_2) - \frac{\hat{\eta}_1\hat{\eta}_2}{\hat{\alpha}} - \hat{\alpha}, \tag{2.19}$$

$$\hat{k}_- = \frac{\hat{\eta}_1\hat{\eta}_2}{\hat{\alpha}}. \tag{2.20}$$

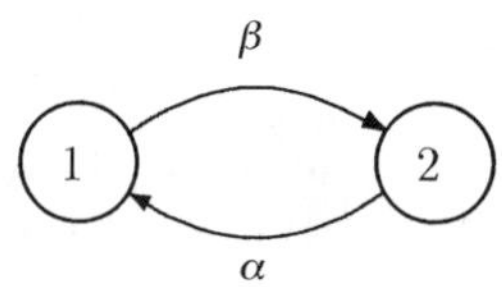

Figure 2.4 The two-state scheme; 1 and 2 respectively denote the state 1 ($u_1=0$) and the state 2. This scheme is usually used in as the Closed–Open scheme. α and β are transition rates.

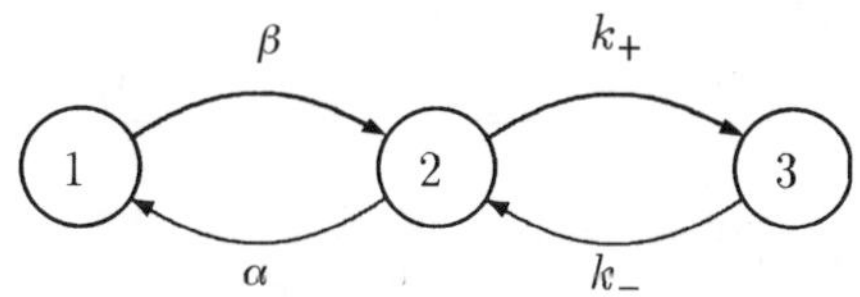

Figure 2.5 The three-state scheme; 1, 2, and 3 respectively designate the state 1 ($u_1=0$), the state 2, and the state 3 ($u_3=0$). This is used as the Closed–Open–Blocked scheme in which the Blocked state is due to substances such as local anaesthetic drugs (Neher and Steinbach, 1978; Colquhoun and Sheridan, 1981). This is also used as the Closed–Open–Inactivated scheme in which the Inactivated state is appeared in sodium channels, etc. α, β, k_+, and k_- are transition rates.

2.4 Computer Simulations

Monte Carlo simulation of ten runs was carried out to assess the validity of the estimation method, and to investigate the accuracy of the parameter estimates against the signal-to-noise ratio SNR (∞ (noise free), 10 dB (noise variance 1/10), and 0 dB (noise variance is the same as that of SICFs)).

SICFs were generated on the basis of two-state and three-state schemes as depicted in Figures 2.4 and 2.5. For these schemes, the parameters were set as follows: data length $N = 81\,920$, sampling instance $\Delta t = 0.5$ ms, and unit current amplitude $u_2 = 2$ pA. The transition rates of the two-state scheme were $\beta = 1$ and $\alpha = 288$, and those of the three-state scheme were $\beta = 1$, $\alpha = 288$, $k_+ = 654$, and $k_- = 737$. The measurement noise, which is the Gaussian process being a fourth-order autoregressive (AR(4)) process with coefficients of $a_1 = -0.492056$, $a_2 = 0.148145$, $a_3 = -0.172044$, and $a_4 = -0.0315019$, was independently added to the generated SICFs. The innovation variance of the AR(4) process, σ_n^2, was determined so that the SNRs were ∞ (noise free), 10 dB (noise variance is 1/10), and 0 dB (noise variance is the same as that of SICFs). For the two-state scheme we have (a) the noise-free situation (the innovation variance of the Gaussian AR(4) process, $\sigma_n^2 = 0$), (b) 10 dB ($\sigma_n^2 = 0.5392749 \times 10^{-24}$), and (c) 0 dB ($\sigma_n^2 = 0.5392749 \times 10^{-23}$), and for the three-state scheme (d) the noise-free situation ($\sigma_n^2 = 0$), (e) 10 dB ($\sigma_n^2 = 0.5390403 \times 10^{-24}$), and (f) 0 dB ($\sigma_n^2 = 0.5392749 \times 10^{-23}$) in which $N_c = 500$. Figure 2.6 depicts an example of realizations in 400 ms length in (a)–(f), showing also the histogram of the noisy SICFs in the left side of SICFs. The shape of these histogram is obviously the asymmetrical distributions possessing non-zero third-order cumulants, except for the cases of SNR = 0 dB. Ten samples of such data were generated for Monte Carlo simulations.

Table 2.1 summarizes the results of the SNR investigation, where the mean, standard deviation (S.D.) and coefficients of variation (c.v.[%]) of the estimated kinetic parameter α in the case of the two-state scheme, and k_+ and k_- in the case

Table 2.1 Theoretical and estimated transition rates (Mean ± S.D. (c.v.%)) in $N_c = 500$ under the assumption that $\alpha \gg \beta$ (10 Monte Carlo runs)

	Theoretical Transition Rates	*SNR* [dB]	*Estimated Mean ± S.D. (c.v.[%])*
		∞	287.966 ± 9.909 (3.441)
Two-state	$\alpha = 288$	10	290.963 ± 15.279 (5.251)
		0	289.439 ± 20.114 (6.950)
		∞	656.430 ± 42.782 (6.517)
	$k_+ = 654$		732.569 ± 101.637 (13.874)
Three-state		10	647.763 ± 66.324 (10.239)
	$k_- = 737$		776.250 ± 167.158 (21.534)
		0	491.664 ± 138.079 (28.084)
			1170.313 ± 453.599 (38.759)

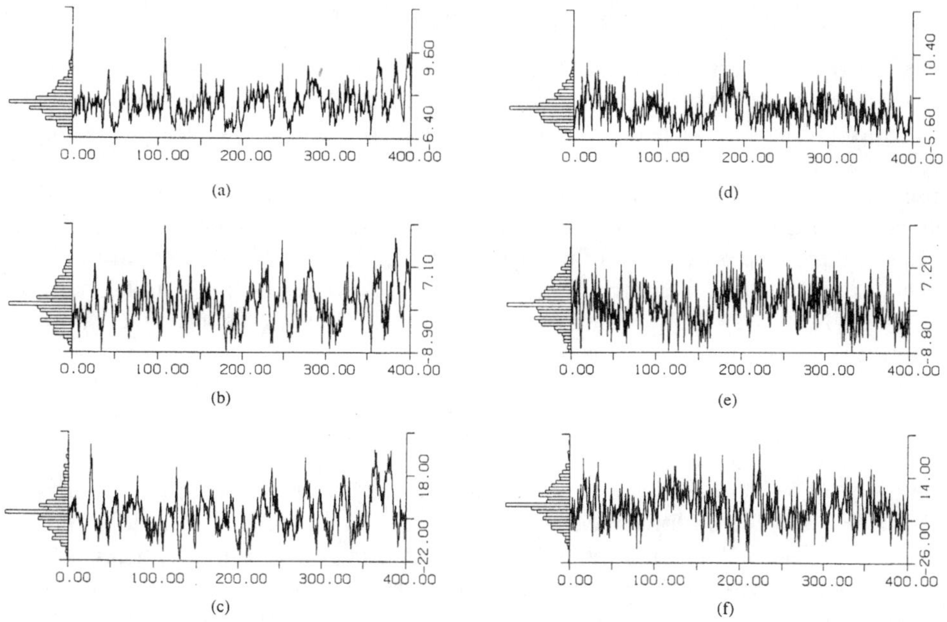

Figure 2.6 A sample path of the SICFs generated under corruption by additive colored Gaussian noise with SNR $=\infty$ ((a) and (b)), 10 dB ((b) and (e)), 0 dB ((c) and (f)) in the C–O scheme ((a), (b) and (c)) and the C–O–B scheme ((d), (e) and (f)) in which $N_c = 500$. The amplitude histogram is depicted on the left side of the data. In all figures, the abscissa is time in [ms] and ordinate is the SICF amplitude in [pA]. (Reprinted from Mino (1993b) with permission.)

of the three-state scheme are shown. The transition rates are estimated according to (2.18)–(2.20) under certain conditions ($\alpha \gg \beta$). Even though the SNR was 10 dB, the mean of estimates was unbiased, while the c.v.'s tend to increase as SNR decreased. This indicates that third-order cumulants are insensitive to additive Gaussian noise. However, we note that in the three-state scheme and SNR $=0$ dB the parameters could not be accurately estimated, because the SICFs corrupted by the Gaussian AR(4) process were almost symmetrically distributed. It follows from the above that the validity and effectiveness of the proposed parametric modeling and estimation method are assessed through Monte Carlo simulations.

3. STATISTICAL METHOD OF ANALYZING NST-ICFs IN THE PRESENCE OF GAUSSIAN NOISE

In this section, a method of estimating the unit current amplitude of single ionic channels, and the number of single ionic channels being activated from observation of the noisy NST-ICFs is presented, and furthermore the validity of the proposed method is assessed through computer simulations.

3.1 Multi-regression Modeling of Noisy NST-ICFs

This section considers modeling of the noisy NST-ICFs on the basis of a combination between the ensemble mean and variance of the noisy NST-ICFs in which NST-ICFs are expressed by an independent superposition of the single ionic channel currents. It is assumed that the experiment can be repeated many times to calculate the ensemble mean and variance of the noisy NST-ICFs for determining u and N_c. It is further assumed that the NST-ICFs have only one state having non-zero amplitude value, and that the probability of being open, which is function of time, has the same time course for each repetition of the experiment. Now, let the ith sample record of the noisy NST-ICFs be represented by $z_i[k]$ $(=z_i(k\Delta t))$ (Δt denotes the sampling instance):

$$z_i[k] = \sum_{j=1}^{Nc} w_i^{(j)}[k] + n_i[k] = w_i[k] + n_i[k], \tag{2.21}$$

where $w_i^{(j)}[k]$, $w_i[k]$ and $n_i[k]$ respectively denote the jth single ionic channel current, NST-ICFs, and the additive measurement noise, and where $k=1,2,\ldots,N_d$ and $i=1,2,\ldots,N_p$. Figure 2.7 shows the NST-ICFs consisting of N_c ionic channel currents, and the noisy NST-ICFs. Here it is assumed that $w_i[k]$ is binomially distributed at an arbitrary time $k\Delta t$. It is further assumed that $n_i[k]$ is a stationary Gaussian process with zero mean and an unknown variance, σ_n^2, and is independent of $w_i[k]$. The assumptions allow us to theoretically derive its mean, $m_z[k]$, and variance, $\sigma_z^2[k]$, which are functions of time because of the non-stationarity of $p[k]$:

$$m_z[k] = E[z[k]] = N_c u p[k] \tag{2.22}$$

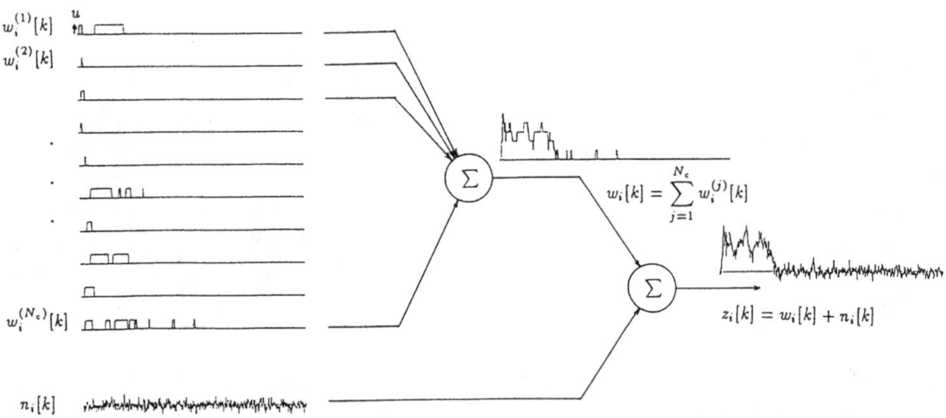

Figure 2.7 A sample record of NST-ICFs, $w_i[k]$, generated by an independent superposition of the single channel current $w_i^{(j)}[k]$, $j=1,2,\ldots,N_c$ having the unit current amplitude u, additive measurement noise $n_i[k]$, and the noisy NST-ICFs $z_i[k]$. (Reprinted from Mino (1993c) with permission.)

and

$$\begin{aligned}\sigma_z^2[k] &= E[(z[k]-m_z[k])^2]\\ &= N_c u^2 p[k](1-p[k])+\sigma_n^2,\end{aligned} \tag{2.23}$$

where $p[k]$, $m_{z(u)}[k]$ and $\sigma_{z(u)}^2[k]$ respectively denote the probability being open at time $k\Delta t$, the mean per unit channel, and the variance per unit channel, and where E[ ] stands for the expectation operation. Substituting (2.22) into (2.23) with respect to $p[k]$ yields a relationship between $m_z[k]$ and $\sigma_z^2[k]$:

$$\sigma_z^2[k] = -\frac{1}{N_c} m_z^2[k] + u m_z[k] + \sigma_n^2, \tag{2.24}$$

where (2.24) obviously represents a parabolic function. The relationship between the mean and variance can be regarded as the following second-order multi-regression expression (Wilks, 1962):

$$\sigma_z^2[k] = b_2 m_z^2[k] + b_1 m_z[k] + b_0 \tag{2.25}$$

with the multi-regression coefficients, b_0, b_1 and b_2. We note not only that the parameters of u and N_c correspond one-to-one with the multi-regression coefficients, but also that the variance of the additive measurement noise is related to one of the coefficients as follows:

$$u = b_1, \tag{2.26}$$

$$N_c = -1/b_2, \tag{2.27}$$

$$\sigma_n^2 = b_0. \tag{2.28}$$

It follows from the above that the parameters of NST-ICFs can be estimated by identifying the multi-regression coefficients, without any interference from the additive measurement noise component, because this multi-regression expression involves not only the NST-ICF component but also the additive measurement noise component, i.e. the effect of the additive measurement noise on the parameter estimates of u and N_c can be canceled by the coefficient b_0. Although σ_n^2 can be simultaneously estimated together with u and N_c, the estimation of σ_n^2 is not essential for the analysis of NST-ICFs. However, the regression coefficient b_0 is necessary for canceling the additive noise component.

3.2 Procedure of Estimating the Parameters of NST-ICFs

In the previous section, we saw that the parameter estimation of NST-ICFs can be regarded as a problem of identifying the multi-regression coefficients. In this section, a procedure for estimating the multi-regression coefficients is described.

Let us consider estimating the multi-regression coefficients, b_0, b_1 and b_2, given the estimated mean $\hat{m}_z[k]$ and the estimated variance $\hat{\sigma}_z^2[k]$, $k=1,2,\ldots,N_d$. Using the matrix form, the multi-regression expression is written as

$$\sigma_z^2 = \mathbf{Mb} + \mathbf{e}, \tag{2.29}$$

where

$$\sigma_z^2 = [\hat{\sigma}_z^2[1], \hat{\sigma}_z^2[2], \ldots, \hat{\sigma}_z^2[N_\mathrm{d}]]^\mathrm{T}, \tag{2.30}$$

$$\mathbf{M} = \begin{bmatrix} 1 & m_z[1] & m_z^2[1] \\ 1 & m_z[2] & m_z^2[2] \\ \vdots & \vdots & \vdots \\ 1 & m_z[N_\mathrm{d}] & m_z^2[N_\mathrm{d}] \end{bmatrix} \tag{2.31}$$

$$\mathbf{b} = [b_0, b_1, b_2]^\mathrm{T} \tag{2.32}$$

and

$$\begin{aligned} \mathbf{e} &= [e[1], e[2], \ldots, e[N_\mathrm{d}]]^\mathrm{T} \\ &= [(\hat{\sigma}_z^2[1] - \sigma_z^2[1]), (\hat{\sigma}_z^2[2] - \sigma_z^2[2]), \ldots, (\hat{\sigma}_z^2[N_\mathrm{d}] - \sigma_z^2[N_\mathrm{d}])]^\mathrm{T}, \end{aligned} \tag{2.33}$$

where $E[\mathbf{e}] = \mathbf{0}$ and $m_z[k]$ is assumed to be approximated by the estimate $\hat{m}_z[k]$. The multi-regression coefficient $\mathbf{b}$ is determined so as to minimize the square value of $\mathbf{e}$ satisfying

$$(\mathbf{M}^\mathrm{T}\mathbf{M})\mathbf{b} = \mathbf{M}^\mathrm{T}\sigma_z^2. \tag{2.34}$$

Thus, the explicit formula for estimating $\hat{\mathbf{b}}$ is given by

$$\hat{\mathbf{b}} = (\mathbf{M}^\mathrm{T}\mathbf{M})^{-1}\mathbf{M}^\mathrm{T}\sigma_z^2, \tag{2.35}$$

where $\hat{\mathbf{b}}$ denotes the estimate of $\mathbf{b}$. We now summarize the estimation procedure as follows:

Step 1: Observe NST-ICFs, $z_i[k]$, $i = 1, 2, \ldots, N_\mathrm{p}$, $k = 1, 2, \ldots, N_\mathrm{d}$.

Step 2: Estimate the ensemble mean, $\hat{m}_z[k]$,

$$\hat{m}_z[k] = \frac{1}{N_\mathrm{p}} \sum_{i=1}^{N_\mathrm{p}} z_i[k] \tag{2.36}$$

and the ensemble variance, $\hat{\sigma}_z^2$,

$$\hat{\sigma}_z^2[k] = \frac{1}{N_\mathrm{p}} \sum_{i=1}^{N_\mathrm{p}} (z_i[k] - \hat{m}_z[k])^2, \tag{2.37}$$

where $k = 1, 2, \ldots, N_\mathrm{d}$.

Step 3: Form vector σ_z^2 and matrix $\mathbf{M}$, using $\hat{\sigma}_z^2[k]$ and $\hat{m}_z[k]$.

Step 4: Estimate u and N_c with (2.26) and (2.27), solving $\hat{\mathbf{b}}$ with (2.35).

3.3 Computer Simulations

In order to assess the validity of the proposed method and the accuracy of its estimation, Monte Carlo simulation is carried out in which the three-state scheme like sodium channels having an inactivation state as shown in Figure 2.5 is treated as a specific example.

Non-stationary single ionic channel currents were generated on the basis of the Closed–Open–Inactivated scheme as is shown in Figure 2.5. These states were the state 1, with the unit amplitude of 0 pA, the state 2, $u=1$ pA, and the state 3, 0 pA, respectively. Initial state probability was set at $p_1[k]=1.0$ for the state 1, $p_2[k]=0.0$ for the state 2, and $p_3[k]=0.0$ for the state 3. The transition rates were as follows: $\beta=550\,\mathrm{s}^{-1}$, $\alpha=368\,\mathrm{s}^{-1}$, $k_+=182\,\mathrm{s}^{-1}$, and $k_-=0.001\,\mathrm{s}^{-1}$. Then, NST-ICFs were generated by an independent superposition of ten single channel currents, i.e., $N_c=10$, with regard to the small channel case. The data length, N_d, and the sampling instance, Δt, were set at 256 and 0.2 ms, respectively. The noise $n_i[k]$, being a Gaussian AR(4) process with $a_1=-0.492056$, $a_2=0.148145$, $a_3=-0.172044$, and $a_4=-0.0315019$, was independently added to NST-ICFs where σ_n^2 was set at 0.0, and 0.3. In this simulation the Gaussian noise was adopted as the additive measurement noise. The number of ensemble records, N_p, was set at 1000. Monte Carlo runs were set at 100. Figures 2.8 and 2.9 show sample paths of the noisy NST-ICFs, mean, and variance in (a), and the parabolic relationship of mean and variance in (b). In Figure 2.8 (b) b_0 is close to zero, while in Figure 2.9 (b) b_0 is close to 0.3, because b_0 corresponds to the variance of the additive measurement noise.

Monte Carlo simulation of 100 runs is carried out, in order to assess the validity of the method. Table 2.2 presents the mean, standard deviation (S.D.) and the coefficients of variation (c.v.[%]) of the estimated parameters u, N_c, and σ_n^2. All the

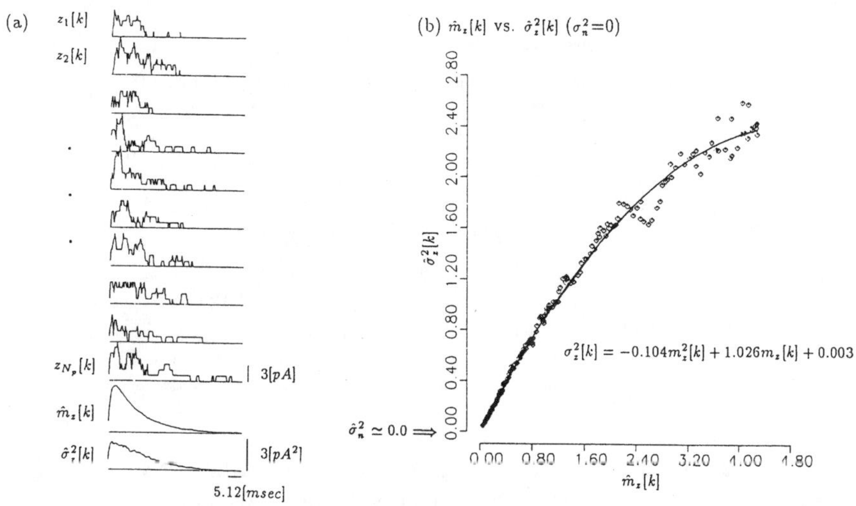

Figure 2.8 (a) Simulated NST-ICFs $z_i[k]$, $i=1,2,\ldots,N_p$, $N_p=1000$, the mean $\hat{m}_z[k]$ and variance $\hat{\sigma}_z^2[k]$ estimated from 1000 sample records in the noise-free case. (b) Plot of $\hat{m}_z[k]$ against $\hat{\sigma}_z^2[k]$ with the square mark. The solid line is the regression curve with the estimated coefficients, $b_2=-0.104$, $b_1=1.026$, and $b_0=0.003$. Note that $b_0\simeq 0$, because $\sigma_n^2=0$ (noise-free case). (Reprinted from Mino (1993c) with permission.)

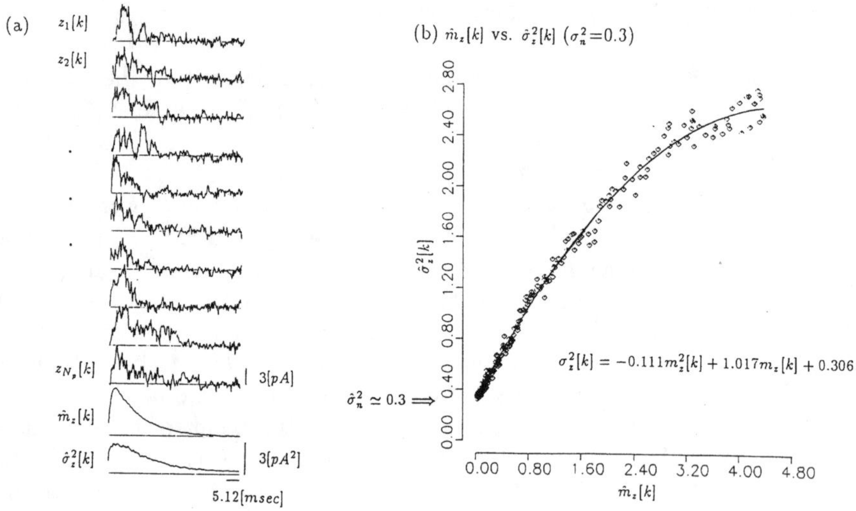

Figure 2.9 (a) Simulated NST-ICFs $z_i[k]$, $i=1,2,\ldots,N_p$, $N_p=1000$, the mean $\hat{m}_z[k]$ and variance $\hat{\sigma}_z^2[k]$ estimated from 1000 sample records in the case of $\sigma_n^2=0.3$. (b) Plot of $\hat{m}_z[k]$ against $\hat{\sigma}_z^2[k]$ with the square mark. The solid line is the regression curve with the estimated coefficients, $b_2=-0.111$, $b_1=1.017$, and $b_0=0.306$. Note that $b_0 \simeq 0.3$, because $\sigma_n^2=0.3$. (Reprinted from Mino (1993c) with permission.)

Table 2.2 Theoretical and estimated parameters (Mean $\pm$ S.D. (c.v.%)) in $N_p=1000$ (100 Monte Carlo runs)

σ_n^2		*Theoretical Values*	*Estimated Mean S.D. (c.v.[%])*
	u	1.000	1.002 ± 0.030 (2.995)
0.0	N_c	10.000	9.953 ± 0.842 (8.461)
	σ_n^2	0.000	-0.001 ± 0.007 (*.***)
	u	1.000	0.989 ± 0.030 (3.092)
0.3	N_c	10.000	10.328 ± 0.980 (9.489)
	σ_n^2	0.300	0.308 ± 0.007 (2.567)

estimated parameters are shown to be unbiased even in the presence of additive measurement noise. The c.v. of u is found to be better than that of N_c, because u is estimated from b_1, which is the coefficient of the first-order function, while N_c is calculated from b_2, which is the coefficient of the second-order function. The c.v. of the estimated parameters tend to increase as σ_n^2 increases, whereas the mean of the estimated parameters is unbiased. Although the c.v.'s tend to decrease as N_p decreases, the unbiasedness still holds.

4. DISCUSSIONS

In this article, two methods have been described for estimating the parameters of ICFs in which the sample realizations of noisy ICFs are available for estimation. The parameters of ICFs estimated by the proposed methods are insensitive to the additive measurement noise, because the third-order cumulants of the noisy SICFs suppress the Gaussian measurement noise component, and the multi-regression model derived for NST-ICFs involves one multi-regression coefficient that cancels the effect of the additive measurement noise. In other words, as long as the proposed methods are used, it is unnecessary to compensate for the additive measurement noise component in spite of the presence of noise, unlike previous methods.

In practical situations of analyzing SICFs, however, we note that higher-order cumulants may be utilized when a large sample size is given, i.e. the estimates of higher-order cumulants usually have large variations, compared with those of the auto-covariance function. To remedy this problem, we adopted the overdetermined MYW method using long lag information of third-order cumulants, since it was observed through an investigation of the auto-covariance function that it improves the accuracy of AR coefficient estimates (Friedlander and Sharman, 1985). In fact, the conventional MYW method of solving $N_s - 1$ equations was not able to provide accurate parameter estimates, although the results were not shown here. Thus, in our case the number of equations to be solved, N_o, was set at 5 and 15, for the two-state and three-state schemes, respectively. N_o was determined according to the results of N_o versus the accuracy of the parameter estimates.

The accuracy of the parameter estimates of SICFs is dependent on the Gaussianity of the noisy SICFs observed, since we saw in computer simulations that the c.v.'s of the parameter estimates tend to increase when SICFs are contaminated by Gaussian measurement noise, e.g. the case of 0 dB, or when the number of contributing channels is large enough to approximate SICFs as the Gaussian process. Thus, for actual data analysis we need to first find non-Gaussianity, for instance, using a histogram of SICFs, before estimating the kinetic parameters. We note again that, if the histogram appears to have an almost symmetrical distribution, the proposed method will have difficulty in estimating the parameters or will no longer apply.

As for NST-ICFs, the proposed method has been derived using a combination of statistical information; the mean and variance. Therefore, alternative combinations can be considered, e.g. the combined mean–skew, mean–kurtosis, and variance–skew methods. For instance, the combined mean–skew method is useful for avoiding interference from additive measurement noise, if the additive measurement noise can be assumed to be a zero mean Gaussian. That is, the combined mean–skew method is also insensitive to measurement noise, since the skew of the Gaussian noise becomes zero. However, higher-order statistics, in general, produce large variations among estimates, and it is recommended that lower-order statistics be utilized in practice, although higher-order statistics convey other information and can be calculated theoretically. Therefore, the method proposed in this chapter utilizes the first two moments, i.e. the combined mean–variance method, in consideration of the accuracy of the estimates.

In the practical analysis of NST-ICFs, it will be noted that parameter estimation is impossible in cases where the assumptions made are not valid, i.e. (i) NST-ICFs at a specific time cannot be approximated by the binomial distribution, (ii) the number of states having non-zero amplitude is not one, and (iii) the mean and variance of the additive noise are not time-invariant. In practical situations, assumption (i) is frequently invalidated by the experimenter, who may use a low-pass filter with a low cut-off frequency to reduce the measurement noise of noisy NST-ICFs. Low-pass filtering distorts the waveform of NST-ICFs, which means that assumption (i) is no longer valid. Hence it is hoped that noisy NST-ICFs will not be low-pass filtered with a low cut-off frequency to avoid distortion, and will be analyzed by the proposed method together with the additive measurement noise component included. However, if the mean and variance of the additive noise vary with time in which the additive noise includes a stationary component with non-stationary components such as currents from other types of channels, and the open channel noise (Heinemann and Sigworth, 1991), the proposed method may not be utilized because the regression coefficient b_0 cannot cancel the time-varying noise component, that is, the parameter estimates of u and N_c may be biased. In this case, we need to construct an advanced model taking the time-varying noise into consideration, using *a priori* information on the additive noise such as the open channel noise. On the other hand, the selection of an interval to be analyzed just for the NST-ICFs component is important for parameter estimation to succeed, because the rising phase of the record is composed of a slight capacitive current, which cannot be fully subtracted, the falling phase of the record is shortened depending on the depolarization level, and so on. It is difficult to determine the optimal interval because of the many factors involved and, this remains a problem for further investigation.

5. CONCLUSIONS AND FUTURE PERSPECTIVES

Recently, the single channel recording technique has reached the status of a routine method, and the parameters of single ionic channels have been analyzed with an aid of maximum likelihood methods given the open and closed times (Sakmann and Neher, 1995), even if brief opening events cannot be detected (Hawkes et al., 1990, 1992; Magleby and Weiss, 1990; Qin et al., 1996; Merlushkin and Hawkes, 1997). Likewise, some software packages for computer simulations have been developed (Pancrazio, 1995; Kukol, 1997). However, as mentioned earlier we often encounter cases where it is not possible to obtain the single channel recording, i.e. the observable data are the ionic channel current fluctuations consisting of multichannels components. In this situation, the ionic channel current fluctuations should be crucial to investigate the underlying mechanism of single ionic channels, so that the methods presented in this article will play a key role even if the methods of analyzing multiple channel currents (Manivannan et al., 1992; Albertsen and Hansen, 1994; Klein et al., 1997) are universally accepted.

Section 3 has described a method of estimating the unit current amplitude and the number of ionic channels being activated without citing estimation methods of kinetic parameters (Baumgartner et al., 1997). Since it is often important to investigate the kinetic properties of voltage gated channels having inactivation processes such as sodium channels, further investigations are necessary for developing alternative methods which have an advantage of the estimation accuracy of kinetic parameters to the previous one (Sigworth, 1981).

In conclusion, the methods presented in this article for the noisy SICFs and NST-ICFs will be extensively utilized in investigating not only the function of ionic channels in physiology, but the effects of substances such as local anaesthetic drugs (Neher and Steinbach, 1978; Colquhoun and Sheridan, 1981) on ionic channels or receptors in pharmacology.

PROBLEMS

2.1. It has been assumed in this Chapter that the ionic channel gating are modeled by a class of Markov processes according to Colquhoun and Hawkes (1977, 1981, 1982). However, it has been reported that the open state of ionic channels may generate an additional noise, so called the "Open Channel Noise" (see Heinemann and Sigworth, 1991). If this additional noise is taken into account, formulate an alternative approach for estimating ionic channel parameters.

2.2. For the analysis of SICFs, the additive measurement noise is assumed to be stationary Gaussian processes so that the third-order cumulants of noisy SICFs are equivalent to those of noise-free SICFs. If this assumption made on the noise property is not validated, develop new methods which do not need a knowledge on the measurement noise property.

2.3. For the analysis of NST-ICFs, the number of ionic channels and the unit current amplitude basically characterize the property of ionic channels, while these would not necessarily express the dynamics, i.e., the transition rates of Markov processes. As a research project attempt the noise-insensitive approaches for estimating the transition rates on the basis of non-stationary third-order cumulants. Note, statistical errors due to estimating higher-order statistics.

REFERENCES

Albertsen, A. and Hansen, U.P. (1994) Estimation of kinetic rate constants from multi-channel recordings by a direct fit of the time series. *Biophys. J.*, **67**, 1393–1403.

Anderson, C.R. and Stevens, C.F. (1973) Voltage clamp analysis of acetylcholine produced end-plate current fluctuations at frog neuromuscular junctions. *J. Physiol.* (*Lond.*), **235**, 655–691.

Anderson, C.R., Cull-Candy, S.G. and Miledi, R. (1978) Glutamate current noise: post-synaptic channel kinetics investigated under voltage clamp. *J. Physiol.* (*Lond.*), **282**, 219–242.

Armstrong, C.M. and Hille, B. (1998) Voltage-gated ion channels and electrical excitability. *Neuron*, **20**, 371–380.

Baumgartner, W., Hohenthanner, K., Hofer, G.F., Groschner, K. and Romanin, C. (1997) Estimating the number of channels in patch-clamp recordings: application to kinetic analysis of multichannel data from voltage-operated channels. *Biophys. J.*, **72**, 1143–1152.

Box, G.E.P. and Jenkins, G.M. (1976) *Time Series Analysis: Forecasting and Control*. California: Holden-Day.

Catterall, W.A. (1993) Structure and modulation of Na^+ and Ca^{++} Channels. *Ann. N. Y. Acad. Sci.: Molecular Basis of Ion Channels and Receptors Involved in Nerve Excitation, Synaptic Transmission and Muscle Contraction*, edited by H. Higashida et al., **707**, 1–19.

Clay, J.R. and DeFelice, L.J. (1983) The relationship between membrane excitability and single channel open-close kinetics. *Biophys. J.*, **42**, 151–157.

Colquhoun, D. and Hawkes, A.G. (1977) Relaxation and fluctuations of membrane currents that flow through drug-operated channels. *Proc. R. Soc. Lond. Ser. B*, **199**, 231–262.

Colquhoun, D. and Hawkes, A.G. (1981) On the stochastic properties of single ion channels. *Proc.R. Soc. Lond. Series B*, **211**, 205–235.

Colquhoun, D. and Hawkes, A.G. (1982) On the stochastic properties of bursts of single ion channel openings and of clusters of bursts. *Phil. Trans. R. Soc. Lond. Ser. B*, **300**, 1–59.

Colquhoun, D. and Sheridan, R.E. (1981) The modes of action of Gallamine. *Proc. R. Soc. Lond. Ser. B*, **211**, 181–203.

Colquhoun, D. and Sakmann, B. (1985) Fast events in single-channel currents activated by acetylcholine and its analogues at the frog muscle end-plate. *J. Physiol. (Lond.)*, **369**, 501–557.

Colquhoun, D. and Ogden, D.C. (1988) Activation of ionic channels in the frog end-plate by high concentrations of acetylcholine. *J. Physiol. (Lond.)*, **395**, 131–159.

Cull-Candy, S.G., Howe, J.R. and Ogden, D.C. (1988) Noise and single channels activated by excitatory amino acid in rat cerebellar granule neurons. *J. Physiol. (Lond.)*, **400**, 189–222.

DeFelice, L.J. (1981) *Introduction to Membrane Noise*. New York: Plenum Press.

Fishman, H.M. (1985) Relaxations, fluctuations and ion transfer across membranes. *Prog. Biophys. Molec. Biol.*, **46**, 127–162.

Friedlander, D. and Porat, B. (1984) The modified Yule-Walker method of ARMA spectral estimation. *IEEE Trans. Aerosp. Electron. Syst.*, **AES-20**, 158–173.

Friedlander, B. and Sharman, K.C. (1985) Performance evaluation of the modified Yule-Walker estimation. *IEEE Trans. Acoust., Speech, Signal Process.*, **ASSP-33**, 719–725.

Friedlander, B. and Porat, B. (1990) Asymptotically optimal estimation of MA and ARMA parameters of non-Gaussian process from higher-order moments, *IEEE Trans. Automat. Contr.*, **AC-35**, 27–35.

Hamill, O.P., Marty, A., Neher, E., Sakmann, B. and Sigworth, F.J. (1981) Improved patch-clamp techniques for high-resolution current recording from cells and cell-free membrane patches. *Pflugers Arch.*, **391**, 85–100.

Hawkes, A.G., Jalali, A. and Colquhoun, D. (1990) The distributions of the apparent open times and shut times in a single channel record when brief events cannot be detected. *Phil. Trans. R. Soc. Lond. A*, **332**, 511–538.

Hawkes, A.G., Jalali, A. and Colquhoun, D. (1992) Asymptotic distributions of apparent open times and shut times in a single channel record allowing for the omission of brief events. *Phil. Trans. R. Soc. Lond. A*, **337**, 383–404.

Heinemann, S.H. and Sigworth, F.J. (1991) Open channel noise VI. Analysis of amplitude histograms to determine rapid kinetic parameters. *Biophys. J.*, **60**, 577–587.

Johnston, D. and Wu, S.M. (1995) *Foundations of Cellular Neurophysiology*, Camb., MA: MIT Press.

Katz, B. and Miledi, R. (1972) The statistical nature of acetylcholine potential and its molecular components. *J. Physiol. (Lond.)*, **224**, 665–699.

Keynes, R.D. (1994) The kinetics of voltage-gated ion channels. *Quart. Rev. Biophys.*, **27**, 339–434.

Kienker, P. (1989) Equivalence of aggregated Markov models of ion channel gating. *Proc. R. Soc. Lond. Series B*, **236**, 269–309.

Kistler, J., Stround, R.M., Klymkowsky, M.W., Lalancette, R.A. and Fairclough, R.H. (1982) Structure and function of an acetylcholine receptor. *Biophys. J.*, **37**, 371–383.

Klein, S., Timmer, J. and Honerkamp, J. (1997) Analysis of multichannel patch clamp recordings by hidden Markov models. *Biometrics*, **53**, 870–884.

Kukol, A. (1997) Ion channels–analysis and simulation embedded in a flexible environment. *Computers & Chemistry*, **21**, 181–184.

Liebovitch, L.S. and Itoh, T.I. (1991) A model of ion channel kinetics using deterministic chaotic rather than stochastic process. *J. Theor. Biol.*, **148**, 243–267.

Liebovitch, L.S. and Todorov, A.T. (1996) Using fractals and nonlinear dynamics to determine the physical properties of ion channel proteins. *Crit. Rev. in Neurobiol.*, **10**, 169–187.

Manivannan, K., Ramanan, S.V., Mathias, R.T. and Brink, P.R. (1992) Multichannel recordings from membranes which contain gap junctions. *Biophys. J.*, **61**, 216–227.

Magleby, K.L. and Weiss, D.S. (1990) Estimating kinetic parameters for single channels with simulation. A general method that resolves the missed event problem and accounts for noise. *Biophys. J.*, **58**, 1126–1411.

Mendel, J.M. (1991) Tutorial on higher-order statistics (spectra) in signal processing and system theory: Theoretical results and some applications. *Proc. IEEE*, **79**, 278–305.

Merlushkin, A. and Hawkes, A.G. (1997) Stochastic behaviour of ion channels in varing conditions. *IMA J. Math. Appl. Med. Biol.*, **14**, 125–149.

Mino, H. (1993a) Estimation Problems in Superposed Markoff Jumping Processes and Their Applications to the Analysis of Membrane Current Fluctuations. PhD. Thesis, Waseda University.

Mino, H. (1993b) A parametric modeling of ionic channel current fluctuations using third-order statistics and its application to estimation of the kinetic parameters of single ionic channels. *IEEE Trans. Biomed. Eng.*, **BME-40**, 970–980.

Mino, H. (1993c) A method of analyzing non-stationary ionic channel current fluctuations in the presence of an additive measurement noise. *IEEE Trans. Biomed. Eng.*, **BME-40**, 222–229.

Neher, E. and Stevens, C.F. (1977) Conductance fluctuations and ionic pores in membranes. *Ann. Rev. Biophys. Bioeng.*, **6**, 345–381.

Neher, E. and Steinbach, J.H. (1978) Local anaesthetics transiently block currents through single acetylcholine-receptor channels. *J. Physiol. (Lond.)*, **277**, 153–176.

Nikias, C.L. and Raghuveer, M.R. (1987) Bispectrum estimation: A digital signal processing framework. *Proc. IEEE*, **75**(7), 869–891.

Ohmori, H. (1981) Unitary current through sodium channel and anomalous rectifier channel estimated from transient current in the tunicate egg. *J. Physiol. (Lond.)*, **311**, 289–305.

Pancrazio, J.J. (1995) Ion channel events simulated with the program SIMSTATE. *Comput. Meth. Prog. Biomed.*, **46** 165–174.

Papoulis, A. (1991) *Probability, Random Variables, and Stochastic Processes*. Singapore: McGraw-Hill, Third Edition.

Qin, F., Auerbach, A. and Sachs, F. (1996) Estimating single-channel kinetic parameters from idealized patch-clamp date containing missed events. *Biophys. J.*, **70**, 264–280.

Sakmann, B. and Neher, E. (1995) *Single-Channel Recording*, Second Edition. New York: Plenum.

Sigworth, F.J. (1980) The variance of sodium current fluctuations at node of Ranvier. *J. Physiol. (Lond.)*, **307**, 97–129.

Sigworth, F.J. (1981) Covariance of nonstationary sodium current fluctuations at the node of Ranvier. *Biophys. J.*, **34**, 111–133.

Sigworth, F.J. (1984) Nonstationary noise analysis of membrane current. In *Membranes, Channel, and Noise*, edited by R.S. Eisenberg, M. Frank and C.F. Stevens. New York: Plenum Press.

Takeuchi, A. and Takeuchi, N. (1959) Active phase of frog's end-plate potentials. *J. Neurophysiol.*, **22**, 203–217.

Traynelis, S.F. and Jaramillo, F. (1998) Getting the most out of noise in the central nervous system. *Trends Neurosci.*, **21**, 137–145.

Tuckwell, H.C. (1989) Stochastic Processes in the Neurosciences. SIAM: Philadelphia.

Wilks, S.S. (1962) *Mathematical Statistics*. New York: John Wiley & Sons.

CHAPTER 3
PHYSIOLOGICAL AND STATISTICAL APPROACHES TO MODELING OF SYNAPTIC RESPONSES

Dennis A. Turner, Ying Chen, Mike West and Howard V. Wheal

1. INTRODUCTION

1.1 Synaptic Function in the CNS

Communication between neurons and peripheral end-organs (such as muscle) is a critical aspect of all nervous system function, and is primarily mediated by chemical neurotransmitter signals released by one cell at synapses and detected by the next. Because of the minute and inaccessible nature of the elements involved, decoding these signals requires often complex interpretation of physiological data to surmise mechanisms of signalling. For this reason, physiological aspects of synaptic communication have been interpreted in a statistical manner, using defined frameworks established in particular at a prototype synapse, the neuromuscular junction. However the neuromuscular junction is optimized for safety, in that an excess of synaptic junctional sites is involved to reliably result in muscle contraction with a wide margin of safety (Del Castillo and Katz, 1954). Contrary to this degree of safety, most CNS excitatory synapses appear to be optimized for insecurity, with often hundreds of synapses required to be concurrently activated to lead to cell discharge. A schematic of each of these synaptic sites is shown in Figure 3.1, with an action potential in the presynaptic axon leading to probabilistic neurotransmitter release (with a certain probability of occurrence, π) and binding of this neurotransmitter to receptors at the postsynaptic site, leading to a response which has an amplitude μ. This insecurity and potential for change leads to a large capacity for processing of incoming signals to individual neurons prior to discharge. Together with mechanisms for both decay and enhancement of responses the large potential for synaptic responses to change presumably allows encoding of very complex cognitive processes, such as memory. However, since many synaptic responses are often stimulated concurrently, the interpretation of responses and their parameters at individual synapses may often require a statistical model, even though such a statistical interpretation may often be obscured by ambient noise from both the neuron and recording conditions (Redman, 1990).

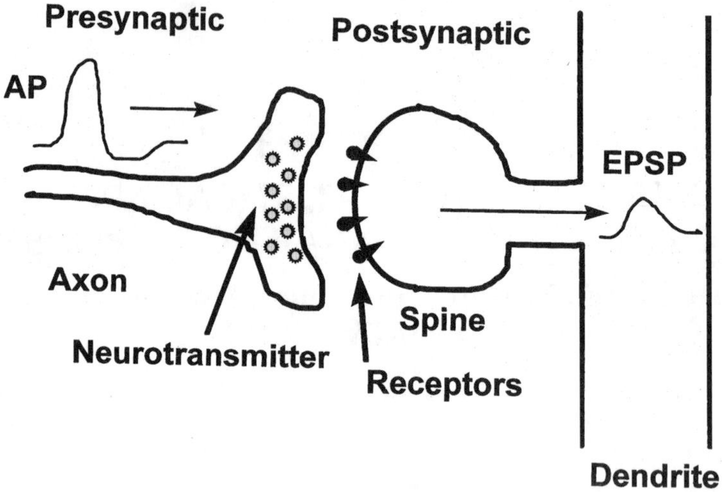

Figure 3.1 *Diagram of Synaptic Site*. This diagram shows the presynaptic axon (to the left) and the presynaptic terminal containing vesicles and neurotransmitter. The presynaptic action potential (AP) conducted along the axon leads to calcium release and vesicle fusion and release occurs, though as a stochastic function. This leads to release probability (π), which is simply the number of releases occurring in comparison to the number of presynaptic action potentials. The neurotransmitter then binds to receptors on the post-synaptic side of the synaptic cleft (usually a dendritic spine in the hippocampus, which is attached to a dendrite), resulting in ionic channel opening and a postsynaptic current or voltage (EPSP) being generated in the dendrite, with peak amplitude μ. On any given neuron there may be multiple such release sites (k), acting in concert, but likely with different parameters. The EPSP signals are summated in time and space from these multiple synapses to determine the postsynaptic activity of the neuron.

Chemical synapses involve common release mechanisms, though the specific parameters of release, type of neurotransmitter elaborated and pre- and post-synaptic receptors involved differ considerably. Such synapses are highly probabilistic with responses from a few synapses showing considerable fluctuation from trial to trial (Bekkers et al., 1990; Bennett, 1995; Del Castillo and Katz, 1954; Faber et al., 1992; Harris and Sultan, 1995; Jack et al., 1994; Korn and Faber, 1991; Redman, 1990; Voronin, 1993). Statistical models of synaptic release mechanisms have been developed to explain the probabilistic release of neurotransmitter and to infer critical changes at individual synaptic release sites when recordings are usually performed from multiple sites at once (Del Castillo and Katz, 1954; Zucker, 1973). These models have been adapted to synapses in the hippocampus for understanding of development, neural plasticity and potentiation. However, many of the basic assumptions of such quantal models have not been tested in CNS neurons. These assumptions include near uniformity of synaptic release amplitudes between sites, and a modest degree of variation in release amplitudes both between sites and also over time at a single site, termed quantal variance. Recently, considerable interest has arisen regarding both heterogeneity between synaptic sites and the degree of intrinsic synaptic amplitude and probability variation over time (Bolshakov and Siegelbaum, 1995; Borst et al., 1994; Brown et al., 1976; Cao and West, 1997; Clements, 1996; Hessler et al., 1993; Rosenmund et al., 1993; Turner et al., 1997; Walmsley et al., 1988). Such heterogeneity requires more complex statistical models to interpret parameter changes at individual synapses than either a simple or compound binomial model allows, yet in many instances data has been analyzed blindly or without regards to such complicating features. One simplifying approach has been to identify or presume a single or unitary connection between cells whenever possible, and using either the rate of failures or the average success (potency) to identify critical changes in release parameter values (Bolshakov and Siegelbaum, 1995; Debanne et al., 1996; Isaac et al., 1995; Liu and Tsien, 1995; Malinow, 1991; Stricker et al., 1996a,b). Thus, most physiological data require a statistical framework in which to interpret the synaptic release parameters, but the framework should be as open and as well defined as possible. Once a framework is decided, critical physiological questions to be addressed include developmental changes in synapses, plasticity and potentiation mechanisms and how synapses are modified on both short-term and long-term time scales.

1.2 Synaptic Release Statistics and Site Heterogeneity

The response to neurotransmitter release in the CNS may be different both between synaptic sites and over time at the same synaptic site because of variable packaging of neurotransmitter, neurotransmitter dynamics in the synaptic cleft, agonist–receptor interactions at the synapse, channel opening properties and postsynaptic modification of receptor proteins (Bekkers et al., 1990; Bennett, 1995; Borst et al., 1994; Buhl et al., 1994; Clements, 1996; Dityatev and Clamann, 1993; Faber et al., 1992; Harris and Sultan, 1995; Hessler et al., 1993; Jack et al., 1994; Korn and Faber, 1991; Lisman and Harris, 1993; Redman, 1990; Rosenmund et al., 1993). Responses may also vary between sites because dendritic location and voltage-dependent dendritic properties and inhibition strongly influence the shape of synaptic events during conduction to summation sites (Magee and Johnston, 1997; Major, 1993; Spruston et al., 1993; Turner, 1984). There are numerous reasons to suspect heterogeneity between synaptic events originating from multiple sites on neurons, and some data confirm this possibility (Bekkers et al., 1990; Bolshakov and Siegelbaum, 1995; Borst et al., 1994; Buhl et al., 1994; Harris and Sultan, 1995; Hessler et al., 1993; Rosenmund et al., 1993; Turner et al., 1997). Two important sources of variability are termed intrasite (or intrinsic) and intersite variability, together often lumped into 'quantal' variance.

There are many sources of variability inherent in synaptic activation. Release is dependent on calcium entry into the presynaptic terminal, often resulting in failures of release. Most neurotransmitters require some uptake mechanism for recycling, as well as a synthetic cycle to redistribute neurotransmitter, resulting in variable neurotransmitter packaging in synaptic vesicles. Once a vesicle is primed and docked for potential release, the duration of docking and amount of neurotransmitter released per evoked action potential is unknown, particularly whether multiple cycles of release can occur from a single vesicle. Once neurotransmitter is released into the synaptic cleft, a receptor–ligand mismatch may result in a variable postsynaptic response from trial to trial by stochastic diffusion and binding (Faber et al., 1992). Depending on the postsynaptic density of ligand-gated receptors and the type of receptor a variable postsynaptic response at the synaptic site may be elicited. The typical postsynaptic response may be further altered by variable dendritic conduction pathways to the soma, and by inhibition or enhancing mechanisms on the conduction path. The large chain of events from presynaptic action potential to postsynaptic response at a soma or summating output site includes many sources of variability. Further, all finite samples of data are subject to sampling variability as well as processes of measurement error, which complicate interpretations.

Quantitative parameters describing synaptic release sites include the number of activated sites on any one trial (k), probability of release (π), and amplitude (μ) and electrotonic conduction from the site of origin to the summation site. The goal of 'quantal' analysis is to infer these parameters at single synapses but usually in the physiological context where a number of release sites may be activated concurrently (Korn and Faber, 1991; Redman, 1990). However, a high level of background synaptic and recording noise often occurs during recordings, obscuring much detail and so requiring statistical inference to deduce the content of the original signal (Wahl et al., 1997). Though a simple quantal–binomial model may hold for inhibitory synapses onto the Mauthner cell (Korn and Faber, 1991) a compound [or complex] binomial distribution has been suggested to be more applicable in the mammalian hippocampus, with varying probability of release [and likely varying amplitude] between synaptic sites required (Hessler et al., 1993; Jack et al., 1994; Rosenmund et al., 1993; Sayer et al., 1990; Stricker and Redman, 1994; Turner et al., 1997). In hippocampal CA1 neurons only a small subset of data ensembles have exhibited clear 'quantal' behavior (Kullmann, 1989; Kullmann and Nicoll, 1992; Larkman et al., 1991; Liao et al., 1992; Malinow, 1991; Manabe and Nicoll, 1994; Manabe et al., 1993). Recent studies of small EPSCs (Bolshakov and Siegelbaum, 1995; Turner et al., 1997) have suggested considerable heterogeneity over time at synaptic sites and between synapses. Thus, techniques to analyze parameters at individual synapses have been limited by adapted statistical frameworks and newer models, including maximum likelihood estimation (MLE; Korn and Faber, 1991; Kullmann, 1989; Kullmann and Nicoll, 1992; Stricker et al., 1996a,b; Stricker and Redman, 1994; Stricker et al., 1994; Walmsley et al., 1988) and Bayesian component (Cao and West, 1997; Escobar and West, 1995; Turner and West, 1993; West and Turner, 1994) and site-directed models (Turner et al., 1997; West, 1997) may provide improved analysis capabilities.

1.3 Binomial Models

Statistical models at the neuromuscular junction centered on binomial models, since it was assumed that vesicles emptied near-identical quantities of neurotransmitter, multiple vesicles were released with each impulse and each vesicle possessed the same likelihood of release (Bennett, 1995; Del Castillo and Katz, 1954; Katz, 1969; Quastel, 1997). With pharmacological manipulation the number of vesicles released could be significantly reduced and the limit of the binomial distribution, a Poisson distribution, could be reached. With the Poisson distribution the binomial parameters n (number) and p (probability) are not separately defined but are merged into a single value, m, applicable only when p is very small compared to n (Brown et al., 1976; Martin, 1966). The adaptation of this framework to central neurons proved to be difficult because of the much higher noise level associated with recording using sharp electrodes, requiring techniques to remove the effects of the noise (Ling and Tolhurst, 1983; Redman, 1990; Wong and Redman, 1980). The MLE approaches were developed to optimize searches for multiple parameters in light of this contaminating noise, using a binomial statistical model extended to a compound binomial allowing differences in probability of release (Bennett, 1995; Kullmann, 1989; Walmsley et al., 1988). Recent MLE approaches have also added estimation of quantal variance and comparison to non-binomial models, though are based on a smoothed function approximation to the original data histogram (Stricker and Redman, 1994; Stricker et al., 1994). Other approaches have included histogram-based smoothing and 'peakiness' functions, using the histogram to identify summation of site responses (termed components) rather than analyzing the primary data (Edwards et al., 1990; Jack et al., 1994; Jonas et al., 1993; Larkman et al., 1991). This approach is severely limited by how the histogram is constructed (bin width, etc.) and very similar data may show highly different histogram appearance randomly (Walmsley, 1995). Thus, approaches based on analyzing the exact, original data are preferred over superficial histogram appearance.

Guidelines for suitability of physiological data sets have also been established, based on simulation approaches (Redman, 1990; Stricker and Redman, 1994; Wong and Redman, 1980). Minimal samples sizes are in the range of 500–750 data points, each from an evoked synaptic response, and a signal to noise ratio of at least 3 to 1 is required for optimal extraction of component locations. Stationarity of the degree of fluctuation of the signal is also required since if the basic synaptic parameters vary a reliable analysis cannot be performed (Brown et al., 1976; Redman, 1990; Turner et al., 1995). Stationarity can usually be assessed by trends of the entire data set or analysis of successive subgroups (Turner et al., 1997). The noise itself can be approximated by either a single or dual Gaussian normal function (σ_{noise}), rather than using the raw data histogram. The results of the MLE analysis are limited in that this particular algorithm often cannot identify a local minimum for parameters and may reach different conclusions based on the starting point used for the initial parameters (Kullmann, 1989; Stricker and Redman, 1994). Thus, multiple initial parameters must be tried and whether the final result is 'optimal' or 'good' in some absolute sense cannot usually be ascertained. Recently, Stricker and Redman (1994) have added a bootstrap technique for data resampling which allows construction of error ranges, and thus some statistical sense of the approximate parameter location. However, this type of solution does not offer a feel for the likelihood that the solution will identify the underlying statistical distribution from which the individual data set has been drawn. Identifying the underlying distribution is a much more difficult task, particularly with a single finite sample, since random samples from any distribution of a finite size may be very unrepresentative in terms of their histogram appearance (Walmsley, 1995). An additional limitation of the current MLE methods is that variability between site amplitudes cannot be detected, even though variability in site probability is allowed with a compound binomial distribution. Together, these limitations severely curtail the number of data sets which can be analyzed to a small fraction of those which are derived from experiments (Kullmann and Nicoll, 1992).

1.4 Bayesian Statistical Models

Approaches using Bayesian site statistical models provide for direct estimation (in terms of full descriptions of uncertainties through posterior probabilities and point estimates) of the number of synaptic sites and their release parameters (Cao and West, 1997; Escobar and West, 1995; Turner and West, 1993; West, 1997; West and Turner, 1994). By comparison with MLE and other pure likelihood-based approaches, Bayesian inference aims to completely explore the regions of parameter spaces supported and indicated by the data, and formally summarize relevant parameter ranges and sets using summary probability distributions rather than just point estimates and error bars. Current standard techniques involve repeated simulation of parameter values jointly from such posterior distributions. Notice that this subsumes and formally manages problems of local maxima in likelihood surfaces, and other such technical problems that often invalidate MLE approaches; the Bayesian strategy is to more thoroughly map out the parameter surfaces via informed navigation of the surface in regions identified as relevant probabilistically. Bayesian site analysis was developed in order to directly represent the underlying synaptic structure in terms of parameters and unobserved/latent variables that describe individual synaptic sites and experimental characteristics (Turner et al., 1997; West, 1997). The resulting general mixture model is an extension of a compound binomial model, but in which the number of sites and the amplitudes and probabilities of the sites are all free variables. An initial statistical outline appears in (West, 1997), though our work has since developed to address more fully estimation of intrinsic variation over time (Turner et al., 1997). We define this approach further below with specific details of both simulations and experimental data sets. Statistical inference is inherent to the Bayesian approach and all parameters are given with appropriate error ranges, including the number of synaptic release sites.

1.5 Developmental and Post-lesion Changes in Synaptic Release Parameters

Systematic changes in the glutamate receptor phenotype (from NMDA to non-NMDA), probability of release and degree of potentiation have been suggested to occur in the hippocampus during development (Bolshakov and Siegelbaum, 1995; Durand et al., 1996). After kainic acid lesions of the hippocampus and denervation of Schaffer collaterals onto CA1 neurons a profound change in the synapses onto these neurons occur, with a significant and unexpected NMDA component (Turner et al., 1995). These post-denervation changes in some ways mirror a developmental re-institution of a large NMDA component, with the likelihood of a later maturation (within weeks of the lesion) of the synaptic glutamate phenotype into a more mature non-NMDA receptor phenotype. This change in receptor phenotype has also been observed at 'new' synapses induced with long-term potentiation (LTP), by observing normally 'silent' NMDA synapses (at resting potential). These 'silent' synapses can be transformed into active non-NMDA synapses by a combination of postsynaptic depolarization and synaptic activation, presumably with receptor modification induced by the calcium influx associated with the NMDA receptor activation (Isaac et al., 1995; Liao et al., 1992). The overall implication of this induction of 'new' synapses is an increased efficacy of synaptic function, but the specific parameters underlying this enhancement remain unresolved. Whether developmental processes continue over the lifespan of an individual remains unknown, particularly whether aging is a continuation of developmental processes.

Other critical physiological situations in which synaptic release parameters may point to mechanisms include short- and long-term plasticity (Andreasen and Hablitz, 1994; Chen et al., 1996; Debanne et al., 1996; Kullmann and Nicoll, 1992; Raastad et al., 1992; Stevens and Wang, 1994; Stricker et al., 1996b; Voronin, 1993). Short-term facilitation is considered to involve primarily changes in presynaptic factors, particularly probability of release, since these changes occur over a few milliseconds (Redman, 1990; Turner et al., 1997). However, long-term potentiation may exhibit many different mechanisms of plasticity, particularly structural changes if the duration is longer than 10–30 minutes. Due to the popularity of the study of long-term potentiation and the wide variety of animal ages and preparations in which this phenomenon is studied, multiple and controversial mechanisms have been suggested (Bolshakov and Siegelbaum, 1995; Debanne et al., 1996; Durand et al., 1996; Malinow, 1991; Stevens and Wang, 1994; Stricker et al., 1996b). Since the evoked response histograms change radically before and after the induction of long-term potentiation various interpretations have included both enhanced probability of release and also increased amplitude of synaptic site responses. However, few of these physiological studies have shown detailed statistical evaluation of the physiological signals which allow for extensive variation between synaptic release sites, other than a compound binomial approach (Kullmann and Nicoll, 1992; Stricker et al., 1996b). Thus, current statistical approaches in the evaluation of long-term potentiation may be premature in that variation between synaptic release sites (and the possible addition of release sites with potentiation) have been insufficiently studied.

1.6 Physiological Systems and Data Requirements

Korn and Faber (1991) and Redman (1990) have previously defined the appropriate physiological circumstances for a proper (and reliable) analysis of underlying synaptic response parameters. These requirements focus on two cell recordings (either using sharp or patch electrodes) where clear, unitary (but usually multi-site) responses can be identified and include:

1. the stimulation of a single presynaptic spike and control of the presynaptic element;
2. optimization of noise level to isolate miniature events to compare to evoked events, with miniature events occurring from same synaptic sites as those from which evoked events emanate;
3. resolution of single site amplitude, with minimum cable or electrotonic filtering (usually requiring a perisomatic or proximal dendritic synaptic location);
4. direct data on variations in quantal amplitude at a single site;
5. morphological identification of number of active sites involved and their dendritic location;
6. independent evidence on uniformity of amplitude and probability between sites.

These fairly rigid criteria are primarily designed for two cell recording and in circumstances where both elements can be stained and then anatomically analyzed. Korn and Faber have extensively performed this type of analysis on goldfish Mauthner cells, which have a glycinergic inhibitory synapse which fulfills these criteria (Korn and Faber, 1991). However, in most other CNS cells, particularly in mammalian systems, unitary recordings are much more difficult to obtain, particularly with evaluation of both physiological responses and the anatomical details of the number and location of synaptic sites between two neurons. Recently, however, the hippocampal slice culture has shown considerable promise, with Debanne et al. (1996) showing dual recordings between CA3 and CA1 neurons with a small number of connections and interesting potentiation properties (Debanne et al., 1996). This same connection in acute *in vitro* slices has shown much less promise due to the small percentage of CA1 and CA3 cells showing excitatory synaptic connections (Malinow, 1991; Redman, 1990; Sayer et al., 1990; Stricker et al., 1996a). Other situations which have been explored include Ia afferent synapses onto motoneurons in the spinal cord (Redman, 1990) and isolation of axonal afferents onto CA1 pyramidal cells using microstimulation techniques (Bolshakov and Siegelbaum, 1995; Manabe and Nicoll, 1994; Turner et al., 1997). All of these excitatory synaptic types, however, show primarily dendritic synapses, which are at some electrotonic distance from the soma and the recording electrode and generally higher noise levels are prevalent, obscuring individual responses. Since there are thousands (25,000) of synapses onto these cells there usually cannot be a rigorous comparison of evoked and spontaneous synaptic currents since the relationship between the origin of the two separate events is unknown.

Skepticism about the value of statistical inference of synaptic responses is widespread amongst physiologists because of the difficulty in assessing whether a solution is either unique or 'good' in an absolute sense and the indirect nature of the analysis, with numerous assumptions often required and the lack of validation of the statistical analysis schemes. On this basis other aspects of synaptic responses have been used, for example in a pharmacological sense, to assess where differences have arisen after a perturbation (Clements, 1990). These techniques include primarily an assessment of coefficient of variation, which is the standard deviation of the ensemble of amplitudes divided by the mean amplitude (Manabe et al., 1993). However, this method provides only an indirect test on the mechanism of changes, for example pre- or postsynaptic location, such as with paired-pulse plasticity and alterations in calcium or magnesium. However, regardless if one wishes to proceed with a formal statistical analysis or instead concentrates on the purely physiological methods, the quality of the data, as stressed by Korn and Faber, is critical and the preparation must be optimized in a number of senses.

All of the statistical techniques require a certain number of evoked response samples to be obtained to be able to perform the analysis satisfactorily. For the initial least mean squares optimization technique employed by Wong and Redman (1980), the minimal critical sample size was approximately 750, assuming a signal to noise ratio of 3 to 1 or better. These samples need to be obtained at a rate which is stable for the preparation and which does not in itself lead to plasticity (either depression or facilitation), which varies from 2 Hz for motoneurons to 0.5–0.1 Hz for many hippocampal preparations. The duration of acquiring such a sample, which is presumably not changing during the acquisition and fulfills the requirements for stationarity, is extensive (up to 40 min at a rate of 0.2 Hz). This property of the ensemble of stability of the fluctuations during the period recorded is termed stationarity, and it implies both the absence of a trend and that the fluctuations during the period are not changing in any essential manner. Stationarity can be partly assessed by a trend plot and also partly by analyzing sequential subgroups of approximately 50 for net significant changes (Turner et al., 1997). A Bayesian time series analysis and formal trend analysis has also been developed to statistically assess stationarity (West and Harrison, 1997).

Other physiological characteristics which are desirable include low noise, particularly achievable with whole cell patch electrodes, where a noise standard deviation less than 1 pA may be achievable, giving a signal to noise ratio of 4 : 1 or better for many excitatory synaptic ensembles. As has been shown in voltage clamp simulations studies, most dendritic synaptic responses show minimal voltage clamp, and there be less error in using current clamp with whole cell patch electrodes (there is no advantage of using voltage clamp) (Major, 1993; Spruston et al., 1993). Having full physiological control of a presynaptic cell or axon is critical, so responses in the postsynaptic cell may be clearly linked to the activity of the presynaptic response, which occurs either spontaneously or may be triggered (by intracellular current or axonal stimulation). As few of synapses as possible should be stimulated, for many of the techniques are limited to less than 8–10 sites due to identifiability of the sites. Other physiological methods have been used to isolate as few of sites as possible, particularly postsynaptic agonists or changes in calcium/magnesium levels to decrease the probability of release (higher magnesium or lower calcium) (Manabe et al., 1993). We have detailed the conditions important for capture and analysis of EPSC data in Turner et al. (1997). Briefly, these conditions included limitation of the synaptic signals to a single glutamate receptor type (non-NMDA), careful activation of a as few of synapses as possible through microstimulation and restriction of the analysis to carefully defined data sets in terms of noise and stationarity.

1.7 Goals of This Chapter

The goal of this Chapter is to discuss both the available statistical models and their degree of applicability to CNS synapses as typically studied, and the physiological framework in which models may be critically assessed. Though a full review of all models used in recent physiological studies is beyond the present scope a pertinent inclusion of a few critical models is crucial to understand current approaches. The types of physiological data requirements to fully derive specific and meaningful statistical inference are also discussed as well as our own laboratory's individual approach to understanding mechanisms of plasticity in CNS neurons. This Chapter of both methods and results with various approaches will then be discussed in terms of present shortcomings in terms of the models and physiological preparations and how additional optimization and mechanisms may be explored. Clearly, the quality of the data is paramount, as stressed by Korn and Faber (1991), as an initial substrate for any statistical modelling application.

2. MODELS FOR SITE SPECIFIC RELEASE CHARACTERISTICS

2.1 Neural Signal and Noise Models

Ensembles of maximum synaptic signal levels (specifically excitatory postsynaptic potentials [EPSPs], excitatory postsynaptic currents [EPSCs], inhibitory postsynaptic currents [IPSCs], etc.) obtained in response to stimulation are assumed to be independently generated according to an underlying stochastic, synaptic site-specific mechanism described as follows. At the identified synaptic connection, the experiment isolates either both the pre- and postsynaptic neurons for complete control of all interactions between the two cells (as shown diagrammatically in Figure 3.1), or a region on the axon branch of a presynaptic neuron leading to activation of only a small number of postsynaptic sites. Stimulation applied to the isolated presynaptic element leads to chemical neurotransmitter release at a random number of available synaptic transmission sites, similar to those shown in Figure 3.1. If release occurs, then each individual site contributes a small electrical signal, which then sums across sites to induce the overall resulting electrical signal in the postsynaptic neuron, which is measured physiologically, though usually in concert with interfering noise. Each of the $s>0$ release sites transmits to a site-specific maximum level independently of all other sites. There is no identification of individual sites (e.g. such as might arise were we able to physically mark sites in the tissue and identify the location of individual transmissions) so we arbitrarily label the sites $1, \ldots, s$. Site by site we assume that, on any trial:

- sites 'release' independently, with individual, site-specific chances of release occurring on any occasion;
- a transmitting site produces a site-specific, fixed packet or 'quantum' of neurotransmitter leading to a characteristic amplitude response at the synaptic site;
- recorded maximum synaptic signal levels represent the sums of signals induced by the various sites randomly transmitting, with additive synaptic and experimental noise.

In any experiment we assume individual transmitter release probabilities and transmission levels to be fixed and with only slight perturbations over time, but no net trend. Changes in these characteristics over time due to forced experimental and environmental changes is of later interest, and evaluating such changes is one of the guiding motivations in developing these models. Symbolically, we represent recorded signal levels as $y=\{y_1,\dots,y_n\}$ where y_i is the level on trial i; the sample size n is typically a few hundreds. Given s sites assumed, site j transmits on any trial with probability π_j, independently of other sites and independently across trials. The level of transmission on occurrence is the site-specific quantity μ_j. Physically, signal levels are measured via either sharp intracellular probes or low-noise patch electrodes as electrical signals induced on cell bodies, and responses are in units of either current (pA) or millivolts (mV). By convention, we change the reading to predominantly positive values so that the μ_j must be non-negative values. The recordings are subject to background synaptic and experimental noise, including both electronic distortions in recording the cellular potential and errors induced in computing the estimated maximum potential levels. These noise sources combine to produce additive errors, assumed to be approximately normally distributed independently across trials, as discussed in the introduction; write v for the variance of these normal errors. Concomitant noise measurements are available to assess these experimental errors and to provide relevant prior information relevant to v; blank signals measured without experimental stimulus provide these measurements. It is stressed that the noise affecting signal recordings has the same distribution, of variance v, as the raw noise recordings obtained without a signal, there being no scientific or experimental reason to assume otherwise. We also model a systematic bias in the signal recordings; although the associated noise measures are typically close to zero mean, the signal recordings are often subject to a small but non-negligible baseline offset induced by the experimental procedures; call this shift m. We then have the following basic signal model.

For $i=1,\dots,n$, the signal observations are conditionally independent,

$$y_i \sim N(y_i|\theta_i, v) \quad \text{with} \quad \theta_i = m + \sum_{j=1}^{s} z_{ij}\mu_j$$

and where the z_{ij} are latent, unobserved, site transmission indicators,

$$z_{ij} = \begin{cases} 1, & \text{if site } j \text{ transmits on trial } i; \\ 0, & \text{otherwise.} \end{cases}$$

Under the model assumptions, these indicators are conditionally independent Bernoulli quantities with $P(z_{ij}=1)=\pi_j$ for each site j and all trials i.

Thus signal data are drawn from a discrete mixture of (at most) $k=2^s$ normal components of common variance, induced by averaging normals determined by all possible combinations of the z_{ij} and weighted by the corresponding chances of those combinations. On any trial, the normal component selected is determined by the column s-vector $z_i=(z_{i1},\dots,z_{is})'$ realized on that trial by individual sites releasing or not; the selected component has mean $\theta_i=m+z_i'\mu$ where $\mu'=(\mu_1,\dots,\mu_s)$, and is selected with chance $\prod_{j=1}^{s}=\pi_j^{z_{ij}}(1-\pi_j)^{1-z_{ij}}$; specifically, and conditional on all model assumptions and parameters, the y_i are conditionally independently drawn from the distribution with density

$$p(y_i|\mu,\pi,m,v)=\sum_{z_i} p(z_i|\pi)\,p(y_i|\mu,z_i,m,v)=\sum_{z_i}\left\{\prod_{j=1}^{s}\pi_j^{z_{ij}}(1-\pi_j)^{1-z_{ij}}\right\}N(y_i|m+z_i'\mu,v),\ (3.1)$$

where the s-vector z_i ranges over all 2^s possible values. If there are common values among the site levels μ_j, then the mixture distribution function will be equivalent to one with fewer than 2^s components. Also, if either $\mu_j=0$ or $\pi_j=0$ then site j disappears from the model and the distribution reduces to a mixture of at most 2^{s-1} components. Additional cases of special interest include:

(a) So-called compound binomial models, in which the site response levels are equal, all at a basic 'quantum' level μ_0, but release probabilities differ. The aggregate response levels then run between m and $m+s\mu_0$ and the mixture of 2^s normal components effectively reduces to a mixture of just $s+1$, with associated weights determined by the discrete compound binomial resulting from the distinct chances $\pi_1,\ldots,\pi_s$ across sites. These kinds of special cases have received considerable attention in the literature (Walmsley et al., 1988; Kullmann, 1989).
(b) Precise quantal–binomial models, in which the site levels are equal and the release probabilities are constant too, $\pi_j=\pi_0$. Now the mixture reduces to one with distinct normal means $m+j\mu_0$ $(j=0,\ldots,s)$, and corresponding binomial weights $\binom{s}{j}\pi_0^j(1-\pi_0)^{s-j}$ (Martin, 1966; Redman, 1990).

Our previous work, following others, modelled synaptic data in the framework of 'standard' mixture models for density estimation, from a particular Bayesian viewpoint; see West and Turner (1994), for example, and Escobar and West (1995) for statistical background. In these models, the component means and weights are essentially unrestricted, rather than being modelled directly as functions of the underlying synaptic parameters μ and π. A significant drawback is that it is then, in general, very difficult to translate posterior inferences about unrestricted normal mixture model parameters to the underlying μ and π parameters of scientific interest, especially in the context of uncertainty about s. One simple example in West and Turner (1994) shows how this can be done; that is a rare example in which the data appear to be consistent with the full quantal hypothesis, so that the convoluted process of backtracking from the mixture model to the underlying site-specific parameters is accessible. However, we have encountered very few data sets in which this is the case, and this inversion process is generally difficult. Hence the direct approach, inferring the neural parameters directly, has been developed.

2.2 Prior Distributions for Synaptic Parameters

Model completion requires specification of classes of prior distributions for the determining parameters μ, π, m and v for any given s. Assessment of reasonable values of s, as well as values of the μ and π quantities for any given s, is part of the statistical inference problem. From a technical viewpoint, uncertainty about s can be formally included in the prior distribution, so as to provide posterior assessment of plausible values. The models below allow this. From the viewpoint of scientific interpretation, however, inference about the site parameters are best made conditional on posited values of s, and then issues of sensitivity to the number of sites explored.

Begin by conditioning on a supposed number of sites s. Current implementation assumes priors for all model parameters which, though rather intricately structured in certain dimensions to represent key qualitative features of the scientific context, are nevertheless inherently 'uniform' in appropriate ways, providing 'reference' initial distributions. Note that other classes of priors may be used, and some obvious variations are mentioned below; however, those used here are believed to be appropriately vague or uninformative in order that resulting posterior distributions provide benchmark or reference inferences that may be directly compared with supposedly objective non-Bayesian approaches (e.g. Kullman, 1989; Redman, 1990; comparative analyses will be reported elsewhere in a further article).

In addressing uncertainty about s alone, we make the observation that models with fewer than s sites are implicitly nested within a model having s sites. To see this, constrain some $r > 0$ (arbitrarily labelled) site levels μ_j to be zero; then (1) reduces to precisely the same form based on $s-r$ sites, whatever the value of the π_j corresponding to the zeroed μ_j. Zeroing some μ_j simply confounds those sites with the noise component in the mixture—a site transmitting a zero level, with any probability, simply cannot be identified. Hence assessing whether or not one or more of the site levels are zero provides assessment of whether or not the data support fewer sites than assumed. This provides a natural approach to inference on the number of active sites, assuming that the model value s is chosen as an upper bound.

Note that a similar conclusion arises by considering $\pi_j = 0$ for some indices j; i.e. an inactive site may have a zero release probability rather than (or as well as) a zero release level. However, the zeroing of a site probability induces a degeneracy in structure of the model and obviates its use as a technical device for inducing a nesting of models with fewer than s sites in the overall model. Hence our models define inactive sites through zeros among the μ_j, restricting the π_j to non-zero values, however small. Thus $\mu_j = 0$ for one or more sites j is the one and only way that the number of active sites may be smaller than the specific s.

2.2.1 *General Structure*

For a specified s, analyses reported below are based on priors with the following structure. Quantities μ, π, (m, v) are mutually independent. We will describe classes of marginal priors for μ and π separately; each will involve certain hyper-parameters that will themselves be subject to uncertainty described through hyper-priors, in a typical hierarchical modelling framework, and these hyper-parameters will also be assumed mutually independent. To anticipate development below, hyper-parameters denoted q and a are associated with the prior for μ, a single quantity b determines the prior for π, and the joint prior is then of the form

$$p(\mu, \pi, m, v, q, a, b) = p(\mu, \pi, m, v|q, a, b)p(q, a, b)$$
$$= p(\mu|q, a)p(\pi|b)p(m, v)p(q)p(a)p(b). \tag{3.2}$$

The component densities here are now described in detail. Two comments on notation: first, conditioning statements in density functions include only those quantities that are required to determine the density, implicitly indicating conditional independence of omitted quantities; second, for any vector of h quantities $x = (x_1, \ldots, x_h)$, for any $j \leqslant h$ the notation x_{-j} represents the vector x with x_j removed, i.e. $x_{-j} = x - \{x_j\} = (x_1, \ldots, x_{j-1}, x_{j+1}, \ldots, x_h)$.

2.2.2 *Priors for μ and Associated Hyper-parameters q, a*

We develop a general classes of priors for μ, and comment on various special cases. The class has the following features:

(a) a component baseline uniform distribution for each μ_j over a prespecified range $(0, u)$, with a specified upper bound u;
(b) components introducing positive prior probabilities at zero for each of the μ_j in order to permit assessment of hypotheses that fewer than the chosen (upper bound) s are actually non-zero, and hence to infer values of the number of active sites;
(c) components permitting exact common values among the elements of μ to allow for the various special cases of quantal transmission, and specifically the questions of whether or not pairs or subsets of sites share essentially the same 'quantal' transmission level.

These are developed as follows. Let $F(\cdot)$ be distribution on $(0, u)$, having density $f(\cdot)$, for some specified upper bound u. Write $\delta_0(x)$ for the Dirac delta function at $x = 0$, and $U(\cdot|a, b)$ for the continuous uniform density over (a, b). Then suppose the μ_j are *conditionally independently* drawn from the model

$$(\mu_j|F, q) \sim q\delta_0(\mu_j) + (1-q)f(\mu_j)$$

for some probability q. Under this prior, the number h of non-zero values among the μ_j—the number of active sites—is binomial, $(h|s, q) \sim Bn(s, 1-q)$, with mean $s(1-q)$, independently of F. If F were uniform, say $U(\cdot|0, u)$, this prior neatly embodies the first two desirable features (a) and (b) described above. Note the two distinct cases:

- Setting $q = 0$ implies that $h = s$ is the assumed number of active sites, so then inference proceeds conditional on $\mu_j > 0$ for each j.
- Otherwise, restricting to $q > 0$ allows for assessment of the number of active sites, subject to the specified upper bound s. In practice we will assign a hyperprior to q in the case $q > 0$. The class of beta distributions is conditionally conjugate, and the uniform prior suggests itself as a reference, $q \sim U(q|0, 1)$. One immediate, and nice, consequence of a uniform prior is that the resulting prior for the number of non-zero values among the μ_j has (averaging the binomial with respect to q) a

discrete uniform prior over $0, 1, \ldots, s$. This is a suitably vague and unbiased initial viewpoint with respect to the number of active sites. Other beta priors may be explored, of course. One specific choice we use in current work is $(q|s) \sim Be(s-1, 1)$; note the explicit recognition of dependence on the specified values of s. The reasoning behind this choice is as follows. First, we are currently focused on experiments designed to isolate rather small numbers of sites, down to just a few, say 1–4 from the viewpoint of scientific intent and expectation. So s values up to 7 or 8 may be explored, but lower values are typical. Whatever value of s is chosen, h is expected to be in the low integers, so guiding choice of the prior for q to induce a prior for h favoring smaller values. Consider values of s in the relevant range $3 \leqslant s \leqslant 8$, or so. Then, integrating $p(h|s, q)$ with respect to the specific prior $(q|s) \sim Be(s-1, 1)$ we obtain a distribution $p(h|s)$ that is almost completely insensitive to s, having a diffuse and decreasing form as h increases, and with $E(h|s) = 1$ for any such s. Thus, this specific beta prior for q has the attractive feature of consistency with a scientifically plausible prior $p(h|s) \approx p(h)$, incorporating the scientific view of likely small numbers of active sites, and almost independently of the upper bound s specified.

This structure provides a baseline uniform prior for release levels, together with the option for allowing a smaller number of sites than the s specified. So far, however, there is no explicit recognition of the special status, in the scientific area, of quantal hypotheses as represented through common values among the μ_j. If $F(\cdot)$ is a continuous distribution, then the prior implies that the non-zero μ_j are distinct. Though this might allow arbitrarily close μ_j values, it is desirable to have the opportunity to directly assess questions about common values, and perhaps subgroups of common values, in terms of posterior probabilities. There is also a significant technical reason, discussed below in connection with issues of parameter identification, that calls for a prior that gives positive probability to exact equality of collections of the non-zero μ_j. We therefore extend the prior structure so far discussed to provide this. We do this using a standard Dirichlet model. Specifically, a Dirichlet process prior for F induces a discrete structure which gives positive prior probability to essentially arbitrary groupings of the set of non-zero μ_j into subsets of common values; this is a general framework that permits varying degrees of partial quantal structure, from the one extreme of completely distinct values to the other of one common value. This structure is most easily appreciated through the resulting set of complete conditional posterior distributions for each of the μ_j given μ_{-j}. These are defined by

$$p(\mu_j|\mu_{-j}, q, a) = q\delta_0(\mu_j) + (1-q)\{r_j U(\mu_j|0, u) + (1-r_j)h_j^{-1} \sum_{iN_j} \delta_{\mu i}(\mu_j)\}, \tag{3.3}$$

where h_j is the number of non-zero elements of μ_{-j} and N_j is the corresponding set of indices, $N_j = \{i|\mu_i > 0, i = 1, \ldots, s; i \neq j\}$, and $r_j = a/(a + h_j)$. The hyper-parameter a is subject to uncertainty and is included in the analysis using existing approaches for inference on precision parameters in Dirichlet models, developed in West (1997) and illustrated in Escobar and West (1995). As shown there, gamma priors for a, or mixtures of gamma priors, are natural choices, and our application currently uses diffuse gammas models.

In summary, (3.3) shows explicitly how site level μ_j may be zero, implying an inactive site, or take a new non-zero value, or be equal to one of the non-zero values of other sites. The roles of hyperparameters q and a are evident in this equation. With this structure we have defined prior components $p(\mu|q,a)p(q)p(a)$ of the full joint prior in (3.2).

2.2.3 *Priors for* π *and Associated Hyperparameter* b

The structure of the prior for π parallels, in part, that of μ in allowing common values. The detailed development for μ can be followed through with the same reasoning about a baseline prior and a structure to induce positive probabilities over subsets of common values. We restrict to positive release probabilities, as discussed earlier, so that the prior structure for the π_j is simpler in this respect. Specifically, we assume the π_j independently drawn from a distribution on $(0,1)$ that is assigned a Dirichlet process prior. We take Dirichlet precision $b>0$ and base measure to be $bU(\cdot|0,1)$. Then, as in the development for μ the full joint prior $p(\pi|b)$ is defined by its conditionals

$$(\pi_j|\pi_{-j},b) \sim wU(\pi_j|0,1)+(1-w)(s-1)^{-1}\sum_{i=1,i\neq j}^{s}\delta_{\pi_i}(\pi_j) \qquad (3.4)$$

where $w=b/(b+s-1)$, for each $j=1,\ldots,s$. As for the site levels, the induced posteriors will now allow inference on which sites may have common release probabilities, as well as on the precise values of such probabilities.

Non-uniform beta distributions might replace the uniform baseline in application, particularly with expected low levels of release probabilities. Our analyses reported below retain the uniform prior, for illustration and so as to avoid question of overtly biasing towards lower or higher values. Note also that we explicitly exclude the possibility of non-zero release probabilities, though some or all may be very small, and so rule out the use of $\pi_j=0$ as a device for reducing the number of active sites; that is completely determined by zero values among the μ_j, as discussed earlier.

As with a in the model for the site levels, the hyper-parameter b will be assigned a prior, and again a diffuse gamma prior is natural. With this structure, we have defined the additional prior components $p(\pi|b)p(b)$ of the full joint prior in (3.2).

2.2.4 *Priors for Noise Moments* m *and* v

In most experiments we anticipate a possible systematic bias m in both noise and signal recordings, induced by the direct measurement of cell membrane potential. This is typically very small relative to induced signal levels, and the raw noise recordings provide data on which to assess this. In addition, the noise measurements inform on background variability v, i.e. they are drawn from the $N(\cdot|m,v)$ distribution. In our analyses, our prior for m and v as input to the signal analysis is simply the posterior from a reference analysis of the noise data alone, i.e. a standard conjugate normal, inverse gamma distribution based on the noise sample mean, variance and sample size. With this structure, we have defined the final component $p(v, m)$ of the full joint prior in (3.2).

2.3 Posterior Distributions and Computation

Calculation of posterior distributions is feasible, as might be expected, via variants of Gibbs sampling (e.g. Gelfand and Smith, 1990, Smith and Roberts, 1993). The specific collection and sequence of conditional posterior distributions used to develop simulation algorithms are briefly summarized here.

A key step is to augment the data with the latent site transmission indicators z; thus the sampling model is expanded as

$$p(y, z|\mu, \pi, m, v, q, a, b) = p(y|\mu, z, m, v)p(z|\pi)$$
$$= \left\{\prod_{i=1}^{n} N(y_i|m + z_i'\mu, v)\right\} \left\{\prod_{j=1}^{s}\prod_{i=1}^{n} \pi_j^{z_{ij}}(1-\pi_j)^{1-z_{ij}}\right\},$$

providing various conditional likelihood functions that are neatly factorized into simple components. The full joint posterior density, for all model parameters together with the uncertain indicators z, has the product form

$$p(\mu, \pi, z, m, v, q, a, b|y) \propto p(\mu|q, a)p(q)p(a)p(z|\pi)p(\pi|b)p(b)\,p(m, v)p(y|\mu, z, m, v), \quad (3.5)$$

where the component conditional prior terms are as described in the previous section.

2.4 Conditional Posteriors and MCMC

The posterior (3.5) yields a tractable set of complete conditional distributions characterizing the joint posterior, and hence leads to implementation of Gibbs sampling based on this structure. Each iteration of this MCMC sampling scheme draws a new set of all parameters and latent variables by sequencing through the conditionals now noted.

1. Sampling site levels μ proceeds by sequencing through $j = 1, \ldots, s$, at each step generating a new value of μ_j given the latest sampled values of μ_{-j} and all other conditioning quantities. For each j, this involves sampling a posterior that is a simple mixture of several point masses with a truncated normal distribution. Sampling efficiency is improved using variations of so-called configuration sampling for the discrete components, as introduced by MacEachern (1994).
2. Sampling site release probabilities π similarly proceeds by sequencing through $j = 1, \ldots, s$, at each step generating a new value of π_j given the latest sampled values of π_{-j} and all other conditioning quantities. For each j, this involves sampling a mixture of discrete components with a beta component. Again, configuration sampling improves simulation efficiency.
3. Sampling indicators transmission z involves a set of n independent draws from conditional multinomial posteriors for the n individual binary 2^s-vectors z_i, $i = 1, \ldots, n$. These simulations are easily performed.

4. Sampling the systematic bias quantity m involves a simple normal draw.
5. Sampling the noise variance v involves sampling an inverse gamma posterior.
6. Sampling the hyper-parameter q is also trivial, simply involving a draw from an appropriate beta distribution.
7. Sampling the hyper-parameters a and b follows West (1997), and Escobar and West (1995), and involves a minor 'augmentation' of the parameter space, with simple beta and gamma variate generations.

Iterating this process produces sequences of simulated values of the full set of quantities $\phi \stackrel{\text{def}}{=} \{\mu, \pi, z, m, v, q, a, b\}$ which represent realizations of a Markov chain in ϕ space whose stationary distribution is the joint posterior (3.5) characterized by the conditionals. The precise sequence (1)–(7) of conditional posteriors detailed above is the sequence currently used in simulation; at each step, new values of the quantities in question are sampled based on current values of all quantities required to determine the conditional in question. Convergence is assured by appeal to rather general results, such as Tierney (1994, especially Theorem 1 and Corollary 2) relevant to the models here; this ensures that successive realizations of $\phi = \{\mu, \pi, z, v, q, a, b\}$ generated by this Gibbs sampling set-up eventually resemble samples from the exact posterior in (3.5). Irreducibility of the resulting chain is a consequence of the fact that all full conditionals defining the chain are everywhere positive. Practical issues of convergence are addressed by experiments with several short runs from various starting values. Important practical issues of convergence is discussed in the next section.

2.5 Parameter Identification and Relabelling Transmission Sites

The model structure as described is not identifiable, in the traditional sense of parameter identification especially associated with mixture models. This is obvious in view of the lack of any physical identification of the neural transmission sites we have arbitrarily labelled $j = 1, \ldots, s$; for example, π_1 is the release probability for the first site in our labelling order, but this could be any of the actual release sites. As a result, the model is invariant under arbitrary permutations of the labels on sites, and likelihood functions are similarly invariant under permutations of the indices of the sets of parameters (μ_j, π_j). Also, as the priors for the μ_j and π_j separately are exchangeable, corresponding posteriors are also invariant under labelling permutations.

To impose identification we need a physically meaningful restriction on the parameters. Perhaps the simplest, and most obvious, constraint is to order either site levels or site release chances. We can do this with the site release levels. In posterior inferences, relabel so that site j corresponds to release level μ_j, the jth largest value in μ. As the priors, hence posteriors, include positive probabilities on common values, we break ties by imposing a subsidiary ordering on the release probabilities for (ordered) sites with common release levels. Thus, if μ_3, μ_4 and μ_5 happened to be equal, the sites relabelled 3–5 would be so chosen in order of increasing values of their chances π_j; were these equal too, as permitted under the priors and hence posteriors here, then the labelling is arbitrary and we have (in this example) three indistinguishable sites: common release levels and common release probabilities. The posterior analysis required to compute corresponding posterior inferences is extremely simple in the context of posterior simulations; each iteration of the Gibbs sampling analysis produces a draw of (μ, π) (among other things). Given a draw, we first create a separate vector μ^* containing the *ordered* values of μ, and a second vector π^* containing the elements of π rearranged to correspond to the ordering of μ to μ^*. If subsets of contiguous values in μ^* are common, the corresponding elements in π^* are then rearranged in increasing order themselves. Through iterations, repeat draws of (μ^*, π^*) are saved, so building up samples from the required posterior distribution that incorporates the identification of sites.

Note that we could alternatively identify the model by a primary ordering on the π_j, rather than the μ_j. In data analyses we routinely explore posterior summaries identified each—i.e. ordered by μ_j and ordered by π_j—as each is only a partial summary of the full posterior. This is relevant particularly in cases when it appears that the data are in conformity, at least partially, with quantal structure.

In connection with this ordering issue, the quantal structure of the priors for μ and π is very relevant, as we alluded to earlier. Consider an application in which the data support common, or very close, values among the μ_j. Suppose that the prior is simply uniform, with no chance that consecutive values of the ordered μ_j are equal. Then imposing the ordering for identification results in a distortion of inferences—values we should judge to be close, or equal, are 'pushed apart' by the imposed ordering. This undesirable effect is ameliorated by the priors we use, allowing exact equality of neighbouring levels. The same applies to the π_j.

Notice that we retain the original, unidentified parametrization for model specification and the simulation analysis operates on the unconstrained posteriors. This has a theoretical benefit in allowing easy and direct application of standard convergence results (Tierney, 1994). More practically, it has been our experience that imposing an identifying ordering on parameters directly through the prior distribution hinders convergence of the resulting MCMC on the constrained parameter space. This arises naturally as the ordering induces a highly structured dependence between parameters, relative to the unconstrained (albeit unidentified) parameter space. In addition, one rather intriguing practical payoff from operating the MCMC on the unrestricted space is an aid in assessing convergence of the simulation. This is due to the exact symmetries in posterior distributions resulting from permutation invariance, partly evidenced through the fact that marginal posteriors $p(\mu_j, \pi_j|y)$ are the same for all j. Hence summaries of marginal posteriors for, say, the μ_j should be indicative of the same margins for all j; the same if true of the π_j.

However, in many applications convergence will be slow. One reason for this is that, in some models and with some observed signal data sets, the posterior will heavily concentrate in widely separate regions of the parameter space, possibly being multi-modal with well separated modes. Such posteriors are notorious in the Markov chain Monte Carlo literature, and the Gibbs sampling routine may get 'stuck' in iterations in just one (or typically several) region of high posterior probability, or around a subset of modes. There is certainly need for further algorithmic research to provide faster alternatives to the raw Gibbs sampling developed here (possibly based on some of the methods mentioned in Besag and Green (1993), or Smith and Roberts (1993), for example). However, this is not a problem as far as the ordered site levels, and the correspondingly reordered release probabilities, are concerned, due to the symmetries in the unconstrained posterior. This is because posterior samples from one region of the (μ, π) space may be reflected to other regions by permuting sites indices, while the resulting ordered values remain unchanged. Hence the unconstrained sampling algorithm may converge slowly as far as sampling the posterior for (μ, π), but the derived sequence for (μ^*, π^*) converges much faster.

Due to the lack of identification, actual output streams from MCMC simulations exhibit switching effects, as parameter draws jump between the modes representing the identification issue. For example, in a study supporting two distinct site release levels with values near 1 and 2 mV, respectively, the simulated series of the unidentified μ_1 (μ_2) will tend to vary around 1 mV (2 mV) for a while, then randomly switch to near 2 mV (1 mV) for a while, and so on. Our experience is that, across several analyses of different data sets, convergence is generally 'clean' in the sense that, following what is usually a rapid initial, burn-in period, the output streams remain stable and apparently stationary between points of switching between these posterior modes. Informal diagnostics based on a small number of short repeat runs from different starting values verify summarized analyses from one final, longer run. In addition, the theoretical convergence diagnostic based on the symmetry of the unidentified posterior is used in final analysis.

2.6 Additional Variance Components—Intrinsic Variance

Recent attention has focussed on questions of additional variability in synaptic signal outcomes due to so-called *intrinsic variability* in release levels of individual synapses (Turner et al., 1997); we describe this concept here, and define an elaborated class of models incorporating it. This concept gives rise to model modifications in which components of the normal mixture representation have variances that increase with level, and this leads to considerable complications, both substantive and technical. The technical complications have to do with developing appropriate model extensions, and associated MCMC techniques to analyze the resulting models; some brief development is mentioned here, with examples. The substantive complication is essentially that competing models with and without intrinsic variance components cannot be readily distinguished on the basis of the observed data alone; an observed data configuration might arise from just one or two sites with significant intrinsic variance, or it might arise from a greater number of sites with low or zero intrinsic variance. In such cases, and especially when inferences about site characteristics are heavily dependent on the number of sites and levels of intrinsic variance, we are left reliant on the opinions of expert neurophysiologists to judge between the models. Unfortunately, in its current state, the field is represented by widely varying expert opinions, from the one extreme of complete disregard of the notion of intrinsic variability, to the other of belief in high levels of intrinsic variability as the norm. In some examples we have studied, including the data sets shown here, this issue is relatively benign, as inferences about release levels and probabilities are relatively insensitive to intrinsic variances. In others cases, it is evidently highly relevant. Further collaborative research to refine knowledge of intrinsic variability effects is part of the current frontiers of the field. Here we give a short discussion of the notion and our current approach to modelling intrinsic variability.

Intrinsic variability refers to variation in levels of neurotransmitter release at specific sites. As developed above, site j has a fixed release level μ_j, and, while these levels may differ across sites, they are assumed fixed for the duration of the experiment under the controlled conditions. However, the mechanism of electro-chemical transmission suggests that this may be an over-simplification. A site transmits by releasing a packet of (many) molecules of a chemical transmitter, and these molecules move across the synaptic 'cleft' to eventually bind to the postsynaptic receptors. The induced signal response is proportional to the number of molecules binding. So, the assumption of fixed μ_j implies that (a) the number of molecules transmitted is constant across occasions, and (b) all ejected molecules are bound to receptors on the postsynaptic cell. Each of these is questionable, and the issue has given rise to the notion of intrinsic variance, i.e. variability in the site-specific level of release across occasions (Faber et al., 1992). Intrinsic variance may also arise when two sites are close to each other and cannot be separated in any statistical sense. Then, it will appear that one site may be fluctuating. Whatever the actual structure of variability, it is apparent in data sets through mixture components having higher variance at higher release levels.

The basic extension of our models to admit the possibility of intrinsic variability involves remodelling the data y_i as coming from the conditional normal model

$$y_i \sim N\left(y_i | m + \sum_{j=1}^{s} z_{ij}\gamma_{ij}, v \right), \tag{3.6}$$

where the γ_{ij} are new site-, and case-, specific release levels and v is the noise variance. Retaining independence across sites and trials, we assume that the γ_{ij} are distributed about an underlying expected release level μ_j—the same site-specific level as before, but now representing underlying average levels about which releases vary. Then the form of the distribution of the γ_{ij} about μ_j represents the intrinsic variability in induced responses due to variation in amounts of neurotransmitter released by site j and also to variation in the success rate in moving the transmitter across the synaptic cleft. Various parametric forms might be considered; our preliminary work, to date, is based on exploration of models in which

$$p(\gamma_{ij}|\mu_j, \tau_j) \propto N(\gamma_{ij}|\mu_j, \tau_j^2 \mu_j^2) I(0 < \gamma_{ij} < u) \tag{3.7}$$

for all trials i and each site j; here u is the earlier specific upper bound on release levels. The new parameters $\tau_j > 0$ measure intrinsic variability of sites $j = 1, \ldots, s$; ignoring the truncation in (3.7), τ_j is an effective constant (though site-specific) coefficient of variation in release levels about the underlying μ_j. This model has been implemented, extending the prior structure detailed in Section 2.2 to incorporate the full set of site-, and trial-, specific release levels $\{\gamma_{ij}\}$ together with the new parameters $\tau_1, \ldots, \tau_s$. Note that, as our model allows $\mu_j = 0$ for inactive sites, we use (3.7) only for $\mu_j > 0$; otherwise, $\mu_j = 0$ implies $\gamma_{ij} = 0$ for each $i = 1, \ldots, n$.

We can see the effects of quantal variability by integrating the data density (3.6) with respect to (3.7). This is complicated due to the truncation to positive values; as an approximation, for the purposes of illustrating structure here, assume this is not binding, i.e. that μ_j and τ_j are such that the mass of the basic normal distribution in (3.7) lies well within the interval $(0, u)$. Then (3.6) and (3.7) combine and marginalize over γ_{ij} to give

$$y_i \sim N\left(y_i | m + \sum_{j=1}^{s} z_{ij}\mu_j, v + \sum_{j=1}^{s} z_{ij}\tau_j^2\mu_j^2 \right). \tag{3.8}$$

The mean here is as in the original formulation, (3.1), with active sites contributing the expected release levels μ_j. But now the variance of y_i is not simply the noise variance v; it is inflated by adding in factors $\tau_j^2\mu_j^2$ for each of the active sites. Hence the feature relevant to modelling increased spread of data configurations at higher response levels.

It should be clear that this extended model can be managed computationally with direct extensions of the MCMC algorithms discussed so far. We are now interested in the full posterior $p(\mu, \pi, z, \gamma, \tau, m, v, q, a, b | y)$, extending the original posterior in (3.5) to include the new quantities $\gamma = \{\gamma_{ij}, i = 1, \ldots, n; j = 1, \ldots, s\}$ and $\tau = \{\tau_1, \ldots, \tau_s\}$. There are difficulties in the posterior MCMC analysis due to the complicated form of (3.7) as a function of μ_j, and also due to the truncation of the basic normal model. These issues destroy part of the nice, conditionally conjugate sampling structure, and are currently handled using some direct analytic approximations and Metropolis-Hasting accept/reject steps in the extended simulation analysis. It is beyond the scope of the current paper to develop the technical aspects of this fully here, but our discussion would be incomplete without raising the issues of intrinsic variability, currently becoming vogue in the field, and without providing some exploratory data analysis. After further technical refinements and practical experience, full modelling and technical details will be reported elsewhere (West, 1997).

3. BAYESIAN ANALYSIS AND EPSC SIMULATIONS

Simulations were performed to accurately guide the definition of the prior parameters under particular conditions. The signal to noise ratio and location and number of sites were variously selected but compared to the actual data after all of the analysis. The format of the simulations was to arbitrarily and randomly generate data sets ($n = 200$ to 500 different ensembles) from a known statistical distribution, but varying the parameters of the random data sets (for example, number of responses in a data set, signal to noise ratio, etc.). The simulated data sets were in general similar to those expected from companion physiological studies, except that the underlying distribution from which the random sample was drawn is well defined for the simulation ensembles. Once these data sets were generated the Bayesian analysis was performed in a manner identical to that used for the physiological data sets. Once optimized, the parameter values (and error ranges) were compared between the results of the analysis and that of the original distribution. If the original distribution parameter values were bracketed by the 95% error ranges of the analyzed results that was counted as a correct parameter value, and an error if the true value was outside of the predicted range. The errors in number of sites and parameters were accumulated and calculated as a percentage of the total number of ensembles which were averaged. This definition of error used the full error estimates available from the Bayesian estimation.

Random ensembles from known distributions were analyzed (Kullmann, 1989), composed of 1–5 sites, using varying signal to noise and ensemble size values as suggested by Stricker and Redman (1994). The raw histograms varied extensively depending on the random sample obtained from the underlying distribution and thus histogram appearance may be very misleading (Larkman et al., 1997; Walmsley, 1995), so the underlying data were used for the analysis rather than a histogram. Numerous simulations were performed using random ensembles drawn from distributions with a variety of site numbers, relative amplitudes (both equal and unequal), signal to noise values, quantal standard deviations and sample sizes to estimate the errors associated with the Bayesian analysis. Each set of simulations included 100–200 separate simulated ensembles, each with $n=200$ to $n=750$ data points, which underwent analysis and comparison to the underlying distribution in the same way as the experimental datasets. Errors in the analysis were defined in two separate ways:

1. the relative error of the standard deviation normalized to the value of the parameter (analyzed as SD/parameter value), similar to that suggested by Stricker and Redman (1994a,b);
2. the percentage of errors per site in which the number of sites, underlying site amplitude or probability did not fall within the 95% confidence interval of the analysis of the data set.

This latter definition shows a more usable definition of error, so that one can estimate how often the analysis will be incorrect taking into account both the limited sample size of the experiment (including sampling errors) and the errors of the analysis. The 95% confidence interval around the median was chosen because the distributions were often asymmetric about the median.

The number of sites (k) derived from the Bayesian analysis reflects partly the prior values, particularly the adherence of the noise in the analysis to the estimated noise standard deviation (σ_{noise}) value. For weak adherence with only a few degrees of freedom ($df=10$) the noise value itself can fluctuate away from the provisional value, leading to fewer sites than in the distribution. Therefore the adherence of the Bayesian analysis to the noise was usually set to a tighter value ($df=n-1$), reflecting a strong faith in the preliminary noise estimate from the experiment. The number of sites with the noise $df=n-1$ evaluated to be either the correct number or larger, in effect to partially overestimate the number of sites to ensure that the site parameters showed the least error. Thus, with the noise set at this level in the analysis the number of sites was highly likely to be at least as large as the number in the simulations. The correct number of sites from the underlying (known) distributions was compared to the predicted number with a cumulative probability of at least 0.5 ($P[k|D] > 0.5$).

3.1 EPSC Simulations with One to Four Synaptic Sites

The summary results of some of these extensive simulations are shown in Figure 3.2 (for 1, 2 and 3 sites) and Figure 3.3 (for 4 sites), and further defined in Turner et al. (1997). The datasets employed in the analysis for Figure 3.2A were defined as having a signal to noise ratio of 3 : 1, with the noise set arbitrarily at $\sigma_{\text{noise}} = 1.0$ pA. There was no quantal or intrinsic variance included. The sample size was varied from 200 to 500, which showed little difference for the analysis of a single site. This analysis was very robust and less than 3% errors were encountered out of 200 datasets analyzed in any of the parameters. The fluctuation noted in Figure 3.2 between the various sample size results represents primarily sampling variation for the random ensembles drawn from the underlying distribution. Figure 3.2B indicates results for 2 sites with signal to noise set at 3 : 1, but with unequal spacing of the sites. This summary shows results for random datasets drawn from a distribution composed of $\mu_1 = 4$ pA ($\pi_1 = 0.4$) and $\mu_2 = 7$ pA ($\pi_2 = 0.25$). Note that the number of errors decreases slightly for sample sizes greater than 200 but the number of errors remains less than 5%. Figures 3.2C and D represent two different random distributions with three sites, with an irregular pattern of sites and a signal to noise of only 2 : 1 for the third site, with equal probabilities: $\mu_1 = 3$ pA ($\pi_1 = 0.4$), $\mu_2 = 3$ pA ($\pi_2 = 0.4$) and $\mu_3 = 5$ pA ($\pi_3 = 0.4$). The primary difference between C and D is that the probability of the sites is changed slightly in D to be different between the sites. Both result show that with irregular sites a much larger sample size is needed to resolve the differences between sites. The overlap between sites also increases the difficulty in resolving correctly the actual number of sites.

Figures 3.3A and C show the parameter errors noted upon varying the characteristics of the distributions with four sites ($\sigma_{\text{noise}} = 1$ pA, 4 sites each with $\mu_i = 3.0$ pA and $\pi_i = 0.4$). Figure 3.3A shows a decline in the errors for the number of sites as the sample size increases, but the amplitude and probability have very few errors, even for small sample sizes. Figure 3.3B shows variations in the signal to noise ratio (site amplitude) with a fixed sample size number = 500. Note that the errors shown in Figure 3.3C are much increased for a signal to noise ratio less than 3 : 1. Again there is considerable variation due to the random sampling of ensembles from the underlying distributions, even for 100–200 ensembles. However, with a reasonable signal to noise of 2 : 1 or greater, there is little sensitivity to sample size in the $n = 200$ to $n = 500$ region and less than 10–15% error in the derived amplitude and probability values.

Figures 3.3B and 3.3D show the relative SD error for four sites, similar to that derived for the bootstrap approach in Stricker and Redman (1994). These values reflect the magnitude of the standard deviation as a function of sample size (Figure 3.3B) and signal to noise ratio (Figure 3.3D), and also indirectly the 95% confidence interval (approximately four times the standard deviation value). Note that these values decline predictably with greater sample size and signal to noise

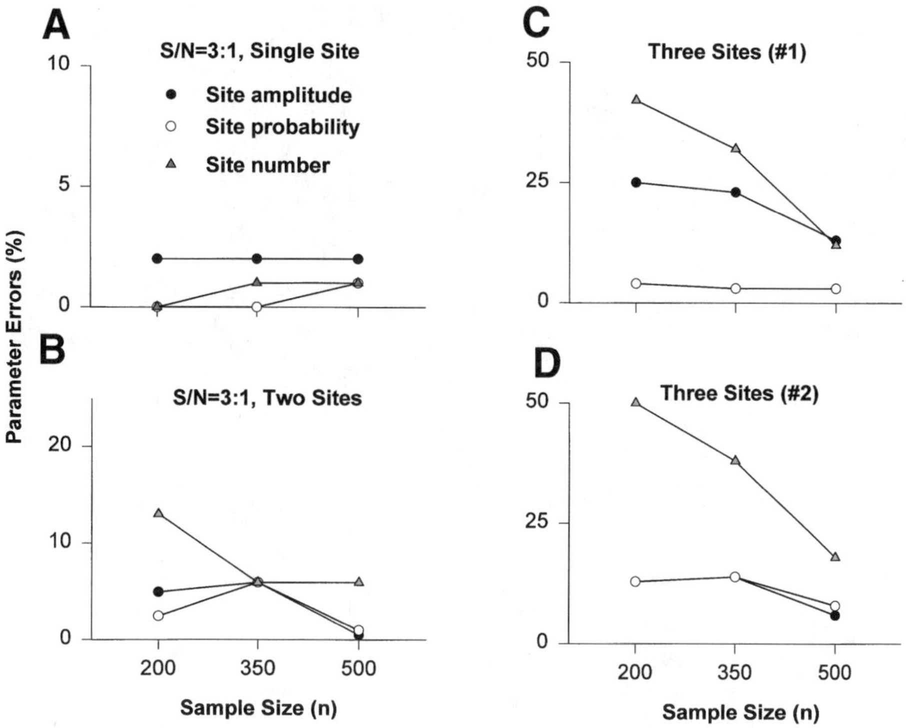

Figure 3.2 *Site Errors from* 1 – 3 *Sites.* Simulations were performed for a large number of randomly sampled ensembles, at various sample sizes (from $n=200$ to $n=500$ responses per ensemble) and signal to noise ratios (noise $\mathrm{SD}=1.0\,\mathrm{pA}$). At each particular point the results of 100 – 200 such simulated ensembles were compiled, in terms of the errors of the analysis, compared to the expected values from the underlying distribution. The error results are shown in Figure 3.2 in terms of the number of absolute errors in synaptic site numbers, amplitude and probability, assessing whether the 'true' underlying value was located within the 95% confidence limits of the analysis. As shown in A there were very few errors with a signal to noise ratio of 3 : 1 and a single synaptic site, with only a random few errors detected. However, there were more errors at low signal to noise ratios (particularly at and below 2 : 1) and for overlapping sites, as shown in C and D, with three sites. The number of errors for the three overlapping sites is larger, particularly for $n=200$ sample size and for the number of sites, but this improves with a larger sample size (particularly around 500).

ratio, but again the sample size range between 200 and 500 and the signal to noise range between 2 : 1 and 5 : 1 appear very similar. However, this value may or may not be helpful in predicting the accuracy of the analysis or particularly the limitations of the sampling error with a small random sample. Thus, the smaller error range may not necessarily reflect the quality of the analysis in terms of the

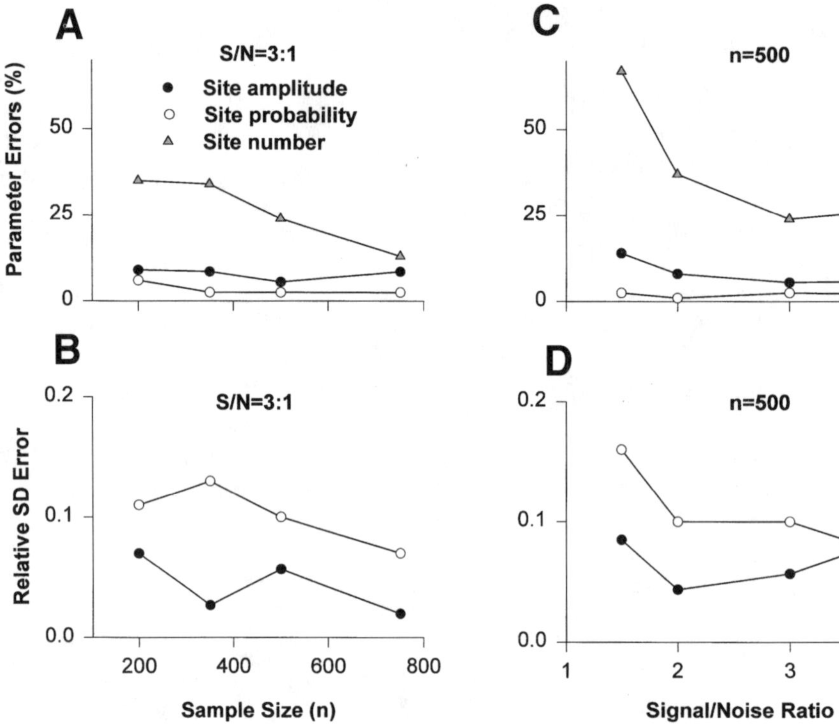

Figure 3.3 *Site Errors from 4 Sites.* This figure shows a summary for the analysis of simulations with four sites and a variable signal to noise ratio and sample number, to estimate the robustness of the Bayesian analysis. Simulations were performed for a large number of randomly sampled ensembles, at various sample sizes (from $n=200$ to $n=750$ responses per ensemble) and signal to noise ratios (noise $SD=1.0\,pA$). At each particular point the results of 100–200 such simulated ensembles were compiled, in terms of the errors of the analysis, compared to the expected values from the underlying distribution. The error results are shown in A and C in terms of the number of absolute errors in synaptic site number, amplitude and probability, assessing whether the 'true' underlying value was located within the 95% confidence limits of the analysis. As shown in A there were few errors with a signal to noise ratio of 3 : 1. However, there were more errors at low signal to noise ratios (particularly at and below 2 : 1 as well as for $n=200$ data points instead of $n=500$). B and D show the relative error, which is derived from the normalized SD value (following division by the site value). This value shows a progressive decrease with both larger sample sizes (B) and improved signal to noise ratios (D), as expected.

percentage errors with respect to the underlying distribution or the actual errors expected for the analysis.

3.2 Simulation and Recovery of Intrinsic ('Quantal') Variance

Additionally, we have simulated data sets (1–4 sites) with varying levels of intrinsic variance, which is directly analyzed in our model (Figure 3.4). Errors remain less than 15% per site up to an intrinsic standard deviation level of 40%, above which there is a sharp increase (to 20% errors at 50% intrinsic standard deviation values; signal to noise = 3 : 1; $n = 200$). The analysis of intrinsic variance shows a linear outcome with the comparison of the sample distribution variance versus the resolved variance, but with the slope showing less variance than expected (0.6). Thus, the analysis can separately resolve intrinsic variance as well as the other site parameters, within limits, though this estimate is linear but currently an underestimate. The use of an error range to approximately locate a site value is highly robust in that if there is less certainty in a particular solution then the error range becomes larger, whereas a high degree of certainty results in a narrow error range. These error ranges naturally impart the desired degree of confidence in the solution, which is a critical aspect of any derived value—how likely is it an answer will be a 'good' solution? These extensive simulations have suggested that the throughput fidelity for even small sampled ensembles is robust using the Bayesian analysis, which includes the clearly identified error ranges.

The Bayesian analysis format shows excellent resolution of synaptic sites, even with $n = 200$ and a signal to noise ratio of 2 : 1, and also reflects the ambiguity of the

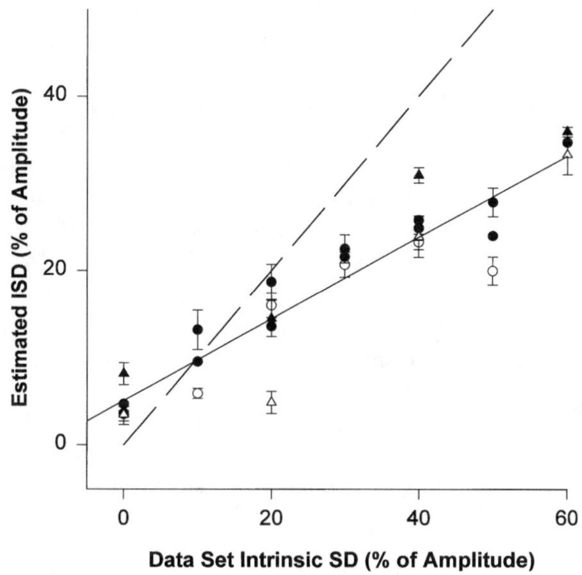

Figure 3.4 *Intrinsic SD Detection.* This plot shows a summary of the value of intrinsic SD from the underlying distributions, from which samples were drawn randomly, plotted against the predicted value for intrinsic SD from the analysis. Multiple data sets were analyzed containing some degree of intrinsic variance, including $n = 200$ and $n = 500$ sample size, and one and four sites. The dashed line shows an ideal curve, with a 45 degree line. However, the solid line is a regression line through the data points, which has a slope of 0.58. This slope suggests that the Bayesian format, as currently implemented, roughly (but linearly) underestimates the degree of intrinsic SD.

analysis in an accurate fashion according to the width of the error ranges for the site values. This format of synaptic site analysis shows improved discrimination of site parameters compared to components analysis, such as the MLE optimization (Stricker and Redman, 1994) and the previous Bayesian components analysis (Turner and West, 1993). However, further simulations (not shown) indicate that unequal site parameters also are resolved nearly as well as these equal site parameters, suggesting that Bayesian site analysis may be helpful in assessing whether or not a binomial model may be appropriate for an individual data set. In addition, intrinsic variation up to approximately 60% may be resolved though with increased errors.

3.3 Example Physiological Data Sets

Physiological data sets were acquired as described in detail in Turner et al. (1997), using minimal stimulation to assure that only a small, stable number of synaptic sites were activated. The postsynaptic responses were recorded as EPSCs using low-noise whole cell patch electrodes, and the ensembles were carefully assessed for stationarity. Figure 3.5A shows a physiological average for a minimal stimulation EPSC ensemble (U2N1), recorded from a CA1 pyramidal neuron under highly controlled conditions, using minimal stimulation just above threshold for the presence of a response with averaging. The peak average was 8.87 ± 7.91 pA ($n=244$) and $\sigma_{noise}=1.22$ pA. Figure 3.5B shows that the noise (dark circles) and signal (light circles) peak data showed neither a trend nor an irregular pattern over the course of the ensemble. The dataset was then analyzed using noise $df=243$, with a resulting probability favoring a single site [$P(1|D)=0.93$], and less than 0.07 likelihood that an additional site was needed. The parameters for the single site (Figure 3.5D) showed $\mu=10.2$ pA and $\pi=0.87$ and a median intrinsic variance of 40%, with the variation around these parameters illustrated as error bars in Figure 3.5D. The raw histogram and calculated Bayesian predictive probability density function (pdf) are shown in Figure 3.5C along with error limits on the pdf as dashed lines. Note the excellent fit of the raw histogram and the pdf. However, the raw histogram was not in any way used in the analysis and is shown only for comparison purposes to the predictive pdf, which was calculated directly from the data. For both of these physiological examples 20,000 Monte Carlo draws were used in the estimation of the parameters.

Figure 3.6 shows another EPSC ensemble (U4N1), again obtained from the CA1 region during a whole cell patch clamp recording from a CA1 pyramidal neuron (Turner et al., 1997). Figure 3.6A illustrates an averaged EPSC trace ($n=386$) and Figure 3.6B shows a trend analysis, with no net changes over the ensemble period. Figure 3.6C shows the comparison of the raw data histogram and the predictive pdf (smooth line). Figure 3.6D shows the two sites which were extracted using the Bayesian analysis: $P(1|D)=0.01$, $P(2|D)=0.99$ and $P(3|D)=0$, with parameters $\mu_1=7.63$ pA ($\pi_1=0.11$), $\mu_2=19.0$ pA ($\pi_2=0.32$) and intrinsic variance of 22%. Note that the two sites do not overlap and both the amplitudes and probabilities are different, suggesting this ensemble to be a complex response, since neither the simple

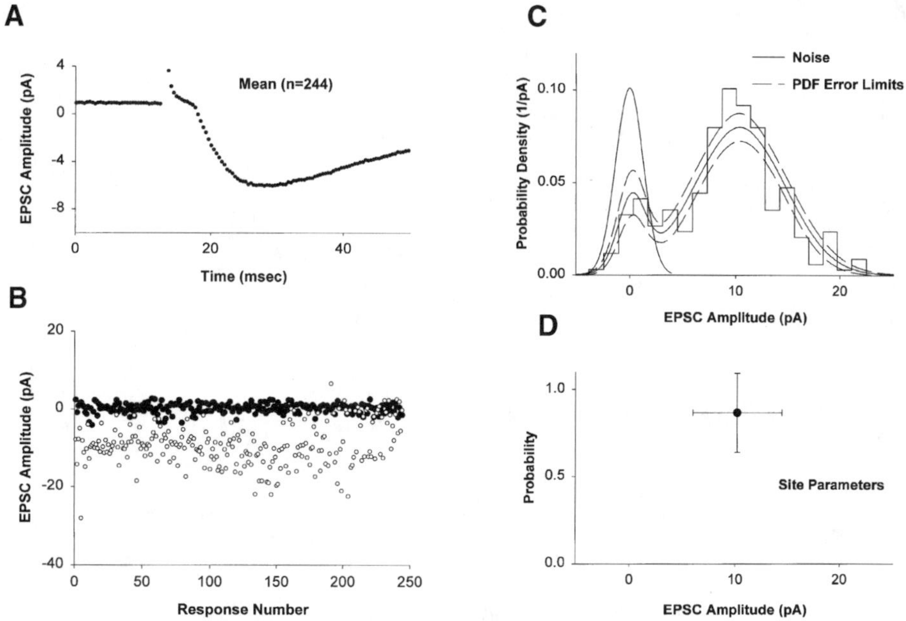

Figure 3.5 *Data Set U2N1 with 1 Site.* A shows the mean trace from U2N1 data set ($n=244$ responses). B shows the scatter plot for stationarity, indicating no change in the degree of fluctuation pattern over the course of the ensemble. The open circles show the responses and the closed circles the noise samples over the course of the ensemble. C illustrates the response histograms and Bayesian analysis results. The smooth function (thick line) in C shows the noise function used in the analysis, overlapping the failure peak. The smooth line indicates the predictive Bayesian pdf and the dotted lines the error ranges around the pdf. D shows the individual synaptic site parameters, with a dual plot comparing site amplitude on the abscissa versus site probability on the ordinate. Note the single site at approximately 10 pA, corresponding well with the histogram in C. This ensemble shows an example of a simple response with either failures or occurrences.

binomial nor the compound binomial would be a suitable fit. Many further simple and more complex responses are shown in Turner et al. (1997), along with the effects of paired pulse plasticity on site parameters. The conclusions from the analysis of these data focus on the considerable between site heterogeneity as well as the moderate level of within site (intrinsic or quantal) variance observed. Less than half of the previously identified and analyzed data sets could be fit by a traditional statistical model, indicating that most responses recorded in the CA1 pyramidal neuron showed unequal amplitudes and probabilities, requiring the more complex model for an adequate fit. Informal assessment of convergence was performed as discussed in Section 2. These summaries reflect the inherent uncertainties in estimating the number of active sites.

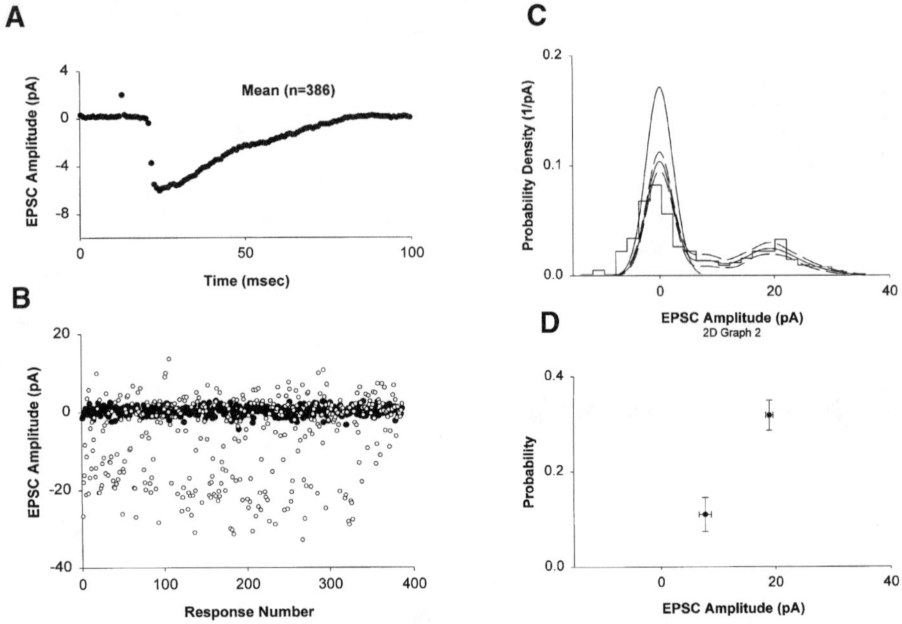

Figure 3.6 *Data Set U4N1 with 2 Sites.* A shows the mean trace from U4N1 data set ($n = 386$ responses). B shows the scatter plot for stationarity, indicating no change in the degree of fluctuation pattern over the course of the ensemble. The open circles show the responses and the closed circles the noise samples over the course of the ensemble. C illustrates the response histograms and Bayesian analysis results. The smooth function (thick line) in C shows the noise function used in the analysis, overlapping the failure peak. The smooth line indicates the predictive Bayesian pdf and the dotted lines the error ranges around the pdf. D shows the individual synaptic site parameters, with a dual plot comparing site amplitude on the abscissa versus site probability on the ordinate. Note the dual sites, with the first at approximately 4 pA and the second at approximately 18 pA, corresponding well with the histogram in C.

4. DISCUSSION

4.1 Comparison of Simulations and Physiological Data Sets

The simulation data sets were purposefully designed to emulate many of the physiological conditions observed, particularly the level of noise, the signal to noise ratio, the level of probability of release and the amplitudes. The variations on signal to noise and sample size also were solidly established to be close to the physiological values. For most of these realistic parameter values the Bayesian Site analysis performed well, though with low signal to noise (less than 2:1), small sample sizes (particularly less than 300) and irregular sites the extraction of parameters was more difficult. The number of parameter errors allows a quantitative estimate of the likelihood of a single random sample leading to the definition of the underlying distribution from which the sample was drawn. This likelihood includes both the effects of a small sample size (particularly when less than 500–750) and the randomness of the subsequent draw and the ability of the analysis technique to resolve the underlying distribution. Since parameter errors have not been estimated in this practical context previously for neurophysiological applications it is difficult to compare this technique to other currently available statistical analysis formats, particularly the bootstrap technique for the MLE algorithm and the compound binomial approach of Stricker and Redman (1994). However, the other estimate of error, that of comparing the standard error, appears comparable for this Bayesian and that of Stricker and Redman (1994). Further simulations, particularly with more sites, may help define the limits of the Bayesian site analysis. Currently the analysis will work with up to 8–10 synaptic release sites, but beyond 4–5 sites practical problems arise with identifying or clustering each individual sites and more site overlap prevents such effective clustering. Thus, analysis of data sets with 1–3 sites may be more robust than those with greater than 4–5 sites and we are particularly cautious regarding a larger of number of sites yet.

Virtually all of the analysis techniques discussed here have limited availability in terms of distribution, and are highly specialized in their application to particular data sets. Kullmann's MLE analysis suite, including the MEND program (Kullmann, 1989), has been available freely by ftp but this older program has been essentially replaced by the more robust MLE algorithm developed by Stricker and Redman (1994). Korn and Faber's binomial analysis program is only applicable under strict conditions and has not become widely available (Korn and Faber, 1991). The Bayesian components (Turner and West, 1993; West and Turner, 1994) and site-directed programs (Turner et al., 1997; West, 1997) are likewise not yet sufficiently robust to become generally available, and require the use of several ranges of parameters to estimate the best solution. In particular, the Bayesian site program does not work well when there are likely multiple sites (greater than 5–6) and there are very few stable data sets currently available which fit the range of limited applicability of this analysis program. Limited usefulness of each type of program has made it difficult to compare efficacy of parameter and distribution detection; it may be better to compare a standard suite of simulation data between labs rather than share such highly specialized programs. However, this comparison is critical and should be performed.

4.2 Analysis of Components in Contrast to Sites

The analysis of sites is a more robust approach than using a histogram approximation procedure, regardless of the particular model and optimization approach utilized. For example, Korn and Faber (1991) have used binomial models which are site-directed, thus ensuring that a low probability tail is adequately represented in the analysis. However, Redman and his group have used a components based approach, beginning with a smoothed histogram but fitting peaks in this smoothed representation of the data (Redman, 1990; Stricker and Redman, 1994; Stricker et al., 1994). The components based approach fits histogram peaks rather than the natural sums and products of underlying sites, and thus the peaks may or may not represent sums of site contributions appropriately. The differences between the two approaches represents an ongoing debate as to the relative usefulness of the two approaches, but clearly the histograms are sums of individual synaptic site contributions, with their own respective amplitudes and probabilities. Whenever there is more than one site, the statistical prediction is that the appropriate sums of the sites (with the probability being a product of the individual values) should be adequately represented, even though on an individual draw the histogram may not be accurate or representative for the distribution (Walmsley, 1995). To study these different approaches we have evolved through both types of analysis.

The Bayesian data analysis has been previously used for analysis of components (peaks in histograms representing sums of synaptic site contributions—Turner and West, 1993; West and Turner, 1994) and data are presented here showing the enhanced usefulness of a site-directed approach. We have performed simulations similar to those illustrated in Figures 1–3 for this older, components-based approach and the results are much less optimal compared to the site-directed approach. If all synaptic sites are assumed equal then a binomial (or in an extreme limiting case, a Poisson) distribution may be appropriate, and the components may be decomposed into their underlying sites. For a binomial distribution this requires estimation of several variables: n, or the number of sites; p, or the probability of release at each site; and the amplitude at each site, q. The histogram is normalized by what is felt to represent the unit amplitude (q) and then integers representing the sums of the individual site contributions are used to assess the probability and number of sites. As shown previously, there is often no unique solution, particularly with the most likely situation that there is contaminating noise and where the first peak may be offset by an arbitrary value (Stricker and Redman, 1994). Estimation of n may be particularly difficult. The situation is highly complicated by the additional likelihood that probability is different between sites, leading to a more complex compound binomial distribution (but which still requires equal site amplitude). When the ambient noise becomes significant, removal of the noise effect also requires an optimization routine, usually termed deconvolution (Redman, 1990). Because of the multiple inferences with this form of deconvolution procedure, the results are difficult to confirm, particularly the likelihood that the solution is 'good' or 'optimal' in any way.

We have developed the Bayesian site method of analysis to overcome many of these problems. By starting with the possibility that sites may not be similar in terms of either amplitude or probability and by not assuming a fixed number of sites, a more flexible and global analysis format is possible. Additionally, as shown in Figures 3.5 and 3.6, the analysis gives error ranges for the predictive pdf and in particular the site parameters, as well as the likelihood for how many sites are required. Though the analysis uses the fitted Gaussian for the noise, this is only a starting point, and depending on the degrees of freedom accompanying the noise value, is more or less strictly used in the analysis. The Bayesian also gives a posterior distribution for the noise variance, indicating how strict this adherence was for a particular data set. Likewise, the Bayesian format is general, and development plans include making the prior distribution for the noise much more flexible, particularly by using a sampled noise distribution rather than any form of fitted smooth distribution. However, this format of analysis begins to give some idea as the 'goodness' of the analyzed fit to the raw data, though of course does not in any way guarantee that the fit is unique or optimal. However, the Bayesian approach uses all of the data for the analysis, as opposed to the more conventional MLE (Kullmann, 1989) and compound binomial (Stricker and Redman, 1994) procedures, which utilize optimized histogram data. Additionally, the Bayesian approach provides a more rigorous and extensive investigation of the data-supported regions of the parameter space than the more restricted MLE solution algorithm (West and Turner, 1994). The direct Bayesian site format may eventually prove to be very helpful to identify changes directly with plasticity and experience rather than the more 'blind' approaches which have conventionally been used.

4.3 Analysis of Physiological Data Sets

The results of the two data sets shown here and the larger group of data sets from Turner et al. (1997) explicitly highlight the considerable heterogeneity of synaptic site parameters in hippocampal neurons during paired-pulse plasticity (Bekkers et al., 1990; Bolshakov and Siegelbaum, 1995; Hessler et al., 1993; Jack et al., 1994; Liu and Tsien, 1995; Rosenmund et al., 1993; Stevens and Wang, 1994). The heterogeneity noted between sites (62% of the mean site amplitude) of non-NMDA responses may be attributable to differences in presynaptic neurotransmitter release, postsynaptic receptor density and electrotonic filtering, but is clearly larger than the identified range of intrasite or intrinsic variability (36% of the mean site amplitude after correction), defining the relative importance of significant alterations in presynaptic release over time or ligand–receptor mismatch (Faber et al., 1992). One strength of the Bayesian site analysis is to resolve these two separate aspects of quantal variance, that due to either differences at a single site over time (here termed intrinsic variance) or that due to differences between site parameters at a single occurrence. The resolution achieved by this analysis suggests that intersite heterogeneity is clearly the rule in CA1 pyramidal neurons, for a variety of expected and important reasons, whereas intrinsic variability at a single site over time is also important and should be explicitly considered. Together these two sources of variability will lead to difficulty with histogram based approaches, since both tend to merge 'peaky' histograms into much smoother and less poorly defined histograms (Turner et al., 1997). This effect of smoothing can easily lead to the two-component histograms defined by earlier studies, in which only a 'failure' and a 'success' component can be identified (Stricker et al., 1994). This result confirms that histogram-based approaches may also mislead interpretations regarding underlying sites due to this smoothing effect and random histogram variations from sampling effects (Walmsley, 1995).

A small subset of histograms showed inflections suggesting approximately equal spacing of synaptic site contributions, but the rule (as observed previously in most reports) is that histogram appearance usually was not 'peaky' and histograms may be unreliable in pointing to the underlying statistical distribution (Larkman et al., 1991; Kullmann and Nicoll, 1992; Walmsley, 1995). Using a paired-pulse stimulation protocol the histogram distributions clearly changed in many ensembles, but there were also data sets which did not exhibit a significant paired pulse potentiation but rather showed depression (Turner et al., 1997). Paired-pulse plasticity appeared to arise primarily from changes in probability of release, as described by many other authors (Andreasen and Hablitz, 1994; Manabe et al., 1993; Redman, 1990). However, the degree of changes in probability appeared to vary between sites, suggesting that past history of activation and/or other factors may be important in the underlying process leading to the short-term changes. The degree and direction of paired pulse plasticity response did correlate with the initial probability of release (for the critical synaptic site underlying most of the change) suggesting that enhanced plasticity may be present if the response starts from a lower initial probability. However, other factors may also be important in the determination of the direction and degree of short-term plasticity, since different sites appear to often respond in varying ways to the same conditioning impulse. Thus, the presence of a response on the conditioning pulse may lead to enhanced plasticity compared to responses which showed a failure on the first response. In contrast, Debanne et al. suggest that an initial response may lead to depression on the test pulse rather than potentiation (Debanne et al., 1996), possibly due to neurotransmitter depletion.

We have also previously evaluated the hypothesis that different release sites over the dendritic surface lead to amplitude and waveform shape differences due to electrotonic filtering of events (Turner, 1984). EPSCs from various data sets were found to contain events with different rise-times, which clearly reflected differences in electrotonic filtering. This electrotonic filtering is likely another source of postsynaptic heterogeneity. These data also leads us to suggest that minimal stimulation may lead to activation of only a few release sites onto individual CA1 cells, often located at different electrotonic locations, and that each site may possess different release probabilities and synaptic site amplitudes. Such variation has also been suggested by anatomical studies showing large variation in synaptic shape and presynaptic densities (Harris and Sultan, 1995). Thus, there is evidence for variable amplitudes of synaptic contributions (as recorded at the soma) due to dendritic spatial filtering, and possibly also due to intrinsic differences between synaptic sites located at similar spatial locations. Additionally, the data presented above indicates that probability of release may vary extensively between synapses, and there does not appear to be a strong rationale for assuming that probability is uniform among any mammalian CNS synapses (Redman, 1990). These findings argue for significant inter-site heterogeneity due to spatial and perhaps inherent site differences as well as likely a certain contribution of intrinsic 'quantal' variability due to changes in release over time. The intrinsic variability can arise from such sources as variable neurotransmitter packaging in vesicles and variable ligand–receptor binding in the synaptic cleft (Faber et al., 1992). The heterogeneity noted at both intrasite and intersite levels of analysis suggests strongly that changes with plasticity may also vary considerably between synapses, possibly as a function of history of use of each synapse. The important role of intrinsic variance is similar to that described by Bolshakov and Siegelbaum (1995; approximately 20–35% in their analyzed data sets) but more than that suggested by Stricker et al. (1996a,b; less than 10% in the analyzable data sets).

5. CONCLUSIONS AND FUTURE PERSPECTIVES

Depending on the suspected number of synaptic release sites there are several applicable statistical models. Because of suspected significant heterogeneity in the hippocampal CA1 neurons, as discussed above, we have applied the Bayesian analysis because it is the most general and unconstrained in terms of site parameter freedom. As described in Turner et al. (1997) the Bayesian complex framework was required for interpretation of the ensemble in a little over half of the datasets, and so clearly allows extension of the number of datasets which can be analyzed. However, limitations include the potential loss of identifiability of data sets with larger numbers of synaptic sites (possibly greater than 5–6 sites), the current lack of realistic noise modelling for complex noise and the lack of any on-line feedback during recordings to show the number of sites activated on a trial by trial basis. Thus, further developments anticipated include simulations to model additional numbers of sites, improved noise characteristics, better characterization of data set validity and stationarity, addition of a dynamic updating to allow some on-line analysis during an experiment and placing the data stream from individual synaptic sites into a time series format of analysis. The latter will allow historical effects to be included, such as potentiation with time, in a dynamic format, and may be more predictive of future trends. Addition of dendritic models to include conduction from dendritic sites to the soma may also enhance the link between synaptic inputs and cell firing capabilities and cell behavior in general.

In addition to the scientifically and statistically live issues arising in connection with investigations of intrinsic variability, several other areas are of current interest and are under initial investigation. As presented, our models make no explicit allowance for outliers in signal recordings, nor for systematic features such as arise when synaptic tissue exhibits inhibitory as well as excitatory responses to stimuli. An example of the latter is discussed in West and Turner (1994), for example. Perhaps the most needed generalization, currently, is to relax the assumption of normally distributed experimental noise. This is currently in development, using noise models based on work in Bayesian density estimation following Escobar and West (1995), and West, Muller and Escobar (1994), and preliminary results will be reported shortly, elsewhere. See Cao and West (1997) for related developments.

One beginning area of research involves issues of dependence in release characteristics over time and across sites. On the first, synaptic responses in general may be viewed as a stream where, at each site, the response may depend on previous occurrences. Current analyses assume independence. The very nature of the paired pulse potentiation experiments, however, explicitly recognize dependence effects, both in terms of response/no response and in terms of response levels, and of the importance of time elapsed between successive stimuli generating the responses. Statistical developments of time series issues are of interest here to explore and, ultimately, model time variation in patterns of synaptic transmission. Interest in transmission dependencies also extends across sites, i.e. to investigations of 'connectivities' between sites that may have physiological interpretation and importance, though serious inroads into study of this issue is some way in the future.

It will be apparent that the technical aspects of simulation based analysis in the novel mixture models here bear further study, from a practical perspective. It is certainly desirable to investigate variations on the basic Gibbs sampling schemes that enhance and improve convergence characteristics. It is possible that hybrid schemes based, in part, on the use of auxiliary variables (e.g. Besag and Green, 1993) will prove useful here; the need for improved computational tools will be more strongly felt in more complex models incorporating non-normal noise components. That said, the current approach described and illustrated here provides a useful basis for exploring and assessing synaptic data in attempts to more directly evaluating the structure and stochastic characteristics of synaptic transmission as it is currently viewed in the neurophysiological community. Our current work on the applications side is focused on refining this basic framework and in exploring ranges of synaptic data sets to isolate and summarize the diversity of observed response mechanisms in mammalian central nervous systems.

PROBLEMS

3.1. How would failures of neural transmission at axon branch points (in addition to those occuring at presynaptic terminals due to changes in calcium fluxes) affect the histogram appearance and analysis of a unitary synaptic event over multiple trials? (Hint: see article by Stricker et al., 1994 on Binomial analysis.)

3.2. What are the problems associated with analysis of histograms alone (such as a 'peakiness' index) in identifying patterns or clusters of synaptic events? (Hint: see the article by Walmsley (1995)).

3.3. Background or ambient noise is a critical problem in the analysis of synaptic events in mammalian neurons, even with low-noise whole cell electrodes. Thus, it is usually assumed that noise occuring during a synaptic event is similar to that recorded in the absence of a synaptic event, to characterize background noise in isolation. What if the ambient noise changes during an actual synaptic event (either less or more)? What are possible physiological reasons for this to occur and how could the analysis be modified to be more realistic?

3.4. If synaptic sites are close together the receptor activation of one site may influence another site in either a negative or positive manner, through direct interactions. What are the physiological possibilities for this type of interaction and how would such an influence be adapted into a framework for analysis of cooperativity? Would such an influence violate the current approaches to synaptic release stsatistics?

3.5. A critical differentiation can be constructed between two scenarios where the histograms may appear superficially similar: (a) a single synaptic site may show a large (greater than 40%) variation in amplitude over time; or (b) there may be two sites which differ slightly and which show only a moderate degree of intrinsic variation over time. How would these two scenarios be differentiated physiologically and statistically?

3.6. As the number of active synaptic sites involved in a response increases, each individual site becomes correspondingly less identifiable, unless there is a large difference between sites in terms of their site parameters. How is the presence of active multiple sites apparent in both histograms and analysis? What if the number of sites changes, as with various forms of potentiation?

ACKNOWLEDGMENTS

Supported by the National Science Foundation (DAT, MW), Department of Veterans Affairs Merit Review (DAT), Wellcome Trust (YC, HVW) and Medical Research Council (HVW).

REFERENCES

Andreasen, M. and Hablitz, J.J. (1994) Paired-pulse facilitation in the dentate gyrus: A patch-clamp study in rat hippocampus *in vitro*. *J. Neurophysiol.*, **72**, 326–335.

Bekkers, J.M., Richerson, G.B. and Stevens, C.F. (1990) Origin of variability in quantal size in cultured hippocampal neurons and hippocampal slices. *Proc. Natl. Acad. Sci.*, **87**, 5359–5362.

Bennett, M.R. (1995) The origin of Gaussian distributions of synaptic potentials. *Prog. Neurobiol.*, **46**, 331–350.

Besag, J. and Green, P.J. (1993) Spatial statistics and Bayesian computation. *J. R. Stat. Soc., Ser. B*, **55**, 25–37.

Bolshakov, V.Y. and Siegelbaum, S.A. (1995) Regulation of hippocampal transmitter release during development and long-term potentiation. *Science*, **269**, 1730–1734.

Borst, J.G.G., Lodder, J.C. and Kits, K.S. (1994) Large amplitude variability of GABAergic IPSCs in melanotropes from Xenopus Laevis: Evidence that quantal size differs between synapses. *J. Neurophysiol.*, **71**, 639–655.

Brown, T.H., Perkel, D.H. and Feldman, M.A. (1976) Evoked neurotransmitter release: statistical effects of nonuniformity and nonstationarity. *Proc. Natl. Acad. Sci.*, **73**, 2913–2917.

Buhl, E.H., Halasy, K. and Somogyi, P. (1994) Diverse source of hippocampal unitary inhibitory post-synaptic potentials and the number of synaptic release sites. *Nature*, **368**, 823–828.

Cao, G. and West, M. (1997) Bayesian analysis of mixtures of mixtures. *Biometrics*, **52**, 221–227.

Clements, J.D. (1990) A statistical test for demonstrating a presynaptic site of action for a modulator of synaptic amplitude. *J. Neurosci. Meth.*, **31**, 75–88.

Clements, J.D. (1996) Transmitter timecourse in the synaptic cleft: its role in central synaptic function. *Trends Neurosci.*, **19**, 163–171.

Debanne, D., Guerineau, N.C., Gahwiler, B.H. and Thompson, S.M. (1996) Paired-pulse facilitation and depression at unitary synapses in rat hippocampus: quantal fluctuations affects subsequent release. *J. Physiol. (Lond.)*, **491**, 163–176.

Del Castillo, J. and Katz, B. (1954) Quantal components of the end-plate potential. *J. Physiol. (Lond.)*, **124**, 560–573.

Dityatev, A.E. and Clamann, H.P. (1993) Limits of quantal analysis reliability: quantal and unimodal constraints and setting of confidence intervals for quantal size. *J. Neurosci. Meth.*, **50**, 67–82.

Durand, G.M., Kovalchuk, Y. and Konnerth, A. (1996) Long term potentiation and functional synapse induction in developing hippocampus. *Nature*, **381**, 71–75.

Edwards, F.A., Konnerth, A. and Sakmann, B. (1990) Quantal analysis of inhibitory synaptic transmission in the dentate gyrus of rat hippocampal slices: a patch-clamp study. *J. Physiol. (Lond.)*, **430**, 213–249.

Escobar, D.M. and West, M. (1995) Bayesian density estimation and inference using mixtures. *J. Am. Stat. Assoc.*, **90**, 577–588.

Faber, D.S., Young, W.S., Legendre, P. and Korn, H. (1992) Intrinsic quantal variability due to stochastic properties of receptor–transmitter interactions. *Science*, **258**, 1494–1498.

Gelfand, A.E. and Smith, A.F.M. (1990) Sampling based approaches to calculating marginal densities. *J. Am. Stat. Assoc.*, **85**, 398–409.

Harris, K.M. and Sultan, P. (1995) Variation in the number, location and size of synaptic vesicles provides an anatomical basis for nonuniform probability of release at hippocampal CA1 synapses. *Neuropharmacology*, **34**, 1387–1395.

Hessler, N.A., Shrike, A.M. and Malinow, R. (1993) The probability of release at a mammalian central synapse. *Nature*, **366**, 569–572.

Isaac, J.T., Nicoll, R.A. and Malenka, R.C. (1995) Evidence for silent synapses: Implications for the expression of LTP. *Neuron*, **15**, 427–434.

Jack, J.J.B., Larkman, A.U., Major, G. and Startford, K.J. (1994) Quantal analysis of the synaptic excitation of CA1 hippocampal pyramidal cells. In, *Molecular and Cellular Mechanisms of Neurotransmitter Release*. Sturme, L., Greengaard, P., Grillner, S. and Hokfelt, T. (Eds). New York: Raven Press, pp. 275–299.

Jonas, P., Major, G. and Sakmann, B. (1993) Quantal components of unitary EPSCs at the mossy fibre synapse on CA3 pyramidal cells of rat hippocampus. *J. Physiol. (Lond.)*, **472**, 615–663.

Katz, B. (1969) *The Release of Neural Transmitter Substances*. Liverpool: Liverpool University Press.

Korn, H. and Faber, D.S. (1991) Quantal analysis and synaptic efficacy in the CNS. *Trends Neurosci.*, **14**, 439–445.

Kullmann, D.M. (1989) Applications of the expectation–maximization algorithm to quantal analysis of post-synaptic potentials. *J. Neurosci. Meth.*, **30**, 231–245.

Kullmann, D.M. and Nicoll, R.A. (1992) Long-term potentiation is associated with increases in quantal content and quantal amplitude. *Nature*, **357**, 240–244.

Larkman, A.U., Jack, J.J.B., and Stratford, K.J. (1997) Assessment of the reliability of amplitude histograms from excitatory synapses in rat hippocampal CA1 *in vitro*. *J. Physiol. (Lond.)*, **505**, 443–456.

Larkman, A., Stratford, K. and Jack, J. (1991) Quantal analysis of excitatory synaptic action and depression in hippocampal slices. *Nature*, **350**, 344–347.

Liao, D., Jones, A. and Malinow, R. (1992) Direct measurement of quantal changes underlying long-term potentiation in CA1 hippocampus. *Neuron*, **9**, 1089–1097.

Ling, L. and Tolhurst, D.J. (1983) Recovering the parameters of finite mixtures of normal distributions from a noisy record: an empirical comparison of differing estimating procedures. *J. Neurosci. Meth.*, **8**, 309–333.

Lisman, J.E. and Harris, K.M. (1993) Quantal analysis and synaptic anatomy-integrating two views of hippocampal plasticity. *Trends Neurosci.*, **16**, 141–147.

Liu, G. and Tsien, R.W. (1995) Properties of synaptic transmission at single hippocampal synaptic boutons. *Nature*, **375**, 404–408.

MacEachern, S.N. (1994) Estimating normal means with a conjugate style Dirichlet process prior. *Commun. Statist: Simulation and Computation*, **23**, 727–741.

Magee, J.C. and Johnston, D. (1997) A synaptically controlled, associative signal for Hebbian plasticity in hippocampal neurons. *Science*, **275**, 209–212.

Major, G. (1993) Solutions for transients in arbitrarily branching cables: III. Voltage clamp problems. *Biophys. J.*, **65**, 469–491.

Malinow, R. (1991) Transmission between pairs of hippocampal slice neurons: quantal levels, oscillations and LTP. *Science*, **252**, 722–724.

Manabe, T. and Nicoll, R.A. (1994) Long-term potentiation: Evidence against an increase in transmitter release probability in the CA1 region of the hippocampus. *Science*, **265**, 1888–1892.

Manabe, T., Wyllie, D.J.A., Perkel, D.J. and Nicoll, R.A. (1993) Modulation of synaptic transmission and long-term potentiation: Effects on paired-pulse facilitation and EPSC variance in the CA1 region of the hippocampus. *J. Neurophysiol.*, **70**, 1451–1459.

Martin, A.R. (1966) Quantal nature of synaptic transmission. *Physiol. Rev.*, **46**, 51–66.

Quastel, D.M.J. (1997) The binomial model in fluctuation analysis of quantal neurotransmitter release. *Biophys. J.*, **72**, 728–753.

Raastad, M., Storm, J.F. and Andersen, P. (1992) Putative single quantum and single fibre excitatory postsynaptic currents show similar amplitude range and variability in rat hippocampal slices. *Eur. J. Neurosci.*, **4**, 113–117.

Redman, S. (1990) Quantal analysis of synaptic potentials in neurons of the central nervous system. *Physiol. Rev.*, **70**, 165–198.

Rosenmund, C., Clements, J.D. and Westbrook, G.C. (1993) Non uniform probability of glutamate release at a hippocampal synapse. *Science*, **262**, 754–757.

Sayer, R.J., Friedlander, M.J. and Redman, S.J. (1990) The time course and amplitude of EPSPs evoked at synapses between pairs of CA3/CA1 neurons in the hippocampal slice. *J. Neurosci.*, **10**, 826–836.

Smith, A.F.M. and Roberts, G.O. (1993) Bayesian computation via the Gibbs sampler and related Markov chain Monte Carol methods. *J. R. Stat. Soc. Ser. B*, **55**, 3–23.

Spruston, N., Jaffe, D.B., Williams, S.H. and Johnston, D. (1993) Voltage- and space-clamp errors associated with the measurement of electrotonically remote synaptic events. *J. Neurophysiol.*, **70**, 781–802.

Stevens, C.F. and Wang, Y. (1994) Changes in reliability of synaptic function as a mechanism for plasticity. *Nature*, **371**, 704–707.

Stricker, C., Field, A.C. and Redman, S.J. (1996a) Statistical analysis of amplitude fluctuations in EPSCs evoked in rat CA1 pyramidal neurones *in vitro*. *J. Physiol. (Lond.)*, **490**, 419–441.

Stricker, C., Field, A.C. and Redman, S.J. (1996b) Changes in quantal parameters of EPSCs in rat CA1 neurones *in vitro* after the induction of long-term potentiation. *J. Physiol. (Lond.)*, **490**, 443–454.

Stricker, C. and Redman, S.J. (1994) Statistical models of synaptic transmission evaluated using the expectation–maximization algorithm. *Biophys. J.*, **67**, 656–670.

Stricker, C., Redman, S.J. and Daley, D. (1994) Statistical analysis of synaptic transmission: Model discrimination and confidence limits. *Biophys. J.*, **67**, 532–547.

Tierney, L.J. (1994) Markov chains for exploring posterior distributions. *Ann. Statist.*, **22**, 1701–1728.

Turner, D.A. (1984) Conductance transients onto dendritic spines in a segmental cable model of CA1 and dentate hippocampal neurons. *Biophys. J.*, **46**, 85–96.

Turner, D.A., Isaac, J., Chen, Y., Stockley, E. and Wheal, H.V. (1995) Analysis of dendritic synaptic sites in hippocampal CA1 pyramidal cells—assessment of variability and uniformity. In *Excitatory Amino Acids and Synaptic Transmission*, Thomson, S.M. and Wheal, H.V. (Eds). San Diego: Academic Press, pp. 173–189.

Turner, D.A., Chen, Y., Isaac, J., West, M. and Wheal, H.V. (1997) Excitatory synaptic site heterogeneity during paired pulse plasticity in CA1 pyramidal cells in rat hippocampus *in vitro*. *J. Physiol. (Lond.)*, **500**, 441–461.

Turner, D.A. and West, M. (1993) Bayesian analysis of mixtures applied to post-synaptic potential fluctuations. *J. Neurosci. Meth.*, **47**, 1–21.

Voronin, L.L. (1993) On the quantal analysis of hippocampal long-term potentiation and related phenomena of synaptic plasticity. *Neuroscience*, **56**, 275–304.

Wahl, L.M., Jack, J.J.B., Larkman, A.U. and Stratford, K.J. (1997) The effects of synaptic noise on measurement of evoked excitatory postsynaptic response amplitudes. *Biophys. J.*, **73**, 205–219.

Walmsley, B. (1995) Interpretation of 'quantal' peaks in distributions of evoked synaptic transmission at central synapses. *Proc. R. Soc. Lond. B*, **261**, 245–250.

Walmsley, B., Edwards, F.R. and Tracey, D.J. (1988) Nonuniform release probabilities underlie quantal synaptic transmission at a mammalian excitatory central synapse. *J. Neurophysiol.*, **60**, 889–908.

West, M. (1997) Hierarchical mixture models in neurological transmission analysis. *J. Am. Stat. Assoc.* **92**, 587–606.

West, M. and Harrison, P.J. (1997) Bayesian Forecasting and Dynamic Models. (2nd Ed.) New York: Springer Verlag.

West, M. and Turner, D.A. (1994) Deconvolution of mixtures in analysis of neural synaptic transmission. *Statistician*, **43**, 31–43.

Wong, K. and Redman, S.J. (1980) The recovery of a random variable from a noisy record with application to the study of fluctuations in synaptic potentials. *J. Neurosci. Meth.*, **2**, 389–409.

Zucker, R.S. (1973) Changes in the statistics of transmitter release during facilitation. *J. Physiol. (Lond.)*, **229**, 787–810.

CHAPTER 4
NATURAL VARIABILITY IN THE GEOMETRY OF DENDRITIC BRANCHING PATTERNS

Jaap van Pelt and Harry B.M. Uylings

1. INTRODUCTION

Dendritic branching patterns are complex and show a large degree of variation in their shapes, within as well as between different cell types and species. This variation is found in typical shape parameters, such as the number, length and connectivity pattern (topological structure) of the segments comprising the dendritic tree, the curved nature of these segments as well as the embedding of the dendrite in three-dimensional (3D) space. Dendritic structure plays an important role in the spatial and temporal integration of postsynaptic potentials, both metrically (e.g. Rall et al., 1992) and topologically (Van Pelt and Schierwagen, 1994, 1995), and is consequently an important determinant for the characteristics of neuronal firing patterns (e.g. Mason and Larkman, 1990; Mainen and Sejnowski, 1996). The implications of dendritic morphological variation for neuronal signal integration is therefore an important topic of research. However, one first needs to know the extent to which individual dendritic trees vary in their morphology, before the range of functional properties can be determined for the existing range of morphological types. Dendritic morphological variation can be expressed by means of distribution functions for detailed shape parameters. For the determination of the functional properties, however, one needs the full dendritic trees themselves. When a sufficient number of original reconstructions is not available, one needs to synthesize representative dendrites from models of dendritic structures. Two approaches can be distinguished in modeling the structure of dendritic trees.

In the *reconstruction model approach* dendritic shapes are reconstructed using empirical distribution functions for the shape parameters. Representative dendrites are obtained by random sampling of these distributions. For instance, Burke et al. (1992) used empirically obtained distribution functions for the length and diameters of dendritic branches and for the diameter relations at bifurcation points. Random dendrites were generated by a repeated process of random sampling of these distributions for deciding whether an elongating neurite should branch and for obtaining the diameters of the daughter branches. The modeled dendrites obtained in this way conform to the original distribution functions for shape characteristics.

An important assumption in this approach is that the shape parameters are independent from each other. Kliemann (1987) considered the segments at a given centrifugal order as individuals of a generation, which may give rise to individuals in a next generation (i.e. by producing a bifurcation point with two daughter segments). Mathematically, such a dendritic reconstruction could be described by a Galton–Watson process, based on empirically obtained splitting probabilities for defining per generation whether or not a segment will be a terminal or an intermediate one. Also in this example it is assumed that the successive choices in the reconstruction of a dendrite are independent. The model's predictability of branching patterns was greatly enhanced by including a systematic component additional to the random one (Carriquiry et al., 1991). Applications of this method to dendritic growth *in vitro* can be found in Uemura et al. (1995). Tamori (1993) used a set of six 'fundamental parameters' to describe dendritic morphology, and introduced a principle of least effective volume to derive dendritic branch angles.

In contrast to the reconstruction model approach, the *growth model approach* is based on modeling dendritic structure from principles of dendritic development. Neuromorphogenesis proceeds via the dynamic behavior of growth cones, specialized structures at the ends of outgrowing neurites, which mediate neuritic elongation, branching, retraction, redirection and synapse formation. Mature neurons have attained their morphological features as a result of this dynamic process. Among the many processes involved in growth cone dynamics are the exploratory interactions of filopodia with the local environment (e.g. Kater et al., 1994), receptor mediated transmembrane signalling to cytoplasmic regulatory systems (e.g. Letourneau et al., 1994) and the formation of the cytoskeleton in the elongating neurite, comprising polymerization of tubulin into microtubules (e.g. Black, 1994). During dendritic development growth cones migrate and branch. Upon branching, a growth cone splits into two, which continue to grow on their own. After a number of such events a branching pattern has been formed in which the number, length and connectivity pattern of the segments is determined by the branching and migration rates of the growth cones. Longitudinal studies on dendritic development *in vivo* are experimentally difficult to perform, because the tissue needs to be processed in order to enable morphological quantification. Tissue culture studies might form an alternative, although, especially in dissociated cultures, neurons develop in an environment, different from their natural one. However, the question how growth cone dynamics result in the typical dendritic shapes can also be approached by means of mathematical modeling. Hypotheses concerning the behavior of growth cones can then be studied for their consequences for dendritic morphology. Because of the large number of processes involved in the behavior of growth cones it is a reasonable assumption to consider the actual behavior of growth cones as outcomes of a stochastic process. Although the probability functions underlying this process may be very complex, it is an interesting question which minimal assumptions are required to obtain a realistic approximation in the modeled dendritic structures.

The *growth model approach* will be described in this chapter. It is based on the assumption that the growth actions are outcomes of a stochastic process. This approach has the advantage that the model parameters allow a direct interpretation in terms of the underlying developmental biological process. Additionally, possible dependencies among morphological parameters, arising during development, are implicitly accounted for. The *growth model approach* has been followed in the past decades by several authors (e.g. Smit et al., 1972; Berry and Bradley, 1976; Van Pelt and Verwer, 1983; Ireland et al., 1985; Horsfield et al., 1987; and Nowakowski et al., 1992). It will be shown that the assumptions of random branching and random elongation of terminal segments (or 'growth cones') are sufficient for generating dendrites with realistic variations in the topological and segment length shape parameters. Because of the structure of the model the variables are directly related to the dynamic behavior of growth cones, which allows for empirical verification. This makes the model also of importance for developmental studies because it provides a tool for analyzing reconstructed dendrites in terms of 'hypothesized' developmental histories, which themselves are often not accessible by empirical techniques.

In the following sections, dendritic shape parameters are introduced and examples are given for the extent of their variations as found in observed dendritic trees. The dendritic growth model is reviewed and discussed. The study of segment length distributions means a new extension to this model and the results are presented here for the first time. The functional implications of dendritic morphological variation are briefly discussed.

2. DENDRITIC SHAPE PARAMETERS

For the characterization of their shapes dendrites are generally simplified to regular geometric structures, which allow easy quantification. An example of such a simplification is given in Figure 4.1 for a cat superior colliculus neuron, in which the segments in the dendrites are stretched and represented by cylinders with a given length and diameter (Schierwagen, 1986).

The embedding in 3D-space and irregularity and curvature of the branches is hereby lost. After such a simplification a dendrite is characterized by the number, length, diameter and connectivity pattern (*topological structure* or *tree type*) of its segments, i.e. the branches between successive bifurcation points and/or end points. *Intermediate segments* (ending in a bifurcation point) and *terminal segments* are distinguished. A given segment can be labeled by its *degree* (denoting the number of terminal segments in its subtree) and/or by its *centrifugal order* (denoting the number of bifurcation points on the path from the root up to the segment). Bifurcation points are labeled by their *partitions*, i.e. the *degrees* of the two subtrees arising from that bifurcation. Dendritic morphological variation is expressed by means of variations in these shape parameters.

2.1 Dendritic Topology

2.1.1 *Connectivity Pattern*

Topological variation arises because of the finite number of *tree types* that are possible for a given number of segments. A summary of all different tree types with up to and including 8 terminal segments is given in Figure 4.2 (taken from Van Pelt

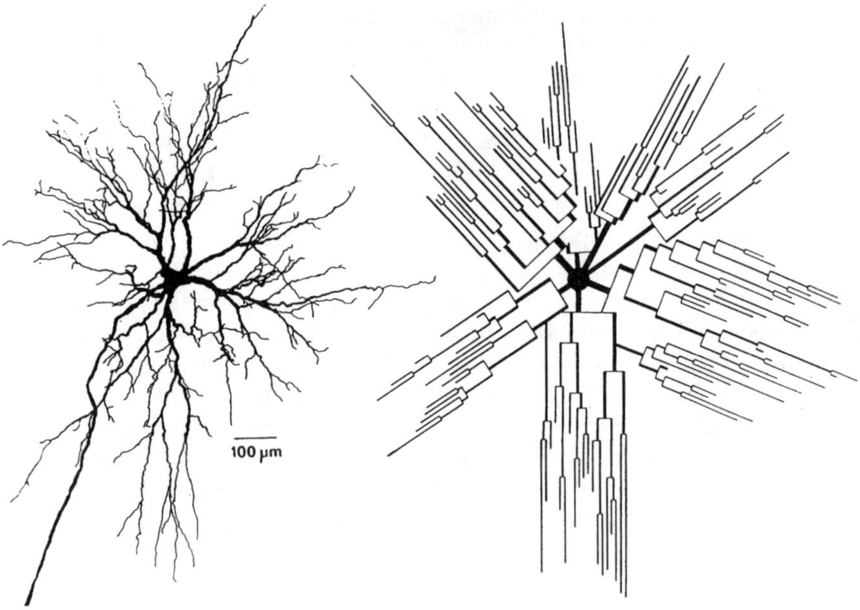

Figure 4.1 Left: Plane reconstruction of the soma-dendritic profile of a tecto-reticulo-spinal neuron of cat superior colliculus. Right: Approximation of neuronal arborizations by regular geometric bodies. (Reproduced from Schierwagen (1986) with permission of Schierwagen and the *J. Hirnforsch.*)

and Verwer, 1983). The trees are labeled by a rank number according to the ranking scheme described in Van Pelt and Verwer (1983). Further reviews of dendritic topological shape parameters are given by Uylings et al. (1989) and Verwer et al. (1992).

For 3D trees there is no distinction between left and right side at bifurcation points. Therefore, the trees could be displayed in a standardized way, such that at each bifurcation point the subtree with the highest rank number is drawn at right side. The number of 3D-tree types with n terminal segments, N_α^n, (α denoting the set of 3D-tree types) is given by the iterative equation

$$N_\alpha^n = \frac{1}{2}\left(\sum_{r=1}^{n-1} N_\alpha^r N_\alpha^{n-r} + (1-\varepsilon(n))N_\alpha^{n/2}\right) \quad \text{and} \quad N_\alpha^1 = 1 \tag{4.1}$$

(Harding, 1971), with

$$\varepsilon(n) = 0 \quad \text{for even } n \quad \text{and} \quad \varepsilon(n) = 1 \quad \text{for odd } n. \tag{4.2}$$

The number of tree types increases rapidly with n and a reasonable approximation for the order of magnitude is given by $N_\alpha^n \approx 2.4^n$ (Van Pelt et al., 1992). [Note, that for 2D branching patterns, like rivers, the left–right distinction is meaningful and

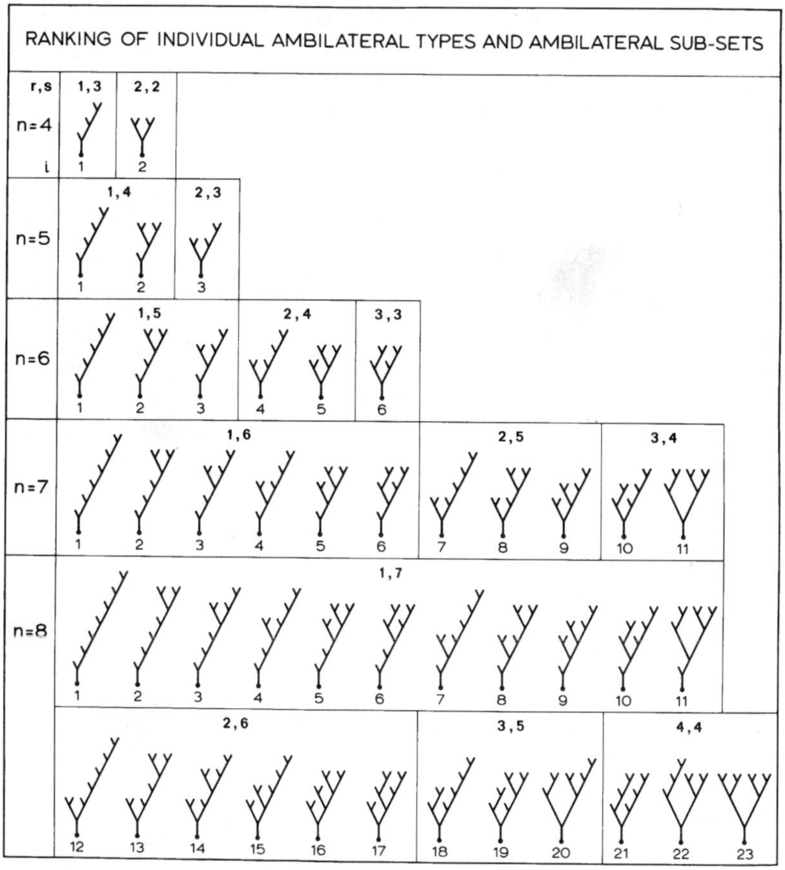

Figure 4.2 Table of tree types of *degree* 4–8. The trees are numbered according to a ranking scheme described in Van Pelt and Verwer (1983). The number pairs above groups of tree types indicate the *degrees* of the first-order subtrees. The tree types are plotted in a standardized way with, at a bifurcation, the subtree with the highest rank number plotted to the right side. (Reproduced from Van Pelt and Verwer, 1983.)

results in a larger number of 2D-tree types N_τ^n (with τ denoting the set of 2D-tree types) which relates to the number of terminal segments n by

$$N_\tau^n = (2n-1)^{-1} \binom{2n-1}{n} \tag{4.3}$$

(Caley, 1859; cf. Shreve, 1966)].

2.1.2 *Tree Asymmetry*

Tree-type frequency distributions become unmanageably large for not too small n necessitating the use of a numerical index for the topological structure to analyze dendritic trees. Among the used topological indices, the tree asymmetry index A_t has proven to be the most discriminative one, i.e. being able to distinguish most (but not all) of the possible tree types (Van Pelt et al., 1989). The tree-asymmetry index A_t of a given tree α^n with n terminal segments is defined by

$$A_t(\alpha^n) = \frac{1}{n-1} \sum_{j=1}^{n-1} A_p(r_j, s_j), \tag{4.4}$$

being the mean value of all the $n-1$ partition asymmetries $A_p(r_j, s_j)$ in the tree, which, at each of the $n-1$ bifurcation points, indicate the relative difference in the number of terminal segments r_j and s_j in the two subtrees emerging from the jth bifurcation point. The partition asymmetry A_p at a bifurcation is defined as

$$A_p(r, s) = \frac{|r-s|}{r+s-2} \tag{4.5}$$

for $r+s>2$, with r and s denoting the number of terminal segments in the two subtrees. $A_p(1,1)=0$ by definition. The discrete values of the tree asymmetry index for trees up to and including *degree* 8 are shown in Figure 4.3.

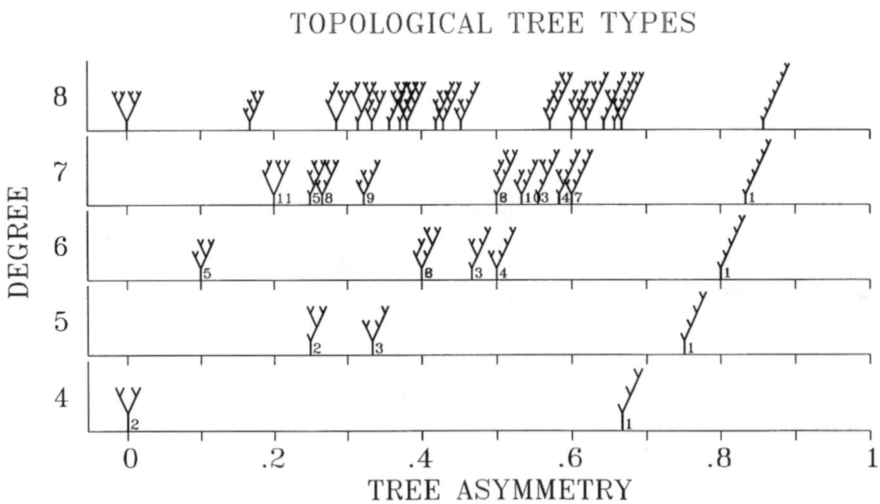

Figure 4.3 Tree-types of *degree* 4–8, plotted versus their tree-asymmetry value. The asymmetry values range from zero for fully symmetric trees (only possible when the *degree* is a power of two) to a maximum value for fully asymmetric trees, which value approaches one when the asymmetric tree becomes larger. The tree numbers correspond to those in Figure 4.2. Some trees have equal values for the asymmetry index and are plotted at the same position. Their numbers overlap like for tree numbers 2 and 6 of degree 6.

Two different ways are used to represent a topological tree by a linear string. The *label array* representation is based on a standardized procedure to trace a tree along its bifurcation points and terminal tips, labeled by 1's and 0's, respectively. The tree is uniquely represented by the sequence of labels (see also Overdijk et al., 1978), which starts at the root segment and, at bifurcations, first follows the right segment, while a label '1' is used for an intermediate segment and a label '0' for a terminal segment. For instance, tree number 11 of *degree* 7 in Figure 4.3 is represented by the array 1110010011000. For a tree of *degree* n the number of labels equals $2n-1$ with $n-1$ '1's and n '0's. Many of the $\binom{2n-1}{n}$ different permutations of such number of labels do not represent a tree. Actually, from each subset of permutations, which transform into each other by rotation, thus containing $2n-1$ permutations, only one of them represents a tree. This makes the total number of different (2D) tree types of degree n equal to $\binom{2n-1}{n}/(2n-1)$, which explains (4.3). The *branching code* representation of a tree indicates iteratively at each bifurcation the number of terminal segments in both subtrees within brackets. The code starts with the number of terminal segments in the complete tree. For instance, the mentioned tree in Figure 4.3 is represented by the branching code 7(3(1 2(1 1)) 4(2(1 1) 2(1 1))). The label arrays and branching codes of the tree types with up to and including 8 terminal segments are given in Table 4.1. Note, that the used ranking scheme (see also Van Pelt and Verwer, 1983) is based on recurrency in the succession of trees. For instance, the sequence of trees of *degree* 6 is repeated as first-order subtrees in trees of *degree* 7.

2.2 Dendritic Metrics

The metrical properties of the simplified dendritic tree are determined by the length and diameter of its segments. Pathlengths are defined by the summed length of the segments between two characteristic points, like for instance between the root point and a terminal tip. The radial distance between two such points is determined by the length of the straight connecting line. The difference between pathlength and radial distance indicates how irregular and curved the branches in the dendrite are.

2.2.1 *Segment Length*

The length of a curved segment is experimentally determined by approximating it by measuring a series of straight pieces.

2.2.2 *Segment Diameter*

Segments are not in all cases smooth structures but may be covered with spines. In such cases, the diameter is an ill-defined quantity. Even in the case of smooth dendrites, the diameter may not be constant along the segment. One speaks of tapering if the diameter gradually changes. Practically, in the case of irregularities the diameter is taken to be an averaged value over several measuring points.

Table 4.1 List of label arrays, branching codes and tree-asymmetry values of tree types with up to and including 8 terminal segments. The asymmetry index has been introduced by Van Pelt et al. (1992) and the given values are calculated according to option 1 for the asymmetry index definition in that paper (the values given by Uylings et al. (1989) have been calculated using option 2 and thus differ from the values listed below). Note, that the branching code of (sub)trees of degree 2 and 3 are expanded in the first two trees only.

Degree	*Number*	*Label array*	*Branching code*	*Asymmetry index*
1	1	0	1	0
2	1	100	2(1 1)	0
3	1	11000	3(1 2(1 1))	0.500
4	1	1110000	4(1 3)	0.667
	2	1100100	4(2 2)	0
5	1	111100000	5(1 4(1 3))	0.750
	2	111001000	5(1 4(2 2))	0.250
	3	111000100	5(2 3)	0.333
6	1	11111000000	6(1 5(1 4(1 3)))	0.800
	2	11110010000	6(1 5(1 4(2 2)))	0.400
	3	11110001000	6(1 5(2 3))	0.467
	4	11110000100	6(2 4(1 3))	0.500
	5	11100100100	6(2 4(2 2))	0.100
	6	11100011000	6(3 3)	0.400
7	1	1111110000000	7(1 6(1 5(1 4(1 3))))	0.833
	2	1111100100000	7(1 6(1 5(1 4(2 2))))	0.500
	3	1111100010000	7(1 6(1 5(2 3)))	0.556
	4	1111100001000	7(1 6(2 4(1 3)))	0.583
	5	1111001001000	7(1 6(2 4(2 2)))	0.250
	6	1111000110000	7(1 6(3 3))	0.500
	7	1111100000100	7(2 5(1 4(1 3)))	0.600
	8	1111001000100	7(2 5(1 4(2 2)))	0.267
	9	1111000100100	7(2 5(2 3))	0.322
	10	1111000011000	7(3 4(1 3))	0.533
	11	1110010011000	7(3 4(2 2))	0.200
8	1	111111100000000	8(1 7(1 6(1 5(1 4(1 3)))))	0.857
	2	111111001000000	8(1 7(1 6(1 5(1 4(2 2)))))	0.571
	3	111111000100000	8(1 7(1 6(1 5(2 3))))	0.619
	4	111111000010000	8(1 7(1 6(2 4(1 3))))	0.643
	5	111110010010000	8(1 7(1 6(2 4(2 2))))	0.357
	6	111110001100000	8(1 7(1 6(3 3)))	0.571
	7	111111000001000	8(1 7(2 5(1 4(1 3))))	0.657
	8	111110010001000	8(1 7(2 5(1 4(2 2))))	0.371
	9	111110001001000	8(1 7(2 5(2 3)))	0.419
	10	111110000110000	8(1 7(3 4(1 3)))	0.600
	11	111100100110000	8(1 7(3 4(2 2)))	0.314
	12	111111000000100	8(2 6(1 5(1 4(1 3))))	0.667
	13	111110010000100	8(2 6(1 5(1 4(2 2))))	0.381
	14	111110001000100	8(2 6(1 5(2 3)))	0.429
	15	111110000100100	8(2 6(2 4(1 3)))	0.452

Table 4.1 (*Continued*)

Degree	*Number*	*Label array*	*Branching code*	*Asymmetry index*
	16	111100100100100	8(2 6(2 4(2 2)))	0.167
	17	111100011000100	8(2 6(3 3))	0.381
	18	111110000011000	8(3 5(1 4(1 3)))	0.619
	19	111100100011000	8(3 5(1 4(2 2)))	0.333
	20	111100010011000	8(3 5(2 3))	0.381
	21	111100001110000	8(4(1 3) 4(1 3))	0.571
	22	111001001110000	8(4(1 3) 4(2 2))	0.286
	23	111001001100100	8(4(2 2) 4(2 2))	0

3. OBSERVED VARIABILITY IN DENDRITIC SHAPE PARAMETERS

3.1 Variation in Topological Structure

The topological variation in a set of dendrites is expressed by the frequencies of the different tree types. Such an outcome is given in Figure 4.4 for basal dendrites of rat cortical pyramidal cells (Uylings et al., 1990). Each panel displays the tree-type frequencies for a subgroup with a given number of terminal segments. The frequency distributions appear to be highly nonuniform, with some tree types abundantly present while others hardly occur. Especially for larger trees many possible tree types

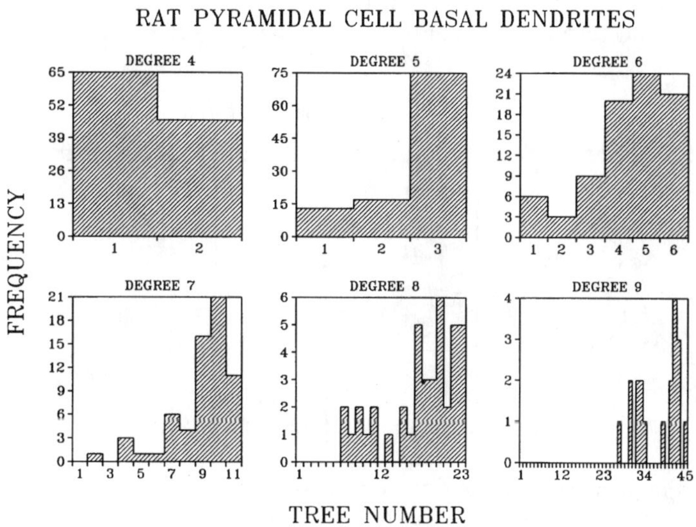

Figure 4.4 Frequency distributions of tree types observed in a set of basal dendrites of rat cortical pyramidal neurons. The tree numbers correspond to those in Figure 4.2. Note, that the distributions are highly nonuniform.

will not occur in a data set resulting in tree-type frequency distributions with many empty classes.

Such distributions become readily unmanageable. Alternatively, the topological variation can be expressed in terms of the tree-asymmetry index, resulting for the whole group of basal dendritic trees of adult rat cortical pyramidal neurons in a mean value of $\bar{A}_t = 0.384$ (SD = 0.217); a mean value of 0.43 (SD = 0.26) for multipolar nonpyramidal neurons, and of 0.492 (SD = 0.020) for rat cerebellar Purkinje cell dendritic trees (see Van Pelt et al., 1992). Dendrites from mature spinal motoneurons in the cat, rat and frog exhibited a narrow range of values for the asymmetry index (0.42–0.47) (Dityatev et al., 1995).

3.2 Variation in the Number of Dendritic Segments

The number of segments in dendritic trees shows large variations. In a recent study on terminal segment number distributions (Van Pelt et al., 1997) it was shown that the number of terminal segments per dendrite ranges from 1 up to about 50 in motoneurons in cat, rat or frog and from 1 up to about 15 in both rat pyramidal, multipolar nonpyramidal and in human dentate granule cells. The shape of the terminal segment number distributions turned out to be highly characteristic for the different cell types (see the figures in Van Pelt et al., 1997). Examples of such distributions for dendrites of 150 days old rat pyramidal and multipolar nonpyramidal cells, and of aged human fascia dentate granule cells are given in Figure 4.5.

3.3 Variation in Segment Length

A large variation has been observed in the length of dendritic segments. A general observation is that terminal segments can be substantially longer than intermediate

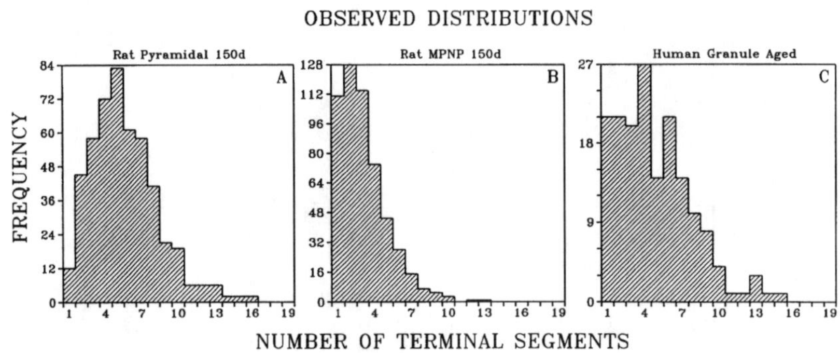

Figure 4.5 Frequency distributions of the number of terminal segments per dendritic tree for 150 days old rat cortical pyramidal (panel A) and multipolar nonpyramidal (MPNP) (panel B) cells (see Uylings *et al.*, 1990) and for human fascia dentate granule cells (panel C) (De Ruiter and Uylings, 1987).

ones as for instance is shown in pyramidal cell basal dendrites of layer V with $\overline{l}_t=$ 117 (SD$=$33) μm for terminal segments and $\overline{l}_i$(*median*)$=$11 μm for intermediate segments (Larkman, 1991), in superior colliculus neurons with $\overline{l}_t=115$ (SD$=$83.3) μm and $\overline{l}_i=80.3$ (SD$=$75.5) μm (Schierwagen and Grantyn, 1986) and in rat Purkinje neurons with terminal segments having a length of about 13 μm or 8 μm and intermediate segments having a length of about 5 μm (Woldenberg et al., 1993). An extensive review of intermediate and terminal segment lengths in dendrites of different cell types and species is given by Uylings et al. (1986, Table II). Additionally, terminal segment length may decrease strongly with increasing centrifugal order as is shown for instance by Uylings et al. (1978). Van Veen and Van Pelt (1993) showed that this decrease in length as well as the difference between intermediate and terminal segment lengths are predicted when the number and the positions of the branch points along the path from the soma to a terminal tip are determined by a Poisson process.

3.4 Variation in Dendritic Diameter

Dendritic segments have small diameters, down to the resolution of optical microscopes. The observed variation in segment diameters may therefore partly be attributed to measurements uncertainties. Nevertheless, a substantial part of the variation is induced by the different positions of the segments in the tree, caused by the correlation that is found between the diameter of a segment and its *degree* (Hillman, 1988). The neuritic cytoskeleton is believed to be a major factor determining the diameter of dendritic segments. Hillman (1979, 1988) assumed a positive correlation between segment diameter and the number of microtubules. This correlation also results in a relation found between the diameters of a parent (d_{p}) and its daughter segments (d_1 and d_2) at a bifurcation, formulated by Rall (1959) as a power law relation

$$d_{\mathrm{p}}^e=d_1^e+d_2^e, \tag{4.6}$$

with e being the branch power parameter. Segment diameters thus decrease across bifurcation points with terminal segments having the smallest diameters. Terminal segments were found to have mean diameters of 1.1 μm for Purkinje cells (Hillman, 1979), 0.7 μm for pyramidal neurons (Hillman, 1979; Larkman, 1991), 0.7 (SD$=$0.3) μm in cat phrenic motoneurons (Cameron et al., 1985) and about 1 μm for cat superior colliculus neurons (Schierwagen and Grantyn, 1986). Values for the branch power were reported of $e=2$ for rat Purkinje and neocortical pyramidal neurons (Hillman, 1979), of $e-1.47$ (SD$-$0.3) (Schierwagen and Grantyn, 1986) for dendrites of cat superior colliculus neurons, of $1.5<e<2$ for rat visual cortex pyramidal neurons (Larkman et al., 1992), and of 3/2 for the distal parts of rat dentate granule cells (Desmond and Levy, 1984). Cullheim et al. (1987) found a value of 1.13 (SD$=$0.34) for the mean ratio of $d_{\mathrm{p}}^{1.5}/(d_1^{1.5}+d_2^{1.5})$ in cat α motoneurons.

4. MODELING DENDRITIC BRANCHING PATTERNS

In the *growth model approach* it is assumed that dendritic growth proceeds by a sequence of branching events. At each branching event a new (terminal) segment is attached to one of the existing segments in the tree, thereby creating a new bifurcation point. Randomness is assumed with respect to (1) the selection of the branching segment from the existing ones, (2) the occurrence of a branching event in time, and (3) the elongation of the segments (because of the migration of growth cones). Thus, the question focusses on whether a simple scheme of probabilities underlying these random choices can be found which will accurately describe variations in observed dendritic trees. In constructing the dendritic growth model the three assumptions of randomness will be implemented separately. It will be shown that the variability in topological tree types can fully be accounted for by an appropriate probability scheme for the selection of the branching segment from the existing ones. The observed variability in the number of terminal segments can accurately be accounted for by an appropriate probability scheme for the occurrences of branching events in the course of time. The variation in segment length can be well described by random growth cone migration additional to the branching process.

4.1 Modeling Topological Variation (*QS*-model)

Topological variability arises when branching events variably occur at different segments in the growing tree. In the so-called *QS*-model, the selection probability of a segment for branching is taken to depend on the type of the segment and on its centrifugal order. The selection probability p_{term} for branching of a terminal segment at centrifugal order γ is defined by $p_{\text{term}}=C_1 2^{-S\gamma}$, with parameter S modulating the dependence on centrifugal order, and C_1 being a normalization constant to make the sum of the branching probabilities of all the segments in the tree equal to one. For $S=0$, all terminal segments have the same probability for being selected for branching. For $S=1$, the branching probability of a terminal segment will decrease by a factor of two for each following centrifugal order. The selection probability of an intermediate segment p_{int} for branching relates to that of a terminal segment of the same order via $p_{\text{int}}=(Q/(1-Q))p_{\text{term}}$, with parameter Q having values between 0 and 1. The parameter Q roughly indicates the total branching probability for all intermediate segments in a tree, with for $Q=0$ branching of terminal segments only, and for $Q=1$ branching of intermediate segments only. With the two parameters Q and S, a range of growth modes can be described, including the well-known random terminal mode of growth (*rtg*) with $(Q,S)=(0,0)$ and the random segmental mode of growth (*rsg*) with $(Q,S)=(0.5,0)$. An accurate description of the topological variability in dendrites from several neuron types and species could be obtained by assuming branching to occur at terminal segments only (i.e. $Q=0$), with possibly a slight dependence of the selection probability on the centrifugal order (small or zero value for S) (Van Pelt and Verwer, 1986; Van Pelt et al., 1992; Dityatev et al., 1995). The *QS*-model reduces to the *S*-model for $Q=0$ and to the

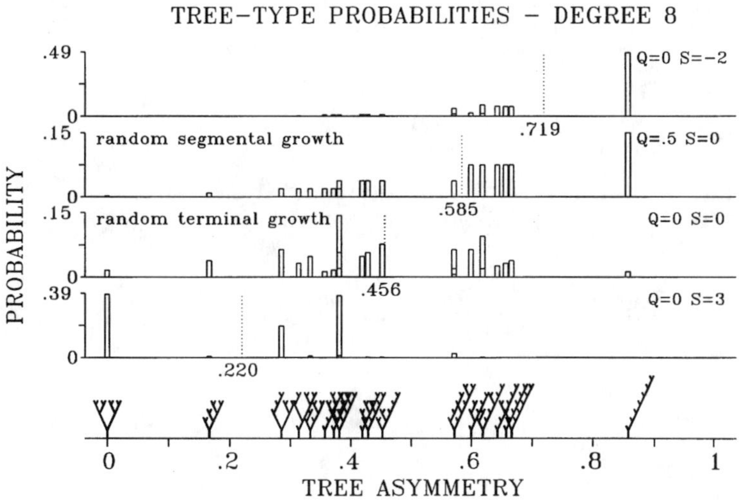

Figure 4.6 Plot of probabilities of occurrence of tree types of *degree* 8. The trees are plotted versus the tree-asymmetry values in the bottom row. The probabilities have been calculated for four different modes of dendritic growth, i.e. for $(Q, S)=(0, -2)$ in the top row, the random segmental mode of growth $(Q, S)=(0, 0.5)$ in the second row, the random terminal mode of growth $(Q, S)=(0, 0)$ in the third row, and for $(Q, S)=(0, 3)$ in the fourth row. The mean value of the tree-asymmetry index for a given mode of growth is indicated by a dotted line. The figures clearly show, that the tree-type probabilities and the tree-asymmetry expectations depend strongly on the mode of growth. Note, that probability bars of trees with equal asymmetry index do overlap.

Q-model for $S=0$. The probabilities of occurrence of tree types appear to depend strongly on the mode of growth. This is clearly shown in Figure 4.6, displaying the probabilities of trees of *degree* 8 for four different modes of growth, including *rtg* and *rsg*.

Analytical expressions for the probabilities and expected values of some shape parameters could be obtained for the Q-model only. Dendritic growth according to the Q-model results in partition probabilities $p(r, n-r|Q)$ (i.e. probabilities for a tree of *degree* n to have first-order subtrees of *degrees* r and $(n-r)$ given by

$$p(r, n-r|Q)=2^{1-\delta(r,n-r)}\left\{1+Q\left(\frac{n(n-1)}{2r(n-r)}-2\right)\right\}\frac{1}{n-1-Q}\prod_{i=1}^{r-1}\frac{1-Q/i}{1-Q/(i+n-r-1)}, \quad (4.7)$$

$$p(1, n-1|Q)=\frac{2+Q(n-4)}{n-1-Q}, \quad (4.8)$$

$$\lim_{n\to\infty} p(1, n-1|Q)=Q, \quad (4.9)$$

$$p(r,n-r|rtg)=\frac{2^{1-\delta(r,n-r)}}{n-1}, \tag{4.10}$$

$$p(r,n-r|rsg)=\frac{N_\tau^r N_\tau^{n-r}}{N_\tau^n}2^{1-\delta(r,n-r)} \tag{4.11}$$

(Van Pelt and Verwer, 1985) with δ denoting the Kronecker δ.

For the random terminal growth mode (*rtg*) the tree-asymmetry expectation is given by

$$E\{A_t^n|rtg\}=\frac{2n}{3(n-1)}\left\{\frac{2-3n_e/n}{4(n_e-1)}-\frac{2}{n_e}+\sum_{k=n_e/2}^{n_e}\frac{1}{k}\right\} \tag{4.12}$$

with $n_e=n-\varepsilon(n)$ and $\varepsilon(n)$ defined in (4.2).

$$\lim_{n\to\infty}E\{A_t^n|rtg\}=-\frac{1}{3}+\sum_{k=1}^{\infty}\frac{1}{(k+1)(2k-1)}=\frac{2}{3}\ln 2=0.4621 \tag{4.13}$$

(Van Pelt et al., 1992).

For the random segmental growth mode (*rsg*) is the tree-asymmetry expectation given by

$$E\{A_t^n|rsg\}=\frac{2}{(n-1)N_\tau^n}\sum_{m=2}^{n}\frac{2(n-m)-1}{m-2}N_\tau^{n-m} \tag{4.14}$$

$$\times\sum_{r-1}^{[m/2]}\{2-\delta(r,m-r)\}(m-2r)N_\tau^r N_\tau^{m-r}$$

(Van Pelt et al., 1992). These expectations are indicated in Figure 4.6 by the dotted lines for the displayed modes of growth.

The *QS*-model has been used to analyze dendrites from a variety of neuronal cell types and species (Van Pelt et al., 1992; Dityatev et al., 1995). In all cases optimized parameter values could be found so as to accurately reproduce the observed topological variation. A given value for the mean asymmetry index, however, is not uniquely associated with a particular combination of *QS* values, but with a contour in *QS*-space. For values of the asymmetry index smaller than about 0.5, the contours remain close to the *S*-axis. All contours start at the *S*-axis, implying that the *S*-axis ($Q=0$) is able to generate a full range of asymmetry expectations (Van Pelt et al., 1992). At present, all reported asymmetry values for neuronal dendrites are smaller than 0.5. We shall therefore assume in the following that $Q=0$, i.e. that branching occurs at terminal segments only, while the probability for a terminal segment to be selected could depend slightly on its position in the tree (i.e. its centrifugal order).

4.2 Modeling the Variation in the Number of Terminal Segments per Dendrite (*BE*-, *BES*- and *BEST*-Model)

Variation in the number of segments per dendritic tree emerges when trees experience a variable number of branching events during outgrowth. In the so-called

BE-growth model (Van Pelt et al., 1997) it is assumed that branching events occur at random points in time and at terminal segments only. To this end, the developmental time period is divided into a number of N time-bins with not necessarily equal durations. In each time-bin i a terminal segment in the tree may branch with a probability given by

$$p_i = B/Nn_i^E, \tag{4.15}$$

with the parameter B denoting the expected number of branching events at an isolated segment in the full period, and the parameter E denoting the dependence of the branching probability of a terminal segment on the total number of terminal segments n_i in the growing tree. Equation (4.15) can also be written as $p_i = Dn_i^{-E}$ with $D = B/N$ denoting the branching probability per time-bin of an isolated terminal segment. The duration of the time-bins is taken sufficiently small (i.e. the number of time-bins sufficiently large) so as to make the branching probabilities p_i much smaller than one, and making the probability for more than one branching event per time-bin in the tree negligibly small. For $E=0$, the branching probability is constant, independent of the number of terminal segments in the tree.

Examples of growth sequences are given in Figure 4.7 for parameter values $B=3$ and $E=0$. The two sequences show the random occurrences of branching events in time as well as the variation in the number of terminal segments in the final trees. Additionally, they show that the number of branching events in the full period can be (much) larger than the value of $B=3$, because of the proliferation in the number of terminal segments during growth. The variation in terminal segment number is most clearly shown in Figure 4.8 in which the distribution functions are displayed of the number of terminal segments per dendritic tree in sets of trees obtained for $B=4$ and for different values of the growth parameter E. For $E=0$, the distribution is monotonously decreasing with increasing number of terminal segments and has a

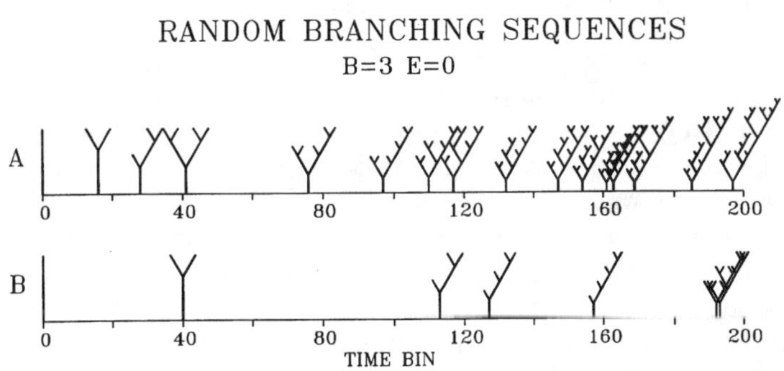

Figure 4.7 Growth of a branching pattern versus a time-bin scale for the parameter values $B=3$ and $E=0$. The full time period is divided into $N=200$ time-bins. At each time-bin, any of the terminal segments in the growing tree may branch with a constant probability. When a branching event has actually occurred, the new tree is plotted at the time-bin of concern.

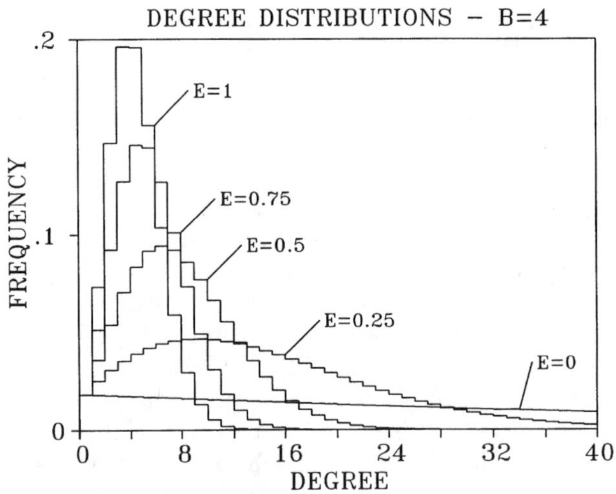

Figure 4.8 *Degree* distributions of trees obtained by growth modes with $B=4$ and different values for the parameter E, i.e. with different dependencies of the branching probability of a terminal segment on the total number of terminal segments in the tree. For $E=0$, the distribution is monotonically decreasing and has a long tail. For increasing values of E, the distributions get a modus and become increasingly narrower.

long tail. For $E>0$, the distributions become unimodal, and increasingly narrower while long tails disappear for increasing values of E.

The distribution of the number of terminal segments in dendritic trees after a period of growth can be calculated by means of the recurrent expression

$$P(n,i)=\sum_{j=0}^{n/2} P(n-j,i-1)\binom{n-j}{j}[p(n-j)]^j[1-p(n-j)]^{n-2j}, \tag{4.16}$$

(Van Pelt et al., 1997) with $P(n,i)$ denoting the probability of a tree of *degree* n at time-bin i with $P(1,1)=1$, and $p(n)$ denoting the branching probability per time-bin of a terminal segment in a tree of *degree* n, with $p(n)=Bn^{-E}/N$. A tree of *degree* n at time-bin i emerges when j branching events occur at time-bin $i-1$ in a tree of *degree* $n-j$. The recurrent equation expresses the probabilities of all these possible contributions from $j=0,\ldots,n/2$. The last two terms express the probability that, in a tree of *degree* $n-j$, j terminal segments will branch while the remaining $n-2j$ terminal segments will not do so. The combinatorial coefficient $\binom{n-j}{j}$ expresses the number of possible ways of selecting j terminal segments from the existing $n-j$ ones.

In the recent study of terminal segment number distributions from dendrites of different cell types and species (Van Pelt et al., 1997), the *BE*-model was able to reproduce accurately the shape of all these distributions. Optimal parameter values were found of $(\bar{B}(\text{sem}), \bar{E}(\text{sem}))=(3.55(0.17), 0.68(0.02))$ for rat pyramidal cell basal dendrites, of $(3.92(0.4), 0.29(0.04))$ for the combined group of motoneurons from cat, rat and frog, of $(\bar{B},\bar{E})=(2.36, 0.38)$ for human dentate granule cells, of $(B,E)=(1.49, 0.42)$ for rat multipolar nonpyramidal cells, and of $(50, 0.54)$ for one month rat cerebellar Purkinje cells.

Including Topological Variation into the *BE*-Model (*BES*-Model)

In the *BE*-model all terminal segments have equal probability for branching and the topological variation, produced by the *BE*-model is similar to that produced by the random terminal growth mode $(Q=0, S=0)$. An accurate account of the topological variability can now be given in a combined *BES*-model by taking the branching probability of a terminal segment per time-bin p_i also dependent on the centrifugal order of the segment, like in the *S*-model, such that

$$p_i = C2^{-S\gamma} B/Nn_i^E, \tag{4.17}$$

with γ denoting the centrifugal order of the terminal segment and $C=n/\sum_{j=1}^{n} 2^{-S\gamma_j}$ being a normalization constant, with a summation over all n terminal segments. The normalization ensures that the summed branching probability per time-bin of all the terminal segments in the tree is independent of the value of S.

Development in Real Time (*BEST*-Model)

The time-bins have been introduced without specifying their durations. To calculate the branching process in real time, the time-bin scale has to be mapped onto the real time scale. The branching probability per time-bin $p_i = Dn_i^{-E}$ then transforms into a branching probability per unit of time $p(t) = Dn_i^{-E}/\Delta T_i$ with ΔT_i being the duration of time-bin i. A linear mapping involves equal durations of the time-bins. A typical example of a growth curve of the terminal segment number per dendrite is given in Figure 4.9. It shows how the mean and standard deviation increase in time. Also the

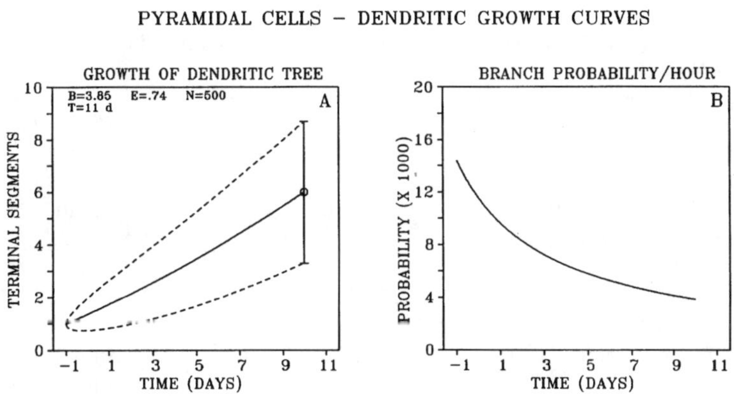

Figure 4.9 Growth curve of the mean terminal segment number in dendritic trees for the growth mode $B=3.85$, $E=0.74$. The dashed lines indicate the standard deviation intervals around the mean. The right panel indicates how the branching probability per hour declines with age because of the increasing number of terminal segments and the nonzero value of E.

growth rate increases with the highest value at the end of the period. The branching probability per unit of time, however, declines in time, because of the increasing number of terminal segments in the growing tree. A nonlinear mapping of the time-bins to real time modifies the shape of the growth curve, but, importantly, does not change the relation between mean and standard deviation. Therefore it offers the possibility to adapt the growth curve to a set of observed values at different points in time for the mean and standard deviation (see Van Pelt et al., 1997 for an example of Purkinje cell development).

4.3 Modeling the Variation in the Length of Dendritic Segments (*BESTL*-Model and Simulation Procedure)

4.3.1 *BESTL Model*

The lengths of segments are determined by both the rate of elongation and branching of terminal segments. In the previous section it has been shown that the branching process can be estimated from the dendritic segment number distribution. Based on this branching process, segment lengths can now be studied by including segment elongation (because of growth cone migration). As an example we will study the growth of dendrites of large layer V rat cortical pyramidal neurons. The relevant empirical findings for these neurons are summarized in Table 4.2. Note, that growth starts one day before birth, and continues with branching up to day 10, while the elongation of segments continues up to day 18 (Uylings et al., 1994).

The topological variation could be well described for the model parameters $Q=0$, $S=0.87$ (Van Pelt et al., 1992), while the terminal segment number distribution was accurately be reproduced by the branching process parameters $B=3.85$ and $E=0.74$ (Van Pelt et al., 1997) (see also Table 4.3). Equal durations of the time-bins will be

Table 4.2 Developmental and shape characteristics of basal dendrites of large layer V rat cortical pyramidal neurons

Onset of branching/elongation	≈ day 1 (24 h) an	Uylings (unpubl. observ.)
Stop of branching	day 10 (240 h) pn	Uylings *et al.*, 1994
Stop of elongation	day 18 (432 h) pn	Uylings *et al.*, 1994
# terminal segments per dendrite	6.0 (SD = 2.7)	Larkman, 1991
Intermediate segment length	11 (median) μm	Larkman, 1991
Terminal segment length	117 (SD = 33) μm	Larkman, 1991
Pathlength to tips	156 (SD = 29) μm	Larkman, 1991
Total dendritic length	777,6 μm	*
Terminal segment diameter	0.8 (SD = 0.2) μm	Larkman, 1991
Branch power	$1.5<e<2$	Larkman *et al.*, 1992
Tree asymmetry	0.38 (SD = 0.22)	Van Pelt *et al.*, 1992

Abbreviations: an—before birth.

* The mean total length per dendritic tree of 777.6 μm is estimated from the total basal dendritic length per pyramidal neuron of 4.510 mm (Larkman and Mason, 1990), and the mean number of basal dendrites per pyramidal neuron of 5.8 (SD = 1.8) (Larkman, 1991).

assumed. Terminal segment elongation will be incorporated into the branching process by first assuming a fixed elongation rate of 0.34 µm/h. This value is estimated from the mean pathlength of 156 µm traversed in the total period of growth of 456 h (19 days). With these model parameters and including different stop times for the branching and elongation process, dendrites were produced with mean values for their intermediate and terminal segment lengths of 23.6 (SD = 18.9) (median = 18.8) and 96.0 (SD = 22.1) µm, respectively. These values appear to be longer, respectively, shorter than the observed values of 11 (median) and 117 (SD = 33) µm. The assumption of constant segment elongation has apparently lead to incorrect results. Branching terminates at 240 h postnatal. Terminal segments become longer when the growth cones propagate faster during the elongation phase. To maintain the same mean elongation rate during the total developmental period this implies a slower rate during the branching phase. In the second simulation run, elongation rates of 0.22 µm/h during the branching phase and of 0.51 µm/h during the elongation phase were assumed, still without variation in these rates. These runs resulted in dendrites with mean (SD) values for the intermediate and terminal segment lengths of 15.2 (SD = 12.2) µm, (with median = 12.1 µm) and 117.4 (SD = 14.2) µm, respectively, and pathlengths of 156 µm without variation. The mean values are now in good agreement with the observed values. Note, that the standard deviations in the segment lengths are solely due to randomness in the branching process, and are smaller than the observed values. In the final simulation run randomness was incorporated in the elongation rate by assigning, at the time of birth of a growth cone after a branching event, a random value from a normal distribution with mean

Table 4.3 Parameter values used for modeling the growth of basal dendrites of large layer V rat cortical pyramidal neurons. The second column indicates which parameters were used for optimizing the shape parameters in the third column to the observed values (indicated as free parameters), and which ones were directly derived from observed data

Parameter	*Use*	*Optimization on*
$S=0.87$	Free	Mean asymmetry
$B=3.85$	Free	*degree* mn, SD
$E=0.74$	Free	*degree* mn, SD
$V_{el}=0.51$ µm/h	Free	Mean terminal segment length
cv in prop. rate = 0.28	Free	SD in pathlength
$V_{br}=0.22$ µm/h	Calculated	*
T(onset) = −24 h	Observed	
T(bstop) = 240 h	Observed	
T(lstop) = 432 h	Observed	

T(onset), *T*(bstop) and *T*(lstop) denote the time of onset of branching and elongation, the stop time of branching, and the stop time of elongation, respectively.
* The mean elongation rate during the branching phase V_{br} is calculated from the optimized value for the elongation rate during the elongation phase V_{el}, the durations of the branching phase $T_{br}=T(\text{bstop})-T(\text{onset})=264$ h and elongation phase $T_{el}=T(\text{lstop})-T(\text{bstop})=192$ h and the mean pathlength $L_p=156$ µm by means of $V_{br}=(L_p-V_{el}*T_{el})/T_{br}$.

0.22 μm/h during the branching phase and 0.51 μm/h during the elongation phase, and with a coefficient of variation (*cv*) of 0.28. With this *cv* value it was found that the model-predicted standard deviation in the pathlength distribution optimally matches the observed value of 29 μm. A summary of the parameter values is given in Table 4.3.

An example of a growing dendrite is given in Figure 4.10. The dendrite is displayed at successive days of its development. Note, that branching terminates 11 days after onset and is followed by a period of elongation only. Typical examples of random full grown dendritic trees, at the end of their developmental period, are shown in Figure 4.11.

These dendrites have intermediate and terminal segment lengths of 15.4 (SD = 13.4) (median 11.6) and 117.1 (SD = 31.4) μm, respectively, and pathlengths of 156 (SD = 29) μm. These outcomes are both in mean and in standard deviations in good agreement with the observed values. For the total dendritic lengths we obtained a value of 774.6 (SD = 342.9) μm. A summary of the shape characteristics for the modeled dendrites is given in Table 4.4, and compared with the observed data. In the last column is indicated whether the agreement is the result of the optimization of the model parameters or an emergent outcome of the model simulations ('prediction'). The agreement between the predicted and observed values underscores the validity of the model and its underlying assumptions. The obtained distributions for several

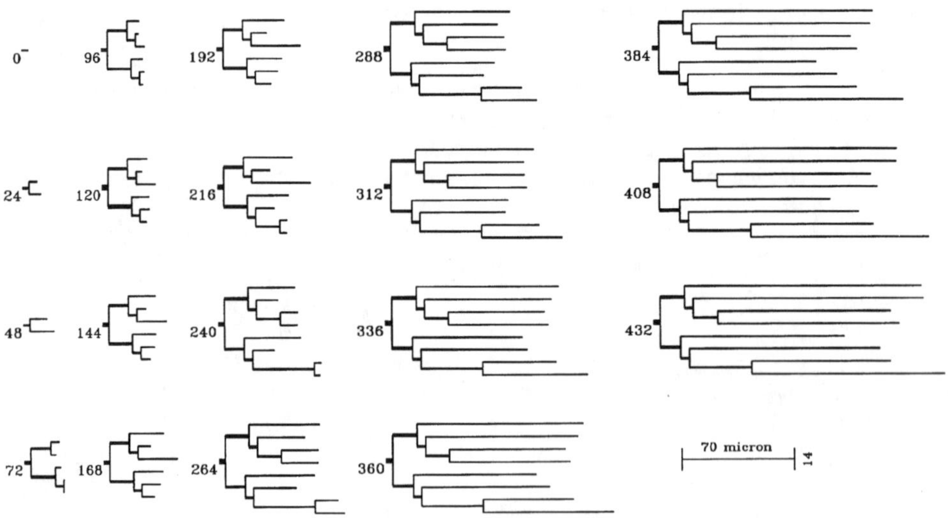

Figure 4.10 Plot of a model dendritic tree at successive days of its development. The growth parameters used were optimized for basal dendrites of large layer V rat cortical pyramidal neurons and given in Table 4.3. A period of branching and elongation from one day before birth up to 10 days after birth (denoted as branching phase) is followed by a period of elongation only up to day 18 (denoted as elongation phase). Different elongation rates are assumed of 0.22 μm/h in the branching phase and 0.51 μm/h in the elongation phase.

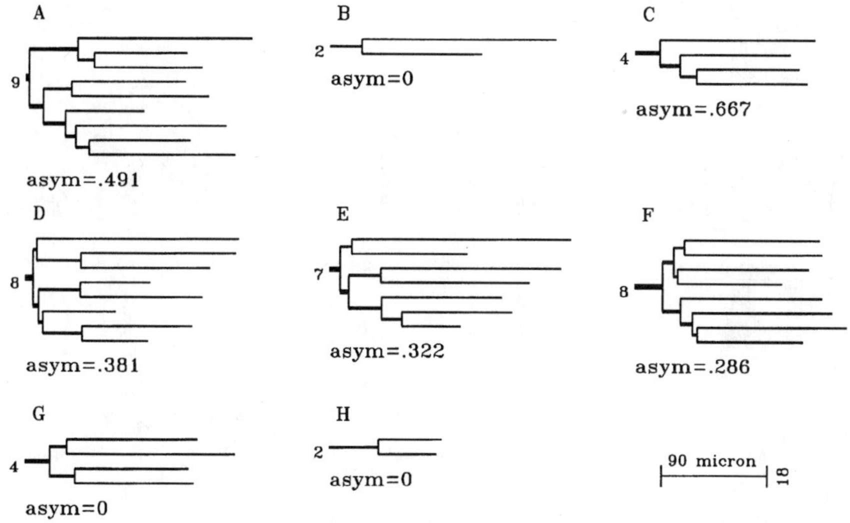

Figure 4.11 Examples of full grown model dendrites obtained for model parameters given in Table 4.3, which were optimized for basal dendrites of large layer V rat cortical pyramidal neurons. For each tree, the number of terminal segments is given as well as the tree asymmetry index (*asym*).

Table 4.4 Shape properties of model dendrites grown by means of the growth model for the parameter values given in Table 4.3. See the text for description of the model. In the last column is indicated which shape variables were matched to the observed values by optimizing the model parameters (see Table 4.3) and which shape variables are just outcomes of the growth model and can be considered as predictions for further empirical verification

Shape variables		*Obs.*	*Model outcomes*	
Degree	mn	6.0	6.0	optimized
	SD	2.7	2.7	optimized
Asymmetry	mn	0.38	0.36	optimized
	SD	0.22	0.20	*prediction*
Centrifugal order	mn		2.26	*prediction*
	SD		1.24	*prediction*
Total length	mn	777.6	774.6	*prediction*
	SD		342.9	*prediction*
Terminal length	mn	117	117.1	optimized
	SD	33	31.4	*prediction*
Intermediate length	mn		15.4	*prediction*
	SD		13.4	*prediction*
	median	11	11.6	*prediction*
Pathlength	mn	156	156.2	estimated
	SD	29	29.2	optimized

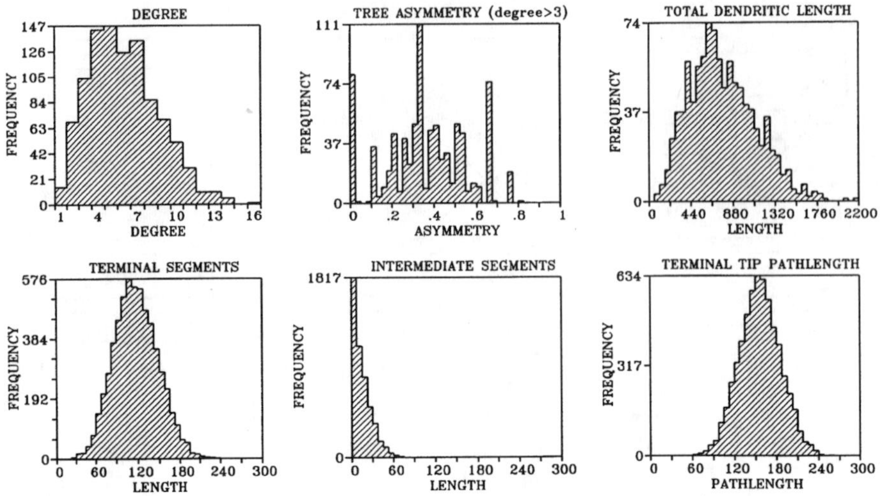

Figure 4.12 Distributions of the shape parameters *degree, tree-asymmetry, total dendritic length, terminal segment length, intermediate segment length and terminal tip pathlength* of model dendrites generated with parameter values given in Table 4.3. The values of the mean (mn) and standard deviation (SD) are given in Table 4.4.

shape parameters are shown in Figure 4.12. The distribution of the tree-asymmetry values shows the irregular pattern, characteristic for the discrete tree types and a mean value of 0.36 that is typical for the $S=0.87$ mode of growth (i.e. random branching of terminal segments with a slight preference for the proximal ones). The simulations without and with variation in elongation rates clearly show which part of the variation in segment lengths and total dendritic lengths is caused by the randomness in branching and which part is due to the randomness in elongation. Note, that the optimized value of 0.28 for the coefficient of variation in the elongation rate is larger than the value of 0.186 (=29/156) for the observed pathlength distribution. The reason is that pathlengths in a dendritic tree are not independent from each other because paths to terminal tips share common intermediate segments (for instance, all paths have the root segment in common). No experimental distributions were available to test the correctness of the shapes of the modeled ones.

4.3.2 *Simulation Procedure*

The simulation of the growth process for the *BESTL*-growth model includes the following steps:

1. Parameter S is estimated from the topological variation in the observed trees (see Van Pelt et al., 1992).
2. Parameters B and E are estimated from the observed *degree* distribution (see Van Pelt et al., 1997).

3. The time-bin scale is mapped onto a real time scale. If the observed dendrites originate from only one point in time, a linear mapping can be used. When dendrites are available at more points in time, a growth curve can be constructed for the mean number of terminal segments. This growth curve can be used to define a (non)linear mapping of the time-bins onto real time (see Van Pelt et al., 1997).
4. Using the bin-to-time mapping, each time-bin corresponds to a time period for which the function $D(t)=B/\Delta T(i)N$ can be calculated, denoting the branching probability per unit of time of an isolated terminal segment.
5. Trees are now generated according to the following iterative algorithm: For a given tree at a given point in time t, the branching probabilities are calculated for all of the $n(t)$ terminal segments with, for $S\neq 0$, the centrifugal order γ being taken into account of each of them, using $p(t)=C(t)2^{-S\gamma}D(t)n(t)^{-E}$. The normalization constant $C(t)$ is obtained via $C(t)=n(t)/\sum_{j=1}^{n(t)}2^{-S\gamma_j}$. Then, using a uniform random number between 0 and 1, it is decided for each terminal segment whether a branching event actually occurs in the present time step (i.e. a branching event occurs when the random number is smaller than or equal to the branching probability for that segment). A branching event implies that the terminal segment becomes an intermediate segment ending in a bifurcation from which two daughter terminal segments, with zero length, arise. If a branching event has occurred, the two daughter segments are assigned an elongation rate by random sampling a normal distribution with mean and standard deviation as given for the elongation rate for that point in time. All terminal segments (including the new ones) are subsequently elongated according to their individual rates. The process starts at the time of onset with a single (root) segment and stops at the last point in time.

4.4 Modeling the Variation in Segment Diameter

No developmental rules have so far been used to generate the diameters of segments. The segment diameters in the displayed dendrograms have therefore been estimated by assuming a power law relation. By the power law relation, the diameter of an intermediate segment (d_s) is determined by the number of terminal segments n_s in its subtree according to $d_s^e=n_s d_t^e$ or $d_s=d_t n_s^{1/e}$. Accordingly, the diameter of a segment in a growing tree is expected to increase with the increasing number of terminal segments. The segment diameters in a tree have been estimated by means of the following procedure. First, terminal segment diameters are estimated by random sampling the observed diameter distribution (or a normal distribution based on the observed mean–SD values). At each bifurcation the diameter of the parent segment is estimated by using a branch power value obtained by random sampling the observed branch power distribution. For the dendrograms in Figures 4.10 and 4.11 the terminal segment diameters were randomly sampled from a *normal distribution* with mean 0.8 µm and standard deviation 0.2 µm (Larkman, 1991). The branch power distribution was assumed to be a normal one with mean 1.6 and standard deviation 0.2 (values estimated from Larkman et al., 1992).

5. DISCUSSION

A growth model for dendritic trees has been described, that is based on randomness in branching and elongation of terminal segments. This model is able to accurately reproduce dendritic complexity with respect to number, length and connectivity of dendritic segments. It is important to note that agreement between model outcomes and empirical data was obtained for a much larger number of shape parameters than the number of optimized model parameters. This finding is a strong support for the assumptions made in the model. Four aspects have been distinguished in dendritic growth, viz., (i) the 'basic' branching process of an individual growth cone, defined by the expected number of branching events B during the developmental period under (hypothetical) isolated conditions; (ii) the 'proliferation' of the number of growth cones in the growing tree; (iii) the 'modulation' of the 'basic' branching probability by the increasing number of growth cones (determined by the parameter E) and the position of the segment in the tree (via parameter S); andc (iv) the elongation of segments by the migration of growth cones.

Because of the structure of the model, both its parameters (i.e. the rates of branching and migration of growth cones) and its outcomes (i.e. the generated dendrites and the distributions of their shape parameters) are directly associated with the biological processes and structures. This reinforces the empirical validation of its assumptions and outcomes. Although the model has shown to accurately describe the results of the dendritic growth process, the actually occurring biological process may be much more complex than captured by the minimal set of rules in the model. The model assumes, for instance, that the growth rules are stable throughout the total period. The behavior of growth cones is highly dynamic, however, involving periods of forward migration as well as retraction. Dendrites may experience phases of remodeling, in which the number of segments is substantially reduced. On a short time scale these dynamic actions are already implicitly accounted for by the stochasticity of the model, but fluctuations on a longer time scale can only be captured by changing the growth rules in the course of time. This happens, for instance, in the case of a nonlinear mapping of the time-bin scale into a real time scale, or when the branching probability is temporary suppressed, immediately following a branching event (Nowakowski et al., 1992). If required, regression can be explicitly incorporated into the model, but with an unavoidable increase in complexity. In a recent study on the impact of pruning on the topological variation of dendritic trees, Van Pelt (1997) showed that under uniform random pruning schemes the topological variance of the pruned trees remained consistent with the mode of initial outgrowth.

Because segment lengths depend on the branching and migration rates of growth cones, their distributions can only be studied by taking both processes into account. It is shown that the segment length distributions in rat cortical pyramidal cell basal dendrites are accurately described by assuming different segment elongation rates during the two phases of development. These findings now await empirical verification.

This chapter has focussed on the morphology of dendrites and on growth models to describe in detail their morphological characteristics, while the variation in the shape parameters and the topological structures are emphasized. By this progress in

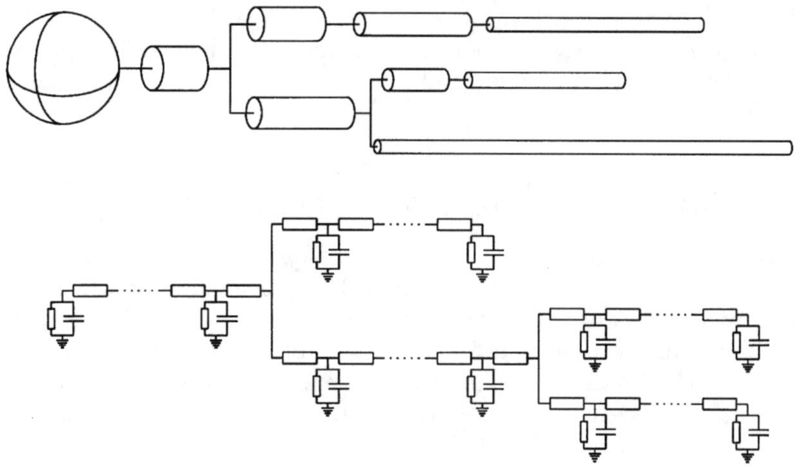

Figure 4.13 Top: Example of a branching cable model in which each segment is modeled as one or more cylindrical membranes. Bottom: a compartmental model in which each segment is modeled as a series of isopotential R–C compartments. (Figure drawn after Figure 3.3 in Koch and Segev (1989).)

modeling dendritic structure it is now possible to study the variations in functional properties induced by realistic morphological variations. Since the pioneering work on cable theory of Rall (see e.g. Rall et al., 1992; Rall, 1995) many methods have been developed for studying the processing of electrical signals in both passive and active dendritic trees. Both the branching cable and the compartmental models (e.g. Koch and Segev, 1989; see also Figure 4.13) have made it clear how signal processing depend on dendritic structure and membrane properties. The impact of topology on input conductance and signal attenuations in passive dendritic trees has first been studied systematically by Van Pelt and Schierwagen (1994, 1995). The use of one single vector equation for the whole dendritic tree, including the Laplace-transformed cable equations of all the individual segments, greatly simplified and accellerated the calculations (Van Pelt, 1992). These studies demonstrated how dendrites with realistic topological structures differ in their passive electrotonic properties from symmetrical trees, and how the natural topological variation in dendritic morphologies contributes to the variation in functional properties.

Dendritic shapes have been studied in this chapter by focussing on the metrics and connectivity of the individual segments (see Section 2). Clearly, dendrites are 3D structures with a particular embedding in 3D space. For the characterization of the 3D geometry, one further needs to know the angles between the segments at the bifurcation points and the curvature of the segments. The 3D dendritic geometry gets functional implications when the extra cellular space participates in the electrical circuitry in an inhomogeneous manner and/or when dendrites are connected to each other via electrical contacts (for instance via gap junctions) (e.g. Poznanski et al., 1995). It is expected that the proximity of dendritic branches becomes an important aspect in dendritic 3D geometry, for which the topological structure and the angles between branches at bifurcation points are crucial parameters. The present equivalent cylinder-, branching cable- and compartmental models, however, do not incorporate these 3D aspects.

6. CONCLUSIONS AND FUTURE PERSPECTIVES

The mathematical modeling of dendritic growth on the basis of random behavior of growth cones has brought many dendritic shape parameters and their variations into a coherent framework. It allows for analysis of dendritic shapes in terms of their developmental histories and for predictions of growth cone behavior. Importantly, the model provides a useful tool for generating sets of randomly varying dendritic trees for use in studies of the functional implications of morphological variations. Both metrical and topological aspects of dendritic shapes have to be incorporated in assessing how the dendrite integrates in space and time the trains of incoming post synaptic potentials resulting in particular series of generated action potentials. Clearly, also membrane properties are subjected to variations as for instance caused by the types and the spatial distributions of the membrane channels, which contribute to the individual functional properties of the particular dendrite. Additionally, a neuron is a highly complex dynamic system and many structural aspects of the system have time dependent properties with large ranges of time constants (like for instance activity-dependent influence on membrane conductances and neurite outgrowth). The integrative properties and firing characteristics are therefore expected to depend on the dynamic history of the neuron in a very complex manner.

It is expected that structural dendritic models become increasingly important in future studies of dendritic structure–function relationships. Structural models will also need to describe dendritic morphologies with increasing natural detail, in order to get better understanding of the rich dynamic repertoire of dendritic operation. A typical example of greater realism is the 3D dendritic geometry and the imbedding in its environmental neural tissue. For the modeling of 3D geometry also the angles between the segments at bifurcation points and the curvature of the segments should therefore be taken into account. For the growth model it would require the migration of growth cones to proceed in 3D as well as a qualification of the 3D environment of the outgrowing neurites. The presented growth model in this chapter for generating branching patterns with realistic topological and metrical variations contribute to this development and allows such future extensions quite naturally.

PROBLEMS

4.1. Real dendrites are structures embedded in 3D space having irregular branches and orientations. How can these morphological aspects be built into the model? To this end the growth process has to be formulated in 3D space, and specific assumptions have to be made for the direction of outgrowth of growth cones.

New possibilities are then offered for validating these assumptions using observed orientation and irregularity properties of dendritic branches.

4.2. The branching process was formulated in real time by transforming the time bins into a real time scale. The elongation process was subsequently formulated on this time scale. A proper transformation can be found when experimental data on the number of segments is available for at least two different points in time. How should one proceed when only one time point is available? One may choose a particular mapping, but how can such a choice be validated?

4.3. A migration rate is assigned to a growth cone at the time of its birth (i.e., after a branching event) by random sampling a given distribution, and remains constant for its life span (i.e., up to its next branching event). Other ways of implementing stochasticity in growth cone migration, however, are also possible (like stochasticity in the actual migration at each time step). Are there other, or even better ways to capture the stochasticity of the underlying biological process, and how to test it?

4.4. A basic assumption in the model is the independency in branching and migration of growth cones. To what extent do possible correlations in the occurrences of branching and elongation events result in altered variations in dendritic shape parameters?

4.5. A thorough experimental test of the model is possible when growth cone behavior is quantified in a longitudinal *in vivo* study of identified neurons in development and the neurons are morphologically reconstructed afterwards. What could be the best experimental approach to this end?

4.6. The power of the growth-model approach is that dendritic geometrical properties and their variations can possibly be captured by a limited set of parameters describing the underlying dendritic growth process. This set of parameters has to be determined for many different neuron types. Such an inventory will give insight in the similarity and dissimilarity of growth actions in the development of different neuron types.

ACKNOWLEDGMENTS

We are very grateful to Dr M.A. Corner for his critical comments and improvements of the manuscript. Part of this work was supported by NATO grant CRG 930426 (JVP).

REFERENCES

Berry, M. and Bradley, P.M. (1976) The application of network analysis to the study of branching patterns of large dendritic fields. *Brain Res.*, **109**, 111–132.

Black, M.M. (1994) Microtubule transport and assembly cooperate to generate the microtubule array of growing axons. In *The Self-Organizing Brain: From Growth Cones to Functional Networks, Progress in Brain Research*: edited by J. van Pelt, M.A. Corner, H.B.M. Uylings and F.H. Lopes da Silva, Vol. 102, pp. 61–77. Amsterdam: Elsevier.

Burke, R.E., Marks, W.B. and Ulfhake, B. (1992) A parsimonious description of motoneuron dendritic morphology using computer simulation. *J. Neurosci.*, **12**, 2403–2416.

Caley, A. (1859) On the analytical forms called trees. *Philos. Mag.*, **18**, 374–378.

Cameron, W.E., Averill, D.B. and Berger, A.J. (1985) Quantitative analysis of the dendrites of cat phrenic motoneurons stained intracellularly with horseradish peroxidase. *J. Comp. Neurol.*, **230**, 91–101.

Carriquiry, A.L., Ireland, W.P., Kliemann, W. and Uemura, E. (1991) Statistical evaluations of dendritic growth models. *Bull. Math. Biol.*, **53**, 579–589.

Cullheim, S., Fleshman, J.W., Glenn, L.L. and Burke, R.E. (1987) Membrane area and dendritic structure in type-identified triceps surae alpha-motoneurons. *J. Comp. Neurol.*, **25**, 68–81.

De Ruiter, J.P. and Uylings, H.B.M. (1987) Morphometric and dendritic analysis of fascia dentate granule cells in human aging and senile dementia. *Brain. Res.*, **402**, 217–229.

Desmond, N.L. and Levy, W.B. (1984) Dendritic caliber and the 3/2 power relationship of dentate granule cells. *J. Comp. Neurol.*, **227**, 589–596.

Dityatev, A.E., Chmykhova, N.M., Studer, L., Karamian, O.A., Kozhanov, V.M. and Clamann, H.P. (1995) Comparison of the topology and growth rules of motoneuronal dendrites. *J. Comp. Neurol.*, **363**, 505–516.

Harding, E.F. (1971) The probabilities of rooted tree-shapes generated by random bifurcation. *J. Appl. Prob.*, **3**, 44–77.

Hillman, D.E. (1979) Neuronal shape parameters and substructures as a basis of neuronal form. In *The Neurosciences*, 4th Study Program, edited by F.O. Schmitt, F.G. Worden, pp. 477–498. Cambridge: MIT Press.

Hillman, D.E. (1988) Parameters of dendritic shape and substructure: Intrinsic and extrinsic determination? In *Intrinsic Determinants of Neuronal Form and Function*, edited by R.J. Lasek, M.M. Black, pp. 83–113. New York: A.R. Liss.

Horsfield, K., Woldenberg, M.J. and Bowes, C.L. (1987) Sequential and synchronous growth models related to vertex analysis and branching ratios. *Bull. Math. Biol.*, **49**, 413–430.

Ireland, W., Heidel, J. and Uemura, E. (1985) A mathematical model for the growth of dendritic trees. *Neurosc. Lett.*, **54**, 243–249.

Kater, S.B., Davenport, R.W. and Guthrie, P.B. (1994) Filopodia as detectors of environmental cues: signal integration through changes in growth cone calcium levels. In *The Self-Organizing Brain: From Growth Cones to Functional Networks*, edited by J. van Pelt, M.A. Corner, H.B.M. Uylings and F.H. Lopes da Silva, Progress in Brain Research, Vol. 102, pp. 49–60. Amsterdam: Elsevier.

Kliemann, W.A. (1987) Stochastic dynamic model for the characterization of the geometrical structure of dendritic processes. *Bull. Math. Biol.*, **49**, 135–152.

Koch, C. and Segev, I. (1989) *Methods in Neuronal Modeling: From Synapses to Networks*. Cambridge: MIT Press.

Larkman, A.U. (1991) Dendritic morphology of pyramidal neurones of the visual cortex of the rat: I. Branching patterns. *J. Comp. Neurol.*, **306**, 307–319.

Larkman, A. and Mason, A. (1990) Correlations between morphology and electrophysiology of pyramidal neurons in slices of rat visual cortex. I. Establishment of cell classes. *J. Neurosc.*, **10**, 1407–1414.

Larkman, A.K., Major, G., Stratford, K.J. and Jack, J.J.B. (1992) Dendritic morphology of pyramidal neurones of the visual cortex of the rat. IV: Electrical geometry. *J. Comp. Neurol.*, **323**, 137–152.

Letourneau, P.C., Snow, D.M. and Gomez, T.M. (1994) Growth cone motility: substratum-bound molecules, cytoplasmic [Ca^{2+}] and Ca^{2+}-regulated proteins. In *The Self-Organizing Brain: From Growth Cones to Functional Networks*, edited by J. Van Pelt, M.A. Corner, H.B.M. Uylings and F.H. Lopes da Silva, Progress in Brain Research, Vol. 102, pp. 35–48. Amsterdam: Elsevier.

Mainen, Z.F. and Sejnowski, T.J. (1996) Influence of dendritic structure on firing patterns in model neocortical neurons. *Nature*, **382**, 363–366.

Mason, A. and Larkman, A. (1990) Correlations between morphology and electrophysiology of pyramidal neurons in slices of rat visual cortex. II. Electrophysiology. *J. Neurosci.*, **10**, 1415–1428.

Nowakowski, R.S., Hayes, N.L. and Egger, M.D. (1992) Competitive interactions during dendritic growth: a simple stochastic growth algorithm. *Brain Res.*, **576**, 152–156.

Overdijk, J., Uylings, H.B.M., Kuypers, K. and Kamstra, A.W. (1978) An economical, semi-automatic system for measuring cellular tree structures in three dimensions, with special emphasis on Golgi-impregnated neurons. *J. Microsc.*, **114**, 271–284.

Poznanski, R.R., Gibson, W.G. and Bennett, M.R. (1995) Electrotonic coupling between 2 CA3 hippocampal pyramidal neurons—A distributed cable model with somatic gap-junction. *Bull. Math. Biol.*, **57**, 865–881.

Rall, W. (1959) Branching dendritic trees and motoneuron membrane resistivity. *Exp. Neurol.*, **1**, 491–527.
Rall, W., Burke, R., Holmes, W., Jack, J., Redman, S. and Segev, I. (1992) Matching dendritic neuron models to experimental data. *Physiol. Rev.*, **72**, 159–186.
Rall, W. (1995) The theoretical foundation of dendritic function. In *Selected papers of Wilfrid Rall with commentaries*, edited by I. Segev, J. Rinzel, G.M. Shepherd. Cambridge: MIT Press.
Schierwagen, A. (1986) Segmental cable modelling of electrotonic transfer properties of deep superior colliculus neurons in the cat. *J. Hirnforsch.*, **27**, 679–690.
Schierwagen, A. and Grantyn, R. (1986) Quantitative morphological analysis of deep superior colliculus neurons stained intracellularly with HRP in the cat. *J. Hirnforsch.*, **27**, 611–623.
Shreve, R.L. (1966) Statistical law of stream numbers. *J. Geol.*, **74**, 17–37.
Smit, G.J., Uylings, H.B.M. and Veldmaat-Wansink, L. (1972) The branching pattern in dendrites of cortical neurons. Acta Morphol. *Neerl.-Scand.*, **9**, 253–274.
Tamori, Y. (1993) Theory of dendritic morphology. *Physical Rev. E*, **148**, 3124–3129.
Uemura, E., Carriquiry, A., Kliemann, W. and Goodwin, J. (1995) Mathematical modeling of dendritic growth *in vitro. Brain Res.*, **671**, 187–194.
Uylings, H.B.M., Kuypers, K. and Veltman, W.A.M. (1978) Environmental influences on the neocortex in later life. In *Maturation of the Nervous System*, edited by M.A. Corner, Progress in Brain Research, Vol. 48, pp. 261–273. Amsterdam: Elsevier.
Uylings, H.B.M., Ruiz-Marcos, A. and Van Pelt, J. (1986) The metric analysis of three-dimensional dendritic tree patterns: A methodological review. *J. Neurosci. Meth.*, **18**, 127–151.
Uylings, H.B.M., Van Pelt, J. and Verwer, R.W.H. (1989) Topological analysis of individual neurons. In *Computer Techniques in Neuroanatomy*, edited by J.J. Capowski, pp. 215–239. New York: Plenum.
Uylings, H.B.M., Van Eden, C.G., Parnavelas, J.G. and Kalsbeek, A. (1990) The prenatal and postnatal development of rat cerebral cortex. In *The Cerebral Cortex of the Rat*, edited by B. Kolb and R.C. Tees, pp. 35–76, Cambridge: MIT Press.
Uylings, H.B.M., Van Pelt, J., Parnavelas, J.G. and Ruiz-Marcos, A. (1994) Geometrical and topological characteristics in the dendritic development of cortical pyramidal and non-pyramidal neurons. In *The Self-Organizing Brain: From Growth Cones to Functional Networks*, edited by J. van Pelt, M.A. Corner, H.B.M. Uylings and F.H. Lopes da Silva, *Progress in Brain Research*, Vol. 102, pp. 109–123. Amsterdam: Elsevier.
Van Pelt, J. and Verwer, R.W.H. (1983) The exact probabilities of branching patterns under segmental and terminal growth hypotheses. *Bull. Math. Biol.*, **45**, 269–285.
Van Pelt, J. and Verwer, R.W.H. (1985) Growth models (including terminal and segmental branching) for topological binary trees. *Bull. Math. Biol.*, **47**, 323–336.
Van Pelt, J. and Verwer. R.W.H. (1986) Topological properties of binary trees grown with order-dependent branching probabilities. *Bull. Math. Biol.*, **48**, 197–211.
Van Pelt, J., Uylings, H.B.M. and Verwer, R.W.H. (1989) Distributional properties of measures of tree topology, *Acta Stereol.*, **8**, 465–470.
Van Pelt, J. (1992) A simple vector implementation of the Laplace-transformed cable equations in passive dendritic trees. *Biol. Cybern*, **68**, 15–21.
Van Pelt, J., Uylings, H.B.M., Verwer, R.W.H., Pentney, R.J. and Woldenberg, M.J. (1992) Tree asymmetry—a sensitive and practical measure for binary topological trees. *Bull. Math. Biol.*, **54**, 759–784.
Van Pelt, J. and Schierwagen, A.K. (1994) Electrotonic properties of passive dendritic trees—effect of dendritic topology. In *The Self-Organizing Brain: From Growth Cones to Functional Networks*, edited by J. van Pelt, M.A. Corner, H.B.M. Uylings and F.H. Lopes da Silva, Progress in Brain Research, Vol. 102, pp. 127–149. Amsterdam: Elsevier.
Van Pelt, J. and Schierwagen, A.K. (1995) Dendritic electrotonic extent and branching pattern topology. In *The Neurobiology of Computation*, edited by J.M. Bower, pp. 153–158. Boston: Kluwer.
Van Pelt, J. (1997) Effect of pruning on dendritic tree topology. *J. Theor. Biol.*, **186**, 17–32.
Van Pelt, J., Dityatev, A.E. and Uylings, H.B.M. (1997) Natural variability in the number of dendritic segments: model-based inferences about branching during neurite outgrowth. *J. Comp. Neurol.*, **387**, 325–340.
Van Veen, M.P. and Van Pelt, J. (1993) Terminal and intermediate segment lengths in neuronal trees with finite length. *Bull. Math. Biol.*, **55**, 277–294.

Verwer, R.W.H., Van Pelt, J. and Uylings, H.B.M. (1992) An introduction to topological analysis of neurones. In *Quantitative Methods in Neuroanatomy*, edited by M.G. Stewart, pp. 295–323. Chicester: John Wiley & Sons.

Woldenberg, M.J., O'Neill, M.P., Quackenbush, L.J. and Pentney, R.J. (1993) Models for growth, decline and regrowth of the dendrites of rat Purkinje cells induced from magnitude and link-length analysis. *J. Theor. Biol.*, **162**, 403–429.

CHAPTER 5
THE MULTIPLE EQUIVALENT CYLINDER MODEL

Jonathan D. Evans

1. INTRODUCTION

The development of neuronal models provides a quantitative framework for studying the spread and summation of synaptic potentials in neurons. Indeed, understanding the passive electrical properties of the soma and dendritic trees of a neuron is an important first step in obtaining a description of a neuron's overall electrical behaviour (Rall et al., 1992).

The objective of neuronal modelling is to construct a mathematical model of synaptic spread and summation, which agrees with existing measurements to within a specified accuracy and can be used with confidence to predict future observations and behaviour. Thus, to pursue neuronal modelling successfully (as in other areas of mathematical modelling) two contrasting but complimentary skills are needed, namely the ability to formulate the problem in appropriate mathematical terms, and then possessing sufficient knowledge (or techniques) to obtain useful information from that mathematical model. The skill in formulation lies in finding a model which is simple enough to give useful information easily, but which is diverse enough to give all the information with sufficient accuracy. In attempting to construct a model, judgement has to be made about which features to include and which to neglect. If an important feature is omitted then the model will not describe the observed phenomenon accurately enough or the model may not be self-consistent. If unnecessary features are included then the model will be more difficult to solve because of its increased complexity. Thus it is advisable to adopt a simple approach with a minimum of features included at the first attempt, and additional features added, if necessary, one by one. It should then be possible to estimate the effect of the additional feature on the results of the original model. It is this approach that has been followed in the development of passive neuron models, where the passive properties of neurons have been investigated first before proceeding to models that include active properties.

The discussion in this article is limited to passive neuron models and concentrates on the recent development of analytical solutions for these models rather than their numerical solutions by, for example, segmental cable analysis or compartmental analysis. Analytical results are of value not only in their own right to reveal trends and provide simple rules of thumb in certain situations, but also as bench marks for confirming the results from numerical approaches which are often of unknown accuracy.

As remarked by Rall (1990), there is no one model for dendritic neurons but rather a variety that have been used to address different problems. Indeed, a model may be appropriate in one set of circumstances but of no value in others because it is either over-complicated or too simple. This is certainly the case with the range of passive cable models that have been developed to investigate, for example, the roles of dendritic branching, dendritic tapering and dendritic spines on the passive voltage spread within the cell. These models are reviewed in the discussion section whilst the text is concerned with the analysis of the multiple equivalent cylinder model, previously called the multi-cylinder somatic shunt cable model (Evans et al., 1992), which is the many equivalent cylinder version of Rall's single equivalent cylinder model of a neuron (Rall, 1962a,b). Before proceeding, it is important to identify two different types of mathematical problems for these passive cable models. The first is termed the *forward problem* and involves solving the model equations for the dependent variable (the voltage) as a function of the independent variables (space and time) for given values of the model parameters. The second is termed the *inverse problem* and involves determining the model parameters for given values of the dependent variable (the voltage) at prescribed independent variable (space and time) values. These two mathematical problems are distinct and generally require very different mathematical techniques to solve them. The purpose of this article is to review and discuss the forward problems only, however, the framework in which we do so allows the conditions for the well-posedness of the inverse problem to be identified.

The mathematical techniques used in the analysis (of the forward problem) of the multiple equivalent cylinder model are equally applicable to the other passive cable models and may be categorized as follows:

(i) *Problem normalization*
A mathematical model can be set in a general framework within which it can be systematically and consistently analyzed. This framework is achieved by non-dimensionalization of the model equations to obtain a dimensionless model in which appear only dimensionless variables (both dependent and independent) and dimensionless parameters. The importance of this process of non-dimensionalisation, called problem normalization, is that when the model equations are expressed in terms of non-dimensional variables (whose typical scale is of order one) the relative magnitude of different physical effects are measured by non-dimensional parameters or dimensionless groups. These dimensionless parameters control the behaviour of the solution to the model and in terms of which the different types of solution behaviour can be classified.

(ii) *Perturbation procedures*
Following non-dimensionalization the simplification of a model obtained by neglecting a term multiplied by a small non-dimensional parameter is the first step in a

systematic procedure for obtaining an asymptotic expansion for the full solution in terms of the small parameter. This perturbation procedure is classified as regular or singular depending upon the limiting behaviour of the full solution with respect to the small parameter. Perturbation procedures in terms of the model's dimensionless parameters provide simplifications of the full solution and often allow progress to be made in situations where numerical schemes can encounter difficulties.

(iii) *Analytical methods*
The passive cable models that we describe in this article are linear systems with governing equations that are parabolic differential equations. As such they are amenable to the techniques of Green's functions and Laplace transforms, using which solutions to the full problem can be obtained.

These techniques provide a coherent methodology for dealing with problems of the type provided by passive cable theory, the techniques (i) and (ii) not being restricted to models with linear equations. Further descriptions of these techniques can be found in, for example, Tayler (1986) and Van Dyke (1964).

2. FORMULATION OF THE MATHEMATICAL MODEL

Rall's model of a nerve cell (Rall, 1962a,b) consists of one single equivalent cylinder representing the whole dendritic tree coupled to a lumped impedance representing the soma (or cell body). This model was originally based on anatomical and neurophysiological information gathered for spinal motoneurons, and reviews for it are given in Jack et al. (1983), Tuckwell (1988) and Rall (1989). The reduction of the entire dendritic structure to an equivalent cylinder is based upon certain assumptions for the dendritic branching pattern which may be summarized as follows:

(i) Each trunk or branch segment has uniform diameter.
(ii) At each dendritic branch point, the diameters of the parent branch d_p [cm] and daughter branches d_k [cm] satisfy

$$d_p^{3/2} = \sum_{k \in dtrs} d_k^{3/2},$$

where the summation is over the set *dtrs* of daughter branches. This expression is often called the 3/2 power law, and it is worth noting that the diameters of the daughter branches need not necessarily be equal. Also worth remarking is that a more general power law is permissible (see Eq. (7.23) in Jack et al., 1983) which includes both electrical and geometrical non-uniformities.
(iii) All dendritic branches in a dendritic tree terminate at the same electrotonic distance from the soma, which is taken to be the electrotonic length of both the dendritic tree and the equivalent cylinder that represents it. For each branch

segment, its electrotonic length L is defined as

$$L = \frac{l}{\lambda} = \frac{l}{d^{1/2}}\left(\frac{4R_\mathrm{i}}{R_m}\right)^{1/2},$$

where l is the branch length [cm] and λ is the space constant [cm], given by

$$\lambda = d^{1/2}\left(\frac{R_\mathrm{m}}{4R_\mathrm{i}}\right)^{1/2},$$

with R_m the membrane specific resistance [$\Omega\,\mathrm{cm}^2$], C_m the membrane specific capacitance [$\mathrm{F/cm}^2$] and R_i the axial resistivity [$\Omega\,\mathrm{cm}$]. The electrotonic length of a path in the dendritic tree is then the sum of the electrotonic lengths of the branch segments composing that path.

(iv) All terminal branches in a dendritic tree end with the same distal boundary condition, which is usually taken to be the sealed end condition (i.e. an "open-circuit" termination).

(v) The values of the electrical parameters, the membrane specific resistance $R_\mathrm{m}[\Omega\,\mathrm{cm}^2]$, the membrane specific capacitance $C_\mathrm{m}[\mathrm{F/cm}^2]$ and the axial resistivity $R_\mathrm{i}[\Omega\,\mathrm{cm}]$, are assumed to be the same for all branches in a particular dendritic tree.

(vi) Each dendritic tree is assumed to have the same electrotonic length.

(vii) If inputs are delivered to dendritic branch locations, then proportional inputs must be delivered into all dendritic branches such that an equal voltage time course would be generated at all electrotonically equivalent dendritic locations.

For ease of generalization it is worth considering the reduction of the entire branching structure to a single equivalent cylinder as a result of two steps: the first being the reduction of each dendritic tree to an equivalent cylinder, which follows if conditions (i)–(v) hold, and the second being the representation of all the equivalent cylinders by a single one, condition (vi). These transformations are summarized in Figure 5.1. Condition (vii) is needed to relate current input into the equivalent cylinder to that in the branched structure it represents, and may be relaxed if input is into the soma.

The Rall model is a particular case of the multiple equivalent cylinder model described later in this section, and consequently, a mathematical statement of the dimensional Rall model can be obtained from the Eqs. (1)–(5) for the particular case in which $N=1$ and the membrane and soma time constants are equal (i.e. $\tau_\mathrm{m}=\tau_\mathrm{s}$, so that the soma resistance is now defined as $R_\mathrm{s}=R_\mathrm{m}/A_\mathrm{s}[\Omega]$).

Analytical solutions for Rall's equivalent cylinder model have been obtained by Rall (1969), Jack and Redman (1971a,b), Iansek and Redman (1973), Redman (1973), Rall (1977), Kawato (1984) and Ashida (1985). A complete analysis and general solution for general current inputs with unrestricted input and recording sites was presented by Bluman and Tuckwell (1987).

A modification of the Rall model is the somatic shunt cable model in which a shunt is introduced at the soma, the entire dendritic branching structure still being represented by a single equivalent cylinder. The effect of a shunt at the soma is

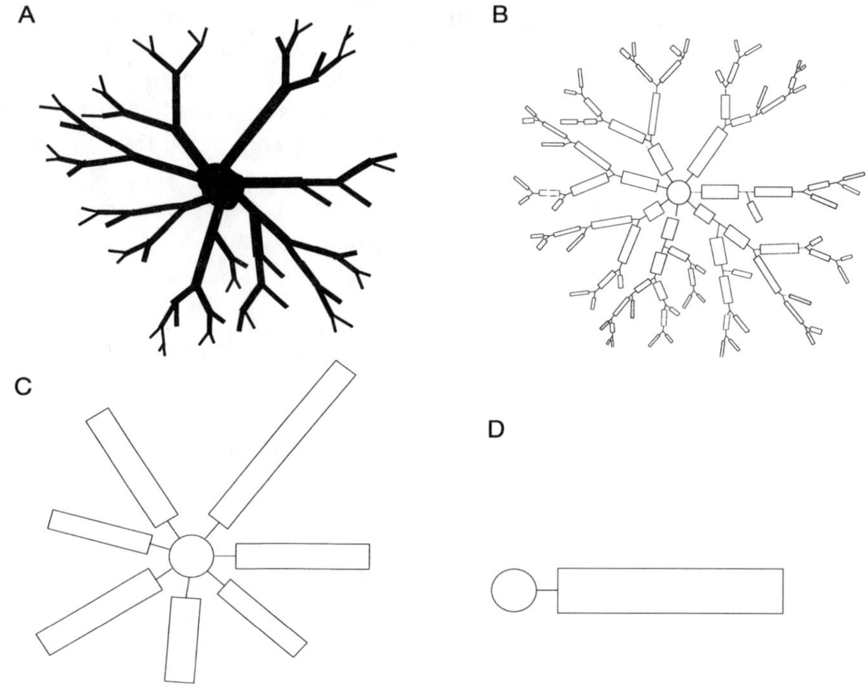

Figure 5.1 Transformations made in the development of models of differing branching complexity. The nerve cell (A) is represented by a model (B) with isopotential soma and each dendritic segement as a cylinder of uniform passive membrane. Under assumptions (i)–(vii) in Section 2, each dendritic tree can be represented by an equivalent cylinder giving the multiple equivalent cylinder model (C). To obtain the Rall model, it is assumed that the individual equivalent cylinders can be combined into a single equivalent cylinder (D).

to produce a lower soma resistance and hence decrease the somatic time constant τ_s and thus accelerate the potential decay as observed experimentally for motoneurons and hippocampal granule cells (Iansek and Redman, 1973; Durand et al., 1983). The shunt is interpreted as being due to either electrode penetration damage or a lower membrane resistance existing at the soma than in the dendrites. The mathematical statement of the dimensional somatic shunt cable model for the single cylinder case is precisely that given by Eqs. (1)–(5) for the multiple equivalent cylinder case with $N=1$ and in this model the somatic time constant τ_s is not necessarily equal to the membrane time constant τ_m. As for the Rall model, analytic solutions for the somatic shunt cable model have been derived by several authors, Iansek and Redman (1973), Kawato (1984), Durand (1984) and Poznanski (1987a). The general analytical solution is discussed in Poznanski (1987b) and parallels that obtained by Bluman and Tuckwell (1987) for the Rall model.

The Rall equivalent cylinder model and its extension, the somatic shunt cable model, require a high degree of symmetry between the branching structures.

The immediate extension to these single equivalent cylinder models is to relax assumption (vi) in the Rall model and consider a model where more than one equivalent cylinder emanates from the soma. The powerful representation of an entire dendritic tree by an equivalent cylinder, as introduced by Rall, is retained but now we allow for more flexibility as individual dendritic trees can be considered (but still not the individual branches within the tree).

The multiple equivalent cylinder model with N equivalent cylinders emanating from the soma with a shunt resistance is illustrated in Figure 5.2, the dimensional mathematical statement for which is the following,

$$\text{in } 0<x_j<l_{\text{cyl}\,j},\quad t>0 \qquad v_j+\tau_{\text{m}}\frac{\partial v_j}{\partial t}-\lambda_j^2\frac{\partial^2 v_j}{\partial x_j^2}=\lambda_j r_{\text{a}j} i_j(x_j,t); \tag{5.1}$$

$$\text{at } x_j=l_{\text{cyl}\,j} \qquad \frac{\partial v_j}{\partial x_j}=0; \tag{5.2}$$

$$\text{at } x_j=0 \qquad v_j=v_{\text{s}}, \tag{5.3}$$

$$v_{\text{s}}+\tau_{\text{s}}\frac{\mathrm{d}v_{\text{s}}}{\mathrm{d}t}-\sum_{j=1}^{N}\frac{R_{\text{s}}}{r_{\text{a}j}}\frac{\partial v_j}{\partial x_j}=R_{\text{s}} i_{\text{s}}(t); \tag{5.4}$$

$$\text{at } t=0,\ 0\leqslant x_j\leqslant l_{\text{cyl}\,j} \qquad v_j=h_j(x_j); \tag{5.5}$$

for $j=1,\ldots,N$, where $v_j(x_j,t)$ is the transmembrane potential in volts at a distance x_j [cm] along the jth equivalent cylinder from the soma ($x_j=0$) and at time t [s], v_{s} is the soma potential in volts, $i_j(x,t)$ is an applied current in amperes to the jth equivalent cylinder, $i_{\text{s}}(t)$ is an applied current in amperes at the soma and $h_j(x_j)$ is the initial voltage distribution in volts in the jth equivalent cylinder. The index j has been introduced as a suffix to the variables and parameters for the jth equivalent

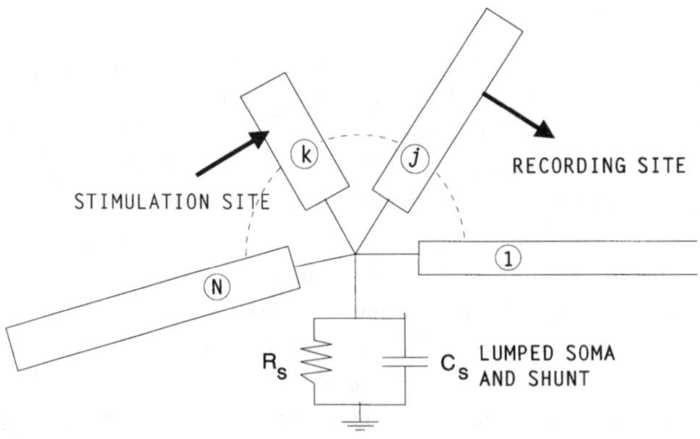

Figure 5.2 Illustration of the multiple equivalent cylinder somatic shunt model with N equivalent cylinders emanating from a common soma.

cylinder where j can range between 1 and N inclusive. It is now worth emphasizing the *core* dimensional parameters,

R_m	membrane specific resistance [$\Omega\,cm^2$],
C_m	membrane specific capacitance [F/cm^2],
R_i	axial resistivity [$\Omega\,cm$],
$d_{cyl\,j}$	physical diameter of the jth equivalent cylinder [cm],
$l_{cyl\,j}$	physical length of the jth equivalent cylinder [cm],
A_s	area of soma [cm^2],
R_{sh}	somatic shunt resistance [Ω],

in terms of which the other more commonly used dimensional parameters can be defined such as,

R_s	resistance of soma [Ω], ($1/R_s = A_s/R_m + 1/R_{sh}$)
$C_s = A_s C_m$	capacitance of soma [F],
$\tau_m = R_m C_m$	membrane time constant of the equivalent cylinder [s],
$\tau_s = R_s C_s$	membrane time constant of the soma [s],
$r_{a\,j} = 4R_i/\pi d_{cyl\,j}^2$	axial resistance per unit length of the jth equivalent cylinder [Ω/cm],
$r_{m\,j} = R_m/\pi d_{cyl\,j}$	membrane resistance of a unit length of the jth equivalent cylinder [$\Omega\,cm$],
$\lambda_j = (r_{m\,j}/r_{a\,j})^{1/2}$	space constant of the equivalent cylinder [cm],
$c_{m\,j} = \pi d_{cyl\,j} C_m$	membrane capacity per unit length of the jth equivalent cylinder [F/cm].

A feature worth emphasizing for this model is that each equivalent cylinder can represent one or more dendritic trees (provided that the dendritic trees that it represents fulfill the necessary conditions for reduction and in particular have the same electrotonic length). The physical diameter $d_{cyl\,j}$ of the jth equivalent cylinder is defined in terms of the diameters of the primary dendrites of the dendritic trees that it represents as follows:

$$d_{cyl\,j} = \left(\sum_{st \in stems\,j} d_{st}^{3/2} \right)^{2/3}, \tag{5.6}$$

where d_{st} is the diameter of a stem segment and the summation is taken over the set stems j of all such stem segments emanating from the soma to be represented by the jth equivalent cylinder. Similarly, the physical length $l_{cyl\,j}$ of the jth equivalent cylinder is defined as

$$l_{cyl\,j} = d_{cyl\,j}^{1/2} \sum_{i \in path\,j} \frac{l_i}{d_i^{1/2}}, \tag{5.7}$$

where l_i and d_i are the length and diameter of a dendritic segment indexed by i and the summation is over the set path j of all dendritic segments which, for any chosen dendritic process being represented by the jth equivalent cylinder, compose the path from the soma to its termination (by assumption all such paths in the dendritic trees being represented by the jth equivalent cylinder have the same electrotonic length). The result (5.7) is derived from the electrotonic length of the jth equivalent cylinder being the same as that of any one of the dendritic trees or processes that it represents, with the electrical parameters R_i and R_m having been removed and leaving the result in terms of the effective length, which for each segment is $l_i/d_i^{1/2}$ and for the jth equivalent cylinder is $l_{\mathrm{cyl}\,j}/d_{\mathrm{cyl}\,j}^{1/2}$. The conditions for the reduction of a dendritic tree to an equivalent cylinder can be analogously stated using effective length in place of electrotonic length and is sometimes used (for example, Clements, 1984; Stratford et al., 1989; Moore and Appenteng, 1991; Rall et al., 1992), the former quantity only depending on morphological parameters.

This model was considered in Evans et al. (1992) with further analysis of the solution behaviour being given in Evans et al. (1995). It involves solving the standard cable equation (5.1) in each equivalent cylinder for appropriate boundary conditions at the ends of the cylinders (5.2) together with conditions of voltage continuity (5.3) and current conservation at the soma (5.4). Indeed, it is the current conservation condition at the soma (5.4) that enforces discontinuity in the derivative of the voltage at the focal point of the equivalent cylinders which makes the problem interesting mathematically. In the following sections a mathematical analysis is presented within a framework well suited to the study of linear models such as those formed by passive cables (branching or otherwise). New results as well as a summary of the important results derived in Evans et al. (1992) and (1995) are presented. Finally, it should be mentioned that Redman (1973), Rinzel and Rall (1974) and Holmes et al. (1992) have obtained analytic or semi-analytic solutions for a multiple equivalent cylinder model with a point soma, which is a particular case of the above model.

3. PROBLEM NORMALIZATION

The dimensional mathematical equations for the model have been presented in the previous section, where the nerve cell is represented by N equivalent cylinders emanating from a uniformly polarized soma.

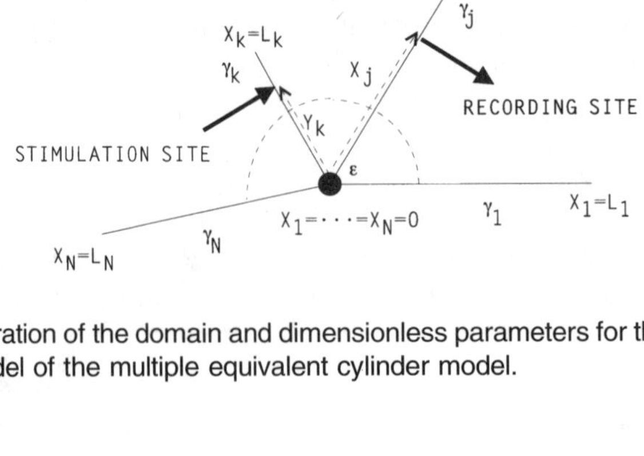

Figure 5.3 Illustration of the domain and dimensionless parameters for the non-dimensional mathematical model of the multiple equivalent cylinder model.

We non-dimensionalize (5.1)–(5.5) with,

$$x_j=\lambda_j X_j,\qquad t=\tau_m T,\qquad v_j=\bar{V}V_j,\qquad v_s=\bar{V}V_s,$$
$$\lambda_j r_{aj} i_j=\bar{V}I_j,\qquad R_s i_s=\bar{V}I_s,\qquad h_j=\bar{V}H_j,$$

where $\bar{V}$ is an appropriate voltage scaling determined by the type of input e.g. for a point charge input of Q_0 coulombs then we may take $\bar{V}=Q_0R_s/\tau_m$. We introduce the non-dimensional parameters,

$$\varepsilon=\frac{\tau_s}{\tau_m},\qquad \gamma_j=\frac{R_s}{\lambda_j r_{aj}},\qquad L_j=\frac{l_{\mathrm{cyl}j}}{\lambda_j},\tag{5.8}$$

for $j=1,\ldots,N$ which gives the non-dimensional mathematical model,

$$\text{in } 0<X_j<L_j,\quad T>0\qquad V_j+\frac{\partial V_j}{\partial T}-\frac{\partial^2 V_j}{\partial X_j^2}=I_j(X_j,T);\tag{5.9}$$

$$\text{at } X_j=L_j\qquad \frac{\partial V_j}{\partial X_j}=0;\tag{5.10}$$

$$\text{at } X_j=0\qquad V_j=V_s,\tag{5.11}$$

$$V_s+\varepsilon\frac{\mathrm{d}V_s}{\mathrm{d}T}-\sum_{j=1}^{N}\gamma_j\frac{\partial V_j}{\partial X_j}=I_s(T);\tag{5.12}$$

$$\text{at } T=0,\ 0\leqslant X_j\leqslant L_j\qquad V_j=H_j(X_j);\tag{5.13}$$

for $j=1,\ldots,N$, illustrated in Figure 5.3.

4. A GENERAL SOLUTION OF THE GOVERNING BOUNDARY-VALUE PROBLEM

The solution to the general problem (5.9)–(5.13) may be expressed as a convolution integral of the unit impulse solution $G_j^k(X_j,Y_k,T)$ with the appropriate input currents and initial voltage distribution in the form

$$\begin{aligned}V_j(X_j,T)=&\sum_{k=1}^{N}\int_0^{L_k}\int_0^{T}G_j^k(X_j,\xi_k,T-u)I_k(\xi_k,u)\,\mathrm{d}u\,\mathrm{d}\xi_k\\&+\sum_{k=1}^{N}\int_0^{L_k}G_j^k(X_j,\xi_k,T)H_k(\xi_k)\,\mathrm{d}\xi_k+\frac{1}{\gamma_j}G_j^j(0,X_j,T)H_j(0)\\&+\frac{1}{\gamma_j}\int_0^{T}G_j^j(0,X_j,T-u)I_s(u)\,\mathrm{d}u,\end{aligned}\tag{5.14}$$

for $j=1,\ldots,N$. The unit impulse solution (or Green's function) $G_j^k(X_j,Y_k,T)$ is defined as the solution of (5.9)–(5.13) with

$$I_k-\delta(T)\delta(X_k-Y_k),\qquad I_j=0,\quad j=1,\ldots,N,\ j\neq k,$$
$$I_s=0,\qquad H_j=0,\quad j=1,\ldots,N;$$

which represents a unit impulse input at a (non-dimensional) distance Y_k from the soma in the kth equivalent cylinder. The unit impulse solution $G_j^k(X_j, Y_k, T)$ so defined represents the response at a point X_j in the jth equivalent cylinder to a unit impulse input at a point Y_k along the kth cylinder. The index k emphasizes that the input or stimulating site Y_k is in the kth cylinder whilst the index j emphasizes that the recording site X_j is in the jth cylinder. Both indexes j and k lie in the range $1, \ldots, N$ and may or may not be equal. For example, $G_2^1(0.5, 0.75, T)$ represents the unit impulse solution recorded at the point $X_2 = 0.5$ in equivalent cylinder 2 for the unit impulse input at $Y_1 = 0.75$ in equivalent cylinder 1. Necessarily, $X_j \in [0, L_j]$ and $Y_k \in [0, L_k]$ and our notation extends to the case where $X_j = 0$ and/or $Y_k = 0$ i.e. recording and/or stimulation is at the soma.

Two useful expressions for the solution G_j^k can be derived, one in the form of an eigenfunction expansion particularly appropriate for $T = O(1)$ whilst another being useful for $T = o(1)$.

The unit impulse solution G_j^k may be expressed as an eigenfunction expansion in the form

$$G_j^k(X_j, Y_k, T) = \sum_{n=0}^{\infty} E_{kn} \psi_{jn}(X_j) \psi_{kn}(Y_k) \mathrm{e}^{-(1+\alpha_n^2)T}, \tag{5.15}$$

where the spatial eigenfunctions are given by

$$\psi_{jn}(X_j) = \frac{\cos[\alpha_n(L_j - X_j)]}{\cos[\alpha_n L_j]}, \qquad \psi_{kn}(Y_k) = \frac{\cos[\alpha_n(L_k - Y_k)]}{\cos[\alpha_n L_k]}, \tag{5.16}$$

the eigenvalues α_n, $n = 0, 1, 2, \ldots$, are the (non-negative) roots of the transcendental equation

$$1 - \varepsilon(1 + \alpha_n^2) = \alpha_n \sum_{j=1}^{N} \gamma_j \tan[\alpha_n L_j], \tag{5.17}$$

and the coefficients E_{kn} are defined as

$$E_{kn} = 2\gamma_k \Bigg/ \left[2\varepsilon + \sum_{j=1}^{N} \gamma_j L_j \left(\frac{\tan[\alpha_n L_j]}{\alpha_n L_j} + \sec^2[\alpha_n L_j] \right) \right]. \tag{5.18}$$

The coefficient E_{k0} needs particular care in the case $\varepsilon = 1$ since the first root of the transcendental equation $\alpha_0 = 0$, but with (5.18) giving the correct limit $E_{k0} = 2\gamma_k / [1 + \sum_{j=1}^{N} \gamma_j L_j]$.

The derivation of this result is discussed in detail in Evans et al. (1992) by two methods; one by the application of the Laplace transform method with complex residues following Bluman and Tuckwell (1987) for the single cylinder case and another by the method of separation of variables with a modified orthogonality condition for this more general multiple equivalent cylinder problem. The former is the preferred method, mainly because of its ease of generalization but also since the Laplace transform solution gives explicit information (in the form of its poles) regarding the eigenvalues of the problem. One of the limitations of the solution expressed in the form of an eigenfunction expansion as in (5.15) is that the eigenvalues α_n for $n=0,1,\ldots,$ must form a complete set, which is more difficult to prove for this N-cylinder problem than in the single cylinder case. Generally, this is the case, however in certain circumstances they do not and a second set of eigenvalues must also be considered. The special circumstances are when the electrotonic lengths of the input and recording cylinders are precisely odd integer multiples of one another, for example $L_k=L_j$, $L_k=3L_j$ or $5L_k=L_j$. The particular case in which the recording cylinder is the input cylinder also being included, then $L_k=L_j$ since $j=k$. Indeed, if we denote this second set of eigenvalues by $\{\beta_n\}_{n=0}^{\infty}$ then they arise only when

$$\cos(\beta_n L_k)=0 \quad \text{and} \quad \cos(\beta_n L_j)=0,$$

i.e. L_k and L_j are in odd integer ratios. In this case the solution (15) is augmented to

$$G_j^k(X_j, Y_k, T)=\sum_{n=0}^{\infty} E_{kn}\psi_{jn}(X_j)\psi_{kn}(Y_k)e^{-(1+\alpha_n^2)T}+\sum_{n=0}^{\infty} D_n\phi_{jn}(X_j)\phi_{kn}(Y_k)e^{-(1+\beta_n^2)T},$$

with

$$\phi_{jn}(X_j)=\cos[\beta_n(L_j-X_j)], \qquad \phi_{kn}(Y_k)=\cos[\beta_n(L_k-Y_k)],$$

and the coefficients D_n are defined as

$$D_n=\begin{cases} -\dfrac{2\gamma_k \operatorname{cosec}[\beta_n L_k]\operatorname{cosec}[\beta_n L_j]}{L_k L_j(\gamma_k/L_k+\sum_{i\in S_n}\gamma_i/L_i)} & j\neq k \\[2ex] -\dfrac{2\gamma_k}{L_k^2(\gamma_k/L_k+\sum_{i\in S_n}\gamma_i/L_i}+\dfrac{2\sin(\beta_n Y_k)\operatorname{cosec}[\beta_n L_k]}{L_k\cos[\beta_n(L_k-Y_k)]}, & j=k,\ Y_k<X_k \end{cases}$$

with the expression for $X_k<Y_k$ in the $j=k$ case being obtained by simply interchanging X_k and Y_k in the expression. The index set S_n is defined as

$$S_n=\{i\in(1,\ldots,N),\ i\neq k: \cos(\beta_n L_i)=0\},$$

and thus includes any cylinder, other than the input cylinder, whose electrotonic length satisfies $\cos(\beta_n L_i)=0$ and thus must be an odd integer multiple of the electrotonic length of the input cylinder. Note that the members of this set may change with n the eigenvalue index. The case in which the recording cylinder is not the input cylinder (i.e. $j\neq k$) is straightforward, with the set S_n necessarily containing the index j for each n and then any other cylinder satisfying the given criteria. The case in which the recording and input cylinder are the same (i.e. $j=k$) needs discussion. In this case the eigenvalues $\{\beta_n\}_{n=0}^{\infty}$ are given by $\beta_n=(2n+1)\pi/2L_k$ with the set S_n containing any cylinder, other than the input cylinder, whose electrotonic length satisfies $\cos(\beta_n L_i)=0$. Note that if this set is empty for any particular n then $D_n=0$ (which can be obtained from the expressions for D_n), and importantly this means that when the electrotonic length of the input cylinder is not an odd integer multiple of the electrotonic lengths of any of the other cylinders then $D_n=0$ for each n.

The full solution has been stated for completeness and is easily derived from the Laplace transform solution (Evans et al., 1992). The two sets of eigenvalues $\{\alpha_n\}_{n=0}^{\infty}$ and $\{\beta_n\}_{n=0}^{\infty}$ arise naturally as the roots ($x \geqslant 0$) of the transcendental equation,

$$\cos[xL_j]\cos[xL_k]\left(1-\varepsilon(1+x^2)-x\sum_{i=1}^{N}\gamma_i\tan[xL_i]\right)=0.$$

The second set of eigenvalues $\{\beta_n\}_{n=0}^{\infty}$ only occur in the certain cases mentioned and are removed by small adjustments e.g. $O(10^{-5})$ to the electrotonic lengths of the cylinders involved, with no discernible effects on the overall resulting waveforms. As such they are less important and the solution (5.15) is adequate, provided these adjustments are made when the electrotonic lengths of the input cylinder and any other cylinder are in odd integer ratios. This point is discussed more fully in Evans et al. (1992) and is a feature which does not arise in the single cylinder case. For the particular case of $\varepsilon=1$ Rall (1969) obtained an expression similar to (5.17) for the eigenvalues in this N cylinder case, whilst Holmes et al. (1992) obtained expressions similar to (5.15) for the particular case of a point soma at which a point charge input occurs.

The case in which electrotonic lengths are equal also needs special mention, since it can be dealt with as just discussed (by small adjustments to the electrotonic lengths) or by combining the cylinders into one. If we assume that the electrotonic lengths of all the cylinders are equal, so that $L_1=L_2=\cdots=L_N=L$ then the Green's function $G(X,Y,T)$ for the 1-cylinder problem satisfies

$$\text{in } 0<X<L, \quad T>0 \qquad G+\frac{\partial G}{\partial T}-\frac{\partial^2 G}{\partial X^2}=\delta(T)\delta(X-Y); \tag{5.19}$$

$$\text{at } X=L \qquad \frac{\partial G}{\partial X}=0; \tag{5.20}$$

$$\text{at } X=0 \qquad G=G_s, \tag{5.21}$$

$$G_s+\varepsilon\frac{dG_s}{dT}-\gamma\frac{\partial G}{\partial X}=0; \tag{5.22}$$

$$\text{at } T=0, \ 0\leqslant X\leqslant L \qquad G=0; \tag{5.23}$$

where $\gamma=\sum_{j=1}^{N}\gamma_j$, $Y=Y_k$, $X=X_j$ and G is related to the Green's function of the N-cylinder problem by,

$$G(X,Y,T)=\frac{\sum_{j=1}^{N}\gamma_j G_j^k(X_j,Y_k,T)}{\sum_{j=1}^{N}\gamma_j}. \tag{5.24}$$

An analogous expression to (5.24) holds when fewer than N cylinders, but all of equal electrotonic length, are combined, the summations now being taken only over the cylinders involved.

Whilst the expression (5.15) is valid for $T \geqslant 0$, its convergence is slow for $T \ll 1$ and the following asymptotic expansion may be obtained (Evans et al., 1992),

$$G_j^k \sim \frac{\gamma_k}{\varepsilon} \sum_{i=1}^{4} \mathrm{e}^{(h\chi_i + h^2 T)} \mathrm{erfc}\left(\frac{\chi_i}{2\sqrt{T}} + h\sqrt{T}\right) + \delta_{jk} \frac{1}{2\sqrt{\pi T}} \sum_{i=1}^{4} (-1)^{i+1} \mathrm{e}^{-(\Omega_i)^2/4T} \quad \text{as } T \to 0, \tag{5.25}$$

where

$$\begin{aligned} \chi_1 &= 2L_k - Y_k + X_j, & \Omega_1 &= X_k - Y_k, \\ \chi_2 &= Y_k + X_j, & \Omega_2 &= 2L_k - X_k - Y_k, \\ \chi_3 &= 2L_k - Y_k + 2L_j - X_j, & \Omega_3 &= 2L_k + Y_k + X_k, \\ \chi_4 &= Y_k + 2L_j - X_j, & \Omega_4 &= 2L_k + Y_k - X_k, \end{aligned} \tag{5.26}$$

with

$$h = \frac{\gamma}{\varepsilon}, \quad \gamma = \sum_{j=1}^{N} \gamma_j.$$

The Kronecker delta δ_{jk} has been introduced for notational conciseness and is defined as,

$$\delta_{jk} = \begin{cases} 1 & \text{if } j = k, \\ 0 & \text{if } j \neq k. \end{cases}$$

The expression (5.25) holds when $j = k$ for $Y_k < X_k$ with the corresponding expression for $X_k < Y_k$ resulting by interchanging X_k and Y_k in (5.25). The expansion (5.25) takes on simplifying forms for the two cases $h = O(1)$ and $h \gg 1$ which are discussed in Evans et al. (1995). The limit $h \to \infty$ (which arises if $\varepsilon \to 0$ or $\gamma \to \infty$) can be obtained from (5.25) but needs care. It is worth mentioning that Abbott et al. (1991) present a trip-based method for determining the Green's function for no soma or shunt in the multiple branching tree case. This can alternatively be derived as is shown in Evans et al. (1995) by retaining higher order terms in the expansion (5.25), with the interpretation of the terms in (5.26) and others of higher order being image or reflection terms. It is noted that in the case of no soma and shunt, more terms in the expansions can be obtained allowing the solution to hold for arbitrarily long times. However, this is not the case once the soma and shunt is introduced, suggesting that in this case trip-based method would not produce a solution valid for arbitrarily large times.

The unit impulse solution may be evaluated in the following manner, as discussed in Evans et al. (1995). The leading order terms are kept for the small time solution (5.25) and the number of terms in the approximation (5.15) is increased until reasonable agreement is obtained between the two results in the form of an overlap region for a range of small T values. A bound on the error involved in truncating the series (5.15) for its numerical calculation can be obtained and is derived in Evans et al. (1995). The error bound explicitly illustrates the fact that as $T \to 0$ more terms of the series need to be retained to achieve a specified accuracy.

Determination of the Green's function allows the voltage response to more general current inputs through use of the convolution integrals in (5.14). These have been explicitly evaluated in Evans et al. (1995) to obtain expressions for the voltage response to a short and long time step current input and for a smooth input current time courses in the form of an alpha-function and a multi-exponential function. Other types of current inputs may be considered as well as multiple input sites.

In this section we have discussed techniques which allow analytical solutions for the unit impulse problem and hence the full dimensionless problem to be obtained. However, this in itself constitutes only part (an important one admittedly) in the understanding of the dimensional mathematical model (5.1)–(5.5). Further understanding is gained by investigating the behaviour of the solution to the model for physically based limits of the dimensional parameters, allowing a perspective on their effect and importance to be inferred. An efficient and consistent way to do this is to consider physically based limits of the dimensionless parameters on the dimensionless model, relating these limits to the appropriate equivalent dimensional problems. This is done in the next section using perturbation techniques, where simplifications of the dimensionless model (5.9)–(5.13) are discussed for the unit impulse input.

5. PERTURBATION ANALYSIS

To understand the behaviour of the solutions to the multiple equivalent cylinder model it is useful to consider physically acceptable based limits of the dimensionless group $(\varepsilon, \gamma_j, L_j) j=1,\ldots,N$ for the unit impulse solution. The relationships between the dimensionless group $(\varepsilon, \gamma_j, L_j) j=1,\ldots,N$ and the primary dimensional parameters (electrical and morphological) $(C_m, R_m, R_i, R_{sh}, d_{cyl\,j}, l_{cyl\,j}, A_s)$ $j=1,\ldots,N$ are

$$\varepsilon = \frac{1}{1+R_m/A_s R_{sh}}, \quad \gamma_j = \frac{\pi d_{cyl\,j}^{3/2} R_m^{1/2}}{A_s R_i^{1/2}(1+R_m/A_s R_{sh})}, \quad L_j = \frac{2 l_{cyl\,j} R_i^{1/2}}{R_m^{1/2} d_{cyl\,j}^{1/2}}, \tag{5.27}$$

for $j=1,\ldots,N$, and the dimensionless parameter ranges are $0 \leqslant \varepsilon \leqslant 1$, $0 \leqslant \gamma_j < \infty$ and $0 \leqslant L_j < \infty$ for $j=1,\ldots,N$.

The dimensionless parameters can each vary independently of one another (as well as dependently) and we now investigate the effect of each separately on the unit impulse solution.

5.1 The Dimensionless Somatic Shunt Parameter ε

The unit impulse solution simplifies for the limits $\varepsilon \to 1$ (somatic (τ_s) and dendritic (τ_m) time constants approximately equal) and $\varepsilon \to 0$ (small somatic time constant relative to the dendritic time constant).

(a) Considering first the limit $\varepsilon \to 0$ with $\gamma_j = O(1)$ and $L_j = O(1)$ for $j=1,\ldots,N$, and seeking regular expansions in powers of ε, it is easily shown that we have $G_j^k \sim \hat{G}_j^k + O(\varepsilon)$ where the leading order term $\hat{G}_j^k$ is the unit impulse solution with $\varepsilon = 0$. Importantly, the relative error $(G_j^k - \hat{G}_j^k)/G_j^k$ is $O(\varepsilon)$ and provides an estimate of the error involved in using the leading order term as an approximation in the limit $\varepsilon \to 0$.

(b) The limit $\varepsilon \to 1$ with $\gamma_j = O(1)$ and $L_j = O(1)$ for $j = 1, \ldots, N$, is discussed in Evans et al. (1995). We have the expansion $G_j^k \sim \hat{G}_j^k + O(\delta)$ as $\delta \to 0$, where $\delta = 1 - \varepsilon$ is a small parameter and the leading order term $\hat{G}_j^k$ is the unit impulse solution with $\varepsilon = 1$. Again, the relative error $(G_j^k - \hat{G}_j^k)/G_j^k \sim O(\delta)$ and provides an estimate of the error involved in using the leading order term as an approximation in the limit $\varepsilon \to 1$.

Example 1

To illustrate the effect of the dimensionless parameter ε on the unit impulse solution we consider a $N = 2$ cylinder problem with the following parameter values:

$$\begin{array}{lll} \text{cylinder 1} & \gamma_1 = 1, & L_1 = 1, \\ \text{cylinder 2} & \gamma_2 = 1, & L_2 = 1.5, \\ \text{soma} & \varepsilon = 0, 0.25, 0.5, 0.75, 1. & \end{array}$$

The input site is in cylinder 1 at $Y_1 = 0.5$ and the recording site is in cylinder 2 at $X_2 = 0.75$. Figure 5.4 shows the voltage response $G_2^1 = G_2^1(0.75, 0.5, T)$ for different

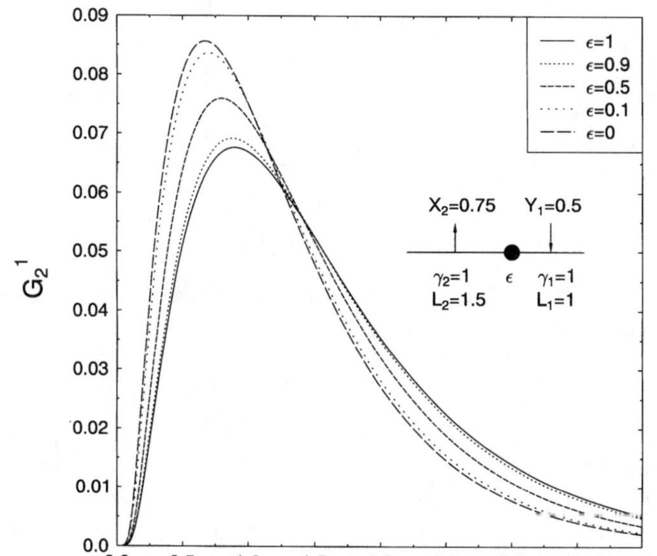

Figure 5.4 Voltage response $G_2^1 = G_2^1(0.75, 0.5, T)$ for an $N = 2$ equivalent cylinder model with parameters as depicted in the diagram, illustrating the effect of varying the nondimensional parameter ε.

values of ε, the parameter ε indeed having the required effect to accelerate potential decay and also potential rise.

5.2 The Dimensionless Parameters γ_j, $j=1,\ldots,N$.

(a) We consider the limit in which $\gamma_j \to 0$ for $j=1,\ldots,M$ (i.e. the first M equivalent cylinders) whilst $\gamma_j = O(1)$ for $j=M+1,\ldots,N$ (necessarily $1 \leqslant M \leqslant N-1$). The electrotonic lengths $L_j = O(1)$ for $j=1,\ldots,N$ and thus this limit corresponds to the dimensional case in which the equivalent diameters of the first M equivalent cylinders are small with corresponding small equivalent lengths. We set,

$$\gamma_j = \delta b_j \quad \text{for } j=1,\ldots,M,$$

where $b_j = O(1)$ and consider the limit $\delta \to 0$. Regular expansions in powers of δ can be obtained for eigenvalues α_n, eigenfunctions ψ_{jn} and expansion coefficients E_{kn} and to leading order we obtain,

$$G_j^k = \begin{cases} O(\delta) & \text{if } k \in (1,\ldots,M), \\ \hat{G}_j^k + O(\delta) & \text{if } k \in (M+1,\ldots,N), \end{cases} \tag{5.28}$$

where

$$\hat{G}_j^k(X_j, Y_k, T) = \sum_{n=0}^{\infty} \hat{E}_{kn} \hat{\psi}_{jn}(X_j) \hat{\psi}_{kn}(Y_k) \mathrm{e}^{-(1+\hat{\alpha}_n^2)T},$$

with

$$\hat{E}_{kn} = 2\gamma_k \Bigg/ \left[2\varepsilon + \sum_{j=M+1}^{N} \gamma_j L_j \left(\frac{\tan[\hat{\alpha}_n L_j]}{\hat{\alpha}_n L_j} + \sec^2[\hat{\alpha}_n L_j] \right) \right],$$

$$\hat{\psi}_{jn}(X_j) = \frac{\cos[\hat{\alpha}_n (L_j - X_j)]}{\cos[\hat{\alpha}_n L_j]}$$

and the eigenvalues $\hat{\alpha}_n$, $n=0,1,2,\ldots$, are the (non-negative) roots of the reduced transcendental equation

$$1 - \varepsilon(1+\alpha^2) = \alpha \sum_{j=M+1}^{N} \gamma_j \tan[\alpha L_j].$$

The case in which $M=N$ needs particular care, and is the case in which all dendritic equivalent diameters are small relative to the length scale given by the value of the parameter $(\pi R_{\mathrm{m}}^{1/2}/2R_{\mathrm{i}}^{1/2}(R_{\mathrm{m}}/R_{\mathrm{sh}}+A_{\mathrm{s}}))^{2/3}$. It is discussed in some detail in Evans et al. (1995), and in agreement with the above result, $G_j^k = O(\delta)$ as $\delta \to 0$ for all input branches.

(b) We now consider the limit in which $\gamma_j \to \infty$ for $j=1,\ldots,M$ (i.e. the first M equivalent cylinders) whilst $\gamma_j = O(1)$ for $j=M+1,\ldots,N$ (necessarily $1 \leqslant M \leqslant N$). The electrotonic lengths $L_j = O(1)$ for $j=1,\ldots,N$ and thus this limit corresponds to the dimensional case of large equivalent diameters of the first M equivalent cylinders with correspondingly large equivalent lengths. We set,

$$\gamma_j = \frac{b_j}{\delta} \quad \text{for } j=1,\ldots,M,$$

where $b_j = O(1)$ and consider the limit $\delta \to 0$. Regular expansions in powers of δ can again be obtained for eigenvalues α_n, eigenfunctions ψ_{jn} and expansion coefficients E_{kn} and to leading order we obtain,

$$G_j^k = \begin{cases} \hat{G}_j^k + O(\delta) & \text{if } k \in (1,\ldots,M), \\ O(\delta) & \text{if } k \in (M+1,\ldots,N), \end{cases} \tag{5.29}$$

where

$$\hat{G}_j^k(X_j, Y_k, T) = \sum_{n=0}^{\infty} \hat{E}_{kn} \hat{\psi}_{jn}(X_j) \hat{\psi}_{kn}(Y_k) \mathrm{e}^{-(1+\hat{\alpha}_n^2)T},$$

with

$$\hat{E}_{kn} = 2b_k \Bigg/ \left[\sum_{j=1}^{M} b_j L_j \left(\frac{\tan[\hat{\alpha}_n L_j]}{\hat{\alpha}_n L_j} + \sec^2[\hat{\alpha}_n L_j] \right) \right],$$

$$\hat{\psi}_{jn}(X_j) = \frac{\cos[\hat{\alpha}_n (L_j - X_j)]}{\cos[\hat{\alpha}_n L_j]},$$

and the eigenvalues $\hat{\alpha}_n$, $n=0,1,2,\ldots$, are the (non-negative) roots of the reduced transcendental equation

$$0 = \sum_{j=1}^{M} b_j \tan[\alpha L_j].$$

The case in which $M=N$ was considered in Evans et al. (1995), and is a particular case which can be obtained from the above result.

Example 2

To illustrate the effect of the dimensionless parameters $\gamma_i (i=1,\ldots,N)$ on the unit impulse solution we consider a $N=2$ cylinder problem with the following parameter values,

cylinder 1	γ_1,	$L_1 = 1$,
cylinder 2	γ_2,	$L_2 = 1.5$,
soma		$\varepsilon = 1$,

with the input site in cylinder 1 at $Y_1 = 0.5$ and the recording site in cylinder 2 at $X_2 = 0.75$.

Figure 5.5(a) shows the voltage response $G_2^1 = G_2^1(0.75, 0.5, T)$ for $\gamma_2 = 1$ and selected values of γ_1, to illustrate the effect of varying the γ parameter for the input cylinder. The voltage responses for non-zero finite values of γ_1 are bounded by those for $\gamma_1 = 0$ and $\gamma_1 = \infty$ (which agree with the limits obtained in (5.28) and (5.29) where $N = 2$, $M = 1$, $k = 1$, $j = 2$).

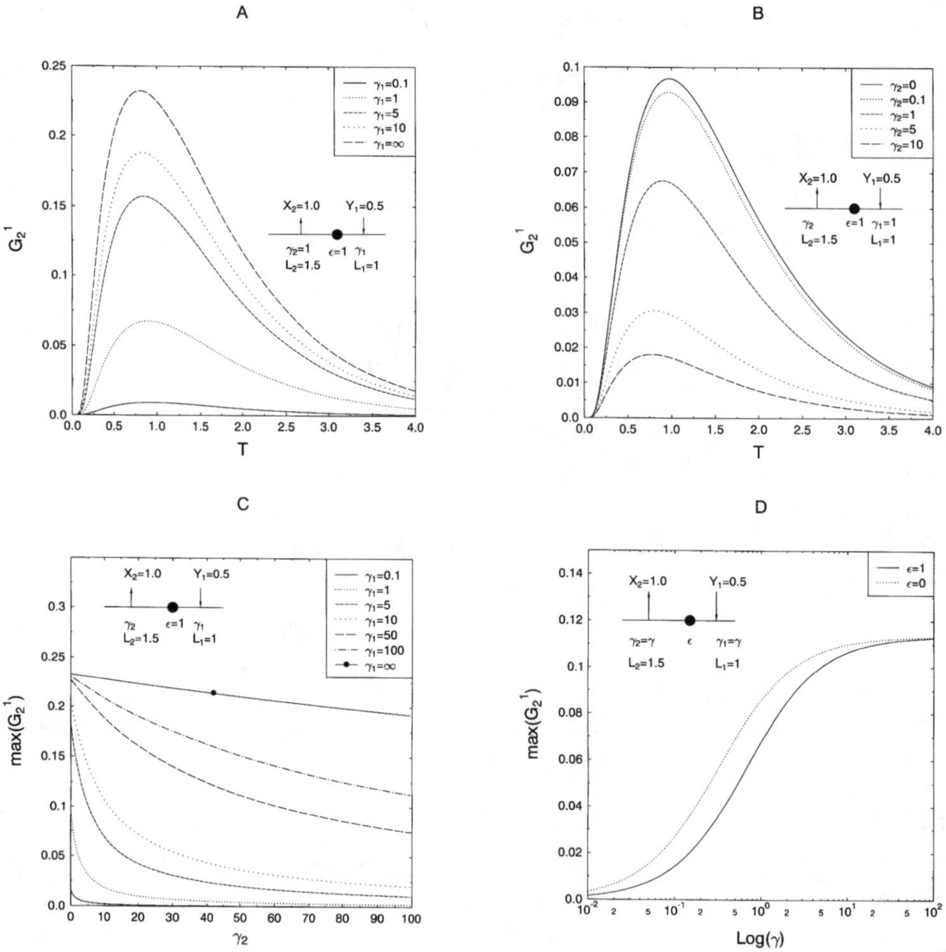

Figure 5.5 Illustrates the effect of the dimensionless parameters γ_1 and γ_2 for the 2 cylinder model described in example 2. (A) shows the voltage response $G_2^1 = G_2^1(1, 0.5, T)$ as the parameter γ_1 of the input cylinder is varied. (B) illustrates the voltage response $G_2^1 = G_2^1(1, 0.5, T)$ as the parameter γ_2 of the recording cylinder is varied. (c) shows the peak voltage, $\max(G_2^1)$, as both γ_1 and γ_2 are varied independently. (D) shows the peak voltage, $\max(G_2^1)$, as both γ_1 and γ_2 are varied simultaneously ($\gamma_1 = \gamma_2 = \gamma$), for $\varepsilon = 1$ and also the case $\varepsilon = 0$ for comparison.

Figure 5.5(b) shows again the voltage response $G_2^1 = G_2^1(0.75, 0.5, T)$ now for $\gamma_1 = 1$ and selected values of γ_2, in order to illustrate the effect of varying the γ parameter for the recording cylinder. The voltage responses for non-zero finite values of γ_2 are again bounded by those for $\gamma_2 = 0$ and $\gamma_2 = \infty$ (which agree with the limits obtained in (5.28) and (5.29) where now we have to interchange the labelling of cylinders 1 and 2 in order to agree with the notation used so that $N = 2, M = 1, k = 2, j = 1$).

Figure 5.5(c) illustrates the peak voltage $\max(G_2^1)$ as γ_2 is varied for selected values of γ_1. For each fixed value of γ_1 the peak voltage occurs for $\gamma_2 = 0$ and steadily falls off for larger values of γ_2, this rate decreasing for larger values of γ_1. The results for the extreme limits of γ_1 and γ_2 are again consistent with those of (5.28) and (5.29).

Figure 5.5(d) considers the peak voltage as both γ_1 and γ_2 are varied at the same rate for not only $\varepsilon = 1$ but also $\varepsilon = 0$. For large values of γ_1, γ_2 the results are consistent with those of (5.29) which are independent of ε, and also for small values of γ_1, γ_2 the results agreeing with those of (5.28).

5.3 Cylinder Reduction

In the multiple equivalent cylinder model we have N equivalent cylinders where in the first instance, each cylinder represents one dendritic tree and thus a neuron with N' dendritic trees will be represented by $N = N'$ equivalent cylinders. However, an equivalent cylinder can represent more than one dendritic tree and thus a number of models can be formed with the number of equivalent cylinders ranging between $N = 1$ (the single cylinder model where one equivalent cylinder represents the whole dendritic structure) and $N = N'$ (where each equivalent cylinder represents one dendritic tree only). An important step in simplifying and understanding the multiple equivalent cylinder model is the reduction of the number of cylinders in the model. This may be achieved in a consistent manner either by cylinder collapse or by cylinder neglect. In this section we consider these two techniques and describe the conditions under which they are applicable.

The results we describe can be interpreted in terms of the membrane surface area of the dendrites and soma which we now state for later reference. The membrane surface area of the jth equivalent cylinder $A_j = \pi d_j l_j$ and the soma A_s may be expressed in the form:

$$A_j = \frac{R_m}{R_s} \gamma_j L_j, \qquad A_s = \frac{R_m}{R_s} \varepsilon. \tag{5.30}$$

1. *Cylinder Neglection*

We begin by considering the neglection of cylinders and the corresponding effect on the voltage response for the unit impulse input. This has been dealt with in Evans et al. (1995) in some detail and here we summarize the important results. We are interested in the removal of cylinders which make a negligible contribution to the voltage response.

(a) We begin by ordering the N cylinders according to the relative sizes,

$$\gamma_1 L_1 \leqslant \gamma_2 L_2 \leqslant \cdots \leqslant \gamma_N L_N.$$

(b) The cylinders are now segregated into two groups, those to be neglected and those to be retained. Let us assume that the first M cylinders are to be neglected and the remaining $N-M$ cylinders are to be retained. The criteria used to determine M (i.e. which cylinders are to be neglected and which are to be retained) is that the following value

$$\theta = \frac{2\sum_{j=1}^{M} \gamma_j L_j}{\varepsilon + \sum_{j=M+1}^{N} \gamma_j L_j}, \tag{5.31}$$

is small i.e. we require $\theta \ll 1$.

(c) If such an M can be found that gives $\theta \ll 1$ then we have the approximation,

$$G_j^k = \hat{G}_j^k + O(\theta), \tag{5.32}$$

where G_j^k is the solution to the full N cylinder problem and $\hat{G}_j^k$ is the solution to the reduced $N-M$ cylinder problem and given by,

$$\hat{G}_j^k(X_j, Y_k, T) = \sum_{n=0}^{\infty} \hat{E}_{kn} \hat{\psi}_{jn}(X_j) \hat{\psi}_{kn}(Y_k) \mathrm{e}^{-(1+\hat{\alpha}_n^2)T}$$

with

$$\hat{E}_{kn} = 2\gamma_k \Bigg/ \left[2\varepsilon + \sum_{j=M+1}^{N} \gamma_j L_j \left(\frac{\tan[\hat{\alpha}_n L_j]}{\hat{\alpha}_n L_j} + \sec^2[\hat{\alpha}_n L_j] \right) \right],$$

$$\hat{\psi}_{jn}(X_j) = \frac{\cos[\hat{\alpha}_n (L_j - X_j)]}{\cos[\hat{\alpha}_n L_j]},$$

and the eigenvalues $\hat{\alpha}_n$, $n = 0, 1, 2, \ldots$, are the (non-negative) roots of the reduced transcendental equation,

$$1 - \varepsilon(1 + \hat{\alpha}_n^2) = \hat{\alpha}_n \sum_{j=M+1}^{N} \gamma_j \tan[\hat{\alpha} L_j].$$

Importantly, the relative error between the solution to the full N cylinder problem G_j^k and the solution to the reduced $N-M$ cylinder problem $\hat{G}_j^k$ is then $O(\theta)$ i.e.

$$\frac{G_j^k - \hat{G}_j^k}{G_j^k} = O(\theta). \tag{5.33}$$

The steps (a)–(c) form a consistent method for cylinder neglection. Note that using (5.30), (5.31) may be expressed as a ratios of areas, and thus the criteria for cylinder neglection, that $\theta \ll 1$, explicitly states that the surface area of the neglected cylinders be far smaller than the surface area of retained cylinders and soma. Moreover, the effect on the voltage response is quantified by the expression (5.32), and shows that the size of the error introduced by neglecting the first M cylinders is $O(\theta)$. It is assumed that the largest value of M is chosen in (5.31) such that the neglection criteria is satisfied, and that cylinders which contain the input or recording sites are not neglected.

Example 3

To illustrate the effect of cylinder neglection on the unit impulse solution we consider a $N=3$ cylinder problem with the following parameter values,

$$\begin{array}{lll} \text{cylinder 1} & \gamma_1=1, & L_1=0.5, \\ \text{cylinder 2} & \gamma_2=1, & L_2=2, \\ \text{cylinder 3} & \gamma_3=10, & L_3=1, \\ \text{soma} & \varepsilon=1, & \end{array}$$

with the input site in cylinder 2 at $Y_2=1$ and the recording site in cylinder 3 at $X_3=0.5$. Thus we have,

$$\gamma_1 L_1=0.5<\gamma_2 L_2=2<\gamma_3 L_3=10$$

and we examine the effect of neglecting cylinder 1. Taking $M=1, N=3$ we have from (5.31) that $\theta=0.077$ which gives an estimate of the relative error on the unit impulse response if cylinder 1 is removed. Figure 5.6 shows the voltage response $G_3^2=G_3^2\,(1, 0.5, T)$ for the full ($N=3$) problem and for the reduced ($N=2$) problem after cylinder 1 is removed. The absolute relative error, defined as $|(G_j^k-\hat{G}_j^k)/G_j^k|$ that for these two solutions is also shown, the maximum value of about 0.07 agreeing well with the θ estimate.

2. *Cylinder Collapse*

We begin by considering the effect on the voltage response for small changes in the electrotonic lengths of the cylinders and then proceed to optimize the changes so that the resulting error in the voltage response is minimized.

For definiteness we order the N equivalent cylinders according to increasing electrotonic length,

$$L_1 \leqslant L_2 \leqslant \cdots \leqslant L_N.$$

The electrotonic lengths are now written in the form:

$$L_j=\hat{L}_j+\theta_j, \quad \text{where } \theta_j \ll 1, \tag{5.34}$$

for $j=1,\ldots,N$ and we now quantify the change in the voltage response in using the approximating electrotonic lengths $\hat{L}_j$. We define the approximating eigenvalues $\hat{\alpha}_n$ as the (non-negative) roots of the transcendental equation,

$$1-\varepsilon(1+\hat{\alpha}^2)=\hat{\alpha}\sum_{j=1}^{N}\gamma_j \tan[\hat{\alpha}\hat{L}_j],$$

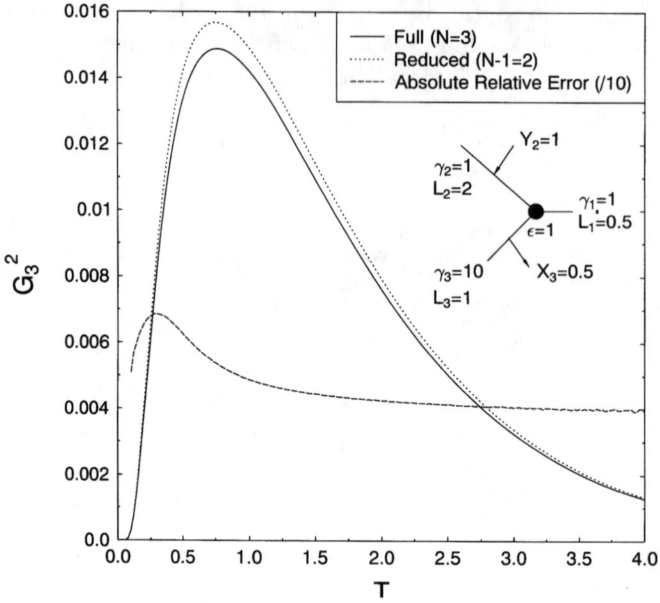

Figure 5.6 Illustration of cylinder reduction by the method of cylinder neglection. The full $N=3$ cylinder model is depicted with the input site in cylinder 2 and recording site in cylinder 3. The voltage response $G_3^2=G_3^2(0.5,1,T)$ is shown for the full model ($N=3$) and the reduced model ($N=2$), where cylinder 1 has been removed. The absolute relative error (divided by 10) for these two voltage responses is shown.

which are related to the true roots α_n by

$$\alpha_n=\hat{\alpha}_n-\delta_n,$$

where

$$\delta_n=\frac{\hat{\alpha}_n\sum_{j=1}^{N}\gamma_j\theta_j\sec^2[\hat{\alpha}_n\hat{L}_j]}{2\varepsilon+\sum_{j=1}^{N}\gamma_j\hat{L}_j(\tan[\hat{\alpha}_n\hat{L}_j]/\hat{\alpha}_n\hat{L}_j+\sec^2[\hat{\alpha}_n\hat{L}_j])}+O(\theta^2),$$

as $\theta\to 0$, where $\theta=\max\{|\theta_j|\}_{j=1}^{N}$. If we choose the approximating lengths $\hat{L}_j$ such that

$$\sum_{j=1}^{N}\gamma_j\theta_j\sec^2[\hat{\alpha}_n\hat{L}_j]=0, \tag{5.35}$$

for each n, then $\delta_n=O(\theta^2)$ and the expansion coefficients E_{kn} can be shown to have the expansion

$$E_{kn}=\hat{E}_{kn}(1+O(\theta^2)),$$

where

$$\hat{E}_{kn}=\frac{2\gamma_k}{2\varepsilon+\sum_{j=1}^{N}\gamma_j\hat{L}_j(\tan[\hat{\alpha}_n\hat{L}_j]/\hat{\alpha}_n\hat{L}_j+\sec^2[\hat{\alpha}_n\hat{L}_j])}.$$

The eigenfunctions have the expansion

$$\psi_{jn}=\hat{\psi}_{jn}+O(\theta),$$

where

$$\hat{\psi}_{jn}=\frac{\cos[\hat{\alpha}_n(\hat{L}_j-X_j)]}{\cos[\hat{\alpha}_n\hat{L}_j]},$$

and the unit impulse solution has the expansion

$$G_j^k=\hat{G}_j^k+O(\theta),$$

where the approximating unit impulse solution is given by

$$\hat{G}_j^k(X_j,Y_k,T)=\sum_{n=0}^{\infty}\hat{E}_{kn}\hat{\psi}_{jn}(X_j)\hat{\psi}_{kn}(Y_k)\mathrm{e}^{-(1+\hat{\alpha}_n^2)T}.$$

Choosing the approximating lengths $\hat{L}_j$ so that (5.35) is satisfied, reduces the error term to $O(\theta^2)$ in the approximation expansions for δ_n and E_{kn} but remains $O(\theta)$ for the solution G_j^k.

Let us now discuss the possibility of satisfying condition (5.35) for each n. We collect the cylinders into P distinct groups where the cylinders in each group are to be collapsed or combined to give one equivalent cylinder. Group (1) is composed of cylinders $1,\ldots,N$, group (2) comprises cylinders $N_1+1,\ldots,N_2$ and so on until the final group (P) which is composed of cylinders $N_{P-1}+1,\ldots,N_P$. Thus a representative group (i), where $i=1,\ldots,P$, is composed of cylinders $N_{i-1}+1,\ldots,N_i$ and has N_i-N_{i-1} cylinders in its group, with the convention that $N_0=0$ and $N_P=N$. Necessarily, $1\leqslant N_i-N_{i-1}\leqslant N$ for each $i=1,\ldots,P$ and $1\leqslant P\leqslant N$.

For the representative group (i), we denote by $L_{(i)}$ the electrotonic length of the collapsed cylinder representing the group and hence set $\hat{L}_{N_{i-1}+1}=\hat{L}_{N_{i-1}+2}=\cdots=\hat{L}_{N_i}=L_{(i)}$. Then, condition (5.36) can be written as,

$$\sum_{i=1}^{P}\left(\sec^2[\hat{\alpha}_n L_{(i)}]\sum_{j=N_{i-1}+1}^{N_i}\gamma_j\theta_j\right)=0,$$

which will be satisfied for each n if

$$\sum_{j=N_{i-1}+1}^{N_i}\theta_j=0, \tag{5.36}$$

for each $i=1,\ldots,P$. Consequently, we obtain

$$\gamma_{(i)}=\sum_{j=N_{i-1}+1}^{N_i}\gamma_j,\qquad L_{(i)}=\frac{\sum_{j=N_{i-1}+1}^{N_i}\gamma_jL_j}{\sum_{j=N_{i-1}+1}^{N_i}\gamma_j}, \tag{5.37}$$

as the dimensionless parameters for the collapsed cylinder representing the cylinders in group (i), with $i=1,\ldots,P$.

For the reduced problem, let the input site be in cylinder (k) at $Y_{(k)}=Y_k$ and the recording site be in cylinder (j) at $X_{(j)}=X_j$. Then the solution is given by

$$\hat{G}^{(k)}_{(j)}(X_{(j)},Y_{(k)},T)=\sum_{n=0}^{\infty}\hat{E}_{(k)n}\hat{\psi}_{(j)n}(X_{(j)})\hat{\psi}_{(k)n}(Y_{(k)})\mathrm{e}^{-(1+\hat{\alpha}_n^2)T}, \tag{5.38}$$

where

$$\hat{E}_{(k)n}=\frac{2\gamma_k}{2\varepsilon+\sum_{i=1}^{P}\gamma_{(i)}L_{(i)}(\tan[\hat{\alpha}_nL_{(i)}]/\hat{\alpha}_nL_{(i)}+\sec^2[\hat{\alpha}_nL_{(i)}])},$$

$$\hat{\psi}_{(j)n}(X_{(j)})=\frac{\cos[\hat{\alpha}_n(L_{(j)}-X_{(j)})]}{\cos[\hat{\alpha}_nL_{(j)}]},$$

and the eigenvalues $\hat{\alpha}_n$ are the (non-negative) roots of the transcendental equation

$$1-\varepsilon(1+\hat{\alpha}^2)=\hat{\alpha}\sum_{i=1}^{P}\gamma_{(i)}\tan[\hat{\alpha}L_{(j)}].$$

The relative error between the solution G^k_j for the full problem and that $\hat{G}^{(k)}_{(j)}$ for the reduced problem is guaranteed to be $O(\theta)$, i.e.

$$\frac{G^k_j-\hat{G}^{(k)}_{(j)}}{G^k_j}=O(\theta), \tag{5.39}$$

where $\theta=\max\,\{\theta_{(i)}\}_{i=1}^{P}$ and $\theta_{(i)}=\max\{|L_j-L_{(i)}|\}_{j=N_{i-1}+1}^{N_i}$.

It is also important to note that in the expression for $\hat{E}_{(k)n}$ its numerator still contains the parameter γ_k of the full problem rather than $\gamma_{(k)}$ of the reduced problem, which we would expect in order to be able maintain proportionately sized inputs to the full and reduced problems.

Thus the N cylinder problem with dimensionless parameters $(\varepsilon,\gamma_j,L_j)$ for $j=1,\ldots,N$ is reduced to a P cylinder problem with dimensionless parameters $(\varepsilon,\gamma_{(i)},L_{(i)})$ for $i=1,\ldots,P$ and it is interesting to notice that the cylinder collapse as defined by (5.37) preserves dendritic membrane surface area.

We do have restrictions on the maximum size of θ that can be used, which we now state and use to provide a useful 'rule of thumb' for cylinder collapse. In order for the above expansions to hold, we require $\alpha_n\theta<0.5$, which, for any given θ, we would expect to hold for $n=0,1,2,\ldots,n_{\max}$. To ensure that the $n_{\max}+1$ terms are sufficient to attain a desired accuracy in the expansion series, we require for a specified T^* that

$$\mathrm{e}^{-(1+\hat{\alpha}^2_{n_{\max}})T^*}=\mu\ll 1,$$

where μ is an accuracy parameter. Then, for $T\geqslant T^*$, the $n_{\max}+1$ terms in the series expansion will give an approximation to the full series with a truncation error of at most $O(\mu)$. Thus for specified T^* and μ we have that

$$\theta \leqslant \frac{0.5}{(-\ln(\mu)/T^*-1)^{1/2}}. \tag{5.40}$$

Once we have an estimate of a suitable θ then in order to ensure that $\theta_j \leqslant \theta$ for each $j=1,\ldots,N$ suggests that the difference between the maximum and minimum electrotonic lengths in each group of cylinders to be collapsed must be at most θ, i.e.

$$L_{N_i} - L_{N_{i-1}+1} \leqslant \theta,$$

for $i=1,\ldots,P$, since the cylinders have already been labelled according to increasing electrotonic length.

Let us see what this means in practice. Suppose we choose $T^*=0.1$ and $\mu=5\times10^{-5}$ then (5.40) gives $\theta\leqslant0.05$. Importantly, this then suggests that cylinders whose electrotonic lengths differ by at most 0.05 can be collapsed together as described above, with the relative error incurred in the collapse being approximately 5% and further more we can be reasonably confident that our results will be valid for at least $T\geqslant T^*=0.1$. The smaller the values we impose on T^* and μ the smaller the value of θ that results and thus the greater the restriction that is placed on the numbers of cylinders that can be collapsed. The above values are only a guide and we may often find that a cylinder collapse performed to set values of θ, T^* and μ will often hold for values of $T<T^*$ and have a relative error less than $O(\theta)$.

Alternatively, suppose that we wish to collapse cylinders whose electrotonic lengths differ by at most 0.1, i.e. we set $\theta=0.1$. Then (5.40) gives

$$-\ln\mu = 26\times T^*,$$

and thus $T^*=0.1, 0.2, 0.3, 0.4$ gives $\mu=7.4\times10^{-2}, 5.5\times10^{-3}, 4.1\times10^{-4}, 3.0\times10^{-5}$, respectively. The value of μ is intended to be a guide to the size of the last term in the series expansion for which the approximations made during the cylinder collapse are expected to hold (i.e. although the perturbation expansions are not valid for subsequent terms of the series, if μ is sufficiently small then we can ensure that their contribution to the solution is negligible).

It is worth noting that whether the collapse of cylinders is done a pair at a time or in larger groups is not important with the same resulting final approximation if it is done consistently as described above.

Example 4

As a first example for cylinder collapse we consider the error involved in collapsing a 2-cylinder problem to a 1-cylinder problem, the recording site thus being in a cylinder that is to be collapsed. We take the following parameter values,

cylinder 1 $\gamma_1=1,\quad L_1=1,$

cylinder 2 $\gamma_2=1,\quad L_2,$

soma $\varepsilon=1,$

with the input site in cylinder 1 at $Y_1=0.75$ and consider the effect of collapse on the unit impulse response as L_2, the electrotonic length of cylinder 2, is varied. The parameters for the reduced 1-cylinder problem are given by

$$\text{cylinder (1)} \quad \gamma_{(1)}=2, \quad L_{(1)}=(L_1+L_2)/2,$$
$$\text{soma} \quad \varepsilon=1,$$

in accordance with (5.37). The input site is taken to be at $Y_{(1)}=0.5$ and if the recording site in the reduced problem is taken to be at $X_{(1)}=0.25$ in the collapsed cylinder, then as discussed in reference to (5.24) we have to consider the two recording sites $X_1=0.25$ and $X_2=0.25$ in the full problem. In Figure 5.7 is shown absolute relative percentage error between the averaged solutions of $G_1^1(0.25, 0.5, T)$ and $G_2^1(0.25, 0.5, T)$ to the full problem and the solution of the reduced problem $\hat{G}_{(1)}^{(1)}(0.25, 0.5, T)$ defined as

$$\left|\frac{(G_1^1+G_2^1)/2-\hat{G}_{(1)}^{(1)}}{\hat{G}_{(1)}^{(1)}}\right| \times 100\%,$$

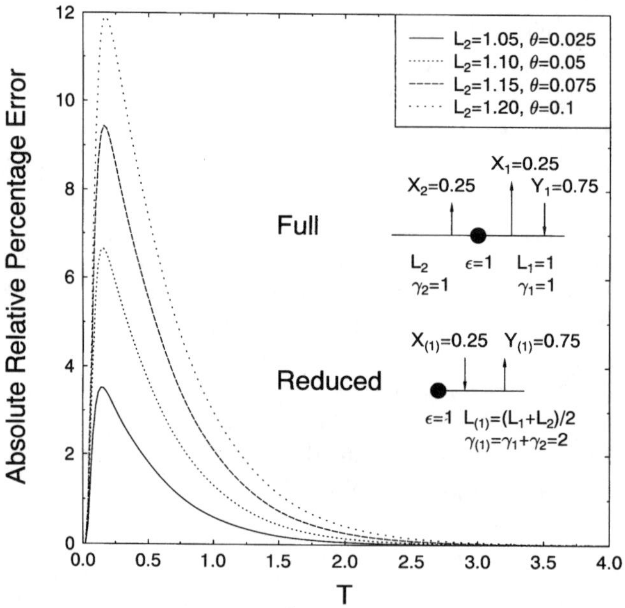

Figure 5.7 Illustration of the method of cylinder collapse for the case when the recording and input sites are in cylinders to be collapsed. The parameters for the full ($N=2$) problem is depicted as are the parameters for the collapsed ($N=1$) problem. Shown is the absolute relative percentage error between the full and collapsed problem as the electrotonic length of cylinder 2, L_2, is varied. For the differing L_2 is stated the corresponding $\theta=L_2-L_{(1)}$ which provides an estimate for the absolute relative error as suggested by (5.40).

for the specified values of L_2. In this example $\theta_2 = -\theta_1 = \theta$ the value of which is also shown for each corresponding L_2 and as predicted from (5.40), providing an estimate of the size of the absolute relative error.

Example 5

As a second example for cylinder collapse we consider the error involved in collapsing a 3-cylinder problem to a 2-cylinder problem, the recording site not being in a cylinder that is to be collapsed cylinder. We take the following parameter values

$$\begin{array}{lll} \text{cylinder 1} & \gamma_1=1, & L_1=1, \\ \text{cylinder 2} & \gamma_2=1, & L_2, \\ \text{cylinder 3} & \gamma_3=1, & L_3=2, \\ \text{soma} & \varepsilon=1, & \end{array}$$

with the input site in cylinder 1 at $Y_1=0.75$ and the recording site in cylinder 3 at $X_3=1$. We consider the effect of collapse on the unit impulse response as L_2, the electrotonic length of cylinder 2, is varied. The parameters for the reduced 2-cylinder problem are given by,

$$\begin{array}{lll} \text{cylinder (1)} & \gamma_{(1)}=2, & L_{(1)}=(L_1+L_2)/2, \\ \text{cylinder (2)} & \gamma_{(2)}=1, & L_{(2)}=2, \\ \text{soma} & \varepsilon=1, & \end{array}$$

in accordance with (5.37), the input site being at $Y_{(1)}=0.5$ and the recording site being at $X_{(2)}=1$ in the reduced problem. In the notation introduced earlier in this section, we are reducing an $N=3$ cylinder problem to a $P=2$ cylinder problem by combining cylinders 1 and 2 in the full problem into a collapsed cylinder (1) in the reduced problem, the cylinder 3 in the full problem remaining uncollapsed and represented by cylinder (2) in the reduced problem. Figure 5.8 shows the absolute relative percentage error, defined as

$$\left|\frac{G_3^1-\hat{G}_{(2)}^{(1)}}{G_3^1}\right| \times 100\%,$$

between the solution to the full problem $G_3^1=G_3^1(1, 0.5, T)$ and the corresponding solution $\hat{G}_{(2)}^{(1)}=\hat{G}_{(2)}^{(1)}(1, 0.5, T)$ of the reduced problem, for selected values of L_2 in the full problem. The corresponding values of θ are also stated and are again representative of the size of the relative error ($\theta \times 100\%$ for the relative percentage error).

Example 4 serves to demonstrate cylinder collapse for the case when the recording site is in one of the cylinders to be collapsed, whilst example 5 demonstrates the case when the recording site is in a cylinder that is not to be collapsed. Both examples demonstrate that θ is a suitable estimate for the relative error, which seems to be greater in the former case than the later. Both these examples have the input site in a cylinder that is to be collapsed, whilst if this is not the case, the relative errors seem to be significantly less than the estimate given by θ.

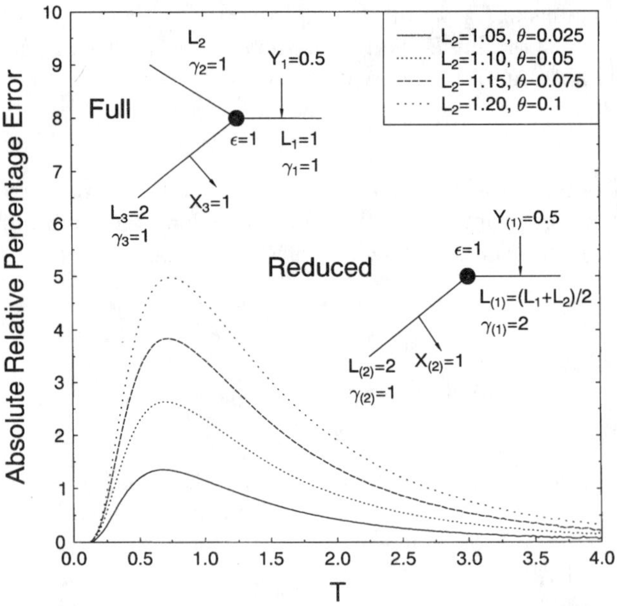

Figure 5.8 Illustration of the method of cylinder collapse for the case when the input site is in a cylinder to be collapsed, but not the recording site. The parameters for the full ($N=3$) problem is depicted as are the parameters for the collapsed ($N=2$) problem. Shown is the absolute relative percentage error between the full and collapsed problem as the electrotonic length of cylinder 2, L_2, is varied. For the differing L_2 is stated the corresponding $\theta = L_2 - L_{(1)}$ which provides an estimate for the absolute relative error as suggested by (5.40).

6. WELL-POSEDNESS OF THE INVERSE PROBLEM

We are now in a position to comment on the inverse problem which in its idealized form may be stated as: Given an entire noise-free voltage response from a model, what are the dimensional parameters for that voltage response? To be more specific, we assume that the model stated in (5.1)–(5.5) describes the passive properties of the nerve cell in which we are interested and we are required to find the values of the dimensional parameters of the model using an experimentally recorded voltage response. We wish to determine the conditions under which this inverse problem is well-posed which for our purposes will be that a solution exists and is unique. The type of current input has an important bearing on our discussion and following the forward problem we discuss the point charge input, other inputs can be treated in a similar manner.

We are given a voltage response from the dimensional model for a point charge input of Q_0 coulombs (assumed known) with input and recording sites at specified fractions (denoted by f_{in} and f_{rec} respectively) of the physical lengths of the appropriate equivalent cylinders. The dimensionless forward problem (5.9)–(5.13) is well-posed and for the unit impulse input at specified input and recording sites, the solution $V=V(T)$ depends uniquely upon the $(2N+1)$ non-dimensional parameters $(\varepsilon, \{\gamma_j\}_{j=1}^N, \{L_j\}_{j=1}^N)$. Introducing the two dimensional scalings a, b for the voltage $v=aV$ and the time scale $t=bT$, respectively, then given the dimensional voltage response the dimensionless inverse problem for the identification of $(\varepsilon, \{\gamma_j\}_{j=1}^N, \{L_j\}_{j=1}^N, a, b)$ has a unique solution so that we can find these and only parameters i.e. the dimensionless inverse problem is well-posed. We emphasize that the dimensionless input site $Y_k=f_{\mathrm{in}}L_k$ and recording site $X_j=f_{\mathrm{rec}}L_j$ depend upon the parameter L_j with the fractions $f_{\mathrm{in}}, f_{\mathrm{rec}}$ known.

The important observation from the well-posedness of the dimensionless inverse problem is that $(2N+3)$ parameters only can determined by fitting the dimensional model to the recorded voltage response. However, the dimensional model has $(2N+5)$ dimensional parameters, namely,

$$R_{\mathrm{m}}, C_{\mathrm{m}}, R_{\mathrm{i}}, \{d_{\mathrm{cyl}\,j}\}_{j=1}^N, \{l_{\mathrm{cyl}\,j}\}_{j=1}^N, R_{\mathrm{sh}}, A_{\mathrm{s}},$$

with the input charge Q_0 known and the input and recording sites given by $y_k=f_{\mathrm{in}}l_{\mathrm{cyl}\,k}$ and $x_j=f_{\mathrm{rec}}l_{\mathrm{cyl}\,j}$, respectively. Thus for the dimensional inverse problem to be well-posed two of the dimensional parameters must be specified (may be determined from independent measurements). If $(2N+4)$ or more dimensional parameters are to be determined then the dimensional inverse problem is ill-posed and the dimensional parameters can trade-off to yield the same underlying dimensionless model. If $(2N+2)$ or less dimensional parameters are to be determined then the values of the assumed known dimensional parameters must be consistent with those of the underlying dimensional model.

As an example, consider the single cylinder $N=1$ somatic shunt model, for which we consider the values of $(\varepsilon, \gamma_1, L_1, a, b)$ known and determined from the fit of the dimensionless model to the dimensional voltage response. The relationship between the dimensional and dimensionless parameters is given by

$$\varepsilon=\frac{1}{1+R_{\mathrm{m}}/A_{\mathrm{s}}R_{\mathrm{sh}}}, \quad \gamma_1=\frac{\pi d_{\mathrm{cyl}\,1}^{3/2}R_{\mathrm{m}}^{1/2}}{A_{\mathrm{s}}R_{\mathrm{i}}^{1/2}(1+R_{\mathrm{m}}/A_{\mathrm{s}}R_{\mathrm{sh}})}, \quad L_1=\frac{2l_{\mathrm{cyl}\,1}R_{\mathrm{i}}^{1/2}}{R_{\mathrm{m}}^{1/2}d_{\mathrm{cyl}\,1}^{1/2}},$$

$$a=\frac{Q_0}{A_{\mathrm{s}}C_{\mathrm{m}}(1+R_{\mathrm{m}}/A_{\mathrm{s}}R_{\mathrm{sh}})}, \quad b=R_{\mathrm{m}}C_{\mathrm{m}}, \tag{5.41}$$

and provide 5 equations for 5 unknown dimensional parameters. Note that we have taken $a=Q_0R_{\mathrm{s}}/\tau_{\mathrm{m}}$. To fix ideas, if we assume that the two dimensional parameters $l_{\mathrm{cyl}\,1}$ and $d_{\mathrm{cyl}\,1}$ are known from morphological measurements then the 5 unknown dimensional parameters $(R_{\mathrm{m}}, C_{\mathrm{m}}, R_{\mathrm{i}}, A_{\mathrm{s}}, R_{\mathrm{sh}})$ are given by

$$R_{\mathrm{m}}=\frac{\pi l_{\mathrm{cyl}\,1}d_{\mathrm{cyl}\,1}ab}{Q_0\gamma_1L_1}, \quad C_{\mathrm{m}}=\frac{Q_0\gamma_1L_1}{\pi l_{\mathrm{cyl}\,1}d_{\mathrm{cyl}\,1}a}, \quad R_{\mathrm{i}}=\frac{\pi d_{\mathrm{cyl}\,1}^2abL_1}{4Q_0l_{\mathrm{cyl}\,1}\gamma_1},$$

$$R_{\mathrm{sh}}=\frac{ab}{Q_0(1-\varepsilon)}, \quad A_{\mathrm{s}}=\frac{\pi\varepsilon l_{\mathrm{cyl}\,1}d_{\mathrm{cyl}\,1}}{\gamma_1L_1}. \tag{5.42}$$

In the Rall single cylinder model we have 6 dimensional parameters (removing the shunt parameter R_{sh}) and 2 dimensionless parameters (γ_1, L_1) since now $\varepsilon = 1$ as $R_{sh} = \infty$. The fit of the dimensionless model to the dimensional voltage response gives the values of the 4 parameters (γ_1, L_1, a, b) which are related to the 6 dimensional parameters $(R_m, C_m, R_i, d_{cyl\,1}, l_{cyl\,1}, A_s)$ through the 4 expressions in (5.41) using $(\varepsilon = 1, R_{sh} = \infty)$. If we assume that the 2 dimensional parameters $l_{cyl\,1}$, and $d_{cyl\,1}$ are known (may be from morphological measurements) then the remaining 4 unknown dimensional parameters (R_m, C_m, R_i, A_s) are given by (5.42).

The inverse problem has received much attention (for example, D'Aguanno et al., 1986; Rose and Dagum 1988; Fu et al., 1989; Stratford et al., 1989; Rall 1990; Holmes and Rall 1992; Rall et al., 1992; White et al., 1992), but remains less well understood than the related forward problem. We do not consider the subject in any more detail here, the intention being only to highlight the importance of the dimensionless model in deciding the well-posedness of the inverse problem.

7. DISCUSSION

The techniques for systematic and consistent analysis of passive cable models are categorized in the introduction and have been elaborated on in detail for the multiple equivalent cylinder model. Problem normalization is an important first step in the analysis of the equations of a model which allows identification of the non-dimensional parameters or dimensionless groups that control the solution behaviour. Perturbation analysis allows the investigation of the solution behaviour to be examined in a limit for one or more of these dimensionless parameters. Since analytical solutions to the passive cable model considered are available in closed form, perturbation analysis can be performed on the full solution expression, rather than on the model equations, as described in section 5.

The multiple equivalent cylinder model involves one or more equivalent cylinders emanating from a common soma, retaining the powerful representation of a dendritic tree by an equivalent cylinder first introduced in the Rall model. Some of the benefits of considering and understanding the solution to this model in detail are (a) that it provides a reference case for other more involved multiple equivalent cylinder models that attempt to accommodate the additional effects of non-uniform electrical parameter values, dendritic spines and tapering of dendritic segments; (b) that the solution in this basic geometry can be used to obtain the solution to arbitrary branching cylinder models by observing that the entire dendritic structure is composed of a multiple cylinder problem at the soma and at each branch point. Importantly, the solution of the multiple equivalent cylinder model is characterized by the $(2N+1)$ dimensionless parameters $(\varepsilon, \gamma_j, L_j)$ for $j = 1, \ldots, N$ where N is the number of equivalent cylinders emanating from the common soma. This set of dimensionless parameters quantifies two important properties in the multiple equivalent cylinder model: (a) The (N) electrotonic lengths L_j, $j = 1, \ldots, N$, which are the ratio of the actual length to the space constant for each equivalent cylinder; (b) The passage of current between cylinders is moderated by the soma where conservation of current introduces the dimensionless parameter ε (the ratio of somatic to dendritic membrane time constants) and the (N) dimensionless parameters γ_j, $j=1,\ldots,N$ (the ratio of dendritic and somatic conductances for each equivalent cylinder). In the single cylinder case ($N=1$), we obtain $(\varepsilon,\gamma_1,L_1)$ as the three controlling dimensionless parameters of the single cylinder somatic shunt model and (γ_1,L_1) as the two dimensionless parameters of the Rall equivalent cylinder model. These dimensionless parameters are typically $O(1)$, although the effect on the solution of certain physically based limits of them have been considered in section 5 which allows an understanding of their effect in the model. Simplification of the model by reduction in the number of cylinders by neglect and collapse has been investigated and criteria developed for systematic application of each method that also allows an estimate of the error involved in the procedure.

As remarked in the introduction, we have addressed the analysis of the forward problem which is well-posed and has a well-defined solution available in closed analytical form. However, practitioners are generally interested in the solution of an inverse problem; the search for model parameters which best describe the physical system being studied. The methodology presented here is in part preparatory to the study of the inverse problem, since a primary requirement before solving an inverse problem is ensuring that it has a unique solution (among other properties), so that it is mathematically well-posed. The search for the solution of the inverse problem may be carried out using dimensional parameters (the dimensional inverse problem) or dimensionless parameters (the dimensionless inverse problem). The process of problem normalization allows the identification of the minimum number of fundamental dimensionless parameters for the model and in terms of which the solution is unique. This number together with the scaling parameters needed to transform the data set to which the model is to be fitted into the required form for working with the dimensionless inverse problem, then sets exactly the number of parameters (dimensional or dimensionless) that can be determined by fitting of the model to the data set. If more than this number of parameters are to be fitted, then the inverse problem is ill-posed but can be made well-posed if the values of the additional parameters can be prescribed (usually by independent means e.g. morphological measurements).

It is straightforward to extend the model to consider inhomogeneous mixed boundary conditions of the form

$$\text{at } x_j=l_{\text{cyl}\,j} \quad a_j v_j + b_j \frac{\partial v_j}{\partial x_j} = c_j, \tag{5.43}$$

for $j=1,\ldots,N$ with a_j, b_j, c_j constants, in place of the sealed end conditions (5.2). These may arise when representing a leakage current between the interior and exterior media (Rall, 1969) or during the analysis for synaptic input located at the terminal end with the synaptic conductance represented by a rectangular pulse (Poznanski, 1987a). This system of equations can be considered with little trouble using the same techniques, the solution for mixed homogeneous boundary conditions being a particular case arising in the transformation of the independent variable for the solution of the modified cable equation representing a tapering core as discussed in Evans and Kember (1997).

The representation of synaptic input as a conductance change, rather than current injection as considered in the text, allows the effects of synaptic excitation and inhibition on the soma-dendritic surface to be included in the model. The derivation of cable theory with synaptic reversal potentials can be found in Rall (1977), and leads to voltage dependent current input terms in (5.9) of the form $I_j(X_j, T; V_j)$. The Green's function G_j^k can be used to reduce the resulting boundary value problem for V_j to a system of volterra integral equations given by (5.14) with $I_k(\xi_k, u)$ now replaced with $I_k(\xi_k, u; V_k)$. The integral equations are linear for the assumed functional forms

$$I_j=(X_j, T; V_j)=\sum_{i=1}^{N_j} g_j^i(T)[V_{\text{rev}\,j}^i - V_j(X_j, T)]\delta(X_j - X_{0j}^i),$$

for $j=1,\ldots,N$ representing in the jth equivalent cylinder N_j synaptic inputs located at the points X_{0j}^i for $i=1,\ldots,N_j$ each with a reversal potential $V_{\text{rev}\,j}^i$ and conductance change $g_j^i(T)$ (usually taken to be an alpha-function). This system has been considered in the single equivalent cylinder case ($N=1$) with a single synaptic input ($N_1=1$) by Poznanski (1987) and in the single tapered equivalent cylinder case ($N=1$) with two synaptic inputs ($N_1=2$) by Poznanski (1990).

The multiple equivalent cylinder model has been extended in several ways to incorporate and investigate more realistic variations in electrical and geometrical non-uniformities, which we now briefly review for completeness.

(a) *Multiple Equivalent Cylinder Models with Non-uniform Electrical Parameters*

Variations in the specific membrane resistivity (R_{m}), the cytoplasmic resistivity (R_{i}) and the specific membrane capacitance (C_{m}) in the cell can occur naturally, from microelectrodes and micropipettes behaving as additional processes when inserted into a cell, or from simplified representations that introduce non-uniformly adjusted values, all of which are discussed in detail in Major and Evans (1994). Consequently, variations in anyone of these parameters leads to differing membrane time constants within the dendritic structure. The multiple equivalent cylinder model may be extended to allow differing membrane time constants for each equivalent cylinder, enabling the investigation of the effect of a variation of the dimensional parameters R_{m}, R_{i} or C_{m} between (but not within) dendritic trees of the neuron. For this case the cable equation (5.1) now becomes,

$$v_j + \tau_{\text{m}\,j}\frac{\partial v_j}{\partial t} - \lambda_j^2 \frac{\partial^2 v_j}{\partial x_j^2} = \lambda_j r_{\text{a}\,j} i_j(x_j, t), \tag{5.44}$$

where $\tau_{\text{m}j}$ emphasizes that the membrane time constants can take different (constant) values in each equivalent cylinder. The model is presented and solved in detail in Evans and Kember (1994), using analogous techniques introduced for the solution of the multiple equivalent cylinder model.

(b) *Multiple Equivalent Cylinder Models with Dendritic Spines*

It is usual to incorporate dendritic spines into the dendritic membrane by a suitable rescaling of either the physical length and diameter or the electrical parameters of the membrane cylinder representing the dendritic segment with its spines (Shelton, 1985; Stratford et al., 1989; Larkman, 1991). However, in general a more careful treatment of dendritic spines is required particularly for those which receive synaptic input so that the effects of spine neck resistances, synaptic conductance changes (Jack et al., 1983; Koch and Poggio, 1983; Brown et al., 1988) and possible active conductances in spine heads (Miller et al., 1985; Perkel and Perkel, 1985) can be investigated.

The first mathematical models of spines (Rall and Rinzel, 1971a, b) and later Wilson (1984) were limited to discrete, passive spine(s) and generalized the original Rall model. Recently, a continuum theory for active spines (i.e. a density of spines with voltage dependent channels), was advanced by Baer and Rinzel (1991). Their continuum model for spines was studied mostly numerically with emphasis on exploring the behaviour of active spines with some analysis of the steady-state, single branch passive neuron model with passive spines. Following Baer and Rinzel (1991) a model for passive spines as a continuum in the multiple equivalent cylinder case was presented and analyzed in Kember and Evans (1995). The essential features of the model are that it couples the cable equation for voltage spread in the dendritic shaft with another representing isopotentiality of the spine head. The standard cable equation (5.1) is replaced with

$$\text{(dendrite)} \qquad v_j + \tau_m \frac{\partial v_j}{\partial t} - \lambda_j^2 \frac{\partial^2 v_j}{\partial x_j^2} + \frac{\bar{N}_j r_{mj}}{R_{ss,j}}(w_j - v_j) = \lambda_j r_{aj} i_{vj},$$

$$\text{(spinehead)} \qquad \tau_{sp} \frac{\partial w_j}{\partial t} + w_j + \frac{R_{sp}}{R_{ss,j}}(v_j - w_j) = R_{sp} i_{wj}, \tag{5.45}$$

where for the jth equivalent cylinder, $v_j(x_j, t)$ is the dendritic membrane potential, $w_j(x_j, t)$ the spinehead membrane potential and i_{vj}, i_{wj} are applied currents to the jth cylinder and to the spinehead membrane in the jth cylinder, respectively. The dimensional parameters introduced for the spines are the spine density $\bar{N}_j$ spines/unit-length (assumed constant), a spine stem resistance $R_{ss,j}$ [Ω], the spine head membrane time constant $\tau_{sp} = R_{sp} C_{sp}$ [s] where R_{sp} [Ω] and C_{sp}[μF] are the spinehead resistance and capacitance respectively. The spine stem resistance can be expressed as $R_{ss,j} = 4R_i l_{ss,j} / \pi d_{ss,j}^2$ where $l_{ss,j}$ and $d_{ss,j}$ are the length and diameter of the spine stems. The other dimensional parameters remain unchanged from their definition in the multiple equivalent cylinder model (5.1)–(5.5). The current conservation condition at the soma can also be modified to include spines.

Equation (5.45) model the dendritic spine system as a continuum, but the spines are electrically independent from each other. The equation for the spinehead in (5.45) contains no term for direct coupling between neighbouring spines with voltage spread along the passive dendrite to which they are attached as the only mechanism for spine interaction. For the multiple equivalent cylinder model the spines on each dendritic tree are assumed to have uniform properties, while the spine density, spine length and spine stem diameter can all vary between dendritic trees. The general solution and its behaviour for this system is presented in Kember and Evans (1995).

(c) *Multiple Equivalent Cylinder Models with Taper*

In the Rall model, the condition that branch segments in dendritic trees have constant diameter is not generally true and neither are the two conditions that the 3/2 power law for branch diameters holds and all branches in a dendritic tree terminate at the same electrotonic distances from the soma (Barrett and Crill, 1974; Turner and Schwartzkroin, 1980, 1983; Moore and Appenteng, 1991). (Although it should be mentioned that the more general 3/2 power law which includes both electrical as well as geometrical non-uniformities, see Jack et al., (1983) Eq. (7.23), may still hold as shown by Poznanski (1992) for starburst amacrine cells.) Thus to obtain a more realistic description of dendritic branching patterns these conditions need to be relaxed which for tapering dendritic segments leads to considering a modified cable equation for the voltage spread (Rall, 1962b, Jack et al., 1983). The standard governing cable equation for jth equivalent cylinder with a tapering core diameter $D_j(x_j)$ now takes the modified form (Evans and Kember, 1997),

$$v_j+\tau_m\frac{\partial v_j}{\partial t}-\lambda_j(0)^2\left(\frac{\partial^2 v_j}{\partial z_j^2}+\frac{1}{F_j}\frac{dF_j}{dz_j}\frac{\partial v_j}{\partial z_j}\right)=\lambda_j r_{aj} i_j(z_j,t), \tag{5.46}$$

where

$$\lambda_j(x_j)=\left(\frac{R_m D_j(x_j)}{4R_i}\right)^{1/2}, \qquad r_{aj}=\frac{4R_i}{\pi D_j(x_j)^2},$$

emphasizes that the diameter D_j of the jth equivalent cylinder is not constant,

$$z_j=\int_0^{x_j}\frac{\lambda_j(0)}{\lambda_j(s)}\,ds$$

is a convenient change of independent variable and $F_j(z_j)=(D_j(z_j)/D_j(0))^{3/2}$ is termed the geometric ratio function for the jth equivalent cylinder.

In the case when the dendritic branch segments of the dendritic trees represented by the jth equivalent cylinder are of constant diameter then (cf. Goldstein and Rall, 1974, Eq. 32)

$$F_j=\frac{\sum_{i=1}^{n_j(x_j)} d_i^{3/2}}{\sum_{i=1}^{n_j(0)} d_i^{3/2}},$$

where x_j is the distance from the soma $x_j=0$ along the dendritic segments, $n_j(x_j)$ represents the number of dendritic segments at the distance x_j from the soma for the dendritic trees represented by the jth equivalent cylinder and d_i represents their diameters. Dendritic segments at any given x_j do not necessarily have the same diameters.

The theory also holds for a more symmetric branching pattern in which the segments taper but segments at equal physical distances from the soma have the same common diameter. The geometric ratio function now takes the form (cf. Poznanski, 1991, Eq. 4),

$$F_j = \frac{n_j(x_j)d_j(x_j)^{3/2}(s_j'(x_j))^{1/2}}{n_j(0)d_j(0)^{3/2}(s_j'(0))^{1/2}},$$

where the diameter $d_j(x_j)$ and the number $n_j(x_j)$ of dendritic segments are functions of the physical distance x_j from the soma and $s_j'(x_j) = [1 + d_j'(x_j)/4]^{1/2}$ with $' \equiv \mathrm{d}/\mathrm{d}x_j$.

The quantity $\sum_{i=1}^{n_j(x_j)} d_i^{3/2}$ is often called the combined dendritic trunk parameter. If $F_j = 1$ then the combined dendritic trunk parameter is constant and the dendritic tree or trees can be transformed into an equivalent cylinder of constant diameter $(\sum_{i=1}^{n_j(0)} d_i^{3/2})^{2/3}$. If $F_j \neq 1$ then the diameter of the equivalent cylinder changes continuously with distance from the soma and is given by $(\sum_{i=1}^{n_j(x)} d_j^{3/2})^{2/3}$, with $F_j > 1$ indicating a profuse branching pattern whilst $F_j < 1$ indicating a paucity of branching compared with that in which the dendritic tree or trees can be reduced to an equivalent cylinder of constant diameter ($F_j = 1$).

In Evans and Kember (1997) the multiple equivalent cylinder model with soma is solved for equivalent cylinders with linear taper and for those with a geometric ratio function that belongs to particular class of six specific functional forms (of which the constant and exponential geometric ratio functions are both members). This class of six geometric ratio functions arise naturally as those functions that allow the modified cable equation (5.46) to be transformed into the standard cable equation (5.1), which has been noted by Kelly and Ghausi (1965) and Poznanski (1991). These analytical results extend those obtained in the single equivalent cylinder case for linear taper by Strain and Brockman (1975), Holmes et al. (1992) and exponential taper considered by Poznanski (1988, 1996). As a final remark, the choice of appropriate geometric ratio function can be obtained from a best fit approach of the function to the measured dendritic trunk parameter, as discussed in Poznanski (1988) for the exponential case. Thus, depending on the choice of the geometric ratio function, each equivalent tapering cylinder can represent different forms of branching and tapering as best fits the dendritic tree or trees that it represents. Indeed, the single tapering equivalent cylinder case has already been shown by Poznanski (1988, 1990, 1992, 1994) and Poznanski and Glenn (1994), to fit a range of dendritic branching patterns that do not follow the criteria for reduction to a single non-tapering equivalent cylinder.

(d) *Arbitrarily Branching Cables*

Dendritic trees in general do not satisfy all of the constraints needed for the formation of the single equivalent cylinder model of Rall nor those for the multiple equivalent cylinder models. As remarked by Rall (1989), these constraints on dendritic branching are not presented as law of nature but are used to define mathematical idealizations that provide useful and valuable reference cases. The tapering multiple cylinder models were developed to accommodate the fact that dendritic trees do not (or only partly, up to a certain electrotonic distance) satisfy the 3/2 power law for their branch diameters due mainly to branches terminating at different electrotonic lengths. The formulation of the tapering multiple equivalent cylinder models in Evans and Kember (1997) allows for the mathematical description of a much wider class of actual dendritic trees than those given by the non-tapering models, and to which real neurons may be fitted with increased accuracy through suitable choices of the geometric ratio functions. However, for a more accurate description of the branching pattern that allows the consideration of branches and segments within dendritic trees and not just between them, branching cylinder models need to be considered.

The system of equations for an arbitrary branching geometry have been considered and solved analytically in Major et al. (1993a, b), the solutions being extensions of those obtained in the multiple equivalent cylinder case. There are simplified representations of the branching structure that are intermediary between the full branching structure and the multiple equivalent cylinder models, where dendritic trees not satisfying the equivalent cylinder criteria are simplified using equivalent dendrite collapses (Stratford et al., 1989). Branching cylinder models allow segments and branches within dendritic trees to be considered and are not restricted to comparisons between dendritic trees as in the multiple equivalent cylinder models. The effects on the voltage response of non-uniform values of the dimensional parameters $R_{\mathrm{m}}, C_{\mathrm{m}}, R_{\mathrm{i}}$ that differ within dendritic trees as well as between, was considered in Major and Evans (1994). The flexibility of the branching models can be seen by their ability to treat tapering branch segments as a series of uniform but different diameter cylinders placed end to end.

However, multiple cylinder models may be perfectly adequate (particularly those with tapering equivalent cylinders) in providing a powerful representation of the passive properties of a nerve cell without the need to complicate the model with complex branching patterns. Indeed, it is these type of intermediary models, which are simplified enough to allow understanding of their behaviour and yet may be complex enough to adequately describe the system they are simulating, which should be developed and investigated fully.

8. CONCLUSIONS AND FUTURE PERSPECTIVES

The methodology presented here for analysis of the the multiple equivalent cylinder model can be applied to linear cable models with arbitrary branching structures and thus can be studied in the same framework. In particular, the perturbation techniques used in Section 5.3 to investigate cylinder collapse can be used to the establish the error involved for collapsing dendritic segments in fully branched structures which do not exactly satisfy the criteria for collapse to an equivalent cylinder. A qualitative description may then be obtained of branching structures that do not satisfy the reduction to equivalent cylinders but nevertheless may be represented by them to within an acceptable error margin (which can be estimated).

The fitting of these models to experimental data necessarily involves parameter identification and thus consideration of the inverse problem. This problem has received a lot more attention in recent years, as difficulties such as ill-posedness and ill-conditionedness have been encountered. The inverse problem has not yet been established, but recently Stratford et al. (1989), Holmes et al. (1992) and Major et al. (1994) have advocated problems associated with parameter estimation. In fact Rall (1990) stated that modelling the complexity of single neurons with multi-compartmental models results in too many degrees of freedom and one way to reduce the degree of freedom is to construct simpler models.

A possible next step in the evolution of these models is to incorporate non-uniform membrane properties introducing time and voltage dependent nonlinearities as originally described by Hodgkin and Huxley (1952). One drawback with this at the moment is that there are few neurons for which the densities, spatial distributions and kinetics of time and voltage dependent conductances are known with any certainty. However, in these cases empirical model results can lend support to initial assumptions and a first step mathematically may be consideration of time varying end conductances.

PROBLEMS

5.1 Extend the multiple equivalent cylinder model (5.1)–(5.5) to consider inhomogeneous mixed boundary conditions of the form given by 5.43. What are the physical applications of boundary conditions of this form? Non-dimensionalise the model and show that the solution $V_j(X_j,T)$ satisfies (5.9)–(5.13) with the boundary condition (5.10) replaced by

$$\text{at } X_j=L_j \qquad A_jV_j+\frac{\partial V_j}{\partial X_j}=B_j;$$

for suitable constants A_j and B_j. Show that the solution can be written in the form

$$V_j(X_j,T)=U_j(X_j,T)+W_j(X_j)$$

where $W_j(X_j)$ may be interpreted as the steady state solution satisfying the inhomogeneous boundary conditions whilst $U_j(X_j,T)$ is the transient solution satisfying a suitable sub-problem of (5.9)–(5.13) with homogeneous mixed boundary conditions at $X_j=L_j$ and an appropriate initial condition. Obtain expressions for the functions W_j and U_j.

5.2 Consider synpatic inputs represented as conductance changes in the form suggested by (5.44) with the time courses given by standard α-functions. Derive a set of integral equations for the voltage response and investigate their properties.

5.3 Discuss the inverse and parameter identification problems for the multiple equivalent cylinder problem, mentioned briefly in section 6.

5.4 Experimental suggestion: As discussed in section 7(c), multiple equivalent cylinder models can be formulated with geometric ratio functions F_j that belong to a class of six specific tapering forms. Determine which functional forms and their parameters from this class that best describe the dendritic trunk parameter for different neuron types (that should be obtained by measurements). The number of equivalent cylinders to be used in the models should also be investigated.

5.5 Formulate and explore suitable non-linear extensions to the linear cable models discussed in this chapter. In particular, consider polynomial membrane current–voltage relations as suggested in Jack et al. (1983).

REFERENCES

Abbott, L.F., Farhi, E. and Gutmann, S. (1991) The path integral for dendritic trees. *Biol. Cybern.*, **66**, 49–60.

Ashida, H. (1985) General solution of cable theory with both ends sealed. *J. Theor. Biol.*, **112**, 727–740.

Baer, S.M. and Rinzel, J. (1991) Propagation of dendritic spikes mediated by excitable spines: A continuum theory. *J. Neurophysiol.*, **65**(4), 874–890.

Barrett, J.N. and Crill, W.E. (1974) Specific membrane properties of cat motoneurones. *J. Physiol.* (Lond.), **239**, 301–324.

Bluman, G.W. and Tuckwell, H.C. (1987) Techniques for obtaining analytical solutions for Rall's model neuron. *J. Neurosci. Meth.*, **20**, 151–166.

Brown, T.H., Chang, V.C., Ganong, A.H., Keenan, C.L. and Kelso, S.R. (1988) Biophysical properties of dendrites and spines that may control the induction and expression of long-term synaptic potentiation. In *Long-Term Potentiation: From Biophysics to Behaviour*, edited by P.W. Landfield and S.A. Deadwyler, New Tork: pp. 201–264. Alan R. Liss.

Clements, J.D. (1984) PhD dissertation. Expt. Neurology Unit, John Curtin School of Medical Research, Australian National University, Canberra.

D'Aguanno, A., Bardakjian, B.J. and Carlen, P.L. (1986) Passive neuronal membrane parameters: Comparison of optimization and peeling methods. *IEEE Trans. Biomed. Eng.*, **33**, 1188–1196.

Durand, D. (1984) The somatic shunt cable model for neurons. *Biophys. J.*, **46**, 645–653.

Durand, D., Carlen, P.L., Gurevich, N., Ho, A. and Kunov, H. (1983) Measurement of the passive electrotonic parameters of granule cells in the rat hippocampus using HRP staining and short current pulses. *J. Neurophysiol.*, **50**, 1080–1096.

Evans, J.D., Kember, G.C. and Major, G. (1992) Techniques for obtaining analytical solutions to the multicylinder somatic shunt cable model for passive neurones. *Biophys. J.*, **63**, 350–365.

Evans, J.D. and Kember, G.C. (1994) Analytical solutions to the multicylinder somatic shunt cable model for passive neurones with differing dendritic electrical parameters. *Biol. Cybern.*, **71**, 547–557.

Evans, J.D., Major, G. and Kember, G.C. (1995) Techniques for the application of the analytical solution to the multicylinder somatic shunt cable model for passive neurones. *Math. Biosci.*, **125**, 1–50.

Evans, J.D. and Kember G.C. (1998) Analytical solutions to a tapering multi-cylinder somatic shunt cable model for passive neurones. *Math. Biosci.* **149**, 137–165.

Fu, P., Bardakjian, B.J., D'Aguanno, A. and Carlen, P.L. (1989) Computation of the passive electrical parameters of neurones using a system model. *IEEE Trans. Biomed. Eng.*, **36**, 55–64.

Goldstein, S.S. and Rall, W. (1974) Changes in action potential shape and velocity for changing core conductor geometry. *Biophys. J.*, **14**, 731–757.

Hodgkin, A.L. and Huxley A.F. (1952) A quantitative description of membrane current and its application to conduction and excitation in nerve. *J. Physiol.* (Lond.), **117**, 500–544.

Holmes, W.R. and Rall, W. (1992) Estimating the electrotonic structure of neurons with compartmental models. *J. Neurophysiol.*, **68**, 1438–1452.

Holmes, W.R., Segev, I. and Rall, W. (1992) Interpretation of time constant and electrotonic length estimates in multicylinder or branched neuronal structures. *J. Neurophysiol.*, **68**, 1401–1420.

Iansek, R. and Redman, S.J. (1973) An analysis of the cable properties of spinal motoneurones using a brief intracellular current pulse. *J. Physiol.* (Lond)., **234**, 613–636.

Jack, J.J.B. and Redman, S.J. (1971a) The propagation of transient potentials in some linear cable structures. *J. Physiol.* (Lond.), **215**, 283–320.

Jack, J.J.B. and Redman, S.J. (1971b) An electrical description of the motoneurone, and its application to the analysis of synaptic potentials. *J. Physiol.* (Lond.), **215**, 321–352.

Jack, J.J.B., Noble, D. and Tsien, R.W. (1983) Electric current flow in excitable cells. Oxford: Clarendon Press.

Kawato, M. (1984) Cable properties of a neuron model with non-uniform membrane resistivity. *J. Theor. Biol.*, **111**, 149–169.

Kember, G.C. and Evans, J.D. (1995) Analytical solutions to a multicylinder somatic shunt cable model for passive neurones with spines. *IMA J. Math. Appl. Med. Biol.*, **12**, 137–157.

Kelly, J.J. and Ghausi, M.S. (1965) Tapered distributed RC networks with similar immittances. *IEEE Trans. Circuit Theory*, **12**, 554–558.

Koch, C. and Poggio, T. (1983) A theoretical analysis of electrical properties of spines. *Proc. R. Soc. Lond. B.*, **218**, 455–477.

Larkman, A.U. (1991) Dendritic morphology of pyramidal neurones of the visual cortex of the rat: III. Spine Distributions. *J. Comp. Neurol.*, **306**, 332–343.

Major, G., Evans, J.D. and Jack, J.J.B. (1993a) Solutions for transients in arbitrarily branching cables: I. Voltage recording with a somatic shunt. *Biophys. J.*, **65**, 423–449.

Major, G., Evans, J.D. and Jack, J.J.B. (1993b) Solutions for transients in arbitrarily branching cables: II. Voltage clamp theory. *Biophys. J.*, **65**, 450–468.

Major, G. and Evans, J.D. (1994) Solutions for transients in arbitrarily branching cables: IV. Nonuniform electrical parameters. *Biophys. J.*, **66**, 615–634.

Major, G., Larkman, A.U., Jonas, P., Sakmann, B. and Jack, J.J.B. (1994) Detailed passive cable models of whole-cell recorded CA3 pyramidal neurons in rat hippocampal slices. *J. Neurosci.*, **14**(8), 4613–4638.

Miller, J.P., Rall, W. and Rinzel, J. (1985) Synaptic amplification by active membrane in dendritic spines. *Brain Res.*, **325**, 325–330.

Moore, J.A. and Appenteng, K. (1991) The morphology and electrical geometry of rat jaw-elevator motoneurones. *J. Physiol.* (Lond.), **440**, 325–343.

Perkel, D.H. and Perkel, D.J. (1985) Dendritic spines: role of active membrane in modulating synaptic efficacy. *Brain Res.*, **325**, 331–335.

Poznanski, R.R. (1987a) Transient response in a somatic shunt cable model for synaptic input activated at the terminal. *J. Theor. Biol.*, **127**, 31–50.

Poznanski, R.R. (1987b) Techniques for obtaining analytical solutions for the somatic shunt cable model. *Math. Biosci.*, **85**, 13–35.

Poznanski, R.R. (1988) Membrane voltage changes in passive dendritic trees: a tapering equivalent cylinder model. *IMA J. Math. Appl. Med. Biol.*, **5**, 113–145.

Poznanski, R.R. (1990) Analysis of a postsynaptic scheme based on a tapering equivalent cable model. *IMA J. Math. Appl. Med. Biol.*, **7**, 175–197.

Poznanski, R.R. (1991) A generalized tapering equivalent cable model for dendritic neurons. *Bull. Math. Biol.*, **53**, 457–467.

Poznanski, R.R. (1992) Modelling the electrotonic structure of starburst amacrine cells in the rabbit retina: A functional interpretation of dendritic morphology. *Bull. Math. Biol.*, **54**, 905–928.

Poznanski, R.R. (1994) Electrotonic length estimates of CA3 hippocampal pyramidal neurons. *Neurosci. Res. Comm.*, **14**, 93 100.

Poznanski, R.R. and Glenn, L.L (1994) Estimating the effective electrotonic length of dendritic neurons with reduced equivalent cable models. *Neurosci. Res. Comm.*, **15**, 69–76.

Poznanski, R.R. (1996) Transient response in a tapering cable model with somatic shunt. *NeuroReport*, **7**, 1700–1704.

Rall, W. (1962a) Electrophysiology of a dendritic neuron model. *Biophys. J.* (*Suppl*), **2**, 145–167.

Rall, W. (1962b) Theory of physiological properties of dendrites. *Ann. N.Y. Acad. Sci.*, **96**, 1071–1092.

Rall, W. (1969) Time constants and electrotonic length of membrane cylinders and neurons. *Biophys. J.*, **9**, 1483–1508.

Rall, W. (1977) Core conductor theory and cable properties of neurons. In *Handbook of Physiology. Section 1. The Nervous System.* Volume 1, edited by E.R. Kandel, pp. 39–97. Bethesda, MD: Am. Physiol. Soc.

Rall, W. (1989) Cable theory for dendritic neurons. In *Methods in Neuronal Modeling: From Synapses to Networks*, edited by C. Koch and I. Segev, pp. 9–62. Cambridge, MA: MIT Press.

Rall, W. (1990) Perspectives on neuron modeling. In *The Segmental Motor System*, edited by M.D. Binder and L.M. Mendell, pp. 129–149. Oxford, UK: Oxford University Press.

Rall, W., Burke, R.E., Holmes, W.R., Jack, J.J.B., Redman, S.J. and Segev, I. (1992) Matching dendritic neuron models to experimental data. *Physiol. Rev.*, **72**, S159–S186.

Rall, W. and Rinzel, J. (1971a) Dendritic spines and synaptic potency explored theoretically. *Proc. Int. Union Physiol. Sci.* (XXV Int. Congr.), **IX**, 466.

Rall, W. and Rinzel, J. (1971b) Dendritic spine function and synaptic attenuation calculations. *Prog. Abstr. Soc. Neurosci.*, **1**, 64.

Redman, S.J. (1973) The attenuation of passively propagating dendritic potentials in a motoneurone cable model. *J. Physiol.* (Lond.), **234**, 637–664.

Rinzel, J. and Rall, W. (1974) Transient response in a dendritic neuron model for current injected at one branch. *Biophys. J.*, **14**, 759–790.

Rose, P.K. and Dagum, A. (1988) Nonequivalent cylinder models of neurons: Interpretation of voltage transients generated by somatic current injection. *J. Neurophysiol.*, **60**, 125–148.

Shelton, D.P. (1985) Membrane resistivity estimated for the Purkinje neuron by means of a passive computer model. *Neuroscience*, **14**, 111–131.

Strain, G.M. and Brockman, W.H. (1975) A modified cable model for neuron processes with non-constant diameters. *J. Theor. Biol.*, **51**, 475–494.

Stratford, K.J., Mason, A.J.R., Larkman, A.U., Major, G. and Jack, J.J.B. (1989) The modelling of pyramidal neurones in the visual cortex. In *The Computing Neuron*, edited by R. Durbin, C. Miall, and G. Mitchison, pp. 296–321. Reading, UK: Addison-Wesley.

Tayler, A.B. (1986) Mathematical models in applied mechanics. Oxford, UK: Oxford University Press.

Tuckwell, H.C. (1988) Introduction to Theoretical Neurobiology. Vol. 1. Linear Cable Theory and Dendritic Structure. Cambridge, UK: Cambridge University Press.

Turner, D.A. and Schwartzkroin, P.A. (1980) Steady-state electrotonic analysis of intracellularly stained hippocampal neurons. *J. Neurophysiol.*, **44**, 184–199.

Turner, D.A. and Schwartzkroin, P.A. (1983) Electrical characteristics of dendrites and dendritic spines in intracellularly stained CA3 and dentate hippocampal neurons. *J. Neurosci.*, **3**, 2381–2394.

Van Dyke, M. (1964) *Perturbation Methods in Fluid Mechanics.* New York: Academic Press.

White, J.A., Manis, P.B. and Young, E.D. (1992) The parameter identification problem for the somatic shunt model. *Biol. Cybern.*, **66**, 307–318.

Wilson, C.J. (1984) Passive cable properties of dendritic spines and spiny neurons. *J. Neurosci.*, **4**, 281–297.

CHAPTER 6
VOLTAGE TRANSIENTS IN MULTIPOLAR NEURONS WITH TAPERING DENDRITES

Loyd L. Glenn and Jeff R. Knisley

1. INTRODUCTION

The physiological significance of dendritic tapering in neurons has long been of interest to neuromorphologists and theorists. One of the first attempts to study tapering was by Rall (Rall, 1962), who modified the cable equation to account for taper or flare. Jack, et al. (1975) expanded this work in their seminal book. Poznanski followed in 1987 with a method of obtaining solutions to the modified cable equation, for a variety of tapers (Poznanski, 1987) that was subsequently published (see Poznanzki, 1991; Schierwagen, 1989). In recent years, Evans of the UK has been at the forefront of the development of the mathematics of tapered cables (see Chapter 5).

All of these groups modified cylindrical cable models in order to incorporate the physical nonuniformities which are evident in almost every study of dendritic morphology or electrotonic conduction. In this chapter, we examine a cable model which incorporates nonuniformities in the dendritic diameter, the membrane resistivity and the electrotonic lengths. In addition, we review the aforementioned work on tapered equivalent cable models.

1.1 Nonuniform Diameter

Tapering is a physical nonuniformity in that diameter varies with the distance from the origin. Tapering can be modelled mathematically by a modified cable equation. The first complete analytical solution to a modified cable equation incorporating a continuously changing diameter was published by Poznanski (1988) using a taper that was electrotonically exponential. Holmes et al. (1992) derived the complete analytic solution for a cable equation with a linear taper, and Poznanski derived solutions for equations which modeled an exponentially-tapered cable with a shunt soma (Poznanski, 1996). All of these derivations focused on either instantaneous or semi-infinite current pulses, but other models have used white noise stimuli (Poznanski and Peiris, 1996), synapse-like membrane conductance changes (Poznanski, 1990), and electrotonic coupling (Poznanski et al., 1991). Recently, Alaburda and Gutman (1996), and Evans and Kember (1998) have derived the cable equation for a cable with a non-uniform cross section.

Only a few studies have been conducted on the frequency, shape and amount of tapering in dendrites. Branch tapering in spinal motoneurons is most prominent in first-order branches at the somatodendritic junction and in the long, thin terminal dendritic branches (Gollvik et al., 1990). Chen and Wolpaw (1994) found the degree of taper in the average branch of a rat spinal motoneuron dendrite was 22% from origin to terminal. Similarly, Rastad et al. (1990) found an average branch taper of 14% for the proximal segments and 22% for the terminal segments in spinal interneurons. In contrast, Ulfhake and Kellerth (1984) found tapering and flaring between branch points of intersegments for feline spinal motoneurons, but the average overall change in the diameter was zero and only the terminal branches were found to taper significantly. Moore and Appenteng (1991) reported extensive tapering and flaring in the branches of brainstem motoneurons, but as in the study by Ulfhake and Kellerth (1984), there was no evidence of a net taper overall. Surkis et al. (1996) found marked non-uniformities in dendritic diameters of laterodorsal tegmental neurons of the pons, with a prevalence of taper but some flaring as well. Feline spinal motoneurons show a net flaring in first-order branches, and to lesser degree, second-order neurons, with an increasing tendency towards tapering as the branch order increases (L.L. Glenn and R.E. Burke, unpublished data). To our knowledge, there have been no systematic studies on the curves which best fit the tapers of individual dendritic branches.

Numeric studies using compartmental models allow more physiological complexity than analytical models. However, little work has been done on numerical methods which incorporate tapering (but see Horwitz, 1983; Holmes, 1986; Stratford et al. 1989; Holmes and Rall, 1992; Pongracz et al., 1993); most numerical investigations used models based on uniform-diameter cables even when working with tapering neurons (see e.g. Ulrich et al., 1994). However, we expect this to change as morphometric technology improves. Later in this chapter, we describe methods for modifying existing numerical methods to accommodate tapers.

1.2 Nonuniform Membrane Resistivity

Another type of nonuniformity that has sparked interest in recent years is nonuniformity in resting membrane resistance. Many recent studies are in accord that distal dendrites have a higher membrane resistance than proximal dendrites or the soma (Fleshman et al., 1988; Clements and Redman, 1989; Nitzan et al., 1990). The first analytic solutions to incorporate nonuniform membrane resistance were those of Durand (1984) and Kawato (1984). This was followed by the work of Major et al. (1993a,b), who extended the models to multipolar models with arbitrary branching. Poznanski (1996) recently derived the solution for a tapering cable model with a shunt soma, which is probably the most advanced solution for this combination of features to date.

1.3 Nonuniform Lengths

A third type of nonuniformity is that of unequal dendritic electrotonic lengths. Single-cable models such as those developed by Rall (1969), Jack and Redman (1971), Durand (1984), Kawato (1984), Ashida (1985) and others, were used almost exclusively between 1970 and 1990 (Burke and ten Bruggencate, 1971; Clements and Redman, 1989; Durand et al., 1983; Glenn et al., 1987; Iansek and Redman, 1973; Tsukahara et al., 1975; Spruston and Johnston, 1992). Glenn (1988) and Rose and Dagum (1988) reported that single-cable models have limitations that lead to incorrect estimates of the electrotonic parameters. In particular, the assumption of uniform dendritic length—which is explicit in single-cable models—has led to an overestimate of neuron dendritic lengths (Glenn and Samojla, 1987; Glenn, 1988). Recently, several others have confirmed or extended the finding on overestimation, including Holmes et al. (1992), Holmes and Rall (1992), Nitzan et al. (1990), and Major et al. (1993). Glenn (1988), Glenn and Samojla (1987), Glenn et al. (1988), and Holmes et al. (1992) derived the solution for equations modelling a multipolar somatic-shunt model with arbitrary dendrite number, diameter, and electrotonic length in which current stimulation and voltage recordings were made in the soma. In an elegant and comprehensive mathematical achievement, Major et al. (1993a) found the solutions for the same models but with arbitrary branching points, diameters, and lengths, and for arbitrary stimulation and recording. Korogod (1996) used such models to study attenuation of axial currents in the distoproximal direction.

The multipolar models of Glenn (1988), Holmes et al. (1992), and Major et al. (1993a,b) have the simplifying assumption of constant diameter branches. This is a potentially important limitation because neither dendrites nor equivalent cable representations of whole dendritic trees have constant diameters. Consequently, a model that combines three basic types of nonuniformities (diameter, length, and time constant) would appear to be at least useful, and may be argued to be the minimal model that should be used to analyze experimental charging curves from neurons. These are the types of models reviewed in this chapter.

The term *tapering* has been used to describe at least two different morphological traits—the tapering of individual dendritic branches, hereafter called branch tapering, and the occurrence of electrotonic mismatches at branchpoints, hereafter called a step change in the equivalent diameter. These step changes are caused by decreases in summed diameter to the three-halves power at branchpoints.

Tapering may refer either to variations in a single dendritic branch or to variations in the summed diameters for whole neurons after the dendrites have been collapsed into a single cone-like structure called an equivalent cable. Dendritic trees are collapsed by summing the diameters (to the three-halves power) at increasing distances from the soma along the dendritic paths. The value of the sum at any particular path distance from the soma is called the dendritic trunk parameter. If this sum is taken to the 2/3 power, it is called the equivalent diameter. In most neurons studied to date, branch tapering, step changes in net diameter, and termination of dendrites all contribute to the (whole neuron) tapering of the equivalent diameter (Cullheim et al., 1987a; Clements and Redman, 1989; Ulfhake and Kellerth, 1984), with dendritic termination predominating.

The difference in the relationship between the form of a taper in electrotonic or dimensionless space (usually denoted by the variable Z) and the form of the taper in metric or dimensioned space (usually indicated by the variable x) has not been widely appreciated. Some have mistakenly supposed that the form of the taper in dimensionless space is the same as the form of the taper in dimensioned space, but the problem of determining the form of the taper in one space given the taper in the other space is not a trivial problem. However, if morphological data are to be applied to solutions that are expressed only in dimensionless variables, it is essential that more of these types of results are discovered and acknowledged.

In contrast to the limited number of studies on the tapering properties of individual branches, there are a large number of studies on whole neuron tapering—that is, tapering of the dendritic trunk parameter. Such studies were motivated by a desire to approximate the whole neuron by a single cable model, or to rule out the possibility that a single constant-diameter cable can be used to model the passive voltage–current responses of the neuron. Two of the earlier studies were by Lux et al. (1970) who used radioactive amino acids as a label and by Barrett and Crill (1974) who used fluorescent compounds to label motoneurons and plot the whole neuron electrotonic profile. Neither of these studies allow full visualization of the entire dendritic tree, but with the advent of biocytin and horseradish peroxidase as intracellular labels, complete staining of dendritic trees became possible in many types of neurons. Using these modern techniques, the profile of the equivalent cable was determined in spinal motoneurons (Ulfhake and Kellerth, 1984; Cullheim et al., 1987b; Cameron et al., 1981; Clements and Redman, 1989; Fleshman et al., 1988), brainstem motoneurons (Moore and Appenteng, 1991; Bras et al., 1987; Nitzan et al., 1990) and hippocampal pyramidal cells (Turner and Schwartzkroin, 1980).

A picture of the relationship between the diameter $D_{\text{taper}} = (\sum d^{3/2})^{2/3}$ and the electrotonic distance Z from the soma has emerged from these studies. For hippocampal pyramidal cells, the D_{taper} as a function of Z is similar in appearance to an exponential decay (as in Poznanski, 1988). However, the diameters of spinal motoneurons do not have this profile. Sometimes the diameters of spinal motoneurons can be approximated by a linear taper (Clements and Redman, 1989), but often a more general curve is necessary (Cullheim et al., 1987b; Ulfhake and Kellerth, 1984). Figure 6.1. shows a plot of the equivalent diameter of spinal motoneurons at different distances from the soma (equivalent distance). The taper begins several hundred μm from the soma rather than at the soma itself as would be required for a linear or exponential taper. Indeed, for the first few hundred μm, there is a slight flaring of the profile (Figure 6.1). Thus, spinal motoneurons collapsed by the $d^{3/2}$ rule should be approximated by two cables—a short cable emanating from the soma with a slight flare followed by a longer cable with a curvilinear taper of diameter with distance. Brainstem motoneurons have shown both of the above types of taper, with jaw-closer motoneurons showing the spinal motoneuron tapering profile (Moore and Appenteng, 1991) and vagal motoneurons showing a profile similar to hippocampal pyramidal neurons (Nitzan et al., 1990).

The curve used in Figure 6.1 to model the taper is a quadratic function of distance. This curve approximates both the proximal flare and the distal taper of spinal motoneurons. If the curve is converted to dimensionless units, specifically to $d^{3/2}$ by electrotonic distance, the curve corresponds exactly to that of Poznanski (1991) and Goldstein and Rall (1974)—a fact which motivates and justifies the study of the models described in the remainder of this chapter.

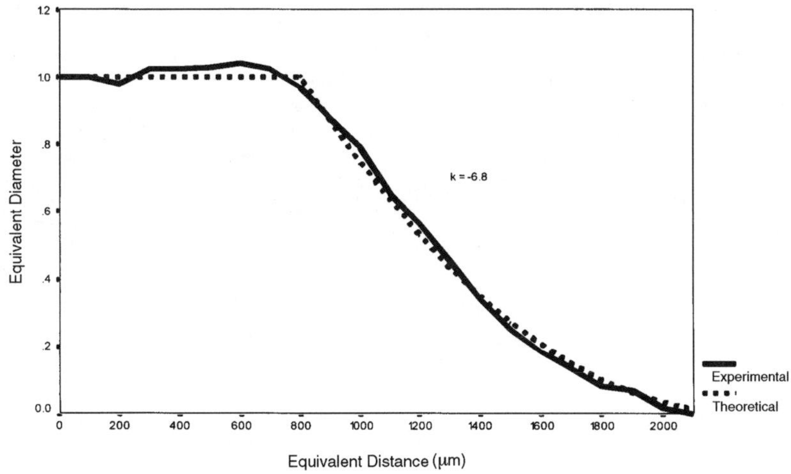

Figure 6.1 Curve fit of a model taper with the morphometric taper of a spinal motoneuron whose collapsed dendrites are collapsed into a single cable by the $d^{3/2}$ rule. The taper form is given by (6.1).

If we write the equation for the rate of a taper as

$$D_{\text{taper}} = D_0(1+kx)^2, \tag{6.1}$$

where D_0 is initial diameter and x is distance from the point at which the initial diameter is measured, then the taper rate k can be calculated from morphometric measurements shown in Figure 6.1. First, the diameter next to the origin of a dendritic branch is measured. Next, the diameter is measured at a given distance (x) from the origin. All measurements are converted to centimeters using $1\,\mu\text{m} = 0.0001\,\text{cm}$ and then the three values are substituted into (6.1) after which we solve for k. If the dendritic branch gets narrower with longitudinal distance, k will be negative (i.e. a taper).

2. A MULTIPOLAR TAPERING CABLE MODEL

Suppose now that a neuron has N dendritic trees originating from the soma, and that the membrane constant for each tree may differ from that of other trees. Then each of the N trees can be collapsed to an exponentially tapering equivalent cable with dimensionless taper constant ($K_j < 0$), membrane time constant τ_j and voltage V_j, $j = 1, \ldots, N$. A schematic diagram of such a neuron is shown in Figure 6.2. Each dendritic tree (all branches that stem from a trunk dendrite which is attached to soma) is modeled by a single equivalent cylinder (Figure 6.3). The justification for collapsing single dendritic trees into an equivalent cylinder is discussed in a later section.

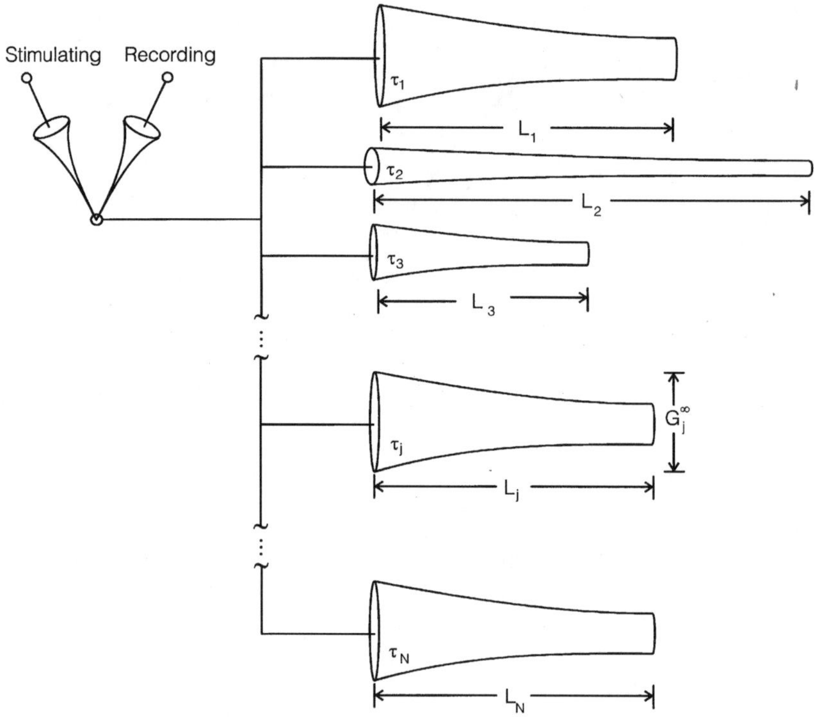

Figure 6.2 Schematic diagram of the multipolar tapered equivalent cable model. Each cable dendrite can have a unique electrotonic length (L), membrane time constant (τ), initial diameter ($d^{3/2}$ is proportional to the G^{∞}), and taper rate (K). Stimulating and recording sites are shown on the soma.

The axial current for the jth equivalent cable is given by

$$I_j(Z_j, t) = G_j^{\infty} \frac{\partial V_j}{\partial Z_j},$$

where

$$G_j^{\infty} = -\frac{\pi D_j^{3/2}(0)}{2\sqrt{R_\mathrm{i} R_\mathrm{m}}},$$

and $D_j(0)$ is the initial diameter of the jth equivalent cable. We assume that current is conserved at the soma,

$$\sum_{j=1}^{N} I_j(0,t) + I_\mathrm{s} = I^{\mathrm{in}}(t),$$

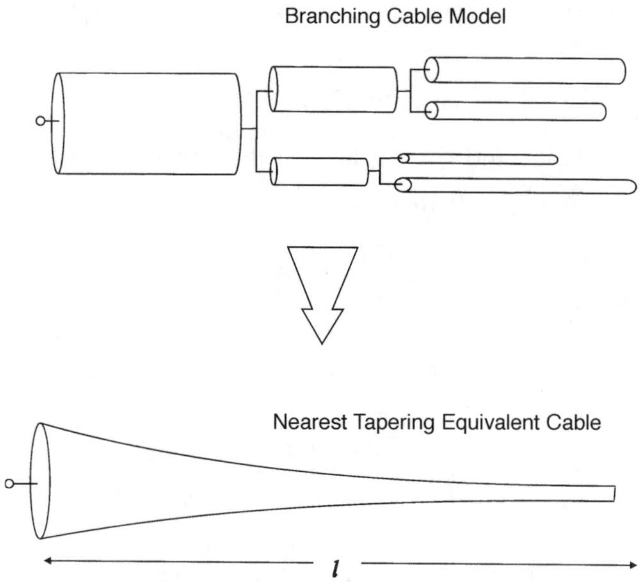

Figure 6.3 Basis of multipolar tapering equivalent cable model. Each cable in the present model represents a dendritic tree, and the dendritic tree is assumed to be collapsible to a tapering equivalent cylinder. Non-equivalent dendritic trees which do not obey the $d^{3/2}$ rule can be approximated by single cylinders. This model is intermediate between full branching models and the unipolar tapering equivalent cylinder models.

where I_s is the somatic current and $I^{\text{in}}(t)$ is the external current stimulus. We also assume that the terminal ends are sealed—i.e. that

$$I_j(L_j, t) = 0.$$

For now, we assume that the soma is the $j = 0$ cable. Later we will show that a lumped soma with a shunt is a special case of our model. As a result, the potential V_j on the jth cable is given by the multipolar model in $N+1$ cables:

$$\frac{\partial^2 V_j}{\partial Z_j^2} - K_j \frac{\partial V_j}{\partial Z_j} - \tau_j \frac{\partial V_j}{\partial t} - V_j = 0, \tag{6.2}$$

$$\frac{\partial V_j}{\partial Z_j}(L_j, t) = 0, \tag{6.3}$$

$$\sum_{j=0}^{N} G_j^{\infty} \frac{\partial V_j}{\partial Z_j}(0, t) = I^{\text{in}}(t), \tag{6.4}$$

$$V_j(0, t) = V_0(0, t), \tag{6.5}$$

where $j=0,\ldots,N$. Since the external stimulus is forcing the system, we can assume that the cell is initially depolarized:

$$V_j(Z_j,0)=0, \quad j=0,\ldots,N. \tag{6.6}$$

2.1 The Steady State and the Laplace Transformation of the Solution

The Laplace transform of (6.2)–(6.6) is given by

$$\hat{V}_j''-K_j\hat{V}_j'=(s\tau_j+1)\hat{V}_j, \tag{6.7}$$

$$\hat{V}_j'(L_j)=0, \tag{6.8}$$

$$\sum_{j=0}^{N}G_j^{\infty}\hat{V}_j'(0)=\hat{I}^{\text{in}}(s), \tag{6.9}$$

$$V_j(0)=V_0(0), \tag{6.10}$$

where $\hat{V}_j(Z_j)=\hat{V}_j(Z_j,s)$ is the Laplace transform of V_j. The solution to (6.7) is given by

$$\hat{V}_j(Z_j)=\mathrm{e}^{K_jZ_j/2}[A_j\cosh(\hat{b}_j(L_j-Z_j))+B_j\sinh(\hat{b}_j(L_j-Z_j))],$$

where $\hat{b}_j=\sqrt{s\tau_j+K_j^2/4+1}$. Application of (8) yields

$$B_j=\frac{-K_j}{2\hat{b}_j}A_j,$$

which in combination with (9) and (10) yields

$$A_j=\frac{\hat{I}^{\text{in}}(s)\hat{H}(s)}{[\cosh(\hat{b}_jL_j)-(K_j/2\hat{b}_j)\sinh(\hat{b}_jL_j)]},$$

where

$$\hat{H}(s)=\left(\sum_{j=0}^{N}\frac{G_j^{\infty}(1+s\tau_j)}{(K_j/2)-\hat{b}_j\coth(\hat{b}_jL_j)}\right)^{-1}.$$

The poles of $\hat{H}(s)$ are defined to be the eigenvalues of the model, denoted α_m^2. By letting $s=-\alpha^2$, we can define α_m to be a solution to the eigenvalue equation:

$$\sum_{j=0}^{N}\frac{G_j^{\infty}(\alpha^2\tau_j-1)}{(K_j/2)-\sqrt{\alpha^2\tau_j-(K_j^2/4)-1}\cot(L_j\sqrt{\alpha^2\tau_j-(K_j^2/4)-1})}=0. \tag{6.11}$$

In Laplace space, the solution is given by

$$\hat{V}_j(Z_j)=\hat{I}^{\text{in}}(s)\hat{H}(s)\mathrm{e}^{K_jZ_j/2}\hat{\phi}_j(Z_j),$$

where

$$\hat{\phi}_j(Z_j)=\frac{\cosh(\hat{b}_j(L_j-Z_j))-(K_j/2\hat{b}_j)\sinh(\hat{b}_j(L_j-Z_j))}{\cosh(\hat{b}_jL_j)-(K_j/2\hat{b}_j)\sinh(\hat{b}_jL_j)}.$$

According to the Final Value Theorem for the Laplace transform, the steady state for the system is given by

$$V^{ss}(Z_j) = \lim_{s \to 0} s\hat{V}_j(Z_j, s). \tag{6.12}$$

If we let

$$I^{\infty} = \lim_{s \to 0} s\hat{I}^{in}(s),$$

then (6.12) can be simplified to

$$V^{ss}(Z_j) = I^{\infty} e^{K_j Z_j/2} \left(\sum_{j=0}^{N} \frac{G_j^{\infty}}{K_j/2 - \sqrt{K_j^2/4+1}\coth(L_j\sqrt{K_j^2/4+1})} \right)^{-1}$$
$$\times \left(\frac{\cosh\left(\sqrt{K_j^2/4+1}(L_j - Z_j)\right) - \dfrac{K_j}{2\sqrt{K_j^2/4+1}}\sinh\left(\sqrt{K_j^2/4+1}(L_j - Z_j)\right)}{\cosh\left(\sqrt{K_j^2/4+1}L_j\right) - \dfrac{K_j}{2\sqrt{K_j^2/4+1}}\sinh\left(L_j\sqrt{K_j^2/4+1}\right)} \right). \tag{6.13}$$

2.2 The Lumped Soma and the Somatic Shunt Solution

In most neuron models, the soma is assumed to be isopotential and is modeled by an RC circuit as shown in Figure 6.4. Often the time constant τ_0 for the soma is assumed to be smaller than the other time constants, in which case the model is said to have a shunt soma and is called a somatic shunt model (Durand, 1984). Such models have had some use in explaining experimental results, but the instability of the inverse problem has limited their usefulness (White et al., 1992).

The somatic shunt model is a limiting case of the multipolar models, as we now illustrate. Let us assume that the somatic cable has a taper of 0. Then the Laplace transform of the solution is given by

$$\hat{V}_j(Z_j) = \hat{I}^{in}(s)\hat{H}(s)e^{-K_j Z_j/2}\hat{\phi}_j(Z_j),$$

where

$$\hat{H}(s) = \left(-G_0^{\infty}\sqrt{s\tau_0+1}\tanh\left(\sqrt{s\tau_0+1}L_0\right) + \sum_{j=1}^{N} \frac{G_j^{\infty}(1+s\tau_j)}{K_j/2 - \hat{b}_j\coth(\hat{b}_j L_j)} \right)^{-1}.$$

We define the constant $\rho = G_0^{\infty} L_0$ so that

$$\hat{H}(s) = \left(-\rho\sqrt{s\tau_0+1}\,\frac{\tanh\left(\sqrt{s\tau_0+1}L_0\right)}{L_0} + \sum_{j=1}^{N} \frac{G_j^{\infty}(1+s\tau_j)}{K_j/2 - \hat{b}_j\coth(\hat{b}_j L_j)} \right)^{-1}.$$

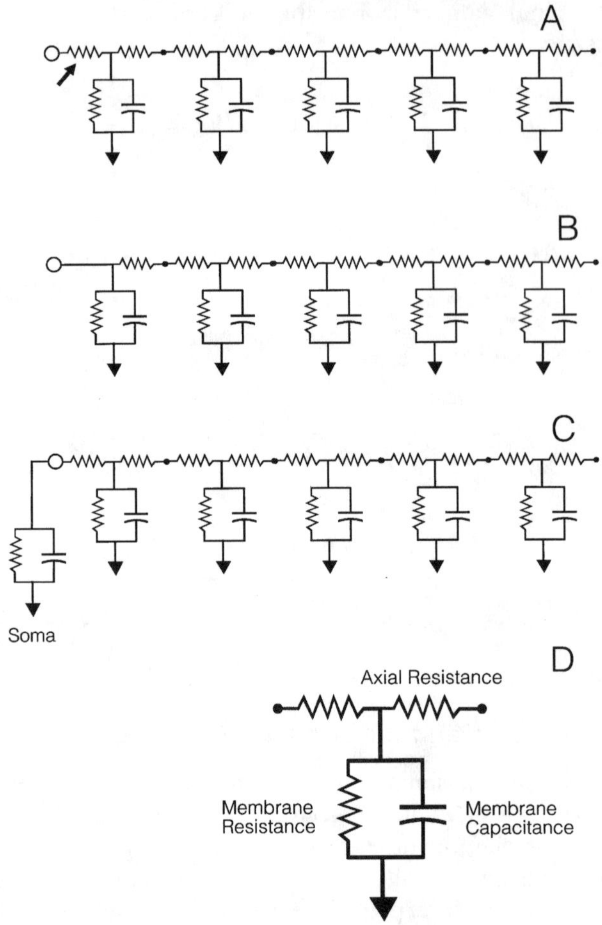

Figure 6.4 Compartmental model of a tapered cable. A: Axial resistor is between stimulus-recording point and the initial compartment. B: No axial resistor proximal to initial compartment. C: Soma compartment incorporated in model. D: A Single compartmental unit.

If L_0 is arbitrarily small (i.e. if we take the limit as $L_0 \to 0$) then

$$\hat{H}(s) = \left(-\rho(s\tau_0 + 1) + \sum_{j=1}^{N} \frac{G_j^{\infty}(1 + s\tau_j)}{K_j/2 - \hat{b}_j \coth(\hat{b}_j L_j)} \right)^{-1},$$

which yields the Laplace transform of the somatic shunt solution since

$$\lim_{L_0 \to 0} \hat{\phi}_0(Z_0) = 1.$$

Correspondingly, the eigenvalue equation reduces to

$$\sum_{j=1}^{N} \frac{G_j^{\infty}(\alpha^2\tau_j - 1)}{K_j/2 - \sqrt{\alpha^2\tau_j - (K_j^2/4) - 1}\,\cot(L_j\sqrt{\alpha^2\tau_j - (K_j^2/4) - 1})} = \rho(1-\tau_0\alpha^2). \quad (6.14)$$

If each of the tapers is identically zero and if

$$\tau_j = \tau_m$$

for all $j = 1, \ldots, N$, then (14) reduces to

$$\rho\frac{1-\varepsilon(1+\beta^2)}{\beta} = \sum_{j=1}^{N} G_j^{\infty}\tan(\beta L_j),$$

where $\beta = \sqrt{\alpha^2\tau_m - 1}$ and $\varepsilon = \tau_0/\tau_m$. This is the same as (6.23) in Major et al. (1993a).

2.3 The Transient Solution for Various Inputs

2.3.1 *The Green's Function and Solutions for Various Inputs*

The Green's function, $U_j(Z_j, t)$, for the model is defined to be the voltage response when $I^{in}(t)$ is given by the delta function. It is useful because complete solutions to a variety of stimuli can often be easily derived once Green's function is known for a particular model. Since the Laplace transform of the delta function is identically 1, it follows that the Laplace transform Green's function is

$$\hat{U}(Z_j, s) = \hat{H}(s)e^{K_jZ_j/2}\hat{\phi}_j(Z_j).$$

Using the method of Poincaré and the theory of residues (Major et al., 1993a), it follows that

$$U_j(Z_j, t) = \sum_{m=0}^{\infty} C_m e^{K_jZ_j/2}\phi_{jm}(Z_j)e^{-\alpha_m^2 t},$$

where $b_{jm} = \sqrt{\alpha_m^2\tau_j - (K_j^2/4) - 1}$ and

$$\phi_{jm}(Z_j) = \frac{\cos(b_{jm}(L_j - Z_j)) - (K_j/2b_{jm})\sin(b_{jm}(L_j - Z_j))}{\cos(b_{jm}L_j) - (K_j/2b_{jm})\sin(b_{jm}L_j)} \quad (6.15)$$

and where

$$C_m = \lim_{s \to -\alpha_m^2}(s + \alpha_m^2)H(s).$$

As a result,

$$\frac{1}{C_m} = \sum_{j=0}^{N} \frac{G_j^{\infty}\tau_j(1-\alpha_m^2\tau_j)([\cot(b_{jm}L_j)/b_{jm}] + L_j\csc^2(b_{jm}L_j))}{(K_j/2 - b_{jm}\cot(b_{jm}L_j))^2}$$
$$- \sum_{j=0}^{N} \frac{2G_j^{\infty}\tau_j}{K_j/2 - b_{jm}\cot(b_{jm}L_j)}.$$

In Laplace Space, we can write $\hat{V}_j(Z_j, s)$ in terms of the Green's function as

$$\hat{V}_j(Z_j, s) = \hat{I}^{\text{in}}(s)\hat{U}_j(Z_j, s).$$

The convolution theorem for the Laplace transform (Carslaw and Jaeger, 1948) implies that

$$V_j(Z_j, t) = \int_0^t U_j(Z_j, t-u) I^{\text{in}}(u)\, \mathrm{d}u.$$

This in turn reduces to

$$V_j(Z_j, t) = \sum_{m=0}^{\infty} C_m \mathrm{e}^{K_j Z_j/2} \phi_{jm}(Z_j) \mathrm{e}^{-\alpha_m^2 t} \int_0^t I^{\text{in}}(u) \mathrm{e}^{\alpha_m^2 u}\, \mathrm{d}u.$$

If $I^{\text{in}}(t)$ is a long step—i.e. if

$$I^{\text{in}}(t) = \begin{cases} I^{\text{stim}} & \text{if } t \geqslant 0, \\ 0 & \text{if } t < 0, \end{cases}$$

then

$$V_j(Z_j, t) = V^{\text{ss}}(Z_j) - \sum_{m=0}^{\infty} \frac{I^{\text{stim}} C_m}{\alpha_m^2} \mathrm{e}^{K_j Z_j/2} \phi_{jm}(Z_j) \mathrm{e}^{-\alpha_m^2 t},$$

where $V^{\text{ss}}(Z_j)$ is the steady-state voltage distribution. For a short step—i.e. for

$$I^{\text{in}}(t) = \begin{cases} I^{\text{stim}} & \text{if } p \leqslant t \leqslant q, \\ 0 & \text{otherwise}, \end{cases}$$

the integration for $t > q$ is given by

$$\int_a^b I^{\text{in}}(u) \mathrm{e}^{\alpha_m^2 u}\, \mathrm{d}u = \frac{I^{\text{stim}}}{\alpha_m^2} (\mathrm{e}^{\alpha_m^2 b} - \mathrm{e}^{\alpha_m^2 a}).$$

Consequently, the solution for $t > q$ is given by

$$V_j(Z_j, t) = \sum_{m=0}^{\infty} \frac{I^{\text{stim}} C_m}{\alpha_m^2} (\mathrm{e}^{\alpha_m^2 q} - \mathrm{e}^{\alpha_m^2 p}) \mathrm{e}^{K_j Z_j/2} \phi_{jm}(Z_j) \mathrm{e}^{-\alpha_m^2 t}.$$

Finally, if $I^{\text{in}}(t)$ is an alpha function—i.e. if

$$I^{\text{in}}(t) = I^{\text{stim}} t \mathrm{e}^{-\gamma t},$$

then the integration is given by

$$\int_0^t u \mathrm{e}^{\alpha_m^2 u} \mathrm{e}^{-\gamma u}\, \mathrm{d}u = I^{\text{stim}} \frac{(\alpha_m^2 - \gamma) t \mathrm{e}^{-t(\gamma - \alpha_m^2)} - \mathrm{e}^{-t(\gamma - \alpha_m^2)} + 1}{(\gamma - \alpha_m^2)^2},$$

so that the solution is finally given by

$$V_j(Z_j, t) = \sum_{m=0}^{\infty} \frac{I^{\text{stim}} C_m}{(\gamma - \alpha_m^2)^2} \mathrm{e}^{K_j Z_j/2} \phi_{jm}(Z_j) ((\alpha_m^2 - \gamma) t \mathrm{e}^{-t\gamma} - \mathrm{e}^{-t\gamma} + \mathrm{e}^{-\alpha_m^2 t}).$$

We note here that such full analytic solutions are not always necessary. Glenn and Knisley (1997) reported a method of obtaining solutions from eigenvalue equations alone, such as (6.11). See also Steinberg (1996) for a novel approach.

Note that we defined $\phi_j(Z_j)$ so that

$$\phi_j(0)=1, \quad j=0,\ldots,N,$$

so that the voltage at the soma is given by

$$V_j(0,t)=\sum_{m=0}^{\infty} C_m \mathrm{e}^{-\alpha_m^2 t}\int_0^t I^{\mathrm{in}}(u)\mathrm{e}^{\alpha_m^2 u}\,\mathrm{d}u.$$

Consequently, the voltage at the soma for a long step stimulus is given by

$$V_j(Z_j,t)=V^{\mathrm{ss}}-\sum_{m=0}^{\infty}\frac{I^{\mathrm{stim}}C_m}{\alpha_m^2}\mathrm{e}^{-\alpha_m^2 t},$$

where

$$V^{\mathrm{ss}}=I^{\mathrm{stim}}\left(\sum_{j=0}^{N}\frac{G_j^{\infty}}{K_j/2-\sqrt{(K_j^2/4)+1}\coth(L_j\sqrt{K_j^2/4+1})}\right)^{-1}. \tag{6.16}$$

2.3.2 *Orthogonality in the Multipolar Model*

An inner product for the multipolar model can be defined, the eigenfunctions of which can be shown to be orthogonal under this inner product. To begin with, we define a state of the multipolar model to be

$$F=(F_0(Z_0),\ldots,F_N(Z_N)),$$

where $F_j(Z_j)$ is the voltage distribution in the jth cable. For example, the steady state of the system is given by

$$V^{\mathrm{ss}}=(V_0^{\mathrm{ss}}(Z_0),\ldots,V_N^{\mathrm{ss}}(Z_N)),$$

and the mth eigenfunction state is given by

$$\phi_m=(\phi_{0m}(Z_0),\ldots,\phi_{Nm}(Z_N)),$$

where ϕ_{jm} is defined in (6.15). We then define the inner product of two states $E=(E_0,\ldots,E_N)$ and $F=(F_0,\ldots,F_N)$ to be

$$\langle E,F\rangle=\sum_{j=0}^{N}G_j^{\infty}\tau_j\int_0^{L_j}E_j(Z_j)F_j(Z_j)\,\mathrm{d}Z_j.$$

A lengthy elementary calculation shows that if $m\neq n$, then

$$\langle\phi_m,\phi_n\rangle=0,$$

when ϕ_m and ϕ_n are eigenfunctions of the multipolar model; that is, the eigenfunctions are orthogonal.

Conversely, when ϕ_m and ϕ_n are distinct states of the somatic shunt model—i.e. when $n \neq m$—then

$$\langle \phi_m, \phi_n \rangle = \varepsilon.$$

That is, the eigenfunctions for a lumped soma are not orthogonal as has been well-documented (Kawato, 1984; Durand, 1984). However, as was shown in Evans et al. (1992), the eigenfunctions do satisfy the modified orthogonality condition

$$B(E, F) = \sum_{j=0}^{N} G_j^{\infty} \tau_j \int_0^{L_j} E_j(Z_j) F_j(Z_j)\, dZ_j + \varepsilon E_j(0) F_j(0).$$

That is, if ϕ_m and ϕ_n are distinct states of the somatic shunt model, then

$$B(\phi_m, \phi_n) = 0.$$

3. RELATION BETWEEN BRANCHING TREE MODELS AND TAPERING EQUIVALENT CABLE MODELS

One of the questions that arise is the degree to which individual dendrites are well-approximated by a single tapered equivalent cylinder. Although there has been a lot of work published on collapsing *all* dendritic trees of an entire neuron into a single equivalent cylinder, there is much less on collapsing *single* dendritic trees, defined as all branches stemming from a *single* dendritic trunk that is attached to a soma (Figure 6.3). We tested this by comparing the responses of branching trees with a single lumped soma. Ten trees were created by the full stochastic growth algorithm (Model II with grandparent branch correction) of Burke et al. (1992). The average tree had 11.0 ± 7.8 S.D. terminal branches with a range of 2–26. The number of parent branches was 10.0 ± 7.9 S.D. parent branches with a range of 1–25, for a total of 21 branches per dendritic tree. The average diameter of the trunk was 7.0 ± 2.4 S.D. μm with a range of 3.0–11.1 μm, and the total cumulative area was $48{,}503 \pm 26{,}110$ S.D. μm^2 S.D. with a range of 12,730–90,320 μm^2. The average path length from soma to terminal was 236 ± 170 μm, and the average maximum path length for each tree was 1891 ± 332 μm. The correlation between the trunk diameter and total membrane area was 0.82. The tree structure compares closely with that of spinal motoneurons (Cullheim et al., 1987a), with some relatively minor discrepancies (see Burke et al., 1992).

The responses of these non-equivalent trees (considered in isolation from the soma using a sealed-end boundary condition) were compared to that of a single tapered (equivalent) cylinder, using a program we developed for modeling branched neurons (ArborVT). The key parameters that required matching were the k and l, which were randomly varied until a best chi-squared fit was obtained between the waveforms. The results are shown in Table 6.1. The best fit was obtained on average when k was -2.4. The fact that the k is meaningful is indicated by the fact that it correlated with both the mean path distance ($r = 0.76$, $p < 0.02$) and with the ratio of the mean path distance to the maximum path distance of the branched dendritic trees ($r = 0.88$, $p < 0.002$) (see Figure 6.5). The best fit for the maximum dendritic path-to-terminal distance was about $l = 3{,}490$ μm, corresponding to the actual of 1891 μm for the fully branched tree model. The poor correspondence between the maximum path length

Table 6.1 Best fit relation between tapered equivalent cylinder and non-equivalent branching dendritic tree model

	Minimum	*Maximum*	*Mean*	*S.D.*
Length (μm)[a]	2,200	6,800	3,490	1,546
Taper (k)[a]	−5.67	2.62	−2.40	2.38
Membrane Area (μm^2)[b]	4,041	23,771	13,276	6,787
Max Path Distance (μm)[b]	1,386	2,574	1,892	333
Number of Branches[b]	3	51	21	16
Mean Length (μm)[b]	236	727	454	170
Mean Length/Max Length[b]	0.14	0.55	0.25	0.13

[a] From single tapering equivalent cylinder model.
[b] From non-equivalent branching tree model.

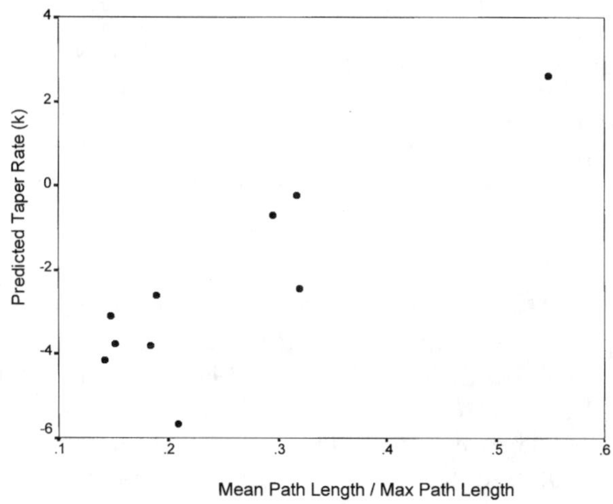

Figure 6.5 Correlation between an index of tree taper and the best fit equivalent cylinder taper. The index is comprised of the mean path distance divided by the maximum path distance where path distance refers to the distance from the soma to a dendritic terminal. Each point represents one dendritic tree and one best fit equivalent cylinder.

of the tree and the l of the single tapered cylinder reflects the relatively rapid effective taper of the tree. As l approaches $-1/k$, which is the case in motoneurons (see earlier section above), it can vary over a wide range without producing any change in the voltage response to a current pulse because of the summed $d^{3/2}$ tapers to a negligible magnitude.

4. NUMERICAL VALIDATION WITH A TAPERING COMPARTMENT MODEL

The solution in section 2.1 was verified by a program to calculate voltage waveforms that result from a step of current in the multipolar tapered model above (TaperMT4

Table 6.2 Comparison of voltage transients computed by analytic and compartmental methods. The model used is a tapered multipolar neuron with three dendrites of unequal τ_m. The parameters for the three dendrites were L of 1,0.65, and 1.19; G^{∞} of 10.0, 17.7, and 21.6 μS; τ_m of 0.010, 0.0021, and 0.0062 msec; and K_j of -1.0, -1.45, and -0.66, respectively. The compartmental solutions are based on models with 100 compartments per dendrite and a calculation time step of 10 μs. The analytic solution computed all 346 $1/\alpha^2$ values (over 10^{-5} ms) and had a tolerance of 10^{-7} ms for $1/\alpha^2$

Time (ms)	*Analytic TaperMT4* (mV)	*Compartmental TSP* (mV)	*Compartmental NEURON* (mV)	*Exact ALL* (mV)
0	0	0	0	0
1	1.1573	1.1572	1.1572	1.163
2	1.6603	1.6586	1.6586	1.666
3	2.0207	2.0179	2.0179	2.026
4	2.3058	2.3021	2.3022	2.311
5	2.5414	2.5370	2.5370	2.547
6	2.7394	2.7341	2.7342	2.745
7	2.9069	2.9012	2.9012	2.912
8	3.0491	3.0429	3.0428	3.054
9	3.1698	3.1632	3.1631	3.175

in Fortran 90 by J.L. Holmes and L.L. Glenn, equivalent program available from the authors). The algorithm for calculating the α_m's and C_m's from the analytic solutions was the same as used by Major et al. (1993a,b). The voltage response was then compared to that of an RC-compartmental model described below. Agreement was good (Table 6.2) between both the analytic solution and compartmental solutions (less than 0.1% difference). Note that the analytic solution, which is comparable in accuracy to the compartmental solution, was produced in 0.065 as much time as required by the NEURON compartmental program of Hines (1989) and 0.070 as much time as that required by our TSP program. Thus, there is a sizable computational speed advantage when analytic methods were used.

Compartmental models are based on multiple ordinary differential equations rather than on a single partial differential equation. Each dendritic segment is modeled by several different compartments and each compartment is represented by an ordinary differential equation. Most compartments use a resistor for the membrane resistance, one or two resistors for the axial resistance, and a capacitor for the membrane capacitance (Figure 6.4). The resulting matrix difference equation is tridiagonal. A finite-analysis program was developed to model multipolar models with tapering dendrites (TSP written by L.L. Glenn and J.L. Holmes in Fortran 90, equivalent available from the authors). The purpose of the model was to have a fast compartmental program that permitted tapering to be specified by the single parameter K. The model could then be used to validate the above equations and to study the multipolar tapered model from a computational approach.

The TSP program uses a typical method of solution. (1) Gauss–Jordan elimination is applied to the tridiagonal matrices representing individual branches in an uncoupled system. (2) Newtonian iteration at the soma-dendritic junction to solve for the boundary condition between the soma and the proximal compartments of each branch. Steps 1 and 2 are alternated until the system converges. (3) The simple Euler method (implicit backward finite difference method) is used to estimate the next time step. The implicit backward finite difference method is commonly used because it provides sufficient accuracy with the short computation times while avoiding the instabilities inherent with forward difference methods.

The execution time of TSP compared favorably to that of the widely used compartmental simulation program named NEURON (Hines, 1989). On an Intel Pentium 90 workstation, voltage responses to a step of current were obtained at a comparable accuracy in 8.3 s for NEURON versus 7.72 s for TSP. The slightly higher performance of TSP was likely due to the use of the following accuracy improvements for compartmental simulations.

We conclude by addressing the estimated error caused by the assumption that dendrites or equivalent cable representations of whole neurons have a uniform diameter. The equations in the previous sections were used to create a 12-cable model that fit the morphometric data of Ulfhake and Kellerth (1983) and the electrotonic data of Fleshman et al. (1988). Each cable had a unique length and diameter, and represented a single collapsed dendritic tree. When the cables were assumed to be untapered, the time constant for the first (slowest) exponential component (τ_0) was 10.0 ms. In comparison, τ_0 was 7.43 ms in the more physiological case. The time required for the signal to decay to 90% of its initial value correlates with the amplitude of the fastest exponential components in a signal. It was 0.30 ms in the untapered case, but 0.41 in the tapered case. The time required for a 50% decay was 5.6 ms in the untapered case and 3.5 in the tapered case. The reduction in both cases is attributed to the lower average L when tapering is incorporated into the model. This elementary change was also noted by Poznanski (1988) and Holmes and Rall (1992) and it was observed in the present model with multiple dendrites and nonuniform tapering.

The consequences of these changes for the estimate of the electrotonic parameters of experimentally-derived transients from neurons is beyond the scope of this article. However, our results in quadriceps (spinal) motoneurons indicate that the average L is between 0.4 and 0.8 for all of the alternative models tested (L.L. Glenn and J.R. Knisley, unpublished findings). The incorporation of some form of tapering, whether by tapering a single cable or varying the lengths of a non-tapering multipolar model (Poznanski and Glenn, 1994), is essential to obtaining accurate estimates of electrotonic parameters.

4.1 Improving Accuracy in Tapering Compartmental Models

Nonuniformity of diameter causes reductions in the accuracy of compartmental models, reductions that are not present when diameters are uniform. These shortcomings can be minimized using the following three methods: adjusting compartment diameters, adjusting compartment width, and optimizing branch length.

4.2 Adjusting Compartment Diameters

In essence, elaborate compartmental models are readily solved by computer because the cable is broken into isopotential branch segments (with discontinuities between the segments) as shown in Figure 6.6. A compartment has a uniform diameter and membrane properties for a given distance (ΔZ) with abrupt changes in diameter between compartments, which corresponds to steps in diameter of a cable representation. This is indicated in Figure 6.6 by the dotted lines. Note that the actual diameter as a function of electrotonic distance is both monotonic and smooth as shown by the solid line in Figure 6.6. As the number of compartments per branch (i.e. steps per cable) increases, the compartmental model more closely approximates the actual taper. The compartments are simpler conceptually and mathematically than smoothly graded changes in physical properties.

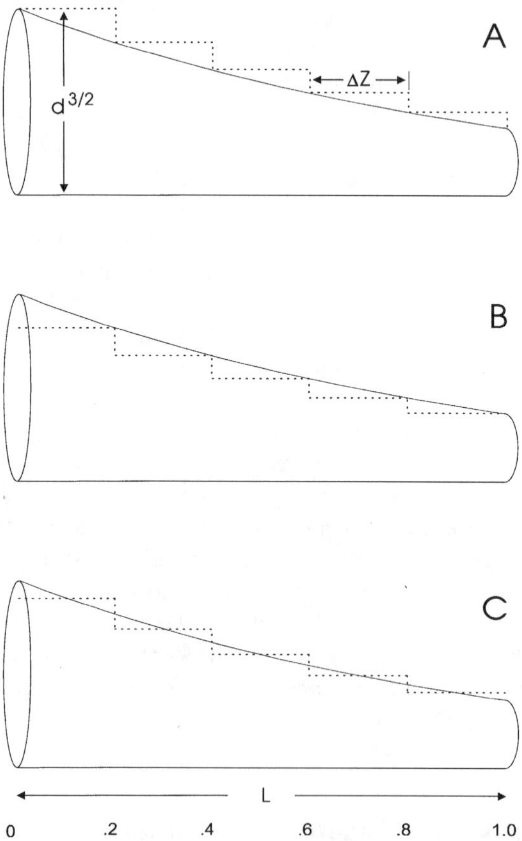

Figure 6.6 Tapering of diameter of a branch in dimensionless units ($d^{3/2} = f(Z)$). Note that $d^{3/2}$ is proportional to G_0^∞. The electrotonic length is 1 and increments correspond to ΔZ, which is the distance from the origin. Dotted lines represent a five-compartment tapering cable model for the branch.

The first problem faced in constructing a compartmental representation of a tapering cable is to set the diameter. Figure 6.6(A) shows the consequence of the most common method of using the beginning diameter of the branch segment as the diameter of the associated compartment. This approach overestimates that cable diameter, and leads to underestimates in the magnitude of the electrical response. An alternative is to used the ending diameter of the segment to be approximated, as in Figure 6.6(B). However, this approach underestimates the cable diameter and leads to overestimates of the magnitude of the electrical response. A better approach, of course, is to use the average diameter of the left and right borders of the branch segment as shown in Figure 6.6(C). However, the average diameter is optimal only if the taper G^{∞} is linear with respect to Z. The model in the present chapter is based on an exponential taper and the optimal diameters would be slightly smaller than the average of the branch segment borders. The exact value for the average diameter for an exponential taper, given here in terms of G^{∞} (proportional to $d^{3/2}$), is

$$G^{\infty}_{\text{avg}} = \frac{G_0}{Z_2 - Z_1} \int_{Z_1}^{Z_2} e^{3KZ}\, dZ = G_0 \frac{e^{3KZ_2} - e^{3KZ_1}}{3K(Z_2 - Z_1)}, \tag{6.17}$$

where Z_1 and Z_2 are the beginning and ending distance of a compartment from the origin in dimensionless units. When the above expression for G^{∞}_{avg} is used to fix the diameter of an RC-compartment, the relative accuracy of a single branch, 10-compartment model with $L=1$ improves from 0.8% to 1.3% over that of using the arithmetic average of the starting and ending diameters. This method was used in all simulations in the present project, and the code for setting these parameters in the NEURON program of (Hines, 1989) is shown in Table 6.2.

4.3 Adjusting Compartment Width

The previous section assumes that all compartments have the same electrotonic width, but this is not necessarily optimal, in particular if the stimulation and recording sites are at the same point, such as at the origin of the branch ($Z=x=0$). For this example, a more accurate response could conceivable be obtained if the compartment width (ΔZ) is smaller for the proximal compartments than the distal compartments. Bush and Sejnowski (1993) showed that an optimum in speed-accuracy curve can obtained if a cable was divided into segments with units of equal xr_a (distance times axial resistance per unit length). Now, if $r_a = 4R_a/\pi D$ and R_a is constant, then each compartment has a uniform x/D. The equation that converts dimensioned (i.e. morphometric) units into dimensionless units for a tapering cable is derived from Goldstein and Rall (1974) for small $(dr/dx)^2$ as

$$Z = \frac{3}{K} \ln(kx+1)$$

and the formula for converting dimensionless into dimensioned units is

$$x = \frac{e^{KZ/3} - 1}{k}.$$

Converting the equal xr_a criterion into dimensionless units and substituting into (6.17) shows that

$$(e^{KZ_2} - e^{KZ_1})/G^{\infty}_{avg} \tag{6.18}$$

should be constant across all compartments of a branch. If we use (6.17) to substitute for G^{∞}_{avg}, then (6.18) reduces to $(Z_2 - Z_1)$. In other words, if the compartment conductance G is calculated by the above equation for G^{∞}_{avg}, then equal increments in Z correspond to equal increments in xr_a, and consequently the optimal strategy is shown to be equal electrotonic widths in the compartments.

4.4 Optimizing L

A more subtle inaccuracy associated with the use of compartmental models—an inaccuracy that is not often appreciated by users of compartmental models programs—is the fact that the origin of a branch model by a series of RC compartments gives different responses for a rise versus a decay waveform. Consider the RC ladder network in Figure 6.4. Note that the two differ by an axial resistor positioned between the stimulus point and the node of the first compartment. When a stimulus is applied and maintained at the left terminal of (A), the above resistor contributes to the voltage division of the signal across the ladder. However, when the stimulus is withdrawn, the same resistor no longer contributes to the circuit, effectively reducing the electrotonic length of the network. The consequence of this asymmetry is that the decay is slower than the rise because the initial rate of decay of a waveform at short times (such as $t < 0.1\tau$) is a function of L.

The problem cannot be solved by removing the initial axial resistor. In the case of Figure 6.4(B), there is no axial resistance between the stimulus terminal and the first node, but this leads to the opposite problem. When the stimulus is applied, the first compartment does not contribute to the electrotonic length of the circuit (each axial and membrane resistor pair contributes to electrotonic length), thereby reducing the length compared to the condition when the stimulus is withdrawn. The addition of a soma, as in Figure 6.4(C) does not ameliorate the asymmetry either, because (C) is just a longer version of (B), and it will also show a faster decay than rise at short times because of the effective differences in electronic length.

A correction for this inaccuracy is to use different circuits for conditions when a stimulus is applied than when a stimulus is withdrawn. When a stimulus is applied, the basic circuit in Figure 6.4(A) was used, but when a stimulus is withdrawn, the value of the resistors and capacitors in the circuit were modified. The modification consisted of recalculating their values after adding 0.5 to the number of compartments in the branch. This compensated for the effective loss of the first resistor after the end of the applied stimulus. Note that the same argument applies to the terminal axial resistor, but such corrections are only necessary if the stimulus is on the terminal axial resistor.

For a three branch model with 100 compartments per dendrite, in which, $\tau_m = 0.005$, $G = 10^{-7}$, $L = 1$ and $K = -1$ for each branch, the accuracy of the voltage

10 ms after the stimulus was withdrawn increased from 1.5% to 0.4% by using the above correction method. The asymmetry problem can be minimized by increasing the number of compartments, however, a comparable increase in accuracy in this case required 800 compartments, which tripled execution time. By making the above correction, an optimal match to theoretical values is obtained for both rises and decays, or synonymously, for calculations of the steady-state and transient cases.

5. DISCUSSION

The present model occupies a position of intermediate complexity and realism, sitting between single (equivalent) models in which an entire neuron is collapsed by the $d^{3/2}$ rule, and fully-branched models, in which the $d^{3/2}$ rule is not used, and each of thousands of cylinders represent a single dendritic branch (i.e. the segment between two adjacent branch points). In the present model, individual dendritic trees are collapsed by the $d^{3/2}$ rule, but there are multiple cylinders, one for each tree. The question arises of whether so many different models are needed, and if so, how to select which of these different models should be applied to a given experimental approach.

Uncollapsed Models. The fully-branched models developed by the J.J.B. Jack and other members of the Oxford group are best suited to neurons for which the complete dendritic architecture (length, diameter, and connectivity of branches) is known. Generally, this knowledge comes from filling neurons with a dye or chromagen-producing substance such as horseradish peroxidase or biotin, and meticulously reconstructing and measuring the branches a series of sections. These models are the most architecturally detailed to date, but also require the most time and work.

Partially Collapsed Models. This model is derived and described in the present chapter (cf. Evans and Kember, 1998), and is best suited for neurons in which the soma diameter and stem diameters of all dendritic trees are known, but little is known about the branching and tapering architecture of individual trees. In this case, each dendritic tree can be assigned a diameter on the basis of direct morphometric measurement, however the length and taper are estimated from the soma diameter. Such estimates require the plausible assumption that the relation between dendritic diameter, length, and taper for the neuron under study is represented by neurons in which such relations have been determined. Limited architectural data is common in morphological studies and arises from the difficulty of obtaining high-contrast, complete fills with chromagen-producing agents.

Fully Collapsed Models. Models of the single equivalent cable type, particularly the tapering type developed by Poznanski at Monash, are optimal when little or nothing is known about the architecture of the experimentally-recorded neurons. It is far more difficult, expensive, and time-consuming to obtain reliable passive response recordings from neurons with a dye-filled microelectrode than from plain microelectrodes. The difficulty level is compounded as the maximal diameter of the neuron becomes lower. Consequently, there will always be experimental problems for which such fully-collapsed models are optimal.

6. CONCLUSIONS AND FUTURE PERSPECTIVES

The primary goal of mathematical models of neural activity is an understanding of the physical mechanisms implied by the equations and their solutions. Computational methods are limited by their inability to establish the fundamental physical mechanisms that produce the output from a given input. Analytical methods cannot be supplanted by computational methods, but rather, computational methods complement analytical work when analytic solutions are cumbersome or unavailable.

The findings in this chapter contribute to the general acceptance of the value and continuing progress of analytical work. In the past five years, we have seen significant mathematical breakthroughs. In particular, we have seen the extension of cylinder cable models to multiple orders of branching dendrites by Evans and coworkers (Evans et al., 1992; Major et al., 1993a,b; Major and Evans, 1994). Also, the ability to analytically manipulate tapers of different type (Poznanski, 1991; Holmes et al., 1992) under different conditions (Poznanski and Peiris, 1996), including the recent ability to add a shunt soma (Poznanski, 1996; Evans and Kember, 1998) adds enormous power to the arsenal of analytical neuroscience. The present study provides a contribution to the modeling of multipolar neurons with unique tapers and time constants. There are still many problems left to solve, however. On the horizon are models with:

(i) arbitrarily branching tapered dendrites;
(ii) smoothly-varying changes in membrane time constant and other membrane properties;
(iii) voltage-dependent conductances; and
(iv) multiple stimulation sites.

With every new model solved, a new tool is available to understand the workings of neurons, neural circuits, and the nervous system. Discovering the underlying mechanisms of neural behavior will continue to be a challenge in analytical neuroscience in the future. Modelers will also be challenged to make their product understandable, and therefore useful, to experimental neurobiologists. For instance, Ohme and Schierwagen (1998) claim to have developed an equivalent cable model with voltage-dependent conductances, but a closer inspection reveals their equivalent cables are not in fact equivalent in the sense that all information is retained, i.e. the relationship between tree and cable is not completely bijective (see Chapter 7).

PROBLEMS

6.1. Major et al. (1993a,b) provided solutions for neuron models with arbitrarily branched dendrites, but the dendritic branches were untapered. Provide the solution for tapered dendrites in an arbitrarily branched model by applying more branching to the model discussed in section 2.

6.2. In the present chapter, the stimulus, $I^{in}(t)$, is applied to the soma ($Z=0$). Derive the solution for a multipolar, unbranched, tapered model where $I^{in}(t)$ is a distance Z^{in} from the soma on any given branch (i.e. on the jth branch for some j).

6.3. Goldstein and Rall (1974) showed that when $(dr/dx)^2$ is small relative to unity, then

$$D_{taper} = D_0 e^{2KZ/3}$$

and also that

$$\frac{dZ}{dx} = \frac{1}{\lambda}\sqrt{\frac{D_0}{D_{tapper}}}$$

Use this to derive the identities in section 4.3 and also to derive (6.1). [Hint: $k = \frac{k}{3\lambda}$].

6.4. Figure 6.1 suggests that the best model for a totally collapsed neuron is a two equivalent cable model in which the first equivalent cable is untapered and the second equivalent cable tapers according to (6.1). We call such a model the *eventually tapered* equivalent cable model of a neuron. Write down the cable equations and boundary conditions for the eventually tapered equivalent cable model with stimulating and recording in the soma, and then determine the steady-state and transient solutions for the somatic potential.

6.5. Under what assumptions on K_j, τ_j and G_j^{∞} could the model in Figure 6.2 be collapsed into a single tapering equivalent cylinder model?

6.6. Experimental suggestion: The following three models have a common property: (a) The multipolar, unbranched, tapered model derived and analyzed in the present chapter; (b) the untapered, arbitrarily branched model of Major et al. (1993a,b); and (c) the tapered equivalent cable models of Poznanski (1988, 1996). The property is that for any two points on the model dendrites or soma, the voltage response produced will be identical regardless of which of the two points is used as the stimulus site and which is used as the recording site. This phenomenon is called *stimulus-response symmetry*. Test whether stimulus-response symmetry is found in mammalian neurons using dual microelectrode penetrations of single neurons. Use signal averaging and small subthreshold current pulses ($V_{ss} < 0.1$ mV) to control nonlinearities.

ACKNOWLEDGMENTS

The authors would like to thank Jeffrey L. Holmes for help with the numerical analysis program. Copies of the source and executables of the TaperMT4 and TSP programs are available from either of the authors on request. Copies of ArborVT are also freely available for computing transients of multipolar branched neurons. This work was supported by NIH grant R01RR08848.

REFERENCES

Alaburda, A. and Gutman, A. (1996) Theory of a cable with an irregular cross section. *Biophys.*, **41**, 727–732.

Ashida, H. (1985) General solution of cable theory with both ends sealed. *J. Theor. Biol.*, **112**, 727–740.

Barrett, J.N. and Crill, W. (1974) Influence of dendritic location and membrane properties on the effectiveness of synapses on cat motoneurones. *J. Physiol.* (*Lond.*), **239**, 325–345.

Bras, H., Gogan, P. and Tyc-Dumont, S. (1987) The dendrites of single brain-stem motoneurons intracellulary labelled with horseradish peroxidase in the cat. Morphological and electrical differences. *Neuroscience*, **22**, 947–970.

Burke, R.E. and ten Bruggencate, G. (1971) Electrotonic characteristics of alpha motoneurones of varying sizes. *J. Physiol.* (*Lond.*), **212**, 1–20.

Burke, R.E. Marks, W.B. and Ulfhake, B. (1992) A parsimonious description of motoneuron dendritic morphology using computer simulation. *J. Neurosci.*, **12**, 2403–2413.

Bush, P.C. and Sejnowski, T.J. (1993) Reduced compartmental models of neocortical pyramidal cells. *J. Neurosci. Meth.*, **46**, 159–166.

Cameron, W.E., Binder, M.D., Botterman, B.R., Reinking, R.M. and Stuart, D.G. (1981) 'Sensory partitioning' of cat medial gastrocnemius muscle by its muscle spindles and tendon organs. *J. Neurophysiol.*, **46**, 32–47.

Carslaw, H.S. and Jaeger, J.C. (1948) *Operational Methods in Applied Mathematics*. Oxford: Oxford University Press.

Chen, X.Y. and Wolpaw, J.R. (1994) Triceps surae motor neuron morphology in the rat: a quantitative light microscopic study. *J. Comp. Neurol.*, **343**, 143–157.

Clements, J.D. and Redman, S.J. (1989) Cable properties of cat spinal motoneurones measured by combining voltage clamp, current clamp, and intracellular staining. *J. Physiol.* (*Lond.*), **409**, 63–87.

Cullheim, S., Fleshman, J.W., Glenn, L.L. and Burke, R.E. (1987a) Membrane area and dendritic structure in type-identified triceps surae alpha motoneurons. *J. Comp. Neurol.*, **255**, 68–81.

Cullheim, S., Fleshman, J.W., Glenn, L.L. and Burke, R.E. (1987b) Three-dimensional architecture of dendritic trees in type-identified alpha motoneurons. *J. Comp. Neurol.*, **255**, 82–96.

Durand, D. (1984) The somatic shunt cable model for neurons. *Biophys. J.*, **46**, 645–653.

Durand, D., Carlen, P.L., Gurevich, N., Ho, A. and Kunov, H. (1983) Electrotonic parameters of rat dentate granule cells measured using short current pulses and HRP staining. *J. Neurophysiol.*, **50**, 1080–1097.

Evans, J.D. and Kember, G.C. (1998) Analytical solutions to a tapering multi-cylinder somatic shunt cable model for passive neurons. *Math. Biosci.* **149**, 137–165.

Evans, J.D., Kember, G.C. and Major, G. (1992) Techniques for obtaining analytical solutions to the multicylinder somatic shunt cable model for passive neurones. *Biophys. J.*, **63**, 350–364.

Fleshman, J.W., Segev, I. and Burke, R.E. (1988) Electrotonic architecture of type-identified alpha motoneurons in the cat spinal cord. *J. Neurophysiol.*, **60**, 60–85.

Glenn, L.L. (1988) Overestimation of the electrical length of neuron dendrites and synaptic electrotonic attenuation. *Neurosci. Lett.*, **91**, 112–119.

Glenn, L.L. and Knisley, J.R. (1997) Use of eigenslope to estimate Fourier coefficients for passive cable models of the neuron. *Neurosci. Res. Comm.*, **21**, 187–194.

Glenn, L.L. and Samojla, B.G. (1987) Three-fold overestimation of dendritic electrotonic length by single equivalent cylinder models. *Soc. Neurosci. Abst.*, **13**, 158.

Glenn, L.L. Samojla, B.G. and Whitney, J.F. (1987) Electrotonic parameters of cat spinal alpha motoneurons evaluated with an equivalent cylinder model that incorporates non-uniform membrane resistivity. *Brain Res.*, **435**, 398–402.

Glenn, L.L. Samojla, B.G. and Rebeta, P.J. (1988) Eigenfunctions of the charging curve in multipolar neurons of nonuniform dendritic length. *Soc. Neurosci. Abst.*, **14**, 248.

Goldstein, S.S. and Rall, W. (1974) Changes of action potential shape and velocity for changing core conductor geometry. *Biophys. J.*, **14**, 731–757.

Gollvik, L., Ornung, G. and Ulfhake, B. (1990) Anatomy of soleus a-motoneurone dendrites in normal cats and in cats subjected to chronic postnatal tenotomy or overload of the soleus muscle. *Exp. Brain Res.*, **80**, 34–43.

Hines, M.L. (1989) A program for simulation of nerve equations with branching geometries. *Int. J. Biomed. Comp.*, **24**, 55–68.

Hines, M.L. and Carnevale, N.T. (1997) The NEURON simulation environment. *Neural. Comput.*, **9**, 1179–1209.

Holmes, W.R. (1986) A continuous cable method for determining the transient potential in passive dendritic trees of known geometry. *Biol. Cybern.*, **55**, 115–124.

Holmes, W.R., Segev, I. and Rall, W. (1992) Interpretation of time constant and electrotonic length estimates in multicylinder or branched neuronal structures. *J. Neurophysiol.*, **68**, 1401–1420.

Holmes, W.R. and Rall, W. (1992) Electrotonic length estimates in neurons with dendritic tapering or somatic shunt. *J. Neurophysiol.*, **68**, 1421–1437.

Holmes, W.R. and Rall, W. (1992) Estimating the electrotonic structure of neurons with compartmental models. *J. Neurophysiol.*, **68**, 1438–1452.

Horwitz, B. (1983) Unequal diameters and their effects on time-varying voltages in branched neurons. *Biophys. J.*, **41**, 51–66.

Iansek, R. and Redman, S.J. (1973) An analysis of the cable properties of spinal motoneurones using a brief intracellular current pulse. *J. Physiol.* (*Lond.*), **234**, 613–636.

Jack, J.J.B., Noble, D. and Tsien, R.W. (1983) *Electric Current Flow in Excitable Cells.* Oxford: Clarendon Press.

Jack, J.J.B. and Redman, S.J. (1971) An electrical description of the motoneurone and its application to the analysis of synaptic potentials. *J. Physiol.* (*Lond.*), **215**, 321–352.

Kawato, M. (1984) Cable properties of a neuron model with non-uniform resistivity. *J. Theor. Biol.*, **111**, 149–169.

Korogod, S.M. (1996) Electro-geometrical coupling in non-uniform branching dendrites. Consequences for relative synaptic reflectiveness. *Biol. Cybern.*, **74**, 85–93.

Lux, H.D., Schubert, P. and Kreutzberg, G.W. (1970) Direct matching of morphological and electrophysiological data in cat spiral motoneurons. In: *Excitatory Synaptic Mechanisms*, Edited by P. Anderson and J.K.S. Jansen Oslo, Universitets forlaget, pp. 189–198.

Major, G. and Evans, J.D. (1994) Solutions for transients in arbitrarily branching cables: IV. Nonuniform electrical parameters. *Biophys. J.*, **66**, 615–634.

Major, G., Evans, J.D. and Jack, J.J. (1993a) Solutions for transients in arbitrarily branching cables: I. voltage recording with somatic shunt. *Biophys. J.*, **65**, 423–449.

Major, G., Evans, J.D. and Jack, J.J. (1993b) Solutions for transients in arbitrarily branching cables: II. voltage clamp theory. *Biophys. J.*, **65**, 450–468.

Moore, J.A. and Appenteng, K. (1991) The morphology and electrical geometry of rat jaw-elevator motoneurones. *J. Physiol.* (*Lond.*), **440**, 325–343.

Nitzan, R., Segev, I. and Yarom, Y. (1990) Voltage behavior along the irregular dendritic structure of morphologically and physiologically characterized vagal motoneurons in the guinea pig. *J. Neurophysiol.*, **63**, 333–345.

Ohme, M. and Schierwagen, A. (1998) An equivalent cable model for neuronal trees with active membrane. *Biol. Cybern.*, **78**, 227–243.

Pongracz, F., McClintock, T.S., Aches, B.W. and Shepherd, G.M. (1993) Signal transmission in lobster olfactory receptor cells: functional significance of electrotonic structure analysed by a compartmental model. *Neuroscience.*, **55**, 325–338.

Poznanski, R.R. (1987) Statistics Research Report No. 159, Monash University.

Poznanski, R.R. (1988) Membrane voltage changes in passive dendritic trees: a tapering equivalent cylinder model. *IMA J. Math. Appl. Med. Biol.*, **5**, 113–145.

Poznanski, R.R. (1990) Analysis of a postsynaptic scheme based on a tapering equivalent cable model. *IMA J. Math. Appl. Med. Biol.*, **7**, 175–197.

Poznanski, R.R. (1991) A generalized tapering equivalent cable model for dendritic neurons. *Bull. Math. Biol.*, **53**, 456–457.

Poznanski, R.R. (1996) Transient Response in a tapering cable model with somatic shunt. *NeuroReport*, **7**, 1700–1704.

Poznanski, R.R., Gibson, W.G. and Bennett, M.R. (1991) Electrotonic coupling between two CA3 hippocampal pyramidal neurons: a distributed cable model with gap-junction. *Bull. Math. Biol.*, **57**, 865–881.

Poznanski, R.R. and Glenn, L.L. (1994) Estimating the effective electrotonic length of dendritic neurons with reduced equivalent cable models. *Neurosci. Res. Comm.*, **15**, 69–76.

Poznanski, R.R. and Peiris, M.S. (1996) Subthreshold response to white-noise current input in a tapering cable model. *IMA J. Math. Appl. Med. Biol.*, **13**, 207–222.

Rall, W. (1962) Theory of physiological properties of dendrites. *Ann. N.Y. Acad. Sci.*, **96**, 1071–1092.

Rall, W. (1969) Time constants and electrotonic length of membrane cylinders and neurons. *Biophys. J.*, **9**, 1483–1508.

Rastad, J., Gad, P., Jankowska, E., McCrea, D. and Westman, J. (1990) Light microscopical study of dendrites and perikarya of interneurones meditating la reciprocal inhibition of cat lumbar alpha-motoneurones. *Anat. Embryol.*, **181**, 381–388.

Rose, P.K. and Dagum, A. (1988) Nonequivalent cylinder models of neurons: interpretation of voltage transients generated by somatic current injection. *J. Neurophysiol.*, **60**, 125–148.

Schierwagen, A.K. (1989) A non-uniform equivalent cable model of membrane voltage changes in a passive dendritic tree. *J. Theor. Biol.*, **141**, 159–179.

Spruston, N. and Johnston, D. (1992) Perforated patch-clamp analysis of the passive membrane properties of three classes of hippocampal neurons. *J. Neurophysiol.*, **67**, 508–528.

Stratford, K., Mason, A., Larkman, Major, G. and Jack, J.J.B. (1989) The modelling of pyramidal neurones in the visual cortex. In, The Computing Neuron (R. Durbin, C. Miall and G. Mitchison, eds.) Addison-Wesley Publishing, New York, chap. 16, pp. 298–321.

Steinberg, I.Z. (1996) On the analytic solution of electrotonic spread in branched passive dendritic trees. *J. Comput. Neurosci.* **3**, 301–311.

Surkis, A., Taylor, B., Peskin, C.S. and Leonard, C.S. (1996) Quantitative morphology of physiologically identified and intracellularly labeled neurons from the guinea-pig laterodorsaltegmental nucleus in vitro. *Neuroscience*, **74**, 375–392.

Tsukahara, N., Murakami, F. and Hultborn, H. (1975) Electrical constants of neurons of the red nucleus. *Exp. Brain Res.*, **23**, 49–64.

Turner, D.A. and Schwartzkroin, P.A. (1980) Steady-state electrotonic analysis of intracellularly stained hippocampal neurons. *J. Neurophysiol.*, **44**, 184–199.

Ulfhake, B. and Kellerth, J.-O. (1984) Electrophysiological and morphological measurements in cat gastrocnemius and soleus alpha motoneurons. *Brain Res.*, **307**, 167–179.

Ulfhake, B. and Kellerth, J.-O. (1983) A quantitative morphological study of HRP-labelled cat alpha-motoneurones supplying different hindlimb muscles. *Brain Res.*, **264**, 1–19.

Ulrich, D., Quadroni, R. and Lüscher, H.-R. (1994) Electrotonic structure of motoneurons in spinal cord slice cultures: a comparison of compartmental and equivalent cylinder models. *J. Neurophysiol.*, **72**, 861–871.

White, J.A., Manis, P.B. and Young, E.D. (1992) The parameter identification problem for the somatic shunt model. *Biol. Cybern.*, **66**, 307–318.

APPENDIX 6.1

The following code can be used to obtain an exponential taper in dimensionless space in dendritic branches using the NEURON v3.0 by John W. Moore and Michael Hines (see also Hines and Carnevale, 1997), which is based on the use of pointers. The taper in the code below is exponential (in dimensionless units), and the input variables are length (in cm), G_m (in S/cm^2), initial diameter d_0, and K (dimensionless). The result of the code is that the resistances and capacitances of the RC-compartments vary along the length of the cable, thereby modeling in accordance with a tapering diameter.

```
create sec[o] for h=0, numdend {
  sec[h]·nseg=ncmpt
  sec[h]·L=leng[h]* 10000
  sec[h]·Ra=200.0
  sec[h]·cm=0.7
  sec[h] insert pas
  sec[h]·e_pas=0
  sec[h]·g_pas=Gmemb[h]// in S/cm 2
  sec[h] for (x) if (x>0 && x<1) {// compute diameter
  tmpl=(d0[h]/2) 0.5
  tmp2=(K[h]*((2*sec[h]·Ra*sec[h]·g_pas) 0.5)*((x-0.5/nseg)*leng[h]))/3
  tempdiam=10 4*2*((tmpl+tmp2) 2)
  diam(x-0.0001:x+0.0001)=tempdiam:tempdiam
  }
}
```

CHAPTER 7
THE LANCZOS PROCEDURE FOR GENERATING EQUIVALENT CABLES

James M. Ogden, Jay R. Rosenberg and Rex R. Whitehead

1. INTRODUCTION

The observed complexity and variety of neuron dendritic geometry has raised interesting questions concerning the role that such structures might play in neuronal signal processing. Much of our present insight into the functional properties of dendritic trees has been obtained through the application of cable theory. However, an extensively branched tree does not lend itself to a simple mathematical analysis which might reveal subtle and important function associated with specific neuronal shapes. One approach to understanding the properties of dendritic geometry involves replacing a branching tree with a simpler, but equivalent, unbranched structure in which one tries to retain the tree's essential electrical characteristics. In this way one avoids the analytical difficulties associated with the complex geometry while still gaining some insight into its function.

Rall (1962) showed how a passive dendritic tree that satisfies certain very restrictive geometrical conditions can be replaced by a single uniform *equivalent cylinder*, attached to the soma. This simplification of dendritic structure is a well established result that has proven effective in guiding the interpretation of experimental results, allowing, for example, electrical parameters of the dendritic membrane to be estimated more accurately than was previously possible (Rall, 1957, 1959, 1960). However, real dendritic trees are unlikely to be equivalent to a cylinder and Rall's result can, at best, only give a good approximation to somatic voltage responses. Rall's equivalent cylinder, by its very nature, ignores potentially interesting properties of tree geometry.

Attempts have been made to extend Rall's result to less restrictive tree geometry. The continuous equivalent cables developed by Rall (1962), Goldstein and Rall (1974) and Poznanski (1988, 1991) are non-uniform cables of specified continuous taper. The corresponding tree exhibits a uniformly spread branching structure, the number of branches defined by a continuous function of distance from the soma. The fractional orders of branching inherent in this approach makes it markedly unphysiological. Nevertheless, these cables have been used to estimate effective electrotonic lengths of dendritic trees. Empirical extensions to Rall's result have also been postulated. The equivalent cables of Fleshman et al. (1988), Clements and Redman (1989), Stratford et al. (1989), Manor et al. (1991) and Burke et al. (1994) may be constructed for dendritic trees of arbitrary geometry. These cables have non-uniform, discontinuous diameters and have been employed, in a similar manner to the equivalent cylinder, to estimate electrical parameters and fit experimental results to theoretical models, as well as to simplify computer modeling of neurons. Burke (1997) discusses several different approaches to empirical equivalent cable construction (lambda, attenuation and delay cables), comparing their usefulness in estimating membrane parameters of motoneurons.

Construction of simplified structures for passive dendritic trees has now been properly generalized with the introduction, by Whitehead and Rosenberg (1993), of the *Lanczos procedure* for generating *equivalent cables*. An equivalent cable—as defined by Whitehead and Rosenberg (1993)—is an unbranched cable with (generally) non-uniform diameter and electrotonic length equal to the total electrotonic length of all the tree's branches (see also Ogden et al., 1997; Lindsay et al., 1997). Dendritic trees of arbitrary geometry may be transformed to equivalent cables. Furthermore, and perhaps most importantly, a two-way *electrical mapping* links a dendritic tree to its equivalent cable. Combined, the cable and mapping preserve all the structural and electrical detail of the tree. The electrical activity at any point on the original dendritic tree may be inferred from the electrical activity over its equivalent cable.

Before a dendritic tree can be transformed to an equivalent cable, a location must be chosen as the *origin*. The origin represents the same point on both tree and cable and the two structures are electrically indistinguishable with respect to this point. To be consistent with Rall's equivalent cylinder the tree-to-soma connection point would be chosen as origin.

An equivalent cable may be made up of several disjoint sections. Only one, called the *connected section*, is actually attached to the origin. The remaining sections, referred to as *disconnected sections*, define, via the electrical mapping, configurations of electrical activity on the dendritic tree that will have no net effect at the origin. The connected section and its mapping contain all information about a dendritic tree that is relevant to the origin. Certain trees will exhibit (possibly obscure) electrical and geometrical symmetries that give rise to some electrical redundancy. Consequently, the connected section will be shorter than the complete equivalent cable, and disconnected sections must exist.

Many tools for investigating electrical activity in passive dendritic trees of arbitrary geometry are currently available. In particular, there are several analytical approaches to solving a set of cable equations, or analysing transient voltage responses (e.g. Butz and Cowan, 1974; Horwitz, 1981; Abbott et al., 1991; Abbott, 1992; Cao and Abbott, 1993; Major et al., 1993a,b; Major, 1993c; Agmon-Snir, 1995; see associated references for further details concerning the computer implementation of these and other schemes). The huge variety of possible configurations of electrical activity over a dendritic tree limit the insight that these approaches can give into the potentially manifold and subtle functional capabilities conferred on a neuron by its dendritic geometry. The equivalent cable, however, through its electrical mapping, constitutes a novel tool that can be used to define and investigate interesting and possibly functional configurations of electrical activity in passive trees that arise solely as a consequence of the tree geometry, and are independent of specific cable electrical parameters. Equivalent cable construction does not involve solving the cable equations, instead it combines and rearranges them.

The main aim of this Chapter is to provide an introduction to the Lanczos procedure for generating equivalent cables. The description given here should be instructive and detailed enough to allow the reader with reasonable mathematical and computational experience to implement an algorithm for themselves. Software for the construction of equivalent cables is available, free of charge, from the authors.

2. THE CABLE EQUATION AND THE MULTI-CYLINDER MODEL OF A DENDRITIC TREE

The branching structure of a passive dendritic tree cannot in general be replaced by a single uniform cylinder as illustrated in Figure 7.1. Rall's equivalent cylinder (see, for example, Rall, 1977; also section 5.2) is the only well known exception. However, uniform cylinders, regarded as building blocks, may be used to represent small sections of a complex tree. An arbitrary number of these cylinders, generally with different diameters, can be placed together to form features such as tapering and branching, allowing a complete dendritic tree to be approximated, as illustrated in Figure 7.2. Smooth changes in diameter along a branch become multiple small abrupt diameter changes when replaced by chains of cylinders. Some overlapping of cylinders may occur where they connect, but this does not cause any mathematical problems in the representation that we use.

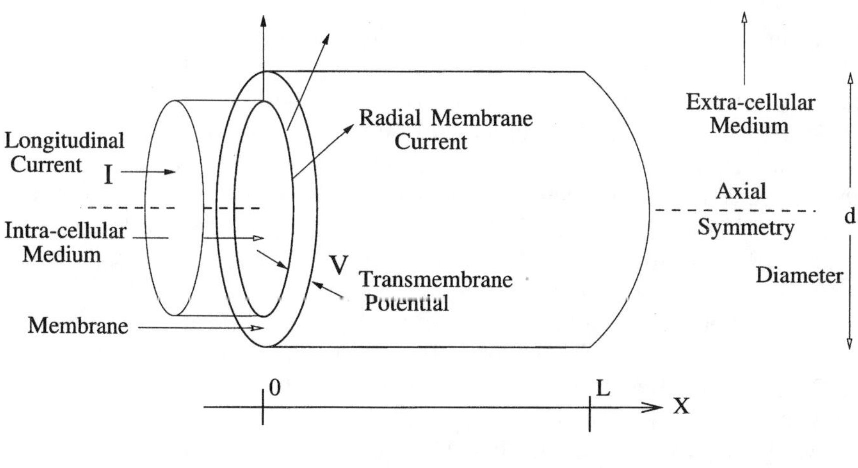

Figure 7.1 Single dendritic cylinder with electrotonic length L and diameter d. In reality, the membrane is very thin compared to the diameter of the cylinder.

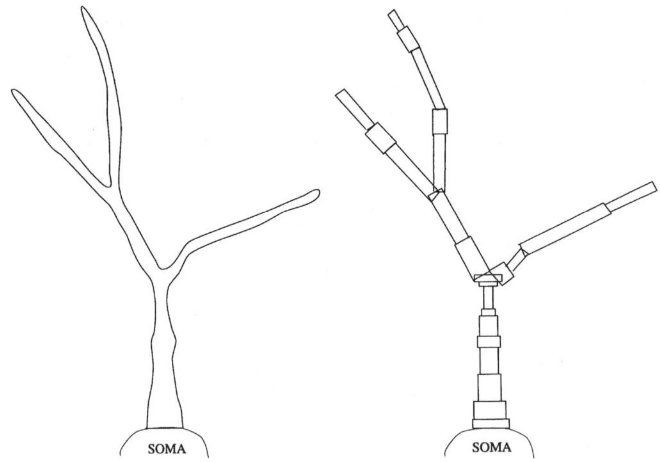

Figure 7.2 Multi-cylinder representation of a dendritic tree.

The electrical characteristics of a single dendritic cylinder are described by the one-dimensional linear cable equation,

$$\frac{r_{\mathrm{m}_i}}{r_{\mathrm{i}_i}}\frac{\partial^2 v_i(x,t)}{\partial x^2} = r_{\mathrm{m}_i} c_{\mathrm{m}_i}\frac{\partial v_i(x,t)}{\partial t} + v_i(x,t) - r_{\mathrm{m}_i} g_i(x,t), \quad 0_i < x < l_i, \quad t>0, \tag{7.1}$$

where, for cylinder i, with diameter d_i, at position x and time $t, v_i(x,t)$ is the transmembrane potential with respect to the tree's uniform resting potential (V), $g_i(x,t)$ is an internally applied current per unit length (input current density) ($\mathrm{A\,cm^{-1}}$), r_{m_i} is the membrane resistance times unit length ($\Omega\,\mathrm{cm}$), r_{i_i} is the internal (axial) resistance per unit length ($\Omega\,\mathrm{cm}^{-1}$), c_{m_i} is the membrane capacitance per unit length ($\mathrm{F\,cm^{-1}}$), and l_i is the length of the cylinder (cm).

Voltage continuity and current conservation conditions between neighbouring cylinders, plus boundary conditions at the end of each terminating cylinder, complete this multi-cylinder representation of a dendritic tree. Electrical activity may give rise to a longitudinal (axial) current in the intra-cellular medium, as well as a radial membrane current due to charge flow across the resistive component of the membrane, and charging/discharging of the capacitive component of the membrane.

Cable theory is well documented and the derivation of the cable equation is set out in several texts (for example, Jack et al., 1983; Tuckwell, 1988a). Several physiologically reasonable assumptions are made in the derivation about electrical and physical properties of the dendritic tree. In the cylinder representation, axial symmetry is satisfied by all branches. Electrical properties of a cylinder are assumed to be independent of any electrical activity in the cylinder—this is a *passive* model. The extracellular medium is perfectly conducting, while membrane thickness, membrane resistivity, internal (cytoplasmic) resistivity, and membrane capacitance per unit area are constant for the whole dendritic tree.

The cable equation can be rewritten in terms of dimensionless, or *electrotonic* units. Electrotonic length is measured in units of the space constant, λ_i, which is actually a function of dendritic diameter, while electrotonic time is measured in units of the membrane time constant, τ, a constant for the tree. The electrotonic units are given by

$$\lambda_i = \sqrt{\frac{r_{m_i}}{r_{i_i}}} = k_2\sqrt{d_i}, \quad \tau = r_{m_i}c_{m_i} = k_1, \tag{7.2}$$

where k_1 and k_2 are dendritic-tree-specific constants determined by various electrical and physical properties.

When expressed in electrotonic units the cable equation for cylinder i becomes

$$\frac{\partial^2 V_i(X,T)}{\partial X^2} = \frac{\partial V_i(X,T)}{\partial T} + V_i(X,T) - \frac{G_i(X,T)}{d_i^{3/2}}, \quad 0_i < X < L_i, \quad T > 0, \tag{7.3}$$

where $X = x/\lambda_i$ is electrotonic distance along the cylinder, $T = t/\tau$ is electrotonic time, $V_i(X,T) = v_i(x,t)$, and $L_i = l_i/\lambda_i$ is the electrotonic length of the cylinder. $G_i(X,T)$ is a scaled internally applied current per unit electrotonic length (scaled electrotonic current density). Because of its behavior, from here on we usually refer to G_i as the applied current term. It is independent of diameter (Tuckwell, 1988a) and, like any current, must be conserved; only relative magnitudes of applied currents will be of interest.

It is essential that electrotonic units are used when constructing equivalent cables since the dimensionless electrotonic length dominates the dendritic tree representation required by the Lanczos procedure. The single space coordinate, X, is measured along paths in the dendritic structure, increasing away from soma towards terminals.

The longitudinal current, I_i for cylinder i, is given by

$$I_i(X,T) = -\frac{1}{\lambda_i r_{i_i}} \frac{\partial V_i(X,T)}{\partial X}. \tag{7.4}$$

By convention, this current is positive where there is a net flow of positive charge in the direction of increasing X. The internal resistance per unit length, like the space constant, only has a diameter dependence—$r_{i_i} \propto d_i^{-2}$ and $\lambda_i \propto d_i^{1/2}$ so $r_{i_i}\lambda_i \propto d_i^{-3/2}$. The longitudinal current can be rewritten as

$$I_i(X,T) = k_3 d_i^{3/2} \frac{\partial V_i(X,T)}{\partial X}, \tag{7.5}$$

where k_3 is another dendritic-tree-specific constant.

Equations (7.1)–(7.5) are equally valid for an equivalent cable if it is represented by uniform cylinders. In the next section we use (7.3) and (7.5) to develop a framework of discretized cable equations that can describe the geometry and boundary conditions of multi-cylinder structures (branching or not). These new equations will allow us to place a complete dendritic tree in a mathematical form that is amenable to transformation to an equivalent cable using the Lanczos procedure.

The constants k_1, k_2, and k_3 are not required when generating an equivalent cable. Provided the original physical lengths and diameters of the dendritic tree cylinders are given, the physical lengths and diameters of the equivalent cable's cylinders, as well as the electrical mapping between the structures, can be correctly determined using arbitrarily chosen constants. If the real constant values (giving correct values for τ and λ_i) are required subsequently, for some reason, an equivalent cable's electrotonic length is readily scaled to the correct value.

3. DISCRETIZATION OF THE CABLE EQUATION AND THE MATRIX REPRESENTATION OF A DENDRITIC TREE

In the multi-cylinder model, the branches of a dendritic tree have been divided up into distinct uniform cylinder sections. When considering how the Lanczos procedure is eventually to be applied, however, we find that a slightly stricter discretisation must be imposed on the structure—all cylinder electrotonic lengths must be integer multiples of one, arbitrarily small, basic electrotonic length. Call this basic length H. This restriction ensures that a tree can be represented by a number of points, or *nodes*, with the internodal electrotonic length a constant throughout the structure (see Figure 7.3). If the internodal electrotonic length is denoted h then $H = zh$ where z is a positive integer.

To indicate structural features, all points where cylinders meet or terminate must be marked by a node. A node is *connected* to its nearest neighbouring node(s) and two connected nodes are always located on the same cylinder (though are not necessarily unique to that cylinder). Only nodes may be chosen as origin, restricting

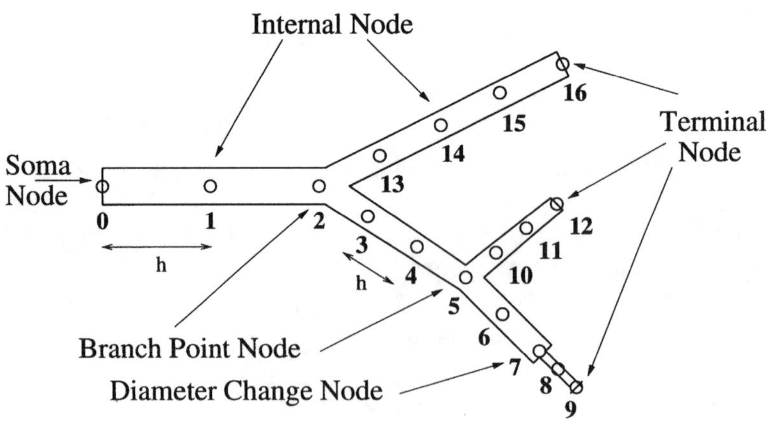

Figure 7.3 Discretized dendritic tree represented by nodes. Constant internodal electrotonic length is h, with basic length $H = h$. The physical lengths of the cylinders are drawn—note that the closer the nodes, the smaller the cylinder diameter. Any node, from 0–16, may be chosen as origin. (The node numbering in this example assumes that none of the terminal nodes have been assigned voltage boundary conditions.)

to a finite number the equivalent cables that can be constructed for a dendritic tree approximated in this way.

If a configuration uses a minimal number of nodes (two per basic length) then $H=h$. However, when this level of discretization is used, ambiguities can arise concerning the correct interpretation of an equivalent cable from the mathematical form generated by the Lanczos procedure. In anticipation of this, a finer discretization may be required, with each basic length H represented by at least three nodes ($H=zh$ where $z\geqslant 2$). The extra level of discretization ensures that we know just enough information about an equivalent cable, prior to its construction, to overcome problems caused by certain boundary conditions and disconnected sections on the cable. These problems are discussed when encountered. We rely on the fact that if the dendritic cylinders have a common basic unit of length $H=zh$, then the same must be true of the equivalent cable cylinders. When moving, for example, from $H=h$ to $H=2h$ in the same multi-cylinder dendritic tree, no new structure is represented by the extra nodes, and consequently no new structure will appear in the equivalent cable. It may be desirable to use a finer discretization anyway, since the Lanczos procedure only generates a discontinuous electrical mapping—from dendritic tree nodes to equivalent cable nodes. (With care, and provided $z\geqslant 2$, it is in fact possible to infer a continuous mapping from the discrete one, though this requires a detailed knowledge of the underlying processes.)

A set of discrete equations is required to describe a dendritic tree in terms of electrical activity at its nodes. A subset of these equations will be equally important for describing the unbranching equivalent cable.

3.1 Node Numbering

The nodes that represent a dendritic tree must be uniquely numbered. A well chosen numbering system makes the Lanczos transformation simpler to implement and computationally more efficient (see also Hines, 1984). The system used (and recommended) by the authors for binary branching trees proceeds as follows:

Node 0 marks the connection of dendritic tree to soma (the soma node) unless this node has been assigned a voltage boundary condition. In fact, any node that marks a point of termination on a dendritic tree and which has been assigned a voltage boundary condition is not included in the numbering process, although such a node is still an essential part of the representation (this is explained in section 3.4).

Each cylinder has a node nearest the soma (near-end node) and a node furthest from the soma (far-end node). Node numbers increase consecutively from near-end to far-end, except where a node has already been numbered (this may occur at a branch point) or must not be numbered (a voltage boundary condition)—in either case the node is skipped and numbering continues from the next (connected) node on the cylinder, if one is available.

The order in which the cylinders are chosen for node numbering is determined by a right-branch-then-left-branch scheme. At a binary branch point, a parent branch has two child branches, right and left. The left branch nodes are only numbered after all the nodes on the adjacent right branch and all nodes on the substructure beyond the right branch have been numbered. At any branch point encountered in the substructure, the child branches are numbered according to this rule. The consecutive numbering of nodes must not be broken.

This scheme is fairly easily visualised through Figures 7.3, 7.6(a) and 7.6(b). A very similar method of assigning cylinder precedence can be used to number dendritic trees with more extensively branched junctions.

Node numbering schemes for branching structures cannot guarantee that all cylinders connected to a branch point have consecutively numbered nodes. This is impossible when a node is common to more than two cylinders. To avoid confusion over which nodes we are referring to, let 1_n indicate an increment of one node from n along a cylinder. Node $(n-1_n)$ is then one node further from n towards the soma, while node $(n+1_n)$ would be one node further away from the soma.

3.2 Discretization of Electrotonic Space Derivatives

Electrotonic space in the dimensionless cable equation (7.3) can be discretized using a central difference approximation for the second-order partial derivative,

$$\frac{\partial^2 V_n}{\partial X^2} = \frac{V_{n+1_n} - 2V_n + V_{n-1_n}}{h^2}, \tag{7.6}$$

where V_n is the transmembrane potential at node n. This is a second-order approximation in h, the internodal electrotonic length. Time dependence can be ignored from here on, though one should consider it implicit in the electrical activity. The construction of equivalent cables is time-independent.

Before proceeding any further, it is convenient for us to define

$$c_i = d_i^{3/2}. \tag{7.7}$$

We refer to c_i as the c-value of cylinder i. It is a frequently occuring parameter in equivalent cable construction and is useful as a replacement measure for cylinder diameters.

Equation (7.6) only applies where n is an internal node of cylinder i, i.e. n must be connected to two other nodes also on cylinder i, as illustrated in Figure 7.4(d). Otherwise, one of the three nodes would either not actually exist or would be located on a different cylinder (with a different diameter) satisfying a separate cable equation. Either case would make the approximation unacceptable, with the dendritic structure not correctly reflected in the equation.

Using (7.6) to replace the second order partial derivative in the dimensionless cable equation (7.3) gives a three term recurrence relation that links an internal node, n, to its two neighbours, $(n-1_n)$ and $(n+1_n)$, namely

$$\frac{V_{n+1_n}}{h^2} - \left(\frac{2}{h^2}+1\right)V_n + \frac{V_{n-1_n}}{h^2} = \frac{\partial V_n}{\partial T} - \frac{G_n}{c_i}, \tag{7.8}$$

where G_n represents the applied current at node n on cylinder i. This is the *discrete cable equation* for internal node n on cylinder i. Changes in dendritic structure must

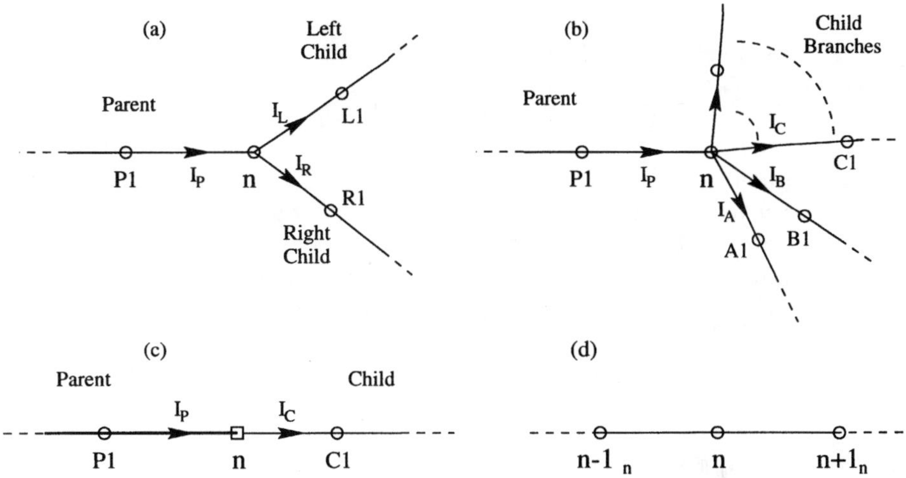

Figure 7.4 Non-terminating structural features found in a cylinder-represented dendritic tree. Cylinders are drawn as lines but have arbitrary diameters. (a) Binary branch point. (b) General branch point. (c) Point of abrupt diameter change. (d) Uniform cylinder section.

be taken into account when constructing discrete cable equations for nodes that link two (diameter change), three (binary branch point) or more (general branch point) cylinders at a common node. We must ensure voltage continuity and current conservation at nodes common to multiple cylinders. From (7.5), the longitudinal current at node n on cylinder i is

$$I_{n_i} = k_3 d_i^{3/2} \left.\frac{\partial V_i}{\partial X}\right|_n = k_3 c_i \frac{\partial V_{n_i}}{\partial X}. \tag{7.9}$$

Expressions for $\partial V_n / \partial X$ are required. They must only involve nodes located on the same cylinder, ensuring independence from adjoining cylinders and no misrepresentation of the cylinder-approximated dendritic geometry. Starting with the central difference approximation for first-order derivatives,

$$\frac{\partial V_n}{\partial X} = \frac{V_{n+1_n} - V_{n-1_n}}{2h}, \tag{7.10}$$

two suitable expressions can be constructed. In one, $(n-1_n)$ and n are the two far-end nodes of a cylinder; in the other, n and $(n+1_n)$ are the two near-end nodes of a cylinder.

Using (7.8), substitute into (7.10) a replacement for V_{n+1_n} to get the 'end of cylinder' expression. Set applied current zero, it will be treated later. If node n is at the far-end of cylinder i,

$$\frac{\partial V_{n_i}}{\partial X} = \frac{1}{2h}\left[(2+h^2)V_n - 2V_{n-1_n} + h^2 \frac{\partial V_n}{\partial T}\right]. \tag{7.11}$$

Similarly, substitute into (7.10) a replacement for V_{n-1_n} to get the 'start of cylinder' expression. If node n is at the near-end of cylinder i,

$$\frac{\partial V_{n_i}}{\partial X} = \frac{1}{2h}\left[2V_{n+1_n} - (2+h^2)V_n - h^2\frac{\partial V_n}{\partial T}\right]. \tag{7.12}$$

Each substitution eliminates a reference to one node. These expressions are correct to second order in h, ensuring that discrete cable equations derived using them are consistent in their order of approximation with the simple discrete cable equation (7.8). The cylinder must be specified since $\partial V/\partial X$ is discontinuous across abrupt changes in diameter. There is no need to distinguish between cylinders when discussing the continuous potentials, V_n.

(Note that (7.11) and (7.12) differ only in the direction electrotonic distance X is measured, towards or away from n. It is useful to have both forms available.)

3.3 Connecting Multiple Cylinders at a Common Node

The most complicated physical feature in the multi-cylinder dendritic tree model is the branch point, where a number of child cylinders diverge from a single parent cylinder, as illustrated in Figure 7.4(b). Properties of each of the cylinders that meet at the junction must be combined into one equation. Voltage continuity at a branch point node is already guaranteed by the discrete model. Conservation of longitudinal current provides the means by which cylinders may be linked in a single discrete cable equation. In this way we can take account of changes in diameter.

From Kirchoff's law, in the absence of an applied current at node n, the total longitudinal current flowing *into* the branch point from the parent cylinder (P) equals the total longitudinal current flowing *out of* the branch point into the child cylinders (j). Thus,

$$I_{n_P} = \sum_j I_{n_j}, \tag{7.13}$$

where n_P and the n_j are all node n, the subscript indicating the cylinder to which the node is considered to belong. If an applied current, I_{A_n}, is also injected *into* the branch point, then

$$I_{A_n} + I_{n_P} = \sum_j I_{n_j}, \tag{7.14}$$

Substituting (7.9) into (7.13) gives

$$\frac{I_{A_n}}{k_3} + c_P\frac{\partial V_{n_P}}{\partial X} = \sum_j c_j\frac{\partial V_{n_j}}{\partial X}. \tag{7.15}$$

Using (7.11) for the parent cylinder, and (7.12) for each of the child cylinders, discretize the first order derivatives in (7.15) to give

$$\frac{2hI_{A_n}}{k_3} + c_P\left[(2+h^2)V_n - 2V_{P1} + h^2\frac{\partial V_n}{\partial T}\right] = \sum_j c_j\left[2V_{j1} - (2+h^2)V_n - h^2\frac{\partial V_n}{\partial T}\right]. \tag{7.16}$$

Node P1 lies on the parent cylinder, while node $j1$ lies on child cylinder j. All are connected to node n. Rearranging gives the discrete cable equation for a general branch point node,

$$-\left(\frac{2}{h^2}+1\right)V_n+\frac{2}{h^2C_n}\sum_i c_iV_{i1}=\frac{\partial V_n}{\partial T}-\frac{2}{C_n}G_n, \tag{7.17}$$

where

$$c_i=d_i^{3/2},\quad C_n=\sum_i c_i,\quad G_n=-\frac{I_{A_n}}{hk_3}. \tag{7.18}$$

Here the summation is over all cylinders (parent and children), i, that meet at node n. The sum of their c-values is C_n. The total applied current at node n is given by G_n. Note that the sum of the coefficients of V_{i1} is $2/h^2$.

Binary branch points, as illustrated in Figure 7.4(a), are usually sufficient for modeling realistic dendritic geometry. Simply set to three the number of cylinders meeting at node n in (7.17). Two child cylinders (left, L, and right, R) diverge from a parent cylinder (P). So,

$$\frac{2c_{\mathrm{P}}}{h^2C_n}V_{\mathrm{P}1}-\left(\frac{2}{h^2}+1\right)V_n+\frac{2c_{\mathrm{R}}}{h^2C_n}V_{\mathrm{R}1}+\frac{2c_L}{h^2C_n}V_{\mathrm{L}1}=\frac{\partial V_n}{\partial T}-\frac{2}{C_n}G_n, \tag{7.19}$$

where

$$c_{\mathrm{P}}=d_{\mathrm{P}}^{3/2},\quad c_{\mathrm{L}}=d_{\mathrm{L}}^{3/2},\quad c_{\mathrm{R}}=d_{\mathrm{R}}^{3/2},\quad C_n=c_{\mathrm{P}}+c_{\mathrm{R}}+c_{\mathrm{L}}. \tag{7.20}$$

A node at a point of abrupt diameter change, as illustrated in Figure 7.4(c), is common to only two cylinders (parent, P, and child, C). From (7.17), the discrete cable equation in this case is

$$\frac{2c_{\mathrm{P}}}{h^2C_n}V_{\mathrm{P}1}-\left(\frac{2}{h^2}+1\right)V_n+\frac{2c_{\mathrm{C}}}{h^2C_n}V_{\mathrm{C}1}=\frac{\partial V_n}{\partial T}-\frac{2}{C_n}G_n, \tag{7.21}$$

where

$$c_{\mathrm{P}}=d_{\mathrm{P}}^{3/2},\quad c_{\mathrm{C}}=d_{\mathrm{C}}^{3/2},\quad C_n=c_{\mathrm{P}}+c_{\mathrm{C}}. \tag{7.22}$$

This is the main equation required for describing unbranching structure, and thus is vital in a discrete equivalent cable representation. Note that the ratio of coefficients of $V_{\mathrm{C}1}$ and $V_{\mathrm{P}1}$ equals the ratio of c-values for the cylinders C and P. The sum of these coefficients must still be $2/h^2$.

Not surprisingly, if $d_{\mathrm{C}}=d_{\mathrm{P}}$, the two adjoining cylinders may be collected together and replaced with one cylinder that has their combined length—node n then becomes an internal node of a cylinder, described by (7.8).

All non-terminal nodes on a dendritic tree have a corresponding discrete cable equation which can be extracted from the general discrete cable equation (7.17).

3.4 Terminal Boundary Conditions

The far-end node of each terminating branch cylinder, along with the soma node, form the set of terminal nodes. A terminal node (except possibly the soma node) is connected to only one other node. In our model, one of two general boundary condition types may be imposed at a terminal node.

(i) Current condition at terminal node n on cylinder i:

$$I_n = k_3 c_i \frac{\partial V_{n_i}}{\partial X} = \kappa_n(T). \tag{7.23}$$

For example, $\kappa_n(T)$ may represent a constant current injection. When $\kappa_n(T)=0$, (7.23) is a *sealed end* condition, which is often taken as the natural terminal condition for dendritic trees. There is no current leak from such a terminal, and it may be treated as a point where the diameter falls to zero.

(ii) Voltage condition at terminal node n:

$$V_n = \phi_n(T). \tag{7.24}$$

Setting $\phi_n(T)$ constant gives a voltage clamp condition. When $\phi_n(T)$ is a *cut end* (or *killed end*) condition. Condition (ii) can arise under experimental conditions. Node n may be treated as a point where the diameter jumps to infinity.

The only node that is exempt from restriction to these two conditions is that chosen as origin. If a terminal node is chosen as origin, it may be assigned any boundary condition that can be incorporated into a discrete cable equation (a lumped soma for example—see Tuckwell, 1988a). Later we show how the nature of the origin boundary condition has no impact on the structure of the equivalent cable. For the sake of simplicity, we assume from now on, without loss of generality, that if a terminal node is chosen as origin then it has a sealed end condition. If a more realistic condition must be imposed at the origin for some reason (for example, we may want to solve the cable equations to give the depolarization at a more realistic soma origin due to transient inputs) then the sealed end condition can be replaced with the desired condition after the Lanczos procedure has been applied. This flexibility makes the soma node an ideal choice of origin, since it may be the only terminal node where (i) is not a generally acceptable condition when mathematically modelling realistic neurons.

We now take each boundary condition type, in turn, and demonstrate how it may be incorporated into an acceptable discrete cable equation.

3.4.1 *Current Conditions*

Representation of condition (i) is a simple task since expressions for $\partial V_{n_i}/\partial X$ are already given. For a current condition at node n, marking the far end of cylinder i, substitute (7.23) into (7.11) and rearrange to give

$$\frac{2}{h^2} V_{n-1_n} - \left(\frac{2}{h^2} + 1\right) V_n = \frac{\partial V_n}{\partial T} - \frac{2\kappa_n}{h k_3 c_i}. \tag{7.25}$$

Similarly, for a current condition at the near end of a cylinder (on the dendritic tree, this can only mean the soma node, $n=0$, under our numbering system), substitute (7.23) into (7.12) to give

$$-\left(\frac{2}{h^2}+1\right)V_n+\frac{2}{h^2}V_{n+1_n}=\frac{\partial V_n}{\partial T}+\frac{2\kappa_n}{hk_3c_i}. \tag{7.26}$$

All non-zero current information at the terminal should be contained in the boundary condition. Note that conventions for direction of current flow imply $\kappa_n=-I_{A_n}$ at a far-end terminal, while $\kappa_n=+I_{A_n}$ at a near-end terminal. There is really nothing to prevent us from regarding the current boundary condition as a sealed end with an additional applied current term, G_n, at the same point, so rewrite (7.25) as

$$\frac{2}{h^2}V_{n-1_n}-\left(\frac{2}{h^2}+1\right)V_n=\frac{\partial V_n}{\partial T}-\frac{2}{c_i}G_n, \tag{7.27}$$

and rewrite (7.26) as

$$-\left(\frac{2}{h^2}+1\right)V_n+\frac{2}{h^2}V_{n+1_n}=\frac{\partial V_n}{\partial T}-\frac{2}{c_i}G_n, \tag{7.28}$$

where, just as before, $G_n=-I_{A_n}/h\kappa_3$. Set $G_n=0$ to yield the discrete cable equations for sealed ends.

Note that (7.27) and (7.28) can also be derived by setting diameter d_C or d_P to zero in discrete cable equations (7.21).

The above two equations assume that only one node is connected to the terminal node. It is possible that multiple cylinders meet at a soma node which has been assigned a current condition. In this case the general discrete cable equation (7.17) applies, with the appropriate number of cylinders and $d_P=0$. Equivalently treat all cylinders, i, as child cylinders.

3.4.2 *Voltage Conditions*

Recall that a terminal node with a voltage boundary condition is skipped during node numbering. Condition (ii) must be incorporated into the discrete cable equation of the single node connected to the terminal node. An acceptable discrete cable equation that links the terminal node to its connected node cannot be constructed for the voltage terminal node itself (consider letting $c_C\to\infty$ in (7.21)). Each discrete cable equation must illustrate the connectivity of nodes in the dendritic tree, otherwise the Lanczos procedure will not work.

The Lanczos procedure will have no problem accepting voltage conditions once they have been treated correctly. If any non-zero voltage conditions are present ($\phi_n\neq 0$), however, the procedure will not transform these conditions in a way that immediately allows their correct interpretation (unlike non-zero current boundary conditions, whose interpretation after transformation will be clear). This is a consequence of the necessary discretization. Incorporation of non-zero voltage boundary conditions initially requires a level of discretization where $H=2h$, or finer, to ensure all required information is transformed. Care must still be taken to properly interpret the mapping of non-zero voltage conditions from dendritic tree to equivalent cable. The correct method of interpretation is outlined in section 4.3.3, however, we allow for non-zero voltage conditions in the equations for completeness rather than in anticipation of a detailed explanation of how to correctly interpret their transformation. Once the reader achieves a solid understanding of the fundamental processes of the Lanczos procedure, the reasoning behind the given interpretation should become clear.

If $H=2h$, each uniform cylinder is represented by at least three nodes. A terminal node with a voltage condition will then only be connected to an internal node on the same cylinder. Only the discrete cable equation for an internal node, n, need be modified to take account of a voltage terminal node connected to n (see Figure 7.5(b)). If the voltage condition is at the far end of cylinder i, set $V_{n+1_n}=\phi_{n+1_n}$ in (7.8) to give

$$\frac{V_{n-1_n}}{h^2}-\left(\frac{2}{h^2}+1\right)V_n=\frac{\partial V_n}{\partial T}-\frac{G_n}{c_i}-\frac{\phi_{n+1_n}}{h^2}. \tag{7.29}$$

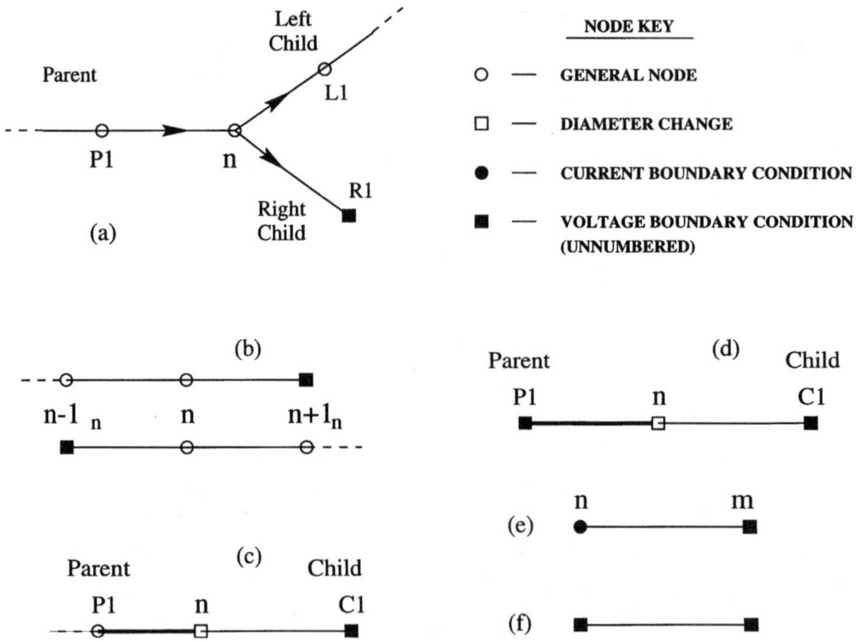

Figure 7.5 Dendritic structures with voltage terminal conditions. (a) Branch point node connected to one cut end terminal node. (b) Two cases of a node with a voltage terminal condition connected to an internal node. (c) One cut end terminal condition connected to a node where there is an abrupt diameter change. (d) Two nodes with cut end terminal conditions connected to a node where there is an abrupt diameter change. (e) Short two-node structure with sealed and cut end terminal conditions. (f) Short two-node structure, both cut ends.

If the voltage condition is at the near-end of cylinder i then set $V_{n-1_n}=\phi_{n-1_n}$ in (7.8) to give

$$-\left(\frac{2}{h^2}+1\right)V_n+\frac{V_{n+1_n}}{h^2}=\frac{\partial V_n}{\partial T}-\frac{G_n}{c_i}-\frac{\phi_{n+1_n}}{h^2}. \tag{7.30}$$

Set $\phi_{n+1_n}=0$ and $\phi_{n-1_n}=0$ to yield the cut end expressions.

If there are no non-zero voltage terminal conditions (i.e. only cut ends and current conditions) on the dendritic tree, the extra discretization may not be necessary. If $H=h$, any type of tree node could be connected to a voltage terminal node, and so any discrete cable equation might have to incorporate a voltage condition. The exact form depends on the number of voltage terminal nodes connected to a node and the structure of the dendritic tree at that node.

For example, in Figure 7.5(a) node n is a binary branch connected to one terminal node, R1, on the right child branch. If $V_{\mathrm{R1}}=0$, a cut end condition, node R1 does not have an associated discrete cable equation and should be ignored in the node numbering. The discrete cable equation for the binary branch point node (7.19) becomes

$$\frac{2c_{\mathrm{P}}}{h^2C_n}V_{\mathrm{P1}}-\left(\frac{2}{h^2}+1\right)V_n+\frac{2c_{\mathrm{L}}}{h^2C_n}V_{\mathrm{L1}}=\frac{\partial V_n}{\partial T}-\frac{2}{C_n}G_n. \tag{7.31}$$

If node n is a point of abrupt diameter change and is connected to one terminal node, C1, with a cut end condition, $V_{\mathrm{C1}}=0$, as in Figure 7.5(c), then (7.21) becomes

$$\frac{2c_{\mathrm{P}}}{h^2C_n}V_{\mathrm{P1}}-\left(\frac{2}{h^2}+1\right)V_n=\frac{\partial V_n}{\partial T}-\frac{2}{C_n}G_n. \tag{7.32}$$

Figure 7.5(d)–(f) illustrate some of the ambiguities and limitations of discrete cable equations associated with certain short dendritic structures when $H=h$. In Figure 7.5(d), node n is connected to two cut end terminal nodes, P1 and C1, on cylinders with different diameters. Only one discrete cable equation exists for this structure. Set $V_{\mathrm{P1}}=0$ and $V_{\mathrm{C1}}=0$ in (7.21) to give

$$-\left(\frac{2}{h^2}+1\right)V_n=\frac{\partial V_n}{\partial T}-\frac{2}{C_n}G_n. \tag{7.33}$$

In Figure 7.5(e), one node, m, has a cut end condition, $V_m=0$, and the other, node n, has a sealed end condition, $\partial V_n/\partial X=0$. Again there is only one discrete cable equation for this structure—(7.28) also collapses to (7.33). The two structures in Figures 7.5(d) and 7.5(e) have different lengths but cannot be distinguished by their discrete cable equations, hence the ambiguity. It is not possible to associate even one discrete cable equation with the two-node structure in Figure 7.5(f), where both nodes are assigned a cut condition.

Dendritic trees are highly unlikely to be represented by such short structures but equivalent cable disconnected sections may well take such forms if $H=h$. If at least three nodes represent each uniform cylinder ($H=2h$) then these structures (hence also the ambiguities) cannot occur.

Note that in all the discrete cable equations given above, the coefficient of V_n is always $-((2/h^2)+1)$. This will emerge as a vital feature for the validity of the Lanczos procedure. The origin coefficient is unaffected by the Lanczos transformation and may in fact take any value, but all others must be the same, hence the restriction to two general boundary condition types except at the origin.

3.5 Matrix Representation of a Dendritic Tree

A complete dendritic tree can now be represented in a simple mathematical form by collecting the discrete cable equations for each numbered node into one matrix equation, like so,

$$A_T v_T = \frac{\partial v_T}{\partial T} + g_T + b_T, \tag{7.34}$$

where A_T is the *Tree Matrix*, the nth row of which contains coefficients of the node potentials from the discrete cable equation for node n. Vector v_T contains the dendritic tree node potentials (V_n). Vector g_T contains applied current terms, including non-zero current boundary condition information ($\bar{G}_n = -2G_n/C_n$). Vector b_T contains non-zero voltage boundary condition information ($\bar{\phi}_n = -\phi_{n\pm 1_n}/h^2$). If a dendritic tree is represented by k nodes, numbered 0 to $(k-1)$, then

$$v_T = \begin{bmatrix} V_0 \\ V_1 \\ \cdot \\ \cdot \\ \cdot \\ V_{k-2} \\ V_{k-1} \end{bmatrix}, \quad g_T = \begin{bmatrix} \bar{G}_0 \\ \bar{G}_1 \\ \cdot \\ \cdot \\ \cdot \\ \bar{G}_{k-2} \\ \bar{G}_{k-1} \end{bmatrix}, \quad b_T = \begin{bmatrix} \bar{\phi}_0 \\ \bar{\phi}_1 \\ \cdot \\ \cdot \\ \cdot \\ \bar{\phi}_{k-2} \\ \bar{\phi}_{k-1} \end{bmatrix}. \tag{7.35}$$

A_T is a square matrix of order k (see examples below). It describes the geometry and boundary condition types (current or voltage) of the dendritic tree and illustrates the connectivity of the nodes. The row and column numbers of a non-zero off-diagonal element indicate nodes that are connected to each other. The numbering scheme used by the authors makes A_T nearly tri-diagonal, with those elements that break the tri-diagonality indicating nodes connected to branch points. The tree matrix is not symmetric, except in the locations of non-zero elements, echoing the simple fact that if one node, i, is connected to node j then j must also be connected to i.

3.5.1 *Matrix Equation Examples*

Given any dendritic tree, it is possible to discretize it to an appropriate resolution and systematically construct a tree matrix equation from discrete cable equations. The two trees in Figure 7.6 are shown in matrix equation form below. In each example the diagonal is $D = -(2+h^2)$ and there are no non-terminal applied current

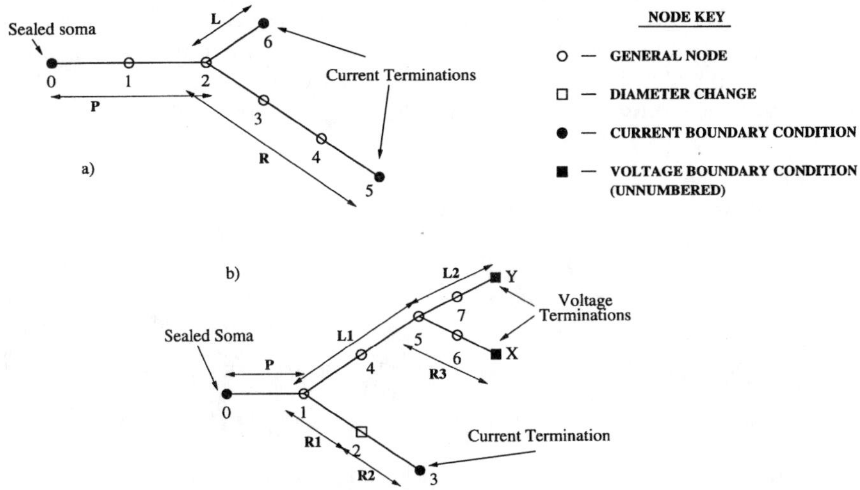

Figure 7.6 (a) Dendritic tree for Example One, consisting of cylinders P, R, and L. (b) Dendritic tree for Example Two, consisting of cylinders P, R1, R2, R3, L1 and L2. Each cylinder has a corresponding arbitrary diameter.

terms ($G_i = 0$ for all i except where i is a terminal node with a non-zero current condition). Example One represents the dendritic tree in Figure 7.6(a) and Example Two represents the dendritic tree in Figure 7.6(b).

Example One

Equation (7.34) becomes

$$\frac{1}{h^2}\begin{bmatrix} D & 2 & 0 & 0 & 0 & 0 & 0 \\ 1 & D & 1 & 0 & 0 & 0 & 0 \\ 0 & \frac{2c_{\mathrm{P}}}{C_2} & D & \frac{2c_{\mathrm{R}}}{C_2} & 0 & 0 & \frac{2c_{\mathrm{L}}}{C_2} \\ 0 & 0 & 1 & D & 1 & 0 & 0 \\ 0 & 0 & 0 & 1 & D & 1 & 0 \\ 0 & 0 & 0 & 0 & 2 & D & 0 \\ 0 & 0 & 2 & 0 & 0 & 0 & D \end{bmatrix}\begin{bmatrix} V_0 \\ V_1 \\ V_2 \\ V_3 \\ V_4 \\ V_5 \\ V_6 \end{bmatrix} = \begin{bmatrix} \partial V_0/\partial T \\ \partial V_1/\partial T \\ \partial V_2/\partial T \\ \partial V_3/\partial T \\ \partial V_4/\partial T \\ \partial V_5/\partial T \\ \partial V_6/\partial T \end{bmatrix} + \begin{bmatrix} 0 \\ 0 \\ 0 \\ 0 \\ 0 \\ -\frac{2G_5}{c_R} \\ -\frac{2G_6}{c_L} \end{bmatrix}, \tag{7.36}$$

where

$$c_{\mathrm{P}} = d_{\mathrm{P}}^{3/2}, \quad c_{\mathrm{R}} = d_{\mathrm{R}}^{3/2}, \quad c_{\mathrm{L}} = d_{\mathrm{L}}^{3/2}, \quad C_2 = c_{\mathrm{P}} + c_{\mathrm{R}} + c_{\mathrm{L}}.$$

The parent, left and right diameters are d_{P}, d_{L} and d_{R}. The soma node is assigned a sealed end condition. The two other terminal nodes have current boundary conditions, G_5 and G_6.

Nodes 0, 5 and 6 have current terminal conditions, described by (7.28) in row 0 and (7.27) in rows 5 and 6 of the tree matrix. Nodes 1, 3, and 4 are cylinder internal nodes described by (7.8) in rows 1, 3 and 4 of the tree matrix. Node 2 is a binary branch point described by (7.19) in row 2 of the tree matrix.

Example Two
Equation (7.34) becomes

$$\frac{1}{h^2}\begin{bmatrix} D & 2 & 0 & 0 & 0 & 0 & 0 & 0 \\ \frac{2c_{\mathrm{P}}}{C_1} & D & \frac{2c_{\mathrm{R1}}}{C_1} & 0 & \frac{2c_{\mathrm{L1}}}{C_1} & 0 & 0 & 0 \\ 0 & \frac{2c_{\mathrm{R1}}}{C_2} & D & \frac{2c_{\mathrm{R2}}}{C_2} & 0 & 0 & 0 & 0 \\ 0 & 0 & 2 & D & 0 & 0 & 0 & 0 \\ 0 & 1 & 0 & 0 & D & 1 & 0 & 0 \\ 0 & 0 & 0 & 0 & \frac{2c_{\mathrm{L1}}}{C_5} & D & \frac{2c_{\mathrm{R3}}}{C_5} & \frac{2c_{\mathrm{L2}}}{C_5} \\ 0 & 0 & 0 & 0 & 0 & 1 & D & 0 \\ 0 & 0 & 0 & 0 & 0 & 1 & 0 & D \end{bmatrix} \begin{bmatrix} V_0 \\ V_1 \\ V_2 \\ V_3 \\ V_4 \\ V_5 \\ V_6 \\ V_7 \end{bmatrix} = \begin{bmatrix} \partial V_0/\partial T \\ \partial V_1/\partial T \\ \partial V_2/\partial T \\ \partial V_3/\partial T \\ \partial V_4/\partial T \\ \partial V_5/\partial T \\ \partial V_6/\partial T \\ \partial V_7/\partial T \end{bmatrix} + \begin{bmatrix} 0 \\ 0 \\ 0 \\ -\frac{2G_3}{c_{\mathrm{R2}}} \\ 0 \\ 0 \\ -\frac{\phi_x}{h^2} \\ -\frac{\phi_y}{h^2} \end{bmatrix},$$

where

$$c_{\mathrm{P}}=d_{\mathrm{P}}^{3/2},\quad c_{\mathrm{R1}}=d_{\mathrm{R1}}^{3/2},\quad c_{\mathrm{L1}}=d_{\mathrm{L1}}^{3/2},\quad c_{\mathrm{R2}}=d_{\mathrm{R2}}^{3/2},\quad c_{\mathrm{L2}}=d_{\mathrm{L2}}^{3/2},\quad c_{\mathrm{R3}}=d_{\mathrm{R3}}^{3/2},$$

$$C_1=c_{\mathrm{P}}+c_{\mathrm{R1}}+c_{\mathrm{L1}},\quad C_2=c_{\mathrm{R1}}+c_{\mathrm{R2}},\quad C_5=c_{\mathrm{L1}}+c_{\mathrm{R3}}+c_{\mathrm{L2}}.$$

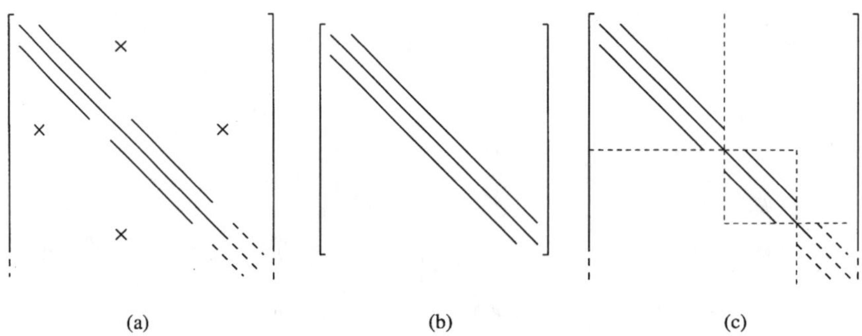

Figure 7.7 (a) The tree matrix for a branching dendritic tree is nearly tri-diagonal, with off-tri-diagonal elements resulting from the branching strucutre. (b) The tree matrix for an unbranching structure is purely tri-diagonal. (c) An equivalent cable matrix is tri-diagonal, but may consist of one or more independent sub-matrices representing disjoint sections of equivalent cable. The first tri-diagonal sub-matrix (upper left) describes the connected section, others (if any) represent disconnected sections. Crosses indicate single non-zero off-tri-diagonal elements while large gaps in the tri-diagonal lines indicate single zeroes.

Again the soma node is assigned a sealed end condition. Node 3 has a current boundary condition, G_3, while nodes x and y have voltage boundary conditions, $V_x = \phi_x$ and $V_y = \phi_y$.

Node 2 is a point of abrupt diameter change described by (7.21) in row 2 of the tree matrix. Nodes 6 and 7 are cylinder internal nodes, their discrete cable equations modified for connection to voltage terminal conditions, as in (7.29), rows 6 and 7 of the tree matrix.

It should be clear that an unbranching dendritic structure, represented by a chain of cylinders with nodes numbered consecutively from soma to far terminal, would have the perfectly tri-diagonal tree matrix structure illustrated in Figure 7.7(b). Each node, n, would only be connected to two others, $(n+1)$ and $(n-1)$, except at the two terminal nodes.

4. THE LANCZOS PROCEDURE

An outline of the Lanczos procedure is given in Whitehead and Rosenberg (1993). Briefly, three matrix transformations (similarity transformations), denoted $\boldsymbol{S}, \boldsymbol{L}$ and $\boldsymbol{X}$, are applied to the tree matrix equation as follows: first the tree matrix is symmetrized ($\boldsymbol{S}$), the result is Lanczos tri-diagonalised ($\boldsymbol{L}$), followed by de-symmetrization ($\boldsymbol{X}$). This sequence of transformations generates a matrix equation that describes an equivalent cable. The new equation carries a form similar to the tree matrix equation, i.e. the matrix elements are coefficients of node potentials in discrete cable equations.

Matrices $\boldsymbol{S}, \boldsymbol{L}$ and $\boldsymbol{X}$ are all nonsingular time-independent square matrices with the same order as $\boldsymbol{A}_\mathrm{T}$. Matrices $\boldsymbol{L}$ and $\boldsymbol{X}$ depend on the node chosen as origin, tree geometry and boundary condition types.

In its simplest incarnation, Lanczos tri-diagonalisation can only be applied to symmetric matrices. A tree matrix, though not symmetric, has just the right form that it can be easily symmetrized using a simple scale transformation. Equation (7.34) then becomes

$$(\boldsymbol{S}\boldsymbol{A}_\mathrm{T}\boldsymbol{S}^{-1})(\boldsymbol{S}\boldsymbol{v}_\mathrm{T}) = \frac{\partial(\boldsymbol{S}\boldsymbol{v}_\mathrm{T})}{\partial T} + (\boldsymbol{S}\boldsymbol{g}_\mathrm{T}) + (\boldsymbol{S}\boldsymbol{b}_\mathrm{T}), \tag{7.37}$$

where $\boldsymbol{A}_\mathrm{TS} = \boldsymbol{S}\boldsymbol{A}_\mathrm{T}\boldsymbol{S}^{-1}$ is the *Symmetric Tree Matrix*.

The Lanczos tri-diagonalisation collapses the branching dendritic structure into an unbranching cable structure (from one symmetric matrix form to another), so that (7.37) becomes

$$(\boldsymbol{L}\boldsymbol{S}\boldsymbol{A}_\mathrm{T}\boldsymbol{S}^{-1}\boldsymbol{L}^{-1})(\boldsymbol{L}\boldsymbol{S}\boldsymbol{v}_\mathrm{T}) = \frac{\partial(\boldsymbol{L}\boldsymbol{S}\boldsymbol{v}_\mathrm{T})}{\partial T} + (\boldsymbol{L}\boldsymbol{S}\boldsymbol{g}_\mathrm{T}) + (\boldsymbol{L}\boldsymbol{S}\boldsymbol{b}_\mathrm{T}), \tag{7.38}$$

where $\boldsymbol{A}_\mathrm{CS} = \boldsymbol{L}\boldsymbol{S}\boldsymbol{A}_\mathrm{T}\boldsymbol{S}^{-1}\boldsymbol{L}^{-1}$ is the tri-diagonal *Symmetric Cable Matrix*.

De-symmetrization of $\boldsymbol{A}_\mathrm{CS}$, using another scale transformation, must be done in such a way that the result is a set of discrete cable equations representing an equivalent cable. The c-values (and thus diameters) of the cylinders that form the equivalent cable, as well as the electrical mapping, can then be determined. After de-symmetrization, (7.38) becomes

$$(\boldsymbol{XLSA}_{\mathrm{T}}\boldsymbol{S}^{-1}\boldsymbol{L}^{-1}\boldsymbol{X}^{-1})(\boldsymbol{XLSv}_{\mathrm{T}})=\frac{\partial(\boldsymbol{XLSv}_{\mathrm{T}})}{\partial T}+(\boldsymbol{XLSg}_{\mathrm{T}})+(\boldsymbol{XLSb}_{\mathrm{T}}), \quad (7.39)$$

or

$$\boldsymbol{A}_{\mathrm{C}}\boldsymbol{v}_{\mathrm{C}}=\frac{\partial \boldsymbol{v}_{\mathrm{C}}}{\partial T}+\boldsymbol{g}_{\mathrm{C}}+\boldsymbol{b}_{\mathrm{C}}, \quad (7.40)$$

where $\boldsymbol{A}_{\mathrm{C}}=\boldsymbol{XLSA}_{\mathrm{T}}\boldsymbol{S}^{-1}\boldsymbol{L}^{-1}\boldsymbol{X}^{-1}$ is the *Cable Matrix*.

The electrical mapping from dendritic tree nodes to equivalent cable nodes is defined by

$$\boldsymbol{M}=\mathrm{XLS}, \quad (7.41)$$

while $\boldsymbol{M}^{-1}=\boldsymbol{S}^{-1}\boldsymbol{L}^{-1}\boldsymbol{X}^{-1}$ defines the mapping from cable to tree. The node potentials for the equivalent cable are given by $\boldsymbol{v}_{\mathrm{C}}=\boldsymbol{Mv}_{\mathrm{T}}$. The reverse node potential mapping from equivalent cable to dendritic tree is given by $\boldsymbol{v}_{\mathrm{T}}=\boldsymbol{M}^{-1}\boldsymbol{v}_{\mathrm{C}}$. Similarly, expressions for the equivalent cable applied current terms and current boundary conditions are given by $\boldsymbol{g}_{\mathrm{C}}=\boldsymbol{Mg}_{\mathrm{T}}$. Vector $\boldsymbol{b}_{\mathrm{C}}=\boldsymbol{Mb}_{\mathrm{T}}$ is the mapping of non-zero voltage boundary conditions and, as stated previously, must be interpreted carefully when non-zero (see section 4.3.3). It is more complex than $\boldsymbol{b}_{\mathrm{T}}$ since, as well as non-zero voltage boundary conditions, it can contain contributions that must be added to non-terminal equivalent cable node potentials. Elements of each cable vector are linear combinations of the original tree vector elements.

The cable matrix is determined by the tree matrix and the location of the origin and, like the tree matrix, defines cylinder-geometry (shape) and boundary condition types. When mapped, tree vectors $\boldsymbol{v}_{\mathrm{T}}$, $\boldsymbol{g}_{\mathrm{T}}$ and $\boldsymbol{b}_{\mathrm{T}}$ together determine electrical activity on the equivalent cable, but cannot influence its shape.

If $\boldsymbol{g}_{\mathrm{T}}$ and $\boldsymbol{b}_{\mathrm{T}}$ are zero (i.e. only sealed and cut end boundary conditions) the eigenvalues of $\boldsymbol{A}_{\mathrm{T}}$ give the characteristic decay rates of voltage distributions over the dendritic tree model nodes (Rall, 1969). If $\bar{\boldsymbol{v}}_i$ is a right eigenvector of $\boldsymbol{A}_{\mathrm{T}}$ and μ_i the corresponding real eigenvalue then, from (7.34),

$$\boldsymbol{A}_{\mathrm{T}}\bar{\boldsymbol{v}}_i=\mu_i\bar{\boldsymbol{v}}_i=\frac{\partial \bar{\boldsymbol{v}}_i}{\partial T}. \quad (7.42)$$

Solving the right-hand equation gives (ignoring the constant of integration)

$$\bar{\boldsymbol{v}}_i=\boldsymbol{k}_i e^{\mu_i T}. \quad (7.43)$$

Eigenvector $\bar{\boldsymbol{v}}_i$ is a characteristic potential distribution over the tree nodes and $\boldsymbol{k}_i$ is this distribution at time $T=0$. The decay time constant for the eigenvector potential distribution $\bar{\boldsymbol{v}}_i$ is given by $\tau_i=-1/\mu_i$, in units of the electrotonic time constant, τ.

Applying the electrical mapping to (7.42) gives

$$\boldsymbol{A}_{\mathrm{C}}(\boldsymbol{M}\bar{\boldsymbol{v}}_i)=\mu_i(\boldsymbol{M}\bar{\boldsymbol{v}}_i)=\frac{\partial(\boldsymbol{M}\bar{\boldsymbol{v}}_i)}{\partial T}. \quad (7.44)$$

The eigenvalues of $\boldsymbol{A}_{\mathrm{C}}$ are the same as those of $\boldsymbol{A}_{\mathrm{T}}$ and the corresponding right eigenvectors for the cable become $\boldsymbol{M}\bar{\boldsymbol{v}}_i$. The tree eigenvector potentials are mapped by $\boldsymbol{M}$ to the equivalent cable eigenvector potentials that will decay at the same characteristic rates. This is an important feature for the equivalence of a dendritic tree and its equivalent cable. If either $\boldsymbol{b}_T$ or $\boldsymbol{g}_T$ are non-zero then the eigenvector solutions are not so simple, although the principle of equivalence is the same. The spectral properties of the tree and cable matrices are discussed in more detail in Whitehead and Rosenberg (1993).

Typical features of the tree and cable matrices are illustrated in Figure 7.7. The practical determination of $\boldsymbol{S}, \boldsymbol{L}, \boldsymbol{X}, \boldsymbol{A}_{TS}, \boldsymbol{A}_{CS}$ and $\boldsymbol{A}_C$ is dependent on several iterative procedures.

4.1 Symmetrizing the Tree Matrix

The symmetrizing scale transformation is an invertible diagonal matrix of scaling factors. If a dendritic tree is represented by k numbered nodes then $\boldsymbol{S}=\delta_{ij}s_i$, where $0 \leqslant i, j \leqslant k-1$, $s_i \neq 0$ and δ_{ij} is the Kronecker delta function. Inversion of such a diagonal matrix is a simple task since $\boldsymbol{S}^{-1}=\delta_{ij}s_i^{-1}$.

If $\boldsymbol{A}_T=(a_{ij})$, then an element in row i and column j of the tree matrix is referenced as a_{ij}. A non-zero element, a_{mn}, where $n>m$, is above the diagonal and has a corresponding element in a symmetric position below the diagonal, a_{nm}. Symmetrization leaves the corresponding elements in $\boldsymbol{A}_{TS}$ equal, and the central diagonal untouched. The diagonal elements (scaling factors) of $\boldsymbol{S}$ are given by

$$s_n = s_m \sqrt{\frac{a_{mn}}{a_{nm}}}. \tag{7.45}$$

The layout and number of off-diagonal elements in the tree matrix ensures that symmetrization uniquely defines the s_i. The scale factors can be initialised by setting $s_0=1$.

The process of determining the s_i can be seen from the following simple example of $\boldsymbol{S}\boldsymbol{A}_T\boldsymbol{S}^{-1}=\boldsymbol{A}_{TS}$,

$$\begin{bmatrix} s_0 & 0 & 0 & 0 & 0 & 0 \\ 0 & s_1 & 0 & 0 & 0 & 0 \\ 0 & 0 & s_2 & 0 & 0 & 0 \\ 0 & 0 & 0 & s_3 & 0 & 0 \\ 0 & 0 & 0 & 0 & s_4 & 0 \\ 0 & 0 & 0 & 0 & 0 & s_5 \end{bmatrix} \begin{bmatrix} D & a_{01} & 0 & 0 & 0 & 0 \\ a_{10} & D & a_{12} & 0 & 0 & 0 \\ 0 & a_{21} & D & a_{23} & a_{24} & 0 \\ 0 & 0 & a_{32} & D & 0 & 0 \\ 0 & 0 & a_{42} & 0 & D & a_{45} \\ 0 & 0 & 0 & 0 & a_{54} & D \end{bmatrix} \begin{bmatrix} \frac{1}{s_0} & 0 & 0 & 0 & 0 & 0 \\ 0 & \frac{1}{s_1} & 0 & 0 & 0 & 0 \\ 0 & 0 & \frac{1}{s_2} & 0 & 0 & 0 \\ 0 & 0 & 0 & \frac{1}{s_3} & 0 & 0 \\ 0 & 0 & 0 & 0 & \frac{1}{s_4} & 0 \\ 0 & 0 & 0 & 0 & 0 & \frac{1}{s_5} \end{bmatrix}$$

$$= \begin{bmatrix} D & a_{01}\frac{s_0}{s_1} & 0 & 0 & 0 & 0 \\ a_{10}\frac{s_1}{s_0} & D & a_{12}\frac{s_1}{s_2} & 0 & 0 & 0 \\ 0 & a_{21}\frac{s_2}{s_1} & D & a_{23}\frac{s_2}{s_3} & a_{24}\frac{s_2}{s_4} & 0 \\ 0 & 0 & a_{32}\frac{s_3}{s_2} & D & 0 & 0 \\ 0 & 0 & a_{42}\frac{s_4}{s_2} & 0 & D & a_{45}\frac{s_4}{s_5} \\ 0 & 0 & 0 & 0 & a_{54}\frac{s_5}{s_4} & D \end{bmatrix} \tag{7.46}$$

Simply equating the two elements in symmetric positions so that

$$a_{ij}\frac{s_i}{s_j} = a_{ji}\frac{s_j}{s_i}, \tag{7.47}$$

gives (7.45).

4.2 Lanczos Tri-Diagonalization

The method of tri-diagonalization we use was introduced by Lanczos (1950). A general discussion and explanation of Lanczos tri-diagonalization and its startling convergence properties can be found in Golub and Van Loan (1990) and Paige (1976, 1980). It has been used extensively in nuclear physics (Whitehead et al., 1977). An outline of the process, with specific application to equivalent cables, and in particular the connected section, is given in Whitehead and Rosenberg (1993). Here we shall give a more thorough introduction to the tri-diagonalization, and indicate how the physical structure of a dendritic tree relates to it. Disconnected sections of equivalent cables are also given greater consideration. Although Lanczos tri-diagonalization does not provide a straight forward and automatic way of naturally generating disconnected sections, it is often possible with a little effort.

(Recently, Ogden et al. (1997) have introduced an alternative method of tri-diagonalization which uses a particular form of Householder transformation (Golub and Van Loan, 1990) that conveniently preserves essential matrix structure. This approach can naturally generate disconnected sections.)

The tri-diagonalizing matrix, $\boldsymbol{L}$, can be constructed simultaneously with $\boldsymbol{A}_{\mathrm{CS}}$. The standard unit vectors, $\boldsymbol{e}_i$, are useful, where

$$\boldsymbol{e}_0 = \begin{bmatrix} 1 \\ 0 \\ 0 \\ \vdots \\ 0 \\ 0 \end{bmatrix}, \quad \boldsymbol{e}_1 = \begin{bmatrix} 0 \\ 1 \\ 0 \\ \vdots \\ 0 \\ 0 \end{bmatrix}, \quad \cdots \quad \boldsymbol{e}_i = \begin{bmatrix} 0 \\ 0 \\ \vdots \\ 1 \\ \vdots \\ 0 \end{bmatrix}, \quad \cdots, \tag{7.48}$$

and $\boldsymbol{e}_i$ represents node i on the dendritic tree. A set of orthonormal *Lanczos vectors*, $\boldsymbol{u}_i$, can be constructed, where $\boldsymbol{u}_i$ represents node i on the equivalent cable.

First, choose an initial, normalised Lanczos vector, $\boldsymbol{u}_0$, which will be the unit vector that represents the origin. The origin node, though not necessarily number zero on the tree, will always be number zero on the cable. If the soma node is taken as origin then $\boldsymbol{u}_0 = \boldsymbol{e}_0$.

After pre-multiplying the first Lanczos vector by the symmetric tree matrix, $\boldsymbol{A}_{\mathrm{TS}}$, the result may be expressed as a linear combination of the first Lanczos vector ($\boldsymbol{u}_0$) and a new normalised Lanczos vector ($\boldsymbol{u}_1$), orthogonal to the first (see below). The process is then repeated, each time pre-multiplying the new Lanczos vector by $\boldsymbol{A}_{\mathrm{TS}}$, and choosing yet another normalised Lanczos vector, orthogonal to all previous Lanczos vectors, to satisfy the following tri-diagonal structure.

$$\begin{aligned}
\boldsymbol{A}_{\mathrm{TS}}\boldsymbol{u}_0 &= \alpha_0\boldsymbol{u}_0 + \beta_0\boldsymbol{u}_1,\\
\boldsymbol{A}_{\mathrm{TS}}\boldsymbol{u}_1 &= \beta_0\boldsymbol{u}_0 + \alpha_1\boldsymbol{u}_1 + \beta_1\boldsymbol{u}_2,\\
\boldsymbol{A}_{\mathrm{TS}}\boldsymbol{u}_2 &= \qquad\quad \beta_1\boldsymbol{u}_1 + \alpha_2\boldsymbol{u}_2 + \beta_2\boldsymbol{u}_3,\\
\boldsymbol{A}_{\mathrm{TS}}\boldsymbol{u}_3 &= \qquad\qquad\qquad \beta_2\boldsymbol{u}_2 + \alpha_3\boldsymbol{u}_3 + \beta_3\boldsymbol{u}_4,\\
\vdots &= \qquad\qquad\qquad\quad \ddots \qquad \ddots \qquad \ddots
\end{aligned}$$

The coefficients of the $\boldsymbol{u}_i$ are the elements of the tri-diagonal symmetric cable matrix. The α_i are all equal, unchanged from the tree matrix diagonals.

The algorithm will terminate when a vector $\boldsymbol{u}_m$ has been generated such that no new vector $\boldsymbol{u}_{m+1}$ is required to represent $\boldsymbol{A}_{\mathrm{TS}}\boldsymbol{u}_m$. Consequently, $\beta_m=0$.

If, when applied to a $k \times k$ symmetric tree matrix, the algorithm terminates when $m=(k-1)$, so that k Lanczos vectors have been produced, then we are finished. If the algorithm terminates prematurely, with fewer than k Lanczos vectors generated, the k-dimensional vector space splits into two or more invariant subspaces. The primary termination of the tri-diagonalization algorithm marks the termination of the connected section of the equivalent cable, described by the first $m \times m$ sub-matrix of $\boldsymbol{A}_{\mathrm{CS}}$.

Further invariant subspaces and associated sub-matrices describe disconnected sections of the equivalent cable and can be generated provided suitable new initial normalised Lanczos vectors, orthogonal to all previous Lanczos vectors, can be found. Unlike $\boldsymbol{u}_0$ for the connected section, the choice of new initial vector for a disconnected section is not immediately clear. A scheme that allows a new Lanczos vector to be chosen is discussed in section 4.2.1. Once a new Lanczos vector has been found, the tri-diagonalization process can be restarted and continued as before. The tri-diagonalisation is complete when k orthonormal Lanczos vectors ($\boldsymbol{u}_0$ to $\boldsymbol{u}_{k-1}$) have been generated, spanning the complete k-dimensional vector space.

The $\boldsymbol{u}_i$ are linear combinations of the $\boldsymbol{e}_i$ and this is the basis of the electrical mapping between the two structures. For example, the potential at a cable node will be a linear combination of potentials at certain tree nodes. The tri-diagonalization rearranges and combines the discrete cable equations of the dendritic tree.

Transformation $\boldsymbol{L}$ is a row-partitioned matrix of transposed orthonormal Lanczos vectors ($\boldsymbol{u}_i^{\mathrm{T}}$), so

$$\boldsymbol{L} = \begin{bmatrix} \boldsymbol{u}_0^{\mathrm{T}} \\ \boldsymbol{u}_1^{\mathrm{T}} \\ \boldsymbol{u}_2^{\mathrm{T}} \\ \cdot \\ \cdot \\ \cdot \\ \boldsymbol{u}_{k-1}^{\mathrm{T}} \end{bmatrix}. \tag{7.49}$$

Orthonormality of the Lanczos vectors implies $\boldsymbol{L}^{\mathrm{T}}=\boldsymbol{L}^{-1}$, allowing simple inversion of $\boldsymbol{L}$.

The Lanczos vectors evolve in a fairly distinct manner, related to the connectivity of the dendritic tree nodes. We say a tree node, i, *contributes* to vector $\boldsymbol{u}_j$ if the ith element of $\boldsymbol{u}_j$ is non-zero.

By construction, the origin always contributes to the first Lanczos vector, $\boldsymbol{u}_0$. Any nodes connected to the origin may contribute to $\boldsymbol{u}_1$. Any nodes connected to those that may contribute to $\boldsymbol{u}_1$ may contribute to $\boldsymbol{u}_2$, and so on until primary termination. If disconnected sections exist, the evolution continues from each new initial vector. Since $\boldsymbol{u}_0$ is always a standard unit vector, orthogonality ensures that the origin never actually contributes to any other vector, thus the equivalent cable structure is independent of the origin nature (though of course it will depend on the location of the origin). The physical structure of a dendritic tree determines whether or not a node that *may* contribute actually *does* contribute to a Lanczos vector.

The evolution process (illustrated by the simple example in section 4.4) ensures that the equivalent cable structure that is j internodal lengths from the origin depends only on dendritic tree structure up to j internodal lengths from the origin. Furthermore, there must be an even number of internodal electrotonic lengths between all nodes contributing to a particular Lanczos vector.

In certain cases, for example where one is interested only in the character of the transmembrane potential seen at the origin (typically the soma), the connected section of an equivalent cable contains all the required information. Denote the matrix of the first m transposed Lanczos vectors by $\boldsymbol{L}_m$, so

$$\boldsymbol{L}_m=\begin{bmatrix} \boldsymbol{u}_0^{\mathrm{T}} \\ \boldsymbol{u}_1^{\mathrm{T}} \\ \cdot \\ \cdot \\ \boldsymbol{u}_{m-1}^{\mathrm{T}} \end{bmatrix}. \tag{7.50}$$

This can be used to generate just the connected section of the equivalent cable, represented by the first $m\times m$ symmetric cable sub-matrix. The required de-symmetrization transformation would then be $\boldsymbol{X}_m$, the first $m\times m$ sub-matrix of $\boldsymbol{X}$. The matrix equation for the connected section is

$$(\boldsymbol{X}_m\boldsymbol{L}_m\boldsymbol{S}\boldsymbol{A}_{\mathrm{T}}\boldsymbol{S}^{-1}\boldsymbol{L}_m^{\mathrm{T}}\boldsymbol{X}_m^{-1})(\boldsymbol{X}_m\boldsymbol{L}_m\boldsymbol{S}\boldsymbol{v}_{\mathrm{T}})=\frac{\partial(\boldsymbol{X}_m\boldsymbol{L}_m\boldsymbol{S}\boldsymbol{v}_{\mathrm{T}})}{\partial T}+(\boldsymbol{X}_m\boldsymbol{L}_m\boldsymbol{S}\boldsymbol{g}_{\mathrm{T}})+(\boldsymbol{X}_m\boldsymbol{L}_m\boldsymbol{S}\boldsymbol{b}_{\mathrm{T}}). \tag{7.51}$$

The electrical mapping from tree to cable connected section is then defined by $\boldsymbol{M}_m=\boldsymbol{X}_m\boldsymbol{L}_m\boldsymbol{S}$. The reverse mapping is defined by $\boldsymbol{M}_m^{-1}=\boldsymbol{S}^{-1}\boldsymbol{L}_m^{\mathrm{T}}\boldsymbol{X}_m^{-1}$.

4.2.1 *Y-junctions and Disconnected Sections*

If the Lanczos tri-diagonalization terminates prematurely, after $m<k$ Lanczos vectors have been generated, then at least one disconnected section will exist.

However, if a complex branched dendritic tree is treated as a single unit, it is not clear exactly how to choose a new vector. For a tree with several orders of branching, there are usually many possible new choices of orthonormal vector, most of which will not allow a proper physical interpretation of the resulting sub-matrix as a disconnected section. A first step towards simplifying the choice is to look at simple terminating substructure. In particular, we are interested in Y-junctions, which consist of two terminating branches that meet at a branch point.

Figure 7.8 illustrates the transformation of a dendritic tree into an equivalent cable, Y-junction by Y-junction. At each step a Y-junction is isolated and detached at a temporary origin point then transformed to an equivalent cable. The connected section is re-attached to the temporary origin point and the disconnected section associated with the Y-junction is kept aside. These disconnected sections play no further role in the subsequent Y-junction transformations. Successively more substructure is incorporated into the transformation until an equivalent cable is fully generated.

The complete Lanczos procedure, including the de-symmetrization described later, must be applied to each Y-junction. (In fact, several computational shortcuts may be taken, however, here we are interested in clarity of description rather than efficiency of algorithm.) Before transformation, the Y-junction's nodes must be renumbered from zero. When the connected section is re-attached, the nodes that form the intermediate dendritic structure must be renumbered to ensure consecutive numbering. It is essential that we keep track of the correct electrical mapping from

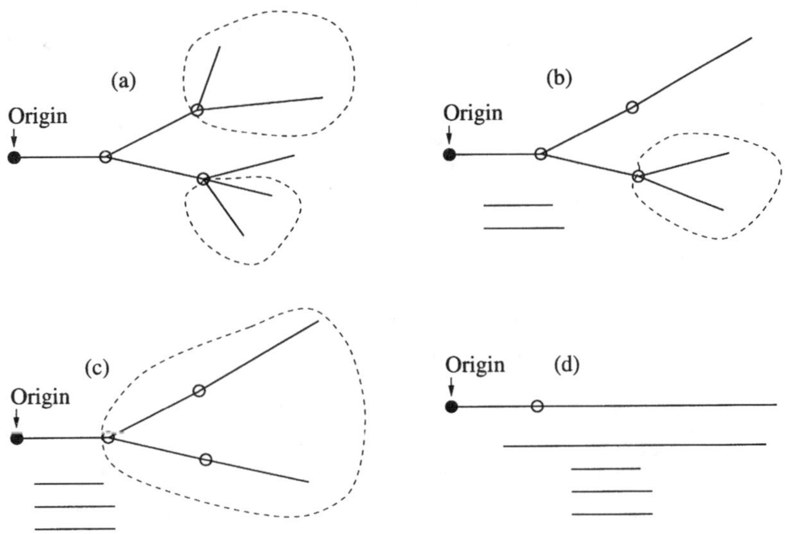

Figure 7.8 The Y-junction approach to transforming a dendritic tree into an equivalent cable. (a)–(d) indicate the repeated isolation and transformation of terminating Y-junctions. Disconnected sections are indicated and carried through each stage.

original tree to final equivalent cable. This is merely a record keeping task, ensuring that we can relate the nodes involved in each electrical mapping through all the intermediate structures. Eventually the matrices $\boldsymbol{X}, \boldsymbol{L}, \boldsymbol{A}_{\mathrm{CS}}$ and $\boldsymbol{A}_{\mathrm{C}}$ for the complete equivalent cable and the mapping from the original tree can be constructed from the matrices generated for each Y-junction.

The Y-junction approach is perfectly acceptable because the branch point nature of the temporary origin (like any origin) does not influence construction of the Y-junction's equivalent cable. The intermediate dendritic structures generated by the process are, as seen by the origin, electrically equivalent.

In fact, any terminating substructure may be isolated, transformed and re-attached in this way. However, the Y-junction is the basic unit of branching dendritic structure and is the key to choosing new Lanczos vectors to restart the tri-diagonalization and generate disconnected sections.

If a Y-junction's equivalent cable has a disconnected section, we need to initialise its construction with a new orthonormal Lanczos vector. It was clear for the connected section which node the initial vector should represent (the origin). We need to choose a new Lanczos vector that will also represent a terminal node on the equivalent cable. In fact, an approach that has proven effective (and which also follows from other transformations, Ogden and co-workers. (1996, 1997)) is to restart where the connected section finished, at the only (non-origin) equivalent cable terminal node we have knowledge of at this stage. It is possible to construct a new Lanczos vector which also represents this node—it incorporates the same boundary condition, but represents structure on the far side of the node, i.e. beyond the connected section.

If a connected section terminated prematurely with cable node $(m-1)$, represented by Lanczos vector $\boldsymbol{u}_{m-1}$, choose a new vector $\boldsymbol{u}_m$ where the only possible non-zero elements are at the same locations as non-zero elements in $\boldsymbol{u}_{m-1}$ but adjusted to ensure orthogonality with $\boldsymbol{u}_{m-1}$ as well as $\boldsymbol{u}_0$ to $\boldsymbol{u}_{m-2}$. So, if

$$\boldsymbol{u}_{m-1} = \begin{bmatrix} p_0 \\ p_1 \\ \cdot \\ \cdot \\ \cdot \\ p_{k-2} \\ p_{k-1} \end{bmatrix} \quad \text{and} \quad \boldsymbol{u}_m = \begin{bmatrix} q_0 \\ q_1 \\ \cdot \\ \cdot \\ \cdot \\ q_{k-2} \\ q_{k-1} \end{bmatrix}, \tag{7.52}$$

then $p_i = 0$ implies $q_i = 0$. The nature of the Lanczos vector evolution ensures that solving for the q_i using

$$\boldsymbol{L}_m \boldsymbol{u}_m = 0 \tag{7.53}$$

will uniquely define orthonormal vector $\boldsymbol{u}_m$. The tri-diagonalization of the Y-junction's symmetric tree matrix can now be restarted.

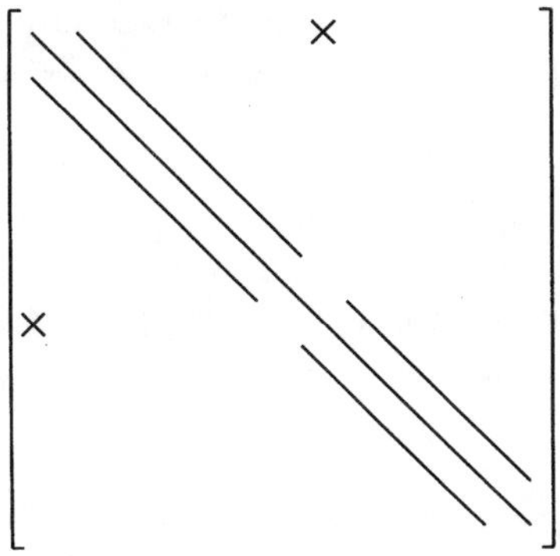

Figure 7.9 General form of the tree matrix for a Y-junction (provided numbering starts from the temporary origin). There are only two off-tri-diagonal elements (the crosses), with two corresponding zero elements on the upper- and lower-diagonals (the gaps).

The typical form of a Y-junction tree matrix is illustrated in Figure 7.9. There are only two off-tri-diagonal elements, corresponding to the single branch point. In fact, if we want an acceptable physical interpretation of equivalent cable sections, the Lanczos procedure cannot generate more than two invariant subspaces for such a matrix. This result relies on the simple structure of the Y-junction tree matrix and the nature of the Lanczos vector evolution. The equivalent cable for a Y-junction will consist of a connected section plus at most one disconnected section. Consequently, the equivalent cable for a dendritic tree can have no more disconnected sections than the number of Y-junctions that must be successively transformed before the equivalent cable is finally generated.

4.2.2 *Computational Lanczos Tri-diagonalization*

In practice, we require a computer implementation of the equivalent cable generation process. The Lanczos tri-diagonalization is the critical transformation, and its numerical stability determines the stability of the whole Lanczos procedure.

Lanczos tri-diagonalization is a powerful tool for finding extremal eigenvalues of large sparse symmetric matrices (Paige, 1980; Golub and Van Loan, 1990), although, due to inherent instability, it is not the preferred choice as a general method of tri-diagonalization. Other techniques (see Golub and Van Loan, 1990) offer more favourable numerical properties. Householder transformations may be used as an alternative method of tri-diagonalization (Ogden et al., 1997), though the speed/storage advantages offered by the Lanczos when constructing connected sections, particularly for realistic trees, are often more appealing.

Round-off errors and loss of orthogonality among the Lanczos vectors are the major concerns. The complexity of a dendritic tree, its boundary conditions, and the number of nodes by which it is represented, as well as the working significance of floating point arithmetic, will determine how much of an equivalent cable can be constructed before arithmetic errors begin to manifest themselves unacceptably. Double floating point arithmetic, at least, should be used when working with complicated structures.

When transforming a complicated tree there may be a loss of acceptable accuracy at distances far (greater than the maximum origin-to-terminal electrotonic distance) from the origin along the connected section, where the equivalent cable structure is usually less significant. A very good approximation to the connected section of an equivalent cable is usually generated.

Similar problems can arise for the Y-junction approach after several orders of branching have been transformed. Disconnected sections associated with the simpler substructures can often be generated, but errors may make it impossible to complete a connected section and thus choose a new Lanczos vector if a Y-junction is highly irregular and represented by a large numbers of nodes.

Four subtly different algorithms, along with an error analysis, are fully discussed by Paige (1972, 1976). In a limited test of complex trees, we have found no single algorithm to give significant advantage over the others when generating equivalent cables. A version of the algorithm Paige calls $A(1, 7)$ is given below. It is modified slightly, to our advantage, for equivalent cable generation since the central diagonal element is always known and need not be computed each iteration. This offsets, to some extent, any numerical instability.

The initial Lanczos vector is $\boldsymbol{u}_0$, and let $\boldsymbol{q}_0 = A_{\mathrm{TS}}\boldsymbol{u}_0$. The repeated central diagonal element is D. For $j = 0$ to termination, do steps (7.54) to (7.58):

$$\alpha_j = D(= \boldsymbol{u}_j^{\mathrm{T}} A_{\mathrm{TS}} \boldsymbol{u}_j), \tag{7.54}$$

$$\boldsymbol{w}_j = \boldsymbol{q}_j - \alpha_j \boldsymbol{u}_j, \tag{7.55}$$

$$\beta_{j+1} = +\sqrt{\boldsymbol{w}_j^{\mathrm{T}} \boldsymbol{w}_j}, \tag{7.56}$$

$$\boldsymbol{u}_{j+1} = \boldsymbol{w}_j / \beta_{j+1}, \tag{7.57}$$

$$\boldsymbol{q}_{j+1} = A_{\mathrm{TS}} \boldsymbol{u}_{j+1} - \beta_{j+1} \boldsymbol{u}_j. \tag{7.58}$$

If a Y-junction is being transformed and a disconnected section exists then, provided errors do not prevent an acceptable choice of new vector, the process can be restarted and the disconnected section generated.

Calculating the central diagonal using (7.54) and comparing to the known value, D, is a useful indicator of where the algorithm loses stability.

Rounding errors can become significant when $\boldsymbol{w}_j$ becomes small due to cancelation. Element β_{j+1} is then small, indicating a sudden narrowing of the cable. This could be considered an approximate termination point if errors are significant enough.

It may be possible to improve the process by re-orthogonalization, i.e. to combat loss of orthogonality, after each iteration the new Lanczos vector, $\boldsymbol{u}_{j+1}$, is forced to be orthogonal to all previous vectors $\boldsymbol{u}_0$ to $\boldsymbol{u}_j$.

4.3 Extracting the Equivalent Cable

If the Lanczos tri-diagonalization has been applied once, to the whole dendritic tree, only the connected section sub-matrix of the symmetric cable matrix will be available for de-symmetrization. If the Y-junction approach has been employed to generate both connected and disconnected sections, the de-symmetrization must be applied to the Y-junction symmetric cable matrix at each stage (computational shortcuts aside).

The de-symmetrizing scale transformation is $\boldsymbol{X}=\delta_{ij}x_i$, where $x_i \neq 0$. Matrices $\boldsymbol{X}$ and $\boldsymbol{A}_{\mathrm{C}}$, can be constructed simultaneously. Consider the symmetrization of $\boldsymbol{A}_{\mathrm{C}}$ to give $\boldsymbol{A}_{\mathrm{CS}}$, the reverse of what we want to do,

$$\boldsymbol{A}_{\mathrm{CS}}=\boldsymbol{X}^{-1}\boldsymbol{A}_{\mathrm{C}}\boldsymbol{X}. \tag{7.59}$$

If $\boldsymbol{A}_{\mathrm{C}}=(\bar{a}_{ij})$, then $\bar{a}_{(i+1)}$ and $\bar{a}_{(i+1)i}$ are two symmetrically located upper- and lower-diagonal elements of $\boldsymbol{A}_{\mathrm{C}}$. Two rules form the basis of the de-symmetrization process:

(i) Recall from (7.47) that the upper- and lower-diagonal elements in a matrix symmetrised by a scale transformation are given by

$$\beta_i=\bar{a}_{i(i+1)}\frac{x_{i+1}}{x_i}=\bar{a}_{(i+1)i}\frac{x_i}{x_{i+1}}. \tag{7.60}$$

Taking the product of the equal terms gives

$$\beta_i^2=\bar{a}_{i(i+1)}\bar{a}_{(i+1)i}. \tag{7.61}$$

Note that the reciprocals of the x_i replace the s_i in (7.47) since $\boldsymbol{X}$ is a de-symmetrizing, not a symmetrizing, transformation.

(ii) Apart from terminal nodes, only discrete cable equation (7.21) is required to represent an equivalent cable in matrix equation form. The sum of the two off-diagonal elements in the same row of the cable matrix must be $2/h^2$ (except possibly at either end of a cable section). This can be seen by summing the coefficients of V_{P1} and V_{C1}.

The de-symmetrization proceeds as follows:

Initialize the rescaling factors by setting $x_0=1$, ensuring consistency with scaling factor s_0. The first upper-diagonal element of the cable matrix, $\bar{a}_{01}$, is required. Effectively, this means the boundary condition type or discrete cable equation at the equivalent cable origin must already be known. The cable origin represents the same point as the dendritic tree origin and, in fact, the cable origin should always have a sealed end condition, so $\bar{a}_{01}=2/h^2$. If the tree origin was a terminal node, it should have been assigned a sealed end boundary condition; if the tree origin was a non-terminal node, the Lanczos automatically treats it as a sealed end condition (a consequence of the sum of elements either side of the diagonal, in the same row of the tree matrix, being $2/h^2$ for both non-terminal node and sealed end node discrete cable equations).

Now, since the symmetric cable matrix elements β_i should be known (at least for the connected section), rule (i) gives the first lower-diagonal element ($\bar{a}_{10}$) on the cable matrix as well as the next rescaling factor (x_1). Then rule (ii) gives the next upper-diagonal element ($\bar{a}_{12}$) and rule (i) gives the next lower-diagonal element ($\bar{a}_{21}$) and the next rescaling factor (x_2). The process continues like this until the connected section (the first $m \times m$ cable sub-matrix) of the equivalent cable has been constructed, along with $\boldsymbol{X}_m$.

The distal (from origin) terminal condition of the connected section will fall out naturally (numerical stability of the tri-diagonalization permitting) when the desymmetrization terminates. It will either be a current condition, described by (7.27), or a voltage condition, described by (7.29) or (7.32). Recall from section 3.4 that if a structure terminates with a voltage boundary condition, the terminal node is unnumbered and the condition is incorporated into the discrete cable equation of the node connected to the terminal node. (In fact we know from section 5.3 what the connected section terminal condition type should be—voltage, if there are any (non-origin) voltage terminal conditions on the dendritic tree; current, if all (non-origin) terminals have current conditions.)

If the cable matrix for a Y-junction is being constructed, and a disconnected section sub-matrix of the symmetric cable matrix is available for de-symmetrization, then the algorithm used to generate the connected section must be restarted. If the Lanczos tri-diagonalization was restarted using the scheme outlined in section 4.2.1, the disconnected section starts with the same condition that the connected section terminated with.

If the disconnected section starts with a current condition, and is not directly connected to a terminal node with a cut end condition, then the first upper diagonal of its sub-matrix, from (7.28), must be $2/h^2$. If the current condition node is connected directly to a cut condition node, (7.33) will apply (just a diagonal element).

If the disconnected section starts with a voltage condition, the first upper diagonal should be $1/h^2$, provided the discretization is set to $H = zh$ where $z \geqslant 2$, as in Eq. (30). If the extra discretization was not in place the first upper diagonal would not be clearly defined since it is determined by the diameters either side of the node connected to the terminal node, as in (7.32).

If a Y-junction's equivalent cable terminates prematurely after m nodes (at cable node $m-1$), restart the de-symmetrization at cable node m with the known upper diagonal of the disconnected section and set $x_m = 1$ (or any non-zero positive value!).

4.3.1 *Equivalent Cable Cylinder Diameters*

Once a section of equivalent cable has been found in its cable matrix form, c-value ratios for neighbouring cylinders in the equivalent cable can be determined. Simply compare the discrete cable equations to known discrete cable equations—most importantly, (7.21) for non-terminal nodes, and also (7.32) for nodes attached to voltage terminal nodes. Discrete cable equations (7.27) and (7.28) for current boundary conditions plus (7.29) and (7.30) for voltage boundary conditions give no c-value information.

We need to know only one cable cylinder c-value so that all others can be determined from the c-value ratios. The c-value of the cylinder connected to the equivalent cable origin is always the sum of c-values of the cylinders that emanate from the dendritic tree origin. This simply expresses the fact that input impedance is the same at both tree and cable origin. We can also justify this c-value by noting that the structure of the dendritic tree just beyond the origin does not violate any of Rall's conditions for an equivalent cylinder (see section 5.2). Only tree structure up to electrotonic distance X from the origin will influence the equivalent cable at electrotonic distance X from the origin. The c-value of the cylinder connected to the cable origin must be the same as that for the equivalent cylinder, i.e. the sum of c-values of the cylinders emanating from the tree origin. The equivalent cable and its electrical mapping will be correct if, for each disconnected section, an arbitrary non-zero c-value is chosen for one cylinder and the rest evaluated from c-value ratios.

It is a simple task to convert electrotonic lengths and times to physical lengths and times using the equivalent cable diameters (from c-values) and (7.2).

We now have the equivalent cable shape (its cylinder lengths and diameters). The electrical mapping ($\boldsymbol{M}=\boldsymbol{XLS}$) completes the equivalent cable.

4.3.2 *Mapping Node Potentials and Applied Currents*

Node potentials are mapped directly by $\boldsymbol{M}$. The same mapping generates simple expressions for the cable applied current terms. Elements of $\boldsymbol{g}_{\mathrm{C}}$ are of the form $2G_n^{\mathrm{C}}/C_n^{\mathrm{C}}$, where G_n^{C} denotes the applied current term at node n on the equivalent cable and C_n^{C} is the c-value sum of cylinder structure that meets at node n. These are linear combinations of similar terms for the original tree, so, if there are k nodes on the tree, and $\boldsymbol{M}=(M_{ij})$, then

$$\frac{G_n^{\mathrm{C}}}{C_n^{\mathrm{C}}}=\sum_{i=0}^{k-1} M_{ni}\frac{G_i}{C_i}, \tag{7.62}$$

from which the G_n^{C} are easily extracted.

An electrical mapping from a disconnected section to the dendritic tree may be scaled by any amount, the magnitude being irrelevant since such configurations of electrical activity will have no net effect at the origin.

4.3.3 *Mapping Voltage Boundary Conditions*

Vector $\boldsymbol{b}_{\mathrm{C}}$ can contain non-zero voltage boundary condition information for unnumbered terminal nodes on the equivalent cable. It can also contain information regarding permanent contributions to the potentials of numbered nodes. These contributions must be added to the corresponding node potential. We just outline the procedure for correct interpretation of $\boldsymbol{b}_{\mathrm{C}}$:

Similarly to the dendritic tree, an equivalent cable terminal voltage condition or permanent potential contribution is incorporated into the discrete cable equation of a node connected to the node where the condition really applies. If a voltage condition appears in equivalent cable discrete cable equation n, then the condition should be re-associated with node $(n+1_n)$ or $(n-1_n)$. We must determine which is the correct node and relocate the voltage condition accordingly. This is possible provided $H=zh$ where $z\geqslant 2$. Contributions from different terminal tree nodes to the same element of $\boldsymbol{b}_{\mathrm{C}}$ should be kept separate, since they might be handled differently. In fact, voltage inputs on non-terminal equivalent cable nodes due to one dendritic tree boundary condition will occur in pairs, on nodes either side of the correct one. If these two conditions appear in the discrete cable equations for nodes $(n+1_n)$ and $(n-1_n)$, replace them with a single condition by shifting both to the correct node, n. Each dendritic tree voltage condition can produce at most one contribution to an equivalent cable terminal voltage boundary condition so simply shift the condition to the unnumbered terminal node.

4.4 Example of the Lanczos Procedure

We illustrate the Lanczos procedure with a simple algebraic example. Consider the Y-junction in Figure 7.6(a), described by matrix equation (7.36). Set $h=1$ for convenience. The branches have a total electrotonic length of 6, and so the equivalent cable must have electrotonic length 6. A discretization level of $H=h$ in this example will allow us to generate the complete equivalent cable. Set $G_i=0$ except for current boundary conditions G_5 and G_6.

The tree matrix is

$$A_{\mathrm{T}}=\begin{bmatrix} D & 2 & 0 & 0 & 0 & 0 & 0 \\ 1 & D & 1 & 0 & 0 & 0 & 0 \\ 0 & \frac{2c_{\mathrm{P}}}{C_2} & D & \frac{2c_{\mathrm{R}}}{C_2} & 0 & 0 & \frac{2c_{\mathrm{L}}}{C_2} \\ 0 & 0 & 1 & D & 1 & 0 & 0 \\ 0 & 0 & 0 & 1 & D & 1 & 0 \\ 0 & 0 & 0 & 0 & 2 & D & 0 \\ 0 & 0 & 2 & 0 & 0 & 0 & D \end{bmatrix}. \tag{7.63}$$

After symmetrization, using (7.45) and (7.47), the tree matrix becomes

$$A_{\mathrm{TS}}=\begin{bmatrix} D & \sqrt{2} & 0 & 0 & 0 & 0 & 0 \\ \sqrt{2} & D & \sqrt{\frac{2c_{\mathrm{P}}}{C_2}} & 0 & 0 & 0 & 0 \\ 0 & \sqrt{\frac{2c_{\mathrm{P}}}{C_2}} & D & \sqrt{\frac{2c_{\mathrm{R}}}{C_2}} & 0 & 0 & \sqrt{\frac{4c_{\mathrm{L}}}{C_2}} \\ 0 & 0 & \sqrt{\frac{2c_{\mathrm{R}}}{C_2}} & D & 1 & 0 & 0 \\ 0 & 0 & 0 & 1 & D & \sqrt{2} & 0 \\ 0 & 0 & 0 & 0 & \sqrt{2} & D & 0 \\ 0 & 0 & \sqrt{\frac{4c_{\mathrm{L}}}{C_2}} & 0 & 0 & 0 & D \end{bmatrix}, \tag{7.64}$$

with symmetrizing scale transformation

$$S=\begin{bmatrix} 1 & 0 & 0 & 0 & 0 & 0 & 0 \\ 0 & \sqrt{2} & 0 & 0 & 0 & 0 & 0 \\ 0 & 0 & \sqrt{\frac{C_2}{c_P}} & 0 & 0 & 0 & 0 \\ 0 & 0 & 0 & \sqrt{\frac{2c_R}{c_P}} & 0 & 0 & 0 \\ 0 & 0 & 0 & 0 & \sqrt{\frac{2c_R}{c_P}} & 0 & 0 \\ 0 & 0 & 0 & 0 & 0 & \sqrt{\frac{c_R}{c_P}} & 0 \\ 0 & 0 & 0 & 0 & 0 & 0 & \sqrt{\frac{c_L}{c_P}} \end{bmatrix}. \tag{7.65}$$

The Lanczos tri-diagonalization can now be applied.
Take the soma, node 0, as the origin. The initial Lanczos vector is $\boldsymbol{u}_0=\boldsymbol{e}_0$, and then

$$A_{TS}\boldsymbol{u}_0=D\boldsymbol{u}_0+\sqrt{2}\boldsymbol{e}_1.$$

Set $\boldsymbol{u}_1=\boldsymbol{e}_1$ to satisfy the tri-diagonal structure and orthonormality of the $\boldsymbol{u}_i$.

$$A_{TS}\boldsymbol{u}_1=\sqrt{2}\boldsymbol{u}_0+D\boldsymbol{u}_1+\sqrt{(2c_P/C_2)}\boldsymbol{e}_2.$$

Set $\boldsymbol{u}_2=\boldsymbol{e}_2$ to satisfy the tri-diagonal structure and orthonormality of the $\boldsymbol{u}_i$.
Non-tri-diagonal structure is now encountered in the matrix, so

$$A_{TS}\boldsymbol{u}_2=\sqrt{(2c_P/C_2)}\boldsymbol{u}_1+D\boldsymbol{u}_2+\sqrt{(2c_R/C_2)}\boldsymbol{e}_3+\sqrt{(4c_L/C_2)}\boldsymbol{e}_6.$$

Choosing $\boldsymbol{u}_3=\sqrt{c_R/(c_R+2c_L)}\boldsymbol{e}_3+\sqrt{(2c_L/(c_R+2c_L))}\boldsymbol{e}_6$ ensures that

$$A_{TS}\boldsymbol{u}_2=\sqrt{(2c_P/C_2)}\boldsymbol{u}_1+D\boldsymbol{u}_2+\sqrt{(2/C_2)(c_R+2c_L)}\boldsymbol{u}_3.$$

$$A_{TS}\boldsymbol{u}_3=\sqrt{(2/C_2)(c_R+2c_L)}\boldsymbol{u}_2+D\boldsymbol{u}_3+\sqrt{c_R/(c_R+2c_L)}\boldsymbol{e}_4.$$

Set $\boldsymbol{u}_4=\boldsymbol{e}_4$.

$$A_{TS}\boldsymbol{u}_4=1\boldsymbol{e}_3+D\boldsymbol{u}_4+\sqrt{2}\boldsymbol{e}_5.$$

This must be rewritten to satisfy symmetry—there must be a term $\sqrt{c_R/(c_R+2c_L)}\boldsymbol{u}_3$.

After a small effort rearranging the terms,

$$A_{TS}u_4=\sqrt{c_R/(c_R+2c_L)}u_3+Du_4+\sqrt{2(c_R+3c_L)/(c_R+2c_L)}u_5,$$

where

$$u_5=\frac{c_L\sqrt{2}}{\sqrt{(c_R+2c_L)(c_R+3c_L)}}e_3+\sqrt{\frac{(c_R+2c_L)}{(c_R+3c_L)}}e_5-\sqrt{\frac{c_Rc_L}{(c_R+2c_L)(c_R+3c_L)}}e_6.$$

At the next iteration the algorithm terminates,

$$A_{TS}u_5=\sqrt{\frac{2(c_R+3c_L)}{(c_R+2c_L)}}u_4+Du_5.$$

No new Lanczos vectors need be generated for the connected section. One more normalized Lanczos vector, orthogonal to all the previous six, must be found so we may complete the symmetric cable matrix. Vector u_6 must be similar to u_5, with coefficients of e_3 and e_6 having the same relative size to ensure orthogonality with u_0 to u_4. Just change the relative size of the coefficient of e_5, so that u_6 can be made orthogonal to u_5. If N is a normalization factor and K is a constant then

$$u_6=N(e_3+Ke_5-\sqrt{(c_R/2c_L)}e_6).$$

The dot product of u_5 and u_6 must be zero so $K=-1/\sqrt{2}$ and, after normalization,

$$u_6=\sqrt{2c_L/(c_R+3c_L)}e_3-\sqrt{c_L/(c_R+3c_L)}e_5-\sqrt{c_R/(c_R+3c_L)}e_6.$$

Finally,

$$A_{TS}u_6=Du_6.$$

The symmetric cable matrix is, from the expressions for $A_{TS}u_i$,

$$A_{CS}=\begin{bmatrix} D & \sqrt{2} & 0 & 0 & 0 & 0 & 0 \\ \sqrt{2} & D & \sqrt{\frac{2c_P}{C_2}} & 0 & 0 & 0 & 0 \\ 0 & \sqrt{\frac{2c_P}{C_2}} & D & \sqrt{\frac{2}{C_2}(c_R+2c_L)} & 0 & 0 & 0 \\ 0 & 0 & \sqrt{\frac{2}{C_2}(c_R+2c_L)} & D & \sqrt{\frac{c_R}{(c_R+2c_L)}} & 0 & 0 \\ 0 & 0 & 0 & \sqrt{\frac{c_R}{(c_R+2c_L)}} & D & \sqrt{\frac{2(c_R+3c_L)}{(c_R+2c_L)}} & 0 \\ 0 & 0 & 0 & 0 & \sqrt{\frac{2(c_R+3c_L)}{(c_R+2c_L)}} & D & 0 \\ 0 & 0 & 0 & 0 & 0 & 0 & D \end{bmatrix}, \tag{7.66}$$

and the matrix of Lanczos vectors is

$$\boldsymbol{L}=\begin{bmatrix} 1 & 0 & 0 & 0 & 0 & 0 & 0 \\ 0 & 1 & 0 & 0 & 0 & 0 & 0 \\ 0 & 0 & 1 & 0 & 0 & 0 & 0 \\ 0 & 0 & 0 & \sqrt{\dfrac{c_\mathrm{R}}{(c_\mathrm{R}+2c_\mathrm{L})}} & 0 & 0 & \sqrt{\dfrac{2c_\mathrm{L}}{(c_\mathrm{R}+2c_\mathrm{L})}} \\ 0 & 0 & 0 & 0 & 1 & 0 & 0 \\ 0 & 0 & 0 & \dfrac{c_\mathrm{L}\sqrt{2}}{\sqrt{(c_\mathrm{R}+2c_\mathrm{L})(c_\mathrm{R}+3c_\mathrm{L})}} & 0 & \sqrt{\dfrac{(c_\mathrm{R}+2c_\mathrm{L})}{(c_\mathrm{R}+3c_\mathrm{L})}} & -\sqrt{\dfrac{c_\mathrm{R}c_\mathrm{L}}{(c_\mathrm{R}+2c_\mathrm{L})(c_\mathrm{R}+3c_\mathrm{L})}} \\ 0 & 0 & 0 & \sqrt{\dfrac{2c_\mathrm{L}}{(c_\mathrm{R}+3c_\mathrm{L})}} & 0 & -\sqrt{\dfrac{c_\mathrm{L}}{(c_\mathrm{R}+3c_\mathrm{L})}} & -\sqrt{\dfrac{c_\mathrm{R}}{(c_\mathrm{R}+3c_\mathrm{L})}} \end{bmatrix}. \tag{7.67}$$

The cable matrix can be extracted using rules (i) and (ii) from section 4.3. The first upper diagonal is 2, from (7.28) for the sealed end at node 0. Set $x_0=1$. The first three nodes on the connected form the same cylinder as those on the original dendritic tree. Cable nodes 2–4 are described by discrete cable equation (7.21). The connected section terminates at node 5, after length 5, with another sealed end, (7.27). The disconnected section must have length 1 for the equivalent cable to have total length 6. The final node must represent a terminal current condition and must be connected to a node with a cut end condition (as in (7.33)). There are no non-zero voltage boundary conditions since $\boldsymbol{b}_\mathrm{T}=0$.

The cable matrix is

$$\boldsymbol{A}_\mathrm{C}=\begin{bmatrix} D & 2 & 0 & 0 & 0 & 0 & 0 \\ 1 & D & 1 & 0 & 0 & 0 & 0 \\ 0 & \dfrac{2c_\mathrm{P}}{C_2} & D & \dfrac{2(c_\mathrm{R}+c_\mathrm{L})}{C_2} & 0 & 0 & 0 \\ 0 & 0 & \dfrac{(c_\mathrm{R}+2c_\mathrm{L})}{(c_\mathrm{R}+c_\mathrm{L})} & D & \dfrac{c_\mathrm{R}}{(c_\mathrm{R}+c_\mathrm{L})} & 0 & 0 \\ 0 & 0 & 0 & \dfrac{(c_\mathrm{R}+c_\mathrm{L})}{(c_\mathrm{R}+2c_\mathrm{L})} & D & \dfrac{(c_\mathrm{R}+3c_\mathrm{L})}{(c_\mathrm{R}+2c_\mathrm{L})} & 0 \\ 0 & 0 & 0 & 0 & 2 & D & 0 \\ 0 & 0 & 0 & 0 & 0 & 0 & D \end{bmatrix}, \tag{7.68}$$

with de-symmetrizing scale transformation

$$
\boldsymbol{X}=\begin{bmatrix}
1 & 0 & 0 & 0 & 0 & 0 & 0\\
0 & \frac{1}{\sqrt{2}} & 0 & 0 & 0 & 0 & 0\\
0 & 0 & \sqrt{\frac{c_P}{C_2}} & 0 & 0 & 0 & 0\\
0 & 0 & 0 & \frac{\sqrt{c_P(c_R+2c_L)}}{\sqrt{2}(c_R+c_L)} & 0 & 0 & 0\\
0 & 0 & 0 & 0 & \sqrt{\frac{c_P}{2c_R}} & 0 & 0\\
0 & 0 & 0 & 0 & 0 & \sqrt{\frac{c_P(c_R+2c_L)}{c_R(c_R+3c_L)}} & 0\\
0 & 0 & 0 & 0 & 0 & 0 & 1
\end{bmatrix}, \qquad (7.69)
$$

The c-value ratio for cylinders either side of a non-terminal equivalent cable node, n, is given by the ratio of elements in row n of the cable matrix, either side of the diagonal. The first c-value is c_P. The remaining cylinder sections are labelled A to D, where

$$
c_A=c_R+c_L, \quad c_B=\frac{c_R(c_R+c_L)}{(c_R+2c_L)}, \quad c_C=\frac{c_R(c_R+3c_L)}{(c_R+2c_L)}, \quad c_D=1. \qquad (7.70)
$$

The electrical mapping is

$$
\boldsymbol{M}=\boldsymbol{XLS}=\begin{bmatrix}
1 & 0 & 0 & 0 & 0 & 0 & 0\\
0 & 1 & 0 & 0 & 0 & 0 & 0\\
0 & 0 & 1 & 0 & 0 & 0 & 0\\
0 & 0 & 0 & \frac{c_R}{(c_R+c_L)} & 0 & 0 & \frac{c_L}{(c_R+c_L)}\\
0 & 0 & 0 & 0 & 1 & 0 & 0\\
0 & 0 & 0 & \frac{2c_L}{(c_R+3c_L)} & 0 & \frac{(c_R+2c_L)}{(c_R+3c_L)} & -\frac{c_L}{(c_R+3c_L)}\\
0 & 0 & 0 & 2 & 0 & -1 & -1
\end{bmatrix}. \qquad (7.71)
$$

The cable node potentials, denoted V_n^C for cable node n, are then

$$
\begin{aligned}
&V_0^C=V_0, \quad V_1^C=V_1, \quad V_2^C=V_2, \quad V_3^C=\frac{c_L}{c_A}V_3+\frac{c_R}{c_A}V_6, \quad V_4^C=V_4,\\
&V_5^C=\frac{2c_L}{(c_R+3c_L)}V_3+\frac{(c_R+2c_L)}{(c_R+3c_L)}V_5-\frac{c_L}{(c_R+3c_L)}V_6, \quad V_6^C=2V_3-V_5-V_6.
\end{aligned} \qquad (7.72)
$$

Again from the mapping, and using (7.62), g_C has non-zero elements at cable nodes 3, 5, and 6, so the non-zero applied current terms given by

$$G_3^C = \frac{2(c_R + c_L)}{(c_R + 2c_L)} G_6, \quad G_5^C = G_5 - \frac{c_R}{(c_R + 2c_L)} G_6, \quad G_6^C = G_5 + \frac{c_R}{c_L} G_6. \tag{7.73}$$

Note that the total applied current, and also total membrane surface area, is conserved between dendritic tree and equivalent cable connected section,

$$G_3^C + G_5^C = G_5 + G_6, \tag{7.74}$$

and

$$c_L + 3c_R = c_A + c_B + c_C, \tag{7.75}$$

while, for each connected section potential, the sum of the coefficients of the tree potentials is 1—a uniform tree potential maps to a uniform connected section potential. These features (characteristic of a sealed tree) can be readily understood from results in the next section.

Figure 7.10 shows the original dendritic tree and the equivalent cable that has just been generated.

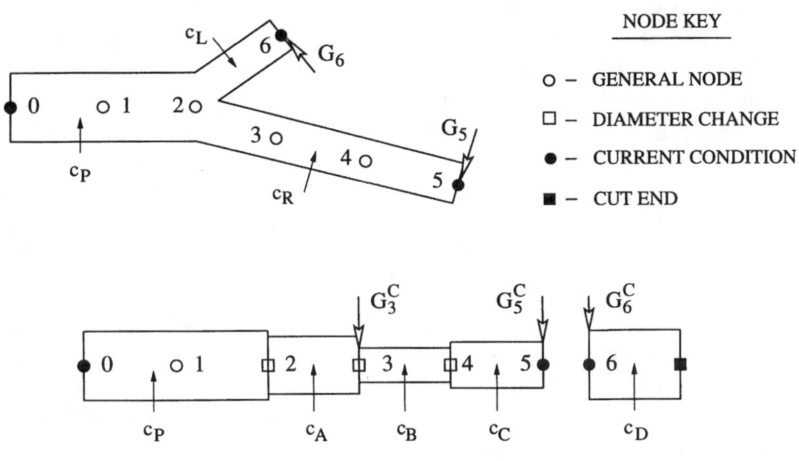

Figure 7.10 The simple Y-junction and its equivalent cable for the example in section 4.4. All lengths are electrotonic. The first cylinder, connected to the origin, is the same in both structures and has electrotonic length 2 and c-value c_P. Three cylinders each of length 1 complete the connected section, which terminates with a current boundary condition. The disconnected section, length 1, starts with a current condition and ends with a cut condition. The diameter of the disconnected section doesn't matter—set it to 1. Cable node 3 has a permanent current input term due to tree node 6. Terminal cable nodes 5 and 6 have current boundary conditions determined by both of the Y-junction's terminal nodes.

5. EXAMPLES OF EQUIVALENT CABLES

We present some examples of equivalent cables for a variety of artificially constructed dendritic trees; some important general features of equivalent cables and electrical mappings are discussed and explained. First we give a brief overview of some aspects of passive electrical signal propagation that will be useful in the discussion.

5.1 Passive Signal Propagation

The cable equation is essentially a linear diffusion equation, describing the spread of charges (and consequently voltage distribution) in a cylinder. Applied currents contribute a forcing term. Electrical input on a passive dendritic tree will, geometry permitting, propagate an effect both towards and away from the soma. These effects are carried by charges, which disperse in an attempt to bring the tree into electrical equilibrium. A transient current input, for example, causes a transient change in dendritic membrane potential at the input location. The magnitude of this effect decays in time, as charge moves both across the membrane (radially) and through the branching tree system (axially), causing changes in the membrane potential at distances further and further away from the input site. Dendritic geometry can have a major influence on dispersal of charge at points such as terminals, branch points and changes in diameter.

Impedance matching is a useful concept, though we shall not give the mathematical detail. (See discussion of input conductance in Rall, 1977.) Two adjacent structures on a dendritic tree are impedance matched if a charge imbalance that has accumulated at the contact point disperses equally into both. For example, any point on a uniform infinite cylinder represents an impedance matched structure and consequently an electrical input generates a voltage distribution that is symmetric about the input site. Impedances are not matched at terminals and abrupt diameter changes. At branch points, where more than two cylinder meet, it is possible that subsets of the cylinders are impedance matched. Where impedances are not matched, which is likely in real neurons, structures with lower impedances (larger diameters) receive more of the accumulated charge.

A cut end terminal may be regarded as a point where cylinder diameter jumps to infinity and consequently there is zero impedance to accumulated charge flowing out of the end. There is *signal loss* at cut ends. A sealed end terminal may be regarded as a point where cylinder diameter falls to zero and therefore has infinite impedance to charge flow. Charges cannot leak from a sealed end; instead they accumulate and propagate a disturbance in the potential away from the terminal. There is no signal loss at sealed ends, but rather *signal reflection.* The corresponding more general non-zero boundary conditions exhibit similar properties, subject to modification by the additional electrical activity at the terminal.

By regarding these particular terminal conditions in this way, we can treat every point on a dendritic tree as a change in diameter of some sort. This is the fundamental unifying feature in the dendritic tree model that allows equivalent cables to exist. This is manifested as the constant diagonal element in tree and cable matrices.

We start with a discussion of Rall's equivalent cylinder in the context of equivalent cables.

5.2 Rall's Equivalent Cylinder

A dendritic tree, represented by uniform cylinders which meet at branch points, is a *Rall tree* if it satisfies Rall's three conditions, namely

1. All (non-soma) terminal boundary conditions are the same.
2. All branches terminate at the same electrotonic distance, L, from the soma.
3. At any branch point the sum of child cylinder c-values equals the c-value of the parent cylinder. This is Rall's 3/2 power law for dendritic cylinder diameters. For the general branch point discrete cable equation (7.17) this condition can be expressed as $C_n = 2c_{\mathrm{P}}$.

Condition (3) is essentially an impedance-matching condition between a parent cylinder and its child cylinders. A Rall tree may be replaced by a uniform cylinder, attached to the soma. This cylinder exhibits the same impedance matching properties as the Rall tree. It has electrotonic length L, the same terminal condition as the tree terminals, and c-value equal to the sum of c-values of the cylinders connected to the soma. An electrical potential difference on a dendritic cylinder, distance X from the soma, is mapped to the same distance on the equivalent cylinder, the magnitude weighted by the ratio of dendritic cylinder c-value to equivalent cylinder c-value. The potential on the equivalent cylinder at electrotonic distance X from the soma is mapped to the tree such that each point at a distance X from the soma has the same potential as that on the cable. An input current on a Rall tree, at distance X from the soma, is also mapped to distance X on the cable, with no weighting.

Rall's equivalent cylinder represents a very restricted subset of equivalent cables, with the soma taken as origin, and disconnected sections (of which there will be at least one) ignored. In fact the number of disconnected sections, as well as their electrotonic structure, is very much dependent upon a Rall tree's branching structure. Rall's equivalent cylinder, while physically equivalent to a Rall tree (as far as the soma is concerned) is not a complete mathematically equivalent structure, whereas the equivalent cable for the Rall tree is both physically and mathematically equivalent.

Electrical inputs on Rall trees exhibit a kind of independence from tree geometry (see Tuckwell, 1988a). Given the response at the soma to an input somewhere on the tree, only the electrotonic distance of the input from the soma may be inferred, not its precise location on the tree. The same current input at two different locations on the tree, equidistant from the soma, will give rise to the same response at the soma. The soma cannot distinguish such inputs, rendering the fine geometrical detail of the tree functionally uninteresting. Branching only serves to provide surface area for synapses.

Rall trees represent the ultimate in symmetry for a dendritic tree, making the equivalent cylinder a useful reference when discussing equivalent cables for dendritic trees that break the generally unrealistic Rall conditions.

An empirical extension of Rall's result to dendritic trees that are not consistent with Rall's conditions has been postulated (Fleshman et al., 1988). The c-value at electrotonic length X from the soma, on a new unbranching structure, is given by a summation of the c-values of all tree branches at electrotonic length X from the soma. The division of the tree into small steps of electrotonic length allows a piecewise construction of the new c-values, moving from soma to the maximum soma-to-tip electrotonic length, where the new structure terminates. Call this structure the *piecewise cable*. Boundary conditions are not taken into account, and the piecewise cable is short compared to the equivalent cable. No electrical mapping can be generated. Later we state a modification to Rall's conditions that must hold for the piecewise cable and equivalent cable connected section to have the same shape.

5.3 Illustrated Examples and General Results

Electrical activity on a dendritic tree will propagate a resultant disturbance in the potential (or in some cases no net disturbance) to the origin. Equivalent cable

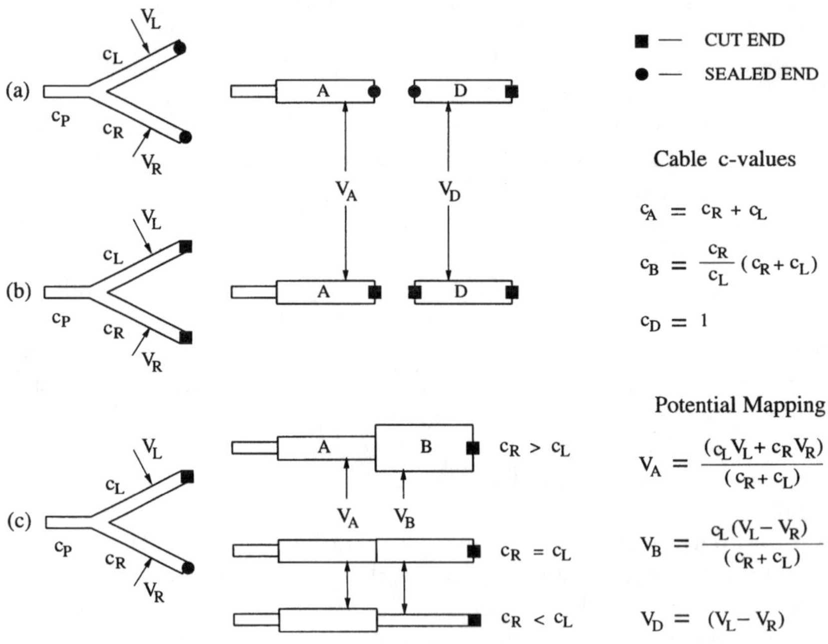

Figure 7.11 Simple algebraically constructed Y-junction examples. Each tree in (a)–(c) consists of one parent and two child branches. Both child branches terminate after electrotonic length 1 (arbitrary unit). The c-values are given and input potentials on the child branches, equidistant from the origin, are mapped on to the equivalent cables. The trees differ only in the boundary conditions assigned to the end of each child cylinder. Each equivalent cable is made up of the parent cylinder and two new cylinders, each with electrotonic length 1.

cylinders must have just the right electrotonic length, diameters and boundary conditions such that the corresponding mapped equivalent cable activity will generate the same effect at the origin. In contrast with Rall's equivalent cylinder, the electrical mapping, in either direction, is generally one-to-many. An input on a dendritic tree will typically map to many inputs on the equivalent cable, and vice-versa.

We shall discuss three sets of illustrated examples (Figures 7.11–7.13) that indicate how violations of Rall's conditions are manifested in equivalent cable structure. Only sealed end and cut end terminal boundary conditions are considered. Non-zero boundary conditions would give rise to the same shape of equivalent cable but with permanent contributions to applied currents and voltages, plus non-zero boundary conditions. The results we obtain for the zero conditions are equally valid for the

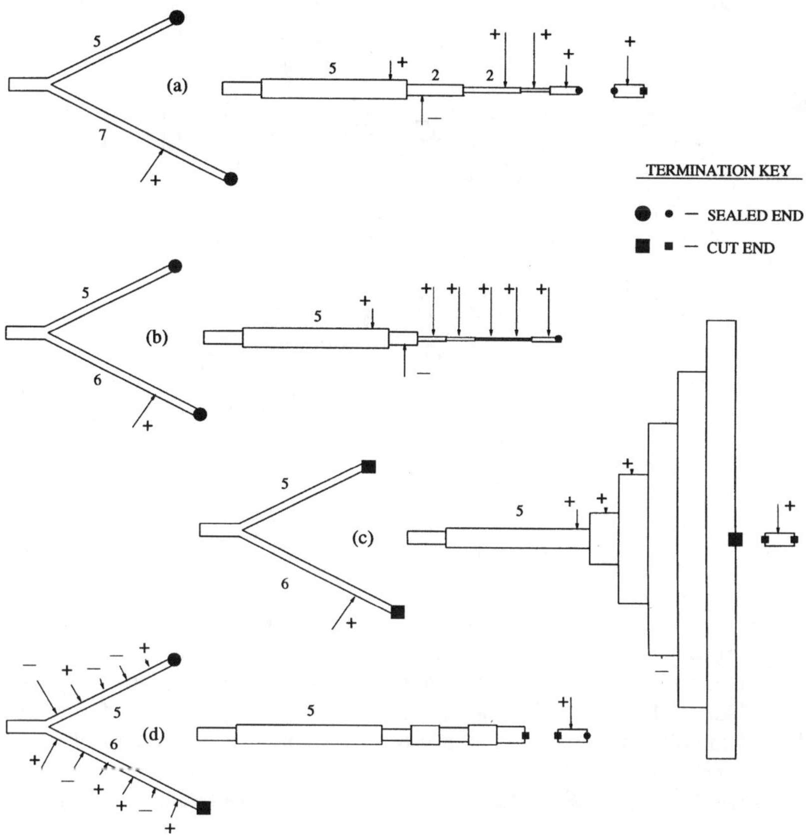

Figure 7.12 Y-junction dendritic trees where the electrotonic lengths of child branches are unequal. Electrotonic lengths are indicated for cylinders longer than electrotonic length 1 (arbitrary unit). Potential mappings between each tree and its equivalent cable are indicated. The sizes of the arrows indicate the strength of the mapping at each point.

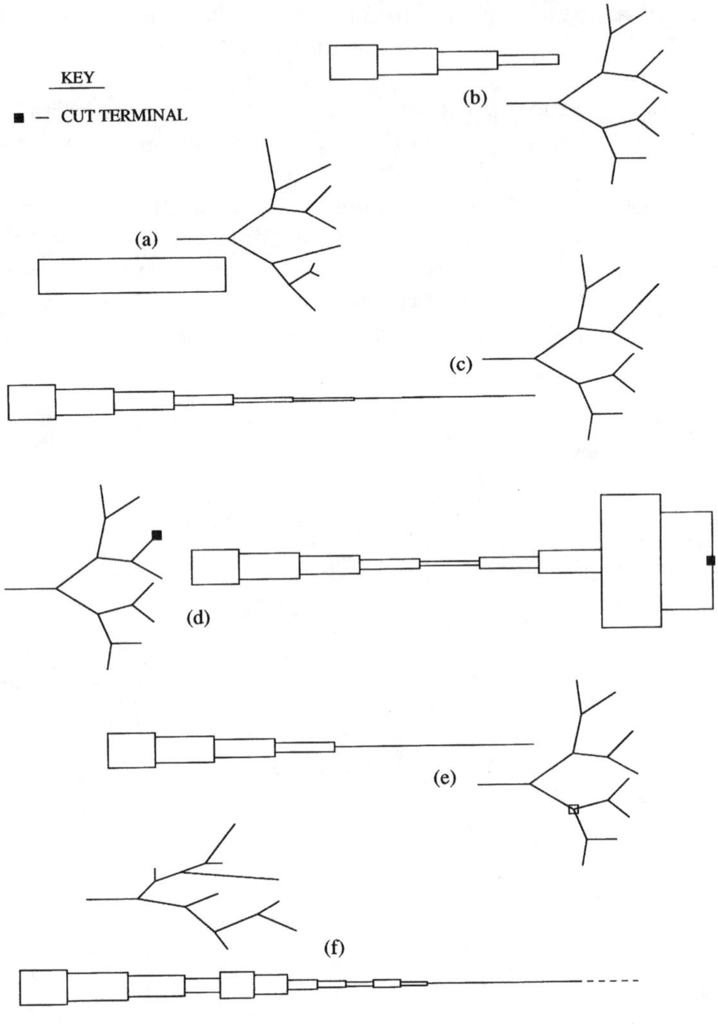

Figure 7.13 The effect of systematic departures from the three Rall conditions and condition (4) on the shape of equivalent cables for dendritic trees with several orders of branching. All terminals are sealed ends unless indicated otherwise. (a) Simple Rall tree. (b) Rall condition (3) is violated, but new condition (4) is still satisfied. (c) Same tree as (b) but Rall condition (2) is violated by one branch. (d) Same tree as (b) but Rall condition (1) is violated at one terminal. (e) Same tree as (b) but new condition (4) is violated at the location indicated. (f) More realistic dendritic tree violating all conditions except (1).

generalized non-zero conditions, subject to some modification of equivalent cable electrical activity. All lengths indicated in the diagrams are integer multiples of some arbitrary unit of electrotonic length. The origin is at the left-hand side of each structure.

5.3.1 *Boundary Condition Asymmetry*

First consider the simple Y-junctions in Figures 7.11(a)–(c) where the left branch and right branch (with potential inputs V_L and V_R equidistant from the origin) have the same electrotonic length. It is easy to apply the Lanczos procedure algebraically to these structures. Expressions for the equivalent cable diameters and potential mapping are also given in Figure 7.11.

Y-junctions 7.11(a) and 7.11(b) both obey Rall conditions (1) and (2). Although condition (3) is violated at the single branch point, for such simple structures this does not affect the part of the equivalent cable that is of particular interest, i.e. that beyond the parent branch. The connected section (parent cylinder plus cylinder A) terminates early, with a sealed end (Figure 7.11(a)) or cut end (Figure 7.11(b)) condition. In both cases a disconnected section completes the equivalent cable. The disconnected section (cylinder D) starts with the condition that the connected section terminated with and in both cases terminates in a cut end. The connected section has a symmetric (additive) potential mapping, weighted by the branch c-values, while the disconnected section has an antisymmetric (subtractive) mapping.

Consider the situation where V_L and V_R are non-zero and $c_L V_L = -c_R V_R$. The potential on the connected section, V_A, will be zero, so there can be no net disturbance in the potential at the origin as a consequence of these inputs. The mapping onto the disconnected section is $V_D = V_L(1 + c_L/c_R)$, which is non-zero. In effect, non-zero inputs that lie just on the disconnected section define non-zero input configurations on the dendritic tree that have no net effect at the origin. In both Y-junctions the two child branches are electrotonically identical. This symmetry ensures that signal dispersal is modified, in the same way, by the two identical boundary conditions and the two potential inputs (with opposite signs) contribute effects of equal magnitude at the origin, thus they cancel. The child cylinder c-values are different, but this merely imposes the weightings on V_L and V_R seen in V_A. In general, mapping a disconnected section potential to the dendritic tree requires the use of $\boldsymbol{v}_T = \boldsymbol{M}^{-1}\boldsymbol{v}_C$.

Y-junction 11(c) breaks the symmetry seen in Figures 7.11(a), (b) by violating Rall condition (1)—the left branch has a cut terminal while the right branch has a sealed terminal. The connected section (parent cylinder plus cylinders A and B) is the complete equivalent cable, and terminates in a cut end. The two boundary conditions affect signal propagation in different ways—there can be signal loss from the cut end and signal reflection from the sealed end. This boundary condition asymmetry allows the anti-symmetric combination of inputs that is ineffective in both 11(a) and 11(b) to be effective in 11(c). The absence of a disconnected section confirms that any non-zero input combination on the tree must have a net effect at the origin.

In Figures 7.11(a)–(c), cylinder A always has the same c-value because the boundary conditions can only influence cable structure at the end of or beyond cylinder A. The disconnected section electrical mappings will always be correct whatever c-value is chosen—multiply the mapping from disconnected section to tree by any real number and the configuration on the tree will still have no effect at the origin. The c-value c_B depends on the c-value ratio of the child cylinders and so dictates the final shape of the equivalent cable in Figure 7.11(c). We can conclude from Figure 7.11 that violation of Rall condition (1) (i.e. different boundary condition types) implies a severe breaking of symmetry in a dendritic tree.

5.3.2 *Branch Electrotonic Length Asymmetry*

The Y-junctions in Figure 7.12 provide further insight into the effects of boundary condition asymmetry and other general features that occur when branch lengths are unequal. In Figures 7.12(a)–(c), a unit potential mapping from each tree to its equivalent cable is indicated. In Figure 7.12(d), a potential on the disconnected section is mapped onto the original tree.

Trees 12(a)–(d) all violate Rall condition (2). It is immediately clear that this results in an equivalent cable with a connected section completely unlike Rall's equivalent cylinder, with multiple changes in diameter, and electrotonic length significantly greater than the maximum origin-to-tip electrotonic length. The length asymmetry of the two child branches ensures that the simple anti-symmetric input combination from Figure 7.11 does not lie on a disconnected section since electrical signals in different child cylinders do not undergo the same sort of modifications at the same distances from the origin. However, for certain trees, complicated input configurations can be found that will not affect the origin and so disconnected sections will still exist (Figure 7.12(d) for example).

The length of the connected section is longer than the maximum origin-to-tip electrotonic length, enabling a boundary condition at electrotonic length X from the origin to influence equivalent cable connected section structure beyond distance X from the origin. Unlike the simple examples in Figures 7.11(a), (b), even when both boundary conditions are the same the cable diameters in Figure 7.12 are highly dependent upon the boundary condition type (cut or sealed). When all terminals are sealed, as in Figures 7.11(a), (b) the equivalent cable connected section tapers with generally decreasing diameters and terminates with a sealed end. When all terminals are cut, as in Figure 7.12(c), moving out from the origin the equivalent cable's diameters become very large and the connected section terminates with a cut end.

In general, a sealed end promotes small diameters while a cut end promotes large diameters. When there is a mixture of boundary conditions, they will exert their influence in the order they are encountered during construction. Example 7.12(d) illustrates this—first a sealed end is incorporated, then, one length unit further from the origin, a cut end. The diameters are reasonably uniform for the whole equivalent cable, the effects of each condition tending to cancel.

The two boundary condition types have extreme and opposite effects on equivalent cable diameters. This can be understood in terms of signal loss and signal reflection at the dendritic tree terminals. The promotion of large diameters by cut conditions is consistent with the fact that accumulated charge leaks from cut terminals. The increasing equivalent cable diameters channel electrical signals away from the equivalent cable origin towards the cut end. The promotion of small diameters by sealed conditions is consistent with the fact that there is no signal loss from such a terminal. The decreasing cylinder diameters channel electrical signals back towards the equivalent cable origin, a result of full reflection at sealed terminals in the original tree.

Two different boundary conditions on the dendritic tree (one cut and one sealed) results in an equivalent cable connected section with a (non-origin) cut end. This ensures that accumulated charge may still leak from the system (other than through the membrane). Only if all non-origin terminals are sealed, ensuring no signal loss from terminals, will the connected section terminate in a sealed end. Imagine holding the entire tree membrane at some uniform non-zero (with respect to resting state) fixed potential (except perhaps in the vicinity of the origin). This is only possible where all (non-origin) terminals are sealed. The corresponding equivalent cable connected section must also have the same uniform potential (ensuring there is still no current flow in the system) and consequently a sealed end. Any (non-origin) cut end conditions disallow such a uniform non-zero distribution.

As illustrated in Figures 12(a)–(c), Y-junctions with different child branch lengths can have quite complicated electrical mappings. A unit positive voltage input at electrotonic distance X on the dendritic tree will map to a scaled input at distance X on the cable, plus multiple positive and negative input potentials further down the cable which describe the modulation of electrical activity at dendritic structure that is not impedance matched. Together, the electrical signals caused by these inputs undergo effectively the same modification in the equivalent cable cylinder-geometry as the electrical signal due to the single input on the original tree undergoes in the dendritic geometry. The effect at the soma is then exactly the same.

There is a clear relationship between equivalent cable cylinder diameter and the relative magnitude of the electrical mapping to that cylinder. For a unit potential input on the dendritic tree at distance X from the origin, invariably, beyond X, there is a large mapping to cylinders with small diameters and a small mapping onto cylinders with large diameters. Again this can be explained in terms of signal loss and signal reflection. Electrical signals undergo more reflections in a dendritic tree with sealed ends and a greater effect is seen at the origin—the strong potential mapping to small diameters mirror these reflections. Conversely, the lack of signal reflection at cut ends reduces the need for strong mapped potentials.

The strength of mapped applied currents will also depend of cylinder diameter. Where there is a large membrane surface area (large diameter), a large current will be required to depolarise it by a small amount, while only small currents are required to produce a significant depolarization across the membrane of thin cylinders. Consequently, the applied current mapping is small to small diameters, large to large diameter.

Obviously, changes in diameter within a dendritic tree will also have effects on equivalent cable structure. Preferential charge flow in directions of lower impedance will have similar, but very much reduced, effects to those due to the two terminal condition types.

From Figures 7.12(a)–(d), we may conclude that even a slight violation of Rall condition (2) implies a severe breaking of dendritic tree symmetry, for Y-junctions at least. The manifestation of the asymmetry in the equivalent cable is determined predominately by boundary condition types.

5.3.3 *Complicated Dendritic Geometry*

Finally, we use the dendritic geometries illustrated in Figure 7.13 to investigate the consequences of Rall condition violations on more complicated trees where there are several orders of branching. Only the connected sections are shown, but disconnected sections will exist for each of the equivalent cables. It is useful here to introduce a fourth condition, in addition to the three Rall conditions given in section 5.2. This condition is automatically satisfied by Rall trees.

4. At every point on a dendritic tree that is equidistant, at electrotonic length X, from the origin, Rall condition (3) is violated by the same fractional amount. Using the notation introduced with the general discrete cable equation (17) the value of c_P/C_n is the same for each tree node, n, located the same number of internodal electrotonic lengths from the origin node.

Figure 7.13(a) illustrates a Rall tree, where conditions (1), (2) and(3) are satisfied. If conditions (1), (2) and (4) are satisfied instead as in Figure 7.13(b), then the tree may still be replaced by a structure that has the same electrotonic length and boundary condition, but with fractional change in c-value at length X on the cable corresponding to the fractional change in c-value on the original tree at electrotonic length X from the origin. Conditions (1), (2) and (4) define where the piecewise cable empirical extension to Rall's conditions is valid—the piecewise cable has the same shape as the connected section of the equivalent cable. It represents a level of symmetry in the dendritic tree that is slightly less restrictive than that required for Rall's equivalent cylinder, and is perhaps a more useful reference when discussing complex trees.

It is quite easy to see why the simple connected section in Figure 7.13(b) is generated if we consider a Y-junction by Y-junction transformation of the dendritic tree. In each successive Y-junction, the two child branches have the same electrotonic length and boundary condition, and have the same fractional changes in diameter at the same distance from the origin because condition (4) is satisfied. The two branches have identical electrical properties. The actual c-values may be different, however this merely imposes a weighting on the relative contributions each branch makes to the electrical mapping, in a similar way to the simple Y-junctions in Figures 7.11(a), (b).

The remaining examples, in Figures 7.13(c)–(f), illustrate how violations of conditions (1), (2), and (4) in the dendritic tree from Figure 7.13(b) will affect the corresponding equivalent cable. In Figure 7.13(c), violation of condition (2) on just one branch results in an extended connected section, in a similar manner to the Y-junctions in Figure 7.12. In Figure 7.13(d), a single boundary condition violation is still a severe breaking of symmetry, with dramatic consequences for the connected section. In Figure 7.13(e), slight violation of new condition (4) has little effect on the connected section other than to extend it with a small insignificant tail. Significant multiple violations of conditions (2) and (4) throughout a tree will have significant consequences for equivalent cable shape, as illustrated in Figure 7.13(f).

These departures from Rall-like structures are more representative of realistic dendritic structure than the idealised cases satisfying conditions (2) and (4). The geometrical independence of inputs that is associated with Rall trees (and in fact is also valid for those trees that may be transformed to piecewise cables) does not hold and, consequently, finer details of dendritic geometry become important when trying to understand the functional properties of the tree.

6. DISCUSSION

We have presented a mathematical procedure that enables a multi-cylinder representation of an arbitrarily branched passive dendritic tree to be transformed to an unbranched, exactly equivalent cable. With respect to a point of origin, an equivalent cable is electrically identical to the original tree. The cable is made up of exactly one connected section plus a set of disconnected sections. The shape of the whole equivalent cable and the number (possibly zero) of disconnected sections depends on the geometry and terminal boundary conditions of the original dendritic tree. The combined electrotonic length of all cable sections equals the combined electrotonic length of all cylinders that form the tree. An electrical mapping relates transmembrane potentials and applied currents between the tree and the cable. This mapping makes the equivalent cable a truly equivalent structure.

In the Lanczos procedure, the one-dimensional linear cable equation, a second order partial differential equation, is employed to devise a set of discrete cable equations which can be used to form a matrix representation of a dendritic tree. Lanczos tri-diagonalization dominates the transformation to equivalent cable and a set of Lanczos vectors is a main component of the electrical mapping (an alternative method of tri-diagonalization has been illustrated by Ogden et al. (1997)). Diameter terms have been replaced with c-values (3/2 power of diameter), which are a more natural measure when dealing with cylinders in a passive electrical context. We have also illustrated some of the general properties of these equivalent cables. The results that have been presented here are also obtainable, although often very painfully, by the application of analytical methods (Ogden and Lindsay, 1996).

Equivalent cables are physically discontinuous structures (abrupt diameter changes), an unavoidable consequence of branching structure where Rall's 3/2 power law for diameters does not hold. This is at odds with the continuous 'equivalent' cables originated by Rall (1962). The discontinuous 'equivalent' cables of Fleshman et al. (1988), Clements and Redman (1989), Stratford et al. (1989) and Manor et al. (1991), are rarely truly equivalent in the sense defined above, yet share the discontinuous nature of the equivalent cable diameters.

The true equivalence of Lanczos-generated equivalent cables is the distinguishing feature that sets them apart from these earlier models. Equally interesting are the disconnected sections which, while in some ways mathematically obvious and essential, are not physically required as far as the origin is concerned. If there are any configurations of electrical input that do not affect the origin then they are implicit in the shape and electrical mapping of the connected section. The disconnected sections and their electrical mappings just define these combinations explicitly.

The description of the Lanczos procedure given here has been geared to the transformation of single dendritic trees. However, multiple trees connected to a single soma may be transformed—all required information and equations have been given. The soma may be represented by a single point, where all dendritic trees meet. Alternatively, the soma may be incorporated directly into the multi-cylinder representation, with appropriate boundary conditions. If the soma is not chosen as origin, it must be assigned one of the two acceptable boundary condition types, however in the majority of situations, it is likely that a soma origin would be most appropriate.

During the transformation, we have assumed a sealed origin, though any condition may be assigned. If we plan to use the matrix representation in further numerical tasks, then a more realistic soma condition may be imposed. Simply take the appropriate expression for the boundary condition and discretise the space derivatives, then rearrange into a discrete cable equation form and incorporate into the matrix representation. For example a lumped soma condition, a somatic shunt, in fact anything that doesn't interfere with the dendritic structure being transformed.

The Lanczos procedure assumes that dendritic tree segments can be described effectively by the one-dimensional linear cable equation. The validity of this model, and the importance of fully understanding the passive case, before proceeding to models which are more complete biophysically, is discussed elsewhere, for example Rapp et al. (1994). At this stage, we sacrifice full realism in an attempt to get a handle on the function of complicated dendritic tree geometry.

The cable equation forms the basis for more sophisticated models of electrical activity in dendritic trees. For example, reversal potentials for different ionic species may be incorporated into the cable equation, along with time-dependent conductance changes in the membrane (see for example Tuckwell, 1988b). The transformation given in this paper, while it may be adapted to incorporate these additional terms, does not allow us to properly interpret the result in terms of an equivalent structure. The introduction of changing membrane conductances means that the time constant will vary spatially. A multi-cylinder model will no longer be described by one spatially invariant differential operator. More complex still are the full non-linear Hodgkin–Huxley equations with active conductances, for which the Lanczos procedure is meaningless. A different approach altogether would have to be sought in order to generate equivalent structures for branching structures described in this way. At this point in our understanding it is not clear if an equivalent structure of some sort exists in these more complex cases.

The foundations of equivalent cable construction have now been laid, though little has been said about specific applications of this work. In fact, any mathematical method applied to a passive dendritic tree model can also be applied to a passive equivalent cable, the lack of branching easing the mathematics somewhat. It is also useful to note that the matrix representation of a tree (or equivalent cable) developed in section 3, is amenable to traditional numerical methods for solving the cable equations (for example, Mascagni, 1989; 1991). Alternatively, eigenvalue methods may be used.

One application, concerning the analysis of voltage transients, has been described by Ogden et al. (1997). A least squares analysis that compares observed transients and theoretical cable voltage responses can be used to estimate the strength, sign and location of equivalent cable connected section inputs that would generate the observed response. The electrical mapping then defines the corresponding tree inputs. Arbitrary mappings from disconnected sections can be added to this configuration with no effect on the response.

The idea of a dendritic sub-unit, as a dendritic region within which local signal processing may take place (Koch et al., 1982), can be adapted to take into account equivalent cable structure and the electrical mapping. A definition of dendritic sub-unit that is independent of dendritic membrane parameters, and can also describe local signal processing within dendritic substructure, is introduced by Lindsay et al. (1997). This work follows, in part, from the observation that singly branched trees (Y-junctions) may be reduced to equivalent cables through a *branch-shifting* process, intermediate steps of which often involve the generation of *equivalent trees* that also have a single branch point.

The equivalent cable and electrical mapping should allow one to directly assess the role of specific passive dendritic geometries, or classes of geometries, in neuronal signal processing.

7. CONCLUSIONS AND FUTURE PERSPECTIVES

Briefly, and not in any order of importance, we suggest several possible equivalent cable applications that might prove fruitful in analysis of dendritic function.

If we take an equivalent cylinder or empirical 'equivalent' cable, used to approximate a passive tree, and replace it with a truly equivalent cable, it may be possible to use present techniques and generate improved estimates of general electrical parameters of the tree.

One implication of the Y-junction approach to transforming dendritic geometry is that a multitude of observably different dendritic trees transform to equivalent cables that will have the same, or approximately the same, connected section. Similarities in connected sections can form the basis for a classification of neuronal types, however the disconnected sections may be different. A further degree of similarity between different trees can be measured, at various levels of geometrical detail, by comparing disconnected sections and electrical mappings associated with particular dendritic substructure.

The electrical mapping should prove useful in the analysis of the effect of synchronisation and the delay of electrical signals in a dendritic tree. Input configurations on the tree defined by inputs on disconnected sections, or narrow cylinders on a connected section, will determine synaptic locations that are most effective at diminishing or enhancing the effect that other synaptic inputs have at the soma. We may gain some insight by relating features of equivalent cables and electrical mappings to the statistical input-output relationships of neuronal spike trains (Rosenberg et al., 1989; Halliday et al., 1995).

One can try to anticipate the consequences of incorporating active membrane properties into the dendritic tree model, and the effects this would have on the equivalent cable, which would then be a dynamic time-dependent structure. We can see from the examples that slight changes in the electrotonic structure of one branch can often lead to dramatic changes in equivalent cable shape. For example, simply increasing the length of one branch can change an equivalent cable with a disconnected section into one without a disconnected section. Variations in membrane conductances can cause transient changes in the effective electrotonic structure of a dendritic tree. The equivalent cable will undergo a corresponding change and, consequently, input configurations that are ineffective at the soma for a passive dendritic tree may influence the soma when parts of the structure becomes active. One might speculate that moment-to-moment changes in membrane conductances will define moment-to-moment input configurations that are effective or not effective at the soma. A consequence of this speculation is that functional properties of neurons are perhaps more subtle and of greater variety than previously expected. The development of techniques for constructing equivalent cables for active cables is part of ongoing work by the authors.

PROBLEMS

7.1. For the purpose of equivalent cable construction, dendritic cylinders are described by the one-dimensional linear cable equation,

$$\frac{\partial^2 V}{\partial X^2}=\frac{\partial V}{\partial T}+V-r_m g. \tag{7.76}$$

Consider instead the cable equation incorporating specific conductances for each ionic species (Tuckwell, 1988b).

$$\frac{\partial^2 V}{\partial X^2}=\frac{\partial V}{\partial T}+V-r_m\sum_i \Delta g_i(V_i-V) \tag{7.77}$$

where g_i is the resting membrane conductance for ionic species i, $\Delta g_i(x,t)$ is the deviation from the resting membrane conductance for species i, and where $r_m=1/\sum_i g_i$.
Following the approach in section 4, derive discrete forms of this cable equation for different dendritic structure, and discuss the problems concerning interpretation that are encountered when trying to use the Lanczos procedure to generate an equivalent unbranched structure. Under what assumptions does this equation reduce to what is essentially the linear cable equation. Discuss the validity of these assumptions for sodium, potassium and chloride ionic currents.

7.2. Construct two dendritic trees which, when reduced to their equivalent cables, have identical connected sections (start with one tree and reduce some of its subtrees or expand terminal branches to generate the second tree). By considering the electrical mappings from disconnected sections for each tree, discuss the differing local operations that may be employed for signal processing within the two trees (which are, as far as the soma is concerned, electrically identical).

7.3. Construct a simple dendritic tree with no more than two or three orders of branching. Assign sealed boundary conditions at all terminals. Choose two non-terminal points on the tree and construct the two equivalent cables where each point is taken as origin (arrange for atleast one disconnected section in one of the cables). Using the bijective electrical mapping, map to each cable a unit potential from the location on the tree not chosen as origin. Compare and discuss the connected sections, the unit potential mappings and mappings from any disconnected sections. Discuss how electrical activity in a region of a tree may be local with respect to structure without the region, but not with respect to the soma.

7.4. Using the matrix equation description of a dendritic tree, write down the general solution of the cable equations for the potential at the origin point in terms of the eigenvalues of the tree matrix, elements of the electrical mapping matrix from tree to equivalent cable connected section, and elements of the diagonalisation matrix for the connected section cable sub-matrix.

7.5. Consider a single junction point where five uniform cylinders, with arbitrary diameters, but equal electrotonic lengths, meet. The equivalent cable for this structure has four disconnected sections. Show how, in whichever order the cylinders are combined in a Y-junction by Y-junction approach, the disconnected sections always define the same configurations of electrical activity over the tree.

7.6. Construct equivalent cables and disconnected sections for dendritic trees satisfying conditions (i), (ii), and (iii) in section 5 (piecewise cables). Compare the properties of the electrical mapping and disconnected sections with the equivalent cables for Rall trees that exhibit similar branching patterns.

ACKNOWLEDGEMENTS

Finally, we would like to note that the work in this Chapter grew out of a question posed by Wilfrid Rall to one of the authors (R.R.W) during a discussion concerning the feasibility of large scale computational modelling of a Purkinje cell: ‘But what would you do with all the numbers?’.

J.M.O. is supported by the Wellcome Trust. This work has also been supported, in part, by the Joint Research Council Cognitive Science and HCI Initiative. We would also like to thank Kenneth A. Lindsay for helpful mathematical discussions on aspects of equivalent cables.

REFERENCES

Abbott, L.F. (1992) Simple diagrammatic rules for solving dendritic cable problems. *Physica A.*, **185**, 343–356.

Abbott, L.F., Farhi, E. and Gutmann, S. (1991) The path integral for dendritic trees. *Biol. Cybern.*, **66**, 49–60.

Agmon-Snir, H. (1995) A novel theoretical approach to the analysis of dendritic transients. *Biophys. J.*, **69**, 1633–1656.

Burke, R.E. (1997) Equivalent cable representations of dendritic trees: Variations on a theme. *Soc. Neurosci. Abstr.*, **23**, 654.

Burke, R.E., Fyffe, R.E.W. and Moschovakis, A.K. (1994) Electrotonic architecture of cat gamma motoneurons. *J. Neurophysiol.* **72**(5), 2302–2316.

Butz, E.G. and Cowan, J.D. (1974) Transient potentials in dendritic systems of arbitrary geometry. *Biophys. J.*, **14**, 661–689.

Cao, B.J. and Abbott, L.F. (1993) A new computational theory for cable theory problems. *Biophys. J.*, **64**, 303–313.

Clements, J.D. and Redman, S.J. (1989) Cable properties of cat spinal motoneurones measured by combining voltage clamp, current clamp and intra-cellular staining. *J. Physiol.* (*Lond.*), **409**, 63–87.

Fleshman, J.W., Segev, I. and Burke, R.E. (1988) Electrotonic architecture of type-identified α-motoneurons in the cat spinal cord. *J. of Neurophysiol.*, **60**(1), 60–85.

Goldstein, S.S. and Rall, W. (1974) Changes is action potential shape and velocity for changing core conductor geometry. *Biophys. J.*, **14**, 731–757.

Golub, G.H. and Van Loan, C.F. (1990) *Matrix Computations* (2nd edn). John Hopkins University Press, Baltimore, Maryland.

Halliday, D.M., Rosenberg, J.R., Amjad, A.M., Breeze, P., Conway, B.A. and Farmer, S.F. (1995) A framework for the analysis of mixed time series/point process data—theory and application to the study of physiological tremor, single motor unit discharges and electromyograms. *Prog. Biophys. molec. Biol*, **64**, 237–278.

Hines, M. (1984) Efficient computation of branched nerve equations. *Int. J. Bio-Med. Comp*, **15**, 69–76.

Horwitz, B. (1981) An analytical method for investigating transient potentials in neurons with branching dendritic trees. *Biophys. J.*, **36**, 155–192.

Jack, J.J.B., Noble, D. and Tsien, R.W. (1983) *Electric Current Flow in Excitable Cells*. Oxford University Press.

Koch, C., Poggio, T. and Torre, V. (1982) Retinal ganglion cells: a functional interpretation of dendritic morphology. *Phil. Trans. R. Soc. Lond.*, **B298**, 227–264.

Lanczos, C. (1950) An iteration method for the solution of the problem of linear differential and integral operators. *J. Res. Natn. Bur. Stand.*, **45**, 255–282.

Lindsay, K.A., Ogden, J.M. and Rosenberg, J.R. (1997) Geometric Basis for Neuronal Signal Processing. Technical report. Division of Neuroscience and Biomedical Systems, University of Glasgow.

Major, G., Evans, D. and Jack, J.J.B. (1993a) Solutions for transients in arbitrarily branching cables: I. voltage recording with a somatic shunt. *Biophys. J.*, **65**, 423–449.

Major, G., Evans, D. and Jack, J.J.B. (1993b) Solutions for transients in arbitrarily branching cables: II. voltage clamp theory. *Biophys. J.*, **65**, 450–468.

Major, G. (1993c) Solutions for transients in arbitrarily branching cables: III. voltage clamp problems. *Biophys. J.*, **65**, 469–491.

Manor, Y., Gonczarowski, J. and Segev, I. (1991) Propagation of action potentials along complex axonal trees: model and implementation. *Biophys. J.*, **60**, 1411–1423.

Mascagni, M.V. (1991) A parallelizing algorithm for computing solutions to arbitrarily branched cable neuron models. *J. Neurosci. Meth.*, **36**, 105–114.

Mascagni, M.V. (1989) Numerical methods for neuronal modeling. In *Methods in Neuronal Modelling*, edited by C. Koch and I. Segev, pp. 255–282. Cambridge, Massachusetts, MIT Press.

Ogden, J.M. and Lindsay, K.A. (1996) Analytical Construction of Equivalent Cables. Technical report, Division of Neuroscience and Biomedical Systems, University of Glasgow.

Ogden, J.M., Lindsay, K.A. and Rosenberg, J.R. (1997) Construction of complete equivalent cables for passive dendritic trees of arbitrary geometry and their application to the analysis of voltage transients. *J. Physiol.*, **501**, 32P.

Paige, C.C. (1972) Computational variants of the Lanczos method for the eigenproblem. *J. Inst., Maths Applics*, **10**, 373–381.

Paige, C.C. (1976) Error analysis of the Lanczos algorithm for tri-diagonalising a symmetric matrix. *J. Inst. Maths Applics*, **18**, 341–349.

Paige, C.C. (1980) Accuracy and effectiveness of the Lanczos algorithm for the symmetric eigenproblem. *Linear Algebra and Appl.*, **34**, 235–258.

Poznanski, R.R. (1988) Membrane voltage changes in passive dendritic trees—a tapering equivalent cylinder model. *IMA J. Math. Appl. Med. Biol.*, **5**, 113–145.

Poznanski, R.R. (1991) A generalized tapering equivalent cable model for dendritic neurons. *Bull. Math. Biol.*, **53**, 457–467.

Rall, W. (1957) Membrane time constants of motoneurones. *Science*, **126**, 454.

Rall, W. (1959) Branching dendritic trees and motoneuron membrane resistivity. *Exp. Neurol.*, **1**, 491–527.

Rall, W. (1960) Membrane potential transients and membrane time constant of motoneurons. *Exp. Neurol.*, **2**, 503–532.

Rall, W. (1962) Theory of physiological properties of dendrities. *Ann. NY Acad. Sci.*, **96**, 1071–1092.

Rall, W. (1969) Time constants and electrotonic length of cylinders and neurons. *Biophys. J.*, **9**, 1483–1508.

Rall, W. (1977) Core conductor theory and cable properties of neurones. In *Handbook of Physiology: the Nervous System*, edited by E.R. Kandel, J.M. Brookhardt and V.B. Mountcastle, Vol. 1, pp. 39–98. Baltimore, Maryland, Williams and Wilkinson.

Rapp, M., Segev, I. and Yarom, Y. (1994) Physiology, morphology, and detailed passive models of guinea-pig cerebellar Purkinje cells. *J. Physiol.*, **474**(1), 101–118.

Rosenberg, J.R., Amjad, A.M., Breeze, P., Brillinger, D.R. and Halliday, D.M. (1989) The fourier approach to the identification of functional coupling between neuronal spike trains. *Prog. Biophys. molec. Biol.*, **53**, 1–31.

Stratford, K., Mason, A., Larkman, A., Major, G. and Jack, J.J.B. (1989) The modeling of pyramidal neurones in the visual cortex. In *The Computing Neurone*, edited by R. Durbin, C. Miall, and G. Mitchison, pp. 296–321. Reading, Massachusetts, Addison-Wesley.

Tuckwell, H.C. (1988a) *Introduction to Theoretical Neurobiology Volume 1—Linear Cable theory and dendritic structure.* Cambridge, Cambridge University Press.

Tuckwell, H.C. (1988b) *Introduction to Theoretical Neurobiology Volume 2—Nonlinear and stochastic theories.* Cambridge, Cambridge University Press.

Whitehead, R.R. and Rosenberg, J.R. (1993) On trees as equivalent cables. *Proc. R. Soc. Lond. B.*, **252**, 103–108.

Whitehead, R.R., Watt, A., Cole, B.J. and Morrison, I. (1977) Computational methods for shell model calculations. *Adv. Nuc. Phys.*, **9**, 123–176.

CHAPTER 8
PARAMETER ESTIMATION ALGORITHMS FOR THE SHUNT CABLE MODEL

Dominique M. Durand, Bassel Tawfik and Jian-Cheng Lin

1. INTRODUCTION

An essential step in data analysis is the formulation of a model capable of explaining or reproducing the experimental data. However, some models require the knowledge of the parameters in order to calculate the output and fit it to the experimental data. The process of obtaining these parameters given the data is known as parameter estimation. Calculation of the model output from a known set of parameters is known as the *forward problem*. The determination of the model parameters from the output data is known as the *inverse problem* and requires the solution to the forward problem. Therefore, parameter estimation is a specialized inverse problem traditionally solved by an optimization algorithm. Typically, an objective function which represents the error between the model output and the experimental data is minimized as in least-squares estimation (such as gradient method) and the unknown parameters are estimated as those which give the best fit to the data.

However, estimation of the unknown coefficients is not always successful. The convergence towards the true solution is affected by such factors as the presence of noise in the data, the number of parameters and the initial choice for the parameters (Bard, 1974). Consequently, the algorithm may not converge or converges towards a local minimum. Although several methods have been designed to escape these local minima, convergence towards the global minimum is not guaranteed.

This local minimum problem usually arises because of the iterative nature of most of the algorithms and could be avoided by using non-iterative procedures such as Linear Associative Memory (LAM). This concept was first introduced by Kohonen for pattern recognition (Kohonen, 1988) and adapted to parameter estimation by Kalaba et al. (1992). The basic principle is to associate a set of inputs (training set of parameters) with another set of model outputs of a given system through a memory matrix (training phase). A new observation is then presented to the memory matrix, which associates it back with the unknown set of parameters (recall phase).

The objective of this Chapter is to present the results of the estimation of the electrotonic parameters of the shunt cable model by three different methods. An iterative gradient method is first applied and tested (Ali-Hassan et al., 1992).

The LAM method is then applied to the same model. The LAM concept is developed and compared to the gradient method (Tawfik and Durand, 1994). The original LAM concept is then modified by adding a weighting procedure (WLAM) (Lin and Durand, 1996). The results of these methods are also compared in the presence and absence of noise. The advantages and disadvantages of each method are discussed.

2. THE SOMATIC SHUNT CABLE MODEL

Under certain conditions, the dendritic tree of a neuron can be approximated by a single equivalent cable. The conditions required for this equivalence are that the branches have approximately the same electrotonic length, with identical termination and that the dendrites diameters satisfy the branching power law (Rall, 1977)

$$d_0^{3/2} = d_1^{3/2} + d_2^{3/2},$$

where d_0, d_1 and d_2 are the parent branch and daughter branches respectively.

Some neurons such as the granule cell of the dentate gyrus satisfy this power law and if the other two conditions can be satisfied the entire dendritic tree can be replaced by an equivalent cable (Durand et al., 1983). The soma is represented by a lumped impedance and the whole neuron is represented by an equivalent model shown in Figure 8.1 (the axon is neglected because of its small diameter). In this

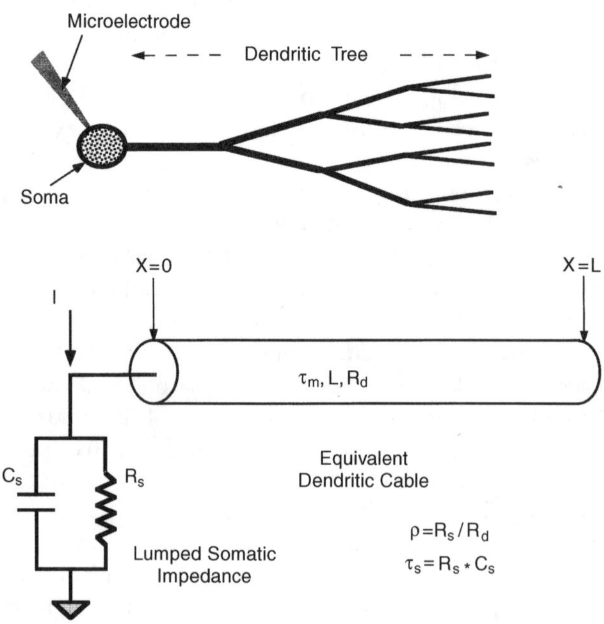

Figure 8.1 The dendritic tree of a neuron is represented by an equivalent cable model. A glass microelectrode injects current pulses at the soma ($X = 0$) while simultaneously measuring the voltage at the same location.

model, the somatic time constant is not assumed to be equal to dendritic time constant in order to account for the shunt which has been found at the soma. This shunt can be generated either by electrode damage or by a lower membrane resistance at the soma (Durand, 1984). The dendritic tree is assumed cylindrical and the tapering of the tree diameter described by Poznanski (1996) is neglected. Moreover, the impedance of the axon is also neglected since it is significantly higher than that of the soma and dendrites (Tranchina and Nicholson, 1986).

The equation describing the electrical behavior of the somatic shunt model is

$$\frac{\partial^2 V(X,t)}{\partial X^2} - V(X,t) - \tau_{\mathrm{m}} \frac{\partial V(X,t)}{\partial t} = 0 \tag{8.1}$$

where $V(X,t)$ is the deviation from the resting membrane voltage as a function of the normalized distance $X(x/\lambda)$ along the cable and time $T(t/\tau_{\mathrm{m}})$. λ and τ_{m} are the space and time constants, respectively. Their values are a function of the electrical properties of the membrane and cytoplasm as well as the geometry of the cell.

Assuming that the voltage at the soma is equal to zero at $t=0$, a sealed end at the end of the cable ($X=L$) and a current $I(t)$ injected at the soma, the following boundary equations and initial conditions must be satisfied (Durand, 1984):

$$V(X,0)=0, \tag{8.2}$$

$$\frac{\partial V(L,t)}{\partial X} = 0, \tag{8.3}$$

$$\frac{\partial V(0,t)}{\partial X} = \frac{\tanh L}{\rho}\left(V(0,t) + \tau_{\mathrm{s}} \frac{\partial V(0,t)}{\partial t} - R_{\mathrm{N}}(\rho+1)I(t) \right), \tag{8.4}$$

where L is the electrotonic length of the equivalent cable, τ_{s} the somatic time constant, R_{N} the input resistance of the cell and $\rho = R_{\mathrm{s}}/R_{\mathrm{d}}$ the ratio of somatic to dendritic resistances R_{s} and R_{d}.

The analytical solution of the model was derived by Durand (1984) and Kawato (1984) for a step and short pulse current inputs. The voltage $V(0,t)$ at the soma was computed as an infinite summation of exponentials with the individual terms obtained from the solution of a transcendental equation. It is also possible to take the Laplace transform of (8.1)–(8.5) with respect to t:

$$\frac{\mathrm{d}^2 \bar{V}(X,s)}{\mathrm{d}X^2} - (\tau_{\mathrm{m}} s + 1)\bar{V}(X,s) = 0, \tag{8.5}$$

$$\frac{\mathrm{d}\bar{V}(L,s)}{\mathrm{d}X} = 0, \tag{8.6}$$

$$\frac{\mathrm{d}\bar{V}(0,s)}{\mathrm{d}X} = \frac{\tanh(L)}{\rho}[(1+\tau_{\mathrm{s}} s)\bar{V}(0,s) - R_{\mathrm{N}}(\rho+1)\bar{I}(s)], \tag{8.7}$$

where $\bar{V}(X,s)$ and $\bar{I}(s)$ are the transforms of $V(X,t)$ and $I(t)$, respectively.

If the voltage response is measured only at the soma then in this special case, the solution of (8.5), (8.6), (8.7) is:

$$\bar{V}(0,s)=\frac{R_{\mathrm{N}}(\rho+1)\bar{I}(s)}{\sqrt{\tau_{\mathrm{m}}s+1}\tanh(L\sqrt{\tau_{\mathrm{m}}s+1})\rho\coth(L)+(1+\tau_{\mathrm{s}}s)}. \qquad (8.8)$$

The five parameters to be estimated are:

1. membrane time constant τ_{m}
2. somatic time constant τ_{S}
3. electrotonic length of the equivalent cable L
4. input resistance of the cell R_{N} ($R_{\mathrm{N}}=R_{\mathrm{s}}R_{\mathrm{d}}/(R_{\mathrm{s}}+R_{\mathrm{d}})$
5. ratio of dendritic to somatic conductance ρ ($\rho=R_{\mathrm{s}}/R_{\mathrm{d}}$).

An equivalent set of parameters is τ_{m}, τ_{S}, L, R_{s} and R_{d}. The equations derived above must be modified when patch-clamp recording is used (cf. Jackson, 1992; Chapter 9) but the estimation procedure is similar.

The voltage responses obtained from (8.8) are converted into the time domain with an FFT algorithm (1024 points, 20 kHz). The time waveform of the voltage response at the soma was simulated using the specified parameter values (Table 8.1) and an input current pulse of 0.5 ms duration and 2 nA amplitude.

3. NOISE AND ARTIFACTS

For the estimation of the electrotonic parameters of neurons, the data to be analyzed is contaminated by noise and stimulation artifacts (Bardakjian, 1990; Ali-Hassan et al., 1992; White et al., 1992; D'Aguanno et al., 1986, 1989; Fu et al., 1989). Experimentally, a current pulse is injected into a nerve cell using a micro-electrode which is also used to record the voltage. Since the current injected into the electrode must pass through the recording electrode, an artifact is generated. A bridge circuit and an artifact rejection circuit in the amplifier can decrease the amplitude of the artifact. However, the time constant of the electrode used for intracellular recording can still generate a significant distortion of the neuronal voltage response. The voltage response is distorted not only during the application of the pulse but also following the termination of the pulse. The effect of the electrode impedance can be minimized by using patch-clamp electrodes (Jackson, 1992), or by including the electrode parameters into the parameter estimation (Wright et al., 1996).

Thermal noise is also generated in the voltage response by the high impedance of the electrode. The RMS (root mean square) value of this noise is given by:

$$V_{\mathrm{RMS}}=(4kRTB)^{1/2},$$

where B is the bandwidth of the signal, R the resistance of the electrode, k, the Boltzmann constant and T the temperature in °K. For an electrode resistance of 100 MΩ and a bandwidth of 10 kHz, one obtains a RMS noise value of 400 μV which gives a peak value of 1.6 mV assuming a crest factor of 4 for thermal noise (Ott, 1976). For patch-clamp recording, the resistance of the electrode is reduced to a few MΩ and the effect of the noise can be reduced in the current clamp mode. In the voltage clamp mode, however, the signal is significantly noisier. Biological noise such as intrinsic membrane noise and synaptic noise can also compromise the signal to noise ratio of the signal. The effect of the noise can be decreased by averaging many waveforms and the amplitude of the noise can be decreased by $n^{1/2}$ where n is the number of waveforms to be averaged. However, the neural tissue response is not time-invariant and significant errors can be introduced by recording the signal for periods long enough to allow the parameters to change. Parameter estimation schemes which use variable frequencies or white noise inputs could be affected by this problem if long time sequences are required.

Therefore, the problem is to estimate accurately the parameters of a neuron model in the presence of significant distortion from artifact and noise. The standard method is to generate an error function between the predicted output and the measured output and to minimize this error.

4. PARAMETER ESTIMATION BY MINIMIZATION OF AN OBJECTIVE FUNCTION

The procedure used to find the parameters using this technique is illustrated in Figure 8.2. The stimulus used in this example is a short current pulse (0.5 ms) but the method is applicable to other stimuli, such as pseudo-random noise sequences or sinusoidal waveforms at various frequencies. Initial guesses for the parameters are used to calculate the voltage output V of the model (forward solution). This output is combined with the experimentally measured response V_m to generate the least-square objective function defined as follows:

$$\Phi = \sum_{k=1}^{N} [V(0,t_k) - V_m(t_k)]^2. \tag{8.9}$$

This function is then minimized with respect to the parameter vector:

$$\mathbf{r} = [\tau_m \tau_s L\, R_s R_d]^T.$$

The minimization can be accomplished with an adaptive, nonlinear, least-squares algorithm gradient method (NL2SOL) (Dennis et al., 1981). The sensitivity of the model output to its parameters $\mathbf{r}$ is described by sensitivity functions:

$$\mathbf{J}_k = \partial V(0,t_k)/\partial \mathbf{r}. \tag{8.10}$$

To obtain the sensitivity functions, the derivatives of the transform solution in the frequency domain are computed with respect to the parameters and then numerically inverted by the FFT (1024 points, 10 kHz). For simultaneous comparison of the model sensitivity to all parameters, a semi-relative sensitivity measure for each parameter (r_i) is used:

$$r_i \cdot [\partial V(0,t)/\partial r_i].$$

The relative sensitivities of V with respect to $\mathbf{r}$ are largest in the region of peak voltage towards the start of the discharging region (Ali-Hassan et al., 1992). For

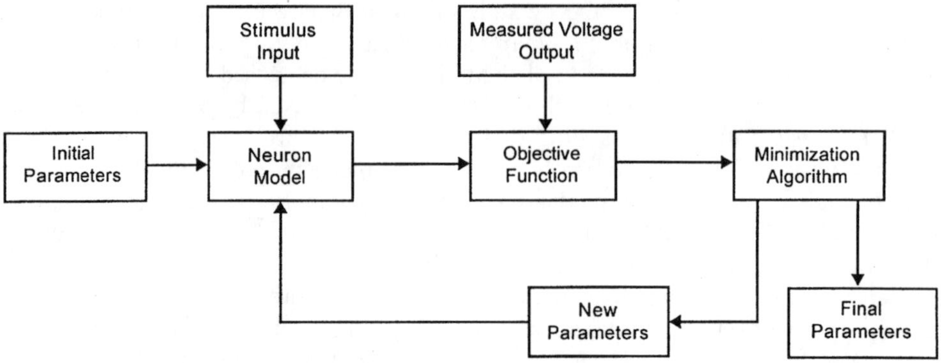

Figure 8.2 Parameter estimation method by error minimization. Initial guesses for the parameters estimated are used to calculate the output of the model. This output is then compared to the experimental output and an error function is generated. An error minimization algorithm then provides new sets of parameters until the error function reaches a minimum. The parameters corresponding to the smallest value of the error function are the estimated parameters.

the least-square objective function, the Hessian can be approximated using the sensitivity functions:

$$\mathbf{H} = \partial^2\Phi/\partial\mathbf{r}^2. \tag{8.11}$$

When the differences between theoretical and experimental outputs are Gaussian with identical variance, then **H** is inversely proportional to the covariance matrix of the parameters:

$$\mathbf{H} \sim 1/\mathrm{cov}(\mathbf{r}). \tag{8.12}$$

Since the confidence region for each parameter (precision) is inversely proportional to the determinant of the covariance matrix, the precision for each parameter is proportional to the det[**H**]. Therefore, in order to minimize the estimation errors the det[**H**] should be maximized (Beck and Arnold, 1977).

The effect of the pulse width on the estimation can then be investigated by plotting the det[**H**] versus the pulse width. It is clear from Figure 8.3(A) that narrow pulses give the maximum precision since the det[**H**] is largest with small pulse width. This can be easily interpreted by the fact that short duration pulses have a wide frequency spectrum when compared to longer pulses. The ideal stimulation waveform is an impulse with flat frequency spectrum, and pulses with short duration will be closest to this ideal. Next, the effect of the artifact can be directly examined by removing the points in the waveforms corresponding the distortion at the onset of the waveform and plotting the det[**H**] with respect to the starting time of data collection. From Figure 8.3(B), it is evident that removing data points close to onset of the pulse significantly decreases the accuracy of the estimation. The duration of the data collection can also be examined by plotting

Table 8.1 Parameters estimated from model-generated data using the error minimization method. The data was generated with a 2 nA, 0.5 ms current pulse with and without noise. The first 1 ms of the data set was ignored

	Parameters		
	True	*Estimated*	
		Zero noise	$\sigma^2=0.1$
τ_m	10.0	10.0	8.92
τ_s	5.0	5.0	8.27
L	1.0	1.0	1.01
R_N	85.7	85.71	86.66
ρ	2.5	2.5	6.31

det[**H**] against the duration of data sample. Figure 8.3(C) shows that there is a region of the voltage in which further data collection is unnecessary since the det[**H**] does not improve by adding more data points.

In order to test the parameter estimation procedure, the model output was calculated with a specified input and a 'true' set of parameter values. Starting with a different set of parameters, an optimization algorithm (Dennis et al., 1981) was applied to minimize the objective function and obtain parameters' estimates that yield the best fit of the model to the data. Pseudo-random white noise with a zero mean and constant variance (0.1 mV) was added to the voltage output.

An estimate of the error in each of the parameters can be obtained by calculating the coefficient of variation defined as

$$CV_i = \sigma_i / \mu_i, \tag{8.13}$$

where σ_i and μ_i are the standard deviation and mean values of the parameters, respectively.

The method was then applied to the voltage responses to short current pulses (0.5 ms, 2 nA) injected into the neuronal model described above (Ali-Hassan et al., 1992) (see Figure 8.4). The optimal estimates and coefficients of variation are given in Table 8.1. With zero noise added to the observation vectors, the error minimization method returns parameters with no error even with data points missing for the first 0.5 and 1 ms. However, when noise is added, the error increases considerably especially when data points for the first 1 ms are missing.

5. PARAMETER ESTIMATION BY LINEAR ASSOCIATIVE MEMORIES

Associative memories were first introduced by Kohonen (1988) in the context of pattern recognition and adapted to parameter association by Kalaba et al. (1992).

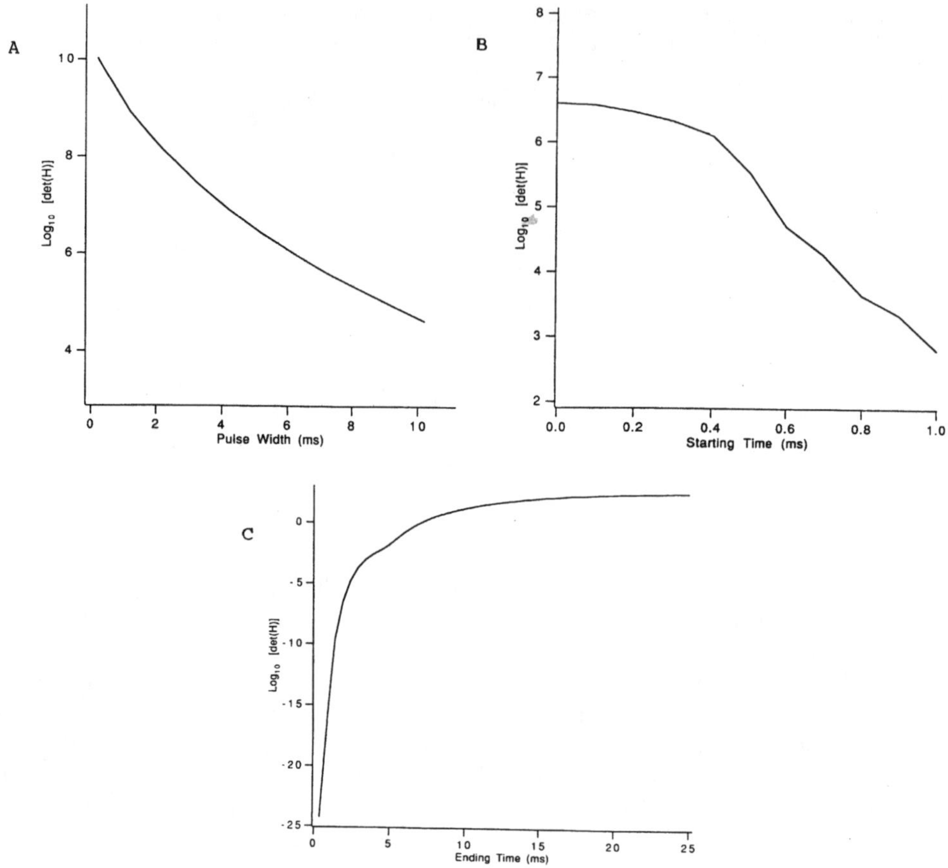

Figure 8.3 Optimization of the experimental procedure by using the determinant of the Hessian. (A) Effect of pulse width variation on the determinant of the Hessian at a constant value of injected charge (5 milliCoulombs). det[**H**] decreases as the pulse width increases indicating that a short pulse width is necessary to obtain more accurate estimates of the parameters. (B) Effect of the starting time of data collection on the determinant of the Hessian. The value of det[**H**] decreases sharply after the end of the input pulse (throughout the discharging region). This indicates that the information content necessary to obtain precise parameter estimates is located near the end of that input pulse. (C) Effect of the ending time of data collection on the determinant of the Hessian. The value of det[**H**] reaches a plateau at 20 ms and data collected after this point does not improve the precision of the estimate. (Modified from Ali-Hassan et al., 1992.)

The objective of LAM is to convert the input/output relationship of a system into a set of simultaneous linear algebraic equations. The estimation procedure is illustrated in Figure 8.5. A training set of parameters is first stored in a matrix R and the corresponding set of output observations in a matrix V. A linear associative

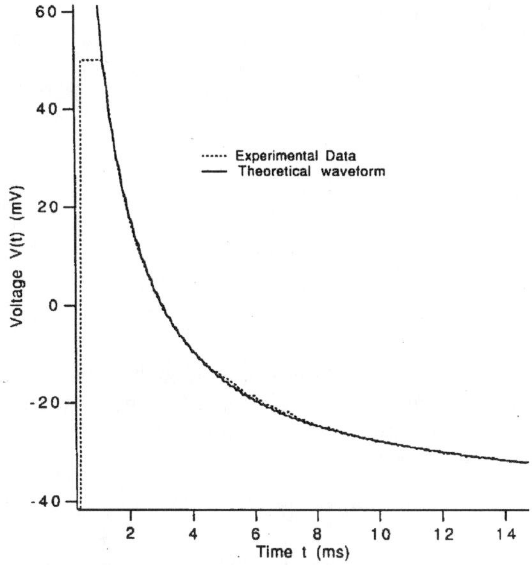

Figure 8.4 Estimation of the electrotonic parameters of a granule cell for the dentate gyrus. A 2 nA, 5 ms pulse was injected into the cell and produced the response shown using the dashed line. The artifact generated by the electrode was cut-off and only the data starting 1.5 ms following the end of the pulse was fitted. The parameters were estimated and used to reconstruct the output (solid line). (Modified from Ali-Hassan et al., 1992.)

memory matrix M is then defined as

$$R = M \cdot V, \tag{8.14}$$

where $\cdot$ is the product between two matrices. In the recall phase, M is used to estimate the values of a parameter vector $\mathbf{r}_e$ from a given observation vector $\mathbf{v}$ by simply reapplying (8.14) in its vector-matrix form:

$$\mathbf{r}_e = M \cdot \mathbf{v}. \tag{8.15}$$

M can be obtained by multiplying R by the inverse of V. In practice, however, the inversion of the observation matrix V is not a straight forward procedure. The columns of V represent repeated output waveforms resulting from different combinations of parameters in the model and are therefore highly correlated. Moreover, the matrix V is usually non-square and the number of equations is less than that of the parameters. Therefore, the system possesses infinitely many solutions. The associative memory matrix M in (8.14) can be expressed in terms of the Minimum Norm Solution (Kohonen, 1988):

$$M = R\{V^{\mathrm{T}}[VV^{\mathrm{T}}]^{-1}\}. \tag{8.16}$$

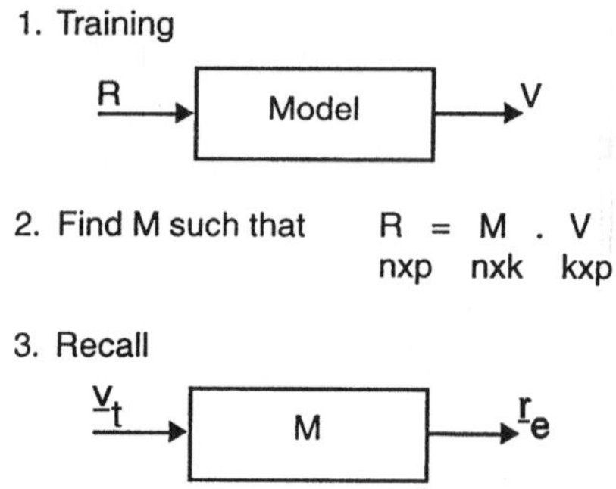

Figure 8.5 Parameter estimation by linear association. In the training phase, a model (or an experiment) is used to obtain a set of outputs v_i from a chosen set of possible parameters r_i. The set of parameters and corresponding outputs are used to construct the matrices R and V. (Modified from Tawfik and Durand, 1994.)

In general, a direct inverse of VV^T does not exist and one must resort to generalized, inverses (Albert, 1972; Ben-Israel and Greville, 1980; Watkins, 1991). One of the widely accepted methods for evaluating the pseudo-inverse of generally non-square matrices such as $[VV^T]^{-1}$ is the singular value decomposition (SVD). This method performs very well when linear dependence is nearly exact (that is when some of the eigenvalues are very close to zero). However, in most cases the eigenvalues are not zero but still small compared to other eigenvalues. In that case the matrix is ill-conditioned and SVD may not always be the best method (Marquardt, 1970; Marquardt and Snee, 1975). This ill-conditioning is a common problem in LAM problems (Eichman and Stojancic, 1987; Homma et al., 1990; Riley, 1956). Another approach has been adopted by Hoerl (1962), and Hoerl and Kennard (1970) for treating ill-conditioned systems by adding a diagonal matrix to VV^T prior to taking its inverse, i.e.,

$$M = RV\{VV^T + \alpha I\}^{-1}, \tag{8.17}$$

where α is a small scalar and I is the identity matrix. The method was called Ridge regression and α the Ridge coefficient. The addition of the αI term gives biased estimates but reduces the mean square error of the estimates (Marquardt, 1970) and improves the stability of the solution (Kalaba et al., 1992). The Ridge regression method is closely related that the method of regularization (Tikhonov and Arsenin, 1977).

In order to apply the method of linear association to the problem of neuronal parameter estimation, a set of parameters is chosen and stored in a matrix R. The corresponding model outputs are stored in V and the matrix M is calculated using the previous equations. A 'true' observation vector, $\mathbf{v}_t$ is then generated from a true set of parameters, $\mathbf{r}_t$, using the equations of the model. Zero-mean, Gaussian-distributed random noise is added to the true observation vector to generate an observation vector **v**. The estimated parameters $\mathbf{r}_e$ are obtained using (8.15) and then entered into the model equation to generate the estimated output (observation) vector $\mathbf{v}_e$. The noise level is determined according to its relative power with respect to that in the recorded signal. The signal-to-noise ratio (SNR) is defined as

$$\text{SNR} = 10\log_{10}[\text{signal power/noise power}]. \tag{8.18}$$

The overall percent relative error due to estimation is defined as

$$v_{\text{err}} = 100 \times \{\|\mathbf{v}_t - \mathbf{v}_e\|\}/\|\mathbf{v}_t\|, \tag{8.19}$$

where $\| \ \|$ is the Euclidean norm and the individual parameter errors are calculated as

$$\mathrm{r}_{\text{err}}(i) = 100 \times [r_t(i) - r_e(i)]/r_t(i), \quad i = 1,2,3,\ldots,n. \tag{8.20}$$

Figure 8.6 shows a typical run where a noisy observation vector $\mathbf{v}_t$ is used to estimate the five electronic parameters of the neuron model using the Ridge method for matrix inversion and with a large amount of noise (SNR set at 46 dB). The solid curve shows the noisy observation vector from the model while the dashed curve represents the

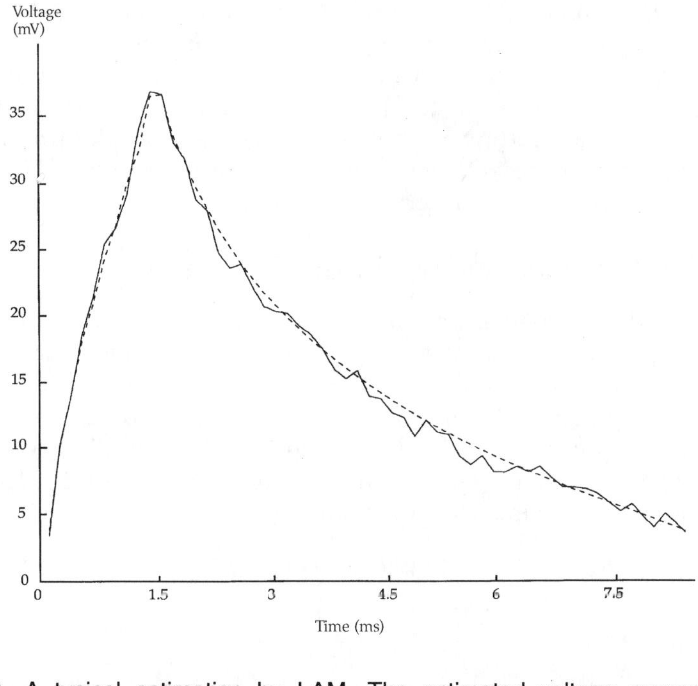

Figure 8.6 A typical estimation by LAM. The estimated voltage response by LAM (interrupted curve) is superimposed on the observed signal v_0 (solid curve) when the latter is corrupted with a 46 dB SNR. Parameter estimation is performed on a 64-point voltage signal using a medium size grid (refer to text) and the Ridge/regularization method. (Modified from Tawfik and Durand, 1994.)

Table 8.2 Parameters estimated from model-generated data using the linear associative method. The data was truncated with 2 ms of data missing and 46 dB of noise was added. The error between the test data and the data reconstructed from the estimated parameters was only 1.85%

	Parameters			
	True	*Estimated*		
		Mean	*SD*	*% Error*
τ_m	7.0	7.57	0.29	8.14
τ_s	5.0	5.03	0.13	0.60
L	1.1	1.21	0.05	10.45
R_s	320.0	321.51	4.14	0.47
R_d	160.0	162.82	1.22	1.76

voltage response calculated from the estimated parameters. An overall relative error of 1.58% is obtained corresponding to individual parameter errors of 21%, 0.9%, 5.6%, 1.5%, and 18% in r_1 through r_5, respectively (Tawfik and Durand, 1994).

The LAM method was tested in the presence of noise and missing data points. The system was trained with truncated and noisy curves. The mean, standard deviation, and percent errors of the parameter estimates for 20 simulations are shown in Table 8.2. A maximum error of 10% with low standard deviations was obtained. The error in the voltage output (1.85%) suggests that this associative method is robust against noise and partial data losses.

6. PARAMETER ESTIMATION USING WEIGHTED LAM

The idea of LAM is to construct an inverse global linear map within a small set of inputs and outputs and extend it to a larger set of inputs and outputs data of a system. LAM parameter estimation consists of two phases: a training phase and a recall phase (or estimation phase). The determination of M is the main task in the training phase. M can be determined by solving (8.14) using SVD (Gruber, 1990; Penrose, 1956) or Ridge regression (Gruber, 1990; Albert, 1972). The latter is essentially equivalent to minimizing the following cost function

$$\Delta = \alpha\|R - MV\|^2 + (1-\alpha)\|M\|^2, \tag{8.21}$$

where

$$\|M\| = \sqrt{\mathrm{Tr}\{M^{\mathrm{T}}M\}} \tag{8.22}$$

is the Euclidean norm of matrix M. A non-zero value of the Ridge parameter α is used to avoid singular behavior of the matrix M.

In the second phase, M is used to provide an approximate representation of the system, so that for a given observation vector $\mathbf{v}$, it generates: $\mathbf{r}_e = M \cdot \mathbf{v}$, where $\mathbf{r}_e$ is an estimate of the 'true' parameter vector $\mathbf{r}_t$.

A careful examination of LAM indicates that the construction of M contains no information about the observation vector. M is built with the expectation that, when supplied with a large number of training parameter vectors, it should be able to locate the position of the true value. However, this method is not economical since typical parameter estimation involves only one observation vector. Moreover it is not accurate, since it does not recognize the possibility that one of the training vectors is actually located at the true value. Therefore the observation vector **v** should be included in building M. A recursive and adaptive determination of M which takes **v** into account has been published (Kalaba and Udwadia, 1993). However, this method is not always efficient since a recursive approach usually starts with a few numbers of training vectors from which one gets a new parameter vector not necessarily closer to the true value. Intuitively, the parameter-response vector pair should play a larger role in constructing M as the response vector gets closer and closer to the observation vector. A formulation involving the observation vector should yield a more efficient and accurate parameter-estimation algorithm. However, this is not generally true and local minima can arise when the model function is not a monotonic function of its parameters. This problem can be partially resolved by implementing an algorithm that allows escaping from such a local minimum. Therefore, a weighted linear associative memory (WLAM) for parameter estimation was developed (Lin and Durand, 1996) and the procedure is shown in Figure 8.7. This is accomplished by introducing a cost function that takes the weighting into

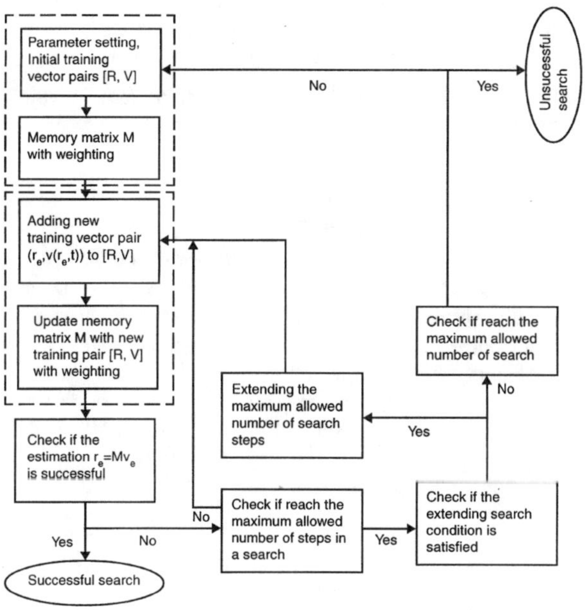

Figure 8.7 Weighted LAM. The LAM procedure was modified to include a recursive algorithm and including a cost function which takes into account a weight associated with the parameters that give a better fit to the data.

account. Moreover, in order to implement the recursive technique in WLAM, the value of Ridge parameter α in the cost function is updated (increased) at each iterative step, yielding a more efficient approach to the nonlinear parameter estimation.

The iterative algorithm for the implementation of the WLAM and procedures for avoiding local minima have been implemented and tested for a highly nonlinear model (Van der Pol equation) as well as the somatic shunt cable model (Lin and Durand, 1996). The results obtained with this WLAM procedure were compared to the LAM procedure and presented in Tables 8.3 and 8.4. In the given example

Table 8.3 Results for the Somatic Shunt Cable Model without noise. Initial Ridge parameter $\alpha = 0.9$; one true value and 6 initial training vectors are selected randomly within the small box (local area around the initial search area). A total of 243 estimations are carried out. Each estimation starts with a different random number generator and gives an error vector. The mean error (standard deviation) vector is the mean value (standard deviation) of the 243 error vectors. The data obtained from LAM estimation are obtained using the same program for WLAM with the weight function replaced by a uniform weight function

	Small box					
Parameter	*Center*	*Width*	*Mean error (%) (WLAM)*	*Standard deviation (%) (WLAM)*	*Mean error (%) (LAM)*	*Standard deviation (%) (LAM)*
τ_m	7	4	0.0058	0.047	0.92	1.01
τ_s	5	4	0.0119	0.100	1.63	1.69
L	1.1	0.8	0.0035	0.028	0.46	0.52
R_s	320	40	0.0128	0.107	1.78	1.83
R_d	160	80	0.0067	0.056	0.98	1.03

Table 8.4 Results for the Somatic Shunt Cable Model with noise. Initial Ridge parameter $\alpha = 0.9$; one true value and 230 initial training vectors are selected randomly within the small box. Each estimation starts with a different random number generator and yields an error vector. The mean error (standard deviation) vector is the mean value (standard deviation) of the 243 error vectors. The data obtained from LAM estimation are obtained using the same program for WLAM with the weight function replaced by a uniform weight function

	Small box					
Parameter	*Center*	*Width*	*Mean error (%) (WLAM)*	*Standard deviation (%) (WLAM)*	*Mean error (%) (LAM)*	*Standard deviation (%) (LAM)*
τ_m	7	4	7.15	6.28	16.02	22.28
τ_s	5	4	5.13	3.75	7.98	7.35
L	1.1	0.8	5.01	4.27	11.82	17.30
R_s	320	40	5.11	4.06	6.31	6.66
R_d	160	80	2.78	2.28	5.37	4.92

without noise, WLAM can increase LAM accuracy by at least two orders of magnitude without increasing the number of iterations. For the given example containing noise at typical experimental level, it can decrease the overall averaged error of parameters from 22.99% to 11.68%. The averaged overall error of observation vectors is also decreased from 3.57% to 2.87% which is very close to the lower bound 2.83% determined by the noise amplitude.

7. DISCUSSION

Three methods for parameter estimation were presented and tested for their ability to provide accurate parameters in the presence of noise and artifacts. The ability of an algorithm to perform well under conditions in which the data is contaminated is known as robustness. Error minimization techniques are currently the method of choice in most parameter estimation problems. These methods are based on the comparison between the error generated by the model output for a given set of parameters and the experimental data. The parameters are updated until the error reaches a minimum. Various techniques for reaching this minimum have been implemented with the gradient-based ones being the most popular. For the neuronal model presented here, the technique chosen (NL2SOL) is gradient-based and has been shown to be capable of solving difficult nonlinear parameter estimation problems (Dennis et al., 1981). NL2SOL gave maximum parameter error less than 15% in the presence of noise and missing data points. With little or no noise, gradient methods gave very accurate results. However, error minimization methods suffer from a major drawback, namely local minima. As various initial guesses are entered into the algorithm, very different estimated parameters can be obtained as the error minimization algorithm gets trapped into a local minimum of the parameter space. This is a serious problem which can be extremely difficult to solve although various methods have been suggested.

The LAM method does not rely on an iterative algorithm to arrive at the estimated parameters but rather on the interpolation between many estimated parameter vectors stored in a matrix. Therefore, the emphasis is not on minimizing the experimental/theoretical errors but rather on finding the set of parameters which gives the best interpolation between an already stored training set of parameter vectors. This method can also give accurate estimates under noisy conditions.

These two methods, error minimization and linear associative memory, were compared directly by applying the same data set to each one. Thirty simulated observation vectors were generated using the same 'true' parameter vector and 30 different noise vectors. The initial guesses for the error minimization method were taken within 20% of the 'true' vector and the parameter space of the matrix method was taken with maximum deviation of 20% of the 'true' parameter vector. The results indicate that the gradient technique is the better method in the absence of noise. With clean data and good initial guesses, this method converges directly to the solution. However, the results also indicate that the gradient method performance degrades quickly when noise is added. Table 8.5 shows that the LAM method gives better results than the error minimization method in the presence of large amount of noise with the largest mean error of 7% compared to 35% for the gradient method. Moreover, the standard deviation of the estimated parameters for LAM is significantly smaller when compared to the error minimization method. This can be attributed to the fact the initial guesses can lead to local minima of the objective function located far away from the true parameters thereby increasing the standard deviation.

The LAM method therefore appears to be very robust against noise and missing points as previously shown (Tawfik and Durand, 1994). Moreover, the main advantage of this method is that it is not an iterative method and does not suffer from the problem of local minima. This represents a significant improvement since local minima can lead to severe errors in parameter estimation problems. However, it is also clear from these studies that LAM is not as accurate as the error minimization method when small amount of noise are present in the data. In the absence of noise, the error minimization method can recover nearly exactly the true parameters. In the absence of noise, the LAM method is not error free since the correlation between the observation vectors can generate some error in the matrix inversion which is required for the algorithm.

The WLAM parameter estimation differs from LAM parameter estimation in two aspects. First, the WLAM algorithm assigns a weight to the error between each model output and the experimental data, and second, it updates the Ridge parameter in the iterative process of the estimation. Both aspects in WLAM contribute to a more efficient search and produce more accurate results than those obtained with LAM in two different estimation problems, the Van der Pol equation and the somatic shunt cable model (see Lin and Durand, 1996 for the Van der Pol results).

One drawback of the WLAM method is the fact that it is an iterative method and in that respect it is similar to gradient methods for parameter estimation. It requires

Table 8.5 Comparison of the parameter estimation results by two different methods: an adaptive gradient search method (NL2SOL) and Linear Associative Memory (LAM). An observation vector was generated from known parameters (true value). Gaussian noise was added before estimation by the two methods. The mean and standard deviation (SD) for 30 different noise vectors are tabulated. The results show that the LAM method gives generally better results with a significantly reduced deviation (F-test, $p < 0.001$)

	Parameters				
	True	*Estimated*			
		Mean *NL2SOL*	*SD* *NL2SOL*	*Mean* *LAM*	*SD* *LAM*
τ_m	7.00	8.50	5.24	7.38	0.30
τ_s	5.0	5.41	2.37	5.18	0.13
L	1.1	1.09	0.17	1.18	0.004
R_s	320.0	355.95	178.57	317.80	2.07
R_d	160.0	181.82	52.54	162.52	1.05

Table 8.6 Relative advantage and disadvantage of the three methods presented in this paper

	Gradient	*LAM*	*WLAM*
Robustness (immunity against noise)	Low	High	High
Accuracy of estimates with low noise data	High	Low	High
Accuracy of estimates with noisy data	Low	High	High
Multiple initial search	No	No	Yes
Sensitivity to local minima problem	Yes	No	Yes

the calculation of the cost function at each step and is subject to the problem of the local minima. However, instead of just using a local property of the model function (the sensitivity function) to generate a new parameter vector in gradient method, WLAM uses both a global property of the model function (the associative memory matrix) and a local property (determined by the weight function) to update a new parameter vector. Because the initial associative memory matrix involves several training parameter vectors and the weight function recognizes the one that is possibly closest to the true parameter vector, WLAM has multiple initial search positions, but only one iteration carries on at an optimal position, suggesting WLAM to be a more efficient estimation method. The various advantages and disadvantages of the three methods are summarized in Table 8.6.

8. CONCLUSIONS AND FUTURE PERSPECTIVES

Matching the shunt cable model with experimental data (cf. Glenn et al., 1987; Rose and Vanner, 1988; Pongracz et al., 1991; Major et al., 1994) requires the electrotonic parameters to be estimated and this often is a difficult problem (Rall et al., 1992). Three methods have been compared to estimate the electrotonic parameters of neurons. The first technique is a gradient technique for minimizing the error between experimental data and the theory. The method works well but is not robust and suffers from the problem of local minima. The LAM method is not an iterative method and is based on the global knowledge of the parameter space. LAM can provide good estimation parameters in the presence of noise and artifact in the data. The method is robust and fast since, once the matrix M has been calculated, a single matrix multiplication is required to obtain the parameters. The method does not suffer from local minima problem but is not as accurate as the standard error minimization method in the presence of low amount of distortion. A final method (WLAM) is an iterative method which combines the advantages of the two previous methods by obtaining knowledge of both the local and global properties of the parameter space. Although still subject to the local minima problem, this method has shown that it is capable of providing both accurate and robust estimates.

Both the LAM and the WLAM are currently in the process of further development. The LAM has been reformulated in order to decrease the order of the matrix inversion required to obtain the parameters. The WLAM has been improved by adding an offset term to the basic formulation (8.14). This term improves the fit of the data to the model formulation and gives more accurate estimates. We are currently testing these various methods on the estimation of parameters using data obtained from dissociated hippocampal neurons.

PROBLEMS

8.1. Modify the shunt cable model to include a parameter K which takes into account the taper of a single equivalent cylinder (Poznanski, 1996). Use the methods provided in this chapter to estimate the six parameters of this model.

8.2. Do a sensitivity analysis of the five electrotonic parameters of the shunt cable model (Verify your solutions with those published in Ali-Hassan et al., 1992). Do a sensitivity analysis of the shunt model with taper.

8.3. The model used in WLAM relies on the following equation for estimating the input output relation $R = MV$. In a one-dimensional system, this would correspond to a straight line with no intercept. This equation was modified to include an intercept and published in Lin and Durand, 1997). Using this method, estimate the six-parameter shunt cable model with taper and compare your results with parameters obtained with other methods.

8.4. The equation $R = MV$ was also modified to reduce the order of the matrix to be inverted (Tawfik and Durand, 1997). Apply this method to the six-parameter shunt cable model and compare your results with those obtained with the WLAM method and the results obtained in Problem 8.3.

8.5. Modify the model to include a multicylinder somatic shunt cable model (cf. Evans et al., 1992 and Chapter 5) and determine the best method to estimate all the parameters of the model.

8.6. Modify the equations of the model response to take into account a voltage clamp input and a current response. Determine which method can best estimate the parameters of the shunt model with and without taper.

8.7. Experimental suggestion: Obtain voltage responses to a short current pulse input in whole-cell patch clamp from granule cells in the dentate gyrus *in vitro*. As determined from the previous sections, apply the 'best' method to estimate the parameters of the shunt cable model or the shunt cable model with taper applied to the experimental data. Determine which model fits the data best.

8.8. Experimental suggestion: Repeat the experiments with a voltage clamp input and current output. Determine the parameters of the model and compare to those obtained in current clamp mode.

ACKNOWLEDGMENTS

This research has been supported by NSF grants # BCS 9315886 and INT 94-17206.

REFERENCES

Albert, A. (1972) *Regression and the Moore-Penrose Pseudo-inverse*. New York: Academic Press.

Ali-Hassan, W.A., Saidel, G.M. and Durand, D. (1992) Estimation of electronic parameters of neurons using an inverse fourier transform technique. *IEEE Trans. Biomed. Eng.*, **39**, 493–501.

Bard, Y. (1974) *Nonlinear Parameter Estimation*. New York: Academic Press.

Bardakjian, B.L. (1990) Characterization of passive electrical properties of excitable cells using optimization methods. *Med. Prog. Tech.*, **16**, 11–17.

Beck, J.V. and Arnold, K.J. (1977) *Parameter Estimation in Engineering and Science*. Chapter 4, pp. 117–129. New York: John Wiley and Sons.

Ben-Israel, A. and Greville, T. (1980) *Generalized Inverses: Theory and Applications*, Robert E. Krieger: Huntington.

Dennis, J.E., David, M.G. and Welsh, R.E. (1981) An adaptive nonlinear least-squares algorithm. *ACM Trans. Math. Softw.*, **7**, 348–368.

D'Aguanno, A., Bardakjian, B.L. and Carlen, P.L. (1986) Passive neuronal membrane parameters: comparison of optimization and peeling methods. *IEEE Trans. Biomed. Eng.*, **33**, 1188–1196.

D'Aguanno, A., Bardakjian, B.L. and Carlen, P.L. (1989) A system model for investigating passive electrical properties of neurons. *Biophys. J.*, **55**, 1169–1182.

Durand, D., Carlen, P.L., Gurevich, N., Ho, A. and Kunov, H. (1983) Electronic parameters of rat dentate granule cells measured using short current pulses and HRP staining." *J. Neurophysiol.*, **50**, 1080–1096.

Durand, D. (1984) The somatic shunt cable model for neurons. *Biophys. J.*, **46**, 645–653.

Eichman, G. and Stojancic, M. (1987) Superresolving signal and image restoration using a linear associative memory. *Appl. Optics*, **26**, 1911–1917.

Evans, J.D., Kember, G.C. and Major, G. (1992) Techniques for obtaining analytical solutions to the multicylinder somatic shunt cable model for passive neurons. *Biophys. J.*, **63**, 350–365.

Fu, P., D'Aguanno, A., Bardakjian, B.L. and Carlen, P.L. (1989) Computation of the passive electrical parameters of neurons using a system model. *IEEE Tran. Biomed. Eng.*, **36**, 55–64.

Glenn, L.L., Samojla, B.G. and Whitney, J.F. (1987) Electrotonic parameters of cat spinal-motoneurons evaluated with an equivalent cylinder model that incorporates non-uniform membrane resistivity. *Brain Res.*, **435**, 398–402.

Gruber, M.H.J. (1990) *Regression Estimators: A Comparative Study*. Boston: Academic Press.

Hoerl, A.E. (1962) Application of Ridge analysis to regression problems. *Chem. Eng. Prog.*, **58**, 54–59.

Hoerl, A.E. and Kennard, R.W. (1970) Ridge regression: biased estimation for non-orthogonal problems. *Technometrics*, **12**, 55–67.

Homma, J., Kosugi, Y. and Sato, T. (1990) Evaluation and improvement of associative memory for pattern recognition. *Appl. Optics*, **29**, 1675–1681.

Jackson, M. (1992) Cable analysis with the whole cell patch clamp: theory and experiment. *Biophys. J.*, **61**, 756–766.

Kalaba, R., Lichtenstein, Z., Simchony, T. and Tesfatsion, L. (1992) Linear and nonlinear associative memories for parameter estimation. *Information Sciences*, **61**, 45–66.

Kalaba, R. and Udwadia, F.E. (1993) Associative memory approach to the identification of structural and mechanical systems. *J. Optim. Theory and Applicat.*, **76**, 207–223.

Kohonen, T. (1988) *Self-Organization and Associative Memory*, pp. 49–50. New York: Springer-Verlag.

Kawato, M. (1984) Cable properties of a neuron model with non-uniform membrane resistivity. *J. Theor. Biol.*, **111**, 149–169.

Lin, J.C. and Durand, D.M. (1997) Non-linear parameter estimation by weighted linear associative memory with non-zero interception. *IEEE Trans. Syst., Man and Cybern.*, **27**, 692–702.

Lin, J.C. and Durand, D.M. (1996) Weighted linear associative memory approach to non-linear parameter estimation. *J. of Optim. Theory and Applications*, **90**, 139–159.

Major, G., Larkman, A.U., Jonas, P., Sakmann, B. and Jack, J.J. (1994) Detailed passive cable models of whole-cell recorded CA3 pyramidal neurons in rat hippocampal slices. *J. Neurosci.*, **14**, 4613–38.

Marquardt, D.W. (1970) Generalized inverses, Ridge regression, biased linear estimation, and nonlinear estimation. *Technometrics*, **12**, 591–612.

Marquardt, D.W. and Snee, R.D. (1975) Ridge regression in practice. *Am. Stat.*, **29**, 3–19.

Ott, H.W. (1976) *Noise Reduction Techniques in Electronic Systems.* New York: Wiley Interscience.

Penrose, R. (1956) On best approximate solution of linear matrix equations. *Proc. Camb. Phil. Soc.*, **52**, 17–20.

Pongracz, F., Firestein, S. and Shepherd, G.M. (1991) Electrotonic structure of olfactory sensory neurons analyzed by intracellular and whole cell patch techniques. *J. Neurophysiol.*, **65**, 747–758.

Poznanski, R.R. (1996) Transient response in a tapering cable model with somatic shunt. *NeuroReport*, **7**, 1700–1704.

Rall, W. (1977) Core conductor theory and cable properties of neurons. In *Handb. Physiol.*, Chapter 3, pp. 39–97. Bethesda, MD: Amer. Physiol. Society.

Rall, W., Burke, R.E., Holmes, W.R., Jack, J.J.B., Redman, S.J. and Segev, I. (1992) Matching dendritic neuron models to experimental data. *Physiol. Rev.*, **72**, S159–186.

Riley, J.D. (1956) Solving systems of linear equations with a positive definite, symmetric, but possibly ill-conditioned matrix, *Math. Tables Aids Comput.*, **9**, 96–101.

Rose, P.K. and Vanner, S.J. (1998) Differences in somatic and dendritic specific membrane resistivity of spinal motoneurons: an electrophysiological study of neck and shoulder motoneurons in the cat *J. Neurophysiol.*, **60**, 149–166.

Tawfik B. and Durand, D.M. (1997) Parameter estimation by reduced order linear associative memory (ROLAM). *IEEE Trans. Biomed. Eng.*, **44**, 297–305.

Tawfik, B. and Durand, D.M. (1994) Non-linear parameter estimation by linear association: application to a five-parameter passive neuron model. *IEEE Trans. Biomed. Eng.*, **41**, 461–469.

Tikhonov, A.N. and Arsenin, V.Y. (1977) *Solutions of Ill-posed Problems*. V.H. Winston and Sons.

Tranchina, D. and Nicholson, C. (1986) A model for the polarization of neurons by extrinsically applied electric fields. *Biophys. J.*, **50**, 1139–1156.

Watkins, D.S. (1991) *Fundamentals of Matrix Computations*, New York: John Wiley and Sons.

White, J.A., Manis, P.B. and Young, E.D. (1992) The parameter identification problem for the somatic shunt model. *Biol. Cybern.*, **66**, 307–318.

Wright, W.N., Bardakjian, B.L., Valiante, T.A., Perez-Velasquez, J.L. and Carlen, P.L. (1996) White noise approach for estimating the passive electrical properties of neurons. *J. Neurophysiol.*, **76**, 3442–3450.

CHAPTER 9
DETERMINATION OF CABLE PARAMETERS FOR NEURONS WITH GAP JUNCTIONS

Roman R. Poznanski and Osamu Umino

1. INTRODUCTION

Electrical coupling occurs between neurons in a number of nuclei in the mammalian brain (for reviews see Korn and Faber, 1979; Dudek et al., 1983; Llinas, 1985; Schuster, 1992). In the adult brain gap-junction connexin proteins have been identified, and associated with purely resistive electrotonic coupling (Dermietzel et al., 1989).

Gap junctions have been observed throughout the nervous system (Dermietzel and Spray, 1993; Spray and Dermietzel, 1996). In particular, they have been known to occur between the dendrites of non-pyramidal neurons (Kosaka and Hama, 1985), dentate gryus granule cells (MacVicar and Dudek, 1982), olivary neurons (Sotelo et al., 1974), olfactory bulb granule cells (Regehr et al., 1991), and dorsal medullary neurons (Dean et al., 1997) just to mention a few. In the retina, gap junctions have been observed in nearly all neuronal types (Vaney, 1994), and in particular the sites of gap-junctions in membranes of bipolar cells have occurred mostly between the dendritic tips (Umino et al., 1994). Indeed, gap junctions in mammals are mostly dendrodendritic and they allow linear transmission of subthreshold potentials (Bennett and Spray, 1987).

The simplified view regarding passive spread of electrical signals in dendrites (see e.g. Spruston et al., 1994) requires change of thought and subsequent modification not only because voltage-dependent membrane properties may come into play, but also because if neurons are found to be electrotonically coupled then their passive electrical constants will undoubtedly be effected by such coupling. This electrical coupling must therefore be taken into account when an attempt is made to determine the passive electrical constants of such neurons. However, few attempts have been made so far to devise models that can quantitatively measure the affect of electrotonic coupling on the evaluation of passive cable parameters of single neurons coupled to other neurons in a network or cluster. Earlier experiments and modeling attempts using single compartmental models of neurons coupled together by a resistance revealed that the input resistance and the time-constant of voltage decay was indeed modified in the presence of electrotonic coupling (see e.g. Getting, 1974). Therefore the aim of this Chapter is to quantitatively characterize electrotonic coupling in dendrites with ligand-gated ionic channels by the use of a mathematical model.

2. CABLE MODEL OF A SINGLE NEURON WITH DENDRODENDRITIC GAP-JUNCTION

A cable model of a single neuron can be represented by a one-dimensional uniform equivalent cylinder attached to a lumped (isopotential) soma (Johnston and Wu, 1995). The combination of an isopotential cell body connected to a single cable is a justifiable first approximation because a reasonable degree of impedance matching does occur between the highly branched structures of the dendrites and the cable (Rall, 1977). In addition, a shunt (i.e. resistance connected to ground) is needed to correspond to the affects of dendrodendritic coupling between coupled neurons and for simplicity this can be attached to the distal end of the dendritic cable, producing a 'leaky' boundary condition.

Investigations using cable models with such 'leaky' boundary conditions began with Rall (1969) and Jack and Redman (1971). With advances in the experimental techniques of recording voltage responses from coupled neurons (see e.g. Dean et al., 1997) new models need to be formulated to adapt to modern electrophysiological techniques (see e.g. Jackson, 1992; Taylor et al., 1996; Wright et al., 1996). A modification to incorporate a more realistic cable with a discretely juxtaposed value of membrane resistivity (R_m) in order to allow for the inclusion of ionic channels (see e.g. Mozrzymas and Bartoszkiewicz, 1993) would be extremely useful, but if the channel distance is assumed to be small then the assumption of the specific membrane resistivity (R_m) being constant and continuously distributed along the dendrites and somata is a reasonable approximation.

The mathematical model of a network of coupled neurons (shown schematically in Figure 9.1) is based on the assumption that the soma of each neuron is isopotential; that the neuronal membrane is passive; that the dendritic and somatic

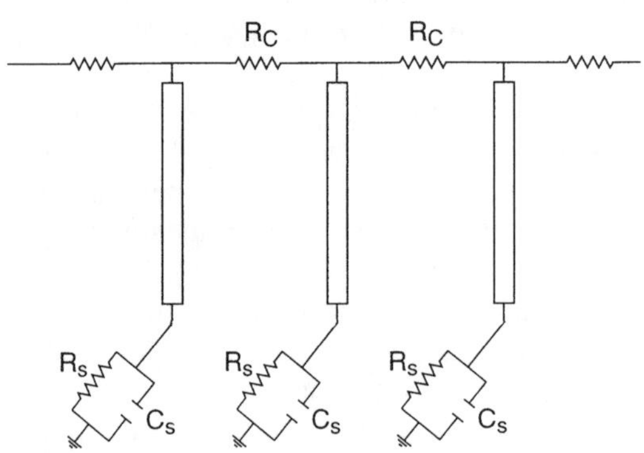

Figure 9.1 A network of electrotonically coupled neurons each represented by an equivalent cylinder representing the dendritic tree and a *RC*-circuit for the soma. Each neuron is coupled to another neuron at the dendritic terminal and the gap-junction is represented by a resistance (R_c).

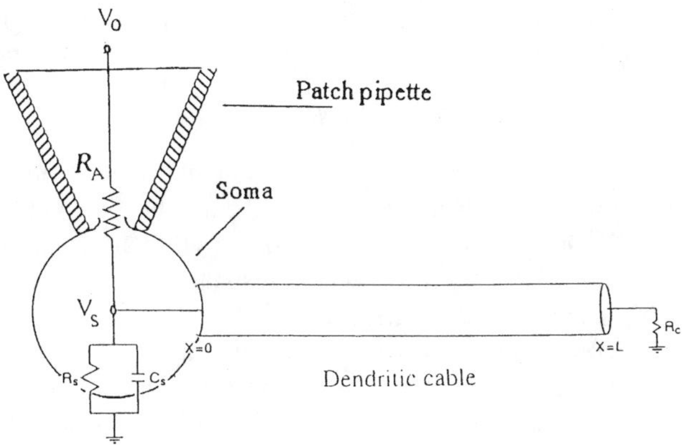

Figure 9.2 A 'ball and stick' cable model of a single neuron with a 'leaky' boundary condition at $X = L$. The soma is voltage clamped in the whole-cell (or perforated) configuration with a patch-pipette. A voltage (V_0) is delivered to the soma across an access resistance (R_A).

membrane resistivity are both identical and continuous; the cytoplasmic resistivity is constant for all the neurons in the cluster; that the somatic membrane resistivity of each neuron is equal; entire dendritic tree can be collapsed into a single equivalent cylinder for each of the neurons in the network. With all the above assumptions the purpose will be to determine the passive cable parameters of single neurons in a network of coupled neurons *in vivo* under a patch-clamp. This will require obtaining expressions for the electrical constants of a cable model with inclusion of leaky boundary conditions to represent a gap-junction in the distal dendrites. Apart from the earlier work of Jack and Redman (1971), and also Winslow and Miller (1989) there appears to be no quantitative method of estimating parameters for cable models with 'leaky' terminations under the patch-clamp recording technique introduced into cable analysis by Jackson (1992) and shown in Figure 9.2.

3. THE FORWARD PROBLEM: ANALYTICAL SOLUTION FOR THE VOLTAGE RESPONSE

The decay of voltage from the steady-state at the soma can be shown to be governed by an infinite series of exponential terms:

$$V_s(t) = C_1 \exp(-t/\tau_1) + C_2 \exp(-t/\tau_2) + C_3 \exp(-t/\tau_3) + \cdots \tag{9.1}$$

where $(C_1, C_2, C_3, \ldots)$ are constants representing each component's amplitude (independent of time) known as the Fourier coefficients (see Appendix 9.1), and $(\tau_1, \tau_2, \tau_3, \ldots)$ represent infinitely many time-constants whose relation to the membrane time-constant can be shown to be:

$$\tau_n = \frac{\tau_m}{(1 + \xi_n^2)}, \quad n = 1, 2, 3, \ldots \tag{9.2}$$

where the roots $\{\xi_n: n=1,2,\ldots\}$ satisfy the following expression:

$$\tan(\xi_n L) = -\xi_n \left\{ \frac{R+\gamma\beta-\xi_n^2}{(R-\xi_n^2)\beta-\gamma\xi_n^2} \right\} \quad (9.3)$$

with $\gamma = R_s/R_{D\infty}$, $\beta = R_{D\infty}/R_c$, and $R = R_s/R_A$. Here L is the electrotonic length of the neuron (dimensionless), D is the dendritic trunk diameter (μm), R_c is the coupling resistance (MΩ), R_s is the somatic input resistance (MΩ), R_D is the input resistance of a semi-infinite cable (MΩ), R_A is the access resistance of the pipette (MΩ).

The above solution exhibits characteristics of the somatic shunt cable model (cf. Redman et al., 1987; Poznanski, 1987, 1996) consisting of an infinite summation of exponential terms of which the largest time-constant of exponential decay (τ_1) is smaller than the membrane time-constant (τ_m) due to the presence of current leakage through the gap-junction.

It is important to mention that the time course of the voltage decay for the coupled system of neurons governed by (9.1) is expected to follow a different time course to that of the uncoupled or single neuron. This is because in the case of no coupling (i.e. $\beta=0$) between all the dendrites, an additional term reflecting the voltage decay with membrane time-constant is needed to be included, and also the eigenvalues expressed by (9.3) will be different. Hence the time course for the uncoupled system follows:

$$V_s(t) = B_1 \exp(-t/\tau_0) + B_2 \exp(-t/\tau_1^*) + \cdots, \quad \text{if } \beta = 0 \quad (9.4)$$

where $\tau_1^*, \tau_2^*, \ldots$ are time-constants whose relation to the membrane time-constant (τ_0) is governed by a different set of eigenvalues than those in (9.3), and this membrane time-constant is not equal to the membrane time-constant in the coupled system (τ_m). In fact, for a cable with no shunt, its membrane time-constant is expected to be greater than the time-constant of a cable with a shunt, since a shunt anywhere along the cable will decrease the time-constant of voltage decay in the whole dendritic tree (Durand, personal communication).

4. THE INVERSE PROBLEM: DETERMINATION OF THE MODEL PARAMETERS

The standard procedure of estimating the membrane time-constant from the slope of the semi-logarithmic plot of the voltage transients at the soma cannot apply if the end-termination of the equivalent cylinder is not sealed (see Johnston and Wu, 1995). Thus, if the specific capacitance of the membrane (C_m) is a biological constant taken to equal 1 μF/cm² (Cole, 1968) then the membrane time-constant (τ_m) can be found from $\tau_m = R_m C_m$. If the neuron showed evidence of membrane folding or any other abnormal membrane morphology this would justify assuming the capacitance was other than 1 μF/cm². The specific membrane resistivity (R_m) can be obtained from the input resistance (R_N) of the neuron measured experimentally from intrasomatic recordings (see Jack et al., 1983: pp. 145–146):

$$R_m = \frac{\pi^2 R_N^2 D^3}{16 R_i} \left\{ 1 + \left[1 + \frac{16 R_i d_{soma}^2}{\pi D^3 R_N} \right]^{1/2} \right\}^2 \quad (9.5)$$

and hence $R_{D\infty} = \{2R_i^{1/2}/(\pi D^{3/2})\} R_m^{1/2}$ and $R_s = R_m/(\pi d_{soma}^2)$ where d_{soma} is the diameter of the soma (μm).

The cytoplasmic resistivity (R_i) is a function of internal electrolyte composition and in most instances is difficult to estimate (see e.g. Schanne, 1969). The value may be increased in the presence of a large concentration of organic solutes and by binding of some of the intracellular organic anions. Alternatively, given R_m the value of R_i can be determined from the following expression:

$$R_i = \frac{R_m \pi^2 R_N^2 D^3}{4[R_m^2 - 2\pi R_N d_{soma}^2 R_m + \pi^2 R_N^2 d_{soma}^4]} \tag{9.6}$$

A plot of the experimentally derived voltage transients on a logarithmic scale, fitted with a regression line to the averaged data, a straight line would show that the charging curve was well approximated by a single exponential (see e.g. Johnston and Wu, 1995), and the negative reciprocal of the slope of this line would yield an estimate of the largest time-constant (τ_1) that is smaller than the membrane time-constant and is expressed by the relation

$$\tau_1 = \frac{\tau_m}{(1+\xi_1^2)}. \tag{9.7}$$

An estimate of the electrotonic length (L) can now be found from the following expression (given that the value of $\beta = R_{D\infty}/R_c$ is unknown and must be treated as a free parameter):

$$L = \begin{cases} \dfrac{1}{\xi_1} \operatorname{atan}\left\{\dfrac{\xi_1(\xi_1^2 - R - \gamma\beta)}{R\beta - \xi_1^2(\gamma+\beta)}\right\}, & R+\gamma\beta < \xi_1^2, \\ \dfrac{1}{\xi_1} \operatorname{atan}\left\{\dfrac{\xi_1(\gamma\beta + R - \xi_1^2)}{R\beta - \xi_1^2(\gamma+\beta)}\right\}, & R+\gamma\beta > \xi_1^2. \end{cases} \tag{9.8}$$

For the multiple cylinder model (see Appendix 9.2) it is possible to isolate L_1 from the transcendental equation (A9.9), but to yield L_1 estimates, L_2 estimates need to be given in advance, and therefore no unique solution for L_1 or L_2 is possible. The dendritic to somatic conductance ratio parameter (ρ) can be obtained from the following relation:

$$\rho = \left(\frac{R_s}{R_{D\infty}}\right)\left\{\frac{1-\exp(-2L)+\beta[1+\exp(-2L)]}{1+\exp(-2L)+\beta[1-\exp(-2L)]}\right\}. \tag{9.9}$$

Finally, the input resistance of the whole neuron (R_N) can be written as a combination of the somatic and dendritic resistance's in parallel:

$$R_N = \frac{R_s R_D}{(R_s + R_D)} \tag{9.10}$$

where $R_D = R_s/\rho$ is the input resistance of the dendritic cable (MΩ).

Since all the above parameters (L, ρ, R_N) are dependent on the unknown coupling resistance (R_c) parameter, one way to estimate this parameter is to plot R_N values generated from (9.10) against a selection of R_c values and subsequently match a corresponding value of R_c that is associated to an R_N value obtained experimentally from the patch-clamp recordings. A check on the appropriateness of the coupling resistance R_c estimate can now be determined from the following expression:

$$R_c = R_{D\infty}\left\{\frac{\xi_1\gamma + (R-\xi_1^2)\tan(\xi_1 L)}{\gamma\xi_1^2\tan(\xi_1 L) + \xi_1(\xi_1^2 - R)}\right\}. \tag{9.11}$$

5. EXPERIMENTALLY MEASURED INPUT RESISTANCE AND TIME CONSTANT OF VOLTAGE DECAY

The Rall cable model has been shown by Durand et al. (1983) and also Turner and Schwartzkroin (1983) to be appropriate for modeling granule cells from the dentate gryus. The novel application of the present chapter is to further extend the model in order to incorporate the leakage of current in the dendrites that may occur as a result of electrical coupling between granule cells as observed by MacVicar and Dudek (1982). With the tight-seal whole-cell configuration an additional resistance of the patch pipette must be included in the model to correspond to the access resistance of the patch pipette tip (cf. Jackson, 1992), and with the perforated patch-clamp technique, the access resistance of the patch pipette is higher due to the fact that the membrane is not ruptured. Spruston and Johnston (1992) had measured the access resistance of the pipette to vary between 30 and 100 MΩ.

It is assumed the data given in Table 9.1 reflects granule cells that are electrically coupled in the distal dendrites, although this assumption was not tested by Staley et al. (1992) or Spruston and Johnston (1992). The equivalence of τ_1 and τ_m holds only if the membrane resistivity (R_m) is uniform or in the absence of dendritic coupling (i.e. $\beta = 0$), so the simple method of equating the largest time-constant of voltage-decay peeled from the semi-logarithmic plot of the somatic potential to the membrane time-constant and thus estimating the membrane resistivity (R_m) from the definition of ($\tau_m = R_m C_m$) as shown by Spruston and Johnston (1992) and Staley et al. (1992) becomes invalid (see also Carlen and Durand, 1981). So τ_1 used here to represent the largest time-constant of voltage decay corresponds to τ_0 referred to by Spruston and Johnston (1992) and Staley et al. (1992).

Table 9.1 Summary of input resistance (R_N) and time-constant (τ_1) values recorded from dentate gryus granule cells with the use of the patch-clamp technique*

Patch-Clamp Method	R_N (MΩ)	τ_1 (ms)
Whole-cell	228 ± 14.2	26.9 ± 1.2
Perforated	446 ± 87	43 ± 4.0

* Values are expressed as averages ± standard error of the average, and have been extracted from the experimental studies of Staley et al. (1992) and Spruston and Johnston (1992).

6. ESTIMATES OF CABLE PARAMETERS FOR THE PASSIVE NEURON MODEL

Based on commonly accepted values of $C_m = 1\,\mu F/cm^2$ and $R_i = 100\,M\Omega$ (used by Spruston and Johnston, 1992), together with the parameters given in Table 9.1 it can readily be shown from (9.5) that $R_m = 28\,k\Omega\,cm^2$ and $103\,k\Omega\,cm^2$ for the whole-cell and perforated patch-clamp, respectively. It follows that $R_{D\infty} = 237\,M\Omega$ (and $455\,M\Omega$), $R_s = 6150\,M\Omega$ (and $22{,}700\,M\Omega$), $\gamma = 26$ (and 50), and from (9.7) and the definition of τ_m, $\xi_1 = 0.18$ (and 1.17) under the whole-cell and perforated patch-clamp techniques, respectively.

Assuming an access resistance of the pipette (R_A) to be 30 and $60\,M\Omega$ for the whole-cell and perforated patch-clamp, respectively, and plotting L governed by (9.8), ρ governed by (9.9), and R_N governed by (9.10) against the unknown parameter (R_c) a possible estimate of these parameters can be found by choosing R_c values that correspond to an R_N value measured experimentally as given in Table 9.1. This procedure is illustrated in Figure 9.3 from which it can be seen that for $R_N = 228$ and $446\,M\Omega$, the approximate value of R_c is 240 and $455\,M\Omega$, respectively. Based on these R_c estimates approximate L and ρ values can be read off as shown in Figure 9.4 and 9.5, respectively. It is clear that for the whole-cell and perforated patch-clamp recordings, the values of L are 1.13 and 0.87, and the values of ρ are 26 and 50, respectively.

A possible discrepancy in the parameter estimates for the perforated and whole-cell patch-clamp technique reflects the recordings made from different species, guinea pigs and rats, respectively.

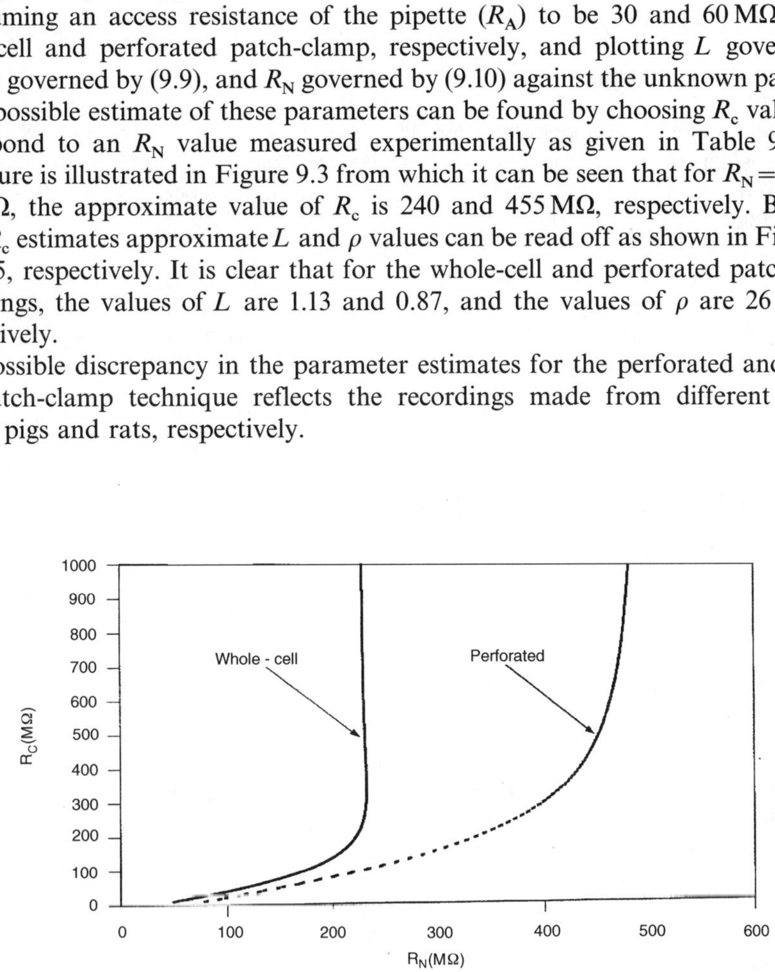

Figure 9.3 A plot of the input resistance (R_N) of the whole neuron governed by (9.10) against the coupling resistance parameter (R_c) varied in the simulation. Experimentally estimated R_N values (see Table 9.1) allow approximate R_c values to be obtained for both whole-cell and perforated patch-clamping techniques. Note $R_N < R_c$.

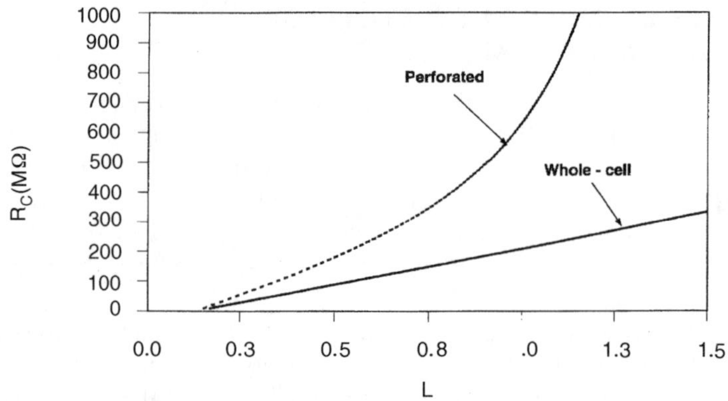

Figure 9.4 A plot of the electrotonic length parameter (L) governed by (9.8) against the coupling resistance parameter (R_c) varied in the simulation. For both the whole-cell and perforated-clamp techniques a specific value of L can be obtained from the R_c estimates found in Figure 9.3.

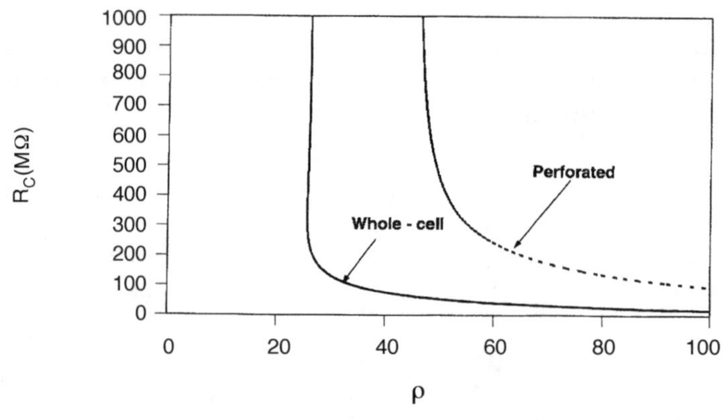

Figure 9.5 A plot of the dendritic-to-somatic conductance ratio (ρ) governed by (9.9) against the coupling resistance parameter (R_c) varied in the simulation. For both the whole-cell and perforated patch-clamp techniques a specific value of ρ is obtained from R_c estimates found in Figure 9.3.

7. DISCUSSION

The effect of a shunt placed along the distal dendrites on the passive membrane parameters was first examined by Carlen and Durand (1981) with a compartmental model. It was shown that the influence of gap-junction coupling decreases the input resistance, decreases the time-constant of decay, and increases the electrotonic length (see also Rörig et al., 1996). Unfortunately, both Carlen and Durand (1981) and Rörig et al. (1996) had equated both the time courses for the voltage transients in the coupled and uncoupled systems, to be identical. Furthermore, a simple expression for the electrotonic length, originally derived by Rall (1969) which applies only to uncoupled neurons with sealed-ends was used. Yet their conclusion on the effects of coupling on the electrotonic length remains valid, even if the membrane time-constants for both systems are not the same, being smaller for the coupled network. This is because electrotonic spread of current is expected to be delayed for the uncoupled system (see Fu et al., 1992).

The largest time-constant of the exponential voltage decay represents the membrane time-constant only if the specific membrane resistivity (R_m) is uniform. Therefore, the largest time-constant of voltage decay in the series (9.1) reflects the membrane time-constant (τ_m^*) only in the absence of dendrodendritic coupling (i.e. $\beta=0$):

$$\tau_0=\tau_m^*. \tag{9.12}$$

Otherwise, the largest time-constant of voltage decay is given from (9.7) which is clear to be smaller, i.e. $\tau_1<\tau_0$ since $\tau_m<\tau_m^*$ and hence the effect of coupling is to decrease the largest time-constant of voltage-decay (cf. Getting, 1974; Carlen and Durand, 1981; Fu et al., 1992).

The influence of gap-junction coupling on L can be shown explicitly by letting $\beta=0$ in (9.8) for $\xi_1^2<\gamma\beta+R$ as it represents the more realistic case, to yield the electrotonic length of an uncoupled neuron (L^*):

$$L^*=\frac{1}{\xi_1^*}\operatorname{atan}\left\{\frac{\xi_1^{*2}-R}{\gamma\xi_1^*}\right\} \tag{9.13}$$

where $\xi_1^*=(\tau_m^*/\tau_1^*-1)^{1/2}$. Unfortunately Spruston and Johnston (1992) did not estimate higher order time-constants (i.e. τ_1^*, τ_2^*, etc.) from their experimental data, and also the uncertainty of the neurons under consideration being coupled, a direct comparison cannot be made. Under the assumption that $\xi_1^*\approx\xi_1$ it follows from (9.8) and (9.13) that $L>L^*$ if $R\beta>\xi_1^2(\gamma+\beta)$ and $\xi_1^2(\gamma^2+\xi_1^2)+R^2>2R\xi_1^2$ which is true for values in Tables 9.1 and 9.2. Hence indicating that coupling can increase the electrotonic length resulting in a less compact neuron. But again this is only speculative since an estimate of ξ_1^* must be found before a firm conclusion can be made.

Table 9.2 Summary of electrotonic parameters influenced by R_c*

	L	ρ
No Coupling†	0.74	20
Coupling	0.87	50

* Based on morphological estimates of $D=2.72\,\mu m$, and $d_{soma}=12\,\mu m$ taken from Spruston and Johnston (1992) and compatible to the morphological estimates of Desmond and Levy (1984).
† Based on values calculated by Spruston and Johnson (1992).

Nevertheless, by adding a gap-junction to the end of a dendritic cable and providing a greater access for current flow, appears to be equivalent to extending the electrotonic length of the dendritic cable.

These results further support the previously found observations of Spruston and Johnston (1992) that the granule cells have a unique electrotonic structure with a high dendritic-to-conductance ratio (ρ). Note that a large ρ represents 'dendritic dominance' or a very small soma attached to a large dendritic tree; such a small size of the granule cell soma has been previously observed (see e.g. MacVicar and Dudek, 1982) and from camera lucida reconstructions (see Turner and Schwartzkroin, 1983). Also, an electrotonic length of approximately 1.0 was observed by MacVicar and Dudek (1982) which falls in the range obtained here for both the perforated and whole-cell patch-clamp techniques.

The theory presented herein is only applicable to dendrites that are passive. For dendrites that are not passive and contain numerous voltage-dependent conductances that are active near the resting membrane potential, the use of electrotonic structure for description of neuronal function is no longer appropriate. However, it is still widely accepted that near the resting potential the electrotonic structure of neurons acts like a "skeleton" and is seen as the scaffolding upon which the active conductances come into action. In dentate gryus granule cells, for example, the non-NMDA receptor channels display a linear I–V relation (see Keller et al., 1991).

8. CONCLUSIONS AND FUTURE PERSPECTIVES

In this paper a theoretical analysis was presented to interpret cable parameters for neurons that are coupled electrotonically via dendrodendritic gap junctions (see Mollgard and Moller, 1975). The equivalent cylinder model was employed for this task since the basic phenomenological behavior of the neuron is preserved at minimal computational expense. It provides a way of interpreting the electrophysiological data that is found in situations when the neuron being studied is coupled with other neurons. A neuron with passive membrane properties, coupled to other neurons via dendrodendritic gap junctions was modeled as a non-sealed-end equivalent cylinder attached to a lumped soma. Both the forward and inverse problems were presented in order to determine the effects of dendrodendritic coupling on the passive cable (electrotonic) parameters. The effect of the leak resistance attached at the end of the dendritic cable on the dendritic-to-somatic conductance ratio (ρ) and the electrotonic length (L) was applied to granule cells in the dentate gyrus of the hippocampus.

The present model assumes that gap junctions are found at all the distal ends of dendrites which is a physiologically restrictive assumption. In fact, a quantitative study of the size and number of gap junctions at various locations (i.e. soma versus distal dendrites) needs to be undertaken to compare the reality of the present model. Indeed, it was shown by Carlen and Durand (1981) that shunts located more proximal to the soma caused greater increases in the electrotonic length (L) than did distally located shunts. Further modifications to the present model will need to be undertaken so that more proximal shunts are distributed. For example, a two-cable model with a shunt (resistance to ground) inserted in the central region of the dendritic tree (see e.g. Major, 1993; Major et al., 1993). If gap junctions are assumed to be located on the somata as well as the dendrites then this would effect the value of R_s by creating a somatic-leakage or pathway for current flow. The present theory can also be extended to incorporate neurons with somatic gap junctions or dendritic trees with taper (Poznanski et al., 1995).

Little conclusive information exists on the strength of gap junctions between neurons in the mammalian brain. The coupling resistance parameter (R_c) needs to be re-evaluated using the dual intracellular recording technique which has been so far only used with intracellular sharp electrodes at the soma of coupled neurons (see e.g. MacVicar and Dudek, 1982). With the recent advances in recording methods it would be interesting to apply a double whole-cell patch-clamp to a pair of isolated coupled cells. This technique could be used on isolated neurons with coupled dendrites to allow for an estimate of R_c based on coupling ratio analysis to be obtained.

Determination of cable parameters for neurons with dendrodendritic gap junctions could have enormous implications in understanding how neurons process information. For example, neurons in the outer retina are known to send graded signals continuously to the next layers of the retina as well as between neurons forming a quasi-syncytial lattice of electrotonic networks (see Chapter 10). Consequently, the possibility of gap junctions shunting the signals in the electrotonic networks could provide a functional explanation for the absence of Na^+ channels and the need for spikes (see for a review Poznanski and Umino, 1997). Furthermore, such shunting effects of gap junctions could be important in dendritic integration as synchronous inputs would be more effective at activating a cluster of cells than would asynchronous inputs, suggesting a possible role of electrotonic coupling in synchronization of electrical activity, particularly in the hippocampus and neocortex (Dudek et al., 1983).

PROBLEMS

9.1. The classical representation of synaptic current first proposed by Eccles is a conductance change which varies in time, with the difference between the membrane potential and the reversal potential: $I(t)=g(t)[V_{REV}-V(t)]$. Prove or refute the theoretical claim by Agmon-Snir and Segev (1993) that the average time of $I(t)$ as it propagates passively from a point on the dendritic tree (i.e. a point on the equivalent cable) is independent of the waveform $I(t)$ by choosing two different representations of $g(t)$:

(i) a Dirac-delta function $\delta(t)=\infty$ for $t=0$ else $\delta(t)=0$,
(ii) α-function of the form $t\alpha^2\exp(-\alpha t)$ where α is a positive constant.

Now by varying the parameter R_c in the model discuss the possible function of electrotonic coupling via dendrodendritic gap junctions in dendritic integration of synaptic signals.

9.2. The 'dumbbell' model consists of a lumped-soma attached to a finite cable terminating in a natural boundary condition (see Chapter 14, Appendix 14.1). By adding a capacitance to the boundary condition at $X = L$ obtain a solution and compare the result with Jackson (1993). How different are the results to the case of a leaky boundary condition used to represent a dendrodendritic gap-junction by a simple resistance? Give an interpretation of the resistive elements by providing suitable values in each case. Derive an analytical solution by attaching a resistance R_L to the soma of the cable model shown in Fig. 9.2. If gap junctions are assumed to be located between the somata of neurons then how would the value of R_A be effected by the inclusion of the additional resistance? Compare your results with the modeling work of Poznanski et al. (1995).

9.3. The present model assumes that gap junctions are found at ALL the distal ends of dendrites which is physiologically unrealistic. Extend the model to include gap junctions at some terminals of the dendritic tree. You may develop models as shown in Chapter 5 by Evans and Chapter 6 by Glenn and Knisley. What effect has branching on the cable parameters and compare the results with those found by Horwitz (1981).

9.4. *Experiment*: Apply a dual whole-cell patch-clamp to a pair of isolated neurons that are electrically coupled between either dendrites and/or somata. After injecting current to the soma of one neuron measure the voltage response at the soma of each neuron and hence estimate R_c using the coupling ratio analysis discussed in this chapter.

9.5. Most gap junctions are electrically linear, however, there are few exceptions like in the Xenopus where the junctional permeability depends on the trans-junctional differences in voltage. As a result, modeling the effect of gap junction nonlinearities in systems of coupled cells have been undertaken by Baigent et al. (1997).

(i) Reformulate the equations to include a non-Ohmic current flow through the gap-junction in the form: $I = f(V)/R_j$ where $f(V) = V - V^2 + (V/2)^3$ and obtain the steady-state solution to the new model. [Hint: Apply Cole's theorem.]
(ii) What happens as R_j approaches infinity?
(iii) Do dendrodendritic gap junctions at distal sites from the soma also have an effect on the electrotonic length (L) and input resistance (R_N) of the neuron?

9.6. Find an analytical solution to the network model shown in Fig. 9.1 for a current step applied to the soma of the central neuron. For a suitable choice of the passive cable parameters find the voltage response at the soma of a neuron other than the central neuron. After how many neurons from the central neuron has the voltage response completely vanished? Now consider the same network model but in order to allow for the inclusion of ionic channels assume discretely juxtaposed values of R_m distributed along each equivalent cable (see Mozrzymas and Bartoszkiewicz, 1993) Check your result by letting the distance between each ionic channel become infinitesimal.

9.7 Repeat problem 9.6 with voltage-dependent ionic channels discretely imposed at specific locations along the dendritic cable. Assume voltage-dependent conductances and compare your findings with those obtained numerically by Yuen and Durand (1991) for a single granule cell. [Hint: Linearize the ionic current and solve the resultant Volterra integral equation using method of successive substitutions.] Now repeat the analysis with the model shown in Figure 9.6 below.

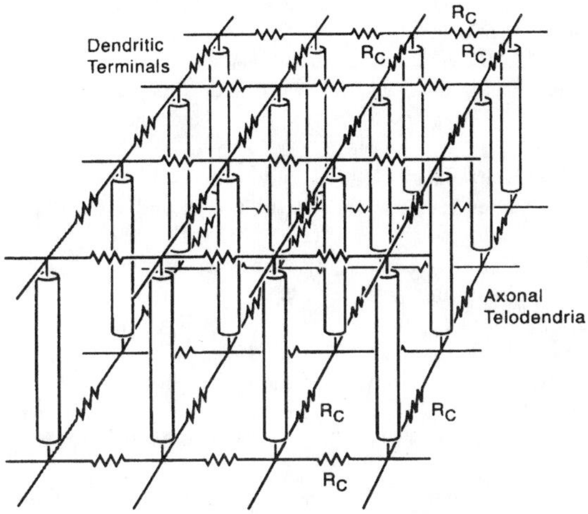

Figure 9.6 Electrical circuit representation of a 3D discrete cable model of a neural syncytium. The electrotonic coupling is represented by a resistance (R_c) shown connecting the cables. (Modified from Poznanski and Umino, 1997).

ACKNOWLEDGMENTS

The authors would like to thank Dominique Durand for reading the manuscript and Dan Johnston for advice on the values for the access resistance used in the perforated patch-clamp experiments. This work was supported by the Ministry of Education, Science, Sports, and Culture (Monbusho). Special thanks to Hiroshi Yamamoto who proved that $\tau_m < \tau_m^*$ holds in general case, and Jonathan Evans for Appendix 9.2.

REFERENCES

Agmon-Snir, H. and Segev, I. (1993) Signal delay and input synchronization in passive dendritic structures. *J. Neurophysiol.* **70**, 2066–2085.

Baigent, S., Stark, J. and Warner, A. (1997) Modelling the effect of gap junction nonlinearities in systems of coupled cells. *J. Theoret. Biol.* **186**, 223–239.

Bennett, M.V.L. and Spray, D.C. (1987) Gap junctions. In Encyclopedia of Neuroscience, edited by G. Adelman, Birkhaüser: Boston.

Carlen, P.L. and Durand, D. (1981) Modelling the postsynaptic location and magnitude of tonic conductance changes resulting from neurotransmitters or drugs. *Neurosci.*, **6**, 839–846.

Cole, K.S. (1968) *Membranes, Ions and Impulses*. Berkeley: University of California Press.

Dean, J.B., Huang, R.-Q., Erlichman, J.S., Southard, T.L. and Hellard, D.T. (1997) Cell-Cell coupling occurs in dorsal medullary neurons after minimizing anatomical-coupling artifacts. *Neurosci.* **80**, 21–40.

Dermietzel, R. and Spray, D.C. (1993) Gap junctions in the brain: where, what type, how many and why? *Trend Neurosci.*, **16**, 186–192.

Dermietzel, R., Traub, O., Kwang, T.K., Beyer, E., Bennett, M.V.L., Spray, D.C. and Willecke, K. (1989) Differential expression of three gap junction proteins in developing and mature brain tissue. *Proc. Natl. Acad. Sci. USA*, **86**, 10148–10152.

Desmond, N.L. and Levy, W.B. (1984) Dendritic caliber and the 3/2 power relationship of dentate granule cells. *J. Comp. Neurol.*, **227**, 589–596.

Dudek, F.E., Andrew, R.D., MacVicar, B.A., Snow, R.W. and Taylor, C.P. (1983) Recent evidence for and possible significance of gap junctions and electrotonic synapses in the mammalian brain. In *Basic Mechanisms of Neuronal Hyperexcitability*, edited by H.H. Jasper and N.M. van Gelder, pp. 31–73. Liss: New York.

Durand, D., Carlen, P.L., Gurevich, N., Ho, A. and Kunov, H. (1983) Electrotonic parameters of rat dentate granule cells measured using short current pulses and HRP staining. *J. Neurophysiol.*, **50**, 1080–1097.

Fu, P., Bardakjian, B.L. and Carlen, P.L. (1992) Ethanol uncouples dentate granule neurons by increasing junctional resistance: a multineuronal system model approach. *Neurosci.*, **51**, 47–54.

Getting, P.A. (1974) Modification of neuron properties by electrotonic synapses. I. Input resistance, time constant, and integration. *J. Neurophysiol.*, **37**, 846–857.

Horwitz, B. (1981) Neuronal plasticity: how changes in dendritic architecture can affect the spread of postsynaptic potentials. *Brain Res.* **224**, 412–418.

Jack, J.J.B., Noble, D. and Tsien, R.W. (1983) *Electric Current Flow in Excitable Cells*. Oxford: Clarendon Press.

Jack, J.J.B. and Redman, S.J. (1971) The propagation of transient potentials in some linear cable structures. *J. Physiol. (Lond.)*, **215**, 283–320.

Jackson, M.B. (1992) Cable analysis with the whole-cell patch clamp: theory and experiment. *Biophys. J.*, **61**, 756–766.

Jackson, M.B. (1993) Passive current flow and morphology in the terminal aborizations of the posterior pituitary. *J. Neurophysiol.* **69**, 692–702.

Johnston, D. and Wu, S. (1995) *Foundations of Cellular Neurophysiology*. Cambridge, MA: MIT Press.

Keller, B.U., Konnerth, A. and Yaari, Y. (1991) Patch clamp analysis of excitatory synaptic currents in granule cells of rat hippocampus. *J. Physiol. (Lond.)*, **435**, 275–293.

Korn, H. and Faber, D.S. (1979) Electrical interactions between vertebrate neurons: field effects and electrotonic coupling. In *Fourth Neurosciences Study Program*, edited by F.O. Schmitt and F.G. Wordon pp. 333–358. Cambridge, MA: MIT Press.

Kosaka, T. and Hama, K. (1985) Gap junctions between non-pyramidal cell dendrites in rat hippocampus (CA1 and CA3 region): A combined Golgi-electron microscopy study. *J. Comp. Neurol.*, **231**, 150–161.

Llinas, R.R. (1985) Electrotonic transmission in the mammalian CNS. In *Gap Junctions*, edited by M.V.L. Bennett and D.C. Spray, pp. 337–353. Cold Spring Harbor Laboratory.

MacVicar, B.A. and Dudek, F.E. (1982) Electrotonic coupling between granule cells of rat dentate gyrus: physiological and anatomical evidence. *J. Neurophysiol.*, **47**, 579–592.

Major, G. (1993) Solutions for transients in arbitrarily branching cables: III Voltage clamp problems. *Biophys. J.*, **65**, 469–491.

Major, G., Evans, J.D. and Jack, J.J.B. (1993) Solutions for transients in arbitrarily branching cables: II Voltage clamp theory. *Biophys. J.*, **65**, 450–468.

Mollgard, K. and Moller, M. (1975) Dendrodendritic gap junctions: a developmental approach. In Advances in Neurology, Vol. 12, edited by G.W. Kreutzberg. pp. 79–89. Raven: New York.

Mozrzymas, J.W. and Bartoszkiewicz, M. (1993) The discrete nature of biological membrane conductance, channel interaction through electrolyte layers and the cable equation. *J. theor. Biol.*, **162**, 371–380.

Poznanski, R.R. (1987) Transient response in a somatic shunt cable model for synaptic input activated at the terminal. *J. Theoret. Biol.*, **127**, 31–50.

Poznanski, R.R. (1996) Transient response in a tapering cable model with somatic shunt. *NeuroReport*, **7**, 1700–1704.

Poznanski, R.R., Gibson, W.G. and Bennett, M.R. (1995) Electrotonic coupling between two CA3 hippocampal pyramidal neurons: a distributed cable model with somatic gap-junction. *Bull. Math. Biol.*, **57**, 865–881.

Poznanski, R.R. and Umino, O. (1997) Syncytial integration by a network of coupled bipolar cells in the retina. *Prog. Neurobiol.*, **53**, 273–291.

Rall, W. (1969) Time constants and electrotonic length of membrane cylinders and neurons. *Biophys. J.*, **9**, 1483–1508.

Rall, W. (1997) Core conductor theory and cable properties of neurons. In *Handbook of Physiology, The Nervous System*, edited by E.R. Kandel, pp. 39–97. Bethesda, MD: American Physiological Society.

Redman, S.J., McLachlan, E.M. and Hirst, G.D.S. (1987) Nonuniform passive membrane properties of rat lumbar sympathetic ganglion cells. *J. Neurophysiol.*, **57**, 633–644.

Reyher, C.H.K., Lubke, J., Larsen, W.J., Hendrix, G.M., Shipley, M.T. and Baumgarten, H.G. (1991) Olfactory bulb granule cell aggregates: morphological evidence for interperikaryal electrotonic coupling via gap junctions. *J. Neurosci.*, **11**, 1485–1495.

Rörig, B., Klausa, G. and Sutor, B. (1996) Intracellular acidification reduced gap junction coupling between immature rat neocortical pyramidal neurones. *J. Physiol.* (*Lond.*), **490**, 31–49.

Schanne, O.F. (1969) Measurement of cytoplasmic resistivity by means of the glass microelectrode. In *Glass Microelectrodes*, edited by M. Lavallee, O.F. Schanne and N.C. Herbert, pp. 299–321. New York: John Wiley & Sons.

Schuster, T.H. (1992) Gap junctions and electrotonic coupling between hippocampal neurons: recent evidence and possible significance-a review. *Concepts in Neurosci.*, **3**, 135–155.

Sotelo, C., Llinas, R. and Baker, R. (1974) Structural study of inferior olivary nucleus of the cat: morphological correlates of electrotonic coupling between neurons. *J. Neurophysiol.*, **37**, 541–559.

Spray, D.C. and Dermietzel, R. (eds.) (1996) Gap Junctions in the Nervous System. Austin: R.G. Landes Publisher.

Spruston, N., Jaffe, D.B. and Johnston, D. (1994) Dendritic attenuation of synaptic potentials and currents: the role of passive membrane properties. *Trends Neurosci.*, **17**, 161–166.

Spruston, N. and Johnston, D. (1992) Perforated patch-clamp analysis of the passive membrane properties of three classes of hippocampal neurons. *J. Neurophysiol.*, **67**, 508–529.

Staley, K.J., Otis, T.S. and Mody, I. (1992) Membrane properties of dentate gyrus granule cells: comparison of sharp microelectrode and whole-cell recordings. *J. Neurophysiol.*, **67**, 1346–1358.

Taylor, W.R., Mittman, S. and Copenhagen, D.R. (1996) Passive electrical cable properties and synaptic excitation of tiger salamander retinal ganglion cells. *Vis. Neurosci.*, **13**, 979–990.

Turner, D.A. and Schwartzkroin, P.A. (1983) Electrical characteristics of dendrites and dendritic spines in intracellularly stained CA3 and dentate hippocampal neurons. *J. Neurosci.* **3**, 2381–2394.

Umino, O., Maehara, M., Hidaka, S., Kita, S. and Hashimoto, Y. (1994) The network properties of bipolar-bipolar cell coupling in the retina of teleost fishes. *Vis. Neurosci.*, **11**, 533–548.

Vaney, D.I. (1994) Patterns of neuronal coupling in the retina. In *Progress in Retinal and Eye Research*, vol. **13**, edited by N.N. Osbourne and G. Chader, pp. 301–355. Oxford: Pergamon Press.

Winslow, R.L. and Miller, R.F. (1989) A theoretical and experimental study of the effects of non-uniform membrane resistance on the shape of single cell charging curves. *Neurosci.*, **29**, 761–771.

Wright, W.N., Bardakjian, B.L., Valiante, T.A., Perez-Velazquez, J.L. and Carlen, P.L. (1996) White noise approach for estimating the passive electrical properties of neurons. *J. Neurophysiol.*, **76**, 3442–3450.

Yuen, G.L.F. and Durand, D. (1991) Reconstruction of hippocampal granule cell electrophysiology by computer simulation. *Neurosci.*, **41**, 411–423.

APPENDIX 9.1 EXPLICIT REPRESENTATION OF THE FOURIER COEFFICIENTS

The Fourier coefficients $(C_1, C_2, C_3, \ldots, C_n)$ can be shown to be expressed by the following expression $(\xi_n \neq 0)$:

$$C_n = \frac{[\gamma V_0 R\xi_n/(R-\xi_n^2)]\{\gamma\xi_n\psi_1 + \beta\gamma\xi_n\psi_2 + (R-\xi_n^2)\psi_3 + \xi_n(\xi_n^2+1)\psi_4\}}{(1+\xi_n^2)(R-\xi_n^2)\psi_5(\psi_6+\psi_7)} \quad \text{(A9.1)}$$

where

$$\psi_1 = \xi_n \sin(\xi_n L) + \sinh(L), \quad \text{(A9.2)}$$

$$\psi_2 = \cosh(L) - \cos(\xi_n L), \quad \text{(A9.3)}$$

$$\psi_3 = \xi_n \cosh(L) - \xi_n \cos(\xi_n L) + \beta[\xi_n \sinh(L) - \sin(\xi_n L)], \quad \text{(A9.4)}$$

$$\psi_4 = \cosh(L) + \beta \sinh(L), \quad \text{(A9.5)}$$

$$\psi_5 = \cosh(L)[1 + R + \gamma\beta] + \sinh(L)[\beta(1+R) + \gamma], \quad \text{(A9.6)}$$

$$\psi_6 = [(\gamma\xi_n)/(R-\xi_n^2)]^2\{L/2 + 1/\gamma + \sin(2\xi_n L)/4\xi_n\}, \quad \text{(A9.7)}$$

$$\psi_7 = L/2 + \gamma \sin^2(\xi_n L)/(R-\xi_n^2) - \sin(2\xi_n L)/4\xi_n. \quad \text{(A9.8)}$$

APPENDIX 9.2 TRANSCENDENTAL EQUATION FOR THE MULTIPLE EQUIVALENT CYLINDER MODEL

For a more realistic model with a multiple number of equivalent cylinders attached to a lumped-soma (see Chapter 5) it can be shown that the transcendental equation becomes:

$$\tan(\xi_n L_1) = -\xi_n \left\{ \frac{R - \gamma\xi_n \tan(\xi_n L_2) + \gamma\beta - \xi_n^2}{[R - \gamma\xi_n \tan(\xi_n L_2) - \xi_n^2]\beta - \gamma\xi_n^2} \right\} \quad \text{(A9.9)}$$

where L_1 is the electrotonic length of cylinder 1 containing a leaky boundary condition, and L_2 is the electrotonic length of cylinder 2 with a sealed-end boundary condition. Note as $L_2 \rightarrow 0$ (A9.9) reduces to (9.3) with $L_1 = L$.

CHAPTER 10
SYNCYTIAL INTEGRATION IN THE OUTER RETINA

Tetsuya Yagi

1. INTRODUCTION

In the neural circuits as well as other excitable cell systems, cells are occasionally coupled by electrical synapses and constitute an electrically continual laminar structure, refered to as a syncytium. The neural syncytium is often found in the developing neocortex and hippocampus, but is also prevalent in a number of nuclei in the mature mammalian brain and especially in the retina. The vertebrate retina is a good model for syncytial integration, i.e. information processing in neural syncytia. Figure 10.1 is a schematic illustration showing a gross structure of the vertebrate retina. The principal cell types are shown with different colors. Each of these principal cell types can be classified into several subtypes, which are not shown in the figure to avoid confusion. In Figure 10.1, the bottom side corresponds to the frontal surface of the retina from which the light comes in through the optical apparatus of the eye. The light passes through the transparent retina to reach the photoreceptor array (gray), where the light signal is transduced to the voltage signal. The voltage signal is transmitted to the horizontal cell (purple) and the bipolar cell (red). The photoreceptors, horizontal cells and bipolar cells interact with each other in the outer plexiform layer (OPL). The bipolar cell transmits the output of the outer retinal circuit to the amacrine cell (white) and the ganglion cell (orange). The bipolar cells, amacrine cells and ganglion cells interact in the inner plexiform layer (IPL). The interplexiform cell (light green) is a unique neuron which provides a feedback pathway from IPL to OPL.

As shown in Figure 10.1, neurons belonging to the same types are arranged in two-dimensional arrays and aggregate in separate layers. It has been well known that the retinal horizontal cells of same subtypes are connected with each other by electrical synapses (Yamada and Ishikawa, 1965; Kaneko, 1971; Kolb, 1977) and possess receptive fields which cover almost the entire retinal surface (Naka and Rushton, 1967; Lamb, 1976; Yagi, 1986). The photoreceptors also constitute a neural syncytium (Cohen, 1965; Baylor et al., 1971; Schwartz, 1976) as well as bipolar cells (Kujiraoka and Saito, 1986; Saito and Kujiraoka, 1988; Umino et al., 1994).

Figure 10.1 Schematic illustration of the vertebrate retina. The principal cell types are distinguished by different colours. See the text for details. Inset: Cross section of the retina. (Reprinted with permission from Matsumoto, T., Kobayashi, H. and Yagi, T. (1993) *Proceedings of IEICE*, **76**(7), 783–791.) (Color plate I)

These syncytia are arranged in multiple layers in the retina and process the visual information hierarchically by chemical synaptic interactions.

Earlier models of chemically interacting networks (of multiple layers) such as those by Tate and Woolfson (1971), Abernethey (1974), and Hara and Takabayashi (1975), did not take into account the syncytial nature of retinal networks (see reviews by Vaney (1994), and Poznanski and Umino, (1997)). On the other hand, models have been proposed to describe the spatio-temporal properties of electrical signals the neural syncytium of the retinal circuit in both discrete (Lamb and Simon, 1976; Torre and Owen, 1983; Torre et al., 1983; Yagi, 1986) and distributed (Naka and Rushton, 1967; Marmarelis and Naka, 1972, Lamb, 1976) forms but did not take into account the multiple layers in the retina. More recently double-layer models by Ohshima et al. (1995) and Yagi et al. (1997) have initiated a move towards combining electrically and chemically interacting networks. These models have shown to provide appropriate descriptions on the spatio-temporal properties of light-induced responses. It can, however, be emphasized that the discrete models reflect finite cell dimensions and cell spacings, which are critical parameters to determine the resolution of visual system.

In this Chapter, analytical solutions are presented for models describing the voltage responses of single layer neural syncytium, and double layer syncytium in which two layers of syncytia are interacting reciprocally. They are shown to provide intuitive understandings on the response properties in terms of physiological parameters, i.e. membrane impedances, strength of chemical synapses and coupling resistances connecting neighbouring cells. Finally, the function of syncytial integration in early vision is characterized by the standard regularization theory.

2. MODEL OF A SINGLE LAYER SYNCYTIUM

In this section, we obtain the analytical solutions describing how electrical signals spread in the single layer syncytium. The retinal rod network is a typical example of such single layer syncytium. Its receptive field size is known to exceed its cell dimension (Schwartz, 1976). The rod horizontal cell, which receives the signal from the rod photoreceptor, is also tightly coupled to neighbors with low resistance gap-junctions and exhibits a large receptive field that encompasses almost the entire retina (Tsukamoto et al., 1987).

2.1 Formulation of the Model

Figure 10.2A illustrates an example of model for the single layer neural syncytium. In this and the following sections, we treat the two-dimensional arrangement of neurons as an one dimensional array, for simplicity. The one-dimensional model can be directly applied to analyse the responses to illuminations which induce homogeneous voltage change in one direction. For example, a long slit of light induces a voltage change which is homogeneous along the long axis and the current spreads only to the right direction of the slit (Lamb, 1976; Detwiler et al., 1978). Although the applications of a one-dimensional model are limited to such stimuli, it is still useful to gain qualitative insights for the properties of voltage responses conducting

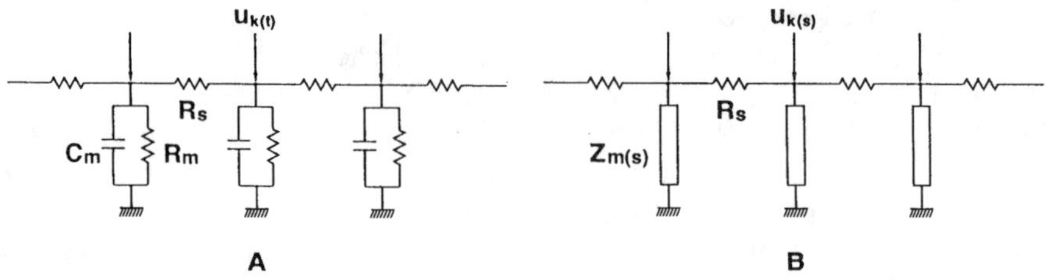

Figure 10.2 A: One-dimensional equivalent electrical circuit model of the single layer syncytium. R_m is a resistance and C_m is a capacitance of the membrane. Each neuron is represented by the RC parallel circuit, which is the simplest version to represent the neuron. R_s represents the resistance of electrical synapse connecting neighboring neurons. $u_k(t)$ is the current induced in kth cell by a stimulus which delivers the current over the syncytium. B: Equivalent circuit model in the frequency domain. Each neuron is represented by the membrane impedance $Z_m(s)$. The voltage response and stimulus current are also expressed in the frequency domain.

two-dimensionally. In Figure 10.2A, each neuron is represented by a RC circuit consisting of a membrane resistance and a membrane capacitance. Neighboring neurons are connected by resistors which represent gap junctions. The gap-junction is a typical structure of the electrical synapse and behaves as a bilateral resistive element with little rectification. Therefore, it can be expressed as a linear resistor in most cases (see Chapter 9).

We consider how the voltage signals spread in the single layer syncytium. Let $v_k(t)$ be the voltage signal of neuron numbered k in the syncytium. Applying Kirchhoff's current law at each node, we obtain

$$\frac{(v_k(t)-v_{k-1}(t))}{R_s}+\frac{(v_k(t)-v_{k+1}(t))}{R_s}+\frac{v_k(t)}{R_m}+C_m\frac{\mathrm{d}v_k(t)}{\mathrm{d}t}=u_k(t),$$

$$(k=-n+1,\ldots,0,\ldots,n-1).$$

When the Laplace transformation is applied, the above equation gives

$$\frac{(v_k(s)-v_{k-1}(s))}{R_s}+\frac{(v_k(s)-v_{k+1}(s))}{R_s}+\frac{v_k(s)}{Z_m(s)}=u_k(s).$$

Here, $v_k(s)$ and $u_k(s)$ are the Laplace transforms of $v_k(t)$ and $u_k(t)$. And

$$Z_m(s)=\frac{R_m}{1+R_mC_ms}$$

corresponds to the membrane impedance of each neuron, when s is substituted by frequency, f as

$$s=2\pi f\mathrm{i} \quad (\mathrm{i}=\sqrt{-1}).$$

Now, it is found to be more convenient and general to apply Kirchhoff's current law in the frequency domain. Figure 10.2B illustrates more general analog circuit model to describe voltage responses in the neural syncytium (Torre and Owen, 1983; Ohshima et al., 1995). As shown in the figure, each neuron is represented by the membrane impedance, $Z_m(s)$, which is not necessary of the RC circuit. The gap-junctional resistance, R_s, is assumed to be constant.

Applying Kirchhoff's current law in the frequency domain, the equation describing the distribution of voltage response is found to be

$$v_{k-1}(s)+c(s)v_k(s)+v_{k+1}(s)=-R_s u_k(s), \tag{10.1}$$

where $c(s)$ is a function of membrane impedance and coupling resistance:

$$c(s)=-\left(2+\frac{R_s}{Z_m(s)}\right).$$

If a symmetry of the stimulus is assumed ($u_{-k}=u_k$), then (10.1) is expressed in matrix form, i.e.

$$\mathbf{Cv}=-R_s\mathbf{u(s)}.$$

Here

$$\mathbf{v}=\begin{bmatrix} v_0(s) \\ v_1(s) \\ \vdots \\ v_n(s) \end{bmatrix}, \quad \mathbf{u}=\begin{bmatrix} u_0(s) \\ u_1(s) \\ \vdots \\ u_n(s) \end{bmatrix}$$

and

$$\mathbf{C}=\begin{bmatrix} c & 2 & 0 & \ldots & 0 & 0 \\ 1 & c & 1 & 0 & \ldots & 0 \\ 0 & 1 & c & 1 & \cdot & \cdot \\ \vdots & \cdot & \cdot & \cdot & \cdot & 0 \\ 0 & \ldots & 0 & 1 & c & 1 \\ 0 & 0 & \ldots & 0 & 1 & c+1 \end{bmatrix}.$$

Since the circuit is linear, we first obtain the point spreading function which is the response to a current stimulus $u_0(s)$ induced in the cell numbered 0 only. The voltage responses to an arbitrary current stimulus are obtained by convolution of the stimulus with the point spreading function. When the stimulus is given at the cell numbered 0, the equation becomes

$$\mathbf{Cv}=-R_s u_0(s)\,\mathbf{e}. \tag{10.2}$$

Here $\mathbf{e}$ is the unit vector. Let us find the general solution using the kth row of (10.2). By assuming the solution of the form,

$$v_k(s)=A(s)\,\rho(s)^k, \tag{10.3}$$

we find the equation for $\rho(s)$ as

$$\rho(s)^2 + c\rho(s) + 1 = 0.$$

Relevant solutions are

$$\rho_1(s) = \frac{-c(s) - \sqrt{c(s)^2 - 4}}{2}, \quad \rho_2(s) = \frac{-c(s) + \sqrt{c(s)^2 - 4}}{2}.$$

Thus, the general solution of $v_k(s)$ is a linear combination of above two solutions, i.e.

$$v_k(s) = A_1(s)\rho_1(s)^k + A_2(s)\rho_2(s)^k.$$

Here, $A_1(s)$ and $A_2(s)$ are functions of frequency and are determined from boundary conditions. Practically, we may think that the number of cell, n, is sufficiently large and $v_n(s) \to 0$ as $n \to \infty$. Therefore, $A_2(s)$ should be 0 because $|\rho_2| \geqslant 1$, and $v_k(s)$ becomes

$$v_k(s) = A_1(s)\rho_1(s)^k.$$

From the first row of (2), $A_1(s)$ is found to be

$$A_1(s) = R_s \frac{u_0(s)}{\sqrt{c^2(s) - 4}}$$

and thus

$$v_k(s) = R_s \frac{u_0(s)}{\sqrt{c^2(s) - 4}} \rho_1(s)^k. \tag{10.4}$$

Equation (10.4) provides intuitive understanding on spatio-temporal properties of the voltage response in the single layer syncytium. Substituting s for $2\pi f$i, we find that an arbitrary frequency component of the voltage response of the kth cell is described by $\rho(2\pi f\text{i})$. We will refer to $\rho(2\pi f\text{i})$ as a decay constant. The modulus of decay constant gives a rate of amplitude decrease as it conducts to the neighbouring neuron. The argument of decay constant gives a phase shift of an arbitrary frequency component of the voltage response during the conduction.

For an arbitrary current stimulus, $u_l(s)(l = -n, \ldots, 0, \ldots, n)$, the voltage response of kth cell is expressed by the spatial convolution of $u_l(s)$ with (10.4):

$$v_k(s) = \frac{R_s}{\sqrt{c^2(s) - 4}} \sum_{l=-\infty}^{\infty} \rho(s)^{|k-l|} u_l(s).$$

The response waveform can be obtained by transforming the above equation into the time domain with the aid of an inverse Fast Fourier Transform (FFT) algorithm.

2.2 Simulation Study on Retinal Rod Syncytium

The electrical signal conducting in the rod syncytium has an interesting property which was referred to as 'a surprising property' (Detwiler et al., 1978). In rods, the time course of voltage response gets faster as the signal conducts the syncytium, which is contradictory to the usual electrotonical signal conduction. In the usual conduction, the time course of signal gets slower as it conducts because of the capacitive delay. To explain this high-pass filtering behavior, Detwiler et al. (1980) added an inductive element in cell membrane which is explained phenomenologically by a partial activation of voltage-dependent conductance. As shown in the Figure 10.3 (inset), each rod is represented by a parallel resistance, r_1 and an inductance l connected in series to a resistance r_2. The membrane capacity is omitted since the light-induced current of rod is so slow that the RC time constant can be neglected. The corresponding membrane impedance is expressed by

$$Z_{\mathrm{m}}(s)=\frac{r_1 r_2 + l r_1 s}{r_1 + r_2 + l s}.$$

The high-pass filtering behaviour of the rod response was calculated with the model and is illustrated in Figure 10.3. We used an empirically obtained equation to simulate the time course of the photo-current (Baylor et al., 1974):

$$u_0(t)=\exp(-\phi t)(1-\exp(-\phi t))^{m-1}.$$

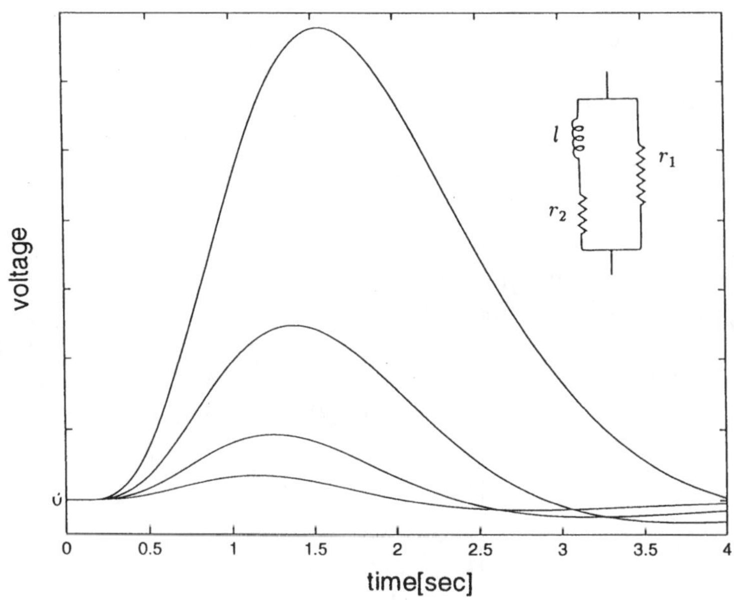

Figure 10.3 A surprising property of rod voltage response calculated from the model. The equivalent circuit of rod membrane is shown in inset. The current calculated with (10.5) is injected at cell numbered 0 and the voltage responses were calculated at cell numbered 0, 2, 4 and 6 with the model.

Here, ϕ and m are constants. The Laplace transform of this equation is

$$u_0(s) = \frac{(m-1)!\phi^m}{(s+\phi)(s+2\phi)\dots(s+m\phi)}. \quad (10.5)$$

In Figure 10.3, the voltage responses to the above current injected in the cell numbered 0 were calculated with (10.4). The voltage responses at different distance from the stimulus are superimposed. As shown in Figure 10.3, the amplitude of response decreases but the time-to-peak becomes shorter as the distance from the stimulus increases. The voltage responses calculated with the analytical solution well reproduce physiological observations (Detwiler et al., 1978, 1980).

3. MODEL OF A DOUBLE LAYER SYNCYTIUM

In the previous section, the mathematical formulation to describe the spatio-temporal properties of electrical signal conducting in the single layer syncytium was explained. In the retina, several types of neurons constitute different layers of neural syncytium and these syncytia are occasionally interacting each other by chemical synapses. The cones are coupled electrically to form the single layer of syncytium (Baylor et al., 1971) as well as the horizontal cells (Naka and Rushton, 1967; Kaneko, 1971). Previous physiological and morphological observations revealed that the cones and the horizontal cells interact by chemical synapses including a negative feedback (Baylor et al., 1971; Fuortes and Simon, 1974; Stell and Lightfoot, 1975). This structure of the retinal circuit suggests that the interaction between neural syncytia plays important roles in visual information processing. In this section, the voltage responses in such reciprocally interacting neural syncytia are studied with a model.

3.1 Formulation of the Model with Chemical Synaptic Interaction

We consider the double layer syncytium in which two layers of syncytium are interacting with reciprocal chemical synapses as shown in Figure 10.4A. The analog electrical circuit model is illustrated in Figure 10.4B. Impedances $Z_{\mathrm{m}1}(s)$ and $Z_{\mathrm{m}2}(s)$ represent membranes of the first and the second layer neurons, respectively. The coupling resistance connecting neighboring neurons of the first layer is represented by $R_{\mathrm{s}1}$ and that of the second layer by $R_{\mathrm{s}2}$. The voltage distributions of the first and the second layer neurons are denoted by $\mathbf{v}_1$ and $\mathbf{v}_2$, respectively:

$$\mathbf{v}_1 = \begin{bmatrix} v_{01}(s) \\ v_{11}(s) \\ \vdots \\ v_{n1}(s) \end{bmatrix}, \quad \mathbf{v}_2 = \begin{bmatrix} v_{02}(s) \\ v_{12}(s) \\ \vdots \\ v_{n2}(s) \end{bmatrix}.$$

Here, $v_{k1}(s)(k=0,\dots,n)$ is the Laplace transform of voltage response of the kth cell in the first layer and $v_{k2}(s)$ in the second layer. We assume the input of current stimulus, $\mathbf{u(s)}$, only in the first layer. This current stimulus corresponds to the

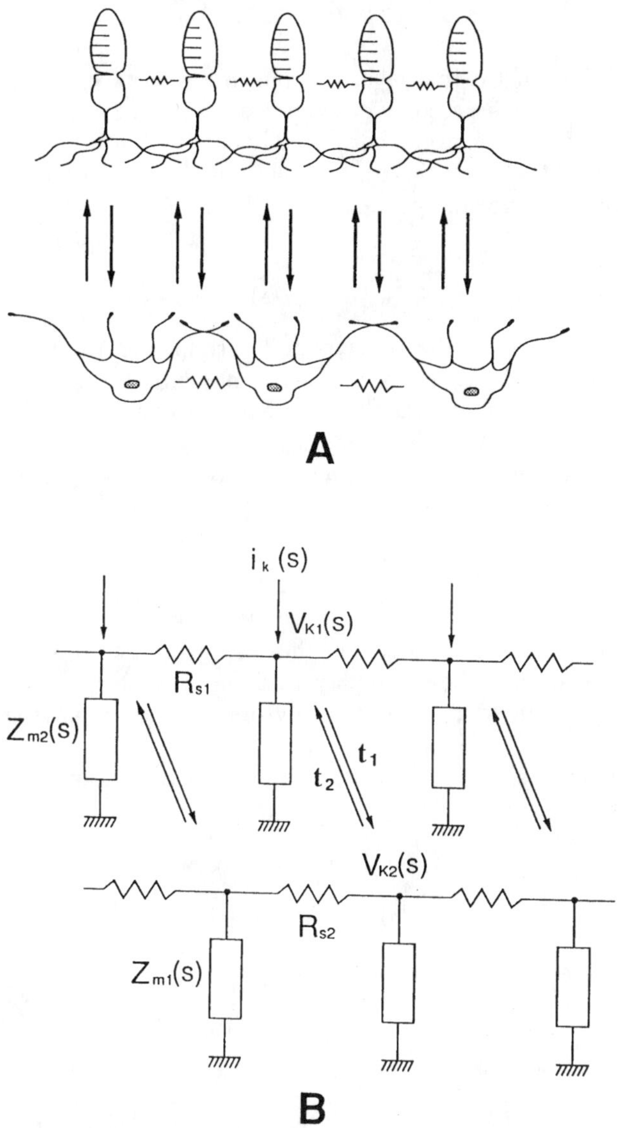

Figure 10.4 A: Schematic drawing of the double layer syncytium. Neurons of the first layer are connected by gap-junctions (indicated by resistances) to constitute a neural syncytium. Neurons of the second layer constitute another syncytium. These two syncytia are interacting with each other by chemical synapses (arrows). B: An analogue circuit model for the double layer syncytium. Each neuron of the first layer is represented by the impedance $Z_{m1}(s)$ and that of the second layer by the impedance $Z_{m2}(s)$. The coupling resistance connecting neighboring cells of the first layer is represented by R_{s1} and that of the second layer by R_{s2}. t_1 is the strength of the synaptic input to the second layer neuron from the neuron of the first layer. The strength of feedback synapse from the second layer neuron to the first layer neuron is represented by t_2. (Reprinted with permission from Ohshima *et al.*, 1995.)

light-induced current, if the first layer neurons are photoreceptors. If the first layer neurons correspond to the second or higher order neurons, the current stimulus corresponds to the postsynaptic current induced by more distal neurons. The strength of synaptic transmission can be represented by a ratio of the postsynaptic current to the presynaptic voltage, i.e.

$$t(s)=\frac{u_{\mathrm{pos}}(s)}{v_{\mathrm{pre}}(s)}.$$

Here, $u_{\mathrm{pos}}(s)$ and $v_{\mathrm{pre}}(s)$ are the postsynaptic current and the voltage response of presynaptic neuron expressed in the frequency domain. The synaptic strength, $t(s)$ has the same dimension as admittance (siemens). In Figure 10.4B, the synaptic input from the first layer neuron to the second layer neuron is expressed by $t_1(s)$ and the feedback synaptic input from the second layer to the first layer is represented by $t_2(s)$. The synaptic strengths are

$$t_1(s)=\frac{u_{k2}(s)}{v_{k1}(s)},\quad t_2(s)=\frac{u_{k1}(s)}{v_{k2}(s)}.$$

Here $u_{k1}(s)$ is the postsynaptic current of the first layer neuron, which is induced by the feedback synapse from the second layer cell. $u_{k2}(s)$ is the postsynaptic current of the second layer neuron which is induced by the synaptic input from the first layer neuron.

Applying Kirchhoff's current law at each node of the circuit, we obtain a set of matrix equations,

$$\mathbf{C}_1\mathbf{v}_1+t_2(s)R_{s1}\mathbf{v}_2=-R_{s1}\mathbf{u(s)}, \tag{10.6}$$

$$t_1(s)R_{s2}\mathbf{v}_1+\mathbf{C}_2\mathbf{v}_2=0. \tag{10.7}$$

Here $\mathbf{C}_1$ and $\mathbf{C}_2$ are

$$\mathbf{C}_1=\begin{bmatrix} c_1 & 2 & 0 & \cdots & 0 & 0 \\ 1 & c_1 & 1 & 0 & \cdots & 0 \\ 0 & 1 & c_1 & 1 & . & \vdots \\ \vdots & . & . & . & . & 0 \\ 0 & \cdots & 0 & 1 & c_1 & 1 \\ 0 & 0 & \cdots & 0 & 1 & c_1+1 \end{bmatrix},\quad \mathbf{C}_2=\begin{bmatrix} c_2 & 2 & 0 & \cdots & 0 & 0 \\ 1 & c_2 & 1 & 0 & \cdots & 0 \\ 0 & 1 & c_2 & 1 & . & \vdots \\ \vdots & . & . & . & . & 0 \\ 0 & \cdots & 0 & 1 & c_2 & 1 \\ 0 & 0 & \cdots & 0 & 1 & c_2+1 \end{bmatrix}.$$

and

$$c_1=-\left(2+\frac{R_{s1}}{Z_{m1}(s)}\right), \tag{10.8}$$

$$c_2=-\left(2+\frac{R_{s2}}{Z_{m2}(s)}\right). \tag{10.9}$$

On combining (10.6) and (10.7), we find equations,

$$(\mathbf{C}_2\mathbf{C}_1 - t_1t_2R_{s1}R_{s2}\mathbf{E})\mathbf{v}_1 = -R_{s1}\mathbf{C}_2\mathbf{u}, \tag{10.10}$$

$$(\mathbf{C}_1\mathbf{C}_2 - t_1t_2R_{s1}R_{s2}\mathbf{E})\mathbf{v}_2 = t_1R_{s2}R_{s1}\mathbf{u}. \tag{10.11}$$

Here, **E** is the identity matrix. As shown previously, we first solve the equation in which only the cell numbered 0 of the first layer is stimulated and the current $u_0(s)$ is induced. Assuming the same form of the solution as (10.3), we find the following equation for $\rho(s)$ from the kth column of (10.10) or (10.11).

$$\rho(s)^4 + (c_1+c_2)\rho(s)^3 + (c_1c_2 + 2 - t_1t_2Z_{m1}Z_{m2})\rho(s)^2 + (c_1+c_2)\rho(s) + 1 = 0. \tag{10.12}$$

Obviously, this can be rearranged to

$$(\rho(s)^2 + 2\alpha(s)\rho(s) + 1)(\rho(s)^2 + 2\beta(s)\rho(s) + 1) = 0. \tag{10.13}$$

Comparing (10.12) with (10.13), we find relations

$$\alpha(s) + \beta(s) = \frac{c_1(s) + c_2(s)}{2}, \quad \alpha(s)\beta(s) = \frac{c_1(s)c_2(s) - t_1(s)t_2(s)Z_{m1}(s)Z_{m2}(s)}{4}.$$

The above relations indicate that $\alpha(s)$ and $\beta(s)$ are solutions of a quadratic equation:

$$x^2 - \frac{c_1(s)+c_2(s)}{2}x + \frac{c_1(s)c_2(s) - t_1(s)t_2(s)R_{s1}R_{s2}}{4} = 0$$

and thus

$$\alpha(s) = \frac{c_1(s)+c_2(s)}{4} + \frac{1}{4}\sqrt{(c_1(s)-c_2(s))^2 + 4t_1(s)t_2(s)R_{s1}R_{s2}},$$

$$\beta(s) = \frac{c_1(s)+c_2(s)}{4} - \frac{1}{4}\sqrt{(c_1(s)-c_2(s))^2 + 4t_1(s)t_2(s)R_{s1}R_{s2}}.$$

Using $\alpha(s)$ and $\beta(s)$ and considering the boundary condition for $n \to \infty$, we find the solutions of (10.13):

$$\rho_1(s) = -\alpha(s) - \sqrt{\alpha(s)^2 - 1}, \tag{10.14}$$

$$\rho_2(s) = -\beta(s) - \sqrt{\beta(s)^2 - 1}. \tag{10.15}$$

Thus, v_{k1} and v_{k2} are expressed as

$$v_{k1}(s) = A_1(s)\rho_1(s)^k + A_2(s)\rho_2(s)^k, \tag{10.16}$$

$$v_{k2}(s) = B_1(s)\rho_1(s)^k + B_2(s)\rho_2(s)^k. \tag{10.17}$$

Equations (10.16) and (10.17) indicate that the amplitude of an arbitrary frequency component of voltage response in the double layer syncytium is characterized by two decay constants, $\rho_1(s)$ and $\rho_2(s)$, for both the first and the second layers.

$A_1(s)$, $A_2(s)$, $B_1(s)$ and $B_2(s)$ are found from the boundary conditions at $k=0$ and $k=1$ which are the first and the second row of (10.10) and (10.11). Solving these boundary conditions, we obtain

$$A_1(s)=\frac{(P_2-c_2Q_2)R_{s1}I_0}{Q_2P_1-Q_1P_2},\quad A_2(s)=\frac{(P_1-c_2Q_1)R_{s1}I_0}{Q_1P_2-Q_2P_1},$$

$$B_1(s)=\frac{t_1R_{s2}R_{s1}Q_2I_0}{Q_2P_1-Q_1P_2},\quad B_2(s)=\frac{t_1R_{s2}R_{s1}Q_1I_0}{Q_1P_2-Q_2P_1}.$$

Here

$$P_1(s)=(c_1c_2+2-t_1t_2R_{s1}R_{s2})+2(c_1+c_2)\rho_1+2\rho_1^2,$$

$$P_2(s)=(c_1c_2+2-t_1t_2R_{s1}R_{s2})+2(c_1+c_2)\rho_2+2\rho_2^2,$$

$$Q_1(s)=(c_1+c_2)+(c_1c_2+3-t_1t_2R_{s1}R_{s2})\rho_1+(c_1+c_2)\rho_1^2+\rho_1^3,$$

$$Q_2(s)=(c_1+c_2)+(c_1c_2+3-t_1t_2R_{s1}R_{s2})\rho_2+(c_1+c_2)\rho_2^2+\rho_2^3.$$

When the retina is illuminated with an image which induces the current $u_1(s)$ in lth neuron of the first layer, $v_{k1}(s)$ and $v_{k2}(s)$ become

$$v_{k1}(s)=A_1(s)\sum_{l=-\infty}^{\infty}\rho_1(s)^{|k-l|}I_l(s)+A_2(s)\sum_{l=-\infty}^{\infty}\rho_2(s)^{|k-l|}I_l(s),$$

$$v_{k2}(s)=B_1(s)\sum_{l=-\infty}^{\infty}\rho_1(s)^{|k-l|}I_l(s)+B_2(s)\sum_{l=-\infty}^{\infty}\rho_2(s)^{|k-l|}I_l(s).$$

3.2 Simulation Study on Red Cone and H1 Horizontal Cell Responses

We calculate the voltage responses of the red cone photoreceptor and the H1 horizontal cell. These neurons constitute the double layer syncytium (Ohshima et al., 1995). Here, the red cone and the horizontal cell are assumed to be represented by membrane impedances:

$$Z_{m1}(s)=\frac{R_{s1}}{1+R_{m1}C_{m1}s},\quad Z_{m2}(s)=\frac{R_{s2}}{1+R_{m2}C_{m2}s},\qquad (s=2\pi f\mathrm{i}).$$

$Z_{m1}(s)$ and $Z_{m2}(s)$ are simple RC circuits since the responses of these neurons do not show prominent high-pass filtering properties. Dynamics of synaptic transmission are also taken into account and the synaptic strengths are described by a first-order low-pass filter (Kamermans et al., 1989), i.e.

$$t_1(s)=\frac{\kappa_1}{1+\tau_1 s},\quad t_2(s)=\frac{\kappa_2}{1+\tau_2 s}.$$

Here, κ_1 and κ_2 are gain, and τ_1 and τ_2 are the time constants.

In Figure 10.5, the spatio-temporal properties of the response calculated from the model were compared with previous physiological experiments. The red cone

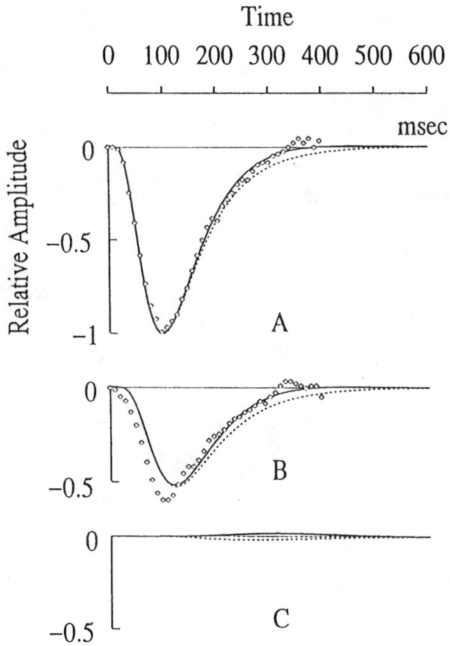

Figure 10.5 Dynamical properties of the red cone response to the slit of light. The current expressed by (10.5) is assumed to be induced in the red cone at the center. ϕ and m are $25\,s^{-1}$ and 6, respectively. The voltage responses were calculated from the cell at 0, 30 and 200 µm and are shown in A, B and C, respectively. Solid lines and dotted lines correspond to $t_2 = -1/(1+\tau_2 s)$ ns and $t_2 = 0$, respectively. $t_1 = -1/(1+\tau_1 s)$ ns in both lines. τ_1 is taken to be 16 ms (Schnapf and Copenhagen, 1982). The time constant of τ_2 is also taken to be 16 ms, for simplicity. Other parameters are: $R_{m1} = R_{m2} = 1\,G\Omega$, $R_{s1} = 30\,M\Omega$, $R_{s2} = 1\,M\Omega$ and $C_{m1} = C_{m2} = 100\,pF$. The responses indicated by ◇ are replotted from Detwiler and Hodgkin (1979). The replotted responses in B and C correspond to −13 µm and −25 µm of the text-figure 10 of their paper. Reprinted with permission from Ohshima et al. (1995).

response to the slit of light is shown by solid lines. In this calculation, only the cell numbered 0 is assumed to be illuminated. The calculated responses in A, B and C are obtained from the cell numbered 0, 3 and 20, respectively. The locations of these cells correspond to the center, 30 and 200 µm, assuming that the cell spacing of the red cone is 10 µm. In this figure, the intracellular responses obained by Detwiler and Hodgkin (1979) are replotted (◇) for comparison. As the distance from slit increases, the response amplitude of red cone decreases and finally the response becomes depolarization due to the negative feedback from H1 horizontal cell (C). The response waveforms were also calculated for the case in which the negative feedback is absent (dotted lines). The waveform of the red cone response becomes slightly slower when the feedback is absent. The responses calculated from the model showed reasonable fits to experimental data.

3.3 Formulation of the Model with Resistive Interaction

Some classes of horizontal cell consist of a soma and an axon terminal which are connected by a slender axon (Boycott and Kolb, 1973; Stell, 1975). The electrical couplings are found between somata as well as between axon terminals. In the fish retina, the soma and the axon terminal are known to be electrotonically interacting by the slender axon (Weiler and Zettler, 1979; Yagi, 1986; Yagi and Kaneko, 1988; Yagi, 1989). Therefore, the horizontal cells in the fish retina form a double layer syncytium with the bilateral interaction through the slender axon (Figure 10.6). We obtain analytical solutions for the voltage response in the two layers. In this analysis, each axon was represented by a pure resistive element, for simplicity. The equations can be obtained in the same way as in the previous section. Relevant equations are

$$\mathbf{D}_1\mathbf{v}_1 + \frac{R_{s1}}{R}\mathbf{v}_2 = -R_{s1}\mathbf{u}(\mathbf{s}), \tag{10.18}$$

$$\frac{R_{s2}}{R}\mathbf{v}_1 + \mathbf{D}_2\mathbf{v}_2 = 0. \tag{10.19}$$

Here $\mathbf{u}(\mathbf{s})$ is the current stimulus induced by the synaptic input from cones. $\mathbf{D}_1$ and $\mathbf{D}_2$ are

$$\mathbf{D}_1 = \begin{bmatrix} d_1 & 2 & 0 & \dots & 0 & 0 \\ 1 & d_1 & 1 & 0 & \dots & 0 \\ 0 & 1 & d_1 & 1 & . & \vdots \\ \vdots & . & . & . & . & 0 \\ 0 & \dots & 0 & 1 & d_1 & 1 \\ 0 & 0 & \dots & 0 & 1 & d_1+1 \end{bmatrix}, \quad \mathbf{D}_2 = \begin{bmatrix} d_2 & 2 & 0 & \dots & 0 & 0 \\ 1 & d_2 & 1 & 0 & \dots & 0 \\ 0 & 1 & d_2 & 1 & . & \vdots \\ \vdots & . & . & . & . & 0 \\ 0 & \dots & 0 & 1 & d_2 & 1 \\ 0 & 0 & \dots & 0 & 1 & d_2+1 \end{bmatrix}.$$

where

$$d_1 = -\left(2 + \frac{R_{s1}}{R} + \frac{R_{s1}}{Z_{m1}(s)}\right), \quad d_2 = -\left(2 + \frac{R_{s2}}{R} + \frac{R_{s2}}{Z_{m2}(s)}\right).$$

Comparing d_1 and d_2 with c_1 and c_2 of (10.8) and (10.9), we find that R_{s1}/R and R_{s2}/R are added to c_1 and c_2 to become d_1 and d_2, respectively. This is because the current at the soma and the axon terminal, leaks not only to the neighbouring cells but also to another syncytium through the axon whose resistance is R. As is easily predicted, (10.18) and (10.19) are analogous to (10.6) and (10.7). Accordingly, the general solutions of (10.18) and (10.19) can be expressed as

$$v_{k1}(s) = A_1(s)\rho_1(s)^k + A_2(s)\rho_2(s)^k,$$

$$v_{k2}(s) = B_1(s)\rho_1(s)^k + B_2(s)\rho_2(s)^k.$$

And

$$\rho_1(s) = -\alpha(s) - \sqrt{\alpha(s)^2 - 1}, \quad \rho_2(s) = -\beta(s) - \sqrt{\beta(s)^2 - 1}.$$

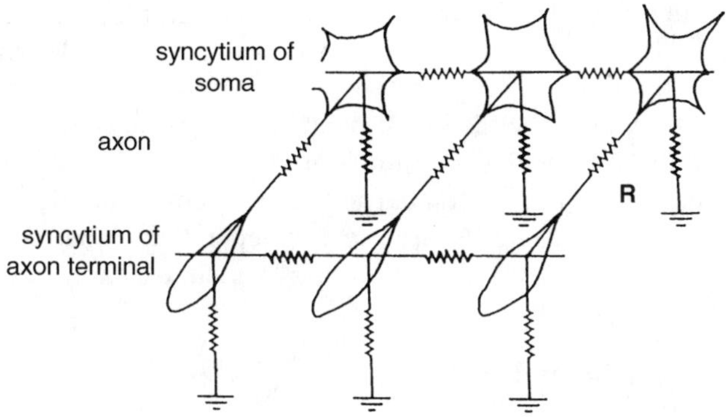

Figure 10.6 The double layer syncytium with resistive interaction. The horizontal cell soma constitutes a neural syncytium. The horizontal cell axon terminal also constitutes another syncytium. Two layers of syncytium are electrotonically interacting through the axons which are expressed by a resistor, *R*.

Here $\alpha(s)$ and $\beta(s)$ are the solutions of the quadratic equation,

$$x^2 - \frac{d_1(s)+d_2(s)}{2}x + \frac{d_1(s)d_2(s) - R_{s1}R_{s2}/R^2}{4} = 0.$$

$A_1(s)$, $A_2(s)$, $B_1(s)$ and $B_2(s)$ are obtained by the boundary conditions, i.e.

$$A_1(s) = \frac{(P_2 - c_2Q_2)R_{s1}u_0}{Q_2P_1 - Q_1P_2}, \quad A_2(s) = \frac{(P_1 - c_2Q_1)R_{s1}u_0}{Q_1P_2 - Q_2P_1},$$

$$B_1(s) = \frac{R_{s2}R_{s1}Q_2I_0}{(Q_2P_1 - Q_1P_2)R}, \quad B_2(s) = \frac{R_{s2}R_{s1}Q_1I_0}{(Q_1P_2 - Q_2P_1)R},$$

where

$$P_1(s) = \left(d_1d_2 + 2 - \frac{R_{s1}R_{s2}}{R^2}\right) + 2(d_1+d_2)\rho_1 + 2\rho_1^2,$$

$$P_2(s) = \left(d_1d_2 + 2 - \frac{R_{s1}R_{s2}}{R^2}\right) + 2(d_1+d_2)\rho_2 + 2\rho_2^2,$$

$$Q_1(s) = (d_1+d_2) + \left(d_1d_2 + 3 - \frac{R_{s1}R_{s2}}{R^2}\right)\rho_1 + (d_1+d_2)\rho_1^2 + \rho_1^3,$$

$$Q_2(s) = (d_1+d_2) + \left(d_1d_2 + 3 - \frac{R_{s1}R_{s2}}{R^2}\right)\rho_2 + (d_1+d_2)\rho_2^2 + \rho_2^3.$$

3.4 How Physiological Parameters Affect Spatial Properties of Voltage Signal

The voltage signal conducting in the single layer syncytium decays monotonically unless negative conductances are included in the circuit and thus it is rather easy to predict how the voltage signal behaves in the single syncytium. However, the voltage signal in the double layer syncytium shows several different modes of conduction, especially when the interaction between the two layers includes negative feedback. In this section, we study how the feedback synapse affects the spatial distribution of voltage signal in the double layer syncytium. Since we consider the signal distribution under steady state conditions, each neuron is represented by the pure resistive element and the synaptic strengths are time independent. Therefore, the decay constants of the syncytium are given by c_1 and c_2 which are expressed as

$$c_1 = -\left(2 + \frac{R_{s1}}{R_{m1}}\right), \quad c_2 = -\left(2 + \frac{R_{s2}}{R_{m2}}\right).$$

Here, R_{m1} and R_{m2} are membranes of the first and the second layers, respectively. Since the spatial properties of voltage responses are characterized by the decay constants, we study how the strength of feedback synapse, t_2 affects their trajectories in the complex plane.

Figure 10.7 illustrates the trajectories of ρ_1 and ρ_2 calculated as t_2 is changed from $-\infty$ to 0. The parameters other than t_2 are estimated from previous physiological observations (Ohshima et al., 1995) and have fixed values in this calculation; ρ_1 and ρ_2 take both real and complex values depending on t_2, although c_1 and c_2 are real in this calculation. ρ_1 and ρ_2 have real values when $-7.0\,\text{ns} \leqslant t_2 \leqslant 0$. In this case, the amplitude of steady state response to the slit of light decays monotonically as the distance from the slit increases. ρ_1 and ρ_2 take identical value (≈ 0.88) when

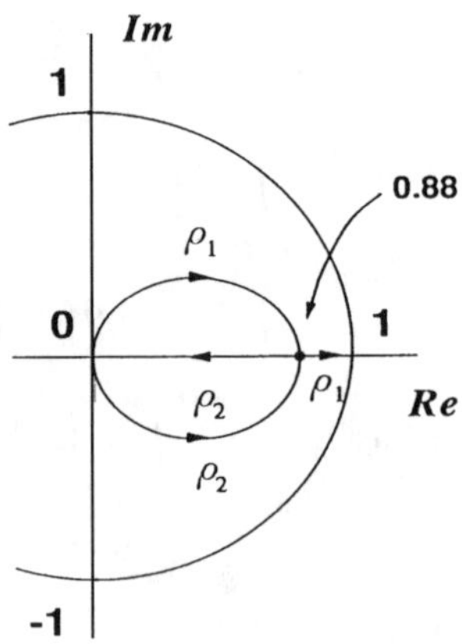

Figure 10.7 Trajectory of decay constant in the complex plane. Decay constants were calculated from (10.14) and (10.15). When t_2 is changed from $-\infty$ to 0, ρ_1 and ρ_2 move to directions indicated by arrows. R_{m1}, R_{m2}, R_{s1}, R_{s2} are same as in Figure 10.5. $t_1 = 1$ ns. (Reprinted with permission from Ohshima *et al.* (1995).)

$t_2 \approx -7.0$ ns. In this case, the response decays with a single exponential in both layers, although the two layers are interacting with each other. As $t_2 \to 0$, which means the strength of negative feedback becomes weaker, $\rho_1 \to 0.84$ and $\rho_2 \to 0.97$. In contrast, when $-\infty < t_2 < -7.0$ ns ρ_1 and ρ_2 take conjugate complex numbers, i.e. $\rho_1 = \rho_2^*$. In this region, the amplitude of the steady-state response decays with the oscillation in the H1 horizontal cell as well as in the red cone. Such oscillatory spatial profile has not been observed in either the red cone nor the H1 horizontal cell responses. Network parameters falling in this region are not appropriate to explain physiological responses.

Figure 10.8 illustrates spatial distributions of the red cone and the H1 horizontal cell responses to the slit of light. In this figure, t_2 is taken to be -1.0 ns. The amplitude of red cone response decreases to half of its maximum at about 30 μm from the edge

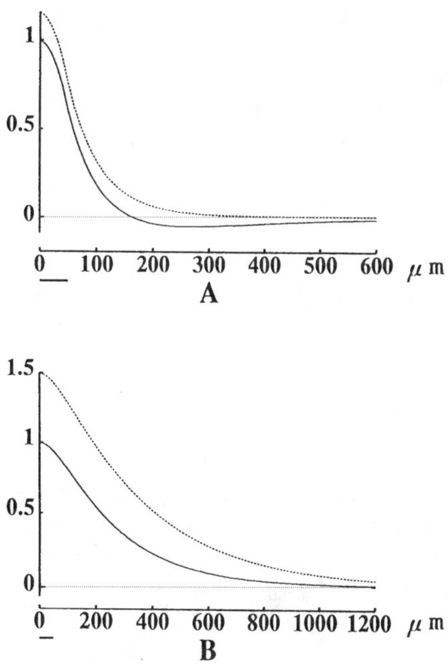

Figure 10.8 Spatial profile of red cone and horizontal cell responses calculated from the model. The response amplitude is plotted as a function of the distance from the cell center. The amplitude is normalized by that obtained in the cell center. The cell spacing of red cone is taken to be 10 μm. Response amplitudes are normalized by the amplitude at 0 μm. The slit of light is assumed to induce a constant current I_0 in red cones located between -50 and 50 μm. The horizontal bars below the abscissa indicate the illuminated region. Dotted lines are calculated with $t_2 = 0$, which corresponds to the absence of negative feedback. Parameters of the model are same as in Figure 10.5. A: Spatial distribution of the red cone voltage. B: Spatial distribution of the H1 horizontal cell voltage. (Reprinted with permission from Ohshima *et al.*, 1995.)

of slit (Figure 10.8A, solid line). This is a reasonable fit to previous physiological experiments (Detwiler and Hodgkin, 1979). The red cone response shows a receptive field surround where a slightly depolarized response due to the negative feedback from the horizontal cell is seen. A half-decay distance of the H1 horizontal cell is about 200 μm (Figure 10.8B, solid line), which is also consistent with previous physiological experiments (Lamb, 1976; Yagi, 1986). When the negative feedback is absent, the receptive field surround of the red cone disappears and the response amplitude increases (dotted line in Figure 10.8A). The increment of amplitude due to the absence of feedback is more prominent in the horizontal cell response (dotted line in Figure 10.8B). The amplitude is about 1.5 times larger at 0 μm when the feedback is absent. The half-decay distance is prolonged about 90 μm in the horizontal cell. Thus, the feedback is found to have larger effects on the H1 horizontal cell response.

4. FUNCTION OF SYNCYTIAL INTEGRATION IN EARLY VISION

In previous sections, spatio-temporal properties of the voltage signal conducting in the neural syncytium are studied with equivalent circuit models. A further step, however, is required to explain what roles the neural syncytium play in visual information processing. It has been pointed out that some of early vision problems can be formulated in terms of standard regularization theory, in which early vision tasks are attributed to minimization of a cost function (Poggio and Koch, 1985; Poggio et al., 1985). This framework is considered to be useful to interpret the function of retinal circuit, since analog electrical circuits are a natural way of solving the minimization problems (Poggio and Koch, 1985; Koch et al., 1986).

4.1 Standard Regularization Theory

Previous theoretical studies pointed out that early vision can be interpreted as a set of ill-posed problems (Poggio and Koch, 1985). Accordingly, some of early vision problems are found to be formulated in terms of the standard regularization theory (Poggio and Koch, 1985; Poggio et al., 1985). For instance, these problems are attributed to minimization of a cost function,

$$G(y) = \|y - d\|^2 + \lambda_r \int \sum_{r=1}^{p} \left(\frac{\mathrm{d}^r y(\xi)}{\mathrm{d}\xi^r} \right)^2 \mathrm{d}\xi, \quad \lambda_r > 0. \tag{10.20}$$

Here, d and y correspond to a given image and a filtered image to be interpolated, respectively. The second term of the cost function ensures uniqueness and convergence of the solution. This choice corresponds to a constraint that the (noiseless) image is smooth because of a band-limited nature of image and has therefore bounded derivatives. In this case, the first term of (10.20) decreases as y becomes closer to d. On the other hand, the second term of (10.20) decreases as y becomes smooth.

Since we described neural syncytia as a discrete network, it is convenient to replace the derivative operations in (10.20) by the difference operations (Kobayashi et al., 1993), i.e.

$$\frac{dy(\xi)}{d\xi} \to y_k - y_{k-1},$$

$$\frac{d^2y(\xi)}{d\xi^2} \to y_{k-1} - 2y_k + y_{k+1}.$$

for $k = 1, \ldots, n$. These operations can be expressed by

$$\frac{dy(\xi)}{d\xi} \to \mathbf{Dy}, \quad \frac{d^2y(\xi)}{d\xi^2} \to \mathbf{Ly}$$

where

$$\mathbf{D} = \begin{bmatrix} 1 & 0 & 0 & \ldots & 0 \\ -1 & 1 & 0 & 0 & \ldots \\ 0 & -1 & 1 & 0 & \ldots \\ \vdots & . & . & . & 0 \\ 0 & \ldots & 0 & -1 & 1 \end{bmatrix}, \quad \mathbf{L} = \begin{bmatrix} -2 & 1 & 0 & \ldots & 0 \\ 1 & -2 & 1 & 0 & \ldots \\ 0 & 1 & -2 & 1 & \ldots \\ \vdots & . & . & . & 1 \\ 0 & \ldots & 0 & 1 & -2 \end{bmatrix}$$

and

$$\mathbf{y} = \begin{bmatrix} y_0 \\ y_1 \\ \vdots \\ y_n \end{bmatrix}, \quad \mathbf{d} = \begin{bmatrix} d_0 \\ d_1 \\ \vdots \\ d_n \end{bmatrix}.$$

Note that

$$\mathbf{DD}^{\mathrm{T}} = -\mathbf{L}.$$

Here T denotes the transpose. Consequently, the regularization problem is formalized in discrete case by the minimization of

$$G(\mathbf{y}) = \|\mathbf{y} - \mathbf{d}\|^2 + \sum_{r=1}^{p} \lambda_r \begin{cases} \|\mathbf{L}^{r/2}\mathbf{y}\|^2 & r\text{: even} \\ \|\mathbf{DL}^{(r-1)/2}\mathbf{y}\|^2 & r\text{: odd,} \end{cases} \tag{10.21}$$

Equation (10.21) is referred to as pth order regularization problem in discrete formulation (Kobayashi et al., 1993). The 2nd order regularization problem, which will be shown to be relevant to the double layer syncytium, is expressed as

$$G(\mathbf{y}) = (\mathbf{y} - \mathbf{d})^{\mathrm{T}}(\mathbf{y} - \mathbf{d}) + \lambda_1(\mathbf{Dy})^{\mathrm{T}}(\mathbf{Dy}) + \lambda_2(\mathbf{Ly})^{\mathrm{T}}(\mathbf{Ly}).$$

The unique solution to minimize above $G(\mathbf{y})$ is obtained by the Euler equation,

$$\mathbf{y} - \mathbf{d} - \lambda_1 L\mathbf{y} + \lambda_2 L^2\mathbf{y} = 0. \tag{10.22}$$

Note that if $G(\mathbf{y})$ has the 2nd order smoothness term, i.e. $\lambda_2 \neq 0$, then (10.22) necessarily contains the 4th order derivative.

4.2 Role in Early Vision

Let us find the quadratic cost function which characterizes the function of the double layer neural syncytium. For an arbitrary input current $\mathbf{u}$, the responses of the first and the second layer syncytium were described by (10.10) and (10.11). We will find the relation between these equations and the Euler equation, (10.22). We use conductances instead of resistances, i.e.

$$g_{m1}=\frac{1}{R_{m1}}, \quad g_{m2}=\frac{1}{R_{m2}}, \quad g_{s1}=\frac{1}{R_{s1}}, \quad g_{s2}=\frac{1}{R_{s2}}.$$

Eliminating the boundary effect and taking into account of the relations:

$$\mathbf{C}_1=\mathbf{L}-\frac{g_{m1}}{g_{s1}}\mathbf{E}, \quad \mathbf{C}_2=\mathbf{L}-\frac{g_{m2}}{g_{s2}}\mathbf{E},$$

we find that the response of the first layer neuron, $\mathbf{v}_1$ holds an equation similar to the Euler equation (10.22). Namely,

$$\mathbf{v}_1-\mathbf{d}_1-\lambda_1\mathbf{L}\mathbf{v}_1+\lambda_2\mathbf{L}^2\mathbf{v}_1=0. \tag{10.23}$$

Here

$$\lambda_1=\frac{g_{m1}g_{s2}+g_{m2}g_{s1}}{g_{m1}g_{m2}-t_1t_2}, \quad \lambda_2=\frac{g_{s1}g_{s2}}{g_{m1}g_{m2}-t_1t_2},$$

and

$$\mathbf{d}_1=(v_1\mathbf{E}\mathbf{u}-v_2\mathbf{L}\mathbf{u}), \tag{10.24}$$

where

$$v_1=\frac{g_{m2}}{g_{m1}g_{m2}-t_1t_2}, \quad v_2=\frac{g_{s2}}{g_{m1}g_{m2}-t_1t_2}.$$

Note that $\mathbf{u}$ is the light-induced current of cones when we consider the interaction between the cone and the horizontal cell. v_1 and v_2 have the dimension of resistance (ohm). Therefore, the first term on the right hand side of (10.24) is the distribution of voltage proportional to the input image, provided that the light-induced current is proportional to the intensity of illumination. Similarly, the second term designates the voltage distribution which is proportional to the second order difference of input image. Note that the second order difference operation enhances the contrast of image as well as the noise. $\mathbf{v}_2$ also satisfies the following equation,

$$\mathbf{v}_2-\mathbf{d}_2-\lambda_1\mathbf{L}\mathbf{v}_2+\lambda_2\mathbf{L}^2\mathbf{v}_2=0. \tag{10.25}$$

λ_1 and λ_2 are same as the case of $\mathbf{v}_1$ and

$$\mathbf{d}_2=v_3\mathbf{E}\mathbf{u}, \tag{10.26}$$

where

$$v_3 = \frac{t_1}{g_{m1}g_{m2} - t_1 t_2}. \tag{10.27}$$

Note that (10.26) does not include the contrast-enhanced image.

Equations (10.23) and (10.25) indicate that the response voltage of the first layer and the second layer of the syncytium distribute in such a way to minimize the cost functions,

$$G_1(\mathbf{v}_1) = (\mathbf{v}_1 - \mathbf{d}_1)^T(\mathbf{v}_1 - \mathbf{d}_1) + \lambda_1(\mathbf{D}\mathbf{v}_1)^T(\mathbf{D}\mathbf{v}_1) + \lambda_2(\mathbf{L}\mathbf{v}_1)^T(\mathbf{L}\mathbf{v}_1),$$

$$G_2(\mathbf{v}_2) = (\mathbf{v}_2 - \mathbf{d}_2)^T(\mathbf{v}_2 - \mathbf{d}_2) + \lambda_1(\mathbf{D}\mathbf{v}_2)^T(\mathbf{D}\mathbf{v}_2) + \lambda_2(\mathbf{L}\mathbf{v}_2)^T(\mathbf{L}\mathbf{v}_2).$$

It is easier to find meanings of above cost functions when we simply rewrite them as follows. For the first layer:

$$G_1(v_{k1}) = \sum_k [v_{k1} - v_1 u_k - v_2(-\Delta U)]^2 + \lambda_1 \sum_k (v_{k1} - v_{k1-1})^2$$

$$+ \lambda_2 \sum_k (v_{k1-1} - 2v_{k1} + v_{k1+1})^2, \tag{10.28}$$

where

$$\Delta U \equiv (u_{k-1} - 2u_k + u_{k+1}).$$

And for the second layer:

$$G_2(v_{k2}) = \sum_k [v_{k2} - v_3 u_k]^2 + \lambda_1 \sum_k (v_{k2} - v_{k2-1})^2 + \lambda_2 \sum_k (v_{k2-1} - 2v_{k2} + v_{k2+1})^2. \tag{10.29}$$

Now it becomes clear to infer the role of the double layer syncytium in early vision. In the first layer neuron, the first term of the right hand side of (10.28),

$$\sum_k [v_{k1} - v_1 u_k - v_2(-\Delta U)]^2,$$

decreases as the voltage distribution of the first layer becomes closer to either the profile of the given image, $v_1 u_k$, or the contrast-enhanced image, $v_2(-\Delta u)$. In the second layer neuron, the first term of the right hand side of (10.29),

$$\sum_k [v_{k2} - v_3 u_k]^2,$$

does not include the contrast enhanced image. Therefore, the voltage distribution of the second layer does not show the contrast-enhancement.

The second and the third term of the cost functions, (10.28) and (10.29) correspond to stabilizers which smooth the image. These stabilizers are important since the input image always contains noise. The degree of smoothing is determined by λ_1 and λ_2. The optimum values of λ_1 and λ_2 are to be determined corresponding to the signal-to-noise ratio of the given image.

5. DISCUSSION

According to previous physiological experiments, the membrane properties of outer retinal neurons, e.g. photoreceptors and horizontal cells, are not linear. The voltage clamp experiments made on isolated horizontal cells revealed several types of nonlinear ionic conductances (Tachibana, 1983; Shingai and Christensen, 1986; Laster, 1986). The nonlinear ionic conductances were also found in photoreceptors (Bader et al., 1982; Barns and Hille, 1989; Yagi and MacLeish, 1994). Computer simulation analyses have been made on the double layer syncytium consisting of the cone and the horizontal cell using models which include these nonlinear ionic conductances (Kamermans et al., 1989; Smith, 1995; Usui et al., 1996). With these nonlinear models, however, the spatio-temporal properties of electrical signals can be evaluated only numerically and it is not easy to characterize how the neural syncytium filters the image with its unique structure. Moreover, the nonlinear properties of membranes affect the voltage response mainly when the intensity of light is bright. For example, the rod response to bright light is known to show a large initial transient (Schwartz, 1976), which is due to the non-specific cation conductance activated by a large hyperpolarization (Bader et al., 1982). However, no transient components are observed in responses to dim or modest light, which are considered to be physiological range of light intensity. Therefore, models presented here in are considered to be adequate and practical to analyse the voltage responses of the outer retinal neurons.

The discrete model has an advantage when we study the electrical properties of the response in relation to cell dimensions and cell spacings, which are critical parameters to determine the spatial resolution of visual system. It is interesting to evaluate the spatio-temporal filtering properties of the retinal circuit with the present formulation, considering the anatomy of the retina as well as the optics of the eye.

The present model was used to characterize the function of neural syncytium with the standard regularization theory, which was proposed to solve early vision problems (Poggio and Koch, 1985; Poggio et al., 1985). This characterization is achieved because linear analog circuits are a natural way of solving the variational principles dictated by standard regularization theory (Poggio and Koch, 1985; Koch et al., 1986). The cost function accompanying the linear circuit model corresponds, in fact, to the total power dissipated as heat from the circuit. The steady state voltages distribute in such a way that the heat dissipated from the linear analog circuit is minimized. Therefore, we are able to infer how the circuit parameters affect the cost function or in other words the function of the double layer syncytium. Concerning the effect of physiological parameters on the double layer syncytium, the conductance of gap-junction connecting the neighbouring horizontal cells, g_{s2}, needs to be elucidated. There are accumulating lines of physiological evidence indicating that the gap-junctional conductance connecting neighboring horizontal cells is controlled by interplexiform cell (Negishi and Drujan, 1979; Teranishi et al., 1983). The cost functions (10.28) and (10.29) indicate that λ_1 and λ_2 increase as g_{s2} increases. This means that the smoothing effect is augmented by g_{s2} in both the cone syncytium and the horizontal cell syncytium.

6. CONCLUSIONS AND FUTURE PERSPECTIVES

Recent physiological and morphological studies have revealed that the electrical coupling exists between bipolar cell (Kujiraoka and Saito, 1986; Saito and Kujiraoka, 1988) as well as between some subtypes of amacrine cells (Naka and Christensen, 1981; Teranishi et al., 1984). These two syncytia are interacting by reciprocal chemical synapses to constitute the double layer syncytium (Tachibana and Kaneko, 1987). These observations indicate that the visual information is processed by more than two layers of syncytium interacting each other through chemical synapses in the retina. The present analysis can be extended to characterize the roles of such hierarchical image processing carried out by the multi-layer syncytium in the retina.

PROBLEMS

10.1. Develop the double layer syncytium model into a multi-layer syncytium and characterize the role of retinal bipolar cells in terms of the standard regularization theory (cf. Yagi et al., 1997).

10.2. Expand the models discussed in this Chapter to two dimensions. [Hint: Formulate equations based on model shown in Figure 9.6, but multi-layered.]

10.3. Experiment suggestion: There are several subtypes of cones and horizontal cells which show different color sensitivity. These subtypes are known to interact with chemical synapses. Study the spatial properties of light-induced responses of these subtypes with electrophysiological experiments and the multi-layer syncytium model.

10.4. Find an algorithm to determine the optimum λs for a given noisy image whose statistical nature is known.

10.5. Elucidate how nonlinear ionic conductances may affect the spatio-temporal properties of the receptive field of outer retinal neurons by incorporating the ionic currents presented in Chapter 11 for the (i) single-layer syncytium model, (ii) double-layer syncytium model.

ACKNOWLEDGMENTS

The author is grateful to Dr S. Ohshima for the valuable discussions.

REFERENCES

Abernethy, J.D. (1974) A dynamic model of a two-synapse feedback loop in the vertebrate retina. *Kybernetik*, **14**, 187–200.

Barnes, S. and Hille, B. (1989) Ionic channels of the inner segment of tiger salamander cone photoreceptors. *J. Gen. Physiol.*, **94**, 719–743.

Baylor, D.A., Fuortes, M.G.F. and O'Bryan, P.M. (1971) Receptive fields of cones in the retina of the turtle. *Journal of Physiol.* (*Lond.*), **214**, 256–294.

Baylor, D.A., Hodgkin, A.L. and Lamb, T.D. (1974) The electrical response of turtle cones to flashes and steps of light. *J. Physiol.* (*Lond.*), **242**, 685–727.

Bader, C.R., Bertrand, D. and Schwartz, E.A. (1982) Voltage-activated and calcium-activated currents studied in solitary rod inner segments from the salamander retina. *J. Physiol.* (*Lond.*), **331**, 253–284.

Boycott, B.B. and Kolb, H. (1973) The horizontal cells of the rhesus monkey retina. *J. Comp. Neurol.*, **148**, 115–140.

Cohen, A.I. (1965) Some electron microscopic observations on inter-receptor contacts in the human and macaque retinae. *J. Anat.*, **99**, 595–610.

Detwiler, P.B. and Hodgkin, A.L. (1979) Electrical coupling between cones in turtle retina. *J. Physiol.* (*Lond.*), **291**, 75–100.

Detwiler, P.B., Hodgkin, A.L. and McNaughton, P.A. (1978) A surprising property of electrical spread in the network of rods in the turtle's retina. *Nature*, **274**, 562–565.

Detwiler, P.B., Hodgkin, A.L. and McNaughton, P.A. (1980) Temporal and spatial characteristics of the voltage response of rods in the retina of the snapping turtle. *J. Physiol.* (*Lond.*), **300**, 213–250.

Fuortes, M.G.F. and Simon, E.J. (1974) Interaction leading to horizontal cell responses in the turtle retina. *J. Physiol.* (*Lond.*), **240**, 177–179.

Hara, K.-I. and Takabayashi, A. (1975) A dynamic model of retinal cells in the vertebrate retina. *Biol. Cybern.*, **20**, 61–67.

Kamermans, M., van Dijk, B.W. and Spekreijse, H. (1989) Lateral feedback from monophasic horizontal cells to cones in carp retina. II. A quantitative model. *J. Gen. Physiol.*, **93**, 695–714.

Kaneko, A. (1971) Electrical connections between horizontal cells in the dogfish retina. *J. Physiol.* (*Lond.*), **213**, 95–105.

Kobayashi, H., Matsumoto, T., Yagi, T. and Shimmi, T. (1993) Image processing regularization filters on layered architecture. *Neural Networks*, **6**, 327–350.

Koch, C., Marroquin, J. and Yuille, A. (1986). Analog 'neuronal' networks in early vision. *Proc. Nat. Acad. Sci. USA*, **83**, 4263–4267.

Kolb, H. (1977) The organization of the outer plexiform layer in the retina of the cat: electron microscopic observations. *J. Neurocytol.*, **6**, 131–153.

Kujiraoka, T. and Saito, T. (1986) Electrical coupling between bipolar cells in the carp retina. *Proc. Nat. Acad. Sci. USA*, **83**, 4063–4066.

Lamb, T.D. (1976) Spatial properties of the horizontal cell in the turtle retina. *J. Physiol.* (*Lond.*), **263**, 239–255.

Lamb, T.D. and Simon, E.J. (1976) The relation between intracellular coupling and electrical noise in turtle photoreceptors. *J. Physiol.* (*Lond.*), **263**, 257–286.

Laster, E. (1986) Ionic currents of cultured horizontal cells isolated from white perch retina. *J. Neurophysiol.*, **55**, 499–513.

Marmarelis, P.Z. and Naka, K.-I. (1972) Spatial distribution of potential in a flat cell. Application to catfish horizontal cell layers. *Biophys. J.*, **12**, 1515–1532.

Naka, K.-I. and Christensen, B.N. (1981) Direct electrical connections between transient amacrine cells in the catfish retina. *Science*, **214**, 462–464.

Naka, K.-I. and Rushton, W.A.H. (1967) S-potential from luminosity units in the retina of the fish (*Cyprinidae*). *J. Physiol.* (*Lond.*), **192**, 437–461.

Negishi, K. and Drujan, B. (1979) Reciprocal changes in center and surrounding *S*-potentials of fish retina in response to dopamine. *Neurochem. Res.*, **4**, 313–318.

Ohshima, S., Yagi, T. and Funahashi, Y. (1995) Computational studies on the interaction between red cone and H1 horizontal cell. *Vis. Res.*, **35**, 149–160.

Poggio, T. and Koch, C. (1985) Ill-posed problems in early vision: from computational theory of analogue networks. *Proc. R. Soc. Lond.*, **B 226**, 303–323.

Poggio, T., Torre, V. and Koch, C. (1985) Computational vision and regularization theory. *Nature*, **317**, 314–319.

Poznanski, R.R. and Umino, O. (1997) Syncytial integration by a network of coupled bipolar cells in the retina. *Prog. Neurobiol.*, **53**, 273–291.

Saito, T. and Kujiraoka, T. (1988) Characteristics of bipolar–bipolar coupling in the carp retina. *J. Gen. Physiol.*, **91**, 275–287.

Schnapf, J.L. and Copenhagen, D.R. (1982) Differences in the kinetics of rod and cone synaptic transmission. *Nature*, **296**, 862–864.

Schwartz, E.A. (1976) Electrical properties of the rod syncytium in the retina of the turtle. *J. Physiol.* (*Lond.*), **257**, 379–406.

Shingai, R. and Christensen, B.N. (1986) Excitable properties and voltage-sensitive ion conductances of horizontal cells isolated from catfish (*Ictalurus punctatus*). *J. Neurophysiol.*, **56**, 32–49.

Smith, R.G. (1995) Simulation of an anatomically defined local circuit: The cone-horizontal cell cell network in cat retina. *Visual Neurosci.*, **12**, 545–561.

Stell, W.K. (1975) Horizontal cell axons and axon terminals in goldfish retina. *J. Comp. Neurol.*, **159**, 503–520.

Stell, W.K. and Lightfoot, D.O. (1975) Color-specific interconnections of cones and horizontal cells in the retina of the goldfish. *J. Comp. Neurol.*, **159**, 473–502.

Tachibana, M. (1983) Ionic current of solitary horizontal cells isolated from goldfish retina. *J. Physiol. (Lond.)*, **345**, 329–351.

Tachibana, M. and Kaneko, A. (1987) γ-Aminobutyric acid exerts a local inhibitory action on the axon terminal of bipolar cells: Evidence for negative feedback from amacrine cells. *Proc. Nat. Acad. Sci. USA*, **84**, 3501–3505.

Tate, C. and Woolfson, M.M. (1971) On modelling neural networks in the retina. *Vis. Res.*, **11**, 617–633.

Teranishi, T., Negishi, K. and Kato, S. (1983) Dopamine modulates *S*-potential amplitude and dye-coupling between external horizontal cells in carp retina. *Nature*, **301**, 234–246.

Teranishi, T., Negishi, K. and Kato, S. (1984) Dye coupling between amacrine cells in carp retina. *Neurosci. Lett.*, **55**, 73–78.

Torre, V. and Owen, W.G. (1983) High-pass filtering of small signals by the network of rods in the retina of the toad, Bufo Marinus. *Biophys. J.*, **41**, 305–324.

Torre, V., Owen, G. and Sandini, G. (1983) The dynamics of electrically interacting cells. *IEEE Trans. syst., man, and cybern.*, **SMC-13**, 757–765.

Tsukamoto, Y., Yamada, M. and Kaneko, A. (1987) Morphological and physiological studies of rod-driven horizontal cells with special reference to the question of whether they have axons and axon terminals. *J. Comp. Neurol.*, **255**, 305–316.

Umino, O., Maehara, M., Hidaka, S., Kita, S. and Hashimoto, Y. (1994) The network properties of bipolar–bipolar cell coupling in the retina of teleost fishes. *Visual Neurosci.*, **11**, 533–548.

Usui, S., Kamiyama, Y., Ishii, H. and Ikeno, H. (1996) Reconstruction of retinal horizontal cell responses by the ionic current model. *Vis. Res.*, **36**, 1711–1719.

Vaney, D.I. (1994) Patterns of neuronal coupling in the retina. *Prog. Ret. Eye Res.*, **13**, 301–355.

Weiler, R. and Zettler, F. (1979) The axon-bearing horizontal cells in the teleost retina are functional as well as structural units. *Vis. Res.*, **19**, 1261–1268.

Yagi, T. (1986) Interaction between the soma and the axon terminal of retinal horizontal cells in *Cyprinus carpio*. *J. Physiol. (Lond.)*, **375**, 121–135.

Yagi, T. and Kaneko, A. (1988) The axon terminal of goldfish retinal horizontal cells: a low membrane conductance measured in solitary preparations and its implication to the signal conduction from the soma. *J. Neurophysiol.*, **59**, 482–494.

Yagi, T. (1989) Dynamics of signal conduction from soma to axon terminal of the teleost retinal horizontal cell: *in vivo, in vitro* and model studies. *Vis. Res.*, **29**, 375–384.

Yagi, T. and Macleish, P.R. (1994) Ionic conductances of monkey solitary cone inner segments. *J. Neurophysiol.*, **71**, 656–665.

Yagi, T., Ohshima, S. and Funahashi, Y. (1997) The role of retinal bipolar cell in early vision: an implication with analogue networks and regularization theory. *Biol. Cybern.*, **77**, 163–171.

Yamada, E. and Ishikawa, T. (1965) The fine structure of the horizontal cells in some vertebrate retinae. *Cold Spring Harbor Symp. Quant. Biol.*, **30**, 383–392.

CHAPTER 11
IONIC CURRENT MODEL OF THE OUTER RETINAL CELLS

Shiro Usui, Yoshimi Kamiyama, Hiroyuki Ishii, Toshihiko Ogura and Akito Ishihara

1. INTRODUCTION

The vertebrate retina has been intensively studied over the past decades using physiological, morphological and pharmacological techniques. Since the retina is a unique and interesting neural system from the viewpoints not only of neuroscience but also of engineering, which has been thought to be a window to the brain because of its accessibility and suitability for the investigation. The retina is an ideal model of neural network system and neural information processing system.

Since the early 1980s, enzymatic dissociation techniques have been employed for the isolation of a variety of retinal cells, including photoreceptors, horizontal cells and bipolar cells (reviewed in Kaneko and Tachibana, 1986; Lasater, 1991). Using isolated cell preparations, membrane ionic currents of solitary retinal cells have been studied in a quantitative fashion using the voltage clamp techniques. These data provide information concerning the functional role of the ionic currents in generating and shaping the light response of the retinal neurons. However, the detailed experimental data alone are not enough to understand how the retinal neurons work. A combination of experimental work and mathematical modeling is necessary to clarify the complicated retinal functions. We believe mathematical modeling and simulation analysis with the advanced computer system have a potential to understand the relationships between the microscopic characteristics and its integrative functions. Models based on the detailed physiology, morphology and pharmacology of the retina illustrate the role of interactions among ionic currents, neurotransmitters and second messengers suggesting further areas of investigation, which may include new experiments.

The retinal outer plexiform layer, which consists of synaptic connections among photoreceptors, horizontal cells and bipolar cells, plays a fundamental role in color- and spatial-information processings. Recent progress in retinal physiology has revealed the detailed characteristics of retinal neurons in terms of ionic currents, neurotransmitters and wiring of the network. Since the ionic mechanism is fundamental for generating the electrical response of neurons, the simulation study based on these characteristics makes it possible to analyze nonlinear dynamical response properties and to understand the role of a particular ionic channel, and thus provides key insights for elucidating the underlying mechanisms in the outer plexiform layer. In this Chapter, models based on the ionic mechanisms are employed to understand individual ionic current dynamics under current clamp conditions and light responses, which cannot be directly measured by the conventional experimental techniques. The models also can easily reveal what happens if a particular current is active or not. From this point of view, we have developed ionic current models of photoreceptors, horizontal cells and bipolar cells to improve our understanding of these retinal neurons.

2. SUMMARY OF IONIC CURRENTS IN THE OUTER RETINA

Figure 11.1 shows an electrical equivalent circuit model of the retinal outer plexiform layer. In the following subsections, we briefly summarize the ionic currents identified in these neurons.

2.1 Ionic Currents in Photoreceptors

There are at least three voltage-dependent ionic currents and two calcium-dependent ionic currents in the photoreceptor inner segments, namely, hyperpolarization activated current (I_h), delayed rectifying potassium current (I_{Kv}), calcium current (I_{Ca}), Ca-dependent potassium current ($I_{K(Ca)}$), and Ca-dependent chloride current ($I_{Cl(Ca)}$). These currents are observed in a variety of rods, salamander cones and lizard cones (Bader et al., 1982; Maricq and Korenbrot, 1988, 1990a,b; Barnes and Hille, 1989). Since both rod and cone inner segments have similar ionic currents (Bader et al., 1982; Barnes and Hille, 1989), we made no distinction between rods and cones when referring to the ionic currents.

2.2 Ionic Currents in Horizontal Cells

Studies of solitary horizontal cells isolated from the retina of lower vertebrates have revealed that the cell soma has highly nonlinear and time-dependent ionic currents (Tachibana, 1981, 1983; Lasater, 1986). On the other hand, the axon terminal of horizontal cells has rather linear and small membrane currents (Yagi and Kaneko, 1988). There are at least five major voltage-dependent currents and one transmitter activated current in the solitary horizontal cell soma, namely sodium current (I_{Na}), calcium current (I_{Ca}), transient outward potassium current (I_A), anomalous rectifying potassium current (I_{Ka}), delayed rectifying potassium current (I_{Kv}), and glutamate-induced current (I_{glu}). Glutamate, which is the most likely candidate for a neurotransmitter of photoreceptors, activates I_{glu} and suppresses I_{Ka} by decreasing the number of active channels (Ishida et al., 1984; Tachibana, 1985; Kaneko and Tachibana, 1985b). Tonic release of glutamate from photoreceptors is mediated by calcium-dependent vesicular process (Copenhagen and Jahr, 1989).

2.3 Ionic Currents in Bipolar Cells

The bipolar cell is morphologically divided into three compartments: dendrite, cell body, and axon terminal. Each compartment contains unique ionic channels.

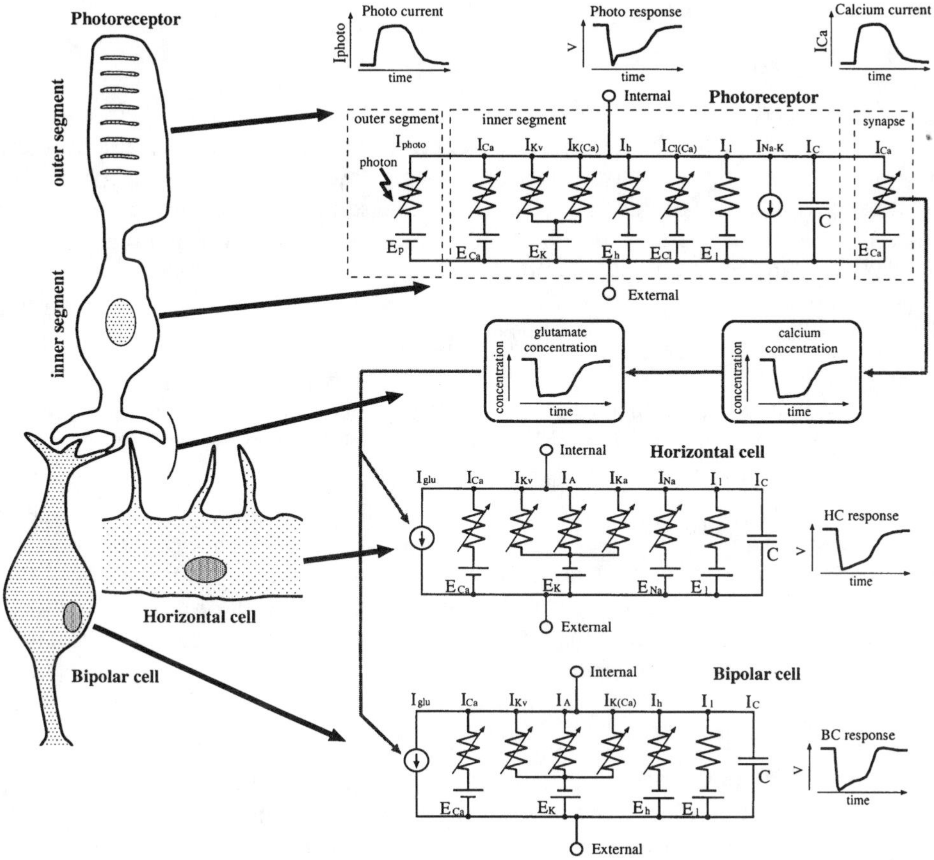

Figure 11.1 Ionic current model of the outer retina.

Channels activated by the transmitter released from photoreceptors are distributed over the dendritic compartment (Nawy and Jahr, 1990, 1991). There are at least five membrane currents in the bipolar cell body, namely, hyperpolarization activated current (I_h), delayed rectifying potassium current (I_{Kv}), transient outward potassium current (I_A), calcium current (I_{Ca}), and voltage- and calcium-dependent potassium current ($I_{K(Ca)}$) (Kaneko and Tachibana, 1985a; Lasater, 1988). Calcium channels are distributed not only over the cell body but also are present in high density over the axon terminal (Heidelberger and Matthews, 1992; Tachibana et al., 1993). The axon terminal also contains channels activated by transmitters from inner retinal neurons (Attwell et al., 1987; Tachibana and Kaneko, 1988; Maguire et al., 1989; Lukasiewicz et al., 1994). It is left as an exercise to include glutamate induced currents in the dendrites of bipolar cells (see Kamiyama et al., 1998), as second-order neurons are interconnected through glutaminergic synapses.

2.4 Description of Ionic Currents

2.4.1 *The Parallel Conductance Model of a Neuron*

Electrical properties of cellular membrane can be modeled by electrical circuits, i.e., the membrane capacitance as a capacitor, membrane permeabilities as conductances, and the equilibrium potential of individual ions as the electrical driving force. The electrical properties of outer retinal cells were described by parallel conductance models as shown in Figure 11.1. The total membrane current in a parallel conductance model can be written as

$$I_{\text{total}} = C\frac{\mathrm{d}V}{\mathrm{d}t} + \sum_{\text{ion}} I_{\text{ion}}. \tag{11.1}$$

We modeled each ionic current using a mathematical formulation similar to the Hodgkin–Huxley equations (Hodgkin and Huxley, 1952). In general, each current is described by an equation of the form

$$I_{\text{ion}}(V,t) = \overline{g_{\text{ion}}} \cdot m^{M}(V,t) \cdot h(V,t) \cdot (V - E_{\text{ion}}) \tag{11.2}$$

where $\overline{g_{\text{ion}}}$ is the maximum conductance, m is the activation variable, h is the inactivation variable, M is the gating exponent and E_{ion} is the reversal potential. Each ionic current was modeled independently from the pharmacologically isolated characteristics under voltage clamp conditions.

2.4.2 *Model Parameters*

The Hodgkin–Huxley type model is based on the idea that the electrical properties of a cell membrane can be modeled by an electrical equivalent circuit, in which each ionic current is expressed by the voltage- and time-dependent conductance and the driving force. Since a particular ionic channel is selectively blocked by the chemical drug (channel blocker) in electrophysiological experiments, the equivalent circuit can be simplified as a single circuit, i.e. each ionic current is analyzed separately. From the engineering point of view, the voltage clamp experiment on the ionic current can be thought of as a process for estimating the model parameters. In this step of modeling, parameter values of each ionic current are estimated independently current by current. In such condition, the steady-state *gate* parameters, $m_\infty(V)$ (activation), $h_\infty(V)$ (inactivation) and their time dependencies can be estimated from the voltage-dependent properties of the ionic conductance measured under several command voltages.

Once values for $m_\infty(V)$, $h_\infty(V)$ and their time-dependent properties have been determined by fitting the conductance data, rate constants, α and β can be expressed by the empirically determined smooth functions. Values of measurable parameters, such as the maximum conductances and the reversal potentials are directly introduced into the model. These procedures are the basic steps for modeling ionic conductance. In addition, in order to solve equations for the estimation of parameters and the simulation, the ordinary differential equations are chosen to be numerically solved on a digital computer (Mascagni, 1989). We have employed a method with adaptive step size. Full descriptions of the proposed model are given in Tables 11.1, 11.2 and 11.3.

2.5 Intracellular Calcium Concentration

Several processes affect the intracellular concentration of calcium ions (Yamada et al., 1989). After calcium ions enter the cell through calcium channels, they diffuse into the cell and bind to various buffers. Specialized channels act as transporters to extrude calcium ions from the cell. The intracellular mechanisms that control calcium ions in the retinal cells are not well understood (but see Von Gersdorff and Matthews, 1996). We thus modeled the calcium system using the fewest mechanisms necessary to reproduce the calcium dependent currents of the cells. Figure 11.2 shows the schematic illustration of the compartment calcium model (Kamiyama et al., 1996; Usui et al., 1996c). The mathematical descriptions and parameters for the photoreceptor model and the bipolar cell model are summarized in Tables 11.4 and 11.5, and Tables 11.6 and 11.7, respectively (Kamiyama et al., 1996; Usui et al., 1996c).

This model is based on the earlier work by DiFrancesco and Noble (1985) in cardiac cells and the recent work on rod outer segments by Torre et al. (1990). Intracellular calcium is controlled by influx via I_{Ca}, by extrusion through the transporters, and by binding and unbinding to high affinity and low affinity internal buffers. Because our primary aim is to model the action of calcium on the calcium-dependent current, we divided the cell space into two compartments, the space just below the cellular membrane and the central space. The calcium concentration just below the membrane ($[Ca^{2+}]_s$) was used for the calcium-dependent currents.

3. SIMULATED RESPONSES OF THE OUTER RETINA

3.1 Photoreceptor

3.1.1 *Voltage Clamp Simulation*

Voltage clamp simulations were made under the conditions identical to electrophysiological experiments by Maricq and Korenbrot (1988). Figure 11.3(A) shows the simulated total ionic current under three different pharmacological conditions corresponding to control, TEA-contained, and TEA- and Co^{2+}-contained solutions (Kamiyama et al., 1996). In these conditions, external solutions always contain Cs^+, and thus we blocked I_h completely. Figure 11.3(B) shows the current–voltage relation measured from the TEA responses. The currents were measured at the peak, 5 ms after the onset of the depolarizing pulse and 135 ms after the onset of the command pulse. The experimental observations are well reproduced by our model.

Each ionic current simulated under control conditions is shown in Figure 11.3(C). From these responses we can explain how the complicated current responses in Figure 11.3(A) are generated. An early inward current elicited by the voltage step to -20 mV is I_{Ca}, and the following outward component is mostly $I_{Cl(Ca)}$ due to the increase of the calcium concentration ($[Ca^{2+}]_s$). The voltage step to 0 and $+20$ mV activates the outward currents, I_{Kv} and $I_{Cl(Ca)}$. The voltage step to $+40$ mV activates

Table 11.1 A mathematical formulation of each ionic current of a photoreceptor

- Delayed rectifying K current

$$\alpha_{mK}=\frac{5(100-V)}{\exp((100-V)/42)-1}$$

$$\beta_{mK}=9\exp\left(-\frac{V-20}{40}\right)$$

$$\frac{dm_K}{dt}=\alpha_{mK}\cdot(1-m_K)-\beta_{mK}\cdot m_K$$

$$\alpha_{hK}=0.15\exp\left(-\frac{V}{22}\right)$$

$$\beta_{hK}=\frac{0.4125}{\exp((10-V)/7)+1}$$

$$\frac{dh_K}{dt}=\alpha_{hK}\cdot(1-h_K)-\beta_{hK}\cdot h_K$$

$$I_{Kv}=\overline{g_{Kv}}\cdot(m_K)^3\cdot h_K\cdot(V-E_K)$$

$$\overline{g_{Kv}}=2.0\,\text{ns}$$

$$E_K=-80\,\text{mV}$$

- Ca current

$$\alpha_{Ca}=\frac{300(80-V)}{\exp((80-V)/25)-1}$$

$$\beta_{Ca}=\frac{1000}{1+\exp((V+38)/7)}$$

$$\frac{dm_{Ca}}{dt}=\alpha_{Ca}\cdot(1-m_{Ca})-\beta_{Ca}\cdot m_{Ca}$$

$$E_{Ca}=12.9\log\left(\frac{[Ca^{2+}]_o}{[Ca^{2+}]_s}\right)$$

$$m_{Cah}=\frac{\exp((40-V)/18)}{1+\exp((40-V)/18)}$$

$$I_{Ca}=\overline{g_{Ca}}\cdot(m_{Ca})^4\cdot m_{Cah}\cdot(V-E_{Ca})$$

$$\overline{g_{Ca}}=1.2\,\text{ns}$$

- Ca-dependent Cl current

$$m_{Cl}=\frac{1}{1+\exp((Cl_h-[Ca^{2+}]_s)/0.09)}$$

$$I_{Cl(Ca)}=\overline{g_{Cl}}\cdot m_{Cl}\cdot(V-E_{Cl})$$

$$\overline{g_{Cl}}=6.5\,\text{ns}$$

$$E_{Cl}=-45\,\text{mV}$$

$$Cl_h=0.37\,\mu\text{M}$$

- Ca-dependent K current

$$\alpha_{Kc}=\frac{15(80-V)}{\exp((80-V)/40)-1}$$

$$\beta_{Kc}=20\exp\left(-\frac{V}{35}\right)$$

$$\frac{dm_{Kc}}{dt}=\alpha_{Kc}(1-m_{Kc})-\beta_{Kc}\cdot m_{Kc}$$

$$m_{Kcl}=\frac{[Ca^{2+}]_s}{[Ca^{2+}]_s+0.3}$$

$$I_{K(Ca)}=\overline{g_{Kc}}\cdot(m_{Kc})^2\cdot m_{Kcl}\cdot(V-E_K)$$

$$\overline{g_{Kc}}=0.5\,\text{ns}$$

- Leakage current

$$I_L=g_l\cdot(V-E_l)$$

$$g_l=0.5\,\text{ns}$$

$$E_l=-55\,\text{mV}$$

- Capacitance

$$C=10\,\text{pF}$$

- Na–K exchanger current

$$I_{Na-K}=9.6\,\text{pA}$$

- Hyperpolarization activated current

$$\alpha_h=\frac{3}{\exp((V+88)/14)+1}$$

$$\beta_h=\frac{20}{\exp(-(V+18)/20)+1}$$

$$M=[C_1\quad C_2\quad O_1\quad O_2\quad O_3]^t$$

$$K=\begin{bmatrix}-4\alpha_h & \beta_h & 0 & 0 & 0\\ 4\alpha_h & -(\beta_h+3\alpha_h) & 2\beta_h & 0 & 0\\ 0 & 3\alpha_h & -(2\beta_h+2\alpha_h) & 3\beta_h & 0\\ 0 & 0 & 3\alpha_h & -(3\beta_h+\alpha_h) & 4\beta_h\\ 0 & 0 & 0 & \alpha_h & -4\beta_h\end{bmatrix}$$

$$\frac{d}{dt}M=KM$$

$$m_h=O_1+O_2+O_3$$

$$I_h=\overline{g_h}\cdot m_h\cdot(V-E_h)$$

$$\overline{g_h}=0.305\,\text{ns}$$

$$E_h=-32\,\text{mV}$$

Table 11.2 A mathematical formulation of each ionic current of a horizontal cell

- Transient outward K current

$$\alpha_{mA}=\frac{40\cdot(140-V_K)}{\exp((140-V_K)/15)-1}$$

$$\beta_{mA}=8\cdot\exp\left(-\frac{V_K}{20}\right)$$

$$m_A=\frac{\alpha_{mA}}{\alpha_{mA}+\beta_{mA}}$$

$$\alpha_{hA}=0.01\cdot\exp\left(-\frac{V_K}{20}\right)$$

$$\beta_{hA}=\frac{2}{\exp((30-V_K)/10)+1}$$

$$\frac{dh_A}{dt}=\alpha_{hA}\cdot(1-h_A)-\beta_{hA}\cdot h_A$$

$$I_A=\overline{g_A}\cdot m_A^3\cdot V_K$$

$$\overline{g_A}=2\,\text{ns}$$

$$V_K=V-E_K$$

- Anomalous rectifying K current

$$m_{Ka}=\frac{1}{1+\exp(V_K/30)}$$

$$I_{Ka}=\overline{g_{Ka}}\cdot m_{Ka}^4\cdot V_K$$

$$\overline{g_{Ka}}=50\cdot[1-0.5\cdot\text{Glu}(t)]\,\text{ns}$$

$$V_K=V-E_K$$

- Na current

$$\alpha_{mNa}=\frac{40\cdot(80-V)}{\exp((80-V)/15)-1}$$

$$\beta_{mNa}=10\cdot\exp\left(-\frac{V}{20}\right)$$

$$\frac{d_{mNa}}{dt}=\alpha_{mNa}\cdot(1-m_{Na})-\beta_{mNa}\cdot m_{Na}$$

$$\alpha_{hNa}=0.01\cdot\exp\left(-\frac{V}{30}\right)$$

$$\beta_{hNa}=\frac{2}{\exp((70-V)/10)+1}$$

$$\frac{d_{hNa}}{dt}=\alpha_{hNa}\cdot(1-h_{Na})-\beta_{hNa}\cdot h_{Na}$$

$$I_{Na}=\overline{g_{Na}}\cdot m_{Na}^3\cdot h_{Na}\cdot(V-E_{Na})$$

$$\overline{g_{Na}}=120\,\text{ns}$$

$$E_{Na}=55\,\text{mV}$$

- Glutamate-induced current

$$I_{glu}=\overline{I_{glu}}\frac{\exp((V+3)/125)-1}{\exp((3-V)/125)+1}$$

$$\overline{I_{glu}}=600\cdot\text{Glu}(t)\,\text{pA}$$

- Delayed rectifying K current

$$\alpha_{mKv}=\frac{10\cdot(120-V_K)}{\exp((120-V_K)/20)-1}$$

$$\beta_{mKv}=5\cdot\exp\left(-\frac{V_K}{20}\right)$$

$$\frac{dm_{Kv}}{dt}=\alpha_{mKv}\cdot(1-m_{Kv})-\beta_{mKv}\cdot m_{Kv}$$

$$I_{Kv}=\overline{g_{Kv}}\cdot m_{Kv}^4\cdot V_K$$

$$\overline{g_{Kv}}=0.25\,\text{ns}$$

$$V_K=V-E_K$$

- Ca Current

$$\alpha_{mCa}=\frac{5\cdot(70-V)}{\exp[(70-V)/28]-1}$$

$$\beta_{mCa}=2\cdot\exp\left(-\frac{V}{15}\right)$$

$$\frac{dm_{Ca}}{dt}=\alpha_{mCa}\cdot(1-m_{Ca})-\beta_{mCa}\cdot m_{Ca}$$

$$h_{Ca}=\frac{K}{K+[\text{Ca}]_i}$$

$$\frac{d[\text{Ca}]_i}{dt}=[\text{Ca}]_{entry}+[\text{Ca}]_{buffer}+[\text{Ca}]_{trans}$$

$$=-\frac{I_{Ca}}{2\cdot F\cdot v}+\frac{d[\text{B}]}{dt}+\frac{[\text{Ca}]_{eq}-[\text{Ca}]_i}{\tau_{trans}}$$

$$\frac{d[\text{B}]}{dt}=0.01\cdot([\text{B}]_{total}-[\text{B}])-10^5\cdot[\text{Ca}]_i\cdot[\text{B}]$$

$$\tau_{trans}=200\cdot\exp\left(\frac{V}{20}\right)$$

$$I_{Ca}=\overline{g_{Ca}}\cdot m_{Ca}^3\cdot h_{Ca}\cdot V_{Ca}$$

$$V_{Ca}=V-E_{Ca}$$

$$E_{Ca}=12.5\log\left(\frac{[\text{Ca}]_o}{[\text{Ca}]_i}\right)$$

$$\overline{g_{Ca}}=135\,\text{ns}$$

$$v=10^{-15}\,\text{m}^3$$

$$F=9.65\times10^4\,\text{C/mol}$$

$$[\text{Ca}]_o=4\,\text{mM}$$

$$[\text{Ca}]_{eq}=0.05\,\text{mM}$$

$$[\text{B}]_{total}=0.03\,\text{mM}$$

$$[\text{K}]=0.002\,\text{mM}$$

- Leakage current

$$I_l=g_l\cdot V_l$$

$$g_l=0.5\,\text{ns}$$

$$V_l=V-E_l$$

- Capacitance

$$C=12.5\,\text{pF}$$

Table 11.3 A mathematical formulation of each ionic current of a bipolar cell

- Delayed rectifying K current

$$\alpha_{mKv}=\frac{400}{\exp(-(V-15)/36)+1}$$

$$\beta_{mKv}=\exp\left(-\frac{V}{13}\right)$$

$$\frac{dm_{Kv}}{dt}=\alpha_{mKv}\cdot(1-m_{Kv})-\beta_{mKv}\cdot m_{Kv}$$

$$\alpha_{hKv}=0.0003\exp\left(-\frac{V}{7}\right)$$

$$\beta_{hKv}=\frac{80}{\exp(-(V+115)/15)+1}+0.02$$

$$\frac{dh_{Kv}}{dt}=\alpha_{hKv}\cdot(1-h_{Kv})-\beta_{hKv}\cdot h_{Kv}$$

$$g_{Kv}=\overline{g_{Kv}}\cdot m_{Kv}^3\cdot h_{Kv}$$

$$I_{Kv}=g_{Kv}\cdot(V-E_K)$$

$$\overline{g_{Kv}}=2.0\,\mathrm{ns}$$

$$E_K=-58\,\mathrm{mV}$$

- Transient outward current

$$\alpha_{mA}=\frac{1200}{\exp(-(V-50)/28)+1}$$

$$\beta_{mA}=6\exp\left(-\frac{V}{10}\right)$$

$$\frac{dm_A}{dt}=\alpha_{mA}\cdot(1-m_A)-\beta_{mA}\cdot m_A$$

$$\alpha_{hA}=0.045\exp\left(-\frac{V}{13}\right)$$

$$\beta_{hA}=\frac{75}{\exp(-(V+50)/15)+1}$$

$$\frac{dh_A}{dt}=\alpha_{hA}\cdot(1-h_A)-\beta_{hA}\cdot h_A$$

$$g_A=\overline{g_A}\cdot m_A^3\cdot h_A$$

$$I_A=g_A\cdot(V-E_K)$$

$$\overline{g_A}=35\,\mathrm{ns}$$

- Capacitance

$$C=0.01\,\mathrm{pF}$$

- Hyperpolarization activated current

$$\alpha_h=\frac{3}{\exp((V+110)/15)+1}$$

$$\beta_h=\frac{1.5}{\exp(-(V+115)/15)+1}$$

$$M=[C_1\quad C_2\quad O_1\quad O_2\quad O_3]^t$$

$$K=\begin{bmatrix}-4\alpha_h & \beta_h & 0 & 0 & 0\\ 4\alpha_h & -(3\alpha_h+\beta_h) & 2\beta_h & 0 & 0\\ 0 & 3\alpha_h & -2(\alpha_h+\beta_h) & 3\beta_h & 0\\ 0 & 0 & 2\alpha_h & -(\alpha_h+3\beta_h) & 4\beta_h\\ 0 & 0 & 0 & \alpha_h & -4\beta_h\end{bmatrix}$$

$$\frac{d}{dt}M=KM$$

$$m_h=O_1+O_2+O_3$$

$$g_h=\overline{g_h}\cdot m_h$$

$$I_h=g_h\cdot(V-E_h)$$

$$\overline{g_h}=0.975\,\mathrm{ns}$$

$$E_h=-17.7\,\mathrm{mV}$$

- Ca current

$$\alpha_{mCa}=\frac{12000(120-V)}{\exp(-(V-120)/25)-1}$$

$$\beta_{mCa}=\frac{40000}{\exp((V+68)/25)+1}$$

$$\frac{dm_{Ca}}{dt}=\alpha_{mCa}\cdot(1-m_{Ca})-\beta_{mCa}\cdot m_{Ca}$$

$$h_{Ca}=\frac{\exp(-(V-50)/11)}{\exp(-(V-50)/11)+1}$$

$$E_{Ca}=12.9\log\left(\frac{[Ca^{2+}]_o}{[Ca^{2+}]_s}\right)$$

$$g_{Ca}=\overline{g_{Ca}}\cdot m_{Ca}^4\cdot h_{Ca}$$

$$I_{Ca}=g_{Ca}\cdot(V-E_{Ca})$$

$$\overline{g_{Ca}}=1.1\,\mathrm{ns}$$

$$[Ca^{2+}]_o=2500\,\mu\mathrm{M}$$

- Ca-dependent K current

$$\alpha_{mKc}=\frac{100(230-V)}{\exp((230-V)/52)-1}$$

$$\beta_{mKc}=120\exp\left(-\frac{V}{95}\right)$$

$$\frac{dm_{Kc}}{dt}=\alpha_{mKc}\cdot(1-m_{Kc})-\beta_{mKc}\cdot m_{Kc}$$

$$m_{Kc1}=\frac{[Ca^{2+}]_s}{[Ca^{2+}]_s+0.2}$$

$$g_{Kc}=\overline{g_{Kc}}\cdot m_{Kc}^2\cdot m_{Kc1}$$

$$I_{K(Ca)}=g_{Kc}\cdot(V-E_K)$$

$$\overline{g_{Kc}}=8.5\,\mathrm{ns}$$

- Leakage current

$$I_l=g_l\cdot(V-E_l)$$

$$g_l=0.23\,\mathrm{ns}$$

$$E_l=-21\,\mathrm{mV}$$

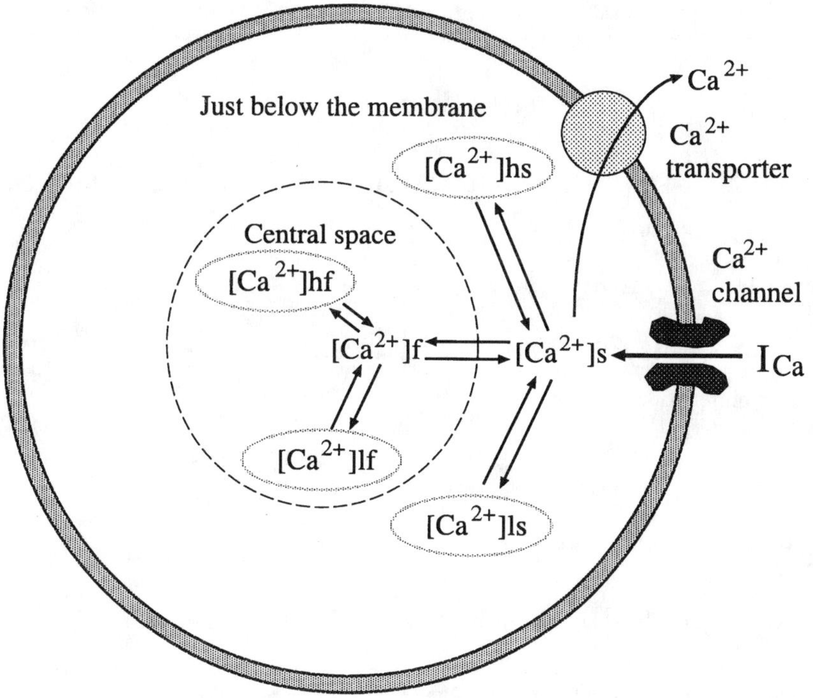

Figure 11.2 Compartment model of intracellular calcium system (Kamiyama et al., 1996; Usui et al., 1996c). The rate of change of internal calcium concentration below the membrane ($[Ca^{2+}]_s$) is determined by influx via I_{Ca}, by extrusion through the transporter, by diffusion into the central space ($[Ca^{2+}]_f$) and by binding and unbinding to high ($[Ca^{2+}]_{ls}$) and low ($[Ca^{2+}]_{ls}$) affinity buffers below the membrane. The calcium ions in the central space ($[Ca^{2+}]_f$) also bind and unbind to the high ($[Ca^{2+}]_{hf}$) and low ($[Ca^{2+}]_{lf}$) affinity internal buffers.

I_{Ca} to a lesser extent than the voltage step to 0 or +20 mV, thus the calcium concentration increases slowly and is reduced compared to these voltage clamp levels. As a result, $I_{Cl(Ca)}$ is smaller and slower compared to the current elicited by 0 and +20 mV voltage steps.

3.1.2 *Current Clamp Simulation*

The ionic current models are derived by the data from voltage clamp experiments, and thus there are no interactions among ionic currents. Here we simulate current clamp responses to validate the proposed model using an independent set of data. The current clamp simulation is a critical test of the ionic current model, since there is no guarantee that the model can reproduce the current clamp responses, in which the simulation has to be made by using the whole ionic components.

Table 11.4 A mathematical description of the calcium system of the photoreceptor

Calcium concentration

$$\frac{d}{dt}[Ca^{2+}]_s = -\frac{I_{Ca}}{2FV_1} - \frac{I_{ex}+I_{ex2}}{2FV_1} - D_{Ca}\frac{S_1}{V_1\delta}([Ca^{2+}]_s - [Ca^{2+}]_f)$$
$$+ Lb_2[Ca^{2+}]_{ls} - Lb_1[Ca^{2+}]_s(BL - [Ca^{2+}]_{ls}) - Hb_1[Ca^{2+}]_s(BH - [Ca^{2+}]_{hs})$$
$$+ Hb_2[Ca^{2+}]_{hs}$$

$$\frac{d}{dt}[Ca^{2+}]_f = D_{Ca}\frac{S_1}{V_2\delta}([Ca^{2+}]_s - [Ca^{2+}]_f) + Lb_2[Ca^{2+}]_{lf} - Lb_1[Ca^{2+}]_f(BL - [Ca^{2+}]_{lf})$$
$$- Hb_1[Ca^{2+}]_f(BH - [Ca^{2+}]_{hf}) + Hb_2[Ca^{2+}]_{hf}$$

Calcium buffer concentration

$$\frac{d}{dt}[Ca^{2+}]_{ls} = Lb_1[Ca^{2+}]_s(BL - [Ca^{2+}]_{ls}) - Lb_2[Ca^{2+}]_{ls}$$

$$\frac{d}{dt}[Ca^{2+}]_{hs} = Hb_1[Ca^{2+}]_s(BH - [Ca^{2+}]_{hs}) - Hb_2[Ca^{2+}]_{hs}$$

$$\frac{d}{dt}[Ca^{2+}]_{lf} = Lb_1[Ca^{2+}]_f(BL - [Ca^{2+}]_{lf}) - Lb_2[Ca^{2+}]_{lf}$$

$$\frac{d}{dt}[Ca^{2+}]_{hf} = Hb_1[Ca^{2+}]_f(BH - [Ca^{2+}]_{hf}) - Lb_2[Ca^{2+}]_{hf}$$

Calcium pump and exchanger

$$I_{ex} = I_{max}\cdot \exp\left(-\frac{V+14}{70}\right)\cdot \frac{([Ca^{2+}]_s - Ca_e)}{([Ca^{2+}]_s - Ca_e) + 2.3}$$

$$I_{ex2} = I_{max2}\cdot \frac{([Ca^{2+}]_s - Ca_e)}{([Ca^{2+}]_s - Ca_e) + 0.5}$$

Maricq and Korenbrot (1988) showed that, in isolated lizard cones, action potentials can be generated by depolarizing current steps. The corresponding spike responses simulated by our model are shown in Figure 11.4. The responses elicited by the depolarizing current under control and TEA conditions are reproduced well. Dotted traces in Figure 11.4(A), (B) are results simulated by blocking $I_{Cl(Ca)}$ in the model. Current dynamics during spike responses are shown in Figure 11.4(C), (D). From these results, we can understand which ionic current participates in the different phases of the spike responses. The membrane depolarization initially elicited by the applied current activates I_{Ca}, and I_{Ca} further depolarizes the cell. I_{Kv} and $I_{Cl(Ca)}$ participate in repolarization of the action potential in phases 1 and 2 respectively. As clearly shown by the dotted traces in Figure 11.4 (A), (B), the membrane potentials remain depolarized without $I_{Cl(Ca)}$, that is, $I_{Cl(Ca)}$ terminates the action potential.

Table 11.5 Parameters of the calcium system of the photoreceptor

F	Faraday constant, 9.649×10^4 C/mol
V_1	volume of submembrane area, 3.812×10^{-13} l
V_2	volume of deep intracellular area, 5.236×10^{-13} l
δ	diffusion distance, 3×10^{-4} cm
S_1	submembrane area, 3.14×10^{-6} cm^2
D_{Ca}	Ca diffusion coefficient, 6.000×10^{-6} cm^2/s
I_{max}	maximum Na^+–Ca^{2+} exchanger current, 20 pA
I_{max2}	maximum Ca–ATPase pump current, 30 pA
Ca_e	minimum intracellular Ca concentration, 0.05 μM
BH	maximum concentration of high-affinity buffer, 500 μM
BL	maximum concentration of low-affinity buffer, 300 μM
Lb_1	on rate constant to low-affinity buffer, 0.4 s^{-1} μM^{-1}
Lb_2	off rate constant to low-affinity buffer, 0.2 s^{-1}
Hb_1	on rate constant to high-affinity buffer, 100 s^{-1} μM^{-1}
Hb_2	off rate constant to high-affinity buffer, 90 s^{-1}

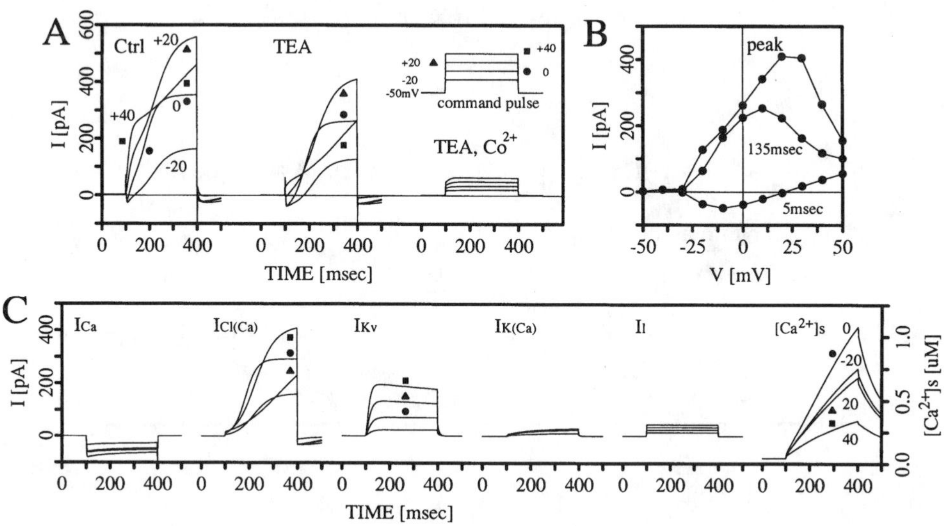

Figure 11.3 Simulated responses of the inner segment model under voltage clamp conditions (Kamiyama et al., 1996). (A) Membrane currents under three different conditions corresponding to control, TEA and TEA–Co^{2+} solutions. The TEA responses were simulated by reducing $\overline{g_{kv}}$ from 2.0 to 0.16 ns. (B) Current–voltage relation at the peak, 5 ms, 135 ms, measured from the TEA responses. (C) Ionic current components and calcium concentration under control condition.

Table 11.6 A mathematical description of the calcium system of the bipolar cell

Calcium concentration

$$\frac{d[Ca^{2+}]_s}{dt} = -\frac{I_{Ca}}{2F \cdot V_s} - \frac{D_{Ca} \cdot S_{sd}}{V_s \cdot d_{sd}}([Ca^{2+}]_s - [Ca^{2+}]_d) - \frac{(I_{ex} + I_{ex2})}{2F \cdot V_s}$$
$$+ \beta_{bl} \cdot [Ca^{2+}]_{bls} - \alpha_{bl} \cdot [Ca^{2+}]_s \cdot ([Ca^{2+}]_{blmax} - [Ca^{2+}]_{bls})$$
$$+ \beta_{bh} \cdot [Ca^{2+}]_{bhs} - \alpha_{bh} \cdot [Ca^{2+}]_s \cdot ([Ca^{2+}]_{bhmax} - [Ca^{2+}]_{bhs})$$

$$\frac{d[Ca^{2+}]_d}{dt} = \frac{D_{Ca} \cdot S_{sd}}{V_d \cdot d_{sd}}([Ca^{2+}]_s - [Ca^{2+}]_d)$$
$$+ \beta_{bl} \cdot [Ca^{2+}]_{bld} - \alpha_{bl} \cdot [Ca^{2+}]_d \cdot ([Ca^{2+}]_{blmax} - [Ca^{2+}]_{bld})$$
$$+ \beta_{bh} \cdot [Ca^{2+}]_{bhd} - \alpha_{bh} \cdot [Ca^{2+}]_d \cdot ([Ca^{2+}]_{bhmax} - [Ca^{2+}]_{bhd})$$

Calcium buffer concentration

$$\frac{d[Ca^{2+}]_{bls}}{dt} = \alpha_{bl} \cdot [Ca^{2+}]_s \cdot ([Ca^{2+}]_{blmax} - [Ca^{2+}]_{bls}) + \beta_{bl} \cdot [Ca^{2+}]_{bls}$$

$$\frac{d[Ca^{2+}]_{bhs}}{dt} = \alpha_{bh} \cdot [Ca^{2+}]_s \cdot ([Ca^{2+}]_{bhmax} - [Ca^{2+}]_{bhs}) + \beta_{bh} \cdot [Ca^{2+}]_{bhs}$$

$$\frac{d[Ca^{2+}]_{bld}}{dt} = \alpha_{bl} \cdot [Ca^{2+}]_d \cdot ([Ca^{2+}]_{blmax} - [Ca^{2+}]_{bld}) + \beta_{bl} \cdot [Ca^{2+}]_{bld}$$

$$\frac{d[Ca^{2+}]_{bhd}}{dt} = \alpha_{bh} \cdot [Ca^{2+}]_d \cdot ([Ca^{2+}]_{bhmax} - [Ca^{2+}]_{bhd}) + \beta_{bh} \cdot [Ca^{2+}]_{bhd}$$

Calcium pump and exchanger

$$I_{ex} = \frac{J_{ex} \cdot ([Ca^{2+}]_s - [Ca^{2+}]_{min})}{[Ca^{2+}]_s - [Ca^{2+}]_{min} + 2.3} \cdot \exp(-(V + 14)/70)$$

$$I_{ex2} = \frac{J_{ex2} \cdot ([Ca^{2+}]_s - [Ca^{2+}]_{min})}{[Ca^{2+}]_s - [Ca^{2+}]_{min} + 0.5}$$

3.1.3 *Simulated Light Responses of the Rod Photoreceptor*

In general, ionic current models are typically used in the context of voltage- and current-clamp experiments. They can also be used in studies investigating several sorts of dynamical behaviors. Here we simulate light responses of rod photoreceptors.

The biophysical processes underlying phototransduction are now well understood (reviewed in Yau and Baylor, 1989; McNaughton, 1990), and a quantitative model

Table 11.7 Parameters of the calcium system of the bipolar cell

F	Faraday constant, 9.649×10^5 C/mol
D_{Ca}	Ca diffusion coefficient, 6.000×10^{-8} dm^2/s
V_s	volume of the submembrane area, 1.692×10^{-13} dm^3
V_d	volume of the deep intracellular area, 7.356×10^{-13} dm^3
S_{sd}	surface area of the submembrane and the deep intracellular area spherical boundary, 4.000×10^{-8} dm^2
d_{sd}	distance between submembrane area and the deep intracellular area, 5.900×10^{-5} dm
$[Ca^{2+}]_{blmax}$	total low-affinity buffer concentration, 400 μM
$[Ca^{2+}]_{bhmax}$	total high-affinity buffer concentration, 300 μM
α_{bl}, β_{bl}	on and off rate constants for the binding of Ca to low-affinity buffer, $0.4\,s^{-1}\,\mu M^{-1}$, $0.2\,s^{-1}$
α_{bh}, β_{bh}	on and off rate constants for the binding of Ca to high-affinity buffer, $100\,s^{-1}\,\mu M^{-1}$, $90\,s^{-1}$
J_{ex}	maximum Na^+–Ca^{2+} exchanger current, 9 pA
J_{ex2}	maximum Ca–ATPase exchanger current, 9.5 pA
$[Ca^{2+}]_{min}$	minimum intracellular Ca concentration for Ca extrusion, 0.05 μM

of phototransduction has been proposed by Torre et al. (1990). In order to simulate the light response of photoreceptors, we introduced the flow of photocurrent into the inner segment as shown in Figure 11.1. In the model of phototransduction (Torre et al., 1990), the cyclic GMP-activated current (J) is expressed as a function of the concentration of cyclic GMP (g). The cyclic GMP-activated current can be measured in an excised patch condition, and it is sufficient to use the current, J, for reproducing the current response in the outer segment. The current–voltage relation of the cyclic GMP-activated current shows an outward rectification reversing near at +10 mV, and thus the photocurrent flow into the inner segment depends on the membrane potential (Baylor and Nunn, 1986). We expressed the voltage dependency of photocurrent (I_{photo}) by multiplying the cyclic GMP-activated current (J) by the electrochemical driving force ($V - E_p$) as shown in Table 11.8.

Figure 11.5 shows the responses of photocurrent (A) and photovoltage (B) to series of light flashes of increasing intensity. The model produces responses highly similar to those measured experimentally (Baylor et al., 1984). In order to assess the importance of each ionic current in shaping the voltage response to light, we estimated the ionic current flow during the light responses as shown in Figure 11.5(C). From these responses, we can explain why the voltage response of a rod to a bright flash is not the mirror image of the photocurrent. In normal conditions, I_h activates slowly as the membrane hyperpolarizes, and thus I_h shapes the initial transient in the voltage response. To confirm the effects of I_h, we simulated the responses by blocking I_h in the model as shown in Figure 11.5(D). The voltage response changes dramatically. The initial transient for bright stimuli disappeared, the time to peak of

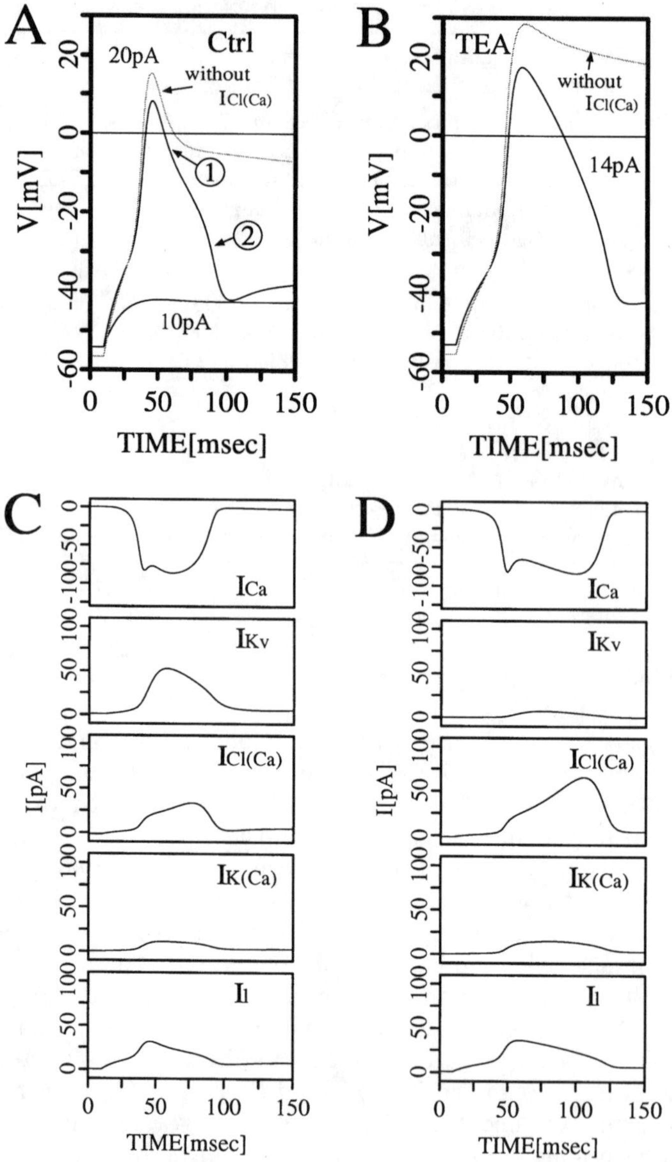

Figure 11.4 Simulations of the current clamp experiments using the inner segment model (Kamiyama et al., 1996). (A) Voltage responses of the model to depolarizing current steps of 10 and 20 pA under control condition. The response without $I_{\mathrm{Cl(Ca)}}$ is plotted by the dotted line. (B) Voltage response of the model to a depolarizing current step of 14 pA under the TEA condition. $\overline{g_{\mathrm{Kv}}} = 0.16\,\mathrm{ns}$ for the TEA condition. Dotted trace was simulated by blocking $I_{\mathrm{Cl(Ca)}}$. (C) (D) Ionic current components under the control and TEA conditions, respectively.

Table 11.8 Model of phototransduction in a rod photoreceptor

Change of intracellular Ca^{2+}

$$\frac{dc}{dt} = b \cdot J - \gamma_{Ca} \cdot (c - c_0) - k_1 \cdot (e_T - c_b) \cdot c + k_2 c_b$$

$$\frac{dc_b}{dt} = k_1 \cdot (e_T - c_b) \cdot c - k_2 \cdot c_b$$

$b = 0.625\,\mu M\,s^{-1}\,pA^{-1}, \quad \gamma_{Ca} = 50\,s^{-1},$

$k_1 = 0.2\,s^{-1}\,\mu M, \quad k_2 = 0.8\,s^{-1}, \quad e_T = 500\,\mu M$

Change of intracellular cGMP

$$\frac{dg}{dt} = A - g \cdot (\bar{V} + \sigma \cdot PDE^*)$$

$$A = \frac{A_{max}}{1 + (c/K_c)^4}$$

$\sigma = 1\,s^{-1}\,\mu M^{-1}, \quad A_{max} = 65.6\,\mu M s^{-1}$

$\bar{V} = 0.4\,s^{-1}, \quad K_c = 0.1\,\mu M$

Mechanism of phototransduction

$$\frac{dRh}{dt} = J_{h\nu}(t) - \alpha_1 Rh + \alpha_2 Rh_i$$

$$\frac{dRh_i}{dt} = \alpha_1 Rh - (\alpha_1 + \alpha_3) \cdot Rh_i$$

$$\frac{dT^*}{dt} = \varepsilon Rh \cdot (T_{Tot} - T^*) - \beta_1 T^* + \tau_2 PDE^* - \tau_1 T^*(PDE_{Tot} - PDE^*)$$

$$\frac{dPDE^*}{dt} = \tau_1 T^* \cdot (PDE_{Tot} - PDE^*) - \tau_2 PDE^*$$

$\alpha_1 = 2.0\,s^{-1}, \quad \alpha_2 = 0.8\,s^{-1}, \quad \alpha_3 = 0.6\,s^{-1},$

$T_{Tot} = 1000\,\mu M, \quad \tau_1 = 0.2\,s^{-1}\,\mu M^{-1}, \quad \tau_2 = 5\,s^{-1}$

$\varepsilon = 1.0\,s^{-1}\,\mu M^{-1}, \quad PDE_{Tot} = 100\,\mu M$

Photosensitive current

$$J = J_{max} \frac{g^3}{g^3 + K^3}$$

$$I_{photo} = K_p \cdot J \cdot (V - E_p)$$

$J_{max} = 5040\,pA, \quad K^3 = 1000\,\mu M, \quad K_p = 2, \quad E_p = 10\,mV$

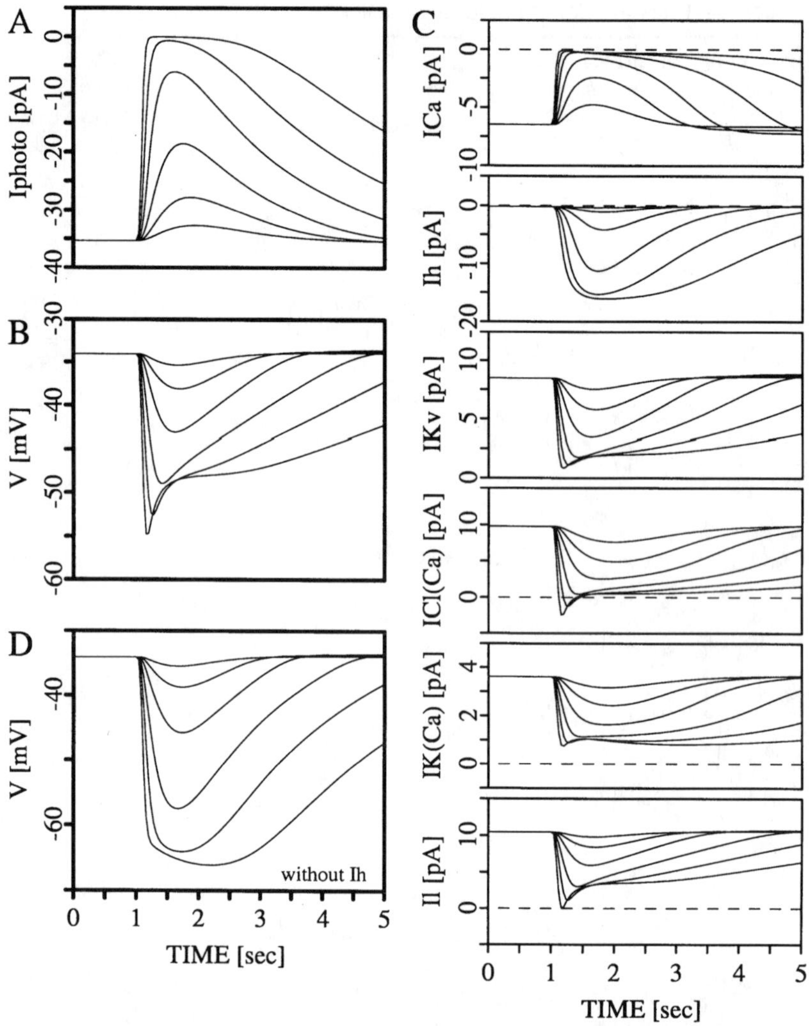

Figure 11.5 Simulated light responses of the photoreceptor (Kamiyama et al., 1996). Stimulus intensities ($J_{hv}(t)$) were 0.03, 0.1, 0.3, 1.0, 3.0, 10.0 $Rh_t^* s^{-1}$. The duration was 50 ms. (A) Photocurrents. (B) Voltage responses. (C) Each ionic current of the inner segment. (D) Voltage responses without $I_h \cdot Cl_h = 0.45\,\mu M$, $A_{max} = 131.2\,\mu M\,s^{-1}$, $\bar{V} = 0.8\,s^{-1}$, $\bar{g}_h = 3.0\,ns$, $\alpha_h = 8/(\exp((V+78)/14)+1)$, $\beta_h = 18/(\exp(-(V+8)/19)+1)$ for simulating the rod responses.

the voltage response was delayed, the amplitude was increased, and the duration of the response was prolonged. These results are consistent with the experimental result by Baylor et al. (1984), who showed that the initial spike characteristics of rod responses were abolished by the I_h blocker, Cs.

3.2 Horizontal Cell

3.2.1 *Voltage and Current Clamp Simulations*

We calculated the *I–V* relationships in the absence (Figure 11.6(A)) and presence (Figure 11.6(B)) of glutamate, the transmitter released from photoreceptors. Each inset in Figures 11.6(A), (B) shows simulated currents when the membrane potential is shifted from the holding potential of −60 to −80 mV, 0 and 20 mV. The simulated responses in the absence of glutamate (Figure 11.6(A)) show the time-dependent,

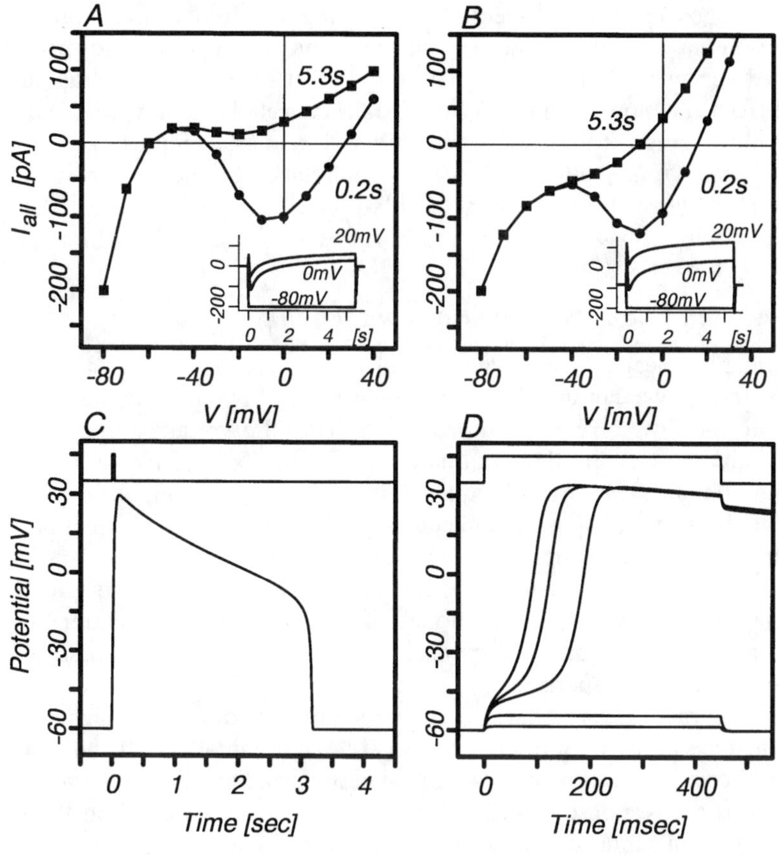

Figure 11.6 The *I–V* characteristics and the action potentials simulated by the model (Usui et al., 1996a). The *I–V* relationship was obtained under the voltage clamp in the absence of glutamate (Glu(t) = 0, (A)) or in the presence of glutamate (Glu(t) = 1.0, (B)). Each inset shows the total ionic currents induced by depolarization to −80, 0 and 20 mV. The *I–V* curves were obtained from the current data measured at 0.2 and 5.3 s after the command onset. The action potentials were generated by injecting current into the model (C) and (D). The magnitude of the current is 50 pA with 45 ms pulse in (C), and 5, 15, 21, 23 and 24 pA with 450 ms steps in (D).

N-shaped I–V characteristics similar to the experimental data (Tachibana, 1983). Since glutamate (denoted as Glu(t)) activates I_{glu} and suppresses I_{Ka}, the negative conductance region in Figure 11.6(B) observed in Figure 11.6(A) disappears in the presence of glutamate, and the cell shows the monotonic I–V relationship when the current amplitude was measured 5.3 s after the onset of voltage pulse. Miyachi and Murakami (1989) explored the I–V characteristics of an *in situ* horizontal cell and found the linear relation which can be explained by the effects of glutamate released from photoreceptors.

Solitary horizontal cells produce a long-lasting action potential when the membrane potential is depolarized by a current injection (Tachibana, 1981). Under current clamp conditions, the model produces action potentials (Figure 11.6(C), (D)) similar to those observed experimentally. The rising dynamics of the action potential depends on the magnitude of injecting current (Tachibana, 1981) and this behavior is simulated very well by the model. Although most of the model parameters were estimated from the slower time resolution data collected under voltage clamp, the faster dynamic change of the membrane potential such as calcium spike is simulated quite well, as shown in Figure 11.6(D). These simulation studies clearly demonstrate that the solitary horizontal cell model is plausible. Thus, using this model we can analyze the mechanisms by which *in situ* horizontal cells produce light responses.

3.2.2 *Inverse Dynamics of the Horizontal Cell*

Since the detailed electrical characteristics of the horizontal cell are incorporated in the present model, we can use the model as an inverse dynamics model. That is, we can estimate the flow of each ionic current during the voltage response recorded experimentally by reversing the dynamics (Usui et al., 1991). This method is similar to the action potential clamp technique (Ibarra et al., 1991), which is applied to the study of the dynamics of specific ionic currents during the action potential. That is, the conventional voltage clamp method is achieved to control membrane potential constant ($dV/dt = 0$) and measures the resulted current. However, from the mathematical point of view with the model, if the membrane is clamped by the experimentally recorded voltage response, then the measured current is the current generating the voltage response.

By using this method, we analyzed the ionic currents during the light responses. Intracellular recordings were made on the L-type horizontal cell in the carp retina. The L-type horizontal cell responds with hyperpolarization to any stimulus condition. However, the response waveform of the cell strongly depends on the stimulus wavelength, size and intensity, and shows nonlinear dynamical behaviors.

Figure 11.7(C) shows the ionic current components for an L-type horizontal cell light response (A) calculated by the ionic current model. I_{Ca} gradually changes by the membrane hyperpolarization. I_{A} is negligible. I_{Ka}, I_{Kv} and I_{l} show the similar waveform to the response and thus these currents are passive components. However, the estimated input current (B), I_{glu}, shows faster dynamics than the voltage response, and note that the inflection like a notch after the peak is more pronounced. This result clearly demonstrates that the L-type horizontal cell membrane acts like a low-pass filter and indicates the advantage for analyzing the input current to the cell response.

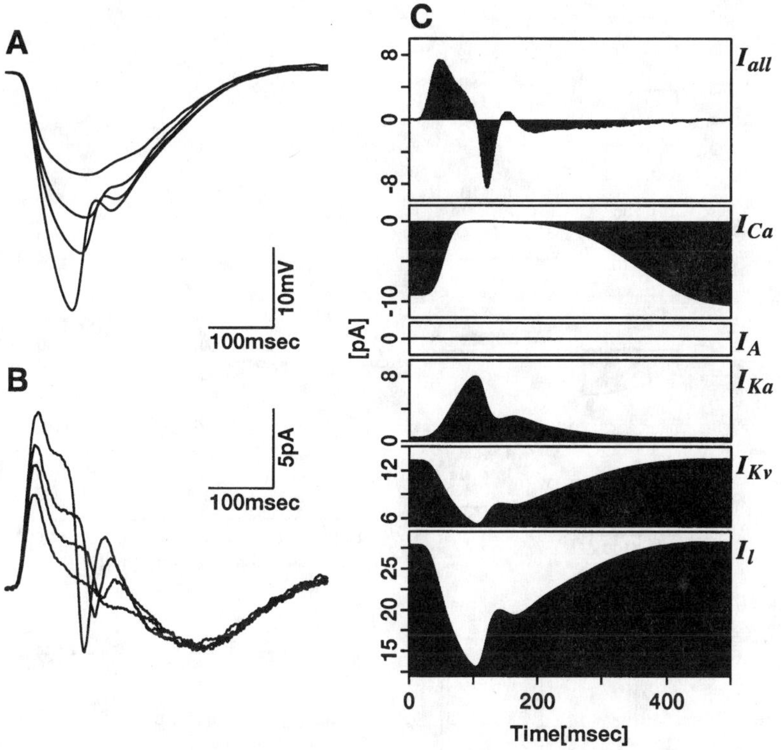

Figure 11.7 Estimated ionic current dynamics in the L-type horizontal cell during the light response (Kamiyama et al., 1993; Usui et al., 1991). Each ionic current component (C) is calculated using the model from the voltage response (A). I_{Ca} slowly changes with hyperpolarization. I_A is almost constant. I_{Ka}, I_{Kv} and I_l flow passively to the response. Note that the estimated input current (B) shows faster dynamics and the result indicated that horizontal cell membrane acts like a low-pass filter.

3.3 Bipolar Cell

3.3.1 *Voltage Clamp Simulation*

Figure 11.8(A) shows the simulated total ionic current in response to depolarizing command pulses to -10, $+10$ and $+30$ mV from the holding potential of -30 mV. The outward current develops with further depolarization and the amplitude peaks near the membrane potential of $+50$ mV (Figure 11.8(C)). The outward current is decreased by depolarization beyond $+50$ mV. These outward currents show a little time dependency. Individual ionic currents and the calcium concentration changes to the depolarizing pulses are shown in Figure 11.8(D). One may expect that since the cytosolic concentration of calcium ions changes slowly, a tail current would appear at the cessation of command pulse. However, the calcium concentration near the ionic channels ($[Ca^{2+}]_s$) returns within a few milli-seconds after the command pulse, and therefore I_{Ca} and $I_{K(Ca)}$ show no tail currents. These results explain why no tail current

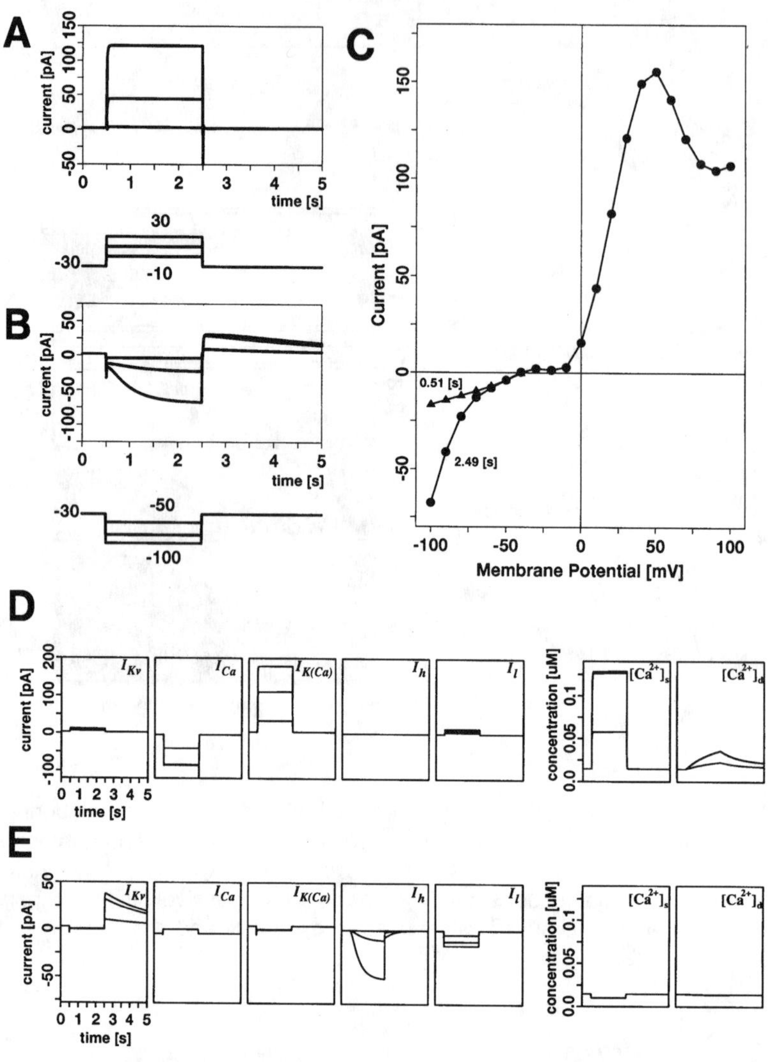

Figure 11.8 Simulated response of the bipolar cell model under voltage clamp conditions (Usui et al., 1996c). The cell was held at -30 mV and 2 s command pulses were applied in 10 mV steps. (A) Simulated total ionic current to depolarizing command pulses to -10, $+10$ and $+30$ mV. (B) Simulated total ionic current to hyperpolarizing command pulses to -50, -80 and -100 mV. (C) I–V relationship under voltage clamp conditions. The I–V curves were obtained from the current data measured at 0.51 s (triangle) and 2.49 s (filled circle). (D) Ionic currents and calcium concentrations under voltage clamp conditions in response to the depolarizing pulses. (E) Ionic currents and calcium concentrations under voltage clamp conditions in response to the hyperpolarizing pulses.

has been observed by the depolarizing pulse (Kaneko and Tachibana, 1985a; Lasater, 1988).

The total current induced by the hyperpolarizing pulses to −50, −80 and −100 mV is shown in Figure 11.8(B). These inward currents show a strong time dependence, and the tail components are observed as in the experimental results of Kaneko and Tachibana (1985a). Figure 11.8(E) provides an interpretation of these current responses to the voltage pulses. The slowly activating inward current results from the time- and voltage-dependent properties of I_h. The tail currents after the hyperpolarizing steps are caused by I_h and I_{Kv} which are activated by the depolarizing pulse of −30 mV from the hyperpolarizing potentials.

3.3.2 *Current Clamp Simulation*

Kaneko and Tachibana (1985a) demonstrated that bipolar cells showed sustained depolarizations to depolarizing current pulses and, in response to hyperpolarizing pulses, showed complicated responses with an initial transient hyperpolarization followed by a smaller plateau hyperpolarized level. These response properties are well reproduced by our model as shown in Figure 11.9. These response characteristics can be explained as follows.

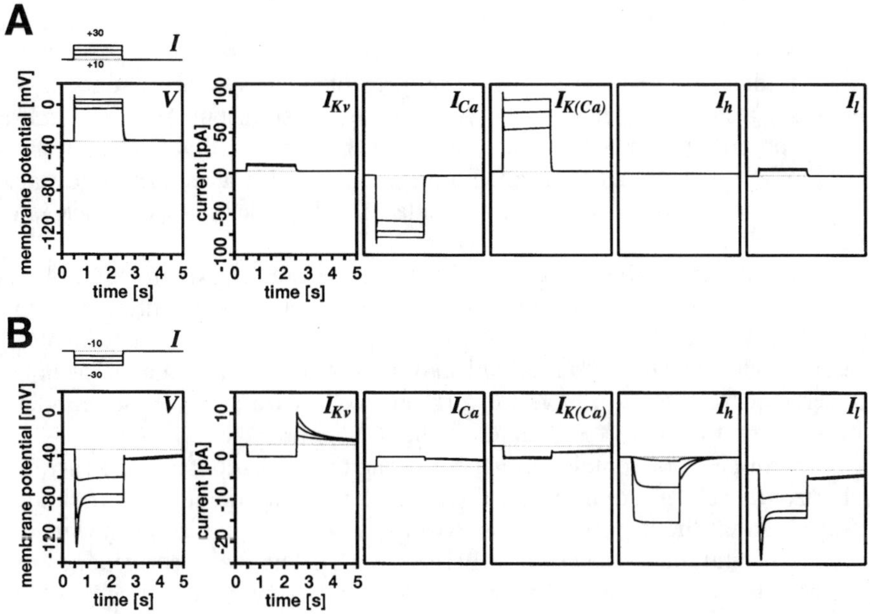

Figure 11.9 Simulated response of the bipolar cell model under current clamp conditions (Usui et al., 1996c). (A) Voltage response and each ionic current to depolarizing current pulses of 10, 20 and 30 pA. (B) Voltage response and each ionic current to hyperpolarizing current pulses of 10, 20 and 30 pA.

As shown in Figure 11.9(A), membrane depolarization gradually reaches a saturated value of around +10 mV as the injecting current is increased. This characteristic is determined by the magnitude of the currents, I_{Ca} and $I_{K(Ca)}$. The initial transient and the subsequent plateau in response to the hyperpolarizing currents in Figure 11.9(B) is caused by activation of I_h, that is, activation of I_h slowly reduces the level of hyperpolarization. The spike-like rebound shown in Figure 11.9(B) seen at the cessation of the hyperpolarizing current pulses and after-hyperpolarization results from I_h and I_{Kv}.

4. DISCUSSION

We have obtained a satisfactory agreement between simulated and experimental responses of rod photoreceptors. The model reproduced the voltage- and current-clamp responses reasonably well and the light response properties observed in the photoreceptors.

Attwell and Wilson (1980) also modeled the time- and voltage-dependent current in rods, and analyzed the electrical behavior of rods using model simulations. They found that the current, I_A, plays a major role in shaping the light responses. Since their I_A is an inward current activated by membrane hyperpolarization, the current partially corresponds to I_h in our model. They did not separate membrane current into each ionic current, because the components were not fully identified in 1980. Our model, however, takes into account many of the ionic current components. This level of details in the model makes it possible to predict the responses under various conditions. The predicted effects of I_h on the light responses suggest that I_h makes the response more transient as the membrane potential hyperpolarizes, which is consistent with the experimental finding by Baylor et al. (1984). We also found that zatebradine, a drug with a specific bradycardiac action, might induce visual symptoms as a side effect by blocking I_h channels in rods (Usui et al., 1996b).

The present model is able to account for many features of experimental observations in photoreceptors but not all the findings. Ionic currents identified in synaptic terminal (Kaneko and Tachibana, 1986) of cone photoreceptors were not involved in the present model. The detailed mechanism of phototransduction in cones is not also well understood yet. We still have to work out to develop a quantitative model of phototransduction in cones to simulate the light responses.

Although solitary horizontal cells generate calcium action potentials (Tachibana, 1981), horizontal cells *in situ* normally do not evoke action potentials. Our simulated result suggests that the membrane property of *in situ* horizontal cells is linearized by the presence of glutamate as shown in Figure 11.6(B). The result corresponds to the measurement by Miyachi and Murakami (1989), who directly examined the linear I–V relation of *in situ* horizontal cells. The horizontal cell also releases GABA in a calcium-independent manner (Schwartz, 1982), and thus the role of the calcium channel in the cell has yet to be clarified. Our result suggests that I_{Ca} in horizontal cell may act as an enhancer for the depolarizing response by its positive feedback characteristics.

The lateral acceleration properties, which were clearly demonstrated in horizontal cells by the stimulus displacement experiment (Usui, 1996a), are also observed in the turtle rod network (Detwiler et al., 1978). The previous study explained the acceleration property by assuming that the cell membrane behaves like an inductance, but the assumption is quite unrealistic for a horizontal cell membrane. Although Detwiler et al. (1980) suggested that if hyperpolarization decreases the potassium conductance, then the membrane shows inductance-like properties, the membrane conductance of the horizontal cell is almost linearized by the effect of glutamate. Therefore, the membrane conductance and capacitance cause the horizontal cell to act as a linear low-pass filter during hyperpolarization, and the current flows proportionally to the membrane voltage (see Figure 11.6(B)). The main factor influencing the hyperpolarizing response is the input signal to the horizontal cell. We conclude that the lateral acceleration for hyperpolarizing response of the horizontal cell was due to the feedback interaction between cone and horizontal cell, that is, the feedback effect causes the high-pass filtering properties in the horizontal cell layer (Usui, 1996a).

A quantitative reconstruction of the membrane ionic currents in the retinal bipolar cells has been realized in the present study. Our model is capable of accurately reproducing the voltage- and current-clamp responses of the bipolar cells. The present model also has an ability to represent the subtype specificities found in bipolar cells. The membrane ionic currents of bipolar cells have been examined not only in lower vertebrates but also in mammals (Kaneko et al., 1989; Karschin and Wässle, 1990). These results show that most of the ionic currents in mammalian bipolar cells are the same kinds as those in the lower vertebrates, although there is a difference in the dynamic properties of the currents. Since our model sufficiently reproduced the different dynamic characteristics of white bass bipolar cells, it should be possible to simulate the mammalian bipolar cell responses as well by re-estimating the model parameters.

Recently, it was found that several ionic channels and receptors on bipolar cells are localized on the dendrite, cell body or axon terminal. Nawy and Jahr (1990) found that cGMP-gated channels closed by glutamate in bipolar cells are responsible for the on-light response. Since the mechanism of the cGMP-mediated pathway is similar to that of phototransduction, it is possible to reconstruct bipolar cell model which produces the light response based on the present model. Tachibana et al. (1993) recently found that most of the calcium channels on bipolar cells are localized at the axon terminal and calcium ions move toward the cell body through the axon by diffusion. Although the present model was able to reproduce the calcium current in bipolar cells, it is important to incorporate the spatial characteristic of the calcium system into our model for the full reconstruction of the bipolar cell. In modeling, we have to introduce more accurate morphological characteristics of bipolar cell (see Poznanski and Umino, 1997).

5. CONCLUSIONS AND FUTURE PERSPECTIVES

A quantitative reconstruction of the membrane ionic currents in the retinal neurons, photoreceptor, horizontal cell and bipolar cell has been realized in this Chapter. The ionic current models are capable of accurately reproducing the voltage- and current-clamp responses of these cells. We believe that our models which replicate the detailed physiological structures and functions of the retinal outer plexiform layer clearly illustrate what we know and what we need to find out.

One still hears the criticism that since our models contain enormous number of parameters, by just changing them one could fit almost anything. This may be critical; however, this is indeed not a serious problem in our case. Our models were constructed based on the voltage clamp measurements, and the voltage clamp experiment itself can be thought of as a process for estimating parameters, since the experimental procedures assume the Hodgkin–Huxley type ionic conductances in the cell membrane (Hille, 1992). In this step of modeling, parameter values of each ionic current were estimated independently current by current, i.e. we do not have to estimate all the model parameters simultaneously. The estimated values of measurable parameter, such as the maximum conductances and the reversal potentials, corresponded well to the experimental data. However, there is no guarantee that the models can reproduce the behavior of the cells under other experimental conditions, since each ionic current was modeled based on pharmacologically isolated conditions. We examined the acceptability of the models by the current clamp experiment, and the results were extremely similar to the experimental data. These results clearly demonstrate the realism of our models.

Our primary aim is to construct a physiologically plausible model which should be distinguished from abstract models that reproduce a particular retinal function (e.g. Mead, 1989). Any new properties and details of the outer retina are discovered may be implemented easily into our dynamical models and examined on it. Thus, these models provide a unified framework for understanding the functional organization of the retinal outer plexiform layer.

PROBLEMS

11.1 The present photoreceptor model consists of two of three functional parts, the inner and outer segments. Extend the model to include the membrane properties of the synaptic terminal.

11.2 In darkness, glutamate, an excitatory neurotransmitter, is tonically released from the synaptic terminal of photoreceptors. Light closes the channels in the outer segment, hyperpolarizes the photoreceptor membrane and lowers the rate of transmitter release. Reduction of transmitter release causes a graded light response in second-order neurons, bipolar and horizontal cells. Discuss the underlying biophysical mechanisms of the signal transmission among photoreceptors and second-order neurons.

11.3 There are C-type (color) horizontal cells which hyperpolarize when the retina is stimulated with a particular wavelength of light and depolarize when another wavelength of light is applied, e.g. the R/G-type cell depolarizes to red light but hyperpolarizes to green light. Discuss the functional role of each ionic current in the C-type horizontal cells.

11.4 Experimental suggestion: Horizontal cells are interconnected by gap junctions. Extend the inverse dynamics method (discussed in section 3.2.2) to estimate the gap junctional conductance. Apply the method to experimental data measured under different light conditions, light adapted and dark adapted retinas.

11.5 There are two types of glutamate receptor in bipolar cells, i.e. ionotropic and metabotropic receptors. The current mediated by the ionotropic receptor flows through cation channels, which is responsible for OFF bipolar cells. On the other hand, the current mediated by the metabotropic receptor flows through the cGMP sensitive cation channels, which is responsible for ON bipolar cells (see section 4). Extend the bipolar cell model to include these glutamate receptors.

11.6 Bipolar cells are the key interneurons which form a signal pathway from the outer retina to the inner retina. Discuss how the visual signals are coded as neural signals in bipolar cell and how the ionic channels in the outer retina contribute to the signal processings.

ACKNOWLEDGMENTS

This work was supported in part by the Human Frontier Science Program (HFSP; P.I.: H. Spekreijse), the Hori Information Science Promotion Foundation and the Grant-in-Aid for Scientific Research (No. 05267104, No. 08279106, No. 08650486 and No. 08750517) from the Ministry of Education, Science, Sports and Culture of Japan.

REFERENCES

Attwell, D., Mobbs, P. and Tessier-Lavigne, M. (1987) Neurotransmitter-induced currents in retinal bipolar cells of the axolotl *ambystoma mexicanum*. *J. Physiol.* (*Lond.*), **387**, 125–161.

Attwell, D. and Wilson, M. (1990) Behaviour of the rod network in the tiger salamander retina mediated by membrane properties of individual rods. *J. Physiol.* (*Lond.*), **309**, 287, 315.

Bader, C.R., Bertrand, D. and Schwartz, E.A. (1982) Voltage-activated and calcium-activated currents studied in solitary rod inner segments from the salamander retina. *J. Physiol.* (*Lond.*), **331**, 253–284.

Barnes, S. and Hille, B. (1989) Ionic channels of the inner segment of tiger salamander cone photoreceptors. *J. Gen. Physiol.* **94**, 719–743.

Baylor, D.A., Matthews, G. and Nunn, B.J. (1984) Location and function of voltage-sensitive conductances in retinal rods of the salamander, ambystoma tigrinum. *J. Physiol.* (*Lond.*), **354**, 203–223.

Baylor, D.A. and Nunn, B.J. (1986) Electrical properties of the light-sensitive conductance of salamander rods. *J. Physiol.* (*Lond.*), **371**, 115–145.

Copenhagen, D.R. and Jahr, C.E. (1989) Release of endogeneous excitatory amino acids from turtle photoreceptors. *Nature*, **341**, 356–359.

DiFrancesco, D. and Noble, D. (1985) A model of cardiac electrical activity incorporating ionic pumps and concentration changes. *Phil. Trans. R. Soc.*, **307**, 353–398.

Detwiler, P.B., Hodgkin, A.L. and McNaughton, P.A. (1978) A surprising property of electrical spread in the network of rods in the turtle's retina. *Nature*, **274**, 562–565.

Detwiler, P.B., Hodgkin, A.L. and McNaughton, P.A. (1980) Temporal and spatial characteristics of the voltage response of rods in the retina of the snapping turtle. *J. Physiol.* (*Lond.*), **300**, 213–250.

Hille, B. (1992) *Ionic Channels of Excitable Membranes.* 2nd edition Sinauer Associates.

Hodgkin, A.L. and Huxley, A.F. (1952) A quantitative description of membrane current and its application to conduction and excitation in nerve. *J. Physiol.* (*Lond.*), **117**, 500–544.

Heidelberger, R. and Matthews, G. (1992) Calcium influx and calcium current in single synaptic terminals of goldfish retinal bipolar neurons. *J. Physiol.* (*Lond.*), **447**, 235–256.

Ibarra, J., Morley, G.E. and Delmar, M. (1991) Dynamics of the inward rectifier K^+ current during the action potential guinea pig ventricular myocytes. *Biophys. J.*, **60**, 1534–1539.

Ishida, A.T., Kaneko, A. and Tachibana, M. (1984) Responses of solitary retinal horizontal cells from *Carassius auratus* to L-glutamate and related amino acids. *J. Physiol.* (*Lond.*), **348**, 255–270.

Kamiyama, Y., Ishihara, A. and Usui, S. (1998) A mathematical model of glutamatergic synapse of the retinal outer plexiform layer. Proc. 2nd RIEC Intern. Sympos. DAIPS, Sendai, Japan, pp. 137–140.

Kamiyama, Y., Ishii, H. and Usui, S. (1993) A method for estimating the neural input to a neuron using the ionic current model. In *Computation and Neural Systems*, edited by F.H. Eeckman and J.M. Bower, Chapter 5, pp. 31–35. Boston: Kluwer Academic Publishers.

Kamiyama, Y., Ogura, T. and Usui, S. (1996) Ionic current model of the vertebrate rod photoreceptor. *Vis. Res.*, **36**, 4059–4068.

Kaneko, A. and Tachibana, M. (1985a) A voltage-clamp analysis of membrane currents in solitary bipolar cell dissociated from *carassius auratus*. *J. Physiol.* (*Lond.*), **358**, 131–152.

Kaneko, A. and Tachibana, M. (1985b) Effects of L-glutamate on the anomalous rectifier potassium current in horizontal cells of *Carassius auratus* retina. *J. Physiol.* (*Lond.*), **358**, 169–182.

Kaneko, A. and Tachibana, M. (1986) Membrane properties of solitary retinal cells. *Prog. Ret. Res.*, **5**, 125–146.

Kaneko, A., Pinto, L.H. and Tachibana, M. (1989) Transient calcium current of retinal bipolar cells of the moise. *J. Physiol.* (*Lond.*), **410**, 613–629.

Karschin, A. and Wässle, H. (1990) Voltage- and transmitter-gated currents in isolated bipolar cells of rat retina. *J. Neurophysiol.*, **63**, 860–876.

Lasater, E.M. (1986) Ionic currents of cultured horizontal cells isolated from white perch retina. *J. Neurophysiol.*, **55**, 499–513.

Lasater, E.M. (1988) Membrane currents of retinal bipolar cells in culture. *J. Neurophysiol.*, **60**, 1460–1480.

Lasater, E.M. (1991) Membrane properties of distal retinal neurons. *Prog. Ret. Res.*, **11**, 215–246.

Lukasiewicz, P.D., Maple, B.R. and Werblin, F.S. (1994) A novel GABA receptor on bipolar cell terminals in the tiger salamander retina. *J. Neurosci.*, **14**, 1202–1212.

Maricq, A.V. and Korenbrot, J.I. (1988) Calcium and calcium-dependent chloride currents generate action potentials in solitaly cone photoreceptors. *Neuron*, **1**, 503–515.

Maricq, A.V. and Korenbrot, J.I. (1990a) Inward rectification in the inner segment of single retinal cone photoreceptors. *J. Neurophysiol.*, **64**, 1917–1928.

Maricq, A.V. and Korenbrot, J.I. (1990b) Potassium currents in the inner segment of single retinal cone photoreceptors. *J. Neurophysiol.*, **64**, 1929–1940.

Mascagni, M.V. (1989) Numerical methods for neuronal modeling. In *Methods in Neuronal Modeling*, edited by C. Koch and I. Segev, Chapter 13, pp. 439–484. Cambridge, MA: MIT Press.

McNaughton, P.A. (1990) The light response of vertebrate photoreceptors. *Physiol. Rev.*, **70**, 847–883.

Maguire, G., Maple, B., Lukasiewicz, P. and Werblin, F. (1989) γ-Aminobutyrate type B receptor modulation of L-type calcium channel current at bipolar cell terminals in the retina of the tiger salamander. *Proc. Natl. Acad. Sci. USA*, **86**, 10144–10147.

Mead, C. (1989) *Analog VLSI and Neural Systems*. Reading, MA: Addison-Wesley.

Miyachi, E.-I. and Murakami, M. (1989) Decoupling of horizontal cells in carp and turtle retinae by intracellular injection of cyclic AMP. *J. Physiol.* (*Lond.*), **419**, 213–224.

Nawy, S. and Jahr, C.E. (1990) Suppression by glutamate of cGMP-activated conductance in retinal bipolar cells. *Nature*, **346**, 269–271.

Nawy, S. and Jahr, C.E. (1991) cGMP-gated conductance in retinal bipolar cells is suppressed by the photoreceptor transmitter. *Neuron*, **7**, 677–683.

Poznanski, R.R. and Umino, O. (1997) Syncytial integration by a network of coupled bipolar cells in the retina. *Prog. Neurobiol.*, **53**, 273–291.

Schwartz, E.A. (1982) Calcium-independent release of GABA from isolated horizontal cells of the toad retina. *J. Physiol.* (*Lond.*), **323**, 211–227.

Shingai, R. and Christensen, B.N. (1983) Sodium and calcium currents measurement in isolated catfish horizontal cells under voltage clamp. *Neuroscience*, **10**, 893–897.

Tachibana, M. (1981) Membrane properties of solitary horizontal cells isolated from goldfish retina. *J. Physiol.* (*Lond.*), **321**, 141–161.

Tachibana, M. (1983) Ionic currents of solitary horizontal cells isolated from goldfish retina. *J. Physiol.* (*Lond.*), **345**, 329–351.

Tachibana, M. (1985) Permeability changes induced by L-glutamate in solitary retinal horizontal cells isolated from *Carassius auratus*. *J. Physiol.* (*Lond.*), **358**, 153–167.

Tachibana, M. and Kaneko, A. (1988) Retinal bipolar cells receive negative feedback input from GABAergic amacrine cells. *Visual Neurosci.*, **1**, 297–305.

Tachibana, M., Okada, T., Arimura, T., Kobayashi, K. and Piccoline, M. (1993) Dihydropyridine-sensitive calcium current mediates neurotransmitter release from bipolar cells of the goldfish retina. *J. Neurosci.*, **13**, 2898–2909.

Torre, V., Forti, S., Menini, A. and Campani, M. (1990) Model of phototransduction in retinal rods. *Cold Spring Harbor Symp. Quant. Biol.*, **LV**, 563–573.

Usui, S., Ishii, H. and Kamiyama, Y. (1991) An analysis of retinal L-type horizontal cell responses by the ionic current model. *Neurosci. Res.*, **Suppl. 15**, S91–S105.

Usui, S., Kamiyama, Y., Ishii, H. and Ikeno, H. (1996a) Reconstruction of retinal horizontal cell responses by the ionic current model. *Vis. Res.*, **36**, 1711–1719.

Usui, S., Kamiyama, Y., Ogura, T., Kodama, I. and Toyama, J. (1996b) Effects of zatebradine (UL-FS 49) on the vertebrate retina. In *Recent Progress in Electropharmacology of the Heart*, edited by J. Toyama, M. Hiraoka and I. Kodama, pp. 37–46. Boca Raton, FL: CRC Press.

Usui, S., Ishihara, A., Kamiyama, Y. and Ishii, H. (1996c) Ionic current model of bipolar cells in the lower vertebrate retina. *Vis. Res.*, **36**, 4069–4076.

Von Gersdorff, H. and Matthews, G. (1996) Calcium-dependent inactivation of the calcium current in synaptic terminals of retinal bipolar neurons. *J. Neurosci.* **15**, 115–122.

Yagi, T. and Kaneko, A. (1988) The axon terminal of goldfish retinal horizontal cells: a low membrane conduction measured in solitary preparations and its implication to the signal conduction from soma. *J. Neurophysiol.*, **59**(2), 482–494.

Yamada, W.M., Koch, C. and Adams, P.R. (1989) Multiple channels and calcium dynamics. In *Methods in Neuronal Modeling*, edited by C. Koch and I. Segev, Chapter 4, pp. 97–133. Cambridge, MA: MIT Press.

Yau, K.W. and Baylor, D.A. (1989) Cyclic GMP-activated conductance of retinal photoreceptor cells. *Annu. Rev. Neurosci.*, **12**, 289–327.

CHAPTER 12
EPHAPTIC INTERACTIONS BETWEEN NEURONS: THE EXAMPLE OF THE HIPPOCAMPUS

Robert Costalat and Bruno Delord

1. INTRODUCTION

It has long been recognized that chemical synaptic transmission is the main way of information transfer between neurons (for a historical perspective, e.g. Shepherd, 1991). However, other mechanisms can contribute to interaction between neurons, and possibly to information transfer, such as: (i) electrotonic coupling *via* gap-junctions, (ii) changes in ionic concentrations, and (iii) electrical field effects (or ephaptic transmission, or extracellular field interaction) (e.g. Dudek and Traub, 1989).

The first documented example of an electrical field effect was the alteration of the excitability of axons by adjacent, stimulated axons within a peripheral nerve (Jasper and Monnier, 1938; Katz and Schmitt, 1940a,b; Arvanitaki, 1942; Marrazzi and Lorente de No, 1944). These experimental studies were soon completed by thorough theoretical work (Lorente de No, 1947). Although the experimental conditions in most of these studies were somewhat artificial, it was strongly suggested that ephaptic transmission was a prominent mechanism of interaction between neurons, even in the central nervous system (CNS). This hypothesis fell in considerable disfavor when chemical synaptic transmission was experimentally demonstrated in the CNS in the early 1950s (see Eccles, 1964; Shepherd, 1991), and few articles on ephaptic transmission were published for some time. Recently, there has been a renewal of interest in the possible involvement of electrical field effects in the physiology and pathophysiology of nervous tissue, both in peripheral nerves and the CNS (see Faber and Korn, 1989; Jeffery's 1995). Electrical field coupling in peripheral nerves has been the subject of several experimental (Kocsis et al., 1982) and theoretical studies (Markin, 1970; Clark and Plonsey, 1970, 1971). Moreover, examples of physiological field effects in the CNS were documented in the early 1960s (Furukawa and Furshpan, 1963). Further experimental work has provided good evidence that electrical field effects can play a role (i) in synchronizing neuronal activity in the CNS, especially during epilepsy, (ii) in cell growth and (iii) in membrane differentiation (Faber and Korn, 1989). Some of these experimental studies were closely correlated with precise modeling of the electrical field within nervous tissue, e.g. in the hippocampus (Dudek and Traub, 1989).

This chapter reviews the main biophysical and mathematical methods used to model electrical field effects. Section 2 is devoted to the derivation of the biophysical equations used to describe electrical field interaction in a nervous tissue, while a simplified model is produced in section 3. Section 4 examines the analytical and numerical methods available for solving these equations. The methods are discussed and examples of their application to real nervous tissues are given in section 5.

2. BIOPHYSICAL BASIS

2.1 Definitions

Stimulated excitable cells generate electrical fields which can have detectable effects on neighboring cells: this effect is known as electrical field effect, electrical interaction, electrical field interaction, or electrical field coupling between neurons. This effect is sometimes referred to as extracellular field interaction, in order to emphasize the importance of extracellular milieu. However, as shown below, the properties of both the intra- and extracellular milieux must be taken into account in any study of electrical field interaction. The term ephaptic transmission was first used in the 1930s to design electrical interaction between two fibers that have been artificially brought together, so as 'to touch' (Arvanitaki, 1942): in Greek, $\varepsilon\phi\alpha\psi\iota\zeta$ is the action of touching. Nowadays, ephaptic transmission is generally used as a synonym for electrical field effects.

2.2 Electromagnetic Analysis of Nervous Tissue

The biophysics of excitable cells are primarily based upon the Maxwell equations:

$$\nabla \cdot \boldsymbol{D} = \rho, \tag{12.1}$$

$$\nabla \times \boldsymbol{E} = -\frac{\partial \boldsymbol{B}}{\partial t}, \tag{12.2}$$

$$\nabla \cdot \boldsymbol{B} = 0, \tag{12.3}$$

$$\nabla \times \boldsymbol{H} = \boldsymbol{j} + \frac{\partial \boldsymbol{D}}{\partial t}, \tag{12.4}$$

where $\boldsymbol{E}$ is the electric field, $\boldsymbol{D}$ is the electric induction (or electric displacement, or electric flux density), $\boldsymbol{B}$ is the magnetic induction or magnetic flux density, and $\boldsymbol{H}$ is the magnetic field. ρ is the spatial charge density and $\boldsymbol{j}$ is the (conduction) current density; ∇ is the del operator (e.g. Shadowitz, 1975). It was unambiguously shown by Lorente de No (1947) and Rosenfalck (1969) that the time derivative $\partial \boldsymbol{B}/\partial t$ can be neglected in excitable cell tissues (note that this does not preclude the possibility of a non-negligible $\boldsymbol{B}$). Hence the curl of the vector field $\boldsymbol{E}$ is zero, so that $\boldsymbol{E}$ derives from a scalar potential V:

$$\boldsymbol{E} = -\nabla V. \tag{12.5}$$

Let us consider the nervous tissue as a domain Ω of R^3 (Figure 12.1) and assume that Ω can be subdivided into (i) intracellular domains Ω_j (cells), $j=1\text{–}m$, with

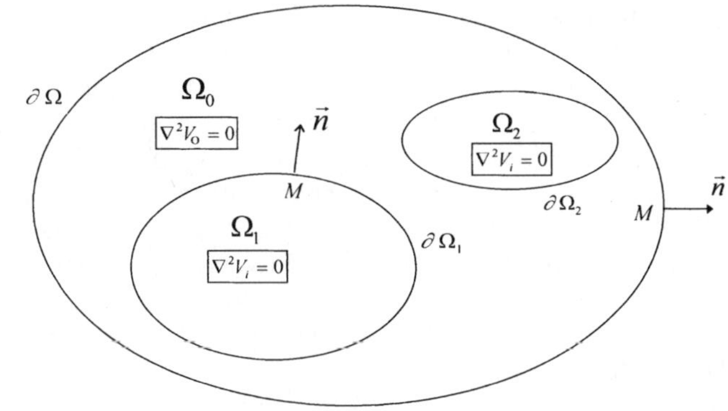

Figure 12.1 Intracellular and extracellular domains and their boundaries. Point M is the generic point at boundaries.

boundaries $\partial\Omega_j$ (infinitely thin cell membranes), and (ii) an extracellular domain Ω_0, with boundaries $\partial\Omega$ (exterior boundary) and $\partial\Omega_j$, $j=1\text{–}m$. We can write $\boldsymbol{E}_\mathrm{i}$, electric field at a point of intracellular space Ω_j, and $\boldsymbol{E}_\mathrm{o}$, electric field at a point of extracellular space Ω_0, as

$$\boldsymbol{E}_\mathrm{i} = -\nabla V_\mathrm{i}, \tag{12.6}$$

$$\boldsymbol{E}_\mathrm{o} = -\nabla V_\mathrm{o}. \tag{12.7}$$

It is generally assumed that capacitive effects are negligible at each point in the intracellular and extracellular domains (although transmembrane capacitive currents do exist): for a discussion based upon experimental results, see Nunez (1981). This means that in the charge conservation equation, which stems from (12.1) and (12.4) (e.g. Shadowitz, 1975):

$$\nabla \cdot \boldsymbol{j} + \frac{\partial \rho}{\partial t} = 0, \tag{12.8}$$

the term $\partial\rho/\partial t$ is zero, so that

$$\nabla \cdot \boldsymbol{j}_\mathrm{i} = 0, \tag{12.9}$$

$$\nabla \cdot \boldsymbol{j}_\mathrm{o} = 0, \tag{12.10}$$

where $\boldsymbol{j}_\mathrm{i}$ and $\boldsymbol{j}_\mathrm{o}$ are the (conduction) current densities in the intracellular and extracellular domains, respectively. It is also generally assumed that intracellular and extracellular milieus behave as linear conductors, so that Ohm's local law is valid (Jack et al., 1983):

$$\boldsymbol{j}_\mathrm{i} = \sigma_\mathrm{i} \boldsymbol{E}_\mathrm{i} = \quad \sigma_\mathrm{i} \nabla V_\mathrm{i}, \tag{12.11}$$

$$\boldsymbol{j}_\mathrm{o} = \sigma_\mathrm{o} \boldsymbol{E}_\mathrm{o} = -\sigma_\mathrm{o} \nabla V_\mathrm{o}, \tag{12.12}$$

σ_i and σ_o being the intracellular and extracellular conductivities; $\sigma_i = 1/\rho_i$ and $\sigma_o = 1/\rho_o$, where ρ_i and ρ_o are the corresponding resistivities. Note that (12.11) and (12.12) may be wrong under some circumstances, especially if the diffusion fluxes of ionic species within the intracellular or extracellular spaces are not negligible. The high-frequency discharge of excitable cells may result in non-uniform spatial distributions of, for instance, the potassium or sodium ion concentrations within a confined extracellular space (Bergman, 1970), so that diffusion of these ions could occur. Hence, ephaptic effects can sometimes be strongly linked to changes in ionic concentrations. This possibility will not be examined in the present study. A recent discussion of the possible effect of ion distribution on membrane excitability behavior can be found in Genet and Cohen (1996).

Now we assume that σ_i and σ_o are constant scalars. Although this latter assumption is generally made (Plonsey, 1964; Clark and Plonsey, 1966, 1968), we must emphasize that it is not trivial, because precise measurements of conductivity are rather difficult. Combining (12.9)–(12.12) we can write the Laplace equations:

$$\nabla^2 V_i = 0, \tag{12.13}$$

$$\nabla^2 V_o = 0, \tag{12.14}$$

which we assume to be valid at every point in the domains Ω_j, $j = 1-m$, and Ω_0. To fully characterize the mathematical problem, we must now write the boundary conditions at boundaries $\partial\Omega_j$ and $\partial\Omega$.

2.3 The Boundary Conditions

The boundary conditions which are generally encountered in excitable cell modeling can be either Dirichlet or Neumann boundary conditions for potential. One says that a Dirichlet boundary condition occurs on a boundary (surface) $\partial\upsilon$ of a domain (volume) υ if the value of the potential V is given at every point of $\partial\upsilon$:

$$\forall M \in \partial\upsilon,\ \forall t,\quad V(M,t) = f_d(M,t), \tag{12.15}$$

where $f_d(M,t)$ is a given function of point $M \in \partial\upsilon$ and time t (note that in many problems of applied mathematics, the Dirichlet boundary condition does *not* depend on time). Furthermore, we shall say that a Neumann boundary condition occurs on a boundary $\partial\upsilon$ of a domain υ if the value of the normal derivative of potential $(\partial V/\partial n)$ is given at every point of $\partial\upsilon$

$$\forall M \in \partial\upsilon,\ \forall t,\quad \frac{\partial V}{\partial n}(M,t) = f_n(M,t), \tag{12.16}$$

where $f_n(M,t)$ is a given function of point M and time.

The choice of the boundary condition for the domain Ω_0 at $\partial\Omega$ (Figure 12.1) depends upon experimental conditions, and can sometimes be non-trivial. Let us consider, for instance, an *in vitro* slice of neural tissue. If a portion of the boundary of the slice is in contact with an electrolyte solution of large enough conductivity and volume, we can assume that the electric potential is the same at every point of this solution and of the boundary surface. If we further assume that the potential of the solution is constant, we can take this potential as zero potential, and the following Dirichlet boundary condition will be fulfilled at every point of the surface:

$$\forall M,\ \forall t,\quad V_{o}(M,t)=0. \tag{12.17}$$

Let us assume now that a portion of the boundary surface is in close contact with an insulator (for instance, a glass surface). Obviously there can be no current orthogonal to the surface (if we disregard any capacitive current), so that the following Neumann boundary condition can be deduced at every point of the surface:

$$\forall M,\ \forall t,\quad \frac{\partial V_{o}}{\partial n}(M,t)=0. \tag{12.18}$$

We now consider *in vivo* measurements. The choice of the boundary condition for the domain Ω_0 is more subtle. First, where is boundary $\partial\Omega$ in the body? Second, what kind of condition can be fulfilled? In most experimental devices, the position of the reference electrode is crucial. If the reference electrode is far enough (from an electrical point of view) from the excitable tissue being studied, it can be assumed that boundary $\partial\Omega$ is at an infinite distance and that the potential V_{o} at infinite distance is zero; this simplifies the mathematical study (see below). Finally, unless the problem of the boundary condition at $\partial\Omega$ is very carefully studied, both theoretically and experimentally, for every experimental device, the resolution of potential equations may lead to nonsense; for further discussion, see Nunez (1981).

The conditions at boundaries Ω_j, $j=1-m$, i.e. at the cell membranes, are generally Neumann boundary conditions. The cell membranes are assumed to be infinitely thin, so that their detailed molecular structure is not taken into account. Current crossing the cell membrane or *transmembrane current* is orthogonal to the membrane (e.g. Johnston and Wu, 1995). Owing to the current density continuity, we can write:

$$\forall M\in\partial\Omega_j,\ \forall t,\quad I_{m}(M,t)=j_{in}(M,t)=j_{on}(M,t), \tag{12.19}$$

where $I_{m}(M,t)$ is the transmembrane current per unit area, $j_{in}(M,t)$ and $j_{on}(M,t)$ are the projections of current density vectors $\boldsymbol{j}_{i}(M,t)$ and $\boldsymbol{j}_{o}(M,t)$ on the normal vector $\boldsymbol{n}$. Using (12.11) and (12.12), we can write

$$I_{m}(M,t)=-\sigma_{i}\frac{\partial V_{i}}{\partial n}(M,t)=-\sigma_{o}\frac{\partial V_{o}}{\partial n}(M,t) \tag{12.20}$$

which is a Neumann boundary condition for V_{i} at boundary $\partial\Omega_j$ and for V_{o} at boundary $\partial\Omega_j$.

Finally, the transmembrane current $I_{m}(M,t)$ must have an explicit form. Equations for membrane current may be either linear ('passive' membrane properties), or nonlinear ('active'); for a full discussion, see Tuckwell (1988a,b). These terms 'passive' and 'active' must not be confused with active and passive transport through the membrane.

Linear membrane properties can be observed (i) in the dendrites, (ii) in the soma or the axon below the firing threshold. Transmembrane current can then be written as

$$I_{\mathrm{m}}(M,t)=C_{\mathrm{m}}\frac{\partial V}{\partial t}(M,t)+G_{\mathrm{m}}V(M,t)=C_{\mathrm{m}}\frac{\partial V}{\partial t}(M,t)+\frac{V(M,t)}{R_{\mathrm{m}}}, \qquad (12.21)$$

where

$$V(M,t)=V_{\mathrm{i}}(M,t)-V_{\mathrm{o}}(M,t)-V_{\mathrm{m,R}}, \qquad (12.22)$$

C_{m} and G_{m} are the membrane capacitance and conductance per unit area, so that $R_{\mathrm{m}}=1/G_{\mathrm{m}}$ is the resistance of a membrane unit area. $V_{\mathrm{m,R}}$ is the membrane resting potential, which is assumed to be constant. $V(M,t)$ is often referred to as the *membrane depolarization*. Equation (12.22) is quite classical in theoretical neurobiology; however, in most neuron models (e.g. in dendrite modeling, see Jack et al., 1983), $V_{\mathrm{o}}(M,t)$ is assumed to be constant, and equal to zero. It is obviously necessary to disregard this latter assumption in order to model properly ephaptic interactions; this results in an important modification of the mathematical problem, as will be shown below.

The membrane properties in the axon, soma (especially during firing) and some regions of dendrites are nonlinear. Equations for transmembrane current are far more complex, and several paradigms have emerged, e.g. the Hodgkin–Huxley equations (Hodgkin and Huxley, 1952) and the FitzHugh–Nagumo equations (FitzHugh, 1961; Nagumo et al., 1962). The (space-clamped) Hodgkin–Huxley equations include at least four dependent variables, while the space-clamped version of the FitzHugh–Nagumo model is two dimensional. The Hodgkin–Huxley equations can be written in their simplest form:

$$I_{\mathrm{m}}(M,t)=C_{\mathrm{m}}\frac{\partial V}{\partial t}(M,t)+\bar{g}_{\mathrm{Na}}[m(M,t)]^3h(M,t)[V(M,t)-V_{\mathrm{Na}}]$$
$$+\bar{g}_{\mathrm{K}}[n(M,t)]^4[V(M,t)-V_{\mathrm{K}}]+\bar{g}_{\mathrm{l}}[V(M,t)-V_{\mathrm{l}}], \qquad (12.23)$$

$$\frac{\partial m}{\partial t}(M,t)=\alpha_m(V(M,t))[1-m(M,t)]-\beta_m(V(M,t))m(M,t), \qquad (12.24)$$

$$\frac{\partial h}{\partial t}(M,t)=\alpha_h(V(M,t))[1-h(M,t)]-\beta_h(V(M,t))h(M,t), \qquad (12.25)$$

$$\frac{\partial n}{\partial t}(M,t)=\alpha_n(V(M,t))[1-n(M,t)]-\beta_n(V(M,t))n(M,t), \qquad (12.26)$$

where $\bar{g}_{\mathrm{Na}}$ and $\bar{g}_{\mathrm{K}}$ are the maximum conductances for sodium and potassium ions, and $\bar{g}_{\mathrm{l}}$ is the leakage conductance. m, h and n are variables which were introduced by Hodgkin and Huxley to describe the activation and inactivation of ionic channels. V_{Na}, V_{K} and V_{l} are the respective Nernst potentials, minus the membrane resting potential; for instance

$$V_{\mathrm{Na}}=\frac{RT}{F}\ln\frac{(\mathrm{Na}^+)_{\mathrm{o}}}{(\mathrm{Na}^+)_{\mathrm{i}}}-V_{\mathrm{m,R}}. \qquad (12.27)$$

$(Na^+)_o$ and $(Na^+)_i$ are the concentrations (more rigorously, the activities) of the sodium ion at the outer and inner face of the membrane. R is the gas constant, T the absolute temperature and F the Faraday constant. α_y and β_y, $y=m, h, n$, are functions of membrane depolarization; they can depend upon T, the absolute temperature, as a parameter. Hodgkin–Huxley equations have undergone many further developments, in order to take into account other types of current, e.g. calcium, persistent sodium or slowly inactivating potassium currents (Hille, 1992; Johnston and Wu, 1995). Equations (12.23)–(12.26) have been written here with all independent variables, namely M and t. It should be noted that any change of outer *or* inner potential alter both the transmembrane driving forces $V(M,t)-V_\eta$, $\eta=\text{Na, K, l}$, and the Hodgkin–Huxley m, h, n variables dynamics.

Finally, synaptic currents can be added to either (12.21) or (12.23). For instance, in the case of a simple, AMPA-like synapse, (12.21) can be replaced by the following (e.g. Johnston and Wu, 1995):

$$I_m(M,t)=C_m\frac{\partial V}{\partial t}(M,t)+G_m V(M,t)+g_{syn}(M,t)[V(M,t)-V_{rev}], \qquad (12.28)$$

where the synaptic conductance $g_{syn}(M,t)$ is a given function of space and time that describes the distribution and dynamics of synaptic channels, and V_{rev} is the synaptic reversion potential (minus membrane resting potential $V_{m,R}$).

3. THE MATHEMATICAL MODEL AND ITS SIMPLIFICATION

3.1 The Mathematical Model

We can now summarize the mathematical problem as follows: we are given ***m*** intracellular domains Ω_j, $j=1-m$, with boundaries $\partial\Omega_j$, and a domain Ω_0, with boundaries Ω_j, $j=1-m$, and $\partial\Omega$, as shown in Figure 12.1. We are looking for:

(i) a function $V_o(x,y,z,t)$, where (x,y,z) is the generic point of the closed domain $\bar{\Omega}_0=\Omega_0\cup\partial\Omega\cup\partial\Omega_1\cup\cdots\cup\partial\Omega_m$. We assume that the function V_o is continuous on $\bar{\Omega}_0$, and we can also assume that function V_o is smooth. Function V_o satisfies the Laplace equation (12.14):

$$\nabla^2 V_o=\frac{\partial^2 V_o}{\partial x^2}+\frac{\partial^2 V_o}{\partial y^2}+\frac{\partial^2 V_o}{\partial z^2}=0 \qquad (12.29)$$

with the following boundary conditions:

$$\forall(x,y,z)\in\partial\Omega_j,\ \forall t,\quad \frac{\partial V_o}{\partial n}(x,y,z,t)=-\frac{1}{\sigma_o}I_m(x,y,z,t), \qquad (12.30)$$

$$\forall(x,y,z)\in\partial\Omega\quad \left(\forall t\ \frac{\partial V_o}{\partial n}(x,y,z,t)=0\right)\ \text{or}\ (\forall t\ V_o(x,y,z,t)=0). \qquad (12.31)$$

(ii) for each $j=1-m$: a function $V_i(x,y,z,t)$, where (x,y,z) is now the generic point of the closed domain $\bar{\Omega}_j=\Omega_j\cup\partial\Omega_j$. We assume that function V_i is continuous and smooth in $\bar{\Omega}_j$. Function V_i satisfies the Laplace equation (12.14):

$$\nabla^2 V_i = \frac{\partial^2 V_i}{\partial x^2} + \frac{\partial^2 V_i}{\partial y^2} + \frac{\partial^2 V_i}{\partial z^2} = 0 \tag{12.32}$$

with the following boundary condition:

$$\forall (x, y, z) \in \partial\Omega_j,\ \forall t,\quad \frac{\partial V_i}{\partial n}(x, y, z, t) = -\frac{1}{\sigma_i} I_m(x, y, z, t). \tag{12.33}$$

The initial conditions for V_o and V_i (as well as for the Hodgkin–Huxley variables m, h and n) must be specified. As noted above, the equations for $I_m(x, y, z, t)$ can be e.g. (12.21), (12.23) or (12.28). Obviously, the Laplace equations (12.29) and (12.32) are linked through the boundary conditions, more precisely, through $I_m(x, y, z, t)$. Conversely, changes in V_o or V_i alter $I_m(x, y, z, t)$.

We must note that $I_m(x, y, z, t)$ must verify a zero flux condition for every cell $j = 1 – m$. Taking into account (12.9) and the Gauss divergence theorem or Green's theorem (e.g. Kellogg, 1929; Shadowitz, 1975):

$$\iiint_{\Omega_j} \nabla \cdot \boldsymbol{j}_i \, dV = \iint_{\partial\Omega_j} \boldsymbol{j}_i \cdot \boldsymbol{n} \, dS, \tag{12.34}$$

we can write

$$\iint_{\partial\Omega_j} \boldsymbol{j}_i \cdot \boldsymbol{n} \, dS = 0 \tag{12.35}$$

or, according to (12.19):

$$\iint_{\partial\Omega_j} I_m(x, y, z, t) \, dS = 0. \tag{12.36}$$

Therefore every choice of $I_m(x, y, z, t)$ must fulfill Eq. (12.36).

To our knowledge, the problem of the existence and uniqueness of the solution of the system of (12.29)–(12.33) and (12.21), (12.23)–(12.26) or (12.28) has not been thoroughly studied. However, in the simplified case of a one-dimensional extracellular space including two axons, it has been shown (Bell, 1980; Bell and Cosner, 1981) that the system admits a unique solution, given certain smoothness and compatibility conditions on the boundary functions.

In the more general case, based on the above-mentioned equations, we can deduce that:

(i) at each time t, there is a unique solution for V_i, more or less an additive arbitrary constant;
(ii) at each time t, there is a unique solution for V_o, provided that the Dirichlet boundary condition (12.17) is fulfilled at least over a part of boundary $\partial\Omega$.

The proof can be found in Courant and Hilbert (1962, p. 329), or in Kellogg (1929, p. 314). The problem of the additive arbitrary constant can generally be solved using biophysical considerations, as will be shown in section 4.

3.2 A Simplification of the Model

The model can be simplified if we consider one-dimensional neurons in a three-dimensional extracellular space, as shown in Figure 12.2 (note that branching and/or tapering of the neurons can be taken into account). This approximation is based upon (i) the fact that the typical diameter of a neuron cross-section (e.g. in the dendritic trees) is generally much smaller than the neuron length; for instance, hippocampal pyramidal cells are about one millimeter long, while the diameter of their dendritic trunk is about one micron; and (ii) calculations taking into account electrophysiological data, e.g. intracellular and membrane resistivities. A theoretical study of the one-dimensional neuron in a three-dimensional space was completed by Pickard (1968, 1969, 1971), who showed that the intracellular potential is almost uniform across a neuron cross-section.

Using these assumptions, Ohm's local law for intracellular fluid (12.11) can be re-written (we assume that the axis of the neuronal segment is the x-axis, see Figure 12.2):

$$j_{\mathrm{i}} = -\sigma_{\mathrm{i}} \frac{\partial V_{\mathrm{i}}}{\partial x} \tag{12.37}$$

or, if the neuron cross-section is circular:

$$i_{\mathrm{a}} = \pi a^2 j_{\mathrm{i}} = -\pi a^2 \sigma_{\mathrm{i}} \frac{\partial V_{\mathrm{i}}}{\partial x} \tag{12.38}$$

where a is the fiber radius, and $i_{\mathrm{a}} = \pi a^2 j_{\mathrm{i}}$ the intracellular axial current of the neuron segment. Note that $r_{\mathrm{a}} = \rho_{\mathrm{i}}/(\pi a^2) = 1/(\sigma_{\mathrm{i}} \pi a^2)$ is the intracellular resistance per unit length. Kirchoff's law of total current conservation leads to

$$\frac{\partial i_{\mathrm{a}}}{\partial x} = -i_{\mathrm{m}} = -2\pi a I_{\mathrm{m}}, \tag{12.39}$$

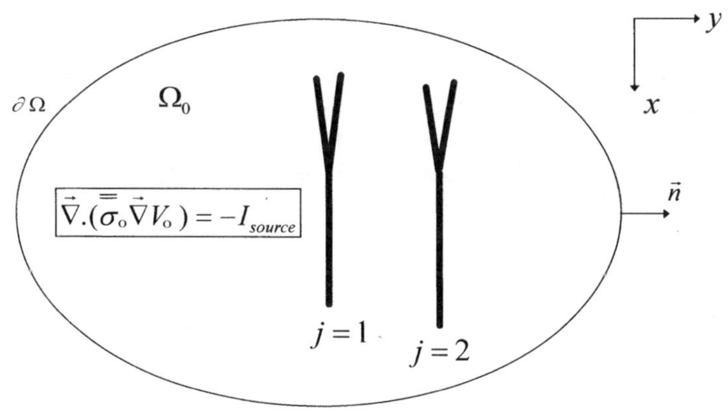

Figure 12.2 Simplified model of unidimensional neurons within an extracellular volume conductor.

where $i_m = 2\pi a I_m$ is the transmembrane current per unit length of fiber. Combining (12.38) and (12.39) gives:

$$\frac{\partial}{\partial x}\left(a^2 \frac{\partial V_i}{\partial x}\right) = \frac{2a}{\sigma_i} I_m. \tag{12.40}$$

If the fiber radius is constant, i.e. there is no tapering, then (12.40) can be rewritten:

$$\frac{\partial^2 V_i}{\partial x^2} = r_a i_m = \frac{2I_m}{a\sigma_i}. \tag{12.41}$$

Since neurons are now considered to be one-dimensional, we assume that the extracellular potential is defined over the whole domain Ω, as well as on the boundary. The domain Ω contains a current source distribution due to neuron membrane currents. More precisely, the extracellular current density can be written (see Mitzdorf, 1985):

$$\boldsymbol{j}_o = \bar{\bar{\sigma}}_o \boldsymbol{E}_o = -\bar{\bar{\sigma}}_o \nabla V_o, \tag{12.42}$$

where $\bar{\bar{\sigma}}_o$ is the conductivity tensor, which is a tensor of order two. It must be emphasized that tensor $\bar{\bar{\sigma}}_o$ is a macroscopic,'tissue' parameter, which includes neuronal tissue anisotropy (Nicholson, 1973). Since neuron spatial domains Ω_j, $j = 1 - m$, are no longer taken into account explicitly, the macroscopic tensor $\bar{\bar{\sigma}}_o$ depends on both the extracellular 'local' conductivity σ_o and on the local geometry of the neurons. For instance, for parallel neurons in a laminar structure (e.g. the hippocampus), the tensor $\bar{\bar{\sigma}}_o$ has different components along the axis of the neurons (x-axis), and along the orthogonal y- and z-axis; using this system of coordinates, the tensor reduces to its three main components, i.e.

$$\bar{\bar{\sigma}}_o = \begin{bmatrix} \sigma_{xx} & 0 & 0 \\ 0 & \sigma_{yy} & 0 \\ 0 & 0 & \sigma_{zz} \end{bmatrix} \tag{12.43}$$

with $\sigma_{yy} = \sigma_{zz}$ in the present case.

An infinitesimal volume element dV in Ω contains a volume current source density I_{source} due to the transmembrane current of the neurons present in dV. Therefore the divergence of $\boldsymbol{j}_o$ is generally different from zero, and can be written in its most general form

$$\nabla \cdot \boldsymbol{j}_o = I_{source}. \tag{12.44}$$

For instance, if the volume element dV contains the membrane area element dS with a transmembrane current in the volume element I_m dS, then the current source density will be written as

$$I_{source} = \frac{I_m \, dS}{dv}. \tag{12.45}$$

If several neurons and/or neuron segments are present in the same volume element, their respective current sources are simply added. Combining (12.42) and (12.44) leads to the Poisson equation:

$$\nabla \cdot (\bar{\bar{\sigma}}_o \nabla V_o) = -I_{\text{source}}. \tag{12.46}$$

Equations (12.40), (12.45) and (12.46) together with the equations for membrane currents (e.g. (12.21), (12.23) or (12.28)) and boundary conditions at boundary $\partial\Omega$ will be referred to as the 'simplified model'. This model is strongly correlated with current source-density analysis (CSDA) (Freeman and Stone, 1969; Nicholson and Llinas, 1971; Mitzdorf, 1985; Kossev et al., 1988). The goal of CSDA is to determine the current source distribution (as a function of space and time) from measurements of extracellular potentials, so that it includes (12.45) and (12.46). CSDA was applied, for instance, to the olfactory bulb (Rall and Shepherd, 1968), to the anuran cerebellum (Nicholson and Freeman, 1975; Freeman and Nicholson, 1975), to pigeon's optic tectum (Stone and Freeman, 1971), to rabbit's visual cortex (Rappelsberger et al., 1981), to cat LGN (Mitzdorf and Singer, 1977), to hippocampal CAI pyramidal neurons (Holsheimer et al., 1982; Leung, 1984; Miyakawa and Kato, 1986; Richardson et al., 1987; Holsheimer, 1987; Herreras, 1990), and to cortically arranged neurons (Klee and Rall, 1977; Chapman and Racine, 1997). The present model differs from CSDA in that it includes explicit equations for intracellular potential and transmembrane current.

To be rigorous, the boundary conditions for intracellular potential, i.e. for (12.40) or (12.41), must not be forgotten, and the corresponding current should be included in the volume current source density I_{source}. A thorough discussion of boundary conditions for intracellular potential can be found in Tuckwell (1988a). Here we can simply assume that a Neumann boundary condition with zero current holds at each neuron extremity, so that no term needs to be added to I_{source}.

If the tissue is isotropic and homogeneous, the tensor $\bar{\bar{\sigma}}_o$ reduces to a constant scalar, so that

$$\nabla^2 V_o = -\frac{I_{\text{source}}}{\sigma_o}. \tag{12.47}$$

If we further assume that a Dirichlet condition (12.17) holds at boundary $\partial\Omega$, and that the boundary $\partial\Omega$ is at infinite distance from the neural tissue (remote reference electrode), then we can simply deduce from classical electrodynamics (Shadowitz, 1975):

$$V_o(N,t) = \frac{1}{4\pi\sigma_o} \iiint_\Omega \frac{I_{\text{source}}(P,t)}{r} \, dV, \tag{12.48}$$

where r is the distance between the generic point $P \in \Omega$ and the observation point N; the integral is computed over the whole neural tissue studied.

Finally, we should like to emphasize the fact that in the CNS, the 'macroscopic' conductivity σ_o can be very low, e.g. $\sigma_o = 0.5\,\text{mS}\,\text{cm}^{-1}$ or even lower (e.g. Faber and Korn, 1989, p. 838). These values strongly differ from the conductivity of an electrolyte Ringer-like solution (about $15\,\text{mS}\,\text{cm}^{-1}$).

4. MATHEMATICAL METHODS FOR SOLVING THE RESULTING PDEs

Both analytical and numerical methods have been used to solve equations of models described in section 3. As in most domains of neuron modeling, analytical methods can be used only for linear membranes and relatively simple geometry: the example of two one-dimensional neurons is given in section 4.1. A method of resolution for models including non-linear equations and three-dimensional lattice is exposed in section 4.3.

4.1 Analytical Methods in Dimension 1: Field Effects in Nerve Trunks and Parallel Dendrites

Let us consider two parallel cylindrical neuron segments, immersed in extracellular space of constant section (Figure 12.3). If we neglect radial fields not only in the intracellular space, but also in the extracellular milieu, we can write

$$\frac{\partial V_{\mathrm{o}}}{\partial x} = -r_{\mathrm{o}} i_{\mathrm{o}} \qquad \frac{\partial V_{\mathrm{i}}^{(j)}}{\partial x} = -r_{\mathrm{a}}^{(j)} i_{\mathrm{a}}^{(j)} \quad \text{for } j = 1, 2, \tag{12.49}$$

where r_{o} is the extracellular resistance per unit length, i_{o} is the extracellular axial current, $r_{\mathrm{a}}^{(j)}$ is the intracellular resistance per unit length and $i_{\mathrm{a}}^{(j)}$ the intracellular axial current of neuron j, $j = 1, 2$. $V_{\mathrm{i}}^{(j)}$ is the intracellular potential of neuron j. Note that $r_{\mathrm{a}}^{(j)} = 1/(\sigma_{\mathrm{i}} \pi (a^{(j)})^2)$, where $a^{(j)}$ is the radius of the neuron j (see (12.38)).

Kirchhoff's law of conservation of total current gives

$$\frac{\partial i_{\mathrm{o}}}{\partial x} = i_{\mathrm{m}}^{(1)} + i_{\mathrm{m}}^{(2)} \qquad \frac{\partial i_{\mathrm{a}}^{(j)}}{\partial x} = -i_{\mathrm{m}}^{(j)} \quad \text{for } j = 1, 2, \tag{12.50}$$

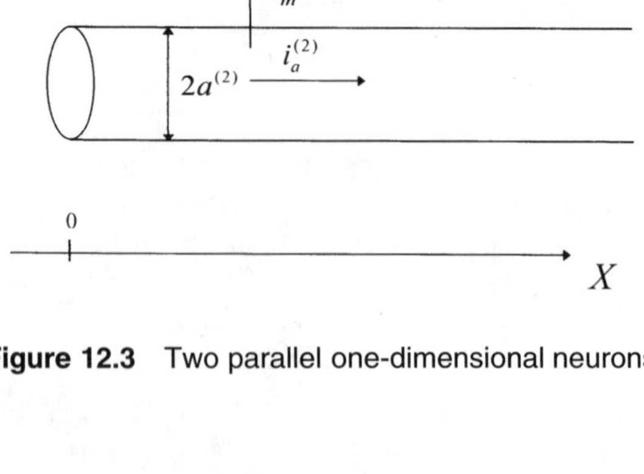

Figure 12.3 Two parallel one-dimensional neurons.

where $i_{\mathrm{m}}^{(j)}$ is the transmembrane current per unit length of neuron j. Combining (12.22), (12.49) and (12.50) and differentiating twice with respect to x, we obtain

$$\frac{\partial^2 V^{(1)}}{\partial x^2} = (r_{\mathrm{a}}^{(1)} + r_{\mathrm{o}})i_{\mathrm{m}}^{(1)} + r_{\mathrm{o}} i_{\mathrm{m}}^{(2)}, \tag{12.51}$$

$$\frac{\partial^2 V^{(2)}}{\partial x^2} = r_{\mathrm{o}} i_{\mathrm{m}}^{(1)} + (r_{\mathrm{a}}^{(2)} + r_{\mathrm{o}})i_{\mathrm{m}}^{(2)}. \tag{12.52}$$

Equations (12.51) and (12.52) can be used for linear and nonlinear membranes; in particular, they can be valid for axons as well as for dendrites. These equations are closely related to the approach of Markin (1970, 1973), Scott and Luzader (1979), Maeda et al. (1980) and Bell (1980, 1981) Bell and Cosner (1981): see for instance (12.1)–(12.3) in Bell's paper (1981). These authors have studied the ephaptic interactions between parallel, unmyelinated axons, some of which are stimulated to carry propagated action potentials. Markin (1970, 1973) and Scott and Luzader (1979) use piecewise linear relations for $i_{\mathrm{ion}}^{(j)} = f(V^{(j)})$, where $i_{\mathrm{ion}}^{(j)}$ is the part of $i_{\mathrm{m}}^{(j)}$ which is due to current through ionic channels. For instance:

$$i_{\mathrm{ion}}^{(j)} = G(V^{(j)} - H(V^{(j)} - a)),$$

where $H(\cdot)$ is the Heaviside step function, a and G are constants and $V^{(j)}$ is properly scaled. Note that this piecewise linear approximation is a further simplification of the FitzHugh–Nagumo model (FitzHugh, 1961; Nagumo et al., 1962). Markin (1970, 1973) then constructed traveling wave solutions, and showed in particular that (i) in the case of two axons having the same geometry and electrophysiological properties, the 'active' (stimulated) axon cannot induce an action potential on the 'inactive' (unstimulated) axon; (ii) however, an unstimulated axon within a fiber bundle can become excited if it is surrounded by a sufficient number of stimulated axons. Markin also showed that pulse trapping or pulse locking is possible: e.g. if a slowly propagating pulse is initiated on a fiber, and a faster pulse is initiated some time later on a second, parallel fiber, then the faster pulse will slow down and both pulses will propagate together. Scott and Luzader (1979) used a structural perturbation technique to study impulse condensation on bundles of parallel fibers. Bell and Cosner (1981) showed that the rest state is locally stable, but a form of threshold behavior occurs which shows that the rest state is not globally stable. Bell (1981) used Hodgkin–Huxley equations and a perturbation theory approach to model pulse 'trapping' due to ephaptic transmission. A somewhat different approach, based on fundamental electromagnetic equations, was developed by Clark and Plonsey (1970) and applied to fiber bundles (1971); this approach met some of the conclusions of Markin's approach, especially the necessity of the presence of several stimulated axons in order to induce an action potential on an unstimulated one; furthermore, Clark and Plonsey pointed out the importance of several parameters, e.g. extracellular resistivity.

The next paragraph is devoted to the case where the neuron membrane properties are linear (e.g. in dendrites).

4.2 Example: Electric Field Effects from Synaptic Potentials

An analytical solution has been derived for semi-infinite neurons with linear membrane properties (Costalat, 1991), in the stationary state and in response to a unit impulse. We shall briefly summarize the response to a unit impulse. The passive membrane properties imply that the membrane current per fiber unit length is

$$i_{\mathrm{m}}^{(j)}=c_{\mathrm{m}}^{(j)}\frac{\partial V^{(j)}}{\partial t}+g_{\mathrm{m}}^{(j)}V^{(j)}=c_{\mathrm{m}}^{(j)}\frac{\partial V^{(j)}}{\partial t}+\frac{V^{(j)}}{r_{\mathrm{m}}^{(j)}},\quad j=1,2, \tag{12.53}$$

where $c_{\mathrm{m}}^{(j)}=2\pi a^{(j)}C_{\mathrm{m}}$ is the capacitance per unit length of fiber, and $g_{\mathrm{m}}^{(j)}=1/r_{\mathrm{m}}^{(j)}=R_{\mathrm{m}}/(2\pi a^{(j)})$ is the corresponding conductance.

Simple transformations then lead to the following system:

$$\frac{\partial^2 V^{(1)}}{\partial X^2}=\frac{\partial V^{(1)}}{\partial T}+V^{(1)}+\beta k\left(\frac{\partial V^{(2)}}{\partial T}+V^{(2)}\right), \tag{12.54}$$

$$\frac{\partial^2 V^{(2)}}{\partial X^2}=k\left(\frac{\partial V^{(1)}}{\partial T}+V^{(1)}\right)+\alpha\left(\frac{\partial V^{(2)}}{\partial T}+V^{(2)}\right), \tag{12.55}$$

where

$$X=\frac{x}{\lambda},\qquad T=\frac{t}{\tau_{\mathrm{m}}},\qquad k=\frac{r_{\mathrm{o}}}{r_{\mathrm{a}}^{(1)}+r_{\mathrm{o}}},$$

$$\alpha=\frac{1/(\sigma_{\mathrm{i}}\pi a^{(2)})+r_{\mathrm{o}}a^{(2)}}{1/(\sigma_{\mathrm{i}}\pi a^{(1)})+r_{\mathrm{o}}a^{(1)}},\quad \beta=\frac{a^{(2)}}{a^{(1)}} \tag{12.56}$$

and $\tau_{\mathrm{m}}=R_{\mathrm{m}}C_{\mathrm{m}}$ is the membrane time constant and λ is defined as

$$\lambda=\sqrt{\frac{r_{\mathrm{m}}^{(1)}}{r_{\mathrm{a}}^{(1)}+r_{\mathrm{o}}}}. \tag{12.57}$$

If the extracellular space resistance is zero, then λ is the first fiber space constant or characteristic length, so that X is simply the space coordinate expressed in units of this space constant. If the radii of the two fibers are equal, obviously $\alpha=\beta=1$.

First it is assumed that the system is closed, so that

$$\forall X\quad i_{\mathrm{a}}^{(1)}+i_{\mathrm{a}}^{(2)}+i_{\mathrm{o}}=0. \tag{12.58}$$

Let us assume that (i) neurons 1 and 2 are semi-infinite, i.e. $0\leqslant X<+\infty$; (ii) the fiber radii are equal, i.e. $a^{(1)}=a^{(2)}$; (iii) depolarizations at infinity are zero, i.e.

$$\lim_{X\to+\infty}V^{(1)}(X,T)=\lim_{X\to+\infty}V^{(2)}(X,T)=0 \tag{12.59}$$

and (iv) fibers depolarization is initially zero, i.e.

$$\forall X\quad V^{(1)}(X,0)=V^{(2)}(X,0)=0. \tag{12.60}$$

We further assume that (v) neuron 1 and/or neuron 2 are stimulated at $T=0$ and $X=0$ by an impulse current, so that

$$i_a^{(1)}(0,T)=Q_0^{(1)}\frac{\delta(T)}{\tau_m},\quad i_a^{(2)}(0,T)=Q_0^{(2)}\frac{\delta(T)}{\tau_m}, \tag{12.61}$$

where $\delta(T)$ is the unit impulsion (the so-called Dirac delta function) and $Q_0^{(j)}$ is the total amount of charge which enters the j neuron; the presence of τ_m in the denominator is simply a consequence of using T (dimensionless) instead of t (ms). $Q_0^{(j)}$ will be either 0 (unstimulated neuron) or a fixed amount Q_0 (stimulated neuron). Equation (12.61) may be interpreted in the following way, for instance: $i_a^{(j)}(0,T)$ can be viewed as a brief synaptic current of fixed magnitude at $X=0$ and $T=0$, followed by a Neumann boundary condition (zero flux) at $X=0$ and $T>0$, which means that there is a sealed end. The response of the system to a more complex stimulus can be derived using convolution (see Jack et al., 1983). With the above assumptions it can be proved using Laplace transforms (Costalat, 1991) that

$$V^{(j)}=V^*(X,T;Q_0^{(j)},k,r_a,\lambda,\tau_m)+V^{**}(X,T;Q_0^{(l)},k,r_a,\lambda,\tau_m),\quad j,l=1,2,\ j\neq l \tag{12.62}$$

with

$$\begin{aligned}V^*(X,T;Q_0^{(j)},k,r_a,\lambda,\tau_m)&=\frac{Q_0^{(j)}\lambda r_a}{2\tau_m}\left(\frac{\sqrt{1+k}}{1-k}\frac{\exp[(-X^2(1+k)-4T^2)/(4T)]}{\sqrt{\pi T}}\right.\\&\quad\left.+\frac{1}{\sqrt{1-k}}\frac{\exp[(-X^2(1-k)-4T^2)/(4T)]}{\sqrt{\pi T}}\right),\\V^{**}(X,T;Q_0^{(l)},k,r_a,\lambda,\tau_m)&=\frac{Q_0^{(l)}\lambda r_a}{2\tau_m}\left(\frac{\sqrt{1+k}}{1-k}\frac{\exp[(-X^2(1+k)-4T^2)/(4T)]}{\sqrt{\pi T}}\right.\\&\quad\left.-\frac{1}{\sqrt{1-k}}\frac{\exp[(-X^2(1-k)-4T^2)/(4T)]}{\sqrt{\pi T}}\right).\end{aligned} \tag{12.63}$$

An example of modeled fiber depolarizations is shown in Figure 12.4, where neuron 1 is stimulated and neuron 2 is unstimulated. The results are compared with the depolarization of one stimulated neuron if there is no ephaptic interaction, i.e. if $k=0$, so that (12.62) obviously reduces to (Hodgkin, in Fatt and Katz, 1951):

$$V(X,T)=\frac{Q_0\lambda r_a}{\tau_m}\frac{\exp[(-X^2-4T^2)/(4T)]}{\sqrt{\pi T}}. \tag{12.64}$$

In this example, ephaptic interactions result in (i) enhanced depolarization of the stimulated neuron, and (ii) a non-monotonic effect on the unstimulated neuron depolarization. More precisely, the unstimulated neuron is depolarized near the 'synaptic region' of the stimulated neuron, and hyperpolarized in the more distal

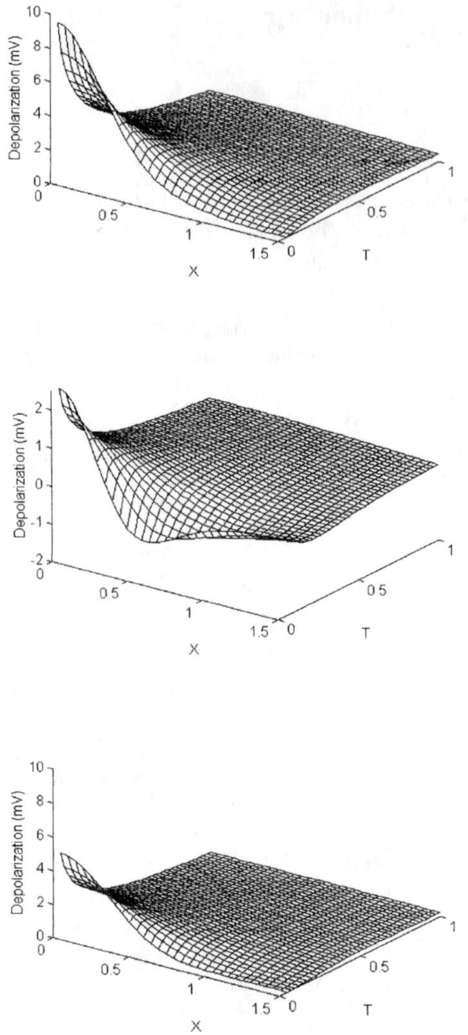

Figure 12.4 Plots of membrane depolarization as a function of space (in units of λ) and time (in units of τ_m). (a) Stimulated neuron, (b) unstimulated neuron, with $k=0.5$ in both cases (cf. (12.62)–(12.63)), (c) stimulated neuron without ephaptic interaction, due to $k=0$ (cf. (12.64)). The injected current is the same in plots (a) and (c).

region. The depolarization front, i.e. the point X where $V^{(2)}$ is zero, propagates with a velocity of

$$v_0 = \frac{1}{2}\sqrt{\frac{\ln((1+k)/(1-k))}{kT}}. \tag{12.65}$$

As expected, k is a prominent parameter in the strength of the coupling between the neurons. An increase in k results in (i) increased stimulated fiber depolarization, (ii) increased absolute values of the depolarization and hyperpolarization of the unstimulated fiber, and (iii) increased v_0 velocity.

The parameter k depends directly on the extracellular volume fraction

$$k=\left(1+\frac{S_o}{\pi a^2}\frac{\sigma_o}{\sigma_i}\right)^{-1}, \tag{12.66}$$

where S_o is the cross-section of the extracellular milieu and σ_o is the extracellular conductivity. Lowering the $S_o/(\pi a^2)$ ratio results in increased ephaptic transmission. These results have been extended to branched dendrites (Costalat, 1991), and an arbitrary number of fibers, some of them stimulated (Costalat and Chauvet, 1998).

4.3 Numerical Methods in Three Dimensions: Field Effects in Populations of Active Neurons

We now consider the numerical resolution of (12.40) or (12.41), (12.45)–(12.46) and (12.21), (12.23)–(12.26) or (12.28). We shall solve the elliptic PDE (12.46) using a finite difference method (see for instance Press et al., 1993). The time evolution will be described using a method we adapted from MacGregor (1987).

The extracellular milieu is first discretized using a three-dimensional grid (Figure 12.5). Points in the extracellular milieu will be referred to using their integer coordinates k, l, m, so that

$$x=k\Delta x,\quad y=l\Delta y,\quad z=m\Delta z,\qquad k=0-k_{\max},\quad l=0-l_{\max},\quad m=0-m_{\max}, \tag{12.67}$$

where Δx, Δy, Δz are the discretization steps along the x, y, and z axes. Assuming that the conductivity tensor does not depend on the space coordinates and can be written along (12.43), we have

$$\bar{\bar{\sigma}}_o \nabla V_o=\begin{bmatrix}\sigma_{xx}\dfrac{\partial V_o}{\partial x}\\[2mm] \sigma_{yy}\dfrac{\partial V_o}{\partial y}\\[2mm] \sigma_{zz}\dfrac{\partial V_o}{\partial z}\end{bmatrix} \tag{12.68}$$

and (12.46) can be written as

$$\sigma_{xx}\frac{\partial^2 V_o}{\partial x^2}+\sigma_{yy}\frac{\partial^2 V_o}{\partial y^2}+\sigma_{zz}\frac{\partial^2 V_o}{\partial z^2}=-I_{\text{source}}. \tag{12.69}$$

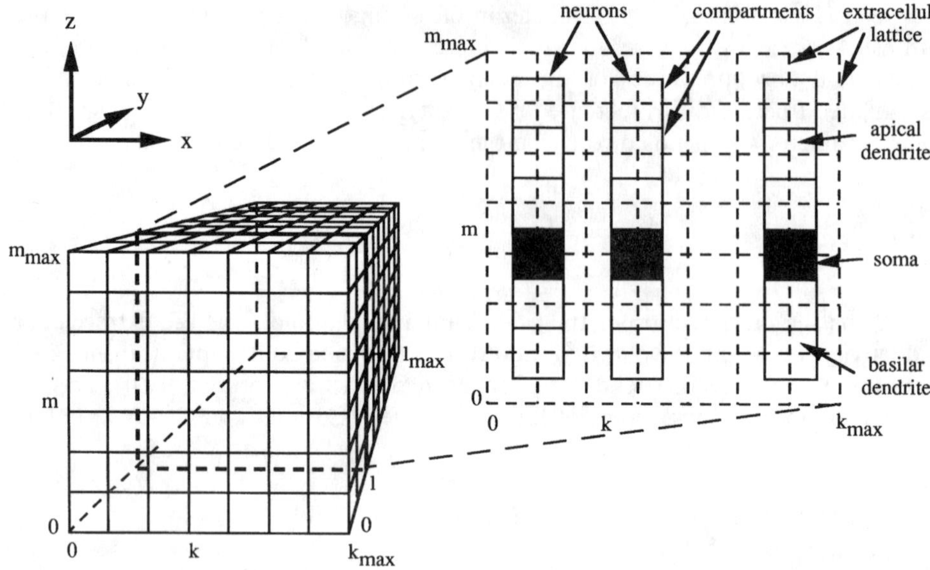

Figure 12.5 Schematic representation of the model structure. The extracellular milieu is modeled as a tri-dimensional (x, y, z) lattice. The model neurons incorporate unbranched apical and basilar dendrites in addition to their soma.

In the finite-difference approximation, this latter equation can be replaced at each point by

$$\sigma_{xx}\frac{V_{o;k+1,l,m}+V_{o;k-1,l,m}-2V_{o;k,l,m}}{\Delta x^2}+\sigma_{yy}\frac{V_{o;k,l+1,m}+V_{o;k,l-1,m}-2V_{o;k,l,m}}{\Delta y^2}$$

$$+\sigma_{zz}\frac{V_{o;k,l,m+1}+V_{o;k,l,m-1}-2V_{o;k,l,m}}{\Delta z^2}=-I_{source;k,l,m}, \tag{12.70}$$

where $V_{o;k,l,m}$ and $I_{source;k,l,m}$ are the extracellular potential and current source term at the point with integer coordinates (k, l, m). Note that both $V_{o;k,l,m}$ and $I_{source;k,l,m}$ depend on time. The current source term $I_{source;k,l,m}$ can be written, as a consequence of (12.45):

$$I_{source;k,l,m}=\frac{I_{m;k,l,m}\Delta S_{k,l,m}}{\Delta x\,\Delta y\,\Delta z} \tag{12.71}$$

where $\Delta S_{k,l,m}$ is the area of the membrane surface contained in the elementary volume $\Delta x\,\Delta y\,\Delta z$; for instance, in the case of one cylindrical neuron placed along the z-axis, we have $\Delta S_{k,l,m}=2\pi a\,\Delta z$. $I_{m;k,l,m}$ is the current which crosses this latter membrane surface (per unit area). Of course $I_{m;k,l,m}$ can be expressed using, for instance, (12.21), (12.23)–(12.26) or (12.28).

Ad hoc boundary conditions must be chosen at the boundaries of the extracellular milieu (see section 2.3). For instance, if a Dirichlet boundary condition (12.17) is fulfilled at every point of $\partial\Omega$, then we can write

$$(k=0 \text{ or } k=k_{\max} \text{ or } l=0 \text{ or } l=l_{\max} \text{ or } m=0 \text{ or } m=m_{\max}) \Rightarrow V_{\mathrm{o};k,l,m}=0. \quad (12.72)$$

Similarly, a Neumann boundary condition can be written at some parts of boundary $\partial\Omega$. For instance, let us assume that boundary condition (12.18) is fulfilled at every point of the surface $m=0$; then we can write using finite-difference approximation for normal derivative:

$$\forall k,\ \forall l,\quad \frac{V_{\mathrm{o};k,l,0}-V_{\mathrm{o};k,l,1}}{\Delta z}=0. \quad (12.73)$$

More elaborate formula for the derivative in (12.73) can be used; see e.g. Press et al. (1993).

We turn now to the discretization of the intracellular milieu. Let us consider a neuron segment, as shown in Figure 12.6. Here we assume that the one-dimensional neuron is placed along the z-axis. Using once more the finite-difference scheme, (12.41) can now be written as

$$\frac{V_{\mathrm{i};k,l,m+1}+V_{\mathrm{i};k,l,m-1}-2V_{\mathrm{i};k,l,m}}{\Delta z^2}=\frac{2I_{\mathrm{m};k,l,m}}{a\sigma_{\mathrm{i}}}. \quad (12.74)$$

Equation (12.74) is clearly derived from Ohm's law and the total current conservation law (Kirchhoff's law). Equations similar to (12.74) can readily be written using Ohm and Kirchoff's laws for particular points, such as branching points in a dendritic tree.

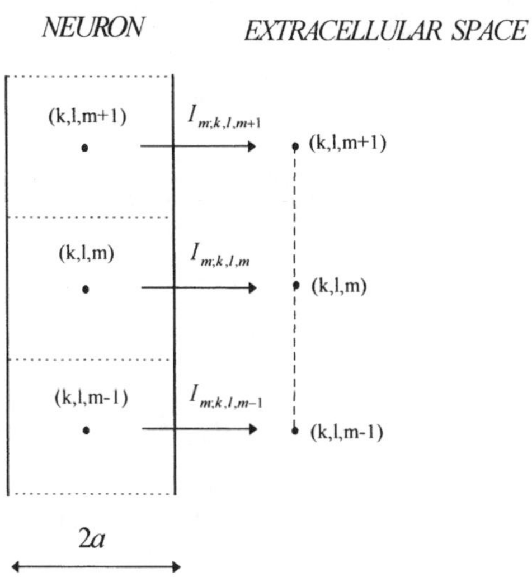

Figure 12.6 Discretization of the intracellular space and connection to extracellular space.

As previously mentioned, the end of the 'fibers', i.e. of one-dimensional neurons, may deserve a careful discussion of the encountered boundary conditions (see for instance Tuckwell, 1988a, pp. 130–9.).

Within the described method, using the formula for transmembrane current ((12.21), (12.23)–(12.26) or (12.28)) is straightforward. For instance, (12.21) can simply be written as

$$I_{\mathrm{m};k,l,m} = C_{\mathrm{m}} \frac{\mathrm{d}V_{k,l,m}}{\mathrm{d}t} + G_{\mathrm{m}} V_{k,l,m}, \tag{12.75}$$

where $V_{k,l,m} = V_{\mathrm{i};k,l,m} - V_{\mathrm{o};k,l,m} - V_{\mathrm{m,R}}$ depends on time.

Similarly, (12.23) would become

$$I_{\mathrm{m};k,l,m} = C_{\mathrm{m}} \frac{\mathrm{d}V_{k,l,m}}{\mathrm{d}t} + \bar{g}_{\mathrm{Na}}(m_{k,l,m})^3 h_{k,l,m}[V_{k,l,m} - V_{\mathrm{Na}}] + \bar{g}_{\mathrm{K}}(n_{k,l,m})^4[V_{k,l,m} - V_{\mathrm{K}}] + \bar{g}_{\mathrm{l}}[V_{k,l,m} - V_{\mathrm{l}}] \tag{12.76}$$

and (12.24)–(12.26) would be replaced by the following:

$$\frac{\mathrm{d}\phi_{k,l,m}}{\mathrm{d}t} = \alpha_{\phi}(V_{k,l,m})[1 - \phi_{k,l,m}] - \beta_{\phi}(V_{k,l,m})\phi_{k,l,m} \tag{12.77}$$

with $\phi \in \{m, h, n\}$, $\phi_{k,,l,m}$ depends on time.

Finally, one has to solve the system consisting of: extracellular milieu Poisson equation (12.70); intracellular milieu Poisson equation (12.74); (12.71), which links the two latter equations; equations such as (12.75) or (12.76)–(12.77), which give explicit form to transmembrane current; *ad hoc* boundary condition, such as (12.73). This system is somewhat difficult to handle, due to the fact that the Poisson equation for intracellular space (12.74) is written in terms of intracellular potential, while equations for membrane current (12.75)–(12.77) are written in terms of membrane depolarization. However, simple algebraic calculus leads to

$$C_{\mathrm{m}} \frac{\mathrm{d}V_{k,l,m}}{\mathrm{d}t} = -\left[G_{\mathrm{m}} + \frac{a\sigma_{\mathrm{i}}}{\Delta z^2}\right] V_{k,l,m} + \left[\frac{a\sigma_{\mathrm{i}}}{2\Delta z^2}(V_{k,l,m-1} + V_{k,l,m+1} + V_{\mathrm{o};k,l,m-1} + V_{\mathrm{o};k,l,m+1} - 2V_{\mathrm{o};k,l,m})\right] \tag{12.78}$$

in the linear case (12.75), and to

$$C_{\mathrm{m}} \frac{\mathrm{d}V_{k,l,m}}{\mathrm{d}t} = -\left[\bar{g}_{\mathrm{Na}}(m_{k,l,m})^3 h_{k,l,m} + \bar{g}_{\mathrm{K}}(n_{k,l,m})^4 + \bar{g}_{\mathrm{l}} + \frac{a\sigma_{\mathrm{i}}}{\Delta z^2}\right] V_{k,l,m} + \left[\bar{g}_{\mathrm{Na}}(m_{k,l,m})^3 h_{k,l,m} V_{\mathrm{Na}} + \bar{g}_{\mathrm{K}}(n_{k,l,m})^4 V_{\mathrm{K}} + \bar{g}_{\mathrm{l}} V_{\mathrm{l}} + \frac{a\sigma_{\mathrm{i}}}{2\Delta z^2}(V_{k,l,m-1} + V_{k,l,m+1} + V_{\mathrm{o};k,l,m-1} + V_{\mathrm{o};k,l,m+1} - 2V_{\mathrm{o};k,l,m})\right] \tag{12.79}$$

in the nonlinear case (12.76).

Equations (12.78) and (12.79) can simply be written

$$\frac{dV_{k,l,m}}{dt} = aV_{k,l,m} + b, \tag{12.80}$$

where coefficients a and b do not depend on $V_{k,l,m}$, while they do depend on: (i) depolarization of adjacent points ('compartments') in the intracellular grid, viz. $V_{k,l,m+1}$ and $V_{k,l,m-1}$, (ii) extracellular potential $V_{o;k,l,m}$, $V_{o;k,l,m+1}$ and $V_{o;k,l,m-1}$, and (iii) maybe other variables such as Hodgkin–Huxley m, h, n variables. At a given time (12.80) can roughly be considered as a linear ordinary differential equation with constant coefficients, the integration of which leads to

$$V_{k,l,m}(t+\Delta t) = V_{k,l,m}(t)e^{a\Delta t} - \frac{b}{a}(1 - e^{a\Delta t}). \tag{12.81}$$

This method of resolution in time is an extension of the method proposed and extensively used by MacGregor (1987) in the case of a neuron in an isopotential extracellular milieu. An alternative way would consist in a discretization of the expression of the time derivative $dV_{k,l,m}/dt$ using one of the classical finite-difference methods for resolution of parabolic PDEs, namely: explicit, implicit or Crank–Nicholson methods.

Finally, we propose an algorithm to solve the resulting system of equations (Costalat, 1991), which can be summarized as follows:

1. Initial values are affected to variables, i.e. to depolarization $V_{k,l,m}$, extracellular potential $V_{o;k,l,m}$ and maybe to other variables such as m, h, n variables.
2. Transmembrane currents are computed along (12.75) or (12.76) (if synaptic current is present, a time-dependent term $g_{syn;k,l,m}(V_{k,l,m} - V_{rev})$ can simply be added).
3. Poisson equation (12.74) is solved using, for instance, a simultaneous over-relaxation technique (Press et al., 1993), so that we obtain new values for extracellular potential.
4. New values for membrane depolarizations $V_{k,l,m}$ are obtained from (12.78) or (12.79), solved along (12.81).
5. New values of m, h, n variables are then computed using (12.77).
6. Finally, new values from transmembrane currents can be obtained: return to 2.

Each iteration corresponds to a time step Δt. In our experience, the numerical method appears to be stable if the time step is about 1–2 μs, using classical values for the Hodgkin–Huxley theory parameters (resulting in an action potential duration about 1 ms).

This method was applied to study the influence of one stimulated neuron upon one unstimulated neuron, assuming active conductances in the soma and axon (Costalat and Chauvet, 1993). From a qualitative point of view, the depolarization profiles were similar to those obtained with the analytical model of section 4.1. However, stimulation of the extremity of the dendrite of one neuron could result in a decrease of the excitability of the other neuron, due to the prolonged hyperpolarization of its soma. Hence this model suggests the possibility of complex changes, not only in the depolarization, but also in the excitability of unstimulated neurons within monolayered structures.

In the case where the neural tissue is homogeneous and isotropic, and a Dirichlet condition (12.17) holds at boundary $\partial\Omega$ situated at an infinite distance, (12.47) and (12.48) are valid, so that extracellular potential can simply be computed using (12.48) (in step 3) of the algorithm). This assumption, which may not be bad *in vivo* (see Nunez, 1981), greatly lowers the calculation time. An example of the solutions obtained using this latter method is displayed in Figures 12.7–12.9. In these figures, the stimulated and unstimulated neurons have a soma embedded

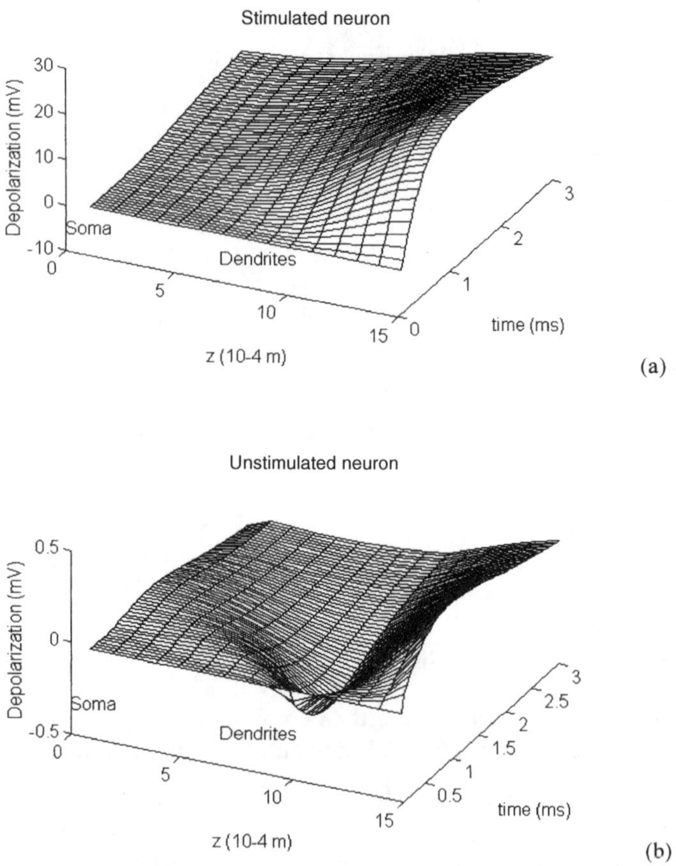

Figure 12.7 Plots of membrane depolarization (mV) as a function of space (in units of 10^4 m) and time (ms). (a) Stimulated neuron, (b) unstimulated neuron. Both neurons are composed of the soma (compartment 1) embedded with Hodgkin–Huxley conductances, and dendrites (compartments 2 to 15) with linear conductance. At the tip of the stimulated neuron (5 last compartments), an excitatory synaptic current is modeled using the formula $\bar{g}_{syn}(\exp(-t/\tau_1)-\exp(-t/\tau_2))(V-V_{rev})$, with $\tau_1=1\,\text{ms}$, $\tau_2=0.1\,\text{ms}$, $V_{rev}=70\,\text{mV}$. $\bar{g}_{syn}$ (maximal synaptic conductance) increases from compartment 11 ($0.28\,\text{mS}\,\text{cm}^{-2}$) to compartment 15 ($1.4\,\text{mS}\,\text{cm}^{-2}$).

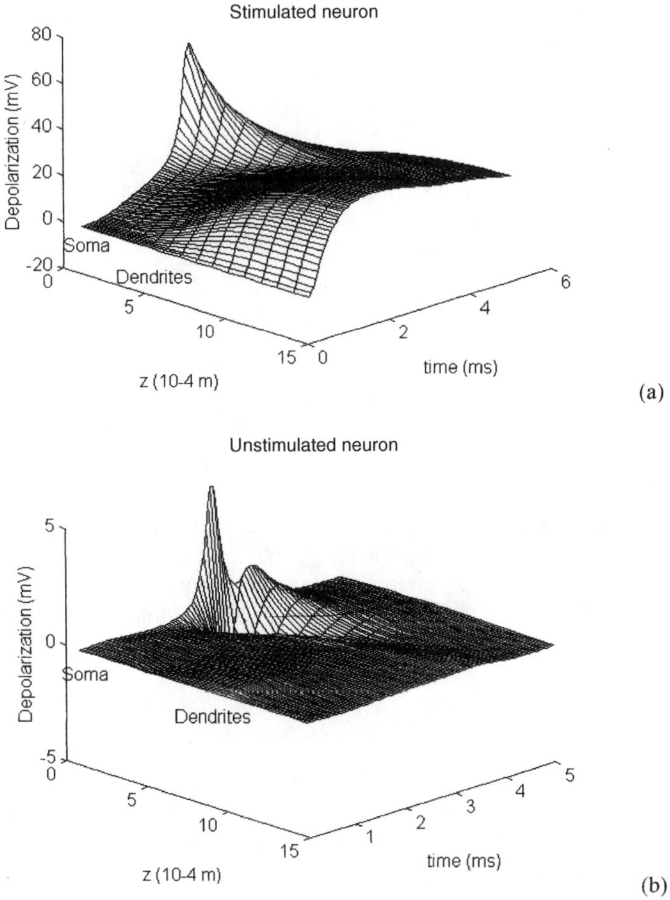

Figure 12.8 Plots of membrane depolarization (mV) as a function of space (in units of 10^{-4} m) and time (ms). (a) Stimulated neuron, (b) unstimulated neuron. The parameter values are the same as in Figure 12.7, with the exception of $\bar{g}_{syn}$ (maximal synaptic conductance) which now increases from compartment 6 (0.28 mS cm^{-2}) to compartment 15 (1.4 mS cm^{-2}). Note the action potential at the stimulated neuron.

with Hodgkin–Huxley conductances and one dendrite with passive, linear conductance. The stimulated neuron is depolarized by a synaptic current (analogous to an AMPA current). When the stimulation effect is below the action potential threshold (Figure 12.7), one can observe an excitatory post-synaptic potential (EPSP) at the stimulated neuron, together with a non-monotonic variation of the unstimulated neuron depolarization; these results can be compared to those obtained using an analytical method in Figure 12.4 (however, note that in Figure 12.4 the first neuron is stimulated by a Dirac-delta function). When the stimulation is increased, for

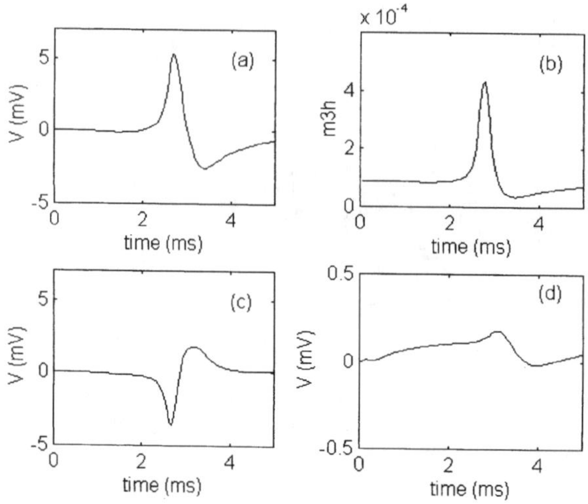

Figure 12.9 Unstimulated neuron. Plots of (a) soma membrane depolarization (mV), (b) soma m^3h product (proportional to Na^+ conductance), (c) second compartment (dendrite) depolarization (d) 12 compartment (dendrite) depolarization. Note that the second compartment is close to the soma, while the 12 is more remote. Parameter and initial condition values are the same as in Figure 12.8.

instance if more synapses are stimulated, so that an action potential is elicited at the stimulated neuron soma, the observed effects in the unstimulated neuron are more complex (Figure 12.8 and Figure 12.9): as in Figure 12.7, a small depolarization is observed at the distal part of the dendrite (Figure 12.9(d)), while a weak hyperpolarization appears at the dendritic compartments close to the soma (Figure 12.9(c)); then the first neuron action potential induces a strong depolarization of the second neuron soma (Figure 12.9(a)). In this example, the behavior of the unstimulated neuron soma potential is polyphasic, due to the superimposition of (i) the dendritic phenomena, which can induce a hyperpolarization at the soma, and (ii) the depolarization and subsequent hyperpolarization induced by the first neuron action potential. This results in non-monotonic changes in the second neuron soma Na conductance, as can be seen on a plot of m^3h versus time (Figure 12.9(b)): in this example, m^3h is first increased, then decreased.

It can be emphasized that Figures 12.7 and 12.8 were obtained with a relatively low value of the 'macroscopic' extracellular conductivity ($\sigma_o = 0.1\,\mathrm{mS\,cm^{-1}}$); quantitatively similar results can be obtained with $\sigma_o = 0.5\,\mathrm{mS\,cm^{-1}}$ and about 5 stimulated neurons around the unstimulated one. It is probable that such values of conductivity can be observed in the CNS (see the end of Section 3.2). This strongly suggests that ephaptic interactions may be important in the CNS, even at the local level.

5. DISCUSSION

5.1 Biophysical and Mathematical Methods

The biophysical basis of ephaptic interaction modeling was established between 1945 and 1970 approximately. Two main assumptions were made that still appear to be valuable: (i) the time derivative of the magnetic induction (Rosenfalck, 1969), and (ii) the variation of the intracellular potential inside a neuron cross-section (Pickard, 1971) are negligible. These hypotheses allow to derive a 3D Poisson equation for the extracellular potential, and a 1D Poisson equation for the intracellular potential. In both cases, the source term is proportional to the transmembrane current, which is generally nonlinear. Analytical solutions are available when membranes are passive and the extracellular space is assumed to be one-dimensional. Although these assumptions are rather strong, this simple case allows to show some general properties of ephaptic interactions, such as the non-monotonic spatial influence of stimulated neurons on the membrane potential of unstimulated neurons. For neurons embedded with active conductances, or when the extracellular space is considered as 3D, numerical methods have to be employed. To our knowledge, no attempt was made to use finite-elements instead of the more classical finite-difference methods. Finite-difference discretization of extracellular space and neurons has been used extensively by Traub et al. (1985a,b; 1987a,b,c). We propose here a relatively simple approach, which is also based on finite-difference discretization in space; however, this method differs from Traub's approach by (i) a time resolution procedure derived from the MacGregor's method (see MacGregor, 1987); (ii) an explicit link to Poisson equation and CSDA. Obviously, the stability of this method, as well as others, should be tested in every concrete case using at least varying time steps.

Ideally, the 3D and nonlinear obtained PDEs should deserve a more complete mathematical study, including existence and uniqueness of solutions, and resolution using for example a finite-element method, together with a study of stability and convergence. However, the resolution would still remain rather cumbersome and time-consuming, especially if one tries to include detailed aspects of neuronal geometry (e.g. dendritic tree branching). Besides, it would be highly desirable to get simplified models of ephaptic transmission. Some models of neural networks include an additional term for diffusion of activity through the extracellular space (see Parodi et al., 1996), which may result from ephaptic transmission. Although interesting, these models are based on *a priori* simplifications and do not include, for instance, the non-monotonic variation of depolarization in the unstimulated neuron, and, more generally, spatial effects observed along neurons. In our view, an alternative way would consist in a careful reduction of the biophysical equations from complete three-dimensional models, in accordance with specific experiments.

5.2 Electrical Field Effects in the Hippocampus and the Cerebral Cortex

Ephaptic interactions have been studied in several nervous tissues (spinal cord, cerebellar cortex, hippocampus), but most of the attention has been paid to the hippocampus, due to historical and experimental reasons. This central nervous structure has long been a major center of interest, because of its functional involvement in learning and memory (Eichenbaum et al., 1992). It is now relatively well known in terms of its connectivity (Rolls, 1989), electrophysiology (Wong and Prince, 1981; Brown and Griffith, 1983), and synaptic plasticity (Bliss and Lynch, 1988).

Clear evidences of ephaptic interactions in the hippocampus were found in the early eighties (Taylor and Dudek, 1982), when spontaneous or evoked synchronized bursts of population spikes were observed in hippocampal slices in which chemical synaptic transmission had been blocked in low-[Ca^{2+}] solution. Differential recordings of intracellular and extracellular potentials clearly showed that field effects could be observed in the slice during the population burst spikes (Taylor and Dudek, 1982). The extracellular potential around the recorded somata was found to decrease transiently during the successive spikes of the bursts, depolarizing the membrane potential by the same amount. The excitability of the unstimulated pyramidal neurons was thus increased since the gating mechanisms of the action potential fast Na^+ channels (as many other types of channels) are sensitive to depolarization of the membrane potential. The instantaneous firing probability of these neurons was consequently increased, eventually leading to their recruitment as additional bursting neurons. These effects were found to remain substantial when normal chemical synaptic transmission was functional (Snow and Dudek, 1986). Ephaptic interactions can thus be a major mechanism of synchronized firing, not only in epileptiform activity but also under physiological circumstances (Richardson et al., 1984).

The modeling effort of Traub et al. (1985a,b; 1987a,b,c; review in Dudek and Traub, 1989) has provided a biophysical theoretical framework for understanding ephaptic interactions in the hippocampus. A model of an array of 2000 pyramidal neurons and their surrounding extracellular space was developed. Each of the model neurons possessed purely passive dendritic processes, including an 8 compartment basilar dendrite, and a 10 compartment apical dendrite. The model neurons also included a somatic compartment, which was embedded with standard Hodgkin–Huxley Na and K conductances. The extracellular space was modeled as a resistive lattice of 19,000 points (with the notations of section 4 and Figure 12.5, $k_{max}=19$, $l_{max}=49$, $m_{max}=18$). Although these authors do not give all the details of the employed method, it clearly appears that it is based on a finite-difference scheme (see section 4). The simulations allowed to reproduce various experimental results, including synchronization properties of the CA3 hippocampal network in the normal condition (excitatory and inhibitory synaptic transmission), low-[Ca^{2+}] condition (blockade of the synaptic transmission), and epileptogenic condition (no inhibitory synaptic transmission).

A hypothetical mechanism of ephaptic interactions in the hippocampus was described (Dudek and Traub, 1989). It took into account several basic anatomical features of this structure: the hippocampus CA1–CA3 contains radially oriented pyramidal neurons, with dendritic trees parallel to each other and perpendicular to the surface; furthermore, the CA1–CA3 can be seen as composed of three principal layers, one layer containing the cell bodies and two layers containing the apical dendritic processes and the basal dendritic processes, respectively. Briefly, the hypothetical scheme can be summarized as follows: the synchronized firing of numerous neurons implies large transient inward (sodium, calcium) ionic currents from the extracellular space into the cell bodies of the discharging cells. A 'current sink' ($I_m < 0$) appears at the cell bodies layer, so that the extracellular potential decreases locally. The unstimulated neurons are consequently locally depolarized at their cell body. As a consequence, a 'current source' ($I_m > 0$) appears in the dendrite layer (see (12.36)). It locally hyperpolarizes unstimulated neurons at their dendritic tip (these findings can be compared with those obtained from the simple model of section 4.1: see Figures 12.4, 12.8 and 12.9). This dual effect has great potential functional implications: it predicts at the same time an increased excitability of neurons, and a more selective dendritic integration of their synaptic inputs. In a neural network, such a tendency may favor a 'winners take all' mode of computation, i.e. a state of increased competition between neurons.

Most of the cited studies were concerned with antidromically evoked electrical fields. What happens when a neuron or a group of neurons are stimulated by a post-synaptic potential, e.g. an EPSP, which can be followed (or not) by an action potential? Yim et al. (1986) proposed a simple model of the unstimulated neuron, consisting of a lumped resistance and capacitance for the neuron membrane, an internal longitudinal cell resistance; the extracellular field was modeled using, e.g., a half sine-wave. They combined this one-neuron model with formula proposed by Patlak (1955) to calculate the extracellular field generated by the synchronous activity of n fibers. Yim et al. (1986) concluded that if the maximum amplitude of the stimulated neurons EPSP is about 30 mV, then the depolarization generated in unstimulated neurons can be about 0.3–1 mV in the hippocampus. They also noted that the action potential of the stimulated neurons can depolarize the unstimulated neuron somata by 5 mV or even more. Our results (Figures 12.4, 12.7, 12.8 and 12.9), although obtained using a different method, are consistent with these latter conclusions; however, they do not preclude the possibility of an inhibition of the unstimulated neuron. Finally, the possible ephaptic effects due to EPSPs in the hippocampus might have amplitudes ranging from 0.1 to 10 mV, depending on the specific situations encountered. Such influences may contribute to integration of synaptic inputs and to synaptic plasticity phenomena.

The question arises whether some functional aspects of other central structures could be influenced by ephaptic interactions. Of special interest is the cerebral cortex, which has been poorly studied from this point of view. This is surprising for several reasons. First, the cerebral cortex is one of the major loci of epileptiform discharges, and ephaptic interaction has been demonstrated to be an important factor in epileptiform discharging in the hippocampus. Second, synchronized discharges of large populations of pyramidal cortical neurons under physiological conditions are now considered to underlie perception processes (Gray and Singer, 1989). Third, several features indicate that ephaptic interactions may be substantial in the cerebral cortex. The anatomical arrangement of the neocortex is relatively similar to that of the hippocampus. It is two-dimensional and contains a dense array of radially oriented cells and dendritic trees. The presence of dendrite bundles implies regions of small extracellular volume and therefore high extracellular resistance (Petsche et al., 1975). The size and shape of the cortical pyramidal neurons are most similar to that of the hippocampal CA1–CA3 pyramidal neurons, which support substantial field effects. Furthermore, the cortical dendritic trees are especially long (0.5–2 mm), favoring the spatial segregation between current 'sinks' and 'sources', and the subsequent impact of field effects. It must be noted that the activity of some neurons can be altered by very small external fields, about a few millivolts or a fraction of a millivolt per centimeter (Bullock, 1997): the possibility of such a phenomenon in the cerebral cortex should be explored. One of the major anatomical features that distinguishes the neocortex from the hippocampus is that the neocortex is constituted of several superimposed layers of neurons whereas the hippocampus is monolayered. The rather simple ephaptic interactions observed in the unstimulated hippocampal neurons (depolarized soma, hyperpolarized dendrites) might not be true anymore for the intricate vertical organization of the cortical tissue. For example, the soma of a neuron in a superficial layer could be both subject to the depolarizing field effect of neurons firing in a near vicinity (of the same layer), while being at the same time under the hyperpolarizing influence of a bundle of dendrites projecting from a group of stimulated, deep layered neurons. This hypothetical example shows how the interplay between dendritic and somatic components of field interactions might be much more complex in the cortex, when compared to the hippocampus. Investigations of field effects in the cortical structure remain to be done and should provide an invaluable contribution to the understanding of information processing and epileptic state of the cortex.

6. CONCLUSIONS AND FUTURE PERSPECTIVES

As was recently emphasized by Bullock (1997), extracellular potentials are not only *signs* of underlying neuronal processes, but also *signals* for neighboring cells. Electric field effects offer several challenges to modelers. The basic biophysical laws appear to be well established, so that writing down the corresponding PDEs can be achieved—at least in a general form—for virtually every concrete system. However, studying concrete systems and/or finding general properties of ephaptic transmission remains a rather difficult issue, due to the relative lack of experimental data, and the complexity of mathematical aspects. Our impression is that future work in this domain should follow two distinct albeit complementary directions. First, the effects of the specific geometry of the nervous tissues studied should be taken into account in a more detailed fashion. This would allow for a deeper understanding of ephaptic interactions for particular neuronal arrangements. One of the major interests of this approach would be to evaluate the contribution of field effects to information transfer between units at various scales (neurons, dendrites, spines). Secondly, carefully simplified models of ephaptic interactions generated from close apposition of single neuronal processess (e.g. dendrites) rather than the synchronous activity of many formal neurons proposed in earlier models (Griffith, 1963, 1965; Wilson and Cowan, 1972; Fischer, 1973; and Nunez, 1974), could be included in neuronal networks models. These models should contribute significantly to the understanding of the dynamics of neuronal populations and to the functional roles they subserve.

PROBLEMS

12.1. The simplified model presented in section 3.2 and the following resolutions given in section 4 assume unbranched one-dimensional neurons. However, considering the case of dendritic tree ramifications appears natural as a further step in the description of electrical field interactions between and within neurons.

(i) Use finite-difference scheme (see Figure 12.6 and (12.74) to write the total current conservation law at a simple branching point (i.e. a main dendrite dividing into two smaller dendrites).
(ii) Derive the expression of the differential equation system describing the evolution of intracellular and extracellular potentials along the neuron.

12.2. Throughout this chapter, ionic concentrations have been taken as constants in intracellular and extracellular milieus. However, reduced extracellular volumes between neurons may allow significative ionic concentration variations. For the following questions, one can build a one-dimensional model, i.e. a "slice"; through somata of adjacent neurons including intracellular and extracellular compartments. Soma diameters and extracellular distances can be respectively taken as 20 μm and 1–10 μm. One will consider action potential (sodium and potassium) conductances and ionic pumps. For estimations concerning channel densities, membranous flows and ionic pumps, see Bergman (1970), Hille (1992), Kabakov (1994) and Genet (1997).

(i) Evaluate extracellular sodium and potassium concentrations in response to spontaneous (1–10 Hz) and repetitive discharges (50–100 Hz). Is it always justified to consider ionic concentrations as constants?
(ii) In the case it is not, determine consequences of these variations on membrane potential.

12.3. Research project. Try to imagine ways to describe ephaptic interactions within an intricated network of neurons.

(i) A preliminary attempt could consist in the description of interactions between neurons which somata are located at different depths (z-dimension, see Figure 12.5).
(ii) A further study could include two (or more) superposed layers of neurons such as found in the entorhinal cortex or different neocortical cortices.

ACKNOWLEDGMENTS

We thank O. Parkes for editing the English text.

REFERENCES

Arvanitaki, A. (1942) Effects evoked in an axon by the activity of a contiguous one. *J. Neurophysiol.*, **5**, 89–108.

Bell, J. (1980) Electrical interactions between parallel nerve fibers: a model and some results. In *Nonlinear Partial Differential Equations in Engineering and Applied Science*, edited by R.L. Sternberg, A.J. Kalinowski, and J.S. Papadakis, pp. 21–32. New York Basel: Marcel Dekker.

Bell, J. (1981) Modeling parallel, unmyelinated axons: Pulse trapping and ephaptic transmission. *SIAM J. Appl. Math.*, **41**, 168–180.

Bell, J. and Cosner, C. (1981) Stability properties of a model of parallel nerve fibers. *J. Diff. Eq.*, **40**, 303–315.

Bergman, C. (1970) Increase of sodium concentration near the inner surface of the nodal membrane. *Pflügers Arch. ges. Physiol.*, **317**, 287–302.

Bliss, T.V.P. and Lynch, M.A. (1988) Long-term potentiation of synaptic transmission in the hippocampus: properties and mechanisms. In *Long-Term Potentiation: From Biophysics to Behavior*, edited by P.W. Landfield and S. Deadwyler, pp. 3–72. New York: Liss.

Brown, D.A. and Griffith, W.H. (1983) Calcium-activated outward current in voltage-clamped hippocampal neurons of the guinea-pig. *J. Physiol.* (*London*), **337**, 287–301.

Bullock, T.H. (1997) Signals and signs in the nervous system: The dynamic anatomy of electrical activity is probably information-rich. *Proc. Natl. Acad. Sci. USA*, **94**, 1–6.

Chapman, C.A. and Racine, R.J. (1997) Converging inputs to the entorhinal cortex from the piriform cortex and medial septum: facilitation and current source density analysis. *J. Neurophysiol.*, **78**, 2602–2615.

Clark, J. and Plonsey, R. (1966) A mathematical evaluation of the core conductor model. *Biophys. J.*, **6**, 95–112.

Clark, J. and Plonsey, R. (1968) The extracellular potential field of a single active nerve fiber in a volume conductor. *Biophys. J.*, **8**, 842–864.

Clark, J.W. and Plonsey, R. (1970) A mathematical study of nerve fiber interaction. *Biophys. J.*, **10**, 937–957.

Clark, J.W. and Plonsey, R. (1971) Fiber interaction in a nerve trunk. *Biophys. J.*, **11**, 281–294.

Costalat, R. (1991) Etude analytique et numérique des interactions neuronales par effet de champ électrique. Application à l'hippocampe. Angers, France: Ph.D Thesis, University of Angers.

Costalat, R. and Chauvet, G. (1993) Electrical interactions in biological neural networks. Portland: World Congress on Neural Networks '93 **II**, 614–617.

Costalat, R. and Chauvet, G. (1998) Basic properties of electrical field coupling between neurons: an analytical approach. *J. Biol. Syst* (submitted).

Courant, R. and Hilbert, D. (1962) *Methods of Mathematical Physics*, Volume 2. New York: John Wiley and Sons.

Dudek, F.E. and Traub, R.D. (1989) Local synaptic and electrical interactions in the hippocampus: experimental data and computer simulations. In *Neural Models of Plasticity. Experimental and Theoretical Approaches*, edited by J.H. Byrne and W.O. Berry. Orlando: Academic Press.

Eccles, J.C. (1964) *The Physiology of Synapses*. Berlin: Springer-Verlag.

Eichenbaum, H., Otto, T. and Cohen, N.J. (1992) The hippocampus: what does it do? *Behav. and Neural Biol.*, **57**, 2–36.

Faber, D.S. and Korn, H. (1989) Electrical field effects: Their relevance in central neural networks. *Physiological Reviews*, **69**, 821–863.

Fatt, P. and Katz, B. (1951) An analysis of the end-plate potential recorded with an intracellular electrode. *J. Physiol.*, **115**, 320–370.

Fischer, B. (1973) A neuron field theory: mathematical approaches to the problem of large numbers of interacting nerve cells. *Bull. Math. Biol.*, **35**, 345–357.

FitzHugh, R. (1961) Impulses and physiological states in theoretical models of nerve membrane. *Biophys. J.*, **1**, 445–466.

Freeman, J.A. and Stone, J. (1969) A technique for current density analysis of field potentials and its application to the frog cerebellum. In *Neurobiology of Cerebellar Evolution and Development*, edited by R. Llinas. Chicago: Am. Med. Assoc., pp. 421–430.

Freeman, J.A. and Nicholson, C. (1975) Experimental optimization of current source-density technique for anuran cerebellum. *J. Neurophysiol.*, **38**, 369–382.

Furukawa, T. and Furshpan, E.J. (1963) Two inhibitory mechanisms in the Mauthner neurons of goldfish. *J. Neurophysiol.*, **26**, 140–176.

Genet, S. and Cohen, J. (1996) The cation distribution set by surface charges explains a paradoxical membrane excitability behavior. *C. R. Acad. Sci. Paris Sciences de la vie/Life sciences*, **319**, 263–268.

Genet S. (1997) Etude expérimentale et théorique des mécanismes d'établissement et de régulation de la différence de potentiel électrique des membranes biologiques. PhD Thesis, University Pierre et Marie Curie.

Gray, C.M. and Singer, W. (1989) Stimulus-specific neuronal oscillations in neuronal columns of cat visual cortex. *Proc. Natl. Acad. Sci. USA*, **86**, 1698–1702.

Griffith, J.S. (1963) A field theory of neural nets: I. Derivation of field equations. *Bull. Math. Biol.* **25**, 111–120.

Griffith, J.S. (1965) A field theory of neural nets: II. Properties of the field equations. *Bull. Math. Biol.* **27**, 187–195.

Herreras, O. (1990) Propagating dendritic action potential mediates synaptic transmission in CAI pyramidal cells in sity. *J. Neurophysiol.* **64**, 1429–1441.

Hille, B. (1992) *Electric Channels of Excitable Membranes*. 2nd edition Sunderland, MA: Sinauer.

Hodgkin, A.L. and Huxley, A.F. (1952) A quantitative description of membrane current and its application to conduction and excitation in nerve. *J. Physiol. Lond.*, **117**, 500–544. (A reprint can be found in *Bull. Math. Biol.*, **52**, 25–71 (1990).)

Holsheimer, J. (1987) Electrical conductivity of the hippocampal CA1 and application to current-source-density analysis. Exp. Brain Res. **67**, 402–410.

Holsheimer, J., Boer, J., Lopes da Silva, F.H., and van Rotterdam, A. (1982) The double dipole model of theta rhythem generation: simulation of laminar field potential profiles in dorsal hippocampus of the rat. *Brain Res.*, **235**, 31–50.

Jack, J.J.B., Noble, D. and Tsien, R.W. (1983) Electric Current Flow in Excitable Cells. Oxford: Clarendon Press.

Jasper, H.H. and Monnier, A.M. (1938) Transmission of excitation between excised non-myelinated nerves. An artificial synapse. *J. Cell. Comp. Physiol.*, **11**, 259–277.

Jefferys, J.G.R. (1995) Nonsynaptic modulation of neuronal activity in the brain: electric currents and extracellular ions. *Physiol. Rev.*, **75**, 689–723.

Johnston, D. and Wu, S. (1995) *Foundations of Cellular Neurophysiology*. Cambridge, MA: MIT Press.

Kabakov A. (1994) The resting potential equations incorporating ionic pumps and osmotic concentrations. J. Theor. Biol., **169**, 51–64.

Katz, B. and Schmitt, O.H. (1940a) Electrical interaction between two adjacent fibers. *J. Physiol. Lond.*, **97**, 471–488.

Katz, B. and Schmitt, O.H. (1940b) A note on the interaction between nerve fibers. *J. Physiol. Lond.*, **100**, 369–371.

Kellogg, O.D. (1929) Foundations of Potential Theory. New York: Frederick Unger Publishing Co.

Klee, M. and Rall, W. (1977) Computed potentials of cortically arranged populations of neurons. *J. Neurophysiol.*, **40**, 647–666.

Kocsis, J.D., Ruiz, J.A. and Cummins, K.L. (1982) Modulation of axonal excitability mediated by surround electric activity: an intra-axonal study. *Exp. Brain Res.*, **47**, 151–153.

Kossev, A., Gydikov, A. and Trayanova, N. (1988) Comparison of the different variants of the current source density analysis methods in neurophysiological studies. *Acta Physiologica et Pharmacologica Bulgarica*, **14**, 75–82.

Leung, L.-W.S. (1984) Model of gradual phase shift of theta rhythm in the rat. *J. Neurophysiol.* **52**, 1051–1065.

Lorente de No, R. (1947) *A Study of Nerve Physiology*. New York: Rockefeller Institute.

MacGregor, R.J. (1987) *Neural and Brain Modeling*. New York: Academic Press.

Maeda, K., Yagi, T., and Noguchi, A. (1980) Induced excitation and synchronization of nerve impulses in two parallel unmyelinated fibers. *IEEE Trans. Biomed. Eng.*, **27**, 139–145.

Marrazzi, A.S., and Lorente de No, R. (1994) Interaction of neighbouring fibres in myelinated nerve, *J. Neurophysiol.*, **7**, 83–101.

Markin, V.S. (1970) Electrical interaction of parallel non-myelinated nerve fibres. I. Change in excitability of the adjacent fibre. *Biophysics*, **15**, 122–133.

Markin, V.S. (1973) Electrical interaction of parallel non-myelinated nerve fibres. III. Interaction in bundles. *Biophysics*, **18**, 324–332.

Mitzdorf, U. (1985) Current source-density method and application in cat cerebral cortex: Investigation of evoked potentials and EEG phenomena. *Physiological Reviews*, **65**, 37–100.

Mitzdorf, U. and Singer, W. (1977) Laminar segregation of afferents to lateral geniculate nucleus of the cat: an analysis of current source density. *J. Neurophysiol.*, **40**, 1227–1244.

Miyakawa, H., and Kato, H. (1986) Active properties of dendritic membrane examined by current source density analysis in hippocampal CA1 pyramidal neurons. *Brain Res.*, **399**, 303–309.

Nagumo, J.S., Arimoto, S. and Yoshizawa, S. (1962) An active pulse transmission line simulating nerve axon. *Proc. IRE.*, **50**, 2061–2071.

Nicholson, C. and Llinas, R. (1971) Field potentials in the alligator cerebellum and theory of their relationship to Purkinje cell dendritic spikes. *J. Neurophysiol.*, **34**, 509–531.

Nicholson, C. (1973) Theoretical analysis of field potentials in anisotropic ensembles of neuronal elements. *IEEE Trans. Biomed. Eng.*, **20**, 278–288.

Nicholson, C. and Freeman, J.A. (1975) Theory of current source-density analysis and determination of conductivity tensor for anuran cerebellum. *J. Neurophysiol.*, **38**, 356–368.

Nunez, P.L. (1974) The brain wave equation: a model for the EEG. *Math. Biosci.*, **21**, 279–297.

Nunez, P.L. (1981) *Electrical Fields of the Brain. The Neurophysics of EEG.* New York, Oxford: Oxford University Press.

Parodi, O., Combe, P. and Ducom, J.-C. (1996) Temporal coding in vision: coding by the spike arrival times leads to oscillations in the case of moving targets. *Biol. Cybern.*, **74**, 497–509.

Patlak, C.S. (1955) Potential and current distribution in nerve: the effect of the nerve sheath, the number of fibers, and the frequency of alternating current stimulation. *Bull. Math. Biophys.*, **17**, 287–307.

Petsche, M., Promaska, O., Rappelsberger, P. and Vollmer, R. (1975) The possible role of dendrites in EEG synchronization. *Adv. Neurol.*, **12**, 53–70.

Pickard, W.F. (1968) A contribution to the electromagnetic theory of the unmyelinated axon. *Math. Biosci.*, **2**, 111–121.

Pickard, W.F. (1969) The electromagnetic theory of electrotonus along an unmyelinated axon. *Math. Biosci.*, **5**, 471–494.

Pickard, W.F. (1971) Electrotonus of a cell of finite dimensions. *Math. Biosci.*, **10**, 201–213.

Plonsey, R. (1964) Volume conductor fields of action currents. *Biophys. J.*, **4**, 317–328.

Press, W.H., Flannery, B.P., Teukolsky, S.A. and Vetterling, W.T. (1993) *Numerical Recipes in Fortran 77*, Second edition (also exists in C, FORTRAN 90 and PASCAL versions). Cambridge, New York: Cambridge University Press.

Rall, W. and Shepherd, G.M. (1968) Theoretical reconstruction of field potentials and dendrodendritic synaptic interactions in olfactory bulb. *J. Neurophysiol.*, **31**, 884–915.

Rappelsberger, P., Pockberger, H. and Petsche, H. (1981) Current source density analysis: methods and application to simultaneously recorded field potentials of the rabit's visual cortex. *Pfuegers Arch.*, **389**, 159–170.

Richardson, T.L., Turner, R.W. and Miller, J.J. (1984) Extracellular fields influence transmembrane potentials and synchronization of hippocampal neuronal activity. *Brain Res.*, **294**, 255–262.

Richardson, T.L., Turner, R.W. and Miller, J.J. (1987) Action-potential discharge in hippocampal CA1 pyramidal neurons: current source–density analysis. *J. Neurophysiol.*, **58**, 981–996.

Rolls, E.T. (1989) Functions of neuronal networks in the hippocampus and neocortex in memory. In *Neural Models of Plasticity*, edited by J.H. Byrne and W.O. Berry. San Diego: Academy Press, pp. 240–265.

Rosenfalck, P. (1969) Intra- and extracellular potential fields of active nerve and muscle fibres. *Acta. physiol. scand. Suppl.*, **321**, 1–168.

Scott, A.C. and Luzader, S.D. (1979) Coupled solitary waves in neurophysics. *Phys. Scripta*, **20**, 395–401.

Shadowitz, A. (1975) *The Electromagnetic Field.* New York: McGraw-Hill. (Reprint by Dover Publications, New York (1988).)

Shepherd, G.M. (1991) *Foundations of the Neuron Doctrine.* New York and Oxford: Oxford University Press.

Snow, R.W. and Dudek, F.E. (1986) Evidence for neuronal interactions by electrical field effects in the CA3 and dentate regions of rat hippocampus slices. *Brain Res.*, **367**, 292–295.

Stone, J. and Freeman, J.A. (1971) Synaptic organization of the pigeon's optic tectum: a Golgi and current source density analysis. *Brain Res.*, **27**, 203–221.

Taylor, C.P. and Dudek, F.E. (1982) Synchronous neural afterdischarges in rat hippocampal slices without active chemical synapses. *Science Wash. DC*, **218**, 810–812.

Taylor, C.P. and Dudek, F.E. (1984) Excitation of hippocampal cells by an electrical field effect. *J. Neurophysiol.*, **52**, 126–142.

Traub, R.D., Dudek, F.E., Taylor, C.P. and Knowles, W.D. (1985a) Simulations of hippocampal afterdischarges synchronized by electrical interactions. *Neuroscience*, **14**, 1033–1038.

Traub, R.D., Dudek, F.E., Snow, R.W. and Knowles, W.D. (1985b) Computer simulations indicate that electrical field effects contribute to the shape of the epileptiform field potential. *Neuroscience*, **15**, 947–958.

Traub, R.D., Knowles, W.D., Miles, R. and Wong R.K.S. (1987a) Models of the cellular mechanism underlying propagation of epileptiform activity in the CA2-CA3 region of the hippocampal slice. *Neuroscience*, **21**, 457–470.

Traub, R.D., Miles, R. and Wong, R.K.S. (1987b) Models of synchronized hippocampal bursts in the presence of inhibition. I. Single population events. *J. Neurophysiol.*, **58**, 739–751.

Traub, R.D., Miles, R., Wong, R.K.S., Schulman, L.S. and Schneiderman, J.H. (1987c) Models of synchronized hippocampal bursts in the presence of inhibition. II. Ongoing spontaneous population events. *J. Neurophysiol.*, **58**, 752–764.

Tuckwell, H.C. (1988a) *Introduction to Theoretical Neurobiology*. Volume 1: Linear Cable Theory and Dendritic Structure. Cambridge: Cambridge University Press.

Tuckwell, H.C. (1988b) *Introduction to Theoretical Neurobiology*. Volume 2: Nonlinear and Stochastic Theories. Cambridge: Cambridge University Press.

Wilson, H.R. and Cowan, J.D. (1972) Excitatory and inhibitory interactions in localized populations of model neurons. *Biophys. J.*, **12**, 1–24.

Wong, R.K.S. and Prince, D.A.J. (1981) Afterpotential generation in hippocampal pyramidal cells. *J. Neurophysiol.*, **45**, 86–97.

Yim, C.C., Krnjevic, K. and Dalkara, T. (1986) Ephaptically generated potentials in CA1 neurons of rat's hippocampus *in situ*. *J. Neurophysiol.*, **56**, 99–122.

CHAPTER 13
NUMERICAL MODELING OF NEOCORTICAL PYRAMIDAL CELLS

Roger D. Orpwood

1. INTRODUCTION

This chapter lays the groundwork for the numerical modelling of the principal neurons of the neocortex, and goes on to show how such numerical models can provide insights into the behaviour of these cells. The models are a mixture of simulations of physiological processes and empirically derived relationships, and the features incorporated are those that appear to be the key ones involved in the information processing capabilities of these neurons. The underlying equations and algorithms are described in sufficient detail to enable such studies to be repeated elsewhere.

The main input/output cells of the cortex are the pyramidal cells (Shepherd, 1979) and these cells are the subject of the work described in this chapter. The pyramidal neuron is primarily involved in receiving information from other cells and making decisions on the basis of that information and its past experience about whether to provide an output or not. It is subject to developmental processes like all cells and the mature configuration of the cell is one to which it adapts following its early experiences, possibly in a Darwinian fashion (Edelman, 1987). However even when the cell has achieved its mature status it is in a constant dynamic state of adapting to its experiences. Its responses to incoming information are not rigid but constantly changing. Any model of the mature pyramidal cell must include all the main factors that enable it to behave in this way.

Neurons, and pyramidal cells in particular, have a very unique structure and there must be an important link between their cellular geometry and the way they behave. The input side of the cell in particular, with its vast arborizations of dendrites providing a very large surface area for interactions with other cells, must be emulated in any model of the cell. In addition many pyramidal cells have a dendritic structure that is divided into two parts. Proximal and basal dendrites are present close to the cell body, echoing the kind of dendritic structure seen in most other neurons, but there are also apical dendrites separated from the cell body by means of a long climbing fibre. The cell appears to be divided into two separate dendritic trees; the proximal dendrites around the cell body, and the distal dendrites, with the apical climbing fibre linking the two. This geometry is a highly distinctive feature of the pyramidal cell and must be included in any model that simulates its behavior.

The adaptive capabilities of pyramidal cells result from the properties of the regions of the cell that receive incoming signals. All facilitatory inputs to the cell occur at dendritic spines. The spine has a very typical structure with a spine head receptor region connected to the dendrite via the spine shaft. Individual receptor sites on dendritic spines have been shown to have the capability to change their sensitivity to inputs, and for the sensitivity changes to remain over a long period (Baudry and Davis, 1991; Madison et al., 1991). It is widely suspected that these long-term changes underlie information storage and retrieval in the cortex, and constitute the biological basis of memory (Bliss and Collingridge, 1993).

Facilitatory signals are transferred from incoming axons to the pyramidal cell by means of the excitatory transmitter glutamate. The glutamate receptors are membrane mounted ion channels in the dendritic spines, and they can be generally classified into two main types. The first (the AMPA receptor), is simply opened by the glutamate transmitter, allowing Na^+ ions through the membrane at the receptor site and causing a local depolarisation. The second kind of receptor (the NMDA receptor) has a more complex response. It is also a membrane mounted ion channel but at resting membrane potential the channel is blocked by Mg^{++} ions. Consequently at resting membrane potentials the glutamate causes no response from the receptor. Only when the local membrane has been depolarized by around 30 mV is the Mg^{++} block removed and the ion channel opened up (Mayer et al., 1984; Nowak et al., 1984). The activated channel then allows an influx of both Na^+ and Ca^{++} ions (Regehr and Tank, 1990), again leading to a local membrane depolarization. The Ca^{++} influx has another effect. It acts as a second messenger within the cell and provides the starting point for a series of intracellular reactions. The end point of this chain of reactions is a change in the sensitivity of the AMPA receptors on (mostly) the same spine (Bekkers and Stevens, 1989).

The receptor mechanisms described above allow the cell to associate one activation with another. If the local membrane potential is depolarized by more than 30 mV then it means that the cell is receiving a large significant input elsewhere. Any local input received at the same time will ultimately lead to the AMPA receptors in the spine increasing their sensitivity. Any subsequent input to the same spine will then lead to a larger depolarization from its AMPA receptors. The ability to associate information in this way is a key aspect of the studies discussed in this chapter.

It appears therefore that there are three important properties of individual pyramidal cells which should be included in any realistic model:

1. The model must include a realistic representation of the cell's geometry.
2. The model must include the mechanisms that exist at individual receptor sites to enable the transformation of the input chemical signal to an electrical one, and include the ability of the sensitivity of the receptor regions to be changed depending on their experiences.
3. The model must include the capabilities of the cell to communicate intracellularly. The communication must enable the activities at a given receptor site to be

transmitted to other receptor regions, and it must enable this activity to be transmitted to the soma of the cell.

The chapter explores all these core aspects of individual pyramidal cells.

The behavior of pyramidal cells has been the subject of enormous amounts of work, both in terms of their single cell behaviour and in terms of the more detailed functioning of their component parts (see Creutzfeldt, 1995). The problem for any modeling study is that not all of the myriad of properties demonstrated can be included and there is always the danger that some key factor will be left out. Nevertheless the models do provide an extension of our cognitive capabilities to mentally model the behaviour of pyramidal cells, given the measured capabilities of their component parts. As long as care is taken over the choice of parameters included in the model the results should provide some insight into properties of the system being modelled and enable experimental work to be more closely focussed.

In the models described in this chapter there is a progression from a sophisticated model using simulations of physiological processes, to a much simpler model that uses empirical equations to describe pyramidal cell behavior. Other workers have followed the same process (Traub et al., 1987; Bush and Sejnowski, 1993; Mainen and Sejnowski, 1996). The more sophisticated models enable the influence of larger numbers of variables to be explored. Once the behavior of these models has been examined, and the key variables affecting this behavior have been identified, it can be expressed in an empirical form. The model can be simplified and checked to make sure the same key properties can still be demonstrated. The simpler models can then be used in studies of larger scale biological systems. It is felt that this process of building up from models of detailed intracellular biological processes to models of complex cellular systems is a useful one in that it enables the key properties of each of the levels of complexity to be elucidated, and ensures that the larger scale models are based on realistic assumptions about the properties of their component elements. As far as the single cell is concerned then the model evolves from a physiological model of the cellular system to a much simpler empirical model that demonstrates the same key properties. The chapter will describe the development of the physiological model and discuss the simplifications embodied in the empirical one. It will compare the performance of the two models to demonstrate that the simpler one is a useful basis on which to explore larger scale cortical simulations.

2. COMPARTMENTALIZATION OF DENDRITES AND DENDRITIC SPINES

2.1 Compartmental Model of Dendrites

Information is communicated at a high speed throughout the cell by means of changes in the membrane potential. This communication is both a passive transfer of potential and an active process, depending on the region of the cell involved (Llinas, 1988). There have been many studies of the way transient membrane potentials are transmitted around neurons, mostly building on the pioneering studies of Rall (1959). The potential distribution over a non-excitable dendrite can be described by the one-dimensional cable equation:

$$\frac{a}{2R_\mathrm{i}}\frac{\partial^2 V}{\partial x^2}=C_\mathrm{m}\frac{\partial V}{\partial t}+\frac{V}{R_\mathrm{m}}, \tag{13.1}$$

where

V = potential difference across the membrane, in terms of the departure from its resting value,
a = radius of dendrite,
R_i = specific resistance of cytoplasm,
x = distance along dendrite,
C_m = membrane capacitance,
R_m = resistance of unit area of membrane.

Analytical solutions to (13.1) have been derived for many simple cellular geometries (Rall, 1977), and simplifications have provided analytical solutions to more complex but basic geometries (Rall and Rinzel, 1973; Stratford et al., 1989; see also Chapter 5). As discussed above the geometry of the pyramidal cell is very distinctive and it is felt that this geometry should be emulated quite closely in any model of its capabilities.

The work described in this chapter uses a compartmental model to simulate the transient potential changes throughout the cell (see Rall, 1964; Dodge and Cooley, 1973; Carnevale and Lebeda, 1987, for discussion). The compartmental approach divides the cell up into a large number of small compartments which are all assumed to be isopotential. The potential difference between compartments and between the compartment and the extracellular space can then be used to calculate the movements of charge between the compartments and out of the cell (see Figure 13.1).

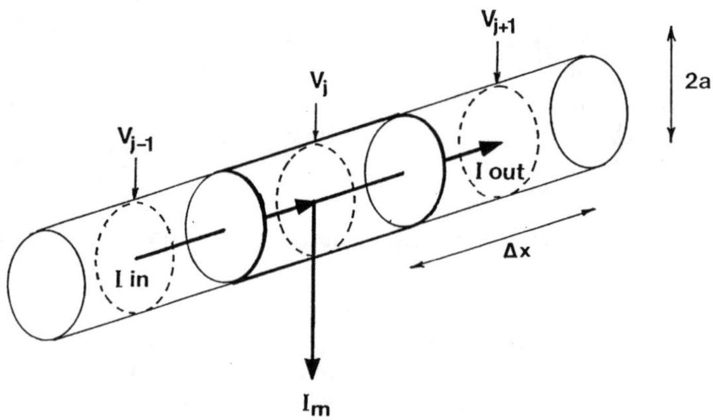

Figure 13.1 Diagram of short length of dendrite illustrating the main variables used in the compartmental model.

A simple forward Euler approximation is used for the second derivative with respect to position:

$$\frac{\partial^2 V}{\partial x^2} \simeq \frac{(V_{j+1}+V_{j-1}-2V_j)}{(\Delta x)^2}. \tag{13.2}$$

Similarly the rate of change of membrane potential can be expressed as a simple approximation of the first derivative with respect to time:

$$\frac{\partial V_j}{\partial t} \simeq \frac{V_j^{t+1}-V_j^t}{\Delta t}, \tag{13.3}$$

where V_j^t is the potential difference at the jth compartment at time t and V_j^{t+1} is the potential difference at time $t+\Delta t$.

The beauty of using these simple approximations (13.2)–(13.3) is that they are very easy to implement and enable an explicit solution to be provided for V^{t+1} in terms of V^t. Unfortunately the solution is unstable unless the following condition applies (Mascagni, 1989):

$$\Delta t \leqslant \frac{R_\mathrm{i} C_\mathrm{m} (\Delta x)^2}{a}. \tag{13.4}$$

This stability criterion (13.4) inevitably leads to small time steps to be necessary and consequently makes the numerical model rather computationally expensive.

So the difference approximations (13.2)–(13.3) can be applied to the cable equation (13.1) to yield a simple expression for the membrane potential of the jth compartment at time $t+\Delta t$:

$$V_j^{t+1} = \left(\frac{\lambda}{\Delta x}\right)_j^2 \frac{\Delta t}{\tau}\left[V_{j+1}^t + V_{j-1}^t - V_j^t\left(2+\left(\frac{\Delta x}{\lambda}\right)_j^2\right)\right] + V_j^t, \tag{13.5}$$

where

$$\lambda^2 = \frac{aR_\mathrm{m}}{2R_\mathrm{i}}, \qquad \tau = R_\mathrm{m} C_\mathrm{m}.$$

Equation (13.5) enables transient potential changes to be plotted along a single cable. The geometry of the neuron is complex with many bifurcations and changes in diameter of dendrites and so similar expressions need to be derived for compartments in different regions of the cell. Figure 13.2 illustrates the main geometries needed for this study and the difference equations that result are provided below:

Figure 13.2(a), different diameters

$$V_j^{t+1} = \left(\frac{\lambda}{\Delta x}\right)_j^2 \frac{\Delta t}{\tau}\left[\alpha V_{j+1}^t + \beta V_{j-1}^t - V_j^t\left(\alpha+\beta+\left(\frac{\Delta x}{\lambda}\right)_j^2\right)\right] + V_j^t, \tag{13.6}$$

where

$$\alpha = \frac{2}{1+(a_j/a_{j+1})^2}, \qquad \beta = \frac{2}{1+(a_j/a_{j-1})^2}.$$

Figure 13.2(b), dendrite ends

$$V_j^{t+1} = \left(\frac{\lambda}{\Delta x}\right)_j^2 \frac{\Delta t}{\tau}\left[\alpha V_{j+1}^t - V_j^t\left(\alpha + \left(\frac{\Delta x}{\lambda}\right)_j^2\right)\right] + V_j^t. \tag{13.7}$$

Figure 13.2(c), bifurcations

$$V_j^{t+1} = \left(\frac{\lambda}{\Delta x}\right)_j^2 \frac{\Delta t}{\tau}\left[\alpha V_{j+1}^t + \gamma_1 V_{(j-1)_1}^t + \gamma_2 V_{(j-1)_2}^t - V_j^t\left(\alpha + \gamma_1 + \gamma_2 + \left(\frac{\Delta x}{\lambda}\right)_j^2\right)\right] + V_j^t, \tag{13.8}$$

where

$$\gamma_1 = \frac{2}{1 + (a_j/a_{(j-1)_1})^2}, \qquad \gamma_2 = \frac{2}{1 + (a_j/a_{(j-1)_2})^2}.$$

Equations (13.6)–(13.8) provide the basis for a numerical model of the potential distribution over a pyramidal cell which takes into account its complex geometry.

2.2 Model of Dendritic Spines

Several studies have explored the significance of dendritic spinal structure for the functioning of the spinal receptors and their communication with the dendrite

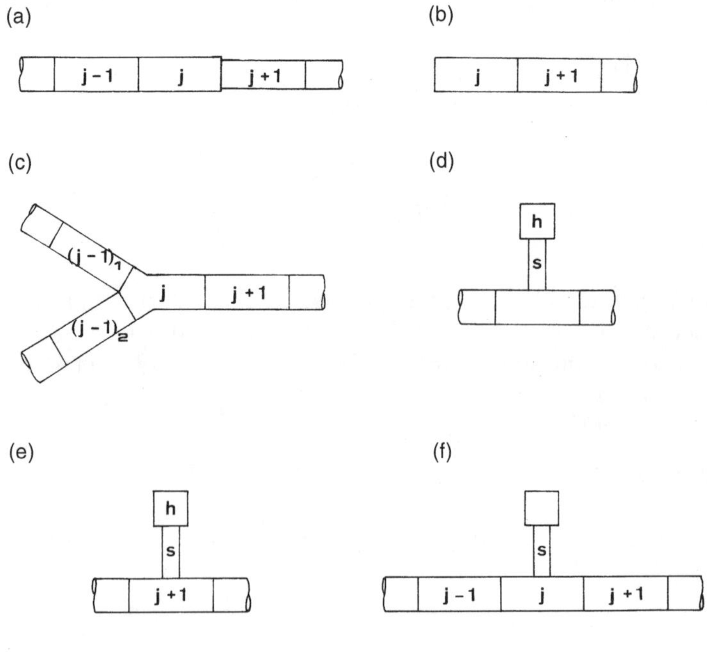

Figure 13.2 The main dendrite geometries for which difference equations are derived for the compartmental model.

(Brown et al., 1988; Zador et al., 1990; Sejnowski and Qian, 1992). When the spine receives an input a large potential difference can be set up between the spine head membrane and the local dendritic membrane. This potential is important for the way that currents are injected into the dendrite and also important for the way the receptor region is able to change its sensitivity depending on prior activity, as will be described below. Consequently each spine that receives an input in the neuron model requires to be modelled explicitly. In the model used here each spine is modelled as two compartments, one representing the spine head and one the spine shaft. These compartments are illustrated in Figure 13.2 and the additional difference equations are as follows:

Figure 13.2(d), spine heads

$$V_{\rm h}^{t+1}=\left(\frac{\lambda}{\Delta x}\right)_{\rm h}^{2}\frac{\Delta t}{\tau}\left[(E-V_{\rm h}^{t})\rho G+\eta_{\rm A}V_{\rm s}^{t}-V_{\rm h}^{t}\left(\rho G+\eta_{\rm A}+\left(\frac{\Delta x}{\lambda}\right)_{\rm h}^{2}\right)\right]+V_{\rm h}^{t}, \qquad (13.9)$$

where

$$\eta_{\rm A}=\frac{2}{1+(a_{\rm h}/a_{\rm s})^{2}(\Delta x_{\rm s}/\Delta x_{\rm h})}, \qquad \rho=\frac{R_{\rm i}\Delta x_{\rm h}}{\pi a_{\rm h}^{2}},$$

G is the total conductance of spine head, as discussed below and E the reversal potential, again discussed below.

Figure 13.2(e), spine shafts

$$V_{\rm s}^{t+1}=\left(\frac{\lambda}{\Delta x}\right)_{\rm s}^{2}\frac{\Delta t}{\tau}\left[\eta_{\rm B}V_{j+1}^{t}+\eta_{\rm C}V_{\rm h}^{t}-V_{j}^{t}\left(\eta_{\rm B}+\eta_{\rm C}+\left(\frac{\Delta x}{\lambda}\right)_{\rm s}^{2}\right)\right]+V_{\rm s}^{t}, \qquad (13.10)$$

where

$$\eta_{\rm B}=\frac{2}{1+(a_{\rm s}/a_{\rm h})^{2}(\Delta x_{\rm h}/\Delta x_{\rm s})}, \qquad \eta_{\rm C}=\frac{2}{1+(a_{\rm s}/a_{j+1})^{2}(\Delta x_{j+1}/\Delta x_{\rm s})}.$$

Figure 13.2(f), parent dendrite

$$V_{j}^{t+1}=\left(\frac{\lambda}{\Delta x}\right)_{j}^{2}\frac{\Delta t}{\tau}\left[\alpha V_{j+1}^{t}+\beta V_{j-1}^{t}+\eta_{\rm D}V_{\rm s}^{t}-V_{j}^{t}\left(\alpha+\beta+\eta_{\rm D}+\left(\frac{\Delta x}{\lambda}\right)_{j}^{2}\right)\right]+V_{j}^{t}, \qquad (13.11)$$

where

$$\eta_{\rm D}=\frac{2}{1+(a_{j}/a_{\rm s})^{2}(\Delta x_{\rm s}/\Delta x_{j})}.$$

If all the spines present on the neuron had to be modelled explicitly it would complicate the model enormously as each cortical pyramidal cell has around 10,000 spines arranged over its dendrites.

Fortunately when current is passing from the dendrite to the spine the situation is simplified by the fact that only a tiny potential difference exists between the dendrite and the spine head (Jack et al., 1983; Segev and Rall, 1988). Consequently the presence of the spines can be modelled implicitly through appropriate changes in the properties of the compartment to which the spines are linked. The method used in this work used changes in the specific membrane properties so as to effectively include the area of the spine in the model (Holmes, 1989):

$$R_{\mathrm{m}}^{*}=\frac{R_{\mathrm{m}}}{F}, \quad C_{\mathrm{m}}^{*}=C_{\mathrm{m}}F, \qquad F=\frac{A_{\mathrm{d}}+A_{\mathrm{s}}}{A_{\mathrm{d}}},$$

where A_{d} is the area of dendrite and A_{s} the area of spines.

The transformed values of R_{m} and C_{m} can then be used in the dendritic compartment to which the spines are attached. The value of F depends on the density of spines existing on the dendritic compartment. The values used in this study took into account the fact that the density varies over the pyramidal cell (Feldman, 1984) and they are indicated in Figure 13.3. The transformation only works, of course, if

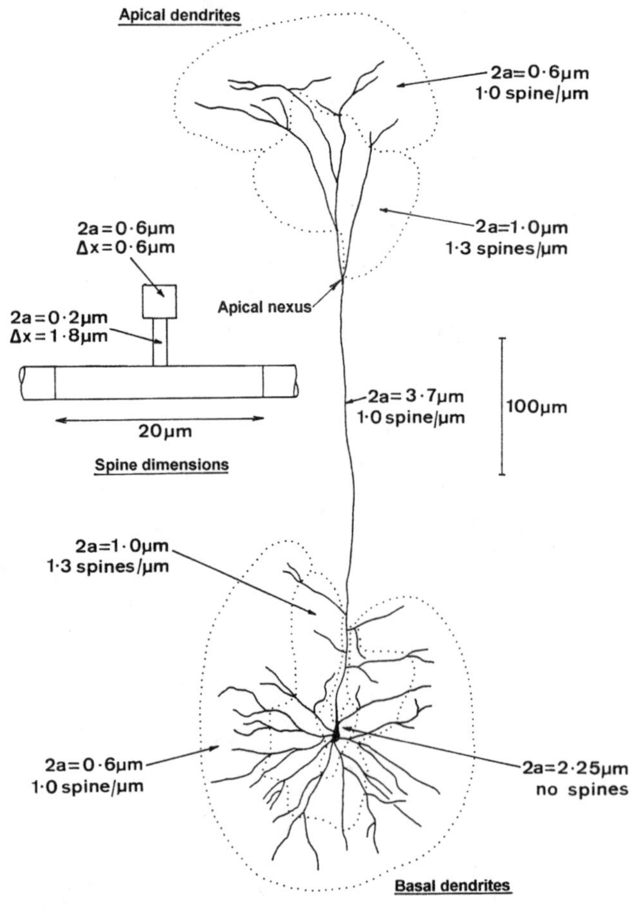

Figure 13.3 Diagram of the pyramidal neuron that was modelled, showing the main geometric variables and their distribution over the cell.

the membrane properties of the dendrite and the spine are the same. Alternative methods for incorporating the spines into the dendritic compartment have been discussed by Stratford et al. (1989), Baer and Rinzel (1991), and Segev et al. (1992).

The values of R_m, R_i and C_m are crucial to the performance of the numerical model. The membrane capacitance has been widely accepted as having a value around $0.01\,F/m^2$ (Cole, 1972) and that value is used in this study. Unfortunately physiological measurements cannot provide unique values for the two resistances. The measured data can be matched to a wide range of pairs of values for R_m and R_i. Stratford et al. (1989) and Major et al. (1994) injected a brief current pulse into a pyramidal cell and then compared the measured voltage response to that predicted by a simplified model of the cell. All the major parameters were allowed to vary within an optimization routine until an optimal least-squares fit was obtained. Cauller and Connors (1992) used a fixed range of R_i values and adjusted the other parameters to fit the observed input resistance and time constant. Their values for R_m and R_i pairs varied from $0.8\,\Omega\,m^2$ and $0.69\,\Omega\,m$, respectively, to $1.73\,\Omega\,m^2$ and $4.0\,\Omega\,m$. The above authors showed that the time courses of potential changes in the soma predicted by compartmental models were not unduly affected by the range of values obtained, except following current injections to the more distal dendrites. Unfortunately inputs to the distal regions are of great interest. As the cell used in this study was the one published by Cauller and Connors (1992) the values finally chosen were a pair from their study, with $R_m = 0.8159\,\Omega\,m^2$ and $R_i = 1.81\,\Omega\,m$. Until further work is completed on exploring more definitive values of R_i for cortical pyramidal cells the uncertainty over these two resistances is a major problem for compartmental modeling of these neurons.

3. RECEPTOR DYNAMICS AND ACTIVE CONDUCTANCES

3.1 Modeling Adaptive Receptors

The main reactions taking place following activation of the glutamate receptor are explored in detail in Orpwood (1988, 1990) where simulations demonstrate the association properties discussed above. For the current work it was felt that the principles demonstrated by these more comprehensive biochemical simulations matched the measured response of receptors well enough for simplifications to be introduced. The receptor dynamics used in these whole cell simulations therefore were empirical in nature as shown below.

The synaptic current that passes into the cell through the receptors is as below:

$$I_h(t) = [V(t) - E]G(t),$$

where $I_h(t)$ is the synaptic current at the spine head, $G(t)$ the membrane conductance caused by the activation of the glutamate receptor, $V(t)$ the local membrane potential and E the reversal potential for the channel given the ion species being admitted.

The actual value for the reversal potential will depend on the Nernst potential for the ion species involved. Most measurements of the reversal potentials for NMDA and AMPA channels indicate values close to zero (Hestrin et al., 1990; Keller et al., 1991) and the values used here were those adopted by Jenson and

Shepherd (1993) of 70 mV relative to resting potential for the AMPA channel and 75 mV for the NMDA. For a more detailed analysis of receptor dynamics see Yamada et al. (1989), and a useful summary of different channel properties is contained in Hille (1992) and Brown et al. (1990).

The channel conductance change, $G(t)$, used the empirically derived time course below (Rall, 1967; Jack et al., 1983):

$$G = \varepsilon t e^{(-t/t_p)}, \tag{13.12}$$

where t_p is the time to the peak conductance change and a value of 0.5 ms is used for the AMPA channel and 2.5 ms for the NMDA. The constant, ε, is set at 5.437×10^{-6} to provide a peak conductance change of 1 nS for the AMPA channel.

If the local membrane depolarization was less than 30 mV then only the AMPA channel is activated and the local depolarization provides an input into the compartmental model and is conducted elsewhere. However if the local membrane is depolarized by more than the crucial 30 mV the NMDA channel is also activated and a sensitivity change in the AMPA channel results. The feedback link between the activation of the NMDA channel and the AMPA sensitivity change was modelled physiologically in Orpwood (1990) but an empirical algorithm was incorporated here. The main ion species admitted by the NMDA receptor are Na^+ and Ca^{++} although the crucial process is the admittance of Ca^{++} as this ion is the second messenger that initiates the AMPA sensitivity changes. It is assumed that conductance time courses for the two ions are the same so that the empirical expression for the conductance change (13.12) can be assumed to reflect the level of activation of the glutamate receptors. Whilst the local membrane was depolarized by more than 30 mV the NMDA receptor membrane current therefore models the rate of admittance of Ca^{++}. The local build up of Ca^{++} is derived in the model by summing the membrane current at each computational time step:

$$[Ca^{++}]_{local} \propto \sum I_{NMDA} \quad \text{while } V_h > 30\,\text{mV}.$$

This is a gross simplification of the mechanisms involved as there are many processes taking place in the cell to remove the free Ca^{++} from the intracellular space (McBurney and Neering, 1987). However it is felt that the proportionality is a valid one. In the model the increase in sensitivity of the AMPA receptor is made proportional to the change in Ca^{++} concentration.

3.2 Modeling Active Conduction

Many authors have shown the existence of actively conducting regions within pyramidal cells (Llinas, 1988; Huguenard et al., 1989; Regehr et al., 1993; Hirsch et al., 1995). Microelectrode recordings (Pockberger, 1991) together with patch clamping (Stuart and Sakmann, 1994) and fluorescence imaging (Yuste et al., 1994) all indicate that the dendrites of neocortical pyramidal cells are capable of supporting action potentials (see Johnston et al. (1996) for a useful review). These studies demonstrate that the dendrites contain one type of Na^+ channel but several types of Ca^{++} channels, with the distribution of the Ca^{++} channels varying over the dendritic tree. Stuart and Sakmann (1994) showed quite clearly the role of Na^+ channels in facilitating active conduction from the soma up to the apical dendritic tuft. In this last study cell firing in the axon hillock was necessary before antidromic Na^+ spikes could be generated to carry the depolarization up into the apical tuft. Rapp et al. (1996) were able to simulate some aspects of these results by stipulating a particular distribution of Na^+ channels.

There are many types of voltage-gated Ca^{++} channels in the dendrites of pyramidal cells, according to studies exploring their responses to pharmacological agents (Johnston et al., 1996). However they can be grouped into two main categories. High voltage activated (HVA) channels require a depolarization from resting of around 20 mV before they will open, whereas low voltage activated (LVA) channels have a much lower threshold and will open with depolarizations of only a few millivolts from resting potential. HVA channels can be activated by the Na^+ action potentials back-propagating into the dendrites. The LVA channels can be activated by EPSPs and appear to be more involved in the forward propagation of activity initiated in the dendrites by synaptic activity. For example, Cauller and Connors (1992) concluded that conduction along the apical dendrite towards the soma was an active process involving Ca^{++} channels, with a trigger region in the apical nexus. They could not account for their measured soma responses without the assumption of this active conduction.

It was important that the pyramidal cell model should include active conduction capabilities but the unknowns about channel types, distribution and dynamics made it difficult to incorporate these active properties using Hodgkin–Huxley kinetics (Hodgkin and Huxley, 1952). It was felt that a reasonable alternative was to use an empirical approach based on published data from recent studies of neocortical pyramidal neurons. It has already been shown that the model used is a hybrid system in that it models the spatial and temporal spread of potential using a model of physiological processes, but models receptor behaviour using empirical equations. It was felt that active conduction could also be included empirically. If the axon hillock region is depolarized below threshold a simulated action potential waveform is generated in the soma:

$$V_{\text{ac.pot.}} = k_1 t \mathrm{e}^{(1-at)} - k_2 t \mathrm{e}^{(1-bt)}. \tag{13.13}$$

The values of the constants can be varied to produce a time course of depolarization that matches published data. For example the Na^+ spikes measured by Stuart and Sakmann (1994) close to the soma can be simulated using the following values: $k_1 = 150$, $k_2 = 10$, $a = 1500$, $b = 600$.

Equation (13.13) simulates the waveform at a given point. So to simulate the active propagation each compartment on the climbing apical dendrite is triggered in turn to follow the same potential time course, up to the apical nexus. In following the results of Stuart and Sakmann (1994) the peak amplitude of the waveform decreases from 90 mV at the soma to 60 mV at the apical nexus. The rate of conduction is also modelled on the measurements of the same authors. It is further assumed that there are trigger regions in the apical dendrites near to the apical nexus which activate Ca^{++} spikes when depolarized below threshold, and in a similar manner to the antidromic spikes, an action potential waveform is simulated and progressively transferred to the soma.

4. PREDICTED RESPONSE TO SIMPLE INPUT PATTERNS

4.1 Structure of Model

It was felt important that the neuron model was based on a real cell. The neuron used was illustrated by Cauller and Connors (1992) and was a layer 5 rat cortical pyramidal cell. The simplified geometry of the cell illustrated by the above authors was digitized as a two-dimensional image and divided up into 280 compartments. Each compartment was 20 μm in length except for the cell body which was pyramidal in shape, 50 μm in length and 8 μm across the base. The dendrites varied in diameter depending on how far away they were from the soma. Figure 13.3 summarises the dimensions used. As discussed above the membrane properties were adjusted for each compartment to take account of the extra membrane area offered by the synaptic spines. For external inputs into the spines a marked potential difference develops between the spine head and the dendrite, and consequently extra compartments were added to the model for each input. A total of 28 inputs were used during the explorations of the model behavior, 16 on the distal dendrites and 12 on the proximal ones. The dimensions of the spinal compartments were held fixed and are shown in Figure 13.3.

The compartmental model was implemented in 'C' language on a fast PC with two forms of output display. The main display consisted of a scale representation of the digitized neuron. Each of the compartments of the model were separately represented and were given false colors to represent the levels of depolarization of the compartment. The colors were arranged on a logarithmic scale and the values used for the color bands are shown on the calibration photo of Figure 13.4. So as the simulation was being run the spatial changes in membrane potential were illustrated in real time on the monitor. In this way the distribution of membrane potential could be monitored from the initial injection of current, until the cell had settled back to its resting state. With the computer system used, 20 ms of cellular activity would take about 1 min of display time. This rate was slow enough to see the details of the changes taking place but quick enough not to be frustrating. In addition to the display of potential distribution the user could call up the time course of potential changes at any point in the cell, and the peak values of potential change at that point. These two display techniques gave the user the bulk of the information needed to explore the response of the cell to inputs.

4.2 Behavior Following Simple Input Patterns

A simple spatial pattern of eight single inputs was presented to the receptor spines and the temporal changes of membrane potential were observed. Activity initiated

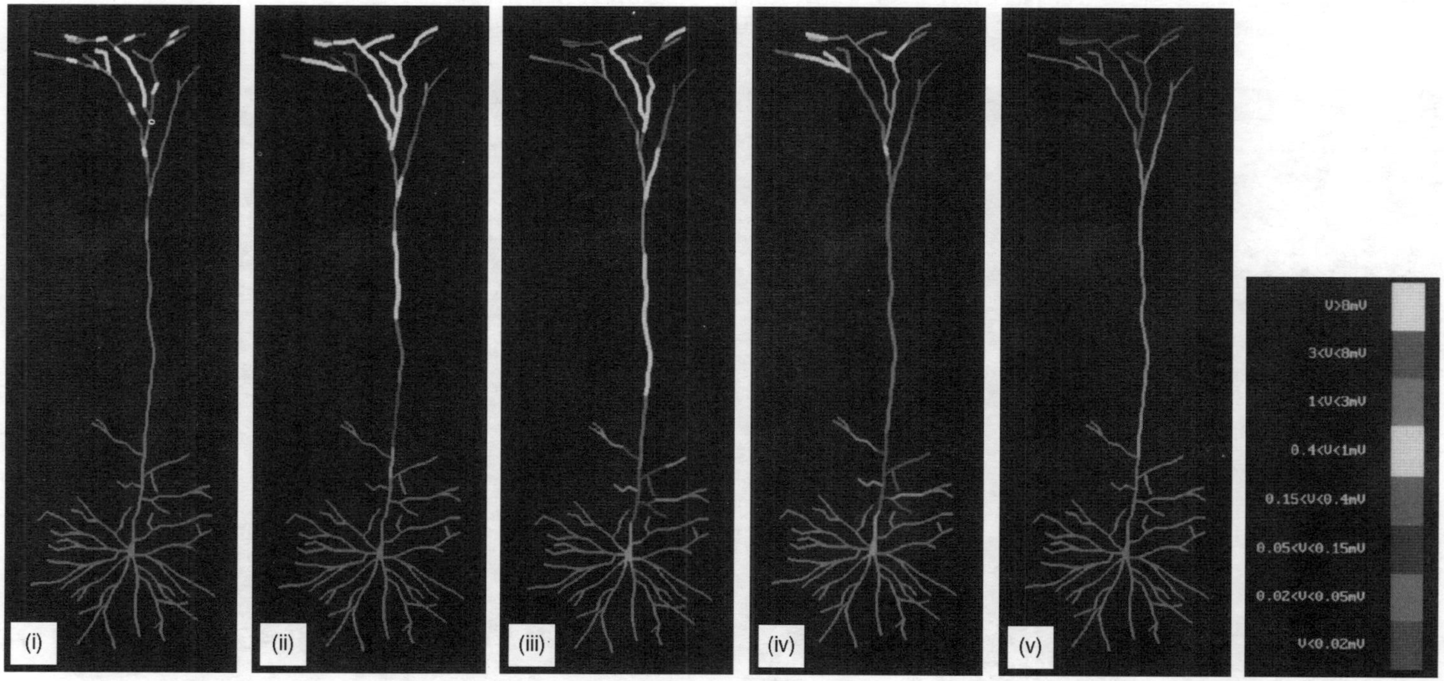

Figure 13.4(a) The distribution of depolarization over the pyramidal cell as a function of time. The colors show the distribution using the log scale illustrated in the calibration. (a) The distribution following eight inputs to the apical dendrites and only passive conduction: (i) 0.4 ms; (ii) 1.0 ms; (iii) 3.5 ms; (iv) 9.0 ms; (v) 18.0 ms. (Color plate II)

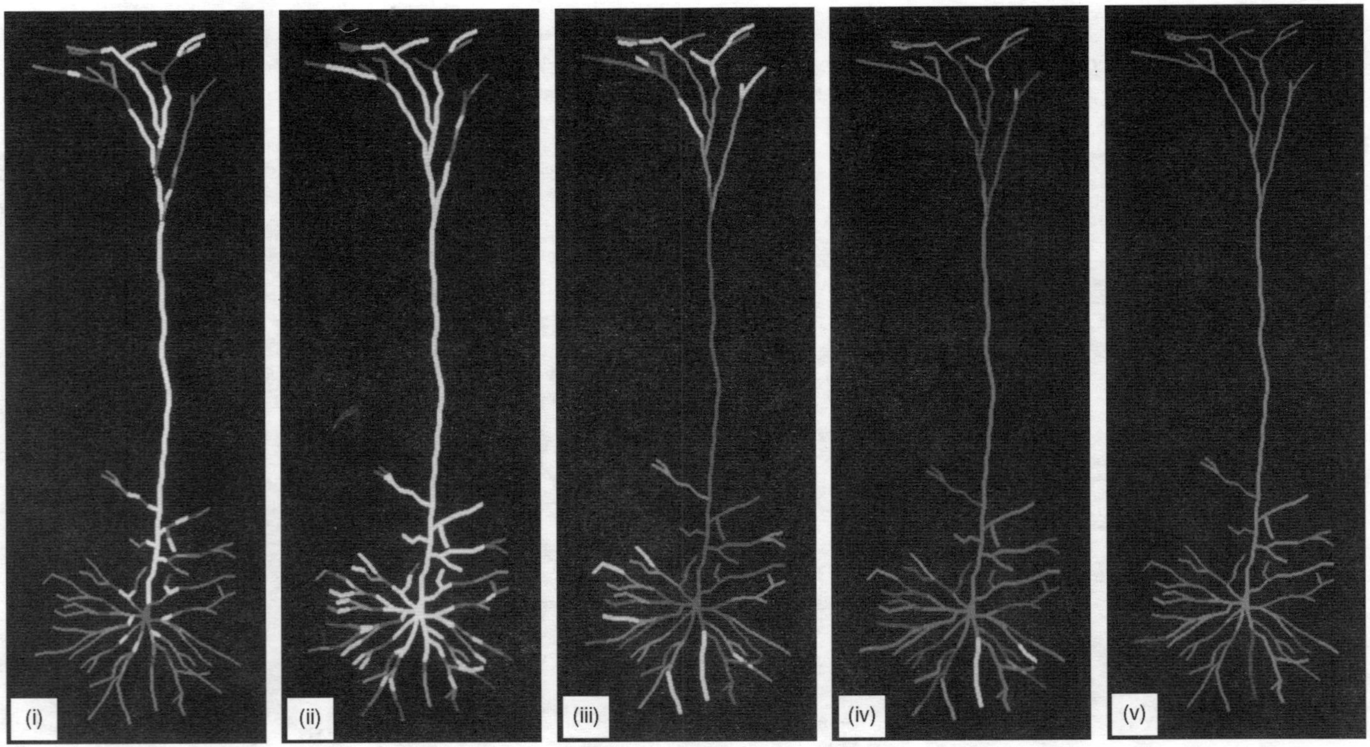

Figure 13.4(b) The distribution following eight inputs to the apical dendrites but with active conduction along the climbing apical dendrite: (i) 2.0 ms; (ii) 4.5 ms; (iii) 11.0 ms; (iv) 18.0 ms; (v) 22.0 ms. The threshold for initiation of active spikes at the apical nexus was set at only 5 mV to allow the small number of inputs to initiate firing. (Color plate III)

in the distal dendrites quickly spread around the dendritic tuft but caused very little activation along the climbing apical dendrite. Depolarizations of the soma were a few microvolts. Figure 13.4(a) shows a series of photos of the potential distribution following such a single input pattern to the apical dendrites. Figure 13.5(a) illustrates the time course of the depolarization achieved, both at one of the spines and at the soma. Similarly activation of the proximal dendrites caused extensive local activation, and depolarized the soma by up to 1 mV, but caused little depolarization up the climbing apical dendrite. These results agree with previous simulations run on simpler cell models (Orpwood, 1994) and with the simulations run by Cauller and Connors (1992), and follow the results of Stratford et al. (1989) and Holmes and Woody (1989). The peak values and time course obtained for soma depolarizations following activation of the proximal dendrites are close to those measured in hippocampal pyramidal cells (Barrionuevo et al., 1986; Anderson, 1990) and a time to peak similar to that measured in neocortical pyramidal cells (Artola et al., 1990). All these results give confidence as to the validity of the model and enable the intended work on pattern recognition to proceed.

4.3 Response to Active Conduction

With the active conduction algorithm in place there was much improved communication between the soma and basal dendrites, and the apical dendrites.

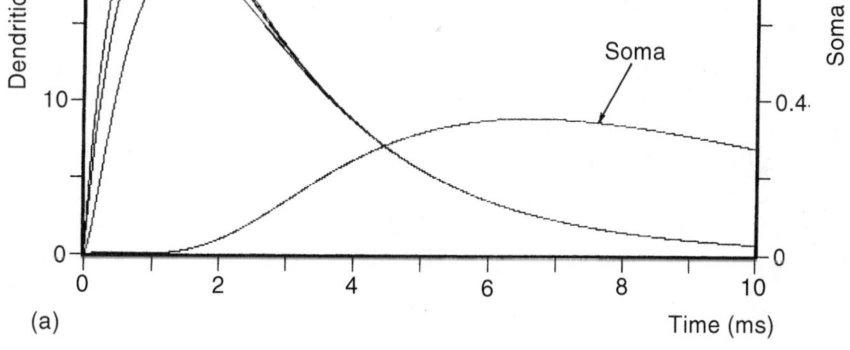

Figure 13.5 Time course of depolarization at several points on the cell following eight inputs to the apical dendrites. (a) Time course assuming passive conduction only (note different scales for soma and dendritic depolarizations). (b) Time course assuming active conduction along climbing apical dendrite.

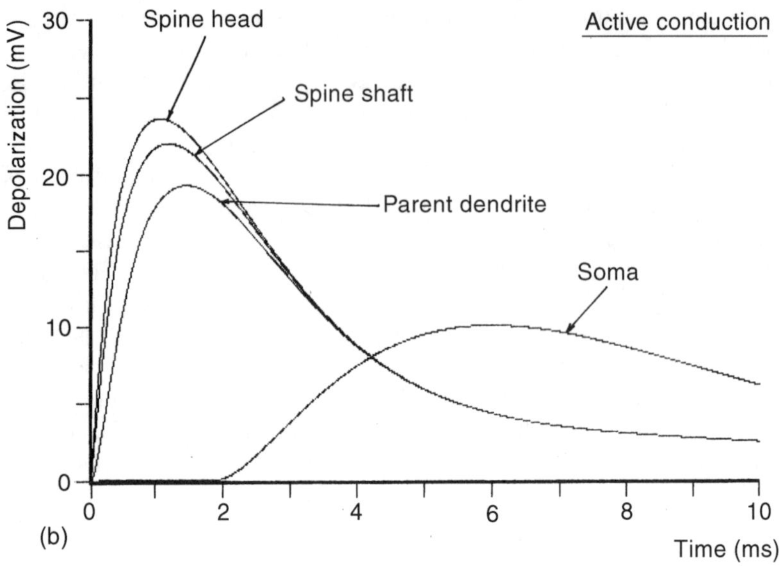

Figure 13.5 (Continued)

Figure 13.4(b) shows the distribution of depolarization at several times following the same input pattern as used in Figure 13.4(a). The active conduction algorithm was used and as can be seen the depolarization soon spread throughout the basal dendritic field. Figures 13.5(a) and 13.5(b) show the time course of soma depolarization following the pattern of eight inputs to the apical dendrites for both passive and active conduction in the climbing apical dendrite. There is no doubt that the incorporation of active conduction provides a much improved communication from the distal apical dendrites to the soma.

Active conduction along the climbing fibre in the antidromic direction was also explored. In the climbing dendrite model an active conduction algorithm was incorporated that could be triggered by firing of the cell. Even though active conduction was provided up to the apical nexus the passive spread from here into the apical dendrites was poor. Figure 13.6 plots the time course of depolarization for the positions in the apical dendrites shown in the diagram. As can be seen the depolarization is rapidly attenuated as it conducts passively into the dendrites. This result has much significance for the pattern associating capabilities of the cell, as is shown below.

4.4 Control of Apical Influence

The climbing apical dendrite is obviously an important channel of communication between the apical dendritic field and the proximal dendrites and the soma. It is also interestingly the site where much of the inhibitory input is received by the cell

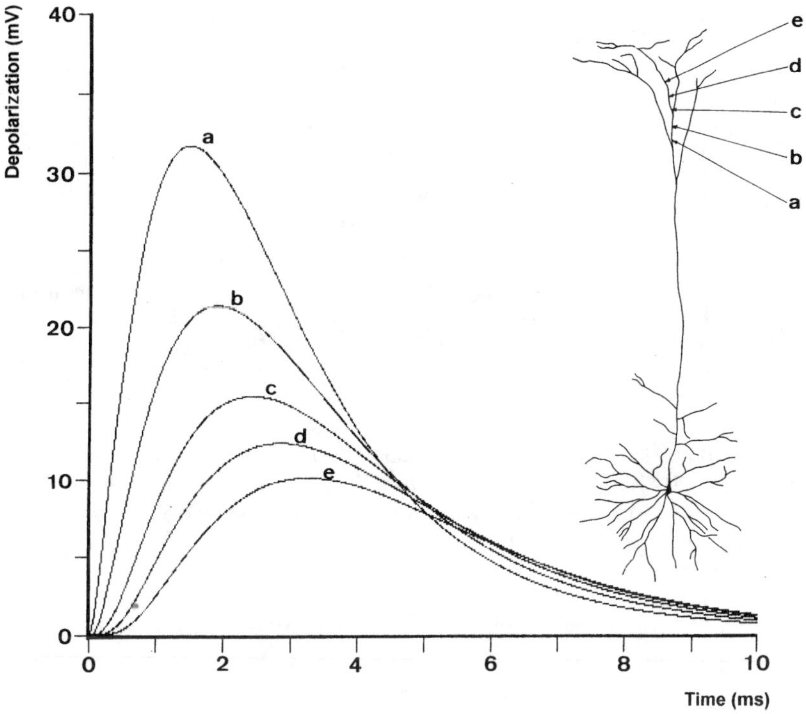

Figure 13.6 Time course of depolarization in the apical dendrites showing the rapid attenuation of the depolarization in the dendritic field where only passive conduction is taking place.

(Shepherd, 1979; Feldman, 1984). This inhibition seems to be ideally located for having a maximal effect on the communication between the two regions of the neuron. If conduction is blocked at the climbing fibre then the fact that input patterns have been recognised in the apical dendrites cannot be communicated to the soma.

The ability for inhibition of the climbing fibre to block the communication was explored with the pyramidal cell model. Inhibition was modelled by providing hyperpolarizing currents in the compartments of the climbing fibre. It was assumed the inhibition was being provided by Cl^- channels and a reversal potential for these inputs was set at -20 mV relative to resting potential with a peak conductance change of 1 nS. The inhibitory inputs were provided at the soma end of the climbing apical dendrite and the cell was provided with active conduction in this fibre.

The results are shown in Figure 13.7 where the soma depolarization following activity triggered at the apical nexus is plotted against time. The inhibition was very effective at reducing the depolarization of the soma. The figure shows the combined response of several tests where the inhibitory input is provided at different delays

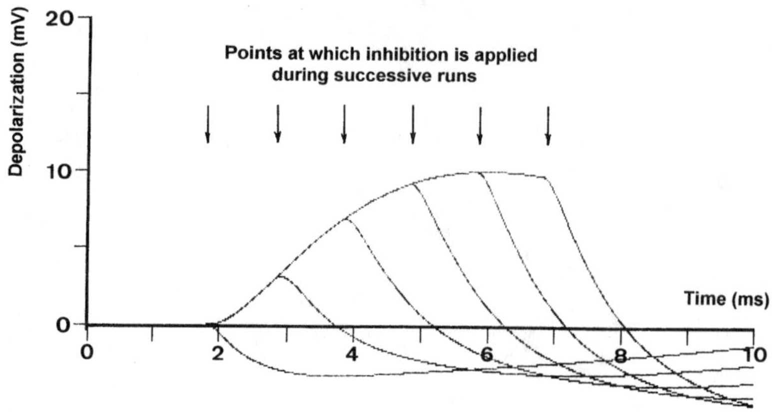

Figure 13.7 The effect on soma depolarization of an inhibitory input to the climbing apical dendrite. The graph shows six successive runs in which the inhibitory input is applied at different times after the arrival of an active spike from the apical dendrites.

relative to the arrival of the depolarizing transient. Even with quite large delays the activity of the soma is severely reduced once the inhibitory input is provided.

5. PREDICTED RESPONSE OF NEURON TO PAIRED PATTERNS: LEARNING IN THE DENDRITES

It was hoped that numerical models of pyramidal cells could provide an indication of the pattern associating capabilities of these neurons. It is felt in this work that even the higher manifestations of cortical function ultimately depend on the ability of cortical networks to process patterns (Orpwood, 1994). However it is also felt that the key to the ability of cortical networks to carry out their information processing lies with the pattern processing capabilities of their individual cells. A few studies have been published on the capabilities of single cortical cells to deal with patterns of information (Orpwood, 1992; Mel, 1992).

For one pattern of inputs to be associated with another, the first pattern must cause a large depolarization in the vicinity of the second. The large local depolarization can be generated in two ways:

1. The depolarizations could be caused by activity in spines adjacent to the one in question. If the dendrites received a spatial pattern of inputs which activated a number of adjacent receptors then the local membrane depolarization would be quite large and would spread to spines in the vicinity.
2. The large depolarization could also arise because the cell has fired. The action potential generated would propagate down the cell's axon but the large depolarization generated in the soma would also transmit antidromically back into the dendrites, causing depolarizations there.

Investigations were carried out to explore the impact on the pyramidal model of both the sources of large depolarization described above. The initial study explored the impact of one pattern of strong inputs on another pattern of weaker ones. The weaker pattern (pattern 'A') consisted of six inputs presented to the same receptors each time. The weaker pattern was always presented together with another six weak inputs chosen at random from the potential 72 receptors in the apical dendrites. So the initial ten runs used inputs to pattern 'A' together with another six random inputs to find a control response for the weak input. The mean response at the apical nexus and its standard deviation are shown in Figure 13.8.

The study was then repeated using the pattern of six strong inputs (pattern 'B') instead of pattern 'A'. Each of the strong inputs was presented to a receptor with a peak conductance of 3 nS. The strong inputs were always presented together with another six weak inputs chosen at random. Another series of ten runs were carried

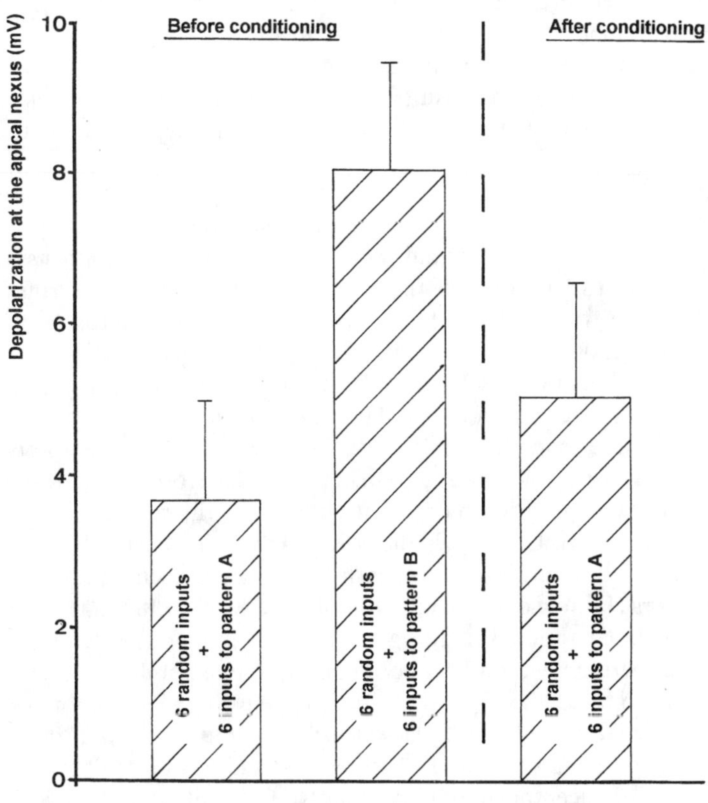

Figure 13.8 The results of the conditioning studies. The mean depolarization achieved at the apical nexus is shown before and after conditioning. The conditioning involved simultaneous exposure to pairs of input patterns, one to a set of weak receptors (pattern A) and one including strong receptors (pattern B), as described in the text.

out and the response at the apical nexus monitored. Figure 13.8 again shows the mean response and standard deviation. As expected the pattern of inputs that included the six strong inputs generated a much larger response.

Following these control studies a process of association was carried out in which weak pattern 'A' was presented together with strong pattern 'B'. Following the mechanisms underlying changes in receptor sensitivity described above, each coincident presentation of the two patterns led to an increase in the sensitivity of the weaker receptors. The coincident presentations were repeated seven times, after which the impact of the association on the cell's response was explored. The initial series of runs was repeated, with pattern 'A' presented together with six random inputs. Because of the change in the sensitivity of the receptors activated by pattern 'A', the response at the apical nexus was increased. Figure 13.8 again illustrates the mean response and standard deviation.

There is no doubt that the apical dendritic region is very able to associate two separate patterns provided at the same time. The depolarization of the apical nexus following only seven presentations of the two patterns shows a very significant increase. The number of receptors activated in these tests is very small and so the resulting depolarizations of the nexus are not large. However with input numbers at a more physiological level the results of the association process illustrated above could make the difference between an input pattern causing active conduction to the soma or not.

The second series of tests used cell firing as a conditioning stimulus. In these tests the cell was fired at the same time as the apical dendrites were presented with pattern 'A', starting with the original receptor sensitivities. These tests also showed an enhancement of the response of the apical dendrites to the set of input patterns. However cell firing was far less effective than the paired patterns used in the first series. The attenuation of depolarization during passive conduction into the apical dendrites that was illustrated in Figure 13.6 means that even active conduction along the climbing fibre causes minimal depolarization of the apical region. Consequently the local membrane potential in the region of the apical receptors was only marginally depolarized following cell firing and therefore much less effective at causing an enhancement in receptor sensitivity, unless the receptor was very close to the apical nexus. Figure 13.9 shows the mean change in sensitivity of the receptors following each conditioning event, for both conditioning techniques.

Several authors (Bunow et al., 1985; Gamble and Koch, 1987) have commented on the fact that depolarization spreading from the soma to the dendrites is rapidly attenuated and therefore would be a poor conditioning stimulus. It has been felt that this result compromised the view that cortical neurons have Hebbian capabilities, because they would be unable to associate pre- and post-synaptic activity and strengthen subsequent responses. It has been suggested that alternative feedback from cell firing to receptors might be involved, possibly biochemical in nature. The results demonstrated here support the difficulty of using cell firing as a conditioning stimulus but illustrate the fact that other patterns of inputs received at the same time can provide a very effective conditioning stimulus.

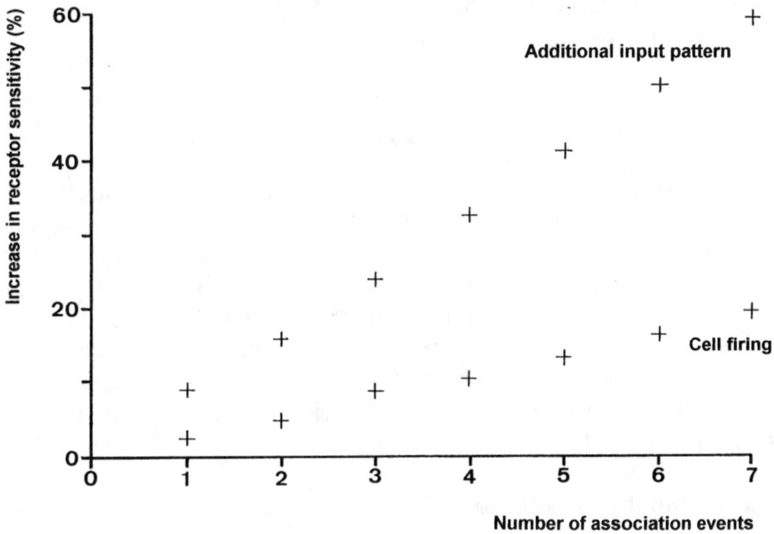

Figure 13.9 Graph showing the mean change in receptor sensitivity for eight receptors following different numbers of association events. When conditioned by cell firing a modest increase in sensitivity results. A larger increase results from using an additional input pattern as the conditioning event.

It would appear therefore that a pattern of inputs received by the apical dendrites is more likely to be associated with another pattern to that region than to the firing of the cell. This is felt to be an important result for the pattern processing capabilities of individual pyramidal cells and also for the pattern processing capabilities of networks of those cells. The result could only easily be demonstrated with models of the kind described.

6. A REDUCED PYRAMIDAL CELL MODEL

6.1 Need for Simple Model

The behavior described above has many implications for the pattern processing capabilities of networks of such pyramidal cells. Feedforward patterns can be recognised by the proximal dendrites and that fact passed onto subsequent layers of cells. Feedback patterns can be recognised by the apical dendrites, and that fact passed onto the cell body but in a conditional way dependant on the level of inhibition in the climbing apical dendrite. Unfortunately large-scale compartmental models such as the one described are fairly computationally intensive and if simple networks are to be explored using such models they would become unacceptably slow if more than a few cells were involved. In order for such network studies to be carried out using reasonably representative neuron models a simplification of the cell model is desired. The subsequent part of this work was therefore to generate such a model with the aim of emulating the behaviour of the full-scale compartmental system.

6.2 Key Features Indicated by Comprehensive Model

The studies described above illustrate the key features that enable the pyramidal neuron to process information. The cell appears to have two separate integrating regions; the apical dendrites at a long distance from the soma, and the proximal and basal dendrites arranged around the cell body. The same conclusion was reached by Cauller and Connors (1992). Both the dendritic regions are able quite independently of each other to recognise patterns amongst their inputs and to learn new patterns through a process of association. The association learning is influenced by cell firing but to a lesser degree. From these studies the firing of the cell seems to be primarily a means of signalling threshold events to other neurons rather than part of the process of internal cell learning. The two cellular regions are linked by the climbing apical dendrite and it was assumed in these studies that active conduction could take place along this dendrite. The role of the dendrite is seen as providing communication between the two regions of the cell and inhibition to this dendrite was very effective in interrupting the channel of communication. The reduced model needed to incorporate these key features.

6.3 Structure of Simple Model

The model that was developed is shown in Figure 13.10. It basically consists of two toroidal structures, representing the apical and proximal dendrites, linked by a channel of communication representing the climbing apical dendrite. For each of the dendritic structures the receptors were arranged over a two-dimensional grid folded to provide a continuous surface in the form of a toroid. Each of the receptor nodes on the grid provided a local depolarization when activated, modelled as a step change in local potential followed by an exponential decay:

$$V = V_0 e^{((t/\Delta t)\ln(\phi))}, \tag{13.14}$$

where V_0 is the step change in local potential at $t=0$ following an input and ϕ the fractional change at each Δt time step.

The exponential decay (13.14) modelled computationally using shift operators. For example if the fractional change at each time step was 0.5 then a right shift of a one bit divides the operand, V, by two at each computational step, and gave a very effective model of the local potential decay which was computationally very fast. Different fractional changes can easily be modelled in a similar fashion (see Appendix 13.1 for a discussion of the use of shift operators). The computational time steps were chosen to allow the time course of the decrement achieved in this way to match that obtained in the full compartmental model. Steps of 1 ms were chosen.

The potentials at each receptor point were communicated to the other receptor points in their vicinity. Several schemes were explored but the one finally chosen is

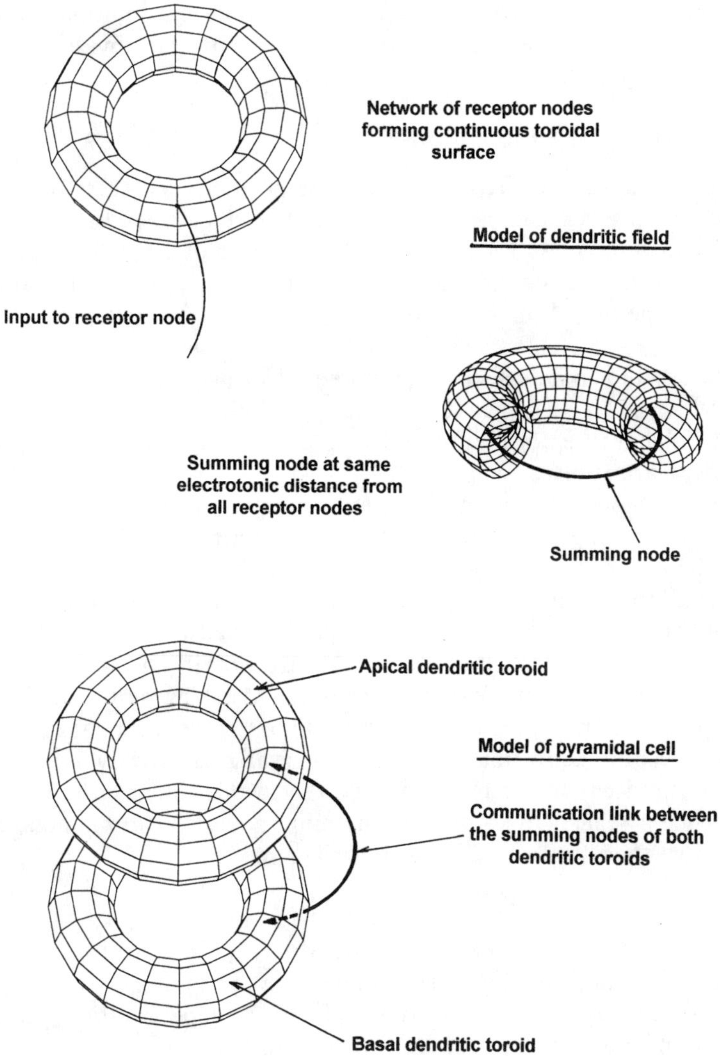

Figure 13.10 Structure of simple pyramidal cell model, showing the toroidal dendritic field and the way two toroids can be linked to provide a complete cell model.

as below:

$$V_{i,j}^{t+1}=(4V_{i,j}^{t}+V_{i,j+1}^{t}+V_{i,j-1}^{t}+V_{i+1,j}^{t}+V_{i-1,j}^{t})/8$$

or using shift operators

$$V_{i,j}^{t+1}=(V_{i,j}^{t}\ll 2+V_{i,j+1}^{t}+V_{i,j-1}^{t}+V_{i+1,j}^{t}+V_{i-1,j}^{t})\gg 3,$$

where $\ll 2$ signifies a left shift of the operand by two bits (i.e. multiplied by 2^2) and $\gg 3$ signifies a right shift of the operand by three bits (i.e. divided by 2^3). See Appendix 13.1 for more details.

6.4 Basic Behavior of Simple Model

An input event received at any of the receptor positions led therefore to a local depolarization that spread across the sheet of dendritic receptors and ultimately decayed away to zero. Figure 13.11 shows the response of a small grid to a single input event. The response is quite crude because of the large computational time step but nevertheless the overall time course is a reasonable representation of the events in the real cell.

Each of the receptor points also incorporated a learning algorithm which ensured that if an input was received when it was locally depolarized below a threshold, then the sensitivity of that receptor point was increased and any subsequent inputs led to larger local depolarizations.

The model also assumed each of the receptor points was at the same electrotonic distance from a summing node. For the proximal dendritic toroid the summing node represented the soma, or more accurately the axon hillock region. For the apical dendritic toroid the summing node represented the apical nexus. As with the physiological neuron model, the summing nodes received the activations generated by the receptors and then leaked the summed activation in an exponential fashion. The decay of depolarization at the summing nodes was achieved by using a fractional change of depolarization at each computational time step, in a similar manner to the decay of the receptor activation. The time interval had already been chosen to provide a realistic decay at the receptor sites. Given this time step the relationship found to provide a realistic decay at the summing node, and to maintain stability, is as below. Note the use of the shift operator again:

$$V_{\text{sum}}^{t+1} = (V_{\text{sum}}^{t} + \sum V_{i,j}^{t}) \gg 2.$$

Figure 13.12 shows the time course of depolarization at a receptor node in the apical toroid and at the basal summing node (representing the soma). A very good match is achieved between this much simplified model and the depolarizations produced by the full compartmental model (compare with Figure 13.5(b)).

6.5 Adaptive Response

A series of three inputs were provided to low sensitivity receptor nodes in the apical toroidal net. At the same time another set of three inputs were provided to receptor nodes that had a sensitivity three times as large. Figure 13.13 shows the activity generated at one receptor following different numbers of these pairings, and as can be seen the response was rapidly enhanced.

The simplified pyramidal cell was exposed to similar procedures to the compartmental model. A series of six low sensitivity inputs was provided to the apical toroid, three random and three to a set pattern. Ten sets of inputs were used and the response of the summing node was monitored. The three random inputs were

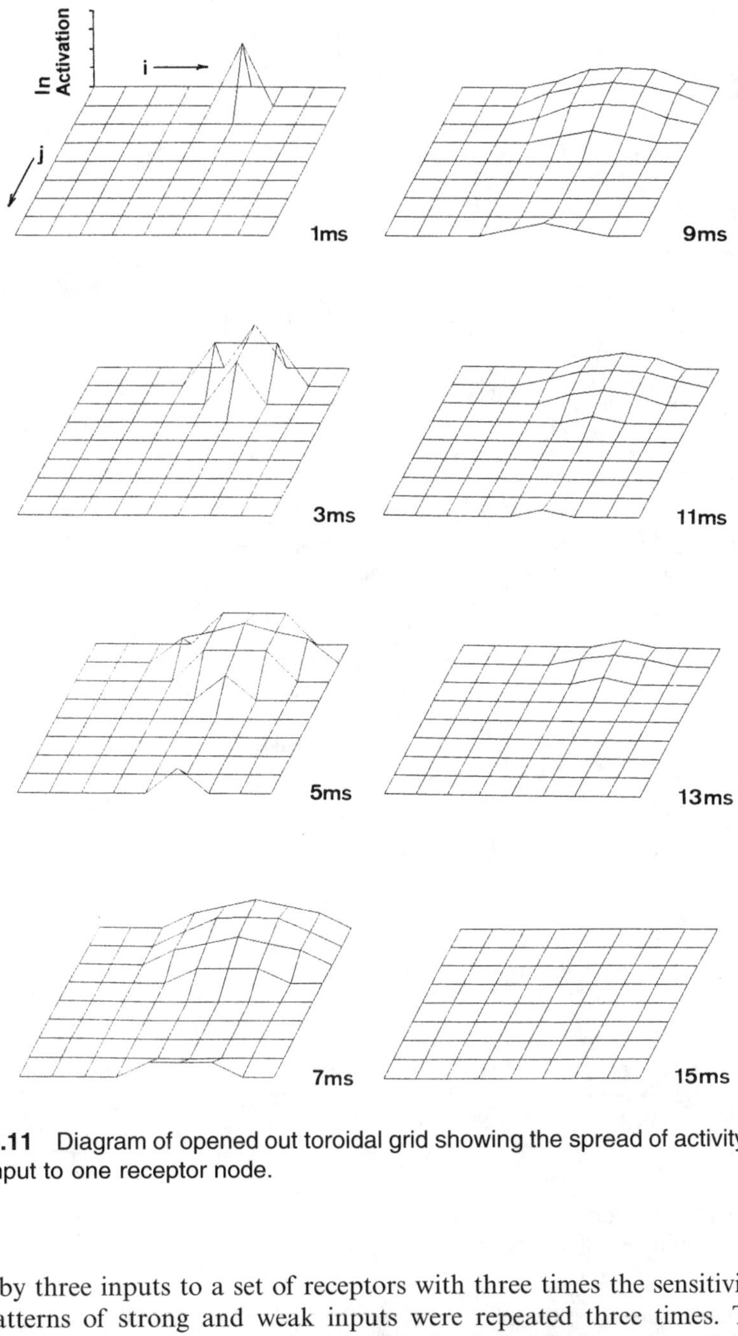

Figure 13.11 Diagram of opened out toroidal grid showing the spread of activity following a single input to one receptor node.

replaced by three inputs to a set of receptors with three times the sensitivity. These paired patterns of strong and weak inputs were repeated three times. The three strong inputs were then replaced by three random ones again and another ten sets of responses were monitored at the summing node. Figure 13.14 compares the responses and as can be seen the association has led to an enhanced response.

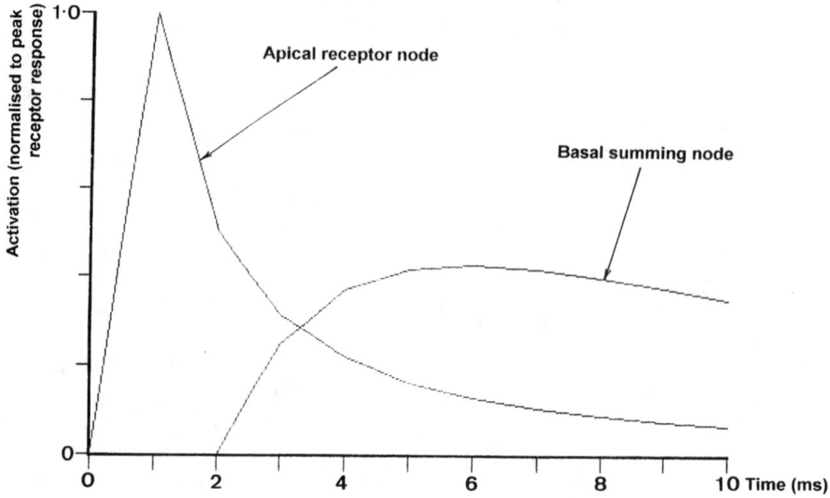

Figure 13.12 Time course of activity for the apical receptor node and the basal summing node (soma) using the simple neuron model.

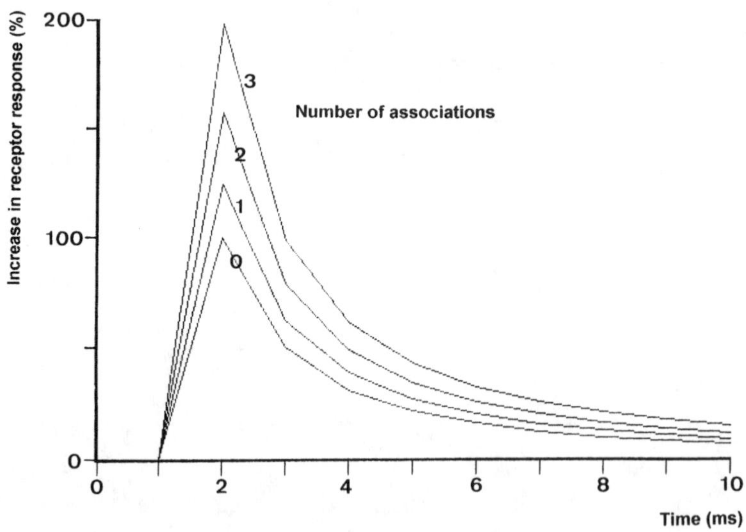

Figure 13.13 Response of one receptor node in the simple neuron model following differing numbers of pattern associations.

The two-dimensional sheet of receptors therefore formed a very effective pattern recognising and learning system. As with the compartmental model the sensitivity of the receptor regions increased and after the conditioning the same patterns of inputs generated a much larger response at the apical nexus.

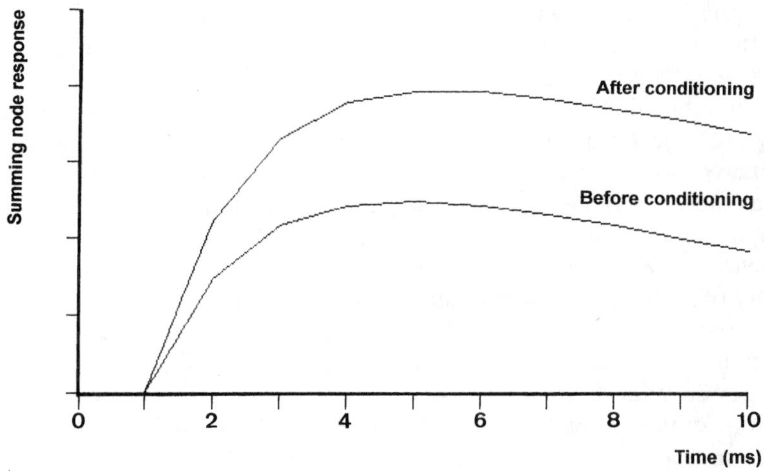

Figure 13.14 The mean response at the apical summing node for ten sets of eight inputs. Activity is shown both before conditioning and after three pattern associations.

6.6 Computation Speeds

It is felt that the simplified pyramidal cell model provides a behavior that emulates that of the full pyramidal cell model quite effectively. It demonstrates the key properties demonstrated by the full model but at a much reduced cost in terms of computing time. For 10 ms of cell activity in the full model the simulation took 25 s with the computer system used. With the same computer system and for a cell with the same number of receptor channels the 10 ms of activity took only 2.5 ms with the simplified model, i.e. four times faster than a real neuron of the same level of complexity. The speeds achieved enable exploration of large networks of such cells to be carried out in the secure knowledge that the individual pattern processing elements used are quite reasonable representations of the pyramidal cells involved in biological networks.

7. DISCUSSION

The aim of the work described in this chapter was to provide a building block that would allow more accurate predictions of cortical network behavior. Much information about cortical structure and properties is now available, but the links between this knowledge and the higher level capabilities of the cortex are difficult to predict. Simulation studies have an important role to play in exploring these links but the network models must include realistic simulations of neuron behavior. As is so obviously indicated by neurobiological studies, the behavior of cortical neurons is very sophisticated. If the predictions of network models are to be believed it is essential that their component neuron models reflect the known properties of neuron behavior. Comprehensive neuron models such as that described in this chapter would require too much computing time to form the basis of a large network model and for this reason the reduced model was developed. Therefore the ultimate aim of the work described is to enable the exploration of cortical networks, but for this to be done in a meaningful way the comprehensive model of the pyramidal cell has to be shown to reflect biological behavior, and the reduced model has to be shown to reflect the core properties of the more comprehensive one. Only in this way can large-scale network models can be generated with confidence.

A major theme of the chapter concerned interactions of activity in different parts of the neuron to enable associative memories to be established. The work predicted the ability of the cell to associate one pattern of input activity with another, and to associate patterns of activity with cell firing. Recent neurobiological work has demonstrated very clearly these association processes taking place. Markram et al. (1997) showed that back-propagating action potentials can cause a long-term potentiation of the response of neocortical cells to a physiological synaptic input. The synaptic input and the action potential both needed to occur within 10 ms of each other for the association to occur and an LTP to take place, and the mechanism involved the NMDA glutamate receptor. This overall mechanism is very similar indeed to that demonstrated by the neuron model described in this chapter. Magee and Johnston (1997) demonstrated a similar role for back-propagating action potentials in causing LTP in hippocampal pyramidal cells. A major prediction of the modeling work has been confirmed by these experiments.

The recent neurobiological work on back-propagating action potentials (Markram et al., 1997; Magee and Johnston, 1997) demonstrates an association between a pattern of inputs and cell firing. The simulations described in the chapter predict this response, but also indicate that a larger association response could arise from one pattern of dendritic inputs associating with another. The reason cell firing was less effective was because back-propagating action potentials did not spread far into the dendritic tree. This poor spread of action potentials was based on experimental data (Stuart and Sakmann, 1994). However the recent work described by Hoffman et al. (1997) shows a much greater and more controlled spread of activity into the dendrites. Their work describes the role of A-type K^+ channels in hippocampal pyramidal cells. They demonstrate that K^+ channels dampen the Na^+ transients, but that this effect could be overridden by local EPSPs. This inactivation of K^+ channels following strong inputs would enable Na^+ action potentials to locally spread much further into the dendrites and could lead to a pronounced association response.

It is clear that active propagation within the pyramidal cell has a profound impact on internal communication. Information arising from activity in the dendritic fields can be communicated actively to the soma, probably though the LVA Ca^{++} channels. Information about the firing of the cell can be communicated back to the dendritic areas through back propagating action potentials using Na^+ channels. In the comprehensive model discussed in this chapter these features were included empirically, using the published results of several authors. It was argued that current understanding about channel types and their distribution was not sufficiently clear to enable their properties to be modelled using Hodgkin–Huxley dynamics, and that empirical techniques were therefore more reliable. This situation is rapidly changing however as more detailed studies of these cells are published. The interaction between Ca^{++} transients and Na^{+} channel activity was demonstrated very clearly *in vivo* by Svoboda et al. (1997). Hoffman et al. (1997) also provided much insight into the interaction between various channel types. The density of A-type K^{+} channels was shown to increase further away from the soma where they were recorded. This distribution was shown to provide an explanation for the soma site of Na^{+} action potential initiation, and for the reducing amplitude of action potentials as they back-propagate into the dendrites. The changing spike amplitude measured during a train of back-propagating action potentials (Callaway and Ross, 1995) was shown by Jung et al. (1997) and Colbert et al. (1997) to be due to an inactivation of the sparse Na^{+} channels in the dendrites. These kinds of data should enable full physiological models of channel activity to be included in future pyramidal cell models.

New experimental data is becoming available all the time and it is important to refine pyramidal models in the light of this new information. The current model places much emphasis on active propagation within the dendrites, and recently Stuart et al (1997) have used multi-patch pipette recordings to explore in much detail the sites of action potential initiation in mature pyramidal cells. They showed Na^{+} action potential initiation in the axon at least 30 μm from the soma. Their work also demonstrated active conduction through Ca^{++} spikes within the distal apical dendrites, but only passive conduction to the soma along the climbing fibre. Using Ca^{++} fluorescence measurements, Schiller et al. (1997) confirmed these results, showing marked Ca^{++} influx in the distal apical dendrites via high threshold channels but little activity along the climbing fibre. This behavior is different to previous assumptions but could easily be incorporated in the full compartmental model described to see what impact it had on communication within the cell.

Modulatory influences could add another dimension to the pattern associating capabilities of pyramidal cells, acting over longer time scales. Hoffman et al. (1997) showed the activity of K^{+} channels to be much influenced by numerous neurotransmitters and other agents. Tsubokawa and Ross (1997) have demonstrated that back-propagating trains of spikes can be influenced by muscarinic receptors in the climbing fibre. A long-standing inclusion in many neural network models that the mean output of a cell remains constant despite changes in input sensitivity has been shown by Turrigiano et al. (1998) to have a biological foundation, at least over 48 hour time scales.

It has already been argued that one of the main roles for models such as those described in this chapter is to provide an extension of our cognitive capabilities. They extend our limited abilities to mentally model the outcome of various ideas and hypotheses. It is encouraging that this role is becoming increasingly recognised even in neurobiological studies. More authors are complimenting such studies with models of neurons or parts of neurons. The models are used in an iterative fashion to suggest further avenues for experiment, or to show the implications of the results obtained. For example Hoffman et al. (1997) demonstrated that blockage of A-type K^{+} channels prevented the amplitude of back-propagating action potentials from decreasing. Using a limited neuron model these authors went on to show that this result is what would be expected. The decrease in action potential amplitude was caused by the larger number of K^+ channels at a distance from the soma, rather than changes in the Na^+ channel properties, as had previously been suggested (Rapp et al., 1996; Mainen et al., 1995).

8. CONCLUSIONS AND FUTURE PERSPECTIVES

This chapter has described some modeling work applied to the most important information processing cell in the cerebral cortex; the pyramidal cell. It has used a straightforward compartmental technique to explore the potential distributions generated over the surface of the cell following input patterns. It has used a hybrid construction where both the receptor dynamics and learning, and the active conduction within the cell, are simply modelled in an empirical fashion. The model is able to demonstrate pattern recognition on the part of the cell. It is also able to demonstrate the ability of the cell to increase its repertoire of learnt patterns through pattern association. Communication between the apical dendrites and the cell body was shown to require active conduction along the climbing apical fibre, and for this region to be ideally placed to allow inhibitory inputs to provide effective control of the communication. Finally a much simplified version of the cell model was described that could demonstrate the same basic behaviour but in a form that increased the speed of computing to a point where large networks of such cells could be efficiently modelled.

Future developments of pyramidal cell models depend on a number of gaps being filled in neurobiological data. Values for membrane resistance are still far from being settled. Information about channel dynamics and distribution inhibits the development of accurate physiologically based models, although as mentioned in the discussion, this problem is rapidly diminishing. And, of course, any properties which are uncovered by cell models need to be verified by experiment before they can be assumed in larger scale models. It would be useful, for example, to have a confirmation of the ability of the cell to associate one pattern of inputs with another. Similarly, a more direct demonstration of the nature of the communication between the apical dendritic tuft and the soma would add confidence to the use of pyramidal cell models. The results of Stuart et al. (1997) described in the discussion are useful indications of the kind of detail about this communication that should be forthcoming. Further modeling refinements probably would involve longer timescale events such as the neuromodulatory effects of some transmitters, and the link between second messenger generation the laying down of changes in synapse sensitivity.

So the basic accuracy of pyramidal cell models is probably close enough for their incorporation in larger scale models. Preliminary work using a cortical network of 400 of the reduced cell models described in this chapter, together with 100 interneurons, has shown interesting attractor behaviour and the ability to generalize categories of input patterns (Orpwood, 1996). It is hoped that the basic pattern processing capabilities of individual pyramidal cells can be used to explore the more sophisticated pattern processing capabilities of networks of such cells, and of hierarchies of such networks. The ultimate aim of such simulations is to attempt to shed some light on the profound high-level capabilities of the most challenging of all biological organs, the cerebral cortex.

PROBLEMS

13.1. Stuart et al. (1997) have demonstrated active propagation within the apical dendritic tuft that is passively conducted down the climbing fibre. What impact might this be expected to have on the behaviour of pyramidal cells?

13.2. Back-propagating action potentials cause both a depolarization of the apical dendritic membrane and an influx of Ca^{++}. The model explores the effect of depolarizations on local changes in receptor sensitivity. How might the changes in Ca^{++} concentration affect sensitivity?

13.3. Use the numerical equations provided in the chapter and the data on K^+ channels published by Hoffman et al (1997) to construct a limited compartmental model of the climbing apical dendrite, and incorporate Hodgkin–Huxley dynamics for the Na^+ and K^+ channels. What distribution of K^+ channels is necessary to emulate the decrement in back-propagating action potentials measured by Stuart and Sakmann (1994)? Can the reducing amplitude of a train of spikes, as shown by Callaway & Ross (1995), be simulated using the Na^+ channel inactivation that is discussed in Jung et al (1997)?

13.4. ***Experimental suggestion.*** Use two separate inputs to the distal apical dendrites of a layer V pyramidal neuron via layer I/II stimulation, without firing the cell. Is one input able to cause long-term changes to the EPSPs generated by the other, and thus demonstrate pattern association in the distal dendrites without cell firing?

13.5. How could the observations of Turrigiano et al. (1998) about the total synaptic strength of a neuron be incorporated in the reduced cell model described in the chapter? Develop appropriate algorithms that would enable the model neuron to behave in this manner.

13.6. Using the principles outlined in section 6, construct a model of an inhibitory interneuron that could be used along side pyramidal cells in a model of a cortical neural network.

REFERENCES

Anderson, P. (1990) Synaptic integration in hippocampal CA1 pyramids. In *Progress in Brain Research*, edited by J. Storm-Mathisen et al., vol. 83, pp. 215–222. Amsterdam: Elsevier.

Artola, A., Brocher, S. and Singer, W. (1990) An *in vitro* model of plasticity in neocortical slices. In *Biology of Memory*, edited by L.R. Squire and E. Lindenlaub, pp. 319–329. Stuttgart: Schattauer Velag.

Baer, S.M. and Rinzel, J. (1991) Propagation of dendritic spikes mediated by excitable spines: A continuum theory. *J. Neurophysiol.*, **65**, 874–890.

Barrionuevo, G., Kelso, S.R., Johnson, D. and Brown, T.H. (1986) Conductance mechanisms responsible for long term potentiation in monosynaptic and isolated excitatory synaptic inputs to hippocampus. *J. Neurophysiol.*, **55**, 540–550.

Baudry, M. and Davis, J.L. (1991) Long-term Potentiation: A Debate of Current Issues. Cambridge, MA: MIT Press.

Bekkers, J.M. and Stevens, C.F. (1989) NMDA and non-NMDA receptors are co-localised at individual excitatory synapses in cultured rat hippocampus. *Nature*, **341**, 230–233.

Bliss, T.V. and Collingridge, G.L. (1993) A synaptic model of memory: Long-term potentiation in the hippocampus. *Nature*, **361**, 31–39.

Brown, D.A., Gahwiler, B.H., Griffith, W.H. and Halliwell, J.V. (1990) Membrane currents in hippocampal neurons. In *Progress in Brain Research*, edited by J. Storm-Mathisen et al., vol. 83, pp. 141–160. Amsterdam: Elsevier.

Brown, T.H., Chang, V.C., Ganong, A.H., Keenan, C.L. and Kelso, S.R. (1988) Biophysical properties of dendrites and spines that may control the induction and expression of long-term potentiation. In *Long-term Potentiation: From Biophysics to Behaviour*, edited by S.A. Deadwyler and P.W. Landfield, pp. 197–260. New York: Liss.

Bunow, B., Segev, I. and Fleshman, J.W. (1985) Modelling the electrical properties of anatomically complex neurons using a network analysis program: Excitable membrane. *Biol. Cyber.*, **53**, 41–56.

Bush, P.C. and Sejnowski, T.J. (1993) Reduced compartmental models of neocortical pyramidal cells. *J. Neurosci. Meth.*, **46**, 159–166.

Callaway, J.C. and Ross, W.N. (1995) Frequency dependent propagation of sodium action potentials in dendrites of hippocampal CA1 pyramidal neurons. *J. Neurophysiol.*, **74**, 1395–1403.

Carnevale, N.T. and Lebeda, F.J. (1987) Numerical analysis of electrotonus in multicompartmental neuron models. *J. Neurosci. Meth.*, **19**, 69–87.

Cauller, L.J. and Connors, B.W. (1992) Functions of very distal dendrites: Experimental and computational studies of layer 1 synapses on neocortical pyramidal cells. In *Single Neuron Computation*, edited by T. McKenna et al., pp. 199–229. San Diego: Academic Press.

Colbert, M.C., Magee, J.C., Hoffman, D.A. and Johnston, D. (1997) Slow recovery from inactivation of Na^+ channels underlies the activity-dependent attenuation of dendritic action potentials in hippocampal CA1 pyramidal neurons. *J. Neurosci.*, **17**, 6512–6521.

Cole, K.S. (1972) *Membranes, Ions and Impulses*. Berkley: University of California Press.

Creutzfeldt, O.D. (1995) Cortex Cerebri: Performance, Structural and Functional Organization of the Cortex. Oxford: Oxford University Press.

Dodge, F.A. and Cooley, J.W. (1973) Action potentials of the motorneuron. *IBM J. Res. Dev.*, **17**, 219–229.

Edelman, G.M. (1987) Neural Darwinism: The Theory of Neuronal Group Selection. New York: Basic Books.

Feldman, M.L. (1984) Morphology of the neocortical pyramidal neuron. In *Cerebral Cortex*, edited by A. Peters and E.G. Jones, vol. 1, pp. 123–200. New York: Plenum Press.

Gamble, E. and Koch, C. (1987) The dynamics of free calcium in dendritic spines in response to repetitive input. *Science*, **236**, 1311–1315.

Hestrin, S., Nicoll, R.A., Perkel, D.J. and Sah, P. (1990) Analysis of excitatory synaptic action in pyramidal cells using whole-cell recording from rat hippocampal slices. *J. Physiol.*, **422**, 203–255.

Hille, B. (1992) *Ionic Channels of Excitable Membranes*. 2nd edition Sunderland, MA: Sinauer.

Hirsch, J.A., Alonso, J. and Clay Reid, R. (1995) Visually evoked calcium action potentials in cat striate cortex. *Nature*, **378**, 612–616.

Hodgkin, A.L. and Huxley, A.F. (1952) A quantitative description of membrane current and its application to conduction and excitation in nerve. *J. Physiol.* (*Lond.*), **117**, 500–544.

Hoffman, D.A., Magee, J.C., Colbert, C.M. and Johnston, D. (1997) K^+ channel regulation of signal propagation in dendrites of hippocampal pyramidal cells. *Nature*, **387**, 869–875.

Holmes, R.W. (1989) The role of dendritic diameter in maximizing the effectiveness of synaptic inputs. *Brain Res.*, **478**, 127–137.

Holmes, W.R. and Woody, C.D. (1989) Effects of uniform and non-uniform synaptic activation distributions on the cable properties of modelled cortical pyramidal cells. *Brain Res.*, **505**, 12–22.

Huguenard, D.R., Hamill, O.P. and Prince, D.A. (1989) Sodium channels in dendrites of rat cortical pyramidal neurons. *Proc. Natl. Acad. Sci.* USA, **86**, 2473–2477.

Jack, J.J.B., Noble, D. and Tsien, R.W. (1983) Electrical Current Flow in Excitable Cells. Oxford: Oxford University Press.

Jensen, R.V. and Shepherd, G.M. (1993) NMDA-activated conductances provide short-term memory for dendritic spine logic computations. In *Computation and Neural Systems*, edited by F.H. Eeckman and J.M. Bower, pp. 425–429. Norwell, MA: Kluwer.

Johnston, D., Magee, J.C., Colbert, C.M. and Christie, R. (1996) Active properties of neuronal dendrites. *Annu. Rev. Neurosci.*, **19**, 165–186.

Jung, H., Mickus, T. and Spruston, N. (1997) Prolonged sodium channel inactivation contributes to dendritic action potential attenuation in hippocampal pyramidal neurons. *J. Neurosci.*, **17**, 6639–6646.

Keller, B.U., Konnerth, A. and Yaari, Y. (1991) Patch clamp analysis of excitatory synaptic currents in Granule cells of rat hippocampus. *J. Physiol.(Lond.)*, **435**, 275–293.

Llinas, R.R. (1988) The intrinsic electrophysiological properties of mammalian neurons: Insights into central nervous system function. *Science*, **242**, 1654–1664.

Madison, D.V., Malenka, R.C. and Nicoll, R.A. (1991) Mechanisms underlying long-term potentiation of synaptic transmission. *Annu. Rev. Neurosci.*, **14**, 379–397.

Magee, J.C. and Johnston, D. (1997) A synaptically controlled, associative signal for Hebbian plasticity in hippocampal neurons. *Science*, **275**, 209–213.

Mainen, Z.F., Joerges, J., Huguenard, J.R. and Sejnowski, T.J. (1995) A model of spike initiation in neocortical pyramidal neurons. *Neuron*, **15**, 1427–1439.

Mainen, Z.F. and Sejnowski, T.J. (1996) Influence of dendritic structure on firing pattern in model neocortical neurons. *Nature*, **382**, 363–366.

Major, G., Larkman, A.U., Jonas, P., Sakmann, B. and Jack, J.J.B. (1994) Detailed passive cable models of whole-cell recorded CA3 pyramidal neurons in rat hippocampal slices. *J. Neurosci.*, **14**, 4613–4638.

Markram, H., Lubke, J., Frotscher, M. and Sakmann, B. (1997) Regulation of synaptic efficacy by coincidence of postsynaptic APs and EPSPs. *Science*, **275**, 213–215.

Mascagni, M.V. (1989) Numerical Methods for Neuronal Modelling. In *Methods in Neuronal Modelling*, edited by C. Koch and I. Segev, pp. 439–484. Massachusetts: MIT Press.

Mayer, M.L., Westbrook, G.L. and Guthrie, P.B. (1984) Voltage dependent block by Mg^{++} of NMDA receptors in spinal cord neurons. *Nature*, **309**, 261–263.

McBurney, R.N. and Neering, I.R. (1987) Neuronal calcium homeostasis. *Trends Neurosci.*, **10**, 164–169.

Mel, B.W. (1992) NMDA-based pattern discrimination in a modelled cortical neuron. *Neural Comp.*, **4**, 502–517.

Nowak, L., Bregestovski, P., Asher, P., Herbert, A. and Prochiantz, A. (1984) Magnesium gates glutamate activated channels in mouse central neurons. *Nature*, **307**, 462–465.

Orpwood, R.D. (1988) Basic module for an adaptive control system based on neuron information processing. *J. Biomed*. Eng., **10**, 201–205.

Orpwood, R.D. (1990) Mechanisms of association learning in the post-synaptic neuron. *J. Theor. Biol.*, **143**, 145–162.

Orpwood, R.D. (1992) Computer simulation of neurone pattern processing. *J. Biomed. Eng.*, **14**, 222–228.

Orpwood, R.D. (1994) A possible neural mechanism underlying consciousness based on the pattern processing capabilities of pyramidal neurons in the cerebral cortex. *J. Theor. Biol.*, **169**, 403–418.

Orpwood, R.D. (1996) A demonstration of core features of perception using networks of cortical pyramidal cell models. *Brain Res. Assoc. Abstr.*, **13**, 78.

Pockberger, H. (1991) Electrophysiological and morphological properties of rat motor cortex neurons *in vivo. Brain Res.*, **539**, 181–190.

Rall, W. (1959) Branching dendritic trees and motoneuron membrane resistivity. *Exp. Neurol.*, **1**, 491–527.

Rall, W. (1964) Theoretical significance of dendritic trees for neuronal input–output relations. In *Neural Theory and Modelling*, edited by R.F. Reiss, pp. 73–97. Stanford University Press: Stanford.

Rall, W. (1967) Distinguishing theoretical synaptic potentials computed for different soma-dendritic distributions of synaptic inputs. *J. Neurophysiol.*, **30**, 1138–1168.

Rall, W. and Rinzel, J. (1973) Branch input resistance and steady attenuation for input to one branch of a dendritic neurone model. *Biophys. J.*, **13**, 648–688.

Rall, W. (1977) Core conductor theory and cable properties of neurons. In *Handbook of Physiology*, Section 1, The Nervous System, edited by E.R. Kandel, vol. 1, Cellular Biology of Neurons, pp. 39–97. Bathesda: American Physiological Society.

Rapp, M., Yarom, Y. and Segev, I. (1996) Modeling back propagating action potentials in weakly excitable dendrites of neocortical pyramidal cells. *Proc. Natl. Acad. Sci* USA, **93**, 11985–11990.

Regehr, W.G. and Tank, D.W. (1990) Postsynaptic NMDA receptor-mediated calcium accumulation in hippocampal CA1 pyramidal cell dendrites. *Nature*, **345**, 807–810.

Regehr, W.G., Kehoe, J., Ascher, P. and Armstrong, C. (1993) Synaptically triggered action potentials in dendrites. *Neuron*, **11**, 145–151.

Schiller, J., Sciller, Y., Stuart, G. and Sakmann, B. (1997) Calcium action potentials restricted to distal apical dendrites of rat neocortical pyramidal neurons. *J. Physiol.*, **505**, 605–616.

Segev, I. and Rall, W. (1988) Computational study of an excitable dendritic spine. *J. Neurophysiol.*, **2**, 499–523.

Segev, I., Rapp, M., Manor, Y. and Yarom, Y. (1992) Analog and digital processing in single nerve cells: Dendritic integration and axonal propagation. In *Single Neuron Computation*, edited by T. McKenna, pp. 173–198. San Diego: Academic Press.

Sejnowski, T.J. and Qian, N. (1992) Synaptic integration by electro-diffusion in dendritic spines. In *Single Neuron Computation*, edited by T. McKenna, pp. 117–139. San Diego: Academic Press.

Shepherd, G.M. (1979) *The Synaptic Organisation of the Brain*. New York: Oxford University Press.

Stratford, K., Mason, A., Larkman, A., Major, G. and Jack, J. (1989) The modelling of pyramidal neurones in the visual cortex. In *The Computing Neuron*, edited by R. Durbin et al., pp. 296–321. Wokingham: Addison-Wesley.

Stuart, G.J. and Sakmann, B. (1994) Active propagation of somatic action potentials into neocortical pyramidal cell dendrites. *Nature*, **367**, 69–72.

Stuart, G., Schiller, J. and Sakmann, B. (1997) Action potential initiation and propagation in rat neocortical pyramidal neurons. *J. Physiol.(Lond.)*, **505**, 617–632.

Svoboda, K., Denk, W., Kleinfeld, D. and Tank, D.A. (1997) *In vivo* dendritic calcium dynamics in neocortical pyramidal neurons. *Nature*, **385**, 161–165.

Traub, R.D., Miles, R. and Wong, R.K.S. (1987) Models of synchronised hippocampal bursts in the presence of inhibition. I. Single population events. *J. Neurophysiol.*, **58**, 752–775.

Tsubokawa, H. and Ross, W.N. (1997) Muscarinic modulation of spike backpropagation in the apical dendrites of hippocampal CA1 pyramidal neurons. *J. Neurosci.*, **17**, 5782–5791.

Turrigiano, G.G., Leslie, K.R., Desai, N.S., Rutherford, L.C. and Nelson, S.B. (1998) Activity-dependent scaling of quantal amplitude in neocortical neurons. *Nature*, **391**, 892–896.

Yamada, W.M., Koch, C. and Adams, P.R. (1989) Multiple channels and calcium dynamics. In *Methods in Neuronal Modelling*, edited by C. Koch and I. Segev, pp. 97–133. MA: MIT Press.

Yuste, R., Gutnick, M.J., Saar, D., Delaney, K.R. and Tank, D. (1994) Ca^{++} accumulations in dendrites of neocortical pyramidal neurons: An apical band and evidence of two functional compartments. *Neuron*, **13**, 23–43.

Zador, A., Koch, C. and Brown, T.H. (1990) Biophysical model of a hebbian synapse. *Proc. Natl. Acad. Sci. USA*, **87**, 6718–6722.

APPENDIX 13.1

This appendix briefly describes the use of shift operators to simulate exponential decays in the simplified neuron model. The value of the local depolarisation is represented in the computer as a binary number. If this number is shifted one bit to the right and the end bit is lost the effect is to divide the binary number by two. Similarly a shift of one bit to the left multiplies the number by two. So if a right shift is used at each computer time step the local depolarisation would decrease by a half each time and an exponential decay can be simulated, with a time course dependent on the time step used. Such operations are computationally very much faster than carrying out floating point divisions or using an exponential function.

In many computer languages the shift operator is signified as follows:

$$x \gg n \text{ is equivalent to } x \text{ shifted right } n \text{ times, i.e. } x/2^n,$$

$$x \ll n \text{ is equivalent to } x \text{ shifted left } n \text{ times, i.e. } x \times 2^n.$$

It is easy to represent different rational changes at each time step. For example,

$$((x \gg 2) + x) \gg 2 = x \times \tfrac{3}{4}.$$

CHAPTER 14
SOME PROBLEMS ARISING IN MODELS OF CONDUCTION IN EXCITABLE DENDRITES

Jonathan Bell

1. INTRODUCTION

A hundred years ago the neuron was established as an independently functioning unit through the anatomical studies of Ramon y Cajal and his mastering of staining techniques. Soon after it was demonstrated that neural membranes could have an all-or-nothing threshold effect. Electrical capacitance was then measured, the squid giant axon was discovered to be an invaluable preparation to work on, and to show that membrane conductance increases during the nerve impulse. Through the development of the voltage-clamp experiment it was determined that transmembrane voltage, rather than current, was a more useful dependent variable to control (see Cole, 1968). All this and more lead to the seminal experimental and theoretical work of Hodgkin and Huxley (1952) (see also Cronin, 1987). One aspect of their work was to show that voltage-clamp experiments gave enough information to develop a model of transmembrane potential, and that such a model would be of use in understanding the behavior of action potentials. It was fortuitous that the squid giant axon had such simple ionic channel dynamics that it allowed for the separate measurement of the two principle independent ionic currents (sodium and potassium). This also greatly facilitated the mathematical modeling of the axon. What had to wait was better techniques to obtain the quality of measurements that would allow better modeling of the channel dynamics and which would work on smaller fibers, including dendritic processes. Important techniques included chemical blockers so that single ionic species channels could be measured, and the patch-clamp method which could isolate a few channels, and sometimes a single channel. This allowed for the environment on both sides of the channel to be controlled, and so careful measurements of channel dynamics could be done (see Hille, 1992). With refinement new classes of channels could be discovered and studied in detail.

Thus, from these various experimental approaches, a pretty good understanding of simple axons could be obtained experimentally and theoretically. However, axons have mostly been considered faithful, though leaky, transmission lines. When threshold conditions for firing an action potential at the axon hillock are achieved, a signal is produced and propagated fast and robustly to all output sites of the axon terminus. Of course, mammalian axons and dendrites show regional specializations in both electrical and morphological properties. For example, axon hillock contains a high density of Na^+ channels, whereas in regions near synaptic release sites voltage-gated Ca^{++} channels are concentrated. There may also be regional differences in resting potential and membrane resistivities. At terminal arborizations axons typically have successive diameter changes.

Classical cable theory provides a framework for understanding propagation of action potentials along an excitable fibers like axons. The theory originally came from the derivation of the cable equation for use in calculations essential to the first trans-Atlantic telegraph cable, around 1855, by Lord Kelvin. In the 1870s Hermann and Weber derived and applied a different partial differential equation to the problem of electric current flow in and around a cylindrical core conductor (model of a nerve axon). Later, around 1900, Hermann and others recognized that when this core conductor equation was reduced to one space dimension, it became equivalent to Kelvin's cable equation (see Scott, 1977).

The basic approach to cable theory is to start with a circuit model for the neural membrane, then apply Kirchhoff principles to the circuit model. Then a current–voltage relation is modeled from experimental measurements. The pioneering work in cable theory applied to neural processes goes back to Rall (1970). To start with, it is generally assumed the excitable membrane locally has a membrane current density representable as a sum of capacitance current (with constant membrane capacity C_m) and a resistive, or ionic, current density I_{ion}; hence, with v being the transmembrane potential,

$$C_m \frac{dv}{dt} + I_{ion} = I_m.$$

The simplest expression for I_{ion} is to write as a sum of individual current densities. In many cases, these currents are associated with sodium, potassium, and calcium ions which pass through the neuronal membrane via ion-specific channels. The biological details are in the description of the dynamics of the various ionic channels. The expression above represents a single circuit element of the membrane model. Applying the conservation of current principle to a circuit chain model and taking a continuum limit, the classical cable equation is derived, namely

$$C_m \frac{\partial v}{\partial t} + I_{ion} = \frac{a}{2R_i} \frac{\partial^2 v}{\partial x^2}.$$

The Hodgkin–Huxley approach has been used to describe the dynamics in many other excitable membrane circumstances, among them myelinated node (Frankenhaeuser and Huxley, 1964), striated muscle fiber (Adrian and Peachey, 1973), barnacle giant muscle (Morris and Lecar, 1981), and cardiac Purkinje fiber (McAllister et al., 1975), to name a few. However, it must be kept in mind that the original formulation of their model was meant for a very simple axon with its principle ionic currents independent of each other, and conductances gated only by potential. There can be cooperative effects among channels. Also, many channels are not very ion selective, and many channels are not of Hodgkin–Huxley's two-state type (one open and one closed state as represented in their activation variable dynamics). It has long been known that sticking to first order kinetics raised to powers in the formulation of the active conductances is phenomenological and rather arbitrary, but the formulation worked well for squid giant axon membrane. (See Cole, 1968; Tuckwell, 1988; Scott, 1977 for further discussion.)

To analytically understand the behavior of various cable models there have been numerous efforts to approximate the models in one way or another. Among the various approaches that could be mentioned, Krinskii and Kokoz (1973), Caston et al. (1975), and others used the disparity in time scales to reduce the number of variables of the Hodgkin–Huxley model or one of its caricatures in one way or another. For example, see Tuckwell (1988), Scott (1977) for some overview. One such model we will mention again below is the Morris–Lecar model for barnacle muscle (Morris and Lecar, 1981), which has been studied most often with a short-time scale process eliminated.

Although most of the mathematical studies have been based on models for neuronal axon, patch-clamp techniques have allowed one to consider the electrophysiological behavior of dendritic processes. Not only has size of the membrane been a challenge to study with dendrites as compared to the giant axons, but also because there are significant nonuniformities. It was evident even in the staining studies of Cajal that dendritic fibers are anything but uniform in cross-section. Not only do the fibers have a lumpiness to them as one moves along the dendritic processes, but branches occur, and cytoplasmic extensions called dendritic spines are evident on CNS nerve fibers. These nonuniformities range from geometric ones like changes in diameter and branching of processes, to electrical ones (changes in axoplasmic resistivity, membrane conductance, membrane resistance), to ionic ones (like spatial differences in various channel densities), as well as often having the presence of dendritic spines.

Dendrites have traditionally been treated as passive cables, conveying synaptic inputs to the cell body for linear summation. However, as experiments improved, it was determined dendritic membranes are not necessarily passive, but contain voltage-dependent membrane conductances (see Adams, 1992). In pyramidal cells, for example, many dendrites have a distribution of both Na^+ and Ca^{2+} channels (Magee and Johnston, 1995). In fact, the distribution of Ca^{2+} ions within dendrites may represent another crucial variable for processing and storing information.

It is still not clear that dendrites can initiate action potentials, but studies show that once initiated in axons, action potentials can actively backpropagate into and within dendritic trees (antidromic spike propagation) (Stuart et al., 1997; Tsubokawa and Ross, 1997). Simultaneous somatic–dendritic recordings have shown action potentials first occur in soma for neocortical and hippocampal pyramidal neurons, cerebellar Purkinje neurons, GABA-releasing neurons, and spinal chord neurons, but that dopamine neurons of substantia nigra might provide an exception where action potentials were first observed in dendrites (Stuart et al., 1997). Vermeulen et al. (1996) modeled the moth olfactory sex-pheromone receptor cell as a single cable and studied the passive backpropagation of action potentials into the dendrite and its influence on the firing frequency of the cell.

From a theoretical modeling point of view we are interested in conditions under which propagation of action potentials can take place, and how such solutions to the model depend on physically relevant parameters. We are particularly interested in situations where it is appropriate to have nonuniformities and consider the effect on conduction of action potentials. Such properties can lead to anomalous behavior. Our purpose here is to review some work which introduces the simplest nonuniformities into classical cable theory, but which shows interesting phenomena can arise due to the presence of the nonuniformity. We also want to point out some questions that should be examined and some approaches to take on certain problems.

We discuss in more detail in succeeding sections a couple of cable scenarios where simple nonuniformities have been introduced. In the next section we review some basic model ideas that we will refer back to throughout the paper. Some details will be confined to Appendix 14.1. In section 3 we consider propagation in a cable model which is uniform except for one abrupt increase in diameter. We will outline previous work covering this case, then discuss where more work needs to done. It is also noted that a steep taper leads to essentially the same results as a jump, and that multiple jumps can also be considered. We will also relate these ideas briefly to the case of modeling branching of daughter processes in dendrites. In section 4 we introduce another fairly simple class of cables which are uniform and passive, and have spines. These models can be considered generalizations of the models discussed in the previous section. We will indicate some previous work done on such models, and bring up some questions which would be interesting to explore. This is an active area of exploration, so the topics and references have to be considered illustrative rather than complete.

2. CABLE MODELING FOR EXCITABLE DENDRITES

As mentioned above, cable theory for excitable membrane systems comes from representing an element of membrane by a circuit model, then considering a limit process to derive a continuum model for the transmembrane potential $v=v(x,t)$. With appropriate scaling, and assuming the fiber diameter, d, is a function of location x, the cable equation takes the form

$$C_{\mathrm{m}}\frac{\partial v}{\partial t}+I_{\mathrm{ion}}=\frac{1}{4R_{\mathrm{i}}d(x)}\frac{\partial}{\partial x}\left[d^2(x)\frac{\partial v}{\partial x}\right], \tag{14.1}$$

where C_{m}, R_{i} and I_{ion} represent, respectively, membrane capacitance, axoplasmic resistivity (which are constant here and below), and ionic current density. We have given a brief description of the equation's derivation in Appendix 14.1. Equation (14.1) expresses a conservation of current principle, so the physiology of the membrane is represented by the form I_{ion} takes on. If the membrane properties of the excitable cell are such that an independence principle between ionic current densities can be used, the form I_{ion} takes on can be represented by

$$I_{\mathrm{ion}}=\sum_j g_j(v-v_j),$$

where the sum is taken over all ionic species associated with the particular membrane (typically calcium, sodium, and potassium), and for each species, v_j and g_j represent, respectively, the reversal potential and conductance. Actually, the specific membrane properties are contained in the form given for the conductances. Based on experiments on squid giant axon, the classical formulation of I_{ion} (as a sum of sodium, potassium, and leakage currents) is the Hodgkin–Huxley (HH) formulation. Because of the interpretation taken of the data, the sodium and potassium conductances are functions of 'activation' variables which satisfy individual voltage-dependent first order dynamics; for details, see, for example, Tuckwell (1988) or Adelman (1971). In the case of diameter $d(x)$ being a constant, a wave (action potential) propagating along the fiber in this case is a function of a single variable, the traveling wave coordinate $s=x/\theta-t$, where θ is a constant wave speed parameter determined by all the membrane and cable parameters. In particular, it is known that θ increases with the square root of the diameter.

To understand the origin of properties observed numerically in the HH model, caricatures of the HH dynamics were developed and examined which were more analytically tractable. The best known of these was the FitzHugh–Nagumo (FN) model dynamics (see Tuckwell, 1988; Rinzel, 1977 and Chapter 16), one representation being

$$I_{\text{ion}}=-f(v)+w, \tag{14.2}$$

where the 'recovery' variable w usually is considered to satisfy an equation of the form

$$\frac{\partial w}{\partial t}=\sigma v-\gamma w. \tag{14.3}$$

Here σ, γ are (constant) recovery rates ($\sigma>0, \gamma\geqslant 0$), and $f(v)$ takes the form shown in Figure 14.1(a). Also, there is no scientific reason one cannot use the FN dynamics with the w equation being

$$\frac{\partial w}{\partial t}=\frac{w_\infty(v)-w}{\tau_w(v)},$$

where w_∞, τ_w are, respectively, the potential-dependent quasi-steady state and rate functions for w. These functions of v would generally have the form given in Figure 14.1(b), and the equation corresponds more closely to a HH first order process.

One can just as well replace the form of f by a smooth polynomial $f(v)=v(1-v)(v-a)$, or a piecewise linear form $f(v)=-1+H(v-a)$ ($0<a<1/2$, $H(\cdot)$ is the usual Heaviside function) and recover most results, at least formally. Since w is a 'slow' variable in FitzHugh's terminology (FitzHugh, 1969), one can consider initiation or threshold questions by leaving w at its rest value, $w\equiv 0$, so the model (with d being a constant) reduces to

$$\tau\frac{\partial v}{\partial t}=\lambda^2\frac{\partial^2 v}{\partial x^2}+f(v), \quad \left(\lambda^2=\frac{dR_{\text{m}}}{4R_{\text{i}}}, \ \tau=R_{\text{m}}C_{\text{m}}\right),$$

a classic reaction–diffusion equation, which is much studied. Here R_{m} is the constant membrane resistivity. We will refer to this model (i.e. with $I_{\text{ion}}=-f(v)$) as the 'reduced' F–N model. For a general background to diffusion equations involving such reaction terms, see Murray (1989).

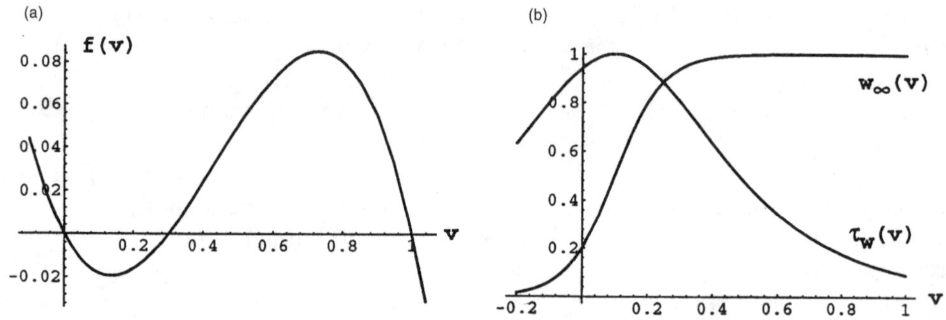

Figure 14.1 (a) The form of the current–voltage relation used in the FitzHugh–Nagumo model. (b) Voltage dependence of the quasi-steady state (w_∞) for the recovery variable, and the recovery rate (τ_w).

A major point to keep in mind, no matter what dynamics to consider for w is that the FN model has no actual biological analog for it to represent. It has been valuable to study the FN problem though because of the insight gained in helping to understand the HH model.

A partial compromise we have found is a model by Morris and Lecar (1981) (see also Rinzel, 1990), which describes potential changes in a barnacle muscle, but is simple enough to be more analytically tractable than HH (but less so than FN). The system consists of Ca^{++} and K^+ channels, neither of which are inactivated appreciably, plus a small ohmic leakage current. There may be a slow Ca^{++}-activated K^+ channel which can produce 'bursting' behavior in certain conditions, but we ignore that mechanism in the model below. We also assume the calcium activation process sufficiently fast so as to consider the conductance in a quasi-steady state. This corresponds to reduction of the model equations by taking advantage of differences in time scales that was mentioned in the introduction. Thus, the form of Morris–Lecar (ML) dynamics in which the fastest time scale has been replaced by a quasi-steady form is

$$I_{\text{ion}} = g_{\text{Ca}} m_\infty(v)[v-1] + g_{\text{K}} w[v-0.7] + g_0[v-v_0] \tag{14.4}$$

where g_{Ca}, g_{K}, g_0, v_0 are constants whose values we have generally taken to be 1, 2, 0.5, and -0.5, respectively, and w is a recovery activation rate for the potassium current. (The first and third terms on the right are, respectively, the calcium and leakage currents.) The dynamics for w is taken to be

$$\frac{\partial w}{\partial t} = \frac{w_\infty(v) - w}{\tau(v)}, \tag{14.5}$$

where w_∞, m_∞ have the form $0.5[1+\tanh((v-v_1)/v_2)]$, and $\tau(v) = \text{sech}((v-v_1)/v_3)$, for appropriate constants v_j.

In the work summarized below various model dynamics possesses nonlinear behavior that can be classified partly by phase plane analysis. For example, one can classify two basic types of excitability when one looks at the phase plane for potential and the fast time activation variable which drives impulse initiation (see Rinzel and Ermentrout, 1989; FitzHugh, 1969). Type 2 excitability in the model dynamics comes from having a single globally stable rest point. Threshold is due to the slow dynamics of a recovery variable, and it is not 'sharp' in the sense that as the initial condition varies, the spike amplitude and threshold response rate varies. An example of a model of this type is the H–H model. Type 1 excitability comes from having three rest points, a stable node and stable node (or spiral) separated by a saddle point with a one-dimensional separatrix serving as a threshold boundary. The threshold is sharp in that spike amplitude is not continuously graded with the initial condition. The dynamics will respond by excitatory behavior, or decaying subthreshold behavior, depending on which side of the separatrix the initial potential resides. Examples of this case (for standard parameter values used) are the Goldstein and Rall model (1974) mentioned below, and the M–L model, among others. With certain parameter changes the M–L model, for example, can be made to have type 1 excitability, so it is not only the model, but the parameter set that determines what type model one is analyzing. In cases below, both with regard to the analysis and computation of situations with reflected wave behavior to be discussed, we consider the case of type 1 excitability. In this case it is easier to analyze the behavior, and the behavior persists over a larger parameter range. But the qualitative results discussed below hold in both cases, and in fact, more work needs to be done to see the extent to which various solution behavior persists.

3. MODELS OF CONDUCTION IN DENDRITES WITH JUMPS IN DIAMETER

Our concern in this section involves simulating simple changes in fiber diameter, leading to either conduction block, successful signal transmission, or transmission with reflecting wave phenomenon. There is a variety of evidence for the existence of partial reflection of impulses, at least in an experimental setting. For example, an antidromically propagating action potential has a delay as it encounters sudden increases in membrane area (Ramon et al., 1975; Khodorov, 1974). Ramon et al. (1975) observed partial reflection in squid axon by using an internal axial wire electrode to alter conduction properties along a segment of fiber. They also performed numerical simulations of delayed conduction and reflection on the H–H model by modeling regions of changing diameter, as well as by altering the local fiber temperature, and by modeling sudden changes in ionic channel density. The computed results may also be relevant to impulse propagation in axon terminal arborizations, and branching of the fiber can be viewed as equivalent to widening of the fiber (see comments below).

Among the earliest numerical studies on the blocking of impulses was done by Berkinblit et al. (1970) and Khodorov et al. (1971). Khodorov and colleagues studied conduction in the H–H model that had an abrupt jump in diameter by factors of $d=3$, 5, 6, 10. That is, they view the fiber as having unit diameter before the jump, and diameter $d>1$ after it. Given a five-fold increase in diameter, an impulse did pass through the discontinuity, although with considerable delay. For a six-fold diameter increase, conduction was blocked. Berkinblit et al. (1970) considered the effect of taper and gave rough conditions for conduction block. Again the H–H dynamics were used. Part of their results are represented in Table 14.1. Khodorov et al. (1971) also considered conduction of a train of impulses along a H–H axon with one abrupt change (increase) in diameter. For example, for a sudden five-fold increase in diameter, only the first impulse successfully propagated past the discontinuity. For a three-fold increase, the first five impulses failed to propagate, but the sixth impulse successfully propagated. More numerics and analysis needs to be done to understand these patterns further.

In a numerical study based on the equivalent cable model, Goldstein and Rall (1974) considered conduction with both abrupt and tapered changes in diameter. They showed when the increase was double, i.e. $d=2$, successful transmission through the jump resulted, but when $d=2.7$, transmission was blocked. This was the earliest study the author knows where reflected impulses (echo waves) were computed, this happening at intermediate values of d, specifically at $d=2.5$. In such cases the geometrical change promotes delay because the spreading longitudinal current from the action potential foot must suddenly excite a much larger membrane area per unit length. The delay in doing this allows for the action potential tail to recharge the membrane behind the action potential to the membrane threshold level, initiating a reflected impulse.

For purposes of discussion below let us consider the simplest conceptual case of a cable with piecewise constant diameter. In particular, for most purposes we consider a single jump in diameter at $x=0$, with unit diameter for $x<0$, and diameter $d>1$ for $x>0$. We also assume propagation of an impulse is from left to right, since we found no interesting phenomena arise for impulses propagating through a fiber decreasing in diameter.

The simplest demonstration of conduction block is provided by the reduced F–N model. If we scale the spatial variable x so that there is no jump in the diffusion term, say by defining $z=x$ for $x<0$, and $z=x/\sqrt{d}$ for $x>0$, then

$$\frac{\partial v}{\partial t}=\frac{\partial^2 v}{\partial z^2}+f(v), \tag{14.6}$$

where f has the cubic form given in Figure 14.1(a). We want continuity of potential at $z=0$, so we must have

$$v(0^-,t)=v(0^+,t). \tag{14.7}$$

Table 14.1

Length of taper	Diameter ratio (d)	
	Blocking	Transmission
0.088	5.5	5
0.785	6	5.5
1.760	8	7
3.810	>10	10

We also do not want to have a current source at $z=0$; that is, by current conservation,

$$\frac{\partial v}{\partial z}(0^-,t)=\frac{\partial v}{\partial z}(0^+,t). \tag{14.8}$$

This can be derived from the equivalent electrical circuit approach (for further comment, see Rinzel, 1977). In the case of no recovery variable in the dynamics (i.e. reduced FN case), we have propagating wave fronts instead of impulses. For $d>d_c$ for some $d_c>1$, one can show there is a steady state solution to (14.6)–(14.8) satisfying $v(z)\to 0$ as $z\to+\infty$, and being monotone decreasing, and using a comparison principle argument, any wave front solution starting sufficiently far to the left and being under the steady state will stay under it. Thus, it cannot propagate toward $z=+\infty$.

Erneux and colleagues investigated the conditions for propagation failure, first in the reduced FN model case (Erneux and Nicolis, 1993), then in the FN case with recovery (Booth and Erneux, 1995), by considering instead the spatially discrete version of the problem. Their model then has the form

$$\frac{\mathrm{d}v_n}{\mathrm{d}t}=d(v_{n+1}-2v_n+v_{n-1})+f(v_n)-w_n, \qquad \frac{\mathrm{d}w_n}{\mathrm{d}t}=\sigma v_n.$$

Here $\sigma>0$ is a constant, $f(v)=v(1-v)(v-a)$, and d, which can be considered a coupling parameter, depends on a. Actually, they took $d=\mathrm{O}(a^2)$, and considered the problem asymptotically for weak threshold, $0<a\ll 1$, and hence weak coupling. Here $v_1(0)=v_0$, $a\ll v_0\ll 1$, $w_1(0)=0$, $v_n(0)=w_n(0)=0$, $n=2, 3, \ldots, N$. In their papers they find, asymptotically, a critical value $d=d_c=d_c(a)$, such that $d>d_c$ implies propagation is possible, and below which propagation is blocked. This results from a slow passage near a limit point for a Riccati equation describing, asymptotically, conduction in each cell. The analysis of the behavior is restricted to the time regime where only the first three cells need be considered. We will return to discussion of discrete dynamics later in the case of ML dynamics, but there has been no comparable work done on obtaining explicit asymptotic conditions for propagation failure in the case of ML dynamics. If Erneux's approach can be extended to the ML model one might gain insight into the nature of reflected wave conditions also. The problem with the FN model is that it appears incapable of more interesting solution behavior like echo waves.

However, Pauwelussen (1982a) studied the propagation failure problem in the FN model from the perspective of the diffusion case with a single jump in diameter, and the use of comparison methods. He scaled his problem to have the form

$$\frac{\partial v}{\partial t}=\delta(x)\frac{\partial^2 v}{\partial x^2}+f(v)-w,$$

$$\frac{\partial w}{\partial t}=\sigma v-\gamma w,$$

where $0 < a < \frac{1}{2}$, $\delta(x) = 1/\varepsilon$ for $x > 0$, and $\delta(x) = 1$ for $x < 0$. He proved that for $\sigma/\gamma < a$, solutions $v(x,t)$ are controlled by a small number in the region $x > 0$, $t > 0$, for $0 < \varepsilon \ll 1$, and so a right moving wave is blocked for ε sufficiently small. Pauwelussen (1982) also generalized his result to more general systems in which the Goldstein–Rall model is a special case. Pauwelussen (1981) also took this model approach regarding a branching fiber, but in general, the most studied modeling approach is to replace a branching dendritic tree by an 'equivalent cylinder' hypothesis, or an 'equivalent profile' approach. In Rall's equivalent cylinder model at a branch point a three-halves rule holds, namely that if a parent trunk fiber of radius d_p branches into a number of daughter branches of diameters d_i's, then $d_p^{3/2} = \sum d_i^{3/2}$. In the 'equivalent profile' this condition is weakened so that the diameter of the profile may vary along the length, giving a cable with the possibility of multiple jumps in diameter. Combining this idea with modeling active spines gives a theoretical situation where one can be in a position to ask various questions, such as how the distribution of active spine heads with both excitatory and inhibitory synapses interact to sum or multiply the incoming inputs. Modeling branching is really too extensive a topic to cover here, but recent work by Manor et al. (1991) Ohme and Schierwagen (1998) are reasonable starting places for background although the latter has severe restrictions imposed on current input.

Motivated by the echo wave phenomena discussed by Goldstein and Rall (1974), the solution behavior exhibited by the M–L dynamics of barnacle muscle, and the wave blocking approach of Pauwelussen (1981), Zhou and Bell (1994) (see also Zhou, 1992) studied the behavior of conduction in a cable model with ML dynamics in the case of changes in fiber diameter modeled by cases given in Figure 14.2. The behaviors isolated were transmission through regions of diameter change with, and without, reflected impulse behavior, and impulse block (see Figure 14.3). For purposes of understanding conditions for having each of these, it did not enlighten the authors to consider taper conditions (Figure 14.2b) versus the abrupt jump (Figure 14.2a), so Zhou and Bell (1994) concentrated on jumps in the diameter. That is, for sharp tapering conditions, all conduction variability was the same as in the case of using a jump in diameter. For a small jump in the diameter, a definition of what is meant for this model to fail to propagate an impulse was defined, and using a weak comparison principle applicable to the ML models with its complicated recovery process, a theorem was established along the philosophical lines of the simple case given above for the reduced FN model of propagation block. Basically what was done was the building of a subsolution and supersolution to the steady state problem which 'pinches together' as $x \to +\infty$, and which brackets any propagating solution for $x \gg 1$. That is, for appropriately constructed functions φ_1, φ_2 which can be made arbitrarily close to each other for large x values, then $\varphi_1(x) \leqslant v(x,t) \leqslant \varphi_2(x)$ for all $t \geqslant 0$. It was also explored when to expect conduction block, and when to expect transmission in the case when there is a finite thickening of the fiber (as shown in Figure 14.2c), since the parameter dependence of fiber thickening length (L) vs diameter increase (D) plays off from one another. In the D–L plane a monotonically decreasing boundary curve between transmission and conduction block behaviors was established. Zhou and Bell (1994) also examined this

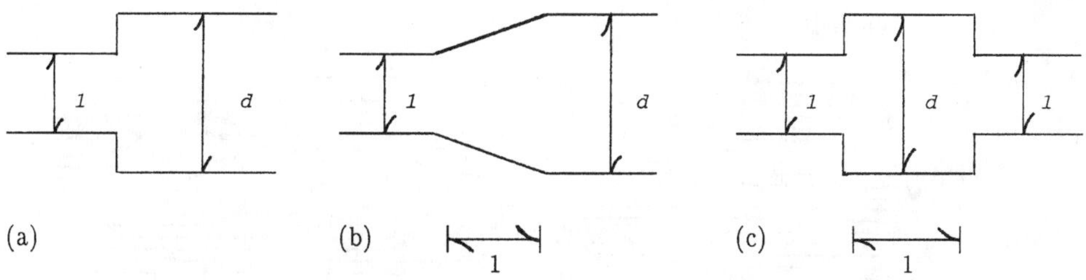

Figure 14.2 (a) Fiber with one abrupt jump in diameter. (b) Fiber with taper. (c) Fiber with an abrupt diameter thickening.

behavior in a model for CNS dendrite with dendritic spines and ML dynamics within the same paper (Zhou and Bell, 1994), more of which will be said in the next section.

Ermentrout and Rinzel (1994) explored reflection in some detail in a model for a single jump in diameter with ML dynamics. For the cable model with one jump, they developed a notation to highlight multiple reflections of impulses. If one indicates by $\ell : r$ the existence of ℓ impulses to the left and r impulses to the right of the jump (all created from a single right moving signal), then simple transmission through the jump, conduction block, and transmission with a single echo wave is labeled 1 : 1, 1 : 0, and 2 : 1 propagation, respectively. Such behavior can be developed in a chain of ML cells with resistive coupling by writing

$$\frac{dv_n}{dt} = -I_{\text{ion}} + I_{\text{app}} + d(v_{n+1} - 2v_n + v_{n-1}), \quad n = 1, 2, \ldots, m-1,$$

$$\frac{dv_m}{dt} = -I_{\text{ion}} + I_{\text{app}} + d(v_{m-1} - v_m)/A,$$

where I_{app} is a constant applied current imposed on all the cells uniformly, and I_{ion} is the ionic current given by the ML dynamics (using (14.4)–(14.5)). Here the parameter A acts like a 'cell enlarging' parameter which takes the place of the diameter parameter. Rinzel (1980) did some numerical studies with $m = 7$, and showed, e.g. when $A = 10$, the last cell failed to fire (conduction block, or the analog of 1 : 0 conduction), when $A = 3.333$ the $(m-1)^{\text{th}}$ cell fired the mth cell with a delay (1 : 1 conduction, interpreted as simple one-way propagation), and when $A = 6.25$, the $(m-1)^{\text{th}}$ cell fired the mth cell, with sufficient recovery in the $(m-1)^{\text{th}}$ cell and delay in the mth cell that the $(m-1)^{\text{th}}$ cell re-fired (hence giving 2 : 1 conduction).

Further work by Ermentrout and Rinzel (1994) showed that the cable model with a single jump produced more complex impulse patterns. For example, the delay could be of sufficient duration that not only would the tail of the impulse recharge the membrane behind it, but that enough charging could be enticed at the jump to cause multiple re-charging of the larger membrane, causing 2 : 2, or 4 : 3 conduction patterns, for example. In fact, given d_* and d^* such that for $d < d_*$ only 1 : 1 conduction is possible, and for $d > d^*$ only 1 : 0 conduction (i.e. conduction block) is

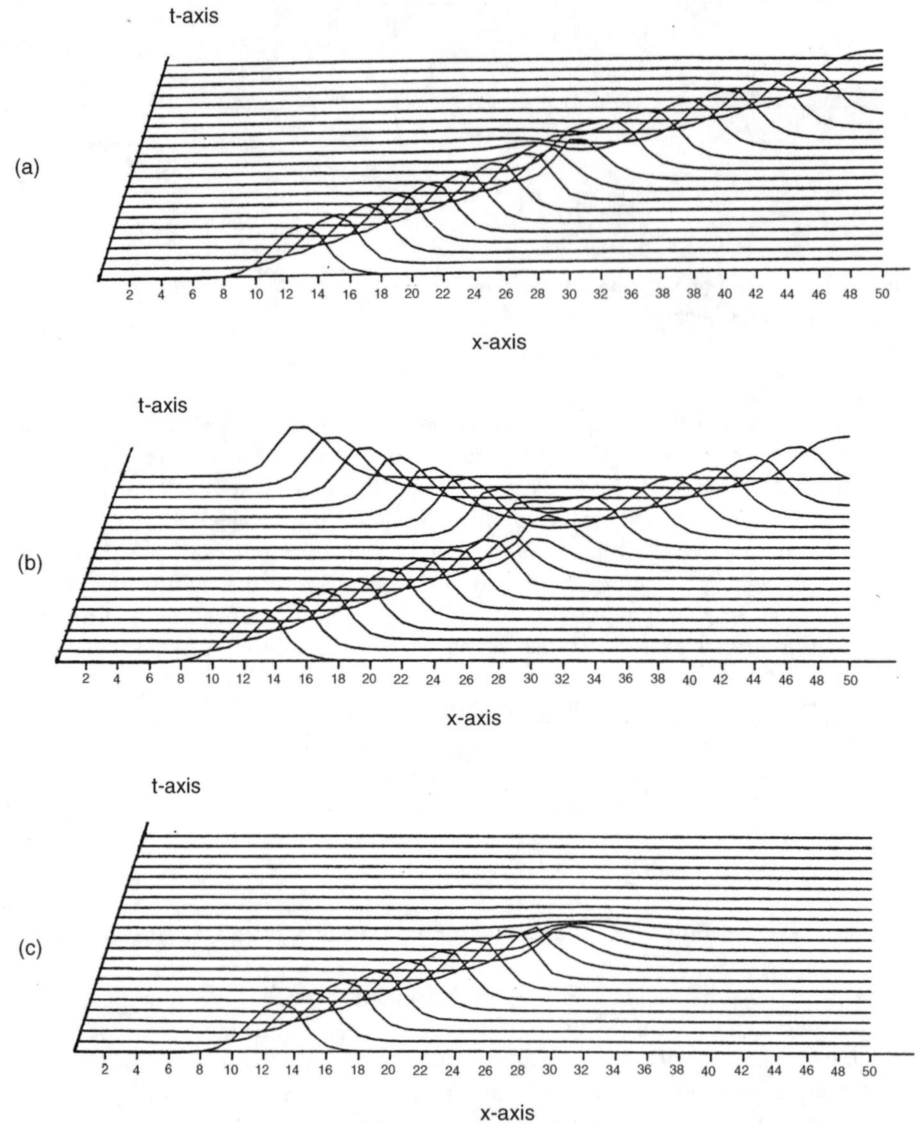

Figure 14.3 Traveling wave for the Morris–Lecar model of an excitable fiber with one abrupt jump in diameter at $x=32$. (This figure is adapted from Zhou, 1992.) (a) Case of successful transmission with diameter ratio $d=1.5$. (b) Case of successful transmission with a reflected (i.e. echo) wave with $d=1.58$. (c) Case of conduction block when $d=1.6$.

possible, then it appears there is a limiting value d_p ($d_* < d_p < d^*$) as a diameter for which there is an infinite number of echo waves for the particular stimulus, and such that as $d \to d_p$ from either the left, or the right, the sequence of conductions have the pattern given in Figure 14.4. That is, as $d \to d_p$ from the right, the conduction pattern seems to be $n+1:n$ with n increasing, while when d approaches d_p from the left the computed conduction pattern is of the form $n:n$. Ermentrout and Rinzel (1994) explored these more complex cases more in the case of a two ML cell model, namely

$$\frac{dv_j}{dt} = -I_{\text{ion}}(v_j, n_j) + I_{\text{app}} + (g/d_j)(v_k - v_j), \quad j, k = 1, 2 \ k \neq j,$$

$$\frac{dn_j}{dt} = (n_\infty(v_j) - n_j)/\tau(v_j),$$

with $d_1 = 1$, $g = 0.16$, then they obtained echo waves when $d_2 = 2$ (2:1 conduction), $d_2 =$ 1.9436 (3:2 conduction), and $d_2 = 1.9433$ (2:2 conduction), for example. The complicated conduction behavior comes about because the firing of the second cell can be arbitrarily delayed. From numerical studies they conjectured it is necessary to have an unstable periodic orbit which will emit alternating spikes between cells. Since Ermentrout and Rinzel (1994) were still dealing with a four-dimensional system, they wanted to build a caricature lower dimensional model that would have a diameter-changing type mechanism that as the parameter is adjusted the presence of a periodic solution would move in such a way as to cause block, transmission, and echo solution phenomena from a single pulse-initiating stimulus. They looked at a pair of coupled phase models of the form

$$\frac{d\theta_1}{dt} = f(\theta_1) + \gamma \sin(\theta_2 - \theta_1),$$

$$\frac{d\theta_2}{dt} = f(\theta_2) + (\gamma/\delta) \sin(\theta_1 - \theta_2), \quad \gamma, \delta > 0 \text{ are constants}.$$

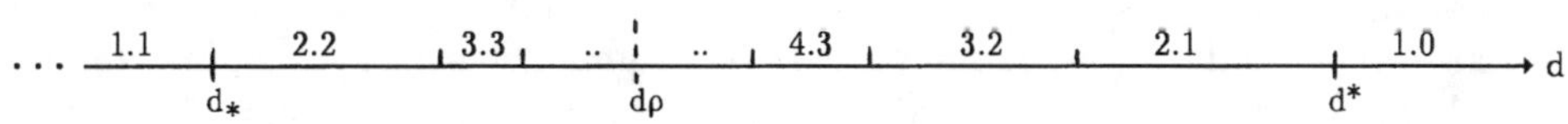

Figure 14.4 Diagram indicating the sequence of reflected wave patterns as the diameter ratio d varies. For $d < d_*$ we have normal transmission, while for $d > d^*$ we have conduction block, but we have reflected waves for $d_* < d < d^*$. See text for further comment. (This diagram is adapted from Ermentrout and Rinzel, 1994.)

Here $f(\theta) = 1 - \alpha\cos(\theta)$, $\alpha > 1$. This function has two zeros in the interval $(0, 2\pi)$, the smaller one (respectively, the larger one) labeled R (resp. T). The system, for $\gamma = 0$, has an asymptotically stable rest point (R, R) and an unstable rest point (T, T), and these persist for γ being nonzero, but small. The system displays transmission block for δ being relatively large, but as δ is decreased the block is removed and an unstable periodic solution emerges due to a saddle-node loop, hence providing a mechanism for echo waves. For details, see Ermentrout and Rinzel (1994).

4. MODELS OF CONDUCTION IN DENDRITES WITH ACTIVE CYTOPLASMIC EXTENSIONS

The existence of dendritic spine has been known since the turn of the century, but interest in these tiny projections has been renewed in the last twenty years by anatomical investigations and interesting questions on spine function and their potential plasticity. Spines are knob-like appendages to the dendritic trunk consisting of a bulbous spine head and a spine stem. They are particularly abundant in pyramidal, Purkinje, and stellate cells in the cortex. In the mammalian nervous system, the most common synaptic connections between two cells is by a presynaptic membrane making contact with a dendritic spine. Berkley (1897) first suggested the idea that spines are discrete areas of synaptic input. Chang (1952) observed that spine stems are often very slender and provide a high ohmic resistance. In fact, spines may vary in shape as one moves along the dendrite, with spine stems getting thinner and longer as one moves away from the cell body (Jones and Powell, 1969). Although it is difficult to link behavior to neurophysiological changes, studies like Coss et al. (1980) prove interesting in seeing changes in spine morphology of calycal interneurons during distinct stages of behavioral development in the honeybee. As the honeybee progresses from newly emerged stage to nurse stage (1–2 weeks), to forager stage (3–5 weeks), the spine stem length decreases and spine head size increases. There is no observed change in spine density, however. Coss et al. (1980) suggest that the spine alterations are in response to increased synaptic activity from greater sensory stimulation in the latter stages of development. According to Rall (1974), the spine stem resistance could be an important variable which might be used physiologically to change the relative weights of synaptic inputs from different afferent sources and could provide a basic cellular mechanism for learning and memory in the nervous system. For more discussion on topic of neural plasticity, see Jack et al. (1975), Rinzel (1982), Coss and Perkel (1985), Fifkova (1985) and Greenough (1975).

In theoretical studies Rall and Rinzel (1971) proposed that spine stem resistance could act as a tunable ‘synaptic weight’ parameter. Miller et al. (1985) showed that an active spine head membrane can provide very significant synaptic amplification of input from many afferent sources (see also Shepherd et al., 1985). Also, in Perkel and Perkel (1985) they show the amplitude of the post-synaptic potential is particularly sensitive to density of active channels and the spine stem resistance. Baer and Rinzel (1991) formulated a cable theory model for which the distribution of spines is treated as a continuum on an electrotonic (i.e. passive) dendritic cable model. In their model the density of spines, spine stem current, and spine head membrane potential are considered distributed variables. Their formulation has the physiologically reasonable feature that there is no direct electrical coupling between neighboring spines; voltage spread along the dendritic cable is the only way for spines to interact.

Another setting where this type of cable theory might apply is at the neural ending of capsulated mechanoreceptors associated with the skin (for sense of touch sensation), as well as other tissues. For example, the fiber inside a Pacinian corpuscle has cytoplasmic extensions, called filopodia, which are structured somewhat like spines (see Bell et al., 1994). It is conjectured that they may be the site of transduction, hence being the principal regions of excitable dynamics. Filopodia are also present in other receptors (Meissner corpuscles, Grandry corpuscles, muscle spindles, etc.).

Leaving aside the specific current–voltage relation, we think of the dendrite as being passive (i.e. linear) cable with current sources coming from each spine. Thus by current conservation for the dendritic membrane potential v, we have (scaling out most constants)

$$\frac{\partial v}{\partial t}+gv=\frac{1}{d(x)}\frac{\partial}{\partial x}\left[d^2(x)\frac{\partial v}{\partial x}\right]+\bar{n}I_{SS}, \tag{14.9}$$

where g is a positive constant conductance (so I_{ion}) has been linearized about the zero rest state), $\bar{n}$ is the density of the spine heads along the fiber, that is $\bar{n}(x)$ is proportional to the number of spines per unit length of membrane centered at x. Here again $d=d(x)$ represents the diameter, so in (14.9) we are allowing for the cable to have variable diameter. Finally, I_{SS} is the spine stem current. The assumptions behind the dendritic spines are: (i) the spine heads are really small so that the membrane potential of the spine head located at x, namely $u(x,t)$, is considered a point source of potential for the nerve; (ii) any spine head is just resistively coupled to the fiber with no other electrical connections to neighboring spines, so that $I_{SS}=(u-v)/R_S$, where R_S is the spine stem resistance; (iii) by current conservation principles, the assumed excitable membrane is, as above, a capacitance current in parallel with a resistive (ionic) current, so that

$$\tau\frac{\partial u}{\partial t}+I_{\text{ion}}+I_{\text{ss}}+I_{\text{syn}}=0, \tag{14.10}$$

where $I_{\text{syn}}=g_{\text{syn}}(x,t)\exp[-t/\tau_{\text{syn}}]$ is the synaptic current at location x, τ is the relative time constant, and τ_{syn} is the synaptic time constant. For the cases below we ignore the synaptic input current and pose an initial boundary value problem to initiate an action potential because our concern is the behavior of the action potential at a discontinuity, not how it was initiated. A special case of (14.9)–(14.10) above is to consider the case when $R_S \to 0$; letting $u=v+R_S v_0+o(R_S)$, then substituting and retaining only O(1) terms, we obtain

$$(1+\tau\bar{n}(x))\frac{\partial v}{\partial t}+gv+\bar{n}(x)I_{\text{ion}}(v)=\frac{1}{d(x)}\frac{\partial}{\partial x}\left[d^2(x)\frac{\partial v}{\partial x}\right].$$

Hence, we recover the former cable equation except for an added linear term in the current–voltage relation, and an augmentation of the membrane capacitance (which may depend on x). Thus the basic spine plus dendrite model of this section has all the freedom of having nonuniformities of the previous cable theory, but with added complications of nonuniformities being introduced in, say the density $\bar{n}(x)$, or the stem resistance $R_S(x)$.

Baer and Rinzel (1991) employed H–H dynamics for the active spine head membrane. In their numerical studies, and our studies (Zhou, 1992; Zhou and Bell, 1994), the principle parameters of concern are the spine density $\bar{n}(x)$ and the spine stem resistance R_S. They have a number of interesting results, including showing that for R_S sufficiently large (but still within the physiologically acceptable range), signal propagation is prevented. Also, they show that successful transmission depends on the spatial distribution of spines, a result we comment more on below.

Zhou and Bell (1994) used ML dynamics with the cable model with spines so that

$$\frac{\partial v}{\partial t}=\frac{1}{d(x)}\frac{\partial}{\partial x}\left[d^2(x)\frac{\partial v}{\partial x}\right]-gv+\bar{n}(x)\left(\frac{u-v}{R_S}\right)$$

$$\frac{\partial u}{\partial t}=-\left(\frac{u-v}{R_S}\right)-m_\infty(u)(u-1)-g_K w(u-u_K)-g_1(u-u_1)$$

$$\frac{\partial w}{\partial t}=\phi\frac{w_\infty(u)-w}{\tau(u)}$$

where m_∞, w_∞, τ are functions of u given in the Section 2. The first question addressed for this model was conditions for propagation of action potentials for the uniform fiber, so that diameter d was a constant, and the spine head density and other parameters were constant. They determined numerically that propagated action potentials only exist in a bounded region in $\bar{n}$, R_S parameter space. Call this region of propagation P. They also determined some relationships between wave speed and $\bar{n}$ and R_S. They looked at cases for fixed $(\bar{n}, R_S)\varepsilon P$ but with d experiencing a single abrupt increase at $x=0$. As expected the model displayed the same three behavior types as discussed in section three, namely conduction block, regular transmission, and transmission with echo waves (though the study was not pushed to show secondary reflections). Undoubtedly this model should have as much solution behavior as the standard cable model, however the calculations would be more costly due to the extra dynamics.

Zhou and Bell (1994) also used a weak comparison theorem to give, as in the classical cable case, a conduction block result for the spine model with ML dynamics. However, in this spine model case we have no analytical classification of block, echo phenomena, or transmission conditions based on jump size, or the value of spine density, or any other parameters.

We have considered simpler dynamics (reduced FN) to get at some analytical work for threshold conditions (Chen and Bell, 1994). In this case the model takes the form

$$\frac{\partial v}{\partial t}=d(x)\frac{\partial^2 v}{\partial x^2}-gv+\bar{n}(x)\rho(u-v),$$

$$\frac{\partial u}{\partial t}=\rho(v-u)+f(u),$$

where d, g and $\rho=1/R_S$ are positive constants, $f(\cdot)$ has the cubic form given in Section 2, and density $\bar{n}(x)$ is positive and bounded. We give conditions under which both subthreshold and superthreshold results can be proved. What works in this case is the ability to construct steady-state solutions with certain characteristics, and the presence of a strong comparison principle for the system (see Chen and Bell, 1994 for details). But it seems difficult to generalize these results to the more complicated ML dynamics case.

More analytical work should be done on asymptotic behavior for any class of models when the density $\bar{n}(x)$ is more complicated. One interesting numerical result of Baer and Rinzel (1991), who synaptically stimulated the spines near $x=0$ in their dendritic spine model with HH dynamics, was to take a fiber of length two and with a fixed number of spines (actually 24 spines) and see what effect clustering them in various ways had on propagation from $x=0$ to $x=2$. Three cases were given. The first is to uniformly distribute the spines (so $\bar{n}=12$), then record the potential in the fiber at $x=0,1,2$. In the three cases think of transmission succeeding if the spine-head potential is above threshold at $x=2$. In the uniformly distributed case transmission fails. However, if the spines are clustered into 8 groups of 3 spines each (cluster width being 0.04, and distance between clusters being 0.24, so that $\bar{n}=75$), then transmission is successful. Hence, the clusters are close enough, and each cluster of spines have enough regenerative capability to sustain propagation of a signal in this case. If, however, the spines were clustered into 4 clusters of 6 spines each (cluster width being 0.08, and distance between clusters being 0.42, so $\bar{n}$ is still 75), transmission failed. Hence, there is an playoff between distance between clusters and cluster strength. Baer and Rinzel (1991) used 'hot' H–H dynamics (the ionic conductances were increased by a constant multiplication factor over the classical values), but it would be interesting to explore this behavior with simpler dynamics to try to understand it better.

Further questions arise when the basic model is interpreted as the nerve fiber model for the Pacinian corpuscle skin receptor. Here a Pacinian corpuscle is a nerve ending surrounded by a capsule constructed of layers of elastic shells separated by fluid layers. The capsule acts as a band-pass filter to pressure stimuli from the skin, and transmits strain to the nerve. This nerve ending has some characteristics of a dendrite (see Bell et al., 1994), so we refer to it as a neurite. The unmyelinated portion of the neurite inside the capsule is covered by cytoplasmic extensions (filopodia), which take the place of the spines in dendrites, and are stimulated through mechanical contact with the surrounding capsule. There are no synapses in the Pacinian corpuscle. Since it is not known whether the current–voltage relation is linear or nonlinear in the filopodial heads, if we initially look at a model with linear dynamics, and take a continuum modeling approach to the filopodia, after some scaling the model has the form

$$\frac{\partial v}{\partial t} = d(x)\frac{\partial^2 v}{\partial x^2} - gv + \bar{n}\rho(u-v),$$

$$\frac{\partial u}{\partial t} = \rho(v-u) - g_1(x,t)u + g_2(x,t),$$

where $g_1(x,t) = \mu(x,t) + b$, and $g_2(x,t) = c\mu(x,t)$, and d, g, $\bar{n}$, ρ, b, and c are initially taken as constants. Hence, the neurite diameter d is constant and scaled out by a change in space scales. Also, $\mu(x,t)$ is a stimulus function associated with local strains imposed on the neurite by the capsule. That is, it is believed that the capsule imposes a strain distribution across the neural membrane which activates strain-gated transducer channels in the filopodia heads (see Bell and Holmes (1992) for details). Initially the fiber is at its rest potential so that $v(x,0) = u(x,0) = 0$. The fiber terminates at $x=0$ with a bead-shaped end, and therefore a 'natural' boundary condition should be imposed (see Tuckwell (1988)). It takes the form (see Appendix 14.1 for comment)

$$\text{at } x=0\text{:} \quad \frac{\partial v}{\partial t} + v - \gamma\frac{\partial v}{\partial x} = 0, \quad t>0.$$

Also, at $x=L$, the unmyelinated portion of the fiber ends at a node (actually called the heminode), and a myelinated neural segment begins and emerges from the capsule. We are interested in whether, and under what conditions, the stimulus represented by $\mu(\cdot,\cdot)$ can get the membrane potential v to the threshold potential at $x=L$. With some assumptions on the dynamics of the heminode, the same boundary condition can be posed at $x=L$ (up to some positive constants), namely

$$\text{at } x=L\text{:} \quad \varepsilon_e\frac{\partial v}{\partial t} + v + \gamma_e\frac{\partial v}{\partial x} = 0, \quad t>0, \tag{14.11}$$

where ε_e, γ_e are dimensionless parameters associated with the imposed end membrane capacitance and conductance at the heminode, respectively. Thus our question transforms to solving the system asking whether $v(L,t) \geqslant V_{\text{thresh}}$ for some finite t. For if this happens, an action potential can be generated and conducted along the myelinated fiber to the spinal chord. In Krzyzanski and Bell (1998) the above model is converted to an integral equation using a Green's function approach, and then a unique solution is established via a contraction mapping argument. By strong maximum principle arguments upper and lower bounds are established for the neuritic potential at the heminode from which conditions for reaching a threshold level can be accessed. Some dependence of the potential on stimulus strength and filopodia density is considered numerically. We also looked at the dependence of the maximum potential at the heminode for various spatial distributions of stimulus and filopodia density, and this suggested an optimization problem of determining if there is a distribution $\bar{n} = \bar{n}(x)$ of filopodia which maximizes the potential response at the heminode. Of course there are natural restrictions to the density function, among them that the number per unit length is bounded above, as well as the total number for the neurite. Given continuity of the functions $\bar{n}(x)$'s, these conditions give us an admissible set of functions A. Then the first problem to consider is the maximizing of the boundary potential in the steady state case. There is a response functional $U: A \to [0, \infty)$ given by $U(\bar{n}) = \bar{u}(L)$, where $\bar{u}$ is the neuritic potential solution to the steady state version of the last u, v system considered above, for the given density $\bar{n}$. Thus, one wants to solve the problem $\sup\{U(\bar{n}) \mid \bar{n} \in A\}$. Then it would be interesting to see how the optimal distribution of filopodia compares to distributions found in freeze-dried experiments on Pacinian corpuscles.

With a slight reformulation of the boundary conditions at $x = L$, another series of questions of physiological importance can be asked. Rewrite the oblique boundary condition (14.11) as

$$v(L, t) = y(t), \qquad \gamma_e \frac{\partial v}{\partial x}(L, t) = -y(t) - \varepsilon_e \frac{dy}{dt}.$$

Then we can ask the following inverse question. Given that we can measure the potential (i.e. $y(t)$) and the current (essentially $z(t) = -y(t) - \varepsilon_e\, dy/dt$) at the heminode via standard experimental methods, can we recover the strain imposed on the nerve by the capsule; that is, can we recover the function $\mu(x, t)$? Given magnitudes on the y, z, can we say anything about bounds on μ? How is the qualitative behavior affected by $\bar{n}$ and R_S? In general the form of μ is too general, and the questions are too broad, but special cases can and will be discussed elsewhere. However, this is just a sample of a whole range of inverse problems that can be considered in the study of nonuniform cable models, but such questions should be closely tied to some experimental findings.

5. DISCUSSION

We have been primarily interested in this Chapter in changing diameter leading to conduction block and reflected wave behavior, whether the fiber is represented by a straight cylinder cable model or one with a continuum of cytoplasmic extensions. There can be significant other inhomogeneities in other parameters which can lead to conduction delays, block, and reflection. In Ermentrout and Rinzel (1994) it was pointed out that in many cardiac preparations conduction may be problematic in localized regions by reduced axial conductance. A model of this situation might be a cable with constant diameter but reduced conductance over some mid-portion of its length. If one starts with (A14.1) in the Appendix 14.1, but instead assumes $d = 2a$ is constant, and $R_i = R_i(x)$, then the scaled cable model to consider has the form

$$C_m \frac{\partial v}{\partial t} + I_{ion} = \frac{d}{4} \frac{\partial}{\partial x}\left[\frac{1}{R_i} \frac{\partial v}{\partial x}\right].$$

Ermentrout and Rinzel (1994) ran a simulation using ML dynamics to show that between 40% and 50% drop in conductance was enough to get reflected wave phenomena.

More recently, Toth and Cunneli (1998) considered a continuously inhomogeneous ion channel distribution with exponentially decaying densities of voltage-activated channels. This assumption is not so realistic in comparison with the inhomogeneity occurring in the discreteness of ion-channel distribution with position along the cable structure as imposed by ionic cable theory (Poznanski, personal communication). It would be interesting based on the examples discussed in this Chapter to recover information about the channel density distribution of say Na^+ given a cable with known stimulus at one end, and voltage response at the other end.

An interesting idea brought up in Goldstein and Rall (1974) is that a pacemaker can be created because of the existence of reflected waves. Consider a dendritic segment with a small diameter central region that is stimulated on the left end to propagate a wave into the thin region. When the wave reaches the right end of the region a reflected wave is generated which propagates back to the left end where a new reflected wave is created. This continues creating a spatially periodic pacemaker. Since, as was pointed out in Ermentrout and Rinzel (1994), that wave reflection occurs in cellular discrete media as well as continuous cables, this may provide another mechanism, and maybe the least restrictive mechanism, for explaining ectopic pacemakers in cardiac tissue.

6. CONCLUSIONS AND FUTURE PERSPECTIVES

We had several objectives here. One objective was that we wanted to point out that even the simplest nonuniformity condition can lead to interesting solution behavior (echo waves and multiple reflections caused by the interaction between dynamic nonlinearities and diffusion) that could be explored further in many directions. We also think there are a number of qualitative questions of interest that could be explored in models with cytoplasmic extensions (spines and filopodia). We also thought it good to bring up the M–L dynamics again because we think it is a worthwhile model to study in its own right. We also want readers aware that oblique boundary conditions can come up in natural settings, and should be used more often, where appropriate, although there is a price to pay in level of work to do to obtain classical results. For example, even for elementary diffusion problems associated with separating variables the set of eigenfunctions is nonorthogonal.

With the ability of nonuniformities in a single fiber causing various impulse delays and reflection phenomena, one can only think that the view of dendritic processes as being just transmission lines is naive, and there is more information processing that can be attributed to single fibers than previously considered. The presence of branching, regional differences in concentrations of various ionic channels, inhomogeneities in electrical and geometric properties, the presence of evaginations, various ionic channels gated by other ions and even membrane strain, it is a wonder any part of the H–H theory remains in the theoretical study of dendritic processes. Yet various aspects of cable theory continue to give us a basis for modeling, simulating, and analyzing numerous neural behavior, and ultimately being inspired by the diversity we observe.

PROBLEMS

14.1. Taking the FitzHugh-Nagumo model (14.1)–(14.3) (or the nonlinear dynamics for w in place of the linear dynamics), and letting $\mathrm{d}(x)=1$ for $x<0$, $\mathrm{d}(x)=1+\varepsilon$ for $x>0$, one would expect successful propagation of a traveling wave through the step if $0<\varepsilon$ is sufficiently small. Can you analytically show this to be the case?

14.2. The work of Zhou (1992), and Zhou and Bell (1994) particularly for the cable model with active spines, concerns conduction block and reflective wave phenomena in the presence of a single wave interacting with a jump in diameter. Extend the work with Morris-Lecar dynamics when a traveling wave train of pulses approaches a jump in diameter. Specific patterns of waves should emerge as a function of d, and this will add to our knowledge about information processing possibilities in neural processes like dendrites.

14.3. As mentioned in section 3, Erneux and colleagues investigated analytically conditions for propagation failure in a simple spatially discrete form of the FitzHugh-Nagumo model. Can you extend their asymptotic approach to a model cable with Morris-Lecar dynamics? Because the latter dynamic supports echo wave phenomena, can the asymptotic approach help establish conditions for the on set of wave reflection?

14.4. Consider the cable model with spines as discussed in section 4, with uniform diameter, but variable spine density. Given a threshold stimulus at the left boundary of the finite length cable, can you establish conditions on the spine density function, $\bar{n}(x)$, for a threshold response to develop at the right boundary in finite time?

ACKNOWLEDGMENT

The author was partially supported by National Science Foundation grant DMS-9404517.

REFERENCES

Adams, P.R. (1992) The platonic neuron gets the hots. *Curr. Biol.*, **2**, 625–627.

Adelman, W.J., Jr. (1971) *Biophysics and Physiology of Excitable Membranes*. New York: Van Nostrand Reinhold.

Adrian, R.H. and Peachey, L.D. (1973) Reconstruction of the action potential of frog sartorius muscle. *J. Physiol.*, **235**, 103–131.

Baer, S.M. and Rinzel, J. (1991), Propagation of dendritic spikes mediated by excitable spines: a continuum theory. *J. Neurophysiol.*, **65**, 874–890.

Bell, J., Bolanowski, S. and Holmes, M.H. (1994) Structure and function of Pacinian corpuscles: a review. *Prog. Neurobiol.*, **42**, 79–128.

Bell, J. and Holmes, M.H. (1992) Model of the dynamics of receptor potential in a mechano receptor. *Math. Biosci.*, **110**, 139–174.

Berkley, H.J. (1897) The intracerebral nerve-fiber terminal apparatus. *John Hopkins Hosp. Rep.*, 89–93.

Berkinblit, M.B., Vvedeenskaya, N.D., Gnedenko, L.S., Kovalev, S.A., Kholopov, A.V., Formin, S.V. and Chailakhyan, L.M. (1970) Computer investigation of the features of conduction of a nerve impulse along fibers with different degree of widening. *Biophysics*, **15**, 1121–1130.

Booth, V. and Erneux, T. (1995) Understanding propagation failure as a slow capture near a limit point. *SIAM J. Appl. Math.*, **55**, 1372–1389.

Caston, R.H., Cohen, H. and Lagerstrom, P. (1975) Perturbation analysis of an approximation to Hodgkin–Huxley theory. *Quart. Appl. Math.*, **32**, 365–402.

Chang, H.T. (1952) Cortical neurons with particular reference to the apical dendrites. *Cold Spring Harbor Symp. Quant. Biol.*, **17**, 189–202.

Chen, P. and Bell, J. (1994) Spine-density dependence of the asymptotic behavior of a model of a nerve fiber with excitable spines. *J. Math. Anal. Appl.*, **187**, 384–410.

Cole, K.S. (1968) *Membranes, Ions, and Impulses.* Berkeley, CA, University of California Press.

Coss, R.G., Brandon, J.G. and Globus, A. (1980) Changes in morphology of dendritic spines on honeybee calycal interneurons associated with cumulative nursing and foraging experiences. *Brain Res.*, **192**, 49–59.

Coss, R.G. and Perkel, D.H. (1985) The function of dendritic spines: a review of theoretical issues. *Behav. Neural Biol.*, **44**, 151–185.

Cronin, J. (1987), *Mathematical Aspects of Hodgkin–Huxley Neural Theory.* Cambridge: Cambridge University Press.

Ermentrout, G.B. and Rinzel, J. (1994) Reflected waves in an inhomogeneous excitable medium. *SIAM J. Appl. Math.*, **56**, 1107–1128.

Erneux, T. and Nicolis, G. (1993) Propagating waves in discrete bistable reaction–diffusion systems. *Physica D*, **67**, 237–244.

Fifkova, E. (1985) A possible mechanism of morphometric changes in dendritic spines induced by stimulation. *Cell Mol. Neurobiol.*, **5**, 47–63.

FitzHugh, R. (1969) Mathematical models of excitation and propagation in nerves. In *Biological Engineering*, edited by H.P. Schwan, New York, McGraw-Hill.

Frankenhaeuser, B. and Huxley, A.F. (1964) The action potential of the myelinated nerve fibre of Xenopus laevis as computed on the basis of voltage clamp data. *J. Physiol.*, **171**, 302–315.

Goldstein, S.S. and Rall, W. (1974) Changes of action potential shape and velocity for changing core conductor geometry. *Biophys. J.*, **14**, 731–757.

Greenough, W.T. (1975) Experimental modification of the developing brain. *Am. Sci.*, **63**, 37–46.

Hille, B. (1992), *Ionic Channels of Excitable Membranes.* 2nd edition. Sinauer Associates, Inc.

Hodgkin, A.L. and Huxley, A.F. (1952) A quantitative description of membrane current and its application to conduction and excitation in nerve. *J. Physiol.* (*London*), **117**, 500–544.

Jack, J.J.B., Noble, D. and Tsien, R.W. (1983) *Electric Current Flow in Excitable Cells.*, Oxford: Clarendon Press.

Jones, E.G. and Powell, T.P.S. (1969) Morphological variations in the dendritic spines of the neocortex. *J. Cell Sci.*, **5**, 509–519.

Khodorov, B.I. (1974) *The Problem of Excitability: Electrical Excitability and Ionic Permeability of the Nerve Membrane.* New York: Plenum.

Khodorov, B.I., Timin, E.N., Pozin, N.V., and Shmelev, L.A. (1971) Theoretical analysis of the mechanisms of conduction of a nerve impulse over an inhomogeneous axon, IV. Conduction of a series of impulses through the fiber portion with increased diameter. *Biophysics*, **16**, 96–104.

Koch, C. (1997) Computation and the single neuron. *Nature*, **385**, 207–210.

Krinskii, V.I. and Kokoz, Y.M. (1973) Analysis of equations of excitable membranes I: Reduction of the Hodgkin–Huxley equations to a second order system. *Biophysics*, **18**, 533–539.

Krzyzanski, W. and Bell, J. (1998) Analysis of a model of membrane potential for a skin receptor. *Math. Biosci.* (submitted).

Magee, J.C. and Johnston, D. (1995) Synaptic activation of voltage-gated channels in the dendrites of hippocampal pyramidal neurons. *Science*, **268**, 301–304.

Manor, Y., Gonczarowski, J. and Segev, I. (1991) Propagation of action potentials along complex axonal trees. *Biophys. J.*, **60**, 1411–1423.

Morris, C. and Lecar, H. (1981) Voltage oscillations in the barnacle giant muscle fiber. *Biophys. J.*, **35**, 193–213.

McAllister, R.E., Noble, D. and Tsien, R.W. (1975) Reconstruction of the electrical activity of cardiac Purkinje fibres. *J. Physiol.*, **251**, 1–60.

Miller, J.P., Rall, W. and Rinzel, J. (1985) Synaptic amplification by active membrane in dendritic spines. *Brain Res.*, **325**, 325–330.

Murray, J.D. (1989) *Mathematical Biology*. Berlin: Springer-Verlag.

Ohme, M. and Schierwagen, A. (1998) An equivalent cable model for neuronal trees with active membrane. *Biol. Cybern.*, **78**, 227–243.

Pauwelussen, J.P. (1981) Waves in Reaction-Diffusion. Mathematical Centre, Amsterdam.

Pauwelussen, J.P. (1982) One way traffic of pulses in a neuron. *J. Math. Biol.*, **15**, 151–171.

Perkel, D.H. and Perkel, D.J. (1985) Dendritic spines: role of active membrane in modulating synaptic efficacy. *Brain Res.*, **325**, 331–335.

Ramon, F., Joyner, R.W. and Moore, J.W. (1975) Propagation of action potentials in inhomogeneous axon regions. *Fed. Proc.*, **34**, 1357–1363.

Rall, W. (1970) Dendritic neuron theory and dendrodendritic synapses in a simple cortical system. In *The Neurosciences: Second Study Program*, edited by F.O Schmitt. New York: Rockefeller University Press.

Rall, W. (1974) Dendritic spines and synaptic potency. In, *Studies in Neurophysiology*. R. Porter (Ed.) pp. 203–209. "Cambridge: Cambridge University Press".

Rall, W. and Rinzel, J. (1971) Dendritic spine function and synaptic attenuation calculations. *Soc. Neurosci. Abst.*, **1**, 64.

Rinzel, J. (1982) Neuronal Plasticity (Learning). In *Some Mathematical Questions in Biology, Neurobiology, Lectures on Mathematics in the Life Sciences*, Vol. 15, edited by R.M. Miura, AMS, Providence, R.I. pp. 7–25.

Rinzel, J. (1977), Repetitive nerve impulse propagation: numerical results and methods, In *Research Notes in Mathematics, Nonlinear Diffusion*, edited by W.E. Fitzgibbon and H.F. Walker. London: Pitman.

Rinzel, J. (1980), Impulse propagation in excitable systems. In *Dynamics and Modeling of Reactive Systems*, edited by W.E. Stewart et al. New York: Academic Press.

Rinzel, J. (1990) Mechanisms for nonuniform propagation along excitable cables. *Annals of NY Acad. Sci.*, **591**, 51–61.

Rinzel, J. and Ermentrout, G.B. (1989) Analysis of neural excitability and oscillations. In *Methods in Neuronal Modeling: From Synapses to Networks*, edited by C. Koch and I. Segev. Cambridge, MA: MIT Press.

Scott, A.C. (1977) *Neurophysics*. New York: John Wiley & Sons.

Shepherd, G.M., Brayton, R.K., Miller, J.P., Segev, I., Rinzel, J. and Rall, W. (1985) Signal enhancement in distal cortical dendrites by means of interactions between active dendritic spines. *Proc. Natl. Acad. Sci. USA*, **82**, 2192–2195.

Stuart, G., Spruston, N., Sakmann, B., and Hausser, M. (1997) Action potential initiation and backpropagation in neurons of the mammalian CNS. *Trends Neurosci.*, **20**, 125–313.

Toth, T.I. and Crunelli, V. (1998) Effects of tapering geometry and inhomogeneous ion channel distribution in a neuron model. *Neuroscience*, **84**, 1223–1232.

Tsubokawa, H. and Ross, W.N. (1997) Muscarine modulation of spike backpropagation in the apical dendrites of hippocampal CA1 pyramidal neurons. *J. Neurosci.*, **17**, 5782–5791.

Tuckwell, H. (1988) *Introduction to Theoretical Neurobiology*, Vol. 1 and 2. Cambridge: Cambridge University Press.

Vermeulen, A., Rospars, J.P., Lansky, P. and Tuckwell, H.C. (1996) Coding of stimulus intensity in an olfactory receptor neuron: role of neuron spatial extent and passive dendritic backpropagation of action potentials. *Bull. Math. Biol.*, **58**, 493–512.

Zhou, Y. (1992) Study of propagation along nonuniform excitable fibers. Ph.D. Thesis, SUNY at Buffalo, Buffalo, N.Y.

Zhou, Y. and Bell, J. (1994) Study of propagation along nonuniform excitable fibers. *Math. Biosci.*, **119**, 169–203.

APPENDIX 14.1

The purpose of this appendix is two-fold. We first want to indicate how to derive a model for an excitable cable as simply as possible with the use of equivalent circuit ideas. Secondly, we want to briefly point out how one thinks about having a 'natural' boundary condition in the excitable cable setting which mathematically gives an oblique boundary condition.

1. The simplest circuit model to consider for an excitable cable membrane model is given in Figure 14.A1, where external resistances to the membrane are considered negligible compared to the internal (axoplasmic) resistivities. By Kirchhoff principles

$$v(x)-v(x+h)=hr_i(x+h)i(x+h),$$

$$i(x)=i(x+h)+hi_m(x).$$

Here $i(x)$, $i_m(x)$, and $r_i(x)$ are longitudinal current, membrane current per unit length, and resistance per unit length, respectively. Making the continuum assumption that all quantities are continuous and differentiable with respect to the space variable, then in the limit as $h \to 0+$, we have

$$-\frac{\partial v}{\partial x}=r_i(x)i(x), \qquad i_m(x)=-\frac{\partial i}{\partial x}.$$

Combining these equations, and making the usual assumption that membrane current is composed of a capacitive current in parallel with a resistive (ionic) current, that is, $i_m = c\,\partial v/\partial t + i_{ion}$, we have

$$c\frac{\partial v}{\partial t}+i_{ion}=\frac{\partial}{\partial x}\left[\frac{1}{r_i(x)}\frac{\partial v}{\partial x}\right]. \qquad \text{(A14.1)}$$

The typical scaling of the problem, which is related to experimental conditions under which current densities are measuring, is

$$C_m = c/2\pi a, \qquad R_i = \pi a^2 r_i, \qquad I_{ion} = i_{ion}/2\pi a,$$

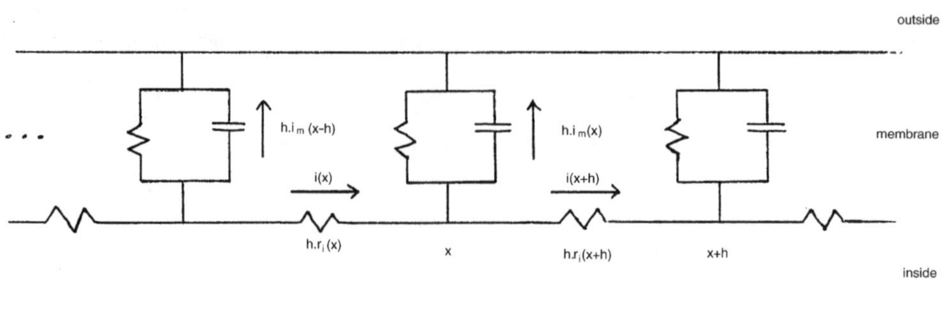

Figure 14.A1 Circuit model of cable theory for an excitable membrane.

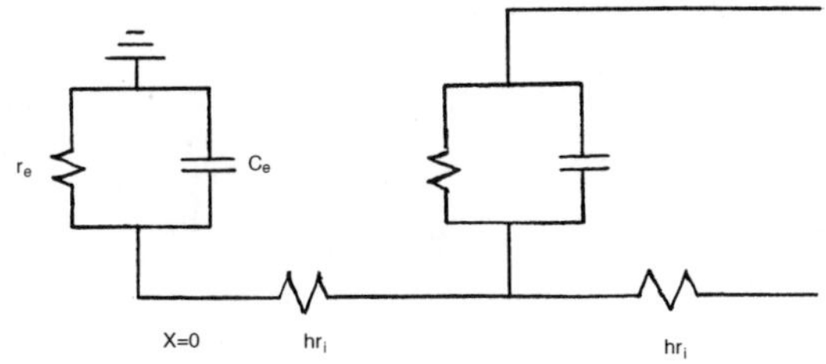

Figure 14.A2 The circuit model for a 'natural' termination to an excitable membrane. Also see Tuckwell (1988) for further comment.

where $a = a(x)$ is the fiber radius. If we assume R_i is constant, and $a = d/2$, then (A14.1) becomes

$$C_m \frac{\partial v}{\partial t} + I_{ion} = \frac{1}{4R_i d} \frac{\partial}{\partial x}\left[d^2 \frac{\partial v}{\partial x} \right].$$

This is expression (14.1) given in the section 2.

2. A natural boundary condition, or termination, in cable theory terms is defined to mean sealing the end of the fiber cylinder with a membrane whose electrical properties are somehow accounted for. In order to keep to a spatial one-dimensional theory, geometric details cannot be included. To do this we impose an endplate circuit model as shown in Figure 14.A2. Using the idea expressed above of the end membrane being a sum of capacitance and resistive currents (here $i_{ion} = v/r_e$), with corresponding electrical constants c_e and r_e. Then, by conservation of current,

$$-\frac{1}{r_i} \frac{\partial v(0,t)}{\partial x} = -\left[c_e \frac{\partial v(0,t)}{\partial t} + \frac{v(0,t)}{r_e} \right].$$

Up to scaling, this is just the boundary condition given in section 4.

CHAPTER 15
SEMI-QUANTITATIVE THEORY OF BISTABLE DENDRITES WITH WIND-UP

Armantas Baginskas, Aron Gutman, Jorn Hounsgaard, Natasha Svirskiene, and Gytis Svirskis

1. INTRODUCTION

In many central neurons, including motoneurons, accommodation of the spike threshold is slow, partial or absent. This means that these cells possess a subthreshold slowly- or non-inactivating inward current (Gutman, 1971). Sodium and calcium currents with these properties have been found in many nerve cells. Direct measurements reveal that slow inward currents are distributed throughout soma-dendritic membrane (Hounsgaard and Kiehn, 1993; Johnston et al., 1996; Llinas and Sugimori, 1980a,b; Llinas et al., 1992; Spruston et al., 1995). They maintain slow dendritic spikes in many CNS neurons, transmitter release from presynaptic dendrites in the brain-stem and dendritic bistability in motoneurons and Purkinje cells. *Dendritic bistability* means that there is a net slow inward current which can cause a steady depolarization (SD) in the distal membrane persisting without stimulation or active synaptic input. The membrane current–voltage characteristic (C–V) is N-shaped and has two stable potential values corresponding to zero current (see Figure 15.1, the solid curve). Earlier theorical work (Butrimas and Gutman, 1978, 1981; Gutman, 1971) stated that voltage-clamped dendrites with a net stationary slow inward current are bistable if the dendrites are electrotonically long. In such neurons the input current–voltage relation and the relation between discharge frequency and stimulus current are hysteretical and possess special features. Bistability of the distal dendritic membrane strongly influences the input current, I_{in}, of the cell measured by clamping soma potential, V_{in}. The input current–voltage relation (C–V_{in}) becomes not only N-shaped but may even split in two branches (see Figure 15.3): there may be two I_{in} values at the same V_{in}. The algebraically larger value corresponds to resting potential (RP) of the distal membrane and the lower value—to its SD. RP or SD of the distal membrane depends on previous activity of the cell and previous V_{in} values. In other words, the two branches of C–V_{in} lead to hysteresis. The hysteresis was predicted *a priori* (Butrimas and Gutman, 1978; Gutman, 1971) and observed experimentally in cat motoneurons (Schwindt and Crill, 1980b; Lee and Heckman, 1996), thus, confirming dendritic bistability. Dendritic SD was observed following direct dendritic excitation in mammalian Purkinje cells and in turtle motoneurons (Hounsgaard and

Kiehn, 1993; Llinas and Sugimori, 1980b). In Purkinje cells the dendritic slow inward current is mediated by P-type Ca-channels and in cat and turtle motoneurons by L-type Ca-channels (Hounsgaard et al., 1988; Hounsgaard and Kiehn, 1989, 1993; Johnston et al., 1996; Lee and Heckman, 1996; Llinas et al., 1992).

The theory of dendritic bistability was developed for the simplest version of time-independent *C–V* (Gutman, 1984, 1991). Later, experiments on interneurons and motoneurons in slices of the turtle spinal cord (Russo and Hounsgaard, 1994; Svirskis and Hounsgaard, 1997) revealed that the bistability develops slowly during depolarization. Indeed, slow activation of the inward current was already observed in previous experiments on cat motoneurons (Hounsgaard et al.,1988; Schwindt and Crill, 1980b). These early observations were not very conclusive since slow development of the observed I_{in} might be explained as a result of slow propagation of the SD along the dendritic cable (Gutman, 1984, 1991). Experiments with repeated application of current or potential pulses or ramps revealed an additional process of a slow facilitation of the slow inward current. Because of this process the slow inward current develops larger and faster during repeated pulses even with complete decay of the electric excitation during the interpulse intervals. Wind-up of spiking response during a series of afferent stimuli (Mendell, 1966; Sivilotti et al., 1993) is at least partly explained by this depolarization-dependent facilitation of inward current. Thus, the depolarization-dependent facilitation is named here *wind-up* of the slow inward current or simply wind-up. This wind-up phenomenon is in accordance with known features of the short-time plasticity of L-type calcium channels (Dolphin, 1996; Kavalali and Plummer, 1994, 1996).

The earlier theory of bistable dendrites excluded the slow kinetics of the inward current in the mathematical analysis. We shall update the theory of bistable dendrites by including wind-up of the inward current. In particular, we shall analyze a cylindrical cable with wind-up of the slow inward current. The relative importance of channel and cable properties for the observed phenomena are considered. Experimental implications of the wind-up in the bistable dendrites are discussed.

The wind-up of inward current *per se* causes a hysteresis of $C\text{–}V_{\text{in}}$ since the *C–V* becomes more deep after depolarization. The wind-up hysteresis interferes with the bistability hysteresis. One must differentiate these phenomena. Here we demonstrate that bistability hysteresis is an all-or-non event while the wind-up hysteresis is not. Wind-up slows propagation of SD many times and, thus, requires voltage-clamp ramps of much longer duration to demonstrate the bistability hysteresis of $C\text{–}V_{\text{in}}$. For the processes which are much slower or much faster than the rate of wind-up the theory of dendritic bistability for the time-independent *C–V* can still provide satisfactory understanding. Thus, the theory of dendrites with wind-up of bistability becomes quite transparent.

2. BISTABLE DENDRITES WITH TIME-INDEPENDENT MEMBRANE CURRENT–VOLTAGE RELATION

The problem of a cable with a slowly changing membrane *C–V* is important for interpretation of the electrical function of nerve cells. In this section we present a short review of the problem as discussed in earlier work by Gutman (1984; 1991) and

Jack et al. (1983). The simplest approximation is a cable with time-independent C–V. The corresponding cable equation differs from the usual Ohmic cable equation by only one term which describes the C–V. The equation in electrotonic units is

$$\frac{\partial^2 V}{\partial x^2}=\frac{\partial V}{\partial t}+f(V), \tag{15.1}$$

where V is the change of transmembrane potential relative to RP; x is electrotonic axial coordinate, expressed in length constants; t is electrotonic time, in membrane time constants; and $f(V)$ is the C–V multiplied by the membrane resistance near the RP, $f(V)\to V$ as $V\to 0$ (see Appendix 15.1 for a list of symbols and abbreviations). Let us analyze the case of a membrane with a net inward current. Such a membrane (or its C–V) is bistable in the following sense: the function $f(V)$ has three points of zero membrane current (Figure 15.1, solid curve): $V=0$ (stable RP), H (SD), and h (unstable depolarization, $0<h<H$). This means that SD may persist without stimulation or synaptic current. The SD of the model is equivalent to the plateau depolarization observed experimentally. The simplest representation of this function is broken-linear (Figure 15.1, solid curve), which we will use for illustration (Butrimas and Gutman, 1981)

$$f(V)=\begin{cases} V, & V\leqslant V_1; \\ k(h-V), & V_1\leqslant V\leqslant V_2; \\ K(V-H), & V\geqslant V_2. \end{cases} \tag{15.2}$$

We use the following parameter values: $V_1=6\,\text{mV}$, $k=1.38$, $h=10.3\,\text{mV}$, $V_2=26\,\text{mV}$, $K=5.7$, $H=30\,\text{mV}$. The values were chosen (see Butrimas and Gutman, 1983) to imitate the C–V_{in} and the relation between discharge frequency and DC stimulus current for the cat motoneuron (Granit et al., 1966; Gustafsson, 1974; Schwindt and Calvin, 1973; Schwindt and Crill, 1977, 1980a).

In order to understand the solutions of (15.1), let us exclude the membrane capacity and consider stationary version of the equation

$$\frac{\mathrm{d}^2 V}{\mathrm{d}x^2}=f(V), \tag{15.3a}$$

which is similar to the dynamic problem of a classical particle. Here the particle moves in a potential well (bottom at $V=h$) between two finite barriers (tops at 0 and H). The phase plane of trajectories, $\mathrm{d}V/\mathrm{d}x(V)$, for the general solution is presented in Figure 15.2(A). The phase trajectories reflect the first integral of the stationary equation, the "energy" conservation,

$$\frac{1}{2}\left(\frac{\mathrm{d}V}{\mathrm{d}x}\right)^2-\int_0^V f(y)\,\mathrm{d}y=E.$$

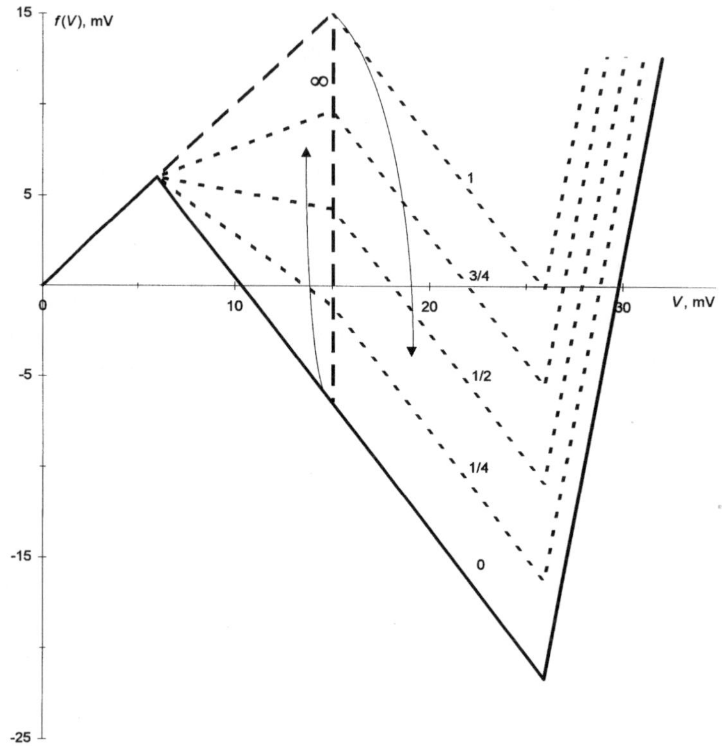

Figure 15.1 Model broken-linear current–voltage characteristics of bistable dendritic membrane, (15.2), (15.9), and (15.10). The lowest solid curve: fully developed wind-up, $N=0$, or time-independent $f(V)$. Stippled curves: incompletely developed wind-up, $N>0$, value of N marked at the curves. The thick stippled line: $f_{\infty}(V)$. For $V<V_S=15\,\mathrm{mV}$ coincides with $f(V)$ before wind-up, $N=1$, then it jumps down to $f(V,N=0)$, after complete wind-up, and coincides with this function for $V \geqslant V_S$. Arrows indicate the change of $C-V$ during moderately fast voltage-ramps when N is decreasing during voltage increase and increasing during voltage decrease.

There are no crossings between phase trajectories because the solutions are unique when the potential and its gradient are defined. Let us consider the boundary condition meaningful for the dendritic cable:

$$V=V_{\text{in}}, \quad x=0; \qquad \frac{\partial V}{\partial x}=0, \quad x=X. \tag{15.3b}$$

Equations (15.3a) and (15.3b) together determine a boundary value problem. Equation (15.3b) describes potential fixation at the "impaled" proximal end and the sealed distant end. The second condition means that the phase trajectory terminates on the abscissa. Thus, there are two types of solutions: (1) monotonic solutions which do

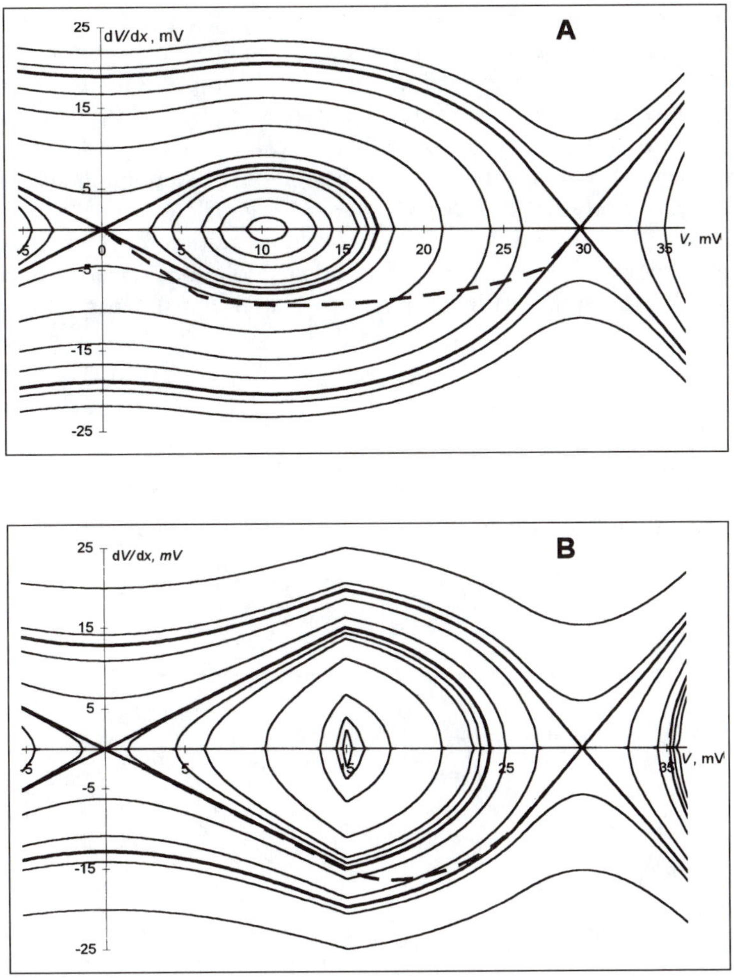

Figure 15.2 Phase trajectories for solutions of the stationary boundary value problem and waves (stippled). Thick lines—separatrixs; their asymptotes, $V \to H + A\exp(\pm x\sqrt{K})$ or $A\exp(\pm x)$, $x \to \pm\infty$. The phase plane is segregated by separatrixs into areas of qualitatively different solutions. (A) for time-independent $C-V$, (15.2); (B) for $C-V_\infty$, (15.10). Note the breaks and the absence of the smooth maxima of dV/dx at V_S in B.

not cross the abscissa, and (2) solutions which may cross the abscissa and oscillate around h.

The boundary value problem (15.3) may have non-unique solutions since the potential V_{in} at the beginning of the cable may correspond to more than one trajectory ending on the abscissa. We are going to define conditions for the non-uniqueness of the solution of this boundary problem. Aiming at finding the possible non-uniqueness we shall analyze solutions near the unstable equilibrium point h. The non-uniqueness of the solution means that the input current–voltage curve of the cable (C–V_{in}), which is equivalent to the dependence of $dV/dx|_{x=0}$ on V_{in}, is also non-unique (Figure 15.3(B)). For electrotonically short cables the C–V_{in} is proportional to the membrane C–V and is unique. In this case the input resistance, $R_{in} = dV_{in}/dI_{in}$, is negative near the point $V_{in} = h$, $I_{in} = 0$. For electrotonically long cables the C–V_{in} deforms and may lose its uniqueness. The non-uniqueness of the stationary solution appears *when and only when a range of positive* R_{in} *between ranges of negative* R_{in} appears in C–V_{in} with the cable prolongation (Figure 15.3). The sufficient condition for the non-uniqueness is: $R_{in} > 0$ at the point $V_{in} = h$, $I_{in} = 0$. The stationary solutions for potentials near the nonstable equilibrium point (Figure 15.2(A)), $[V(x) - h] \approx 0$, satisfy the following equation:

$$\frac{d^2V}{dx^2} = k(h - V) = \frac{1}{L^2}(h - V), \qquad (15.4)$$

where $k = -df(V)/dV|_h$. Here $f(V)$ is for bistability in general, not only for the broken-linear C–V of (15.2).

The corresponding harmonic solution of the boundary value problem (15.3) is

$$V = h + (V_{in} - h)\frac{\cos[(x - X)/L]}{\cos(X/L)}. \qquad (15.5)$$

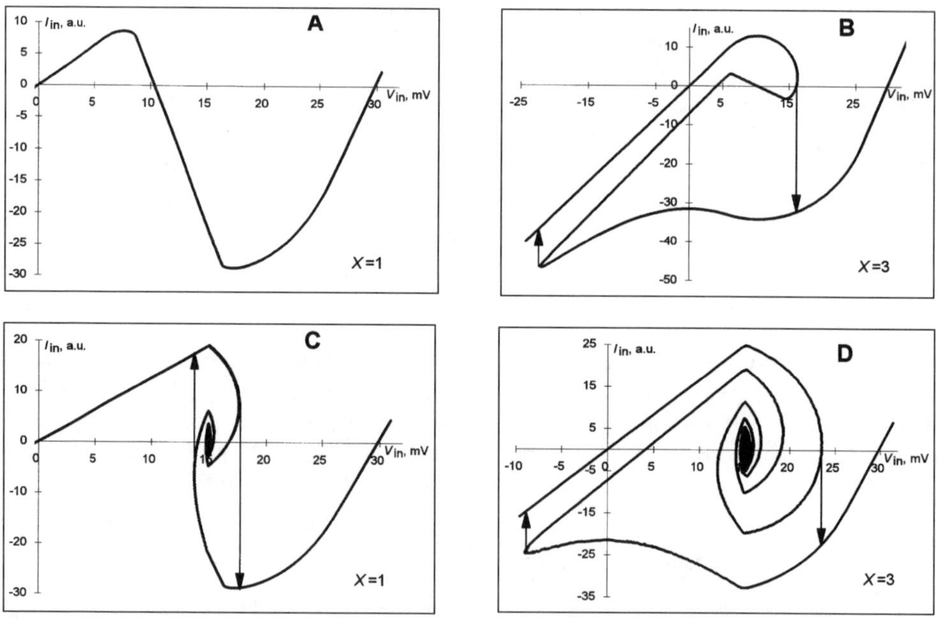

Figure 15.3 Input current–voltage curves for cylindrical cables. (A) and (B) for time-independent C–V, (15.2). (C) and (D) for C–V_{∞}, (15.10). An infinite number of oscillating solutions in (C) and (D) merge in a patch. The electrotonic length, X, is given to the lower right. All stable and unstable solutions are presented. Input current in arbitrary units.

If $X < \pi L/2$, then R_{in} is negative at $V_{in} = h$, $I_{in} = 0$ (Figure 15.3(A)). If $\pi L/2 < X < \pi L$, R_{in} becomes positive at this point and C–V_{in} surely loses uniqueness. In case of the C–V described by (15.2), $\pi L/2 = 1.34$. Thus, when X exceeds the critical cable length $\pi L/2$, there are several I_{in} values in some range of fixed V_{in}.

Now we shall write out some results relating the number and stability of stationary solutions to the boundary value problem (15.3). The rather lengthy proof of these results has been published earlier (Gutman,1984). These results (Figure 15.3(A) and (B)) are the following:

1. There are at least $(2n+1)$ solutions, if $\pi L(n-1/2) < X < \pi L(n+1/2)$.
2. The solution with the largest positive value I_{in} monotonously changes from V_{in} to $V(X)$ which is close to the RP. This RP-solution is stable. The RP-solution can be determined if $-\infty < V_{in} < V_{RP}$. At V_{RP}, the largest possible V_{in} of RP-solutions, the input resistance, $dV_{in}/dI_{in}|_{V_{RP}} = 0$. V_{RP} is larger than h.
3. The solution with the largest negative value I_{in} monotonously approaches SD at the distal end. This SD-solution is also stable. The SD-solution exists for all $V_{SD} < V_{in} < \infty$. At V_{SD}, the smallest possible V_{in} of the SD-solutions, the input resistance, $dV_{in}/dI_{in}|_{V_{SD}} = 0$. V_{SD} is smaller than h.
4. All other $(2n-1)$ solutions with the medial values of I_{in} are unstable. Almost all of them oscillate along the cable.
5. With the increase of cable length, X, the value of V_{RP} is shifted rightward and the value of V_{SD} leftward.
6. The monotonous stable solutions are qualitatively similar. Their quantitative difference depends on a C–V parameter, $u = \int_0^H f(V)\,dV$. With u increasing, V_{RP} and V_{SD} increase. In case of a semi-infinite cable and positive u, the RP-solution is absolutely stable, $V_{RP} = \infty$, while V_{SD} is determined by an equation $\int_{V_{SD}}^H f(V)\,dV = 0$. For $X = \infty$ and negative u the SD solution is absolutely stable, $V_{SD} = -\infty$, while V_{RP} is determined by $\int_0^{V_{RP}} f(V)\,dV = 0$. The C–V of (15.2) gives an example of $f(V)$ with negative u. In case of the semi-infinite cable, $I_{in}(V_{SD}) = I_{in}(V_{RP}) = 0$.

Thus, there may be two stable I_{in} at the same fixed V_{in}. In other words, the cable may be bistable in voltage-clamp conditions. We shall use the notion of the cable bistability only in this sense. There exists a hysteresis of its C–V_{in} for $V_{SD} < V_{in} < V_{RP}$ (Figure 15.3(B)). When V_{in} exceeds V_{RP} or decreases below V_{SD} there must be jumps of I_{in}. This reflects changes from the RP-solutions to SD-solutions, in other words the distal potential shifts from values close to RP to values close to SD, and *vice versa*. Experimental observations in cat motoneurons confirm this hysteresis (Schwindt and Crill, 1980b; Lee and Heckman, 1996). Voltage-clamp of RP after a strong depolarization reveals an inward current which does not decay, at least for hundreds milliseconds. C–V_{in} of a branching dendrite or a cell with several dendrites has several jumps reflecting different V_{RP} and V_{SD} and abrupt changes of potential distribution in individual dendrites or even their branches (Figure 15.4) (Baginskas and Gutman, 1990).

The next problem is to analyze the process of switching between RP- and SD-solutions (Nagumo et al., 1965). In an infinite cable the switch proceeds as a wave,

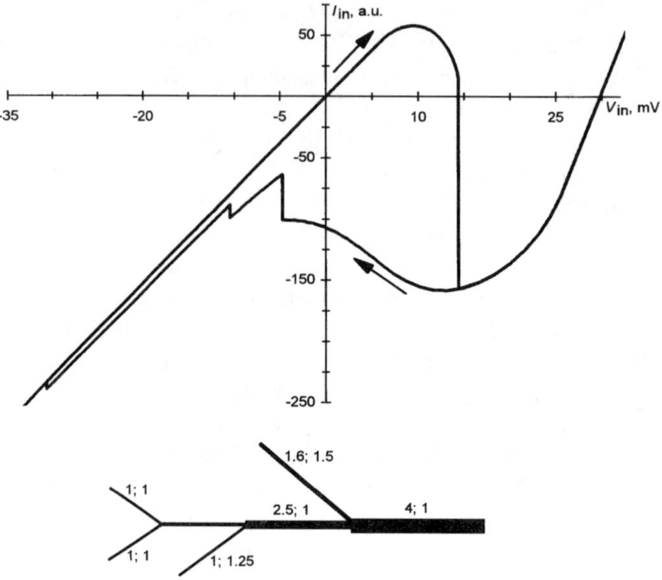

Figure 15.4 Input current–voltage curve of stable solutions for a branching dendrite in case of time-independent C–V. The dendrite is presented below. Diameters and electrotonic lengths, X, of the branches are marked in the following way: diameter; X. There is a single jump in I_{in} at depolarization, $V_{RP} \approx 14\,\text{mV}$. This occurs because the jump to SD-solution of the proximal branch induces the jumps of the more distal branches. There are three independent jumps at hyperpolarization, $V_{SD} \approx -5$, -10, and $-30\,\text{mV}$, for the proximal, intermediate and distal branches, respectively. This difference between jumps from RP to SD and from SD to RP reflects considerable negativity of the parameter u: negative current at $h<V<H$ is much larger than positive current at $0<V<h$. Thus, the SD-solutions are more stable than the RP-solutions.

$V(x,t) = V(x-ct)$. Thus, (15.1) simplifies to

$$\frac{d^2V}{dx^2} = -c\frac{dV}{dx} + f(V). \tag{15.6}$$

In the case of absolute stability of the SD-solution, $u<0$, the boundary conditions for the wave traveling from $-\infty$ to ∞ are

$$V|_{x=\infty} = 0, \qquad V|_{x=-\infty} = H. \tag{15.7}$$

In the phase plane (15.6) and (15.7) transform to:

$$\frac{1}{2}\frac{d}{dV}\left(\frac{dV}{dx}\right)^2 + c\frac{dV}{dx} = f(V), \quad \left.\frac{dV}{dx}\right|_{V=0 \text{ or } H} = 0. \tag{15.8}$$

In the equivalent dynamic problem the wave speed c is the coefficient of viscous friction. One must adjust its value numerically in order to satisfy the boundary conditions of (15.8). The mechanical analogy of the adjustment means that the particle falls without initial speed from the top of the high barrier, $V = H$, and stops exactly at the top of the low barrier, $V = 0$ (Figure 15.2(A)). In case of the broken-linear C–V, (15.2), the wave has the following numerical values of parameters (Appendix 15.2): $c = 0.63$; length of the wave front, defined as, $\Lambda = H/(\mathrm{d}V/\mathrm{d}x)|_{V=h}$, equals 2.7; duration of the wave front, $T = \Lambda/c = 4.3$.

Thus, in addition to the electrotonic length and time constants (both equal 1 in equations used here), the bistable cable with stationary C–V is characterized by additional parameters L, c, Λ and T. If $X < \pi L/2$, the cable is short with respect to bistability and there is no hysteresis of its C–V_{in} (Figure 15.3(A)). If $X > \Lambda$, the cable is very long. The wave of switching from RP to SD may develop fully in such long cables. If the potential, V_{in}, at the proximal end of the long cable is fixed at the SD, at least for a time $\approx T$, then the cable recharges to SD (there is sufficient time for the wave to develop).

The C–V_{in} of a cell determines its discharge qualities. We have modeled cell discharge in response to DC current (Butrimas and Gutman, 1983). The trigger zone confined to soma is described by the Hodgkin–Huxley equations with addition of a post-spike hyperpolarization. There are two events in bistable dendrites (Figure 15.5).

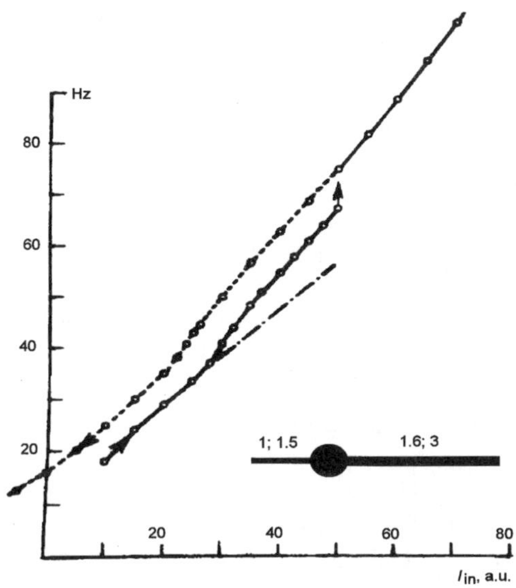

Figure 15.5 Discharge frequency–stimulus current curve for a model neuron (insert) with time-independent C–V. The soma is electrotonically compact and connected with two dendrites, $X = 1.5$ and 3. The diameter of the long dendrite is 1.6 times larger than the diameter of the short dendrite. The soma sodium and potassium channels are described by Hodgkin–Huxley equations. The jump-and-hysteresis reflects the SD of the long dendrite. The smooth increase in rise of frequency is due to the part-time SD of the short dendrite during interspike interval (see text).

In long dendrites, $X \gg L$, the potential of the distal membrane jumps to SD and remains steadily depolarized in spite of the post-spike hyperpolarization. The discharge frequency–stimulus current relation, F–C, has a jump-and-hysteresis, reflecting the jump-and-hysteresis of the dendritic C–V_{in}. In accordance with observations on cat motoneurons (Hounsgaard et al., 1984, 1988; Schwindt and Crill, 1977, 1980a) the frequency hysteresis ends at small depolarizing or even hyperpolarizing currents. In a cell with many dendrites the individual jumps may be small. In not so long dendrites, but still $X > \pi L/2$, the SD of the distal membrane cannot resist maximal values of the post-spike hyperpolarization and there is a part-time SD and part-time RP during the interspike interval. In some range of the current strength F–C rises more steeply than for Ohmic dendrites but jump-and-hysteresis is absent. The ranges of steep rise and small jumps may explain the phenomenon of two linear ranges of F–C (Granit et al., 1966; Gustafsson, 1974; Schwindt and Calvin, 1973). The primary range for small suprathreshold currents is linear and ends with a break which is followed by a more steep linear secondary range for larger currents. In our model the primary range corresponds to the Ohmic part of dendritic C–V and the secondary range consists of small jumps and steep rises determined by the dendritic bistability. The frequency jump-and-hysteresis was observed in cat motoneurons (Hounsgaard et al., 1988).

Thus, bistable dendrites with time-independent C–V explain important experimental observations on C–V_{in} and discharge properties of real neurons.

3. WIND-UP OF SLOW INWARD CURRENT

The inward current develops slowly with a characteristic time of hundreds milliseconds or even seconds (Figure 15.6(A)) in neurons of the turtle spinal cord slices (Russo and Hounsgaard, 1994; Svirskis and Hounsgaard, 1997) and in cat motoneurons (Hounsgaard et al., 1988; Schwindt and Crill, 1980b). The early observations were not very conclusive since slow development of the observed I_{in} might be explained as a result of slow propagation of the SD along the dendritic cable (Gutman, 1984; 1991). Only experiments with repeated application of current or potential pulses or ramps revealed that there is an additional process named wind-up. Indeed, after decay of I_{in} or V_{in} produced by a current pulse, the slow inward current develops larger and faster during the next pulse. Under some conditions wind-up results in an SD which persists beyond the train of current pulses. This observation means that C–V has a net inward current, as in Figure 15.1 and (15.2), after development of the wind-up.

Formally, wind-up may be described as reverse channel accommodation, i.e. slow inward current inactivation at RP or during hyperpolarization and de-inactivation during considerable depolarization. We decided to imitate wind-up as a slow shift of the C–V from an initial "shallow" to the final "deep" shape. In other words, the potential-sensitivity of the inward current is enhanced and shifted leftward during the wind-up. This assumption is in agreement with tail-current development in parallel with wind-up (Russo and Hounsgaard, 1994; Svirkis and Hounsgaard, 1997). The leftward shift is expressed as a continuous shift of the conductance at $V_1 < V < V_S$

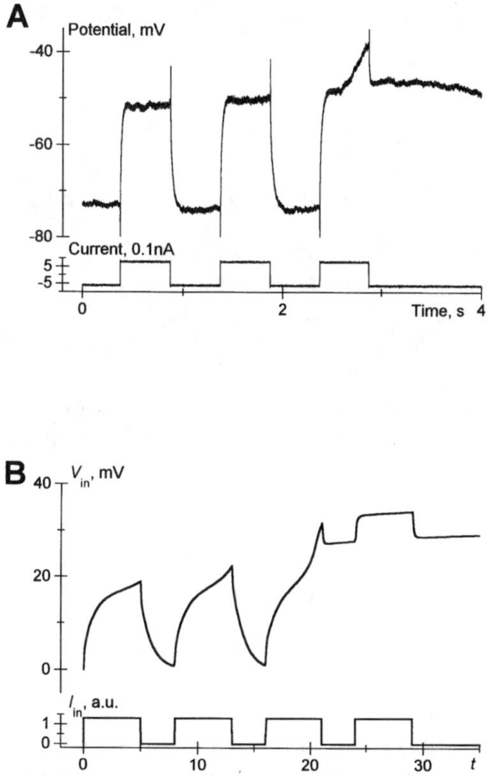

Figure 15.6 Wind-up induced by current pulses. (A) experimental curve, motoneuron from turtle spinal cord slice, (B) model, cable length, $X=1$. Note: the wind-up time constant, θ, is too small for detailed imitation of experimental curves.

from the initial Ohmic to the negative final value of (15.2) (Figure 15. 1): $\mathrm{d}f/\mathrm{d}V = -k(1-N)+N$. Here $N(t)$ is the variable of the wind-up process. There is no wind-up when $N=1$, wind-up is completely developed and C–V corresponds to (15.2) when $N=0$. Obviously, the C–V during wind-up becomes more deep at $V>V_S$. For illustrations we choose the C–V of the completely developed wind-up to be identical to the stationary function of (15.2). Thus, we choose the simplest description of the wind-up as follows (Figure 15.1, stippled curves):

$$f(V,N) = \begin{cases} V, & V \leqslant V_1; \\ V-(1+k)(1-N)(V-V_1), & V_1 \leqslant V \leqslant V_S; \\ V-(1+k)(1-N)(V-V_1)-(1+k)N(V-V_S), & V_S \leqslant V \leqslant V_2; \\ V-(1+k)(1-N)(V-V_1)-(1+k)N(V-V_S)+(k+K)(V-V_2), & V \geqslant V_2. \end{cases} \tag{15.9}$$

The variable of the wind-up is described by first order kinetics:

$$\frac{dN}{dt}=\frac{1-N}{\theta};\quad V<V_S;$$
$$\frac{dN}{dt}=-\frac{N}{\theta};\quad V\geqslant V_S; \tag{15.10}$$
$$N(t)=1,\quad \text{if } V(a)<V_S \text{ at all } a\leqslant t.$$

The process has a threshold V_S. The return of the C–V to a more shallow shape takes place at potential values $V<V_S$ as the potential is decreasing after a large depolarization, $V>V_S$, which have caused wind-up (Figure 15.1, arrows). The time-constant of wind-up, θ, in our calculations equals 10 and the threshold, V_S = 15 mV, coincides with the maximum of C–V before the wind-up starts.

Let us consider a stationary potential distribution along the cable. In this case, $t=\infty$, thus, $N=0$ or 1 and C–V (C–V_∞) corresponds to the following function (Figure 15.1, stippled thick curve):

$$f_\infty(V)=\begin{cases} V, & V<V_S; \\ k(h-V), & V_2\geqslant V\geqslant V_S; \\ K(V-H), & V_2\leqslant V. \end{cases} \tag{15.11}$$

Equation (15.11) differs from (15.2) by the Ohmic segment extending to V_S and a jump of current at this threshold potential of the wind-up (Figure 15.1). In our model $f_\infty(V_S+0)=f(V_S,N=0)<0$. The unstable zero-current point is now V_S, not h, and $\mathrm{d}f_\infty/\mathrm{d}V|_{V_S}=-\infty$. Thus, the length parameter for bistable cables ((15.4) and (15.5)), $L_\infty=0$. This means that even the shortest cable is bistable. Of course, in more realistic and complicated models wind-up should depend continuously on the membrane potential value in some range. The threshold of the wind-up is an oversimplification which, however, facilitates understanding. Correspondingly, in such a model there will be a smooth transition of N from 1 to 0, i.e. a smooth transition from Ohmic to negative current in f_∞. But still the parameter k must be larger than in case of stationary $f(V)\equiv f(V,N=0)$. For the time- and potential-dependent wind-up the theory of stationary bistable cables is useful as a marginal approximation. Bistability at very slow ramps of V_{in} depends on C–V_∞, e.g., (15.11), while during the relatively fast ramps the wind-up parameter N remains approximately constant. In the latter case the bistability is determined by $\mathrm{d}f/\mathrm{d}V$ at the medial zero point of $f(V,N)$ (Figure 15.1, stippled curves). The corresponding critical length parameter, $L(N)$, is not smaller than L for a stationary bistable cable, (15.4): $L(N)\geqslant L$. In general, bistability is more probable for slow potential ramps. We may conclude that even in realistic models the criterion for cable bistability is different for extremely slow and relatively fast potential ramps.

The bistability of infinitely short cables may be seen in the phase plane of the stationary equation, $\mathrm{d}^2V/\mathrm{d}x^2=f_\infty(V)$ (Figure 15.2(B)). $\mathrm{d}^2V/\mathrm{d}x^2$ has a jump at V_S due to the jump of $f_\infty(V)$. This jump is reflected by breaks of the phase trajectories (Figure 15.2(B)). The jump at the point of unstable equilibrium, V_S, means that $\mathrm{d}^2V/\mathrm{d}x^2\neq 0$ also for the trajectories that are very close to the point, $V=V_S$, $I=0$, in contrast to trajectories with continuous $f(V)$. Thus, oscillating solutions (the phase trajectory crosses the abscissa one or more times before ending on this axis) with infinitely small $|V-V_S|$ and $|dV/dx|$ are possible for infinitely small cable length X. The oscillating solutions are unstable and a hysteresis around V_S exists for infinitely short cable. The physical meaning of this hysteresis may be explained by using a simple two-compartment approximation (Booth and Rinzel, 1995; Gutman, 1984) of a short cable. When the input potential is rising and $V_{in}= V_S+0$, wind-up develops in the proximal compartment while still absent in the distal compartment. At the same V_{in} wind-up is developed in both compartments when returning from larger depolarizations. Thus, I_{in} is different at the same V_{in}. A similar situation exists for $V_{in}=V_S-0$. In a short cable, the stationary hysteresis reflecting the bistability is approximately as narrow as the potential difference, $[V_{in}-V(X)]$, between the ends of the cable. In short cables with sealed ends this difference rises proportionally to X^2, thus, the width of hysteresis increases considerably with X. An example of C–V_{in} hysteresis for a short cable in case of $f_\infty(V)$ is presented at Figure 15.3(C). There is no hysteresis for this cable in the case of stationary $f(V)$ or $f(V,N=0)$, Figure 15.3(A).

4. CABLE WITH WIND-UP. MODELING AND DISCUSSION

Our model imitates basic experimental phenomena of wind-up when the cell is stimulated with current pulses (Figure 15.6(B)). The electrotonic response is followed by a delayed rise of V_{in} which increases in amplitude and begins earlier with each pulse in the series. Finally, this increased potential may persist at SD even after the current has been switched off. The wind-up is a prerequisite for bistability. The apparent difference between experimental and model curves is due to a large difference in time-scale (compare Figure 15.6(A) and (B)). The time-constant for wind-up in the model, θ, is too small and, thus, the electrotonic component of the response is over-expressed. The smaller value of θ in the model was chosen for convenience in calculations.

Now, we may advance to the more complex problem of defining C–V_{in} in case of wind-up. The basic observation comes from the experiment with application of a train of voltage-clamp ramps (Figure 15.7(A)). C–V_{in} becomes deeper and separates down from the straight Ohmic line at lower depolarization during the second application of the voltage ramp. There is a hysteresis loop in C–V_{in} but in contradiction with the theory of cables with time-independent C–V it is not accompanied by a potential jump. This observation deserves more analysis.

While V_{in} is moving from small depolarizations and becomes larger than V_S, wind-up C–V corresponds to $f(V,N\approx 1)$ (Figure 15.1, upper curves). During the return from large depolarizations, $\approx$SD or more, to values below V_S, C–V corresponds to $f(V,N\approx 0)$ (Figure 15.1, lower curves). This means that a relatively fast ramp gives two types of hysteresis curves. One type is for and only for long cables, $X>\pi L/2$, here L is defined according to (15.4) and (15.5) for $f(V,N\approx 0)$. This hysteresis is essentially similar to the hysteresis described for cables with time-independent bistable C–V (Figure 15.7(C)). Indeed, the potential distribution at the minimum of the lower hysteresis loop of Figure 15.7(C) corresponds to the potential

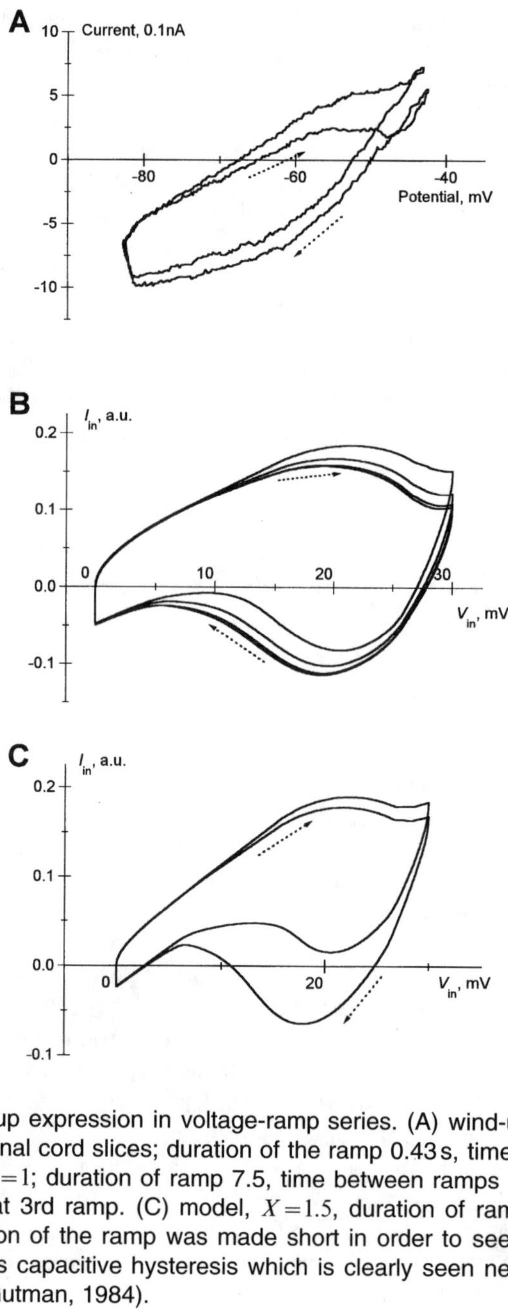

Figure 15.7 Wind-up expression in voltage-ramp series. (A) wind-up observed in turtle motoneuron from spinal cord slices; duration of the ramp 0.43 s, time between two ramps 0.57 s. (B) model, $X = 1$; duration of ramp 7.5, time between ramps 3. Note saturation of the model wind-up at 3rd ramp. (C) model, $X = 1.5$, duration of ramp 15, time between ramps 7. The duration of the ramp was made short in order to see the development of wind-up. This causes capacitive hysteresis which is clearly seen near the RP (Butrimas and Gutman,1980; Gutman, 1984).

at the distal end of the cable being close to SD. The second type of hysteresis, with fast voltage-ramps, has no jumps and threshold potentials V_{RP} or V_{SD}. It may not be bistability hysteresis and may be expressed even in the proximal membrane because N changes during the V_{in}-ramp (Figure 15.7(B)).

In contrast, very slow ramps, that pass the negative resistance region slow compared to θ, always give rise to bistability hysteresis in our model (Figure 15.3(C)). In more realistic models the bistability hysteresis in relatively short cables is also more probable with slow ramps, because of more steep negative resistance range in C–V_∞. Both types of hysteresis are illustrated in Figure 15.8. As the maximal V_{in} gradually shifts to more depolarized values with every ramp in the train, there are narrow hysteresis loops, at first, due to change of N (Figure 15.8(B), three upper loops; Figure 15.8(C), two upper loops; Figure 15.8(D), two upper loops). When V_{in} exceeds some threshold value there is a dramatic increase in the magnitude of the loop (Figure 15.8(B), two lower loops; Figure 15.8(D), three lower loops). Thus, two families of hysteresis are clearly distinguished in the case of long cables or very slow ramps. It is difficult to distinguish them in case of short cables and moderately slow ramps (Figure 15.8(C)). The family of narrow loops corresponds to the wind-up of membrane C–V while the family of broad loops corresponds to the bistability of the cable according to the shape of C–V_∞ (Figure 15.8(D)) or $f(V, N \approx 0)$ (Figure 15.8(B)). This conclusion is confirmed

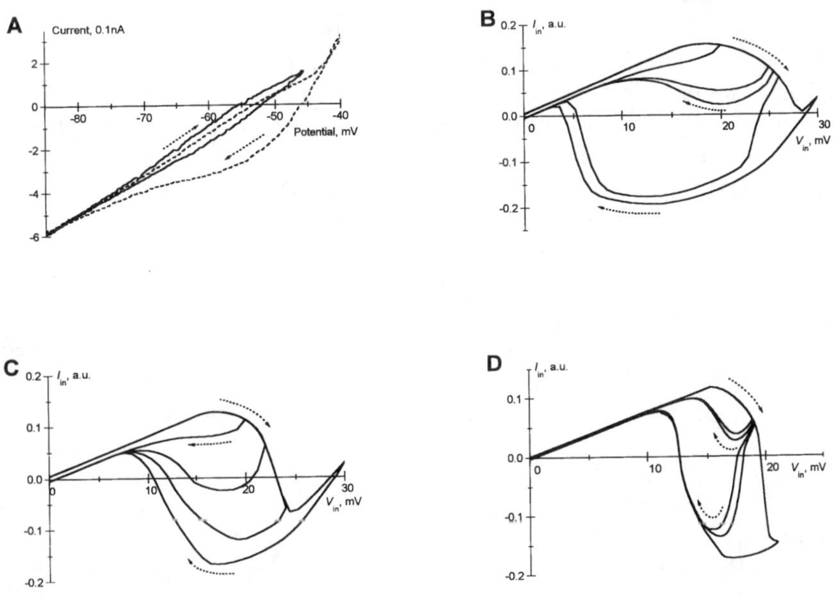

Figure 15.8 Dependence of hysteresis on maximal V_{in} in voltage-ramp. (A) experimental hysteresis loops with two values of maximal clamped potential, turtle motoneuron. Duration of ramp—4 s. (B), (C) and (D) model: (B) $X = 1.5$, duration of the ramp 100; (C) $X = 1$, duration of ramp 100; (D) $X = 1$, duration of ramp 390.

by the distribution of the potential along the cable (not shown). In order to check the possibility of this phenomenon we suggest to investigate experimentally how the hysteresis loop depends on maximal V_{in} and on the duration of the potential ramp. In the model, bistability hysteresis is apparent with ramp durations of few tens θ. In turtle motoneurons wind-up is initiated within 1 s during depolarization (Figures 15.6(A), 15.7(A) and 15.8(A)). Thus, the duration of the ramp ought to be about 10 s or more for experimental detection of possible dendritic bistability. There is one additional criterion for discrimination between wind-up and bistability hysteresis. When hysteresis is caused only by wind-up all loops return to the straight Ohmic lines at the same point (Figure 15.8(B), three upper loops; Figure 15.8(C); Figure 15.8(D), two upper loops). According to Eq. (10), $N(t)$ decays independently of its maximal value. The loops of bistability hysteresis return to Ohmic C–V_{in} at distinctly different point (Figure 15.8(B), two lower loops; Figure 15.8(D), three lower loops). This point may change slightly if SD is not completely established at the distal end of the cable (Figure 15.8(B), compare two lower loops). Thus, if experimental loops are merging with the Ohmic C–V_{in} at the same point then the hysteresis is predominantly caused by wind-up. If these loops end at clearly different points then they may be caused by dendritic bistability.

A different manifestation of those types of hysteresis may be the tail-current observed by Schwindt and Crill (1980b). The experimental paradigm was the following: first, the soma potential was clamped at RP, then at high depolarization, then again at RP or low depolarization. Two types of tail-currents were observed on return to RP or low depolarization: a non-decaying inward current or a slowly decaying inward current. The latter decayed an order of magnitude slower than simple electrotonic decay. We have imitated this observation. The non-decaying current corresponds to a depolarization that is strong enough to induce SD at the distal end of the cable (Figure 15.9, dashed curve). A slightly lower depolarization, which cannot induce the SD, is followed by a tail-current that decays in accordance with the time-constant of wind-up (Figure 15.9, solid curve).

Finally, let us consider the wave of SD in a long cable in case of wind-up (Figure 15.10). The analytical solution of the wave equation in a semi-infinite cable is presented in Appendix 15.3. The important result is the very slow speed of the wave, $c=0.023$. This wave proceeds 27 times slower than in the cable with stationary C–V. The wave length, $\Lambda = H/(dV/dx)|_{V_s} \approx 2$, calculated at the point of unstable equilibrium, is as for time-independent C–V. The length of the wave front in a cable with wind-up is a little shorter than the wave in a cable with time-independent C–V. Correspondingly, the duration of the wave in the case of wind-up becomes almost 20 times longer, $T \approx 80$. Taking into account that wind-up in our model is too fast compared with experimentally observed (Figure 15.6), we suppose that the speed of the real wave could be much slower and its duration much longer. The wave development may be tested using the same paradigm as in the experiments of Schwindt and Crill (1980b). If the wave has not developed completely in the long cable, then the return of the V_{in} to RP from moderately long lasting fixation of SD may block the advance of the wave, and the whole cable repolarizes (Figure 15.10(B)). The critical duration of SD fixation is shorter for shorter cables (Figure 15.10(B)). For moderate durations of the SD fixation in the very long cable the wave occupies only the proximal part of the cable.

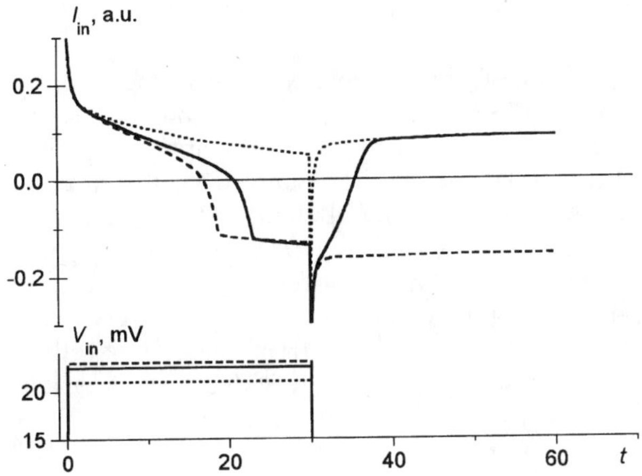

Figure 15.9 Model tail currents at clamped 10 mV following large depolarization. Dotted line: almost Ohmic tail-current; solid line: the slow decaying tail-current of the wind-up; stippled line: the non-decaying tail-current of bistability. $X = 1.5$.

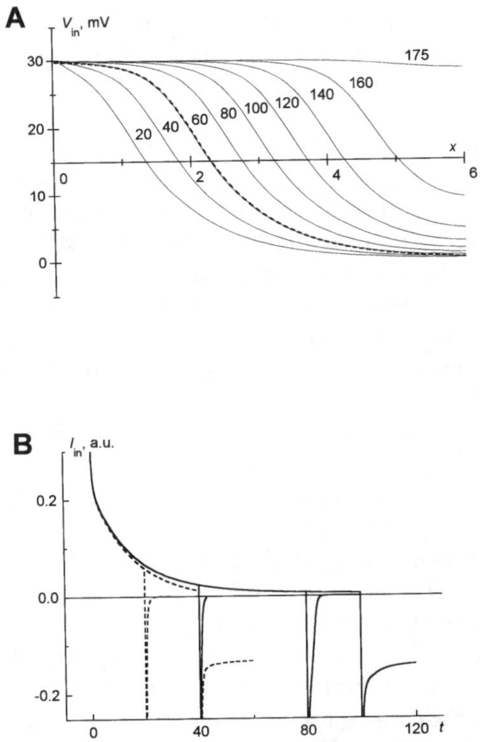

Figure 15.10 Wave of SD development in long cable with wind-up. (A) fronts of propagating wave in very long cable, $X = 6$; time indicated at the curves. Stippled line: analytical solution (Appendix C) for semi-infinite cable. (B) critical duration of fixed depolarization step for the non-decaying tail-currents. Solid lines: $X = 6$, stippled: $X = 3$.

Thus, following the clamp of the proximal end at RP, the wave is loaded from the both ends of the cable and may collapse. In not so long cables the wave can reach the distal membrane of the cable with the same duration of the SD clamp. Thus, following the fixation of RP, it is loaded only from the proximal end and may survive. The slow development of SD at the distal end of the cable, particularly prominent in the long cables, explains why the bistability hysteresis is not accompanied by a jump in current during moderately slow ramps. The jump simply requires long time to develop (compare Figure 15.8(B) and (C)).

Therefore, cables with wind-up differs from the cables with time-independent bistable C–V in much slower establishment of the bistability and shorter critical length for cable bistability.

5. CONCLUSIONS AND FUTURE PERSPECTIVES

The theory of bistable dendrites with time-independent C–V explains important electrophysiological features of some central neurons including bistable behavior and discharge pattern evoked by depolarizing current. However, it fails to account for the slow development, wind-up, of the inward current responsible for bistability.

We have introduced the wind-up of the membrane C–V in the model by means of a threshold of depolarization-dependent slow de-inactivation of the inward current. Now, there are two types of hysteresis of the cable C–V_{in} during voltage-ramps. A small hysteresis loop corresponds to the de-inactivation–inactivation sequence during the ramp. It develops if the ramp duration is at least comparable to the time-constant for the de-inactivation–inactivation but vanishes for very long ramps. A larger loop corresponds to the bistability of the distal membrane of the cable. The latter bistability should be considered in two contexts. There is bistability in a cable with an approximately constant level of inactivation and in a cable with completely developed inactivation–de-inactivation at every point corresponding to stationary potential distribution along the cable. The latter develops much slower and may be established in shorter cables.

We consider an additional possibility for experimental investigation of the properties of cells with wind-up. The two types of hysteresis may be discriminated if the experimenter uses voltage-ramps of different duration and step-by-step shifting of the maximal potential from test to test.

Future theoretical tasks include a more realistic imitation of the wind-up phenomenon and solving cable problems more relevant for real neurons. The revision of previous results on fine structure of the C–V_{in} of branched dendrites and the dependence of the discharge frequency on stimulus current is needed.

Dendritic bistability brings about neuronal multi-stability or at least bistability. There are few trivial implications for modeling of neural networks with the multi-stable cells (Gutman, 1984). First, long dendritic branches may accomplish all logical functions. In this way, neuron becomes an integrating circuit. Secondly, a sheet of non-interacting multi-stable cells may serve as a memory screen. Such a screen imitates iconic memory of phenomenal human beings.

More interesting is a non-trivial relation of multi- or bistable cells (Gutman, 1994, Gutman and Svirskis, 1997) to a neurodynamical principle postulated by Gelfand and Tsetlin for CNS (Tsetlin, 1969) in accordance with games of automata. They stated that afferentation (in general sense) is minimized, i.e., afferentation is meant to be punishment. Neural nets with bistable cells and balanced excitation and inhibition may implement the minimization of the afferent input (Gutman, 1994; Gutman and Svirskis, 1997) because: (a) they change their state from the depolarization plateau to the rest and vice versa under transient afferent command; (b) a stable state is achieved if afferentation becomes small enough when the network with bistable neurons functions properly.

The Gelfand–Tsetlin principles probably reflects the real mechanisms of CNS functioning. Macefield et al. (1991) recorded afferent activity in humans during long and strong contraction. In contradiction with the servo-motor theory, they found that Ia-afferentation declines with time in spite of fatigue. If the same contraction is repeated several times then the starting afferent flow becomes smaller and smaller. The last observation seems to be a general phenomenon: the firing is siginificant during learning and declines during memory consolidation at least in 'networks' involved in the motor control (Kandel et al., 1991).

Gelfand–Tsetlin principle is a meta-theory for two existing theories of the motor control. Firstly, the Feldman's λ-model of muscle length control (Feldman and Levin, 1995). The main gist of the model is stable equilibrium of the loaded muscle length determined by a neural command. The net with bistable cells has a stable equilibrium at the minimal afferentation and, thus, Gelfand–Tsetlin principle includes the λ-model. Secondly, Hogan–Flash principle of minimal jerks (Hogan and Flash, 1987). A smooth movement between fixed start and finish points during time T corresponds to $\min H$, where $H \sim \int_0^T (\mathrm{d}F/\mathrm{d}t)^2\,\mathrm{d}t$ and F is force. F is approximately proportional to the number of active motoneurons, n, if agonists and antagonists are not active at the same time, thus: $H \sim \int_0^T (\mathrm{d}n/\mathrm{d}t)^2\,\mathrm{d}t$. For bistable motoneurons $\mathrm{d}n/\mathrm{d}t$ is a measure of afferentation, then $H \sim \int_0^T (\text{afferentation})^2\,\mathrm{d}t$ and minimization of H is an example of the Gelfand–Tsetlin principle.

Bistable neurons can be introduced in the concept of the classical quasi-neuron. If a classical quasi-neuron has an excitatory suprathreshold autapse then the cell becomes functionally bistable: transient excitation evokes and transient inhibition stops steady firing. We hope that Gelfand–Tsetlin principle together with the concept of neuronal multi- or bistability may be exploited in neural network modeling.

PROBLEMS

15.1. C–V_{in} measurement using slow voltage ramp in cat and turtle motoneurons and mammalian Purkinje cells where dendritic bistability is shown. For discrimination between wind-up and bistability hysteresis one should use step by step change of the maximal voltage value or other parameters.

15.2. The application of the same experimental approach in CNS neurons where dendritic bistability was not looked for.

15.3. Modeling of C–V_{in} and wave propagation in cylindrical cable with more realistic wind-up imitation.
15.4. Analysis of electrophysiological phenomena in model cells based on morphological reconstruction.
15.5. Discharge frequency–current strength relation analysis using wind-up of membrane C–V_{in}.
15.6. Investigation in functional possibilities of neural nets with bi- or multistable neurons.

REFERENCES

Baginskas, A. and Gutman, A. (1990) Dependence of the excitatory synaptic currents on the clamped potential of the soma in a model of a neuron with non-linear dendrites. *Biophysics*, **35**, 495–502.

Booth, V. and Rinzel, J. (1995) A minimal, compartmental model for a dendritic origin of bistability of motoneuron firing pattern. *J. Comput. Neurosci.*, **2**, 299–312.

Butrimas, P. and Gutman, A. (1978) Theoretical analysis of an experiment with voltage clamping in the motoneurone. Proof of the *N*-shape pattern of the steady voltage–current characteristic of the dendrite membrane. *Biophysics*, **23**, 897–904.

Butrimas, P. and Gutman, A. (1980) Theoretical input voltage–current characteristic of the neuron with even shift of the clamped potential of the soma. RC-dendrites. *Biophysics*, **25**, 1105–1109.

Butrimas, P. and Gutman, A. (1981) Theoretical input voltage–current characteristic of the neurone with NC-dendrites on uniform shift of the clamped potential of the soma. *Biophysics*, **26**, 340–345.

Butrimas, P. and Gutman, A. (1983) Theoretical analysis of impulse activity of a neuron with stationary N-shaped current–voltage characteristic of dendritic membrane. Model considerations. *Neurofiziologija*, **15**, 404–411 (in Russian).

Dolphin, A.C. (1996) Facilitation of Ca^{2+} current in excitable cells. *Trends NeuroSci.*, **19**, 35–43.

Feldman, A.G. and Levin, M.F. (1995) The origin and use of positional frames of reference in motor control. *Behav. Brain Sci.*, **18**, 723–806.

Granit, R., Kernell, D. and Lamarre, Y. (1966) Synaptic stimulation superimposed on motoneurones firing in the secondary range to injected current. *J. Physiol. (London)*, **187**, 401–415.

Gustafsson, B. (1974) Afterhyperpolarization and the control of repetitive firing in spinal neurones of the cat. *Acta Physiol. Scand. Suppl.*, **416**, 1–47.

Gutman, A. (1971) Further remarks on the effectiveness of the dendritic synapses. *Biophysics*, **16**, 131–138.

Gutman, A. (1984) Nerve Cell Dendrites. Theory, Electrophysiology, Function. Vilnius, Lithuania: Mokslas (in Russian).

Gutman, A. (1991) Bistability of dendrites. *Internatl. J. Neural Syst.*, **1**, 291–304.

Gutman, A. (1994) Gelfand-Tsetlin principle of minimal afferentation and bistability of dendrites. *Internatl. J. Neural Syst.*, **5**, 83–86.

Gutman, A. and Svirskis, G. (1997) Bistability of motoneurons and minimization of afferent flow in motor control. In Proc. Internatl. Symp. "Brain and Movement", edited by V.S. Gurfinkel and Yu.S. Levik. Institute of Information Transmission Problems. Russian Academy of Science, Moscow, pp. 88–89.

Hogan, N. and Flash, T. (1987) Moving gracefully: quantative theories of motor co-ordination. *Trends Neurosci.*, **10**, 170–174.

Hounsgaard, J., Hultborn, H., Jespersen, B. and Kiehn, O. (1984) Intrinsic membrane properties causing bistable behavior of α-motoneurones. *Exp. Brain Res.*, **55**, 391–394.

Hounsgaard, J., Hultborn, H., Jespersen, B. and Kiehn, O. (1988) Bistability of α-motoneurones in the decerebrate cat and in the acute spinal cat after intravenous 5-hydroxytryptophan. *J. Physiol. (London)*, **405**, 345–367.

Hounsgaard, J. and Kiehn, O. (1989) Serotonin-induced bistability of turtle motoneurones caused by a nifedipine-sensitive calcium plateau potential. *J. Physiol.*, **414**, 265–282.

Hounsgaard, J. and Kiehn, O. (1993) Calcium spikes and calcium plateaux evoked by differential polarization in dendrites of turtle motoneurones *in vitro*. *J. Physiol.* (*London*), **468**, 245–259.

Jack, J.J.B., Noble, D. and Tsien, R.W. (1983) Electric current flow in excitable cells. Clarendon Press, Oxford.

Johnston, D., Magee, J.C., Colbert, C.M. and Christie, B.R. (1996) Active properties of neuronal dendrites. *Annu. Rev. Neurosci.*, **19**, 165–186.

Kandel, E.R., Schwartz, J.H. and Jessell, T.M. (1991) Principles of Neural Science. Prentice-Hall International, London.

Kavalali, E.T. and Plummer, M.R. (1994) Selective potentiation of a novel calcium channel in rat hippocampal neurones. *J. Physiol.* (*London*), **480**, 475–484.

Kavalali, E.T. and Plummer, M.R. (1996) Multiple voltage-dependent mechanisms potentiate calcium channel activity in hippocampal neurons. *J. Neurosci.*, **16**, 1072–1082.

Lee, R.H. and Heckman, C.J. (1996) Influence of voltage-sensitive dendritic conductances on bistable firing and effective synaptic current in cat spinal motoneurons *in vivo*. *J. Neurophysiol.*, **76**, 1967–1970.

Llinas, R. and Sugimori, M. (1980a) Electrophysiological properties of *in vitro* Purkinje cell somata in mammalian cerebellar slices. *J. Physiol.* (*London*), **305**, 171–196.

Llinas, R. and Sugimori, M. (1980b) Electrophysiological properties of *in vitro* Purkinje cell dendrites in mammalian cerebellar slices. *J. Physiol.* (*London*), **305**, 197–214.

Llinas, R., Sugimori, M., Hillman, D.E. and Cherskey, B. (1992) Distribution and functional significance of the P-type, voltage-dependent Ca^{2+} channels in the mammalian central nervous system. *Trends NeuroSci.*, **15**, 251–355.

Macefield, G., Hagbarth, K.-E., Gorman, R., Gandevia, S.C. and Burke, D. (1991) Decline in spindle support to α-motoneurones during sustained voluntary contractions. *J. Physiol.* (*London*), **440**, 497–512.

Mendell, L.M. (1966) Physiological properties of nonmyelinated fibre projection to the spinal cord. *Exp. Neurol.*, **16**, 316–332.

Nagumo, J., Yoshizawa, S. and Arimoto, S. (1965) Bistable transmission lines. *IEEE Trans. CT*, **12**, 400–412.

Russo, R.E. and Hounsgaard, J. (1994) Short-term plasticity in turtle dorsal horn neurons mediated by L-type Ca^{2+} channels. *Neuroscience*, **61**, 191–197.

Schwindt, P.C. and Calvin, W.H. (1973) Nature of conductance underlying rhythmic firing in cat spinal motoneurones. *J. Neurophysiol.*, **36**, 955–973.

Schwindt, P. and Crill, W.E. (1977) A persistent negative resistance in cat lumbar motoneurons. *Brain Res.*, **120**, 173–178.

Schwindt, P. and Crill, W.E. (1980a) Role of persistent inward current in motoneuron bursting during spinal seizures. *J. Neurophysiol.*, **43**, 1296–1318.

Schwindt, P. and Crill, W.E. (1980b) Properties of a persistent inward current in normal and TEA-injected motoneurons. *J. Neurophysiol.*, **43**, 1700–1724.

Sivilotti, L.G., Thompson, S.W.N. and Woolf, C.J. (1993) Rate of rise of the cumulative depolarization evoked by repetitive stimulation of small-caliber afferents is a predictor of action potential windup in rat spinal neurons *in vitro*. *J. Neurophysiol.*, **69**, 1621–1631.

Spruston, N., Schiller, Y., Stuart, G. and Sakmann, B. (1995) Activity-dependent action potential invasion and calcium influx into hippocampal CA1 dendrites. *Science*, **268**, 297–300.

Svirskis, G. and Hounsgaard, J. (1997) Depolarization-induced facilitation of a plateau-generating current in ventral hour neurons in the turtle spinal cord. *J. Neurophysiol.*, **78**, 1740–1742.

Tsetlin, M.L. (1969) Research in Theory of Automata and Modeling of Biological Systems. Nauka: Moscow (in Russian).

APPENDIX 15.1: LIST OF SYMBOLS AND ABBREVIATIONS

$C\text{–}V$	stationary membrane current–voltage relation
$C\text{–}V_{\text{in}}$	cable input current–voltage relation
$C\text{–}V_{\infty}$	membrane current–voltage relation for steady potential value in case of wind-up
$F\text{–}C$	discharge frequency as function of stimulus current
RP	resting membrane potential
RP-solution	solution approaching RP at the distal end of the cable
SD	steady depolarization
SD-solution	solution approaching SD at the distal end of the cable
c	wave velocity in an infinite bistable cable, electrotonic units
$f(V)$	$C\text{–}V$ multiplied by membrane resistance, mV
$f(V,N)$	$f(V)$ in case of wind-up
$f_{\infty}(V)$	$f(V)$ for steady potential value in case of wind-up
h	potential of unstable equilibrium depolarization, medial zero-point of $f(V)$
H	potential of steady depolarization
I_{in}	input current
k	$=-(\mathrm{d}f/\mathrm{d}V)\vert_h$, or negative membrane conductance at $V=\text{h}$, normalized to conductance at RP
K	$=(\mathrm{d}f/\mathrm{d}V)\vert_H$, membrane conductance at SD normalized to conductance at RP
L	$=1/\sqrt{k}$, length parameter for oscillatory solution around h, in length constant units
$N(t)$	variable of wind-up, dimensionless
R_{in}	$=(\mathrm{d}V_{\text{in}}/\mathrm{d}I_{\text{in}})$, differential input resistance of the cable
t	electrotonic time, in membrane time constant units
u	$=\int_0^H f(V)\,\mathrm{d}V$, $C\text{–}V$ parameter
V	transmembrane potential change relative to the resting potential
V_{in}	$=V(x=0)$, input potential at the proximal end of the cable
V_{RP}	maximal V_{in} possible for RP-solution
V_{S}	threshold of wind-up
V_{SD}	minimal V_{in} possible for SD-solution
V_1	potential at current maximum of stationary bistable $C\text{–}V$
V_2	potential at current minimum of linear-broken $C\text{–}V$
x	cable electrotonic axial coordinate, in length constant units
X	cable electrotonic length
Λ	electrotonic length of wave front in infinite bistable cable
θ	wind-up time constant
T	electrotonic duration of wave front in infinite bistable cable

APPENDIX 15.2: WAVE PROPAGATION IN INFINITE BISTABLE DENDRITES WITH TIME-INDEPENDENT *C–V*

The numerical solution of the wave propagation in bistable dendrite with time-independent *C–V* was already done (Butrimas and Gutman, 1981; Gutman, 1984, 1991). Here we shall do the same analytically. The wave solution of (15.6) and (15.7) for the broken-linear *C–V* (15.2) is presented as an ensemble of solutions for the three linear segments of $f(V)$; for the Ohmic segment near the RP, the negative resistance segment near h, and for the Ohmic segment near SD:

$$\frac{d^2V}{dx^2}+c\frac{dV}{dx}-V=0,\quad V\leqslant V_1,\quad V\to 0 \text{ as } x\to\infty;\quad \text{let } V(0)=V_1;$$

thus,

$$V=V_1\exp\left[-x\left(\frac{c}{2}+\sqrt{\frac{c^2}{4}+1}\right)\right];$$

$$\frac{d^2V}{dx^2}+c\frac{dV}{dx}+k(V-h)=0,\quad V_1\leqslant V\leqslant V_2; \tag{A15.1}$$

thus,

$$V=h+\frac{V_1-h}{\cos\varphi}\exp\left(-\frac{cx}{2}\right)\cos\left(x\sqrt{k-\frac{c^2}{4}}+\varphi\right);$$

$$\frac{d^2V}{dx^2}+c\frac{dV}{dx}-K(V-H)=0,\quad V\geqslant V_2,\quad V\to H \text{ as } x\to-\infty;\quad \text{let } V(-l)=V_2;$$

thus,

$$V=H+(V_2-H)\exp\left[(x+l)\left(\sqrt{K+\frac{c^2}{4}}-\frac{c}{2}\right)\right].$$

The potential continuity at V_1 is provisioned in (A15.1). Continuity of the axial current along the wave front at potentials V_1 and V_2 and the potential continuity at V_2 give the required equations for establishing of the parameters l, φ and c:

$$-V_1\left(\sqrt{1+\frac{c^2}{4}}+\frac{c}{2}\right)=(h-V_1)\left(\operatorname{tg}\varphi\cdot\sqrt{k-\frac{c^2}{4}}+\frac{c}{2}\right);$$

$$V_2=h+\frac{V_1-h}{\cos\varphi}\exp\left(\frac{cl}{2}\right)\cos\left(l\sqrt{k-\frac{c^2}{4}}+\varphi\right); \tag{A15.2}$$

$$\frac{h-V_1}{\cos\varphi}\exp\left(\frac{cl}{2}\right)\left[\sqrt{k-\frac{c^2}{4}}\sin\left(l\sqrt{k-\frac{c^2}{4}}+\varphi\right)+\frac{c}{2}\cos\left(l\sqrt{k-\frac{c^2}{4}}+\varphi\right)\right]$$
$$=(V_2-H)\left(\sqrt{K+\frac{c^2}{4}}-\frac{c}{2}\right).$$

The obvious relations from (15.2) were used for (A15.2): $h=(1+1/k)V_1$ and $H=V_2(1+k/K)-V_1(1+k)/K$.

APPENDIX 15.3: WAVE PROPAGATION IN INFINITE DENDRITES WITH WIND-UP OF BISTABLE C–V

Analytical solution of the wave propagation is possible when wind-up is considered as a threshold phenomena. Wind-up develops along the wave-front at points behind the point with the potential, $V=V_S$. Thus, for the wave-front segment ahead of this point, i.e., for $V<V_S$, the Ohmic solution and corresponding continuity conditions of (A15.1) and (A15.2) are valid:

$$V=V_S\exp\left[-x\left(\sqrt{1+\frac{c^2}{4}}+\frac{c}{2}\right)\right],\quad V<V_S. \tag{A15.3}$$

Here, $V(0)=V_S$ and, thus, (15.3) is valid for $x\geqslant 0$. The wind-up (15.10) holds for all $V\geqslant V_S$, i.e., for points $x\leqslant 0$:

$$N=\exp\left(-\frac{t}{\theta}\right)=\exp\left(\frac{x}{c\theta}\right),\quad V\geqslant V_S. \tag{A15.4}$$

Thus, according to (15.9):

$$\frac{d^2V}{dx^2}+c\frac{dV}{dx}+k(V-h)-[V_S(1+k)-kh]\exp\left(\frac{x}{c\theta}\right)=0;$$

$$V(0)=V_S,\quad \left.\frac{dV}{dx}\right|_{V_S}=-V_S\left(\sqrt{1+\frac{c^2}{4}}+\frac{c}{2}\right);\quad V_S\leqslant V\leqslant V_2; \tag{A15.5}$$

$$\frac{d^2V}{dx^2}+c\frac{dV}{dx}-K(V-H)-[V_S(1+k)-kh]\exp\left(\frac{x}{c\theta}\right)=0;$$

$$V(-\infty)=H;\quad V\geqslant V_2.$$

Continuity of potential and its derivative at $V=V_2$ for both solutions of (A15.5) is assumed. These solutions may be expressed analytically. For $V_S\leqslant V\leqslant V_2$ we get:

$$V=h+\frac{V_S(1+k)-kh}{[1/(c\theta)^2]+(1/\theta)+k}\exp\left(\frac{x}{c\theta}\right)$$

$$+\left[V_S-h-\frac{V_S(1+k)-kh}{[1/(c\theta)^2]+(1/\theta)+k}\right]\frac{\exp(-xc/2)}{\cos\varphi}\cos\left(x\sqrt{k-\frac{c^2}{4}}+\varphi\right);$$

$$\tan\varphi=\frac{V_S(\sqrt{1+(c^2/4)}+(c/2))+[(V_S(1+k)-kh)/((1/c\theta)+c+c\theta k)]}{[V_S-h-(V_S(1+k)-kh)/((1/(c\theta)^2)+(1/\theta)+k)]\sqrt{k-(c^2/4)}}-\frac{c}{2\sqrt{k-(c^2/4)}}. \tag{A15.6}$$

For $V \geqslant V_2$:

$$V=H-\frac{V_S(1+k)-kh}{[1/(c\theta)^2]+(1/\theta)-K}\exp\left(\frac{x}{c\theta}\right)+u\exp\left[x\left(\sqrt{K+\frac{c^2}{4}}-\frac{c}{2}\right)\right]. \quad \text{(A15.7)}$$

Now we have three equations for establishing parameters c, u and l. At the point, $x=l$, solutions of both (A15.6) and (A15.7) equal V_2 and their derivatives are equal.

CHAPTER 16
BIFURCATION ANALYSIS OF NEURONAL EXCITABILITY AND OSCILLATIONS

Shinji Doi, Junko Inoue, Shunsuke Sato and Charles E. Smith

1. INTRODUCTION

The Bonhoeffer–van der Pol (BVP) or the FitzHugh–Nagumo (FHN) neuronal model (FitzHugh, 1961; Nagumo et al., 1962; Barnes and Grimshaw, 1997) is well known to be a simplified and a tractable model of the Hodgkin–Huxley (HH) model (Hodgkin and Huxley, 1952), and preserves several neuronal properties such as an excitability, a refractoriness, a repetitive activity by an external current. Much work has been done on the response characteristics of the HH model to sinusoidal input (e.g. Holden, 1976a; Aihara et al., 1984), to periodic pulse trains (e.g. Guevara et al., 1983; Takabe et al., 1988), as well as on the response characteristics of the BVP model to sinusoidal input (Baer and Tier, 1986; Rajasekar and Lakshmanan, 1988a,b; Wang, 1989; Alexander et al., 1990; Braaksma and Grasman, 1993; Yasin et al., 1993; Rajasekar, 1996), to periodic pulse trains (Yoshizawa et al., 1982; Guevara et al., 1983; Sato and Doi, 1992; Nomura et al., 1993, 1994a; Othmer and Watanabe, 1994; Doi and Sato, 1995; Kaplan et al., 1996; Xie and Othmer, 1996), and also on the response characteristics of the squid giant axon to periodic current pulses (Takahashi et al., 1990; Kaplan et al., 1996), to sinusoidal input (Matsumoto et al., 1984). Other references on the response characteristics of different neuron models are included in Doi and Sato (1995).

Neuron models are mathematically described by a set of nonlinear parabolic differential equations and are considered to be a nonlinear dynamical system (Guckenheimer and Holmes, 1983; Degn et al., 1987). The influence of noise on nonlinear dynamical systems has been an object of intense investigations (Moss and McClintock, 1987) since real systems, especially biological systems, always have noise. The phenomenon of transitions induced by external noise has led to a revival of interest for the role of fluctuations in physical systems (Horsthemke and Lefever, 1984). For instance, in a multi-stable system which possesses several competing states of local stability, noise can be responsible for transitions between these states. Recently noisy systems have also received considerable attention within the context of stochastic resonance (Moss et al., 1993; Mannella and McClintock, 1995; Wiesenfeld and Moss, 1995).

Stochastic neuron models have a long history (Holden, 1976b; Ricciardi, 1977; Tuckwell, 1988, 1989) since random fluctuations of membrane voltage, which may originated from the random opening and closing of ion channels, are considered to be inevitable for a neuron (DeFelice, 1981; see also Chapter 2). Noise, as expected, washes out the detailed response characteristics of a neuron (Glass et al., 1980) and linearizes the nonlinear relation between an input current amplitude and the output frequency of a repetitive activity of a neuron (Clay, 1976; Yu and Lewis, 1989). Noise can also considerably change the propagation of impulses on a nerve fiber (Skaugen and Walloe, 1979; Horikawa, 1991, 1992, 1993; Fox, 1997).

In order to study the noise effects on neurons, it is necessary to consider stochastic (nonlinear) differential equations which are difficult to solve analytically and even are time-consuming to simulate numerically! Grasman and Roerdink (1989) analyzed the van der Pol relaxation oscillator with additive noise and reduced the problem of examining the period of the noisy oscillator to the analysis of the time necessary for a one-dimensional stochastic process to reach a boundary for the first time (a first-passage time). Tateno et al. (1995) extended this first-passage-time approach to the present case of a sinusoidal input and proposed a numerical framework which can be done with considerably short computation time and can analyze detailed noise effects on the periodically forced relaxation oscillator.

Using the method of Tateno et al. (1995), we study the noise effects on the (piecewise-linearized) Bonhoeffer–van der Pol neuronal model in the relaxation limit. The neuron model forced by a sinusoidal input displays various coupling patterns between the sinusoidal input and the neuron if some control parameters such as the amplitude and/or period of the periodic input are changed. The neuron may be sometimes 'entrained' or 'phase-locked' to the input so that for each m cycles of the input the neuron generates n spikes at fixed phase (or phases) of the input ($m:n$ phase locking). A phase-locking pattern is considered to be 'bifurcated' from another phase-locking pattern as a control parameter (bifurcation parameter) changes. A bifurcation is a typical phenomenon in nonlinear dynamical systems and a bifurcation analysis is a useful method to analyze nonlinear dynamical systems (Guckenheimer and Holmes, 1983). Thus we investigate the noise effects on the bifurcation phenomena in the neuron model.

Section 2 presents the BVP neuronal model and briefly explains the neuronal feature (excitability, refractoriness, repetitive activity, etc.) of the model. Section 3 considers the model in the (singular) limit of a parameter and studies the effects of a sinusoidal input on the singular model. The sinusoidally forced BVP model is analyzed using a one-dimensional (return) mapping of a 'phase' variable and examples of (deterministic) phase lockings and bifurcations are presented. The effects of the asymmetry of the model on the bifurcation structures are also analyzed. Section 4 analyzes the forced (singular) BVP model in the presence of noise using the method proposed by Tateno et al. (1995). In the presence of noise, the one-dimensional map is extended to a stochastic kernel function, and the iteration of the map is extended to an iteration of an operator defined by the stochastic kernel function. Stochastic phase lockings and stochastic bifurcations are analyzed in terms of the sequence of probability density functions produced by the operator. A new framework which analyzes the stochastic bifurcations in detail from the viewpoint of the spectral properties (eigenvalues and eigenfunctions) of the operator, is also proposed.

2. A SIMPLE MODEL IN THE CONTEXT OF A SPIKING NEURON AND A NEURONAL OSCILLATOR

The Bonhoeffer–van der Pol (BVP) model is derived from the well-known van der Pol equation by FitzHugh (1961) which mimics well the qualitative behavior such as excitability, refractoriness, repetitive activity of the Hodgkin–Huxley model. The BVP model is also called as the FitzHugh–Nagumo (FHN) model because the model is equivalent to the electronic circuit with a tunnel diode proposed by Nagumo et al. (1962).

The (space-clamped) BVP model is expressed by the following set of differential equations:

$$\dot{x} = c(x - x^3/3 + y + z), \tag{16.1a}$$

$$\dot{y} = -(x + by - a)/c, \tag{16.1b}$$

where $\dot{x}$ means $\mathrm{d}x/\mathrm{d}t$. The variable x denotes the membrane potential of the neuron with reverse sign, y is a 'recovery' variable which corresponds to the combination of Na^+ inactivation and K^+ activation of the HH model, and z the current stimulus applied externally. For the equations to mimic the behavior of a neuron, the parameters a, b and c are set in some range (FitzHugh, 1961; Nagumo et al., 1962; Hadeler et al., 1976), and the conventional values are $a = 0.7$, $b = 0.8$, $c = 3.0$.

In the following, we consider the much simpler version of the BVP model (16.1):

$$\varepsilon\dot{x} = f(x, y) \equiv y - x + \tfrac{5}{6}\{|x+1| - |x-1|\}, \tag{16.2a}$$

$$\dot{y} = g(x) \equiv -x + a, \tag{16.2b}$$

where the cubic polynomial in the original BVP model (16.1a) is approximated by a piecewise linear equation and the parameters are also simplified. Note that the externally applied constant current z of (16.1) is now considered to be included in the new parameter a of (16.2). We consider that this simplified BVP model still preserves the neuronal features of the original BVP model and of the HH model although this equation is extremely simpler than (16.1). Throughout, (16.2) will be referred to as the BVP model unless otherwise specified in text.

Let us explain the 'neuronal' features of the BVP model (16.2). Figure 16.1 shows the x–y phase plane of (16.2). The N-shaped line is the x-nullcline on which $\dot{x} = 0$ and orbits (solutions) of (16.2) move vertically, and the vertical line is the y-nullcline ($\dot{y} = 0$). So the intersection point P of the two nullclines is an equilibrium or a fixed point of (16.2) which corresponds to the resting state of a neuron. Typical orbits $(x(t), y(t))$ with different initial values of (16.2) are also displayed. If the neuron is in its resting state and a pulse like current is injected, the neuron responds and settles

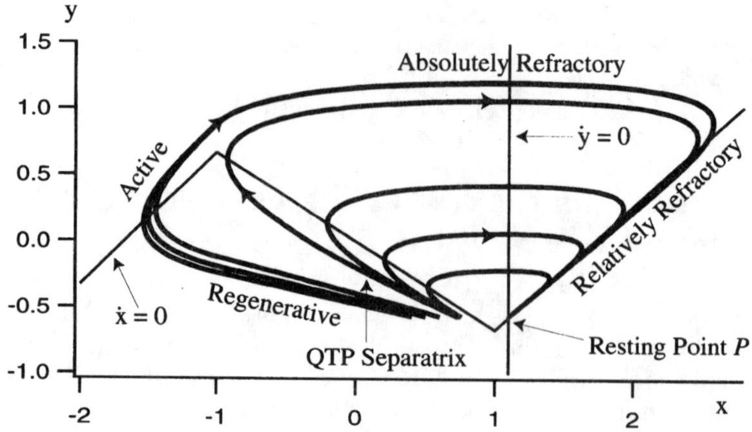

Figure 16.1 The x–y phase plane of (16.2) with the N-shaped x-nullcline ($\dot{x}=0$) and the linear y-nullcline ($\dot{y}=0$). The parameters are $\varepsilon=0.1$, $a=1.1$.

down back to the resting state. For example, an orbit starts at the equilibrium or the resting point P, and a large perturbation in the negative (depolarizing) direction of the x-axis displaces it leftwards, it goes through both the regenerative and active regions, finally comes back to P. If the perturbation is small, the orbit does not enter the regenerative region, i.e. the BVP neuron model is not excited. As is seen, a phase-plane analysis is very useful so that we can get the intuitive understanding of the total behavior of the model without need to make brute-force computer simulations of the model changing initial values.

The response of a BVP model to a single pulse perturbation is not all-or-none (neurons *in vivo* are not all-or-none also). There are neutral orbits which we cannot decide whether they excite or not. The orbits around QTP (quasi-type) separatrix are such type's orbits. In this region, the orbit changes its amplitude dramatically (and continuously) depending on its initial value. Thus, the QTP separatrix is considered as a (quasi-) threshold of the neuron model. Discussions on the geometric property of this threshold curve and on the excitability of the BVP neuron model (16.1) are made by Okuda (1981). The time waveforms $-x(t)$ of the orbits shown in this figure are plotted in Figure 16.2. Depending on the strength of input pulse current, various amplitude of responses are generated.

Figure 16.3 shows a response characteristics of the BVP neuron model to a single pulse stimulus. The abscissa shows the amplitude of such a depolarizing stimulus, and the ordinate shows the response amplitude (the maximum values of $-x(t)$) of the neuron model. In the case of $\varepsilon=0.1$ (solid curve), the response amplitude grows rapidly in the range [0.3, 0.4] of the abscissa. So the BVP model is considered to have a certain (quasi-) threshold property. If the parameter ε decreases to 0.05 (dashed curve), this threshold property becomes more strict; the response amplitude changes abruptly (but still continuously) near a value 0.28. In the limit case of $\varepsilon=0$, the BVP

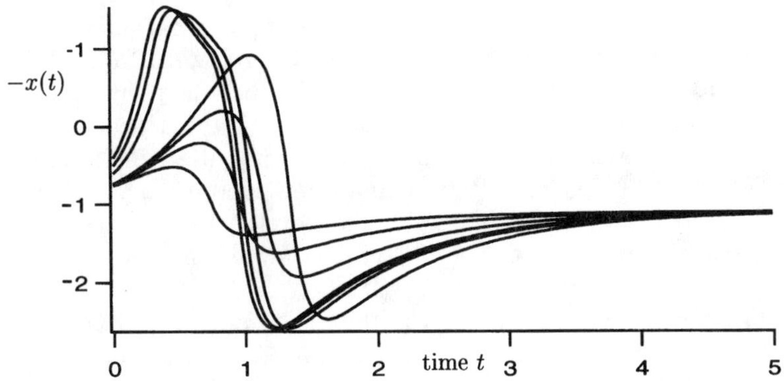

Figure 16.2 Response waveforms $-x(t)$ to various amplitude of single pulses. The time wave forms of various orbits which are shown in Figure 16.1 are plotted.

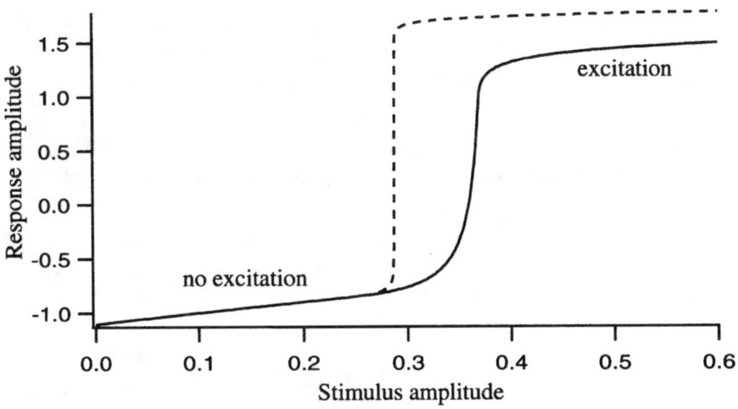

Figure 16.3 Response characteristics of the BVP model (16.2) to a single pulse input, with parameter values $\varepsilon=0.1$ (solid curve) and $\varepsilon=0.05$ (dashed curve). The amplitudes (the maximum values) of $-x(t)$ are plotted against the amplitudes of single pulse inputs.

model responds ideally (all-or-none) and the response characteristics become a step function.

Let $(x^*, y^*)=\left(a, a-\frac{5}{3}\right)$ denote the resting point P. This equilibrium point is stable when $|a|>1$ and unstable when $|a|<1$. In the latter case, (16.2) has a stable limit cycle and the BVP model excites repetitively. So, we can consider the BVP model (16.2) with $|a|>1$ as a spiking neuron (non-oscillatory) model, and one with $|a|<1$ as a neuronal oscillator (pacemaker cell) model.

A spiking neuron shows a repetitive activity when an external constant current is applied. Mathematically this repetitive activity corresponds to a limit cycle solution of a model. The HH model and the original BVP model (16.1) have such a limit cycle which is 'bifurcated' from an equilibrium point (a resting state) by the Hopf bifurcation (Hadeler et al., 1976; Hassard, 1978; Rinzel, 1978). No such bifurcation occurs in the piecewise-linearized BVP model (16.2) since the right hand side (r.h.s.) of (16.2) at the equilibrium point P is not differentiable at the critical parameter value $a=1$. This difference between (16.1) and (16.2) does not affect the total behavior of the model: The model (16.2) has a stable equilibrium point when $|a|>1$ and has both an unstable equilibrium point and a stable limit cycle when $|a|<1$. The amplitude of the limit cycle grows as the value of $|a|$ decreases away from unity. This qualitative behavior of (16.2) is the same as that of (16.1).

Figure 16.4 shows a repetitive activity of the BVP model with different parameter values $a=0.0$ (dashed curve) and $a=0.9$ (solid curve). As the value of a increases, the asymmetry of the waveform increases and the depolarized ($x(t)<0$) fraction of time in one cycle (period) decreases. This fraction of time is called an activity of a neuron model and the importance of asymmetry is already noted in the context of coupled oscillators (Kepler et al., 1990; Meunier, 1992). In the figure, the parameter ε is set much smaller than that used in Figures 16.1 and 16.2. As the value of ε approaches zero, the waveform of $-x(t)$ loses its smoothness.

As stated above, the parameter a may be considered as an 'external' current applied to a neuron. If we regard a as an 'intrinsic' property of a neuron, then the BVP model (16.2) with parameter $|a|<1$ oscillates spontaneously and thus we consider the neuron to be a neuronal oscillator or a pacemaker cell.

Note that the BVP model (16.2) with $a=0$ is equal to the piecewise-linearized version of the van der Pol oscillator (van der Pol, 1926; Grasman and Jansen 1979; Xu and Jiang, 1996).

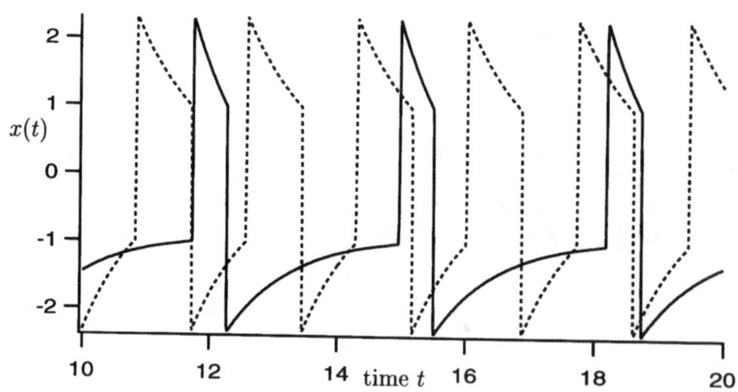

Figure 16.4 Repetitive activities of the BVP model (2) with parameter values: $a=0.0$ (dashed curve) and $a=0.9$ (solid curve). The parameter ε is set to 0.001 which is much smaller than that used in Figure 16.1 and the waveforms have a sharp shape. As the a value increases, the fraction time such that $-x(t)>0$ decreases and the waveform becomes asymmetric (neuron-like).

3. SINGULAR VERSION OF THE MODEL AND ITS RESPONSE TO SINUSOIDAL INPUT

In the following, we consider the BVP model (16.2) in the limit of $\varepsilon=0$. Figure 16.5 shows the x–y phase plane of (16.2) with $\varepsilon=0$. The parameter is set as $|a|<1$, so the equilibrium point is unstable and the model shows repetitive activity. The rectangle ABCD is a stable limit cycle.

In the limit of $\varepsilon=0$, the orbit with an initial value which is not on the x-nullcline ($\dot{x}=0$) instantaneously jumps to the x-nullcline in the horizontal direction since the velocity $\dot{x}=f(x,y)/\varepsilon$ becomes infinite unless $f(x,y)=0$. Thus, all orbits are considered to move on the x-nullcline. On the right (left) branch of the N-shaped x-nullcline, an orbit moves toward the point B (D, resp.). At the point B (D), the orbit instantaneously jumps to the point C (A, resp.), since we consider the limit of $\varepsilon=0$. This limit case is called a singular limit since orbits are not differentiable at such a jump point.

Note that all orbits move on the x-nullcline even for the case of a non-oscillatory parameter value $|a|>1$ although all orbits tend to equilibrium (resting) point after a suitable time in this case.

Let us introduce a sinusoidal external input to the BVP model (16.2):

$$\varepsilon\dot{x}=f(x,y)=y-x+5/6\{|x+1|-|x-1|\}, \tag{16.3a}$$

$$\dot{y}=g(x)+v(t)\equiv -x+a+A\sin\{2\pi(t/T+\theta_0)\}, \tag{16.3b}$$

where A, T, θ_0 are the amplitude, the period, the initial phase of the sinusoidal input, respectively.

Note that the external stimulus $v(t)$ appears on the r.h.s. of $\dot{y}$ rather than the r.h.s. of $\dot{x}$. It is usual to apply the stimulus to the r.h.s. of $\dot{x}$ since x is considered as a voltage

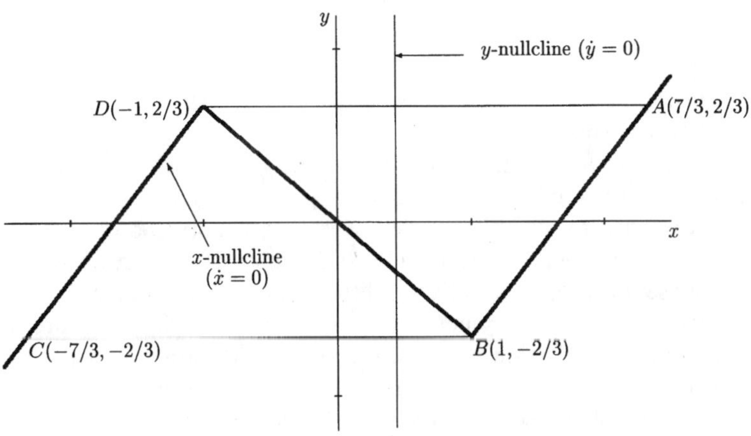

Figure 16.5 The x–y phase plane of (16.2) in the limit of $\varepsilon=0$ and the N-shaped x-nullcline ($\dot{x}=0$). The closed orbit ABCD is a limit cycle. The parameter a shifts the y-nullcline in the horizontal direction ($a=0.4$ in this figure).

and the dimension of the r.h.s. of $\dot{x}$ corresponds to a current. Mathematically, however, the application of a current stimulus to the r.h.s. of $\dot{y}$ is equivalent to the r.h.s. of $\dot{x}$ because (Alexander et al., 1990): If $(x(t), y(t))$ is a solution of (16.3) and define $\xi(t) \equiv x(t)$ and $\eta(t) \equiv y(t) - \int v(t)\,dt$, then $(\xi(t), \eta(t))$ is a solution of

$$\varepsilon\dot{\xi} = f(\xi, \eta) + \int v(t)\,dt,$$

$$\dot{\eta} = g(\xi).$$

Thus the addition of $v(t)$ to the r.h.s. of $\dot{y}$ is equal to that of $\int v(t)\,dt$ to the r.h.s. of $\dot{x}$.

In the singular limit, (16.3) is reduces to

$$\dot{y} = g(x) + v(t), \; f(x, y) = 0$$

or more explicitly, to

$$\dot{x} = \dot{y} = -x + a + A\sin\{2\pi(t/T + \theta_0)\}, \tag{16.4a}$$

$$y = \begin{cases} x - \frac{5}{3} & (x \geqslant 1), \\ x + \frac{5}{3} & (x \leqslant -1). \end{cases} \tag{16.4b}$$

This differential equation is a one-dimensional (piecewise) linear equation. Thus we can obtain its analytical solution piecewise (in each region $x \geqslant 1$ or $x \leqslant -1$). However, a total solution which connects two solutions of both region $x \geqslant 1$ and $x \leqslant -1$ cannot be obtained since we have to solve a transcendental equation which includes both exponential functions and trigonometric functions.

Note that all orbits move on the x-nullcline of the x–y phase plane (see Figure 16.5) even in the case of the presence of the sinusoidal input; the sinusoidal input $v(t)$ changes the velocity of orbits on the x-nullcline.

3.1 Phase Lockings

Figure 16.6 shows examples of responses of the BVP model (16.3) with $\varepsilon = 0.001$ to a sinusoid. Part (a) corresponds to a neuronal oscillator ($a = 0.9$) and (b)–(d) to a (non-oscillatory) neuron ($a = 1.1$). In (a) and (b), the neuron model excites once during one period of a sinusoid. This response pattern is called a 1 : 1 phase locking. Generally, a response in which m cycles of an input correspond to n spikes of a neuron is called an $m:n$ phase locking. Parts (c) and (d) show a 2 : 1 phase locking and a 1 : 2 phase locking, respectively. Note that we do not see any essential difference between an oscillatory neuron (a) and a non-oscillatory neuron (b).

In order to study the relation between the sinusoidal forcing and the forced BVP model, let us observe the sinusoidal input whenever the orbit (x, y) visits the point A of the x–y phase plane (see Figure 16.5). If the parameter a is set in the range of $|a| > 1$ (non-oscillatory neuron), orbits will stay near the resting point P and may not visit the point A without some input $v(t)$. If the amplitude of $v(t)$ is sufficiently large, an orbit can excite (move to point C) and then visit the point A even in the case of non-oscillatory neuron.

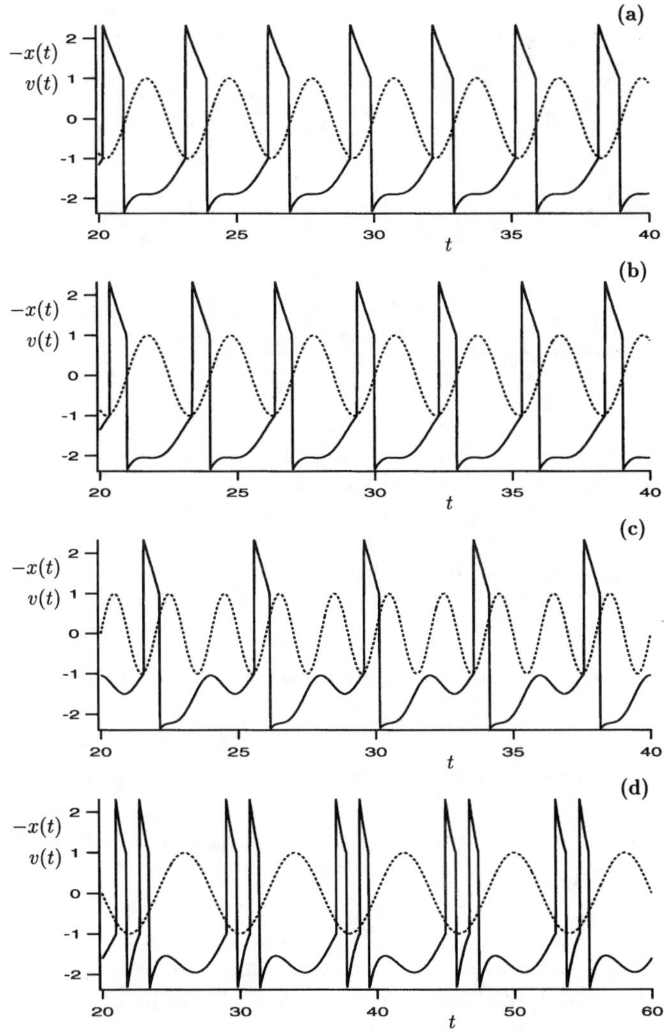

Figure 16.6 Examples of phase lockings of the forced BVP model (3). The amplitude A of the sinusoidal input $v(t)$ is set to 1.0 and other parameters are: (a) $a=0.9$, $T=3.0$ (b) $a=1.1$, $T=3.0$ (c) $a=1.1$, $T=2.0$ (d) $a=1.1$, $T=8.0$. The solid curves are waveforms of $-x(t)$ and dashed curves are that of $v(t)$.

Figure 16.7 is an example of waveforms of both $v(t)$ and $x(t)$. The point A of the x–y phase plane corresponds to the peaks of the waveform $x(t)$. The phase of the sinusoidal input (at the point A) is a number between 0 and 1 normalized by the period T. Note that this terminology 'phase' is irrelevant to the 'phase' of the x–y phase plane. Suppose that the orbit starts at the point A of the x–y phase plane with an initial phase θ_0 of the sinusoidal input and that the orbit returns to the point A

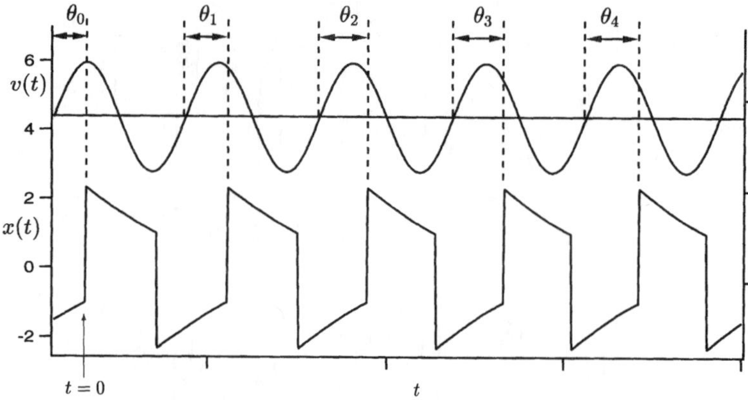

Figure 16.7 Phase of the sinusoidal input (a 1 : 1 phase locking). Waveforms of both $x(t)$ and the input $v(t)$ are plotted ($v(t)$ is shifted upwardand its scale is arbitrary). $A=0.3$, $T=1.5$. The peaks (maximum values) of $x(t)$ correspond to the point A of the x–y phase plane (Figure 16.5). At a time $t=0$, the initial phase of the sinusoidal input is θ_0 and the phase is changed to θ_1, $\theta_2, \ldots$ as the time elapses. In this case, the sequence of the phases converges to a certain fixed value θ^*; a 1 : 1 phase locking occurs.

again after a time t_1. Then the second phase θ_1 is obtained by

$$\theta_1 = p(\theta_0) \stackrel{\text{def}}{=} \theta_0 + t_1/T \pmod 1. \tag{16.5}$$

The relation $p(\theta_0)$ between θ_0 and θ_1 is called a return map or a phase transition curve (PTC).

Using the return map, phases $\theta_2, \theta_3, \ldots$ are recursively defined by the one-dimensional mapping:

$$\theta_{n+1} = p(\theta_n), \quad n = 0, 1, 2, \ldots \tag{16.6}$$

The sequence $\{\theta_n\}$ generated by this equation is also called the orbit of (16.6).

Thus, in order to analyze the phase lockings of the forced BVP model (16.3), we investigate the asymptotic properties of the sequence (orbit) $\{\theta_n\}$ of phases. In Figure 16.7, the sequence $\{\theta_n\}$ asymptotically approaches certain values θ^*; in this case a 1:1 locking occurs. Figure 16.8 shows an example of the return map $p(\theta)$ and its orbit $\{\theta_n\}$ which corresponds to Figure 16.7. The graph of $p(\theta)$ intersects with a diagonal line at two points (indistinguishable in this figure) and the lower intersection point θ^* is a stable fixed point of (16.6). Thus an orbit θ_0, $\theta_1, \ldots$, with any initial phase θ_0 asymptotically converges to this fixed point θ^*, which shows a 1:1 phase locking occurs.

Note that we cannot obtain the explicit form of $p(\theta)$ analytically by the same reason as stated in the beginning of section 3. So, we must numerically calculate $p(\theta)$ by solving transcendental equations by the Newton method. The computation time to yield $p(\theta)$ is very short due to the convergence of the Newton method, and is much faster than the numerical simulation of differential equations.

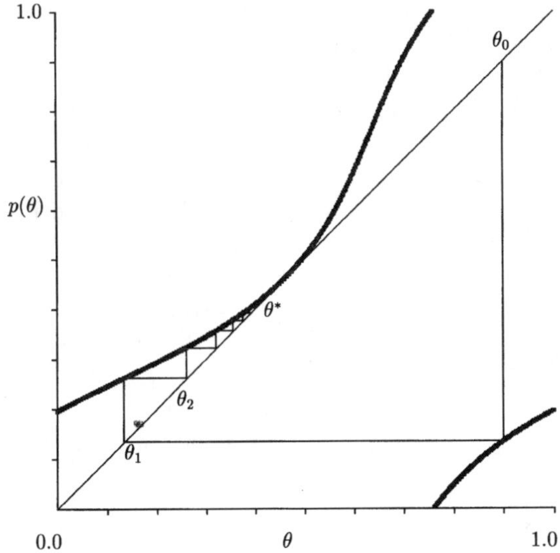

Figure 16.8 Return map $p(\theta)$ (thick curve) and its orbit $\{\theta_n\}$ (thin lines). $A=0.3$, $T=1.5$. The orbit θ_0, $\theta_1,\ldots$ asymptotically converges to a fixed point θ^* which corresponds to a 1:1 phase locking of Figure 16.7.

Generally, in the case of $m:n$ locking, an orbit or sequence $\{\theta_n\}$ asymptotically approaches to an n-periodic sequence:

$$\theta^{(1)}, \theta^{(2)}, \ldots, \theta^{(n)}, \theta^{(1)}, \ldots$$

3.2 Bifurcation Diagrams

Patterns of phase lockings depend on both the amplitude and the period of the sinusoidal input. One phase-locking pattern is considered to be bifurcated from the other phase-locking pattern as a (bifurcation) parameter is changed. A bifurcation diagram is a very useful tool to systematically investigate the effect of a parameter change on a system's behavior.

Figure 16.9 shows how the asymptotic value(s) of the sequence $\{\theta_n\}$ is changed depending on the input period T. Part (a) is a bifurcation diagram of $a=0.0$ (original van der Pol oscillator). For example, in the $T=2$ case, only one dot is plotted in the vertical direction ($\theta^* \approx 0.14$); a 1:1 locking occurs. In the $T=7$ case, three dots are plotted and a 1:3 locking occurs; one cycle of the sinusoidal input synchronizes with three cycles of the oscillator. In the $T=4$ case, many points are plotted; no locking occurs (in this case, the sequence $\{\theta_n\}$ approaches a so-called quasi-periodic sequence rather than a periodic sequence). In a quasi-periodic case, the oscillator is not phase-locked to the input and the phase of the input takes infinitely many values between 0 and 1 although the oscillator seems to be locked for a short time range (quasi-periodic orbits are different from chaotic ones).

In part (b), the value of a is set to 0.5. As shown in section 2, the asymmetry of the waveform increases if the absolute value of a increases. The larger $|a|$ becomes, the narrower the T range of quasi-periodic orbits becomes, although the total bifurcation structure (1 : 1, 1 : 3 lockings of (a)) does not change much. From this observation, we can see that the asymmetry of an oscillator increases the tendency to phase lockings. In parts (c) and (d), the value $|a|$ is increased more. We can still see the similarity in all the figures (a)–(d) although the bifurcation diagrams seem to be prolonged to the right direction as $|a|$ increases. A notable point is that we can see no essential difference between (c) and (d) although the part (c) corresponds to a neuronal oscillator while the part (d), to a non-oscillatory neuron.

We note that the intrinsic period T_a^{Int} of the singular BVP model which has a repetitive activity ($|a|<1$) without the sinusoidal input ($v(t)\equiv 0$) is given by

$$T_a^{\mathrm{Int}} = \ln \frac{(7/3)^2 - a^2}{1 - a^2}$$

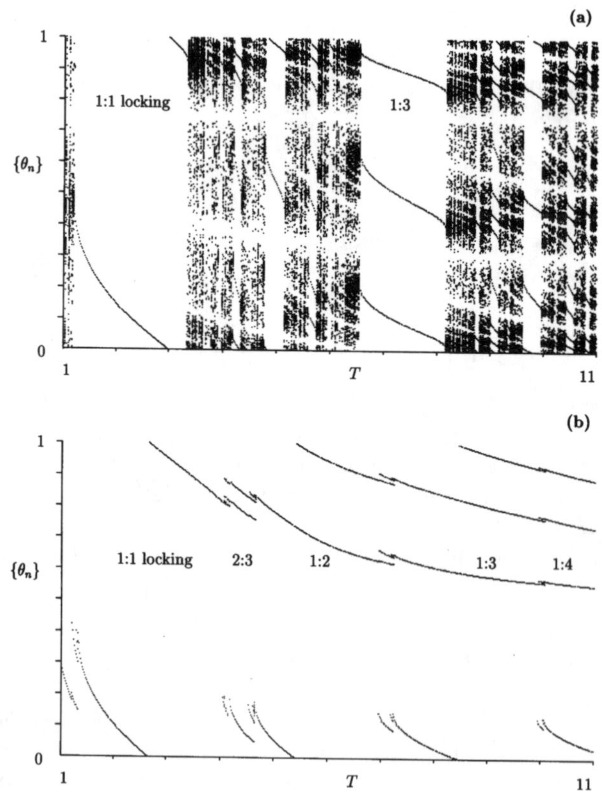

Figure 16.9(a) and (b)

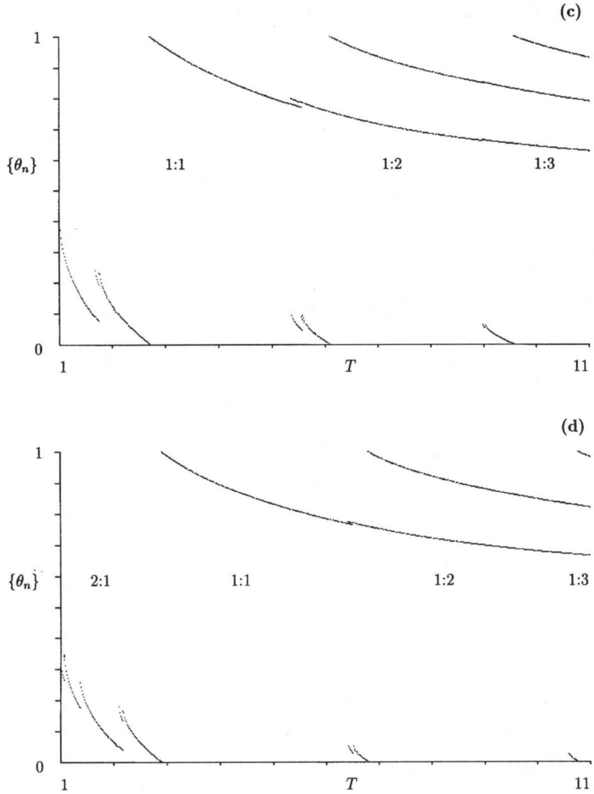

Figure 16.9 Bifurcation diagrams of the return map $p(\theta)$. $A=1.0$. A stationary (or asymptotic) sequence after some transient initial sequence $\{\theta_n\}$, $n=101,\ldots,600$ produced by (16.6) were plotted for each of 600 equally spaced T values on the interval [1, 11]. (a) $a=0$ (the van der pol oscillator) (b) $a=0.5$ (the BVP oscillator) (c) $a=0.9$ (the BVP oscillator) (d) $a=1.1$ (non-oscillatory neuron).

and does not change much for a parameter range $0 \leqslant |a| \leqslant 0.5$:

$$T^{\text{Int}}_{0.0}=1.69,\ T^{\text{Int}}_{\pm 0.3}=1.77,\ T^{\text{Int}}_{\pm 0.5}=1.94,\ T^{\text{Int}}_{\pm 0.9}=3.19.$$

Thus the difference between (a) and (b) is caused not by the difference of intrinsic period but by the asymmetry.

4. A SINGULAR VERSION OF THE MODEL WITH ADDITIVE NOISE

Real neurons are noisy since the opening and closing of ion channels are associated with fluctuations (see Chapter 2). Thus let us study the noise effects on the forced

BVP model (16.3) and thus we consider a stochastic differential equation:

$$\varepsilon\dot{x} = y - x + \frac{5}{6}\{|x+1| - |x-1|\}, \tag{16.7a}$$

$$\dot{y} = -x + a + A\sin\{2\pi(t/T + \theta_0)\} + \sigma\frac{\mathrm{d}W(t)}{\mathrm{d}t}, \tag{16.7b}$$

where $W(t)$ is the standard Wiener process or the Brownian motion (Gardiner, 1983; Tuckwell, 1989) and the term $\sigma\,\mathrm{d}W(t)/\mathrm{d}t$ denotes a Gaussian white noise with noise intensity σ. We also suppose $\varepsilon = 0$ and then this equation is reduced to

$$\dot{x} = -x + a + A\sin\{2\pi(t/T + \theta_0)\} + \sigma\frac{\mathrm{d}W(t)}{\mathrm{d}t}, \tag{16.8a}$$

$$y = \begin{cases} x - \frac{5}{3} & (x \geqslant 1), \\ x + \frac{5}{3} & (x \leqslant -1). \end{cases} \tag{16.8b}$$

In order to investigate the noise effects in detail, the parameter a of the BVP model (16.7) or (16.8) is set to zero in the following numerical examples. Thus we are specifically considering the noise effects on the forced (piecewise-linearized) van der Pol oscillator. All the following analysis, however, does not depend on this specific choice of the parameter value a, and is valid for all parameter values a, even for the case $|a| > 1$ (a non-oscillatory neuron).

4.1 Stochastic Phase Lockings

In the presence of noise, both variables θ_0, θ_1 defined in (16.5) fluctuate by noise and are thus random variables Θ_0, Θ_1. We extend the deterministic map $p(\theta)$ to the case with noise as follows. Define a kernel function $g(\theta_0, \theta_1)$ using conditional probability density functions:

$$g(\theta_0, \theta_1)\mathrm{d}\theta_1 = \Pr\{\theta_1 \leqslant \Theta_1 \leqslant \theta_1 + \mathrm{d}\theta_1 | \Theta_0 = \theta_0\}. \tag{16.9}$$

The function $g(\theta_0, \theta_1)$ can be calculated numerically without simulations of the stochastic differential equations (16.7) or (16.8) (Tateno et al., 1995).

Figure 16.10 is an example of the kernel function g which corresponds to the deterministic return map of Figure 16.8. The function g takes relatively high values along the curves of the return map p. Thus g can be considered as the stochastic extension of the deterministic return map. The heights and widths of the peaks of g depend on the values of (θ_0, θ_1) and thus we can see that the effects of blurring of the deterministic map by noise are not uniform.

Using the kernel function g, we extend the system (16.6) to the noisy case as follows. Let S denote a unit interval $[0,1]$ and $\mathscr{D}$ the set of absolutely integrable functions with a unit L^1 norm on S. An operator $\mathscr{P}$ on $\mathscr{D}$ is defined by

$$\mathscr{P}h(\theta) = \int_S g(\theta_0, \theta)h(\theta_0)\mathrm{d}\theta_0, \quad h \in \mathscr{D}. \tag{16.10}$$

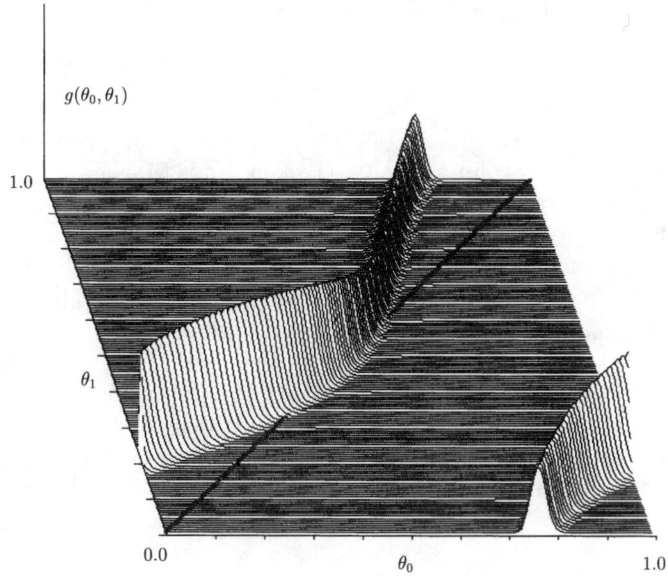

Figure 16.10 Stochastic kernel function $g(\theta_0, \theta_1)$. $a=0.0$, $A=0.3$, $T=1.5$, $\sigma=0.03$. This figure corresponds to the deterministic return map of Figure 16.8.

This operator is called a Markov operator since it transforms a probability density function to an another probability density function (Lasota and Mackey, 1994). Let $h_0(\theta)\in\mathscr{D}$ denote the probability density function of the initial phase Θ_0 when an orbit starts at the point A of the x–y phase plane. Then the density function of Θ_1 when the orbit returns to the point A again is obtained by $h_1(\theta)=\mathscr{P}h_0(\theta)$. Thus, the deterministic mapping (16.6) is extended to the system

$$h_{n+1}(\theta)=\mathscr{P}h_n(\theta), \quad n=0,1,2,\ldots \tag{16.11}$$

Thus we investigate the asymptotic behavior of the sequence $\{h_n(\theta)\}$ of probability density functions rather than the sequence $\{\theta_n\}$ of phases.

Let us list several preliminary definitions (Lasota and Mackey, 1994). A function $h^*(\theta)$ is called the *invariant density* function of an operator $\mathscr{P}$ if the relation $\mathscr{P}h^*=h^*$ holds. The invariant density is *asymptotically stable* if for any initial density function $h_0\in\mathscr{D}$

$$\lim_{n\to\infty} \|\mathscr{P}^n h_0 - h^*\| = 0,$$

where $\|\cdot\|$ is the L^1 norm on S.

As is easily seen from its definition, the function g has the property

$$g(\theta_0, \theta_1)\geqslant 0, \quad \int_S g(\theta_0, \theta_1)\mathrm{d}\theta_1 = 1$$

and is called a stochastic kernel. The inequality

$$\int_S \inf_{\theta_0} g(\theta_0, \theta_1) d\theta_1 > 0 \tag{16.12}$$

is assumed to hold (Tateno et al., 1995). The operator with this property has a unique asymptotically stable invariant density (Lasota and Mackey, 1994). Thus the sequence $\{h_n(\theta)\}$ produced by the operator $\mathscr{P}$ asymptotically approaches a unique invariant density as $n \to \infty$. The invariant density is considered as the stationary probability density function of the phase Θ_n.

Figure 16.11 is an example of the density sequence $\{h_n(\theta)\}$. The uniform initial density function $h_0(\theta)$ changes its shape and approaches an invariant density function h^* ($\approx h_{30}$) with one sharp peak, which shows a 1 : 1 phase locking does occur in a stochastic sense. The peak of the density functions $\{h_n(\theta)\}$ is highest near $n=7$ and is relatively lower in the invariant density h^*, which means that fluctuations of phases are bigger in the stationary state than in the transient state. This phenomenon is due to the fact that the peaks of the kernel function $g(\theta_0, \theta_1)$ near $\theta_0 = \theta_1 = \theta^*$ are relatively lower than the other (cf. Figure 16.10 and Figure 16.8).

Figure 16.12 (a) is also an example of the density sequence in the case of stochastic 5 : 3 phase locking. The initial density function $h_0(\theta)$ has a high mode near $\theta = 0.42$ which is one of the three phases to which oscillator is locked in the noise-free case. The functions h_1 and h_2 have a high mode near the other two phases of the

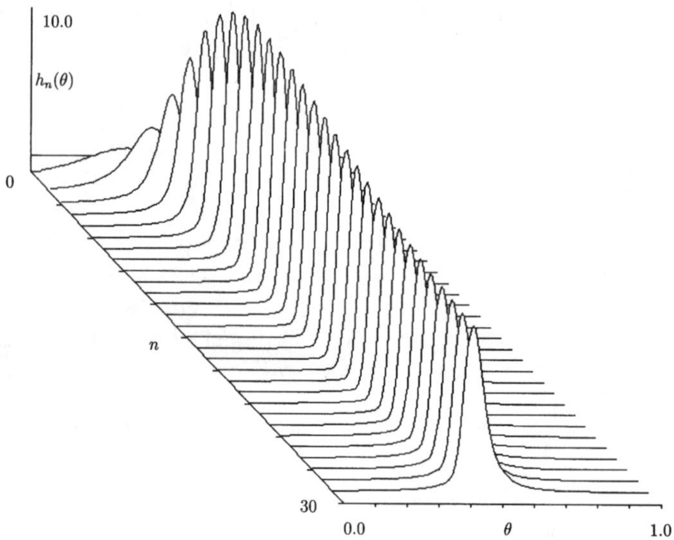

Figure 16.11 Density evolution (sequence) $\{h_n(\theta)\}$ of a stochastic 1 : 1 phase locking. $A=0.3$, $T=1.5$, $\sigma=0.03$. A uniform initial density function evolves to an invariant density function $h^* \approx h_{30}$.

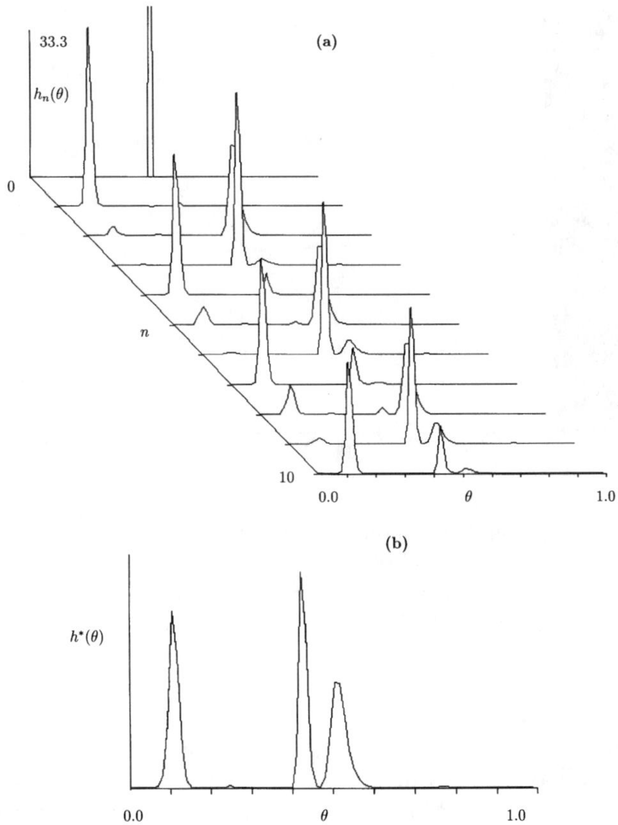

Figure 16.12 (a) Density evolution (sequence) $\{h_n(\theta)\}$ and (b) its invariant density $h^* \approx h_{200}$ of a stochastic 5 : 3 phase locking. $A=2.0$, $T=0.85$, $\sigma=0.02$.

three phases. This sequence seems to initially vary with period 3, but finally converges to an invariant density function (part (b)) with three sharp peaks, which shows a 5 : 3 locking occurs in a stochastic sense.

In the noise-free case, the amplitude A of the sinusoid should be sufficiently large in order to define the deterministic map $p(\theta)$ in the case of $|a|>1$ (non-oscillatory neuron), since the neuron may not excite (and thus orbits of the BVP model may not visit the point A of the x–y phase plane) otherwise. Note that this condition is not necessary in the noisy case since there are always positive possibilities (probabilities), for all values of a, that a non-oscillatory neuron excites owing to noise.

4.2 Stochastic Bifurcation Diagrams

Patterns of deterministic phase lockings depend on both the amplitude A and the period T of the input. Figure 16.13 shows a deterministic bifurcation diagram of the

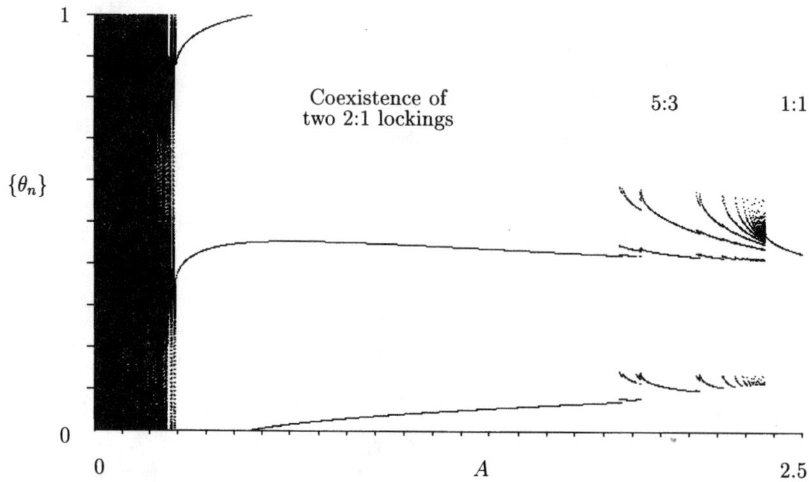

Figure 16.13 Deterministic bifurcation diagram of the return map $p(\theta)$. The bifurcation parameter is the amplitude A rather than the period T, which is fixed to $T=0.85$. A sequence $\{\theta_n\}$, $n=101,\dots,600$ produced by Eq. (6) were plotted for each of 600 equally spaced A values on the interval [0, 2.5].

return map p for the noise-free ($\sigma=0$) case with a bifurcation parameter A rather than T (cf. Figure 16.9). In this case, the period of the sinusoidal input is about half of the intrinsic period of the van der Pol oscillator. So, for a wide range of the amplitude A, 2 : 1 phase lockings occur (two cycles of the input synchronize with one cycle of the oscillator). In particular, two different patterns of 2 : 1 phase lockings coexist; the pattern depends on both the initial phase of the input and the initial state of the oscillator.

Various phase-locking patterns are bifurcated as the value of A varies. If A is large enough ($A>2.4$), one cycle of the input can synchronize with the one cycle of the oscillator although the intrinsic periods differ twofold.

Figure 16.14 is a stochastic or a density bifurcation diagram which corresponds to the deterministic one (Figure 16.3). Invariant density functions are plotted with various value of A for two different noise intensities (a) $\sigma=0.05$ and (b) $\sigma=0.2$. In part (a), corresponding to the deterministic bifurcation diagram, the modes and the shape of the invariant density are changed as the bifurcation parameter A varies. Particularly, in the range of $1.8<A<2.4$, the invariant densities have several peaks and complicated shapes although the shapes are not so complicated as the deterministic one.

In the case of large noise (part (b)), the stochastic bifurcation diagram is very simple; the invariant densities have at most two peaks. The shapes of invariant densities do not depend on the amplitude A; only the heights of the two peaks depend on A. Thus noise washes out the dependency of the density shapes (the number of modes) on the amplitude of the input.

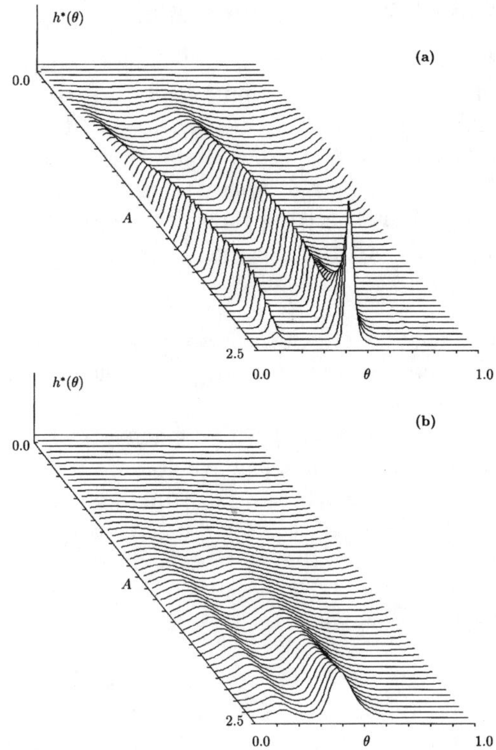

Figure 16.14 Stochastic (invariant density) bifurcation diagram. $T=0.85$. Invariant densities $h^*(\theta)$ were plotted for each of 50 equally spaced A values on the interval [0, 2.5]. (a) $\sigma=0.05$ (b) $\sigma=0.2$.

In the deterministic dynamical systems, the word 'bifurcation' means qualitative change (number, stability) of solutions of a system. As stated above, the equation

$$\mathscr{P}h^* = h^*$$

which the invariant density satisfies, always has a unique asymptotically stable solution and thus no bifurcation of this equation occurs in such a sense. So what is a stochastic bifurcation? This is a newly developing field (Arnold, 1995). One (classical) definition of stochastic bifurcation is based on the 'shape' of invariant density functions. For example, near $A=0.29$ of Figure 16.13, two 1-periodic orbits (2:1 phase lockings) of the mapping (16.6) bifurcate from a quasi-periodic orbit. In the presence of noise, the density functions always have two peaks near $A=0.29$ (cf. Figure 16.14); the densities do not change their shape. So, in the presence of noise, we cannot see any stochastic bifurcations of a 2:1 phase locking. In the next section, we will give an alternative approach which analyzes stochastic bifurcations more quantitatively.

4.3 Spectral Analysis of Stochastic Bifurcations

The linear operator $\mathscr{P}$ has all information about the dynamics of the forced neuron model (16.7) in the limit case of $\varepsilon=0$. Numerically, this operator $\mathscr{P}$ (kernel g) is discretized and is expressed by a matrix (the density functions $\{h_n(\theta)\}$ are vectors). This section analyzes the spectral properties (eigenvalues and eigenfunctions) of the matrix. The operator $\mathscr{P}$ is positive (all elements) and is a stochastic matrix (the sum of every column is unity). Many spectral properties are known about such a positive (or non-negative) matrix (theorem of Perron–Frobenius) and are summarized in (Gantmacher, 1959):

(ia) An irreducible non-negative (stochastic) $n \times n$ matrix A has a real eigenvalue 1 which is a simple root of characteristic equation and the moduli of other eigenvalues do not exceed 1.
(ib) The 'maximal' eigenvalue 1 has a positive eigenvector and there are no other (linearly independent) non-negative eigenvectors.
(ic) If A has k eigenvalues $\lambda_0=1, \lambda_1, \ldots, \lambda_{k-1}$ of modulus unity, these numbers are all distinct roots of $\lambda^k-1=0$ and the whole spectrum $\lambda_0=1, \lambda_1, \ldots, \lambda_{n-1}$ of A are invariant under a rotation with the angle $2\pi/k$ of the complex plane.
(ii) Moreover, if all elements of A are positive, the maximal eigenvalue 1 exceeds the moduli of all other eigenvalues.

The operator $\mathscr{P}$ with small noise is practically considered to be a non-negative matrix (some elements might be zero practically) although it should be positive theoretically. So it is useful to keep both properties of non-negative matrices (i) and positive matrices (ii) in mind. We note that the eigenfunction (eigenvector) which belongs to the maximal eigenvalue 1 is the invariant density function h^* of the operator $\mathscr{P}$ and that the eigenvalue with largest modulus other than 1 governs the convergence speed of the sequence $\{h_n\}$ to this invariant density.

Figure 16.15 shows examples of the whole spectra (eigenvalues) of operators $\mathscr{P}$ with different values of A. In all figures, we can see that the operator has a unique maximum eigenvalue 1. Part (a) corresponds to the deterministic (noise-free) quasi-periodic case (cf. Figure 16.13). We can see many complex eigenvalues between zero and unity eigenvalues. Part (b) and (c) correspond to the deterministic 2:1 phase locking. Comparing (a)–(c), we can see that eigenvalues change their values from complex to real in order of the modulus, which represents a bifurcation from quasi-periodic to 2:1 phase locking in a stochastic sense. Part (d) corresponds to the deterministic 5:3 phase locking (3-periodic orbit of (16.6)). We can see that the whole spectrum is roughly invariant under an angle $2\pi/3$ rotation of complex plane which is a sign of a stochastic 5:3 phase locking.

In Figure 16.16, the modulus and the angles of the second eigenvalue of the operator $\mathscr{P}$ are plotted against A values. Note that a bifurcation from quasi-periodic orbits to 2:1 phase lockings occur in the corresponding deterministic case (cf. Figure 16.13). Near $A=0.35$, the second eigenvalue changes its value from complex to real abruptly and the modulus also abruptly changes from decreasing to increasing. Thus we can say that this point is a bifurcation point of a 2:1 phase

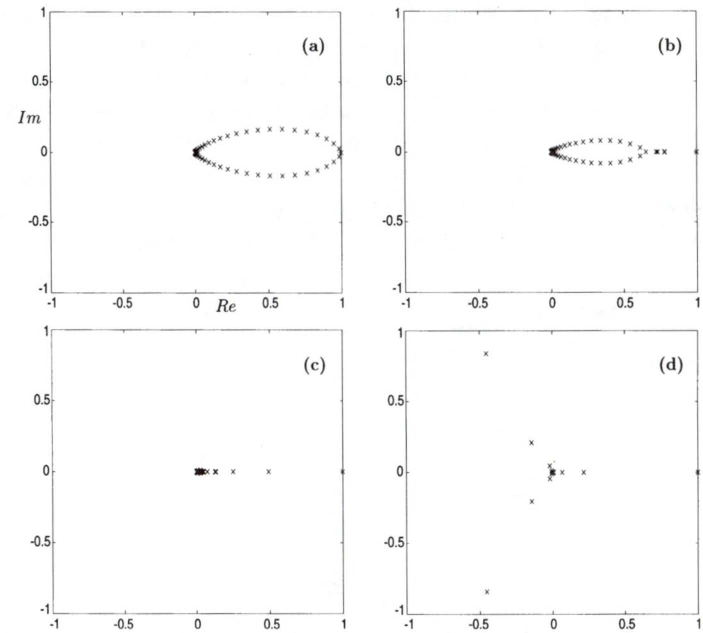

Figure 16.15 Examples of spectra (eigenvalues) of operators $\mathscr{P}$. $T=0.85$, $\sigma=0.02$: (a) $A=0.28$, quasi-periodic case; (b) $A=0.6$, 2 : 1 locking case; (c) $A=1.0$, 2 : 1 locking case; (d) $A=2.0$, 5 : 3 locking case. The abscissa and the ordinate are the real part and imaginary part of the eigenvalues, respectively.

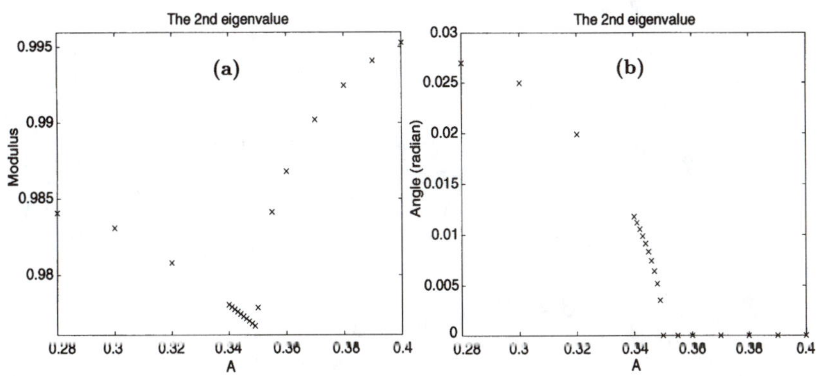

Figure 16.16 Spectral bifurcation diagram (2 : 1 locking). (a) Modulus and (b) angles of the second largest eigenvalues of the operator $\mathscr{P}$ are plotted with various values of the bifurcation parameter $A\in[0.28, 0.40]$. $T=0.85$, $\sigma=0.02$. This figure corresponds to the deterministic bifurcation diagram of Figure 16.13.

locking in a stochastic sense. We can convince ourselves of the validity of defining this point as a stochastic bifurcation if we also see the eigenfunction of the second eigenvalue (Doi et al., 1998). Note that all invariant density functions have the same topological shape (two modes) in this parameter range $0.28 < A < 0.4$. So, we can observe no stochastic bifurcation in a classical sense (Arnold, 1995).

Figure 16.17 is also a spectral bifurcation diagram corresponding to a deterministic 5:3 phase locking. In this figure, sixth eigenvalues are plotted rather than second ones. In the range of $1.98 \leqslant A \leqslant 2.04$, the eigenvalue takes real values. This corresponds to a

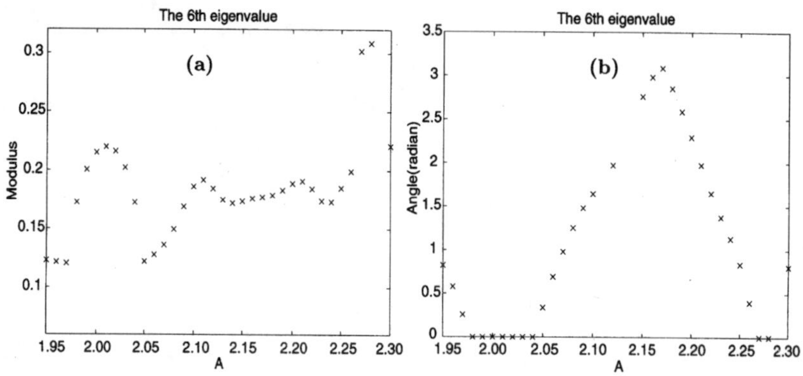

Figure 16.17 Spectral bifurcation diagram (5:3 locking). (a) Modulus and (b) angles of the sixth largest eigenvalues of the operator $\mathscr{P}$ are plotted with various values of the bifurcation parameter $A \in [1.95, 2.30]$. $T = 0.85$, $\sigma = 0.02$.

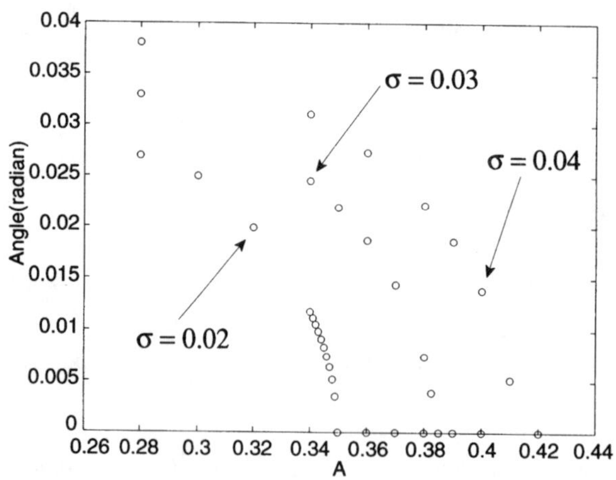

Figure 16.18 Shift of a stochastic 2:1 bifurcation point. Angles of the second eigenvalues of the operator $\mathscr{P}$ are plotted with different noise intensities $\sigma = 0.02, 0.03, 0.04$ in the range of $0.26 < A < 0.44$. $T = 0.85$, $\sigma = 0.02$.

stochastic bifurcation of 5 : 3 phase locking. Note that no clear change is observed in the other (from second to fifth) eigenvalues. Not only is the second eigenvalue, but also the other eigenvalues, are important to analyze a stochastic bifurcation.

In Figure 16.18 angles of the second eigenvalues are plotted in the range of $0.26 < A < 0.44$ with different noise intensities. This corresponds to the spectral bifurcation diagram of Figure 16.16. The stochastic bifurcation point in the sense of our definition (the point in which the eigenvalue changes its value from complex to real) is shifted rightward as the noise intensity increases.

5. DISCUSSION

We have considered the BVP or FHN neuronal model in its simplified form: (1) the cubic polynomial is piecewise linearized and (2) the number of parameters are reduced. This simplified BVP model still keeps the neuronal features (excitability, refractoriness, repetitive activity) although this seems to be closer to the van der Pol model than to the original BVP model.

We have analyzed the response characteristics of the model to sinusoidal stimuli in the singular limit of $\varepsilon = 0$. This numerical analysis reveals that there are no essential differences between the response characteristics of a (non-oscillatory) neuron model and that of a neuron model with repetitive activities (neuronal oscillator).

The singular limit reduces the dynamics of the neuron model to one dimension, and thus we can analyze the dynamics by a one-dimensional mapping called the phase transition curve (PTC). If the externally applied stimulus is rectangular pulse then the PTC can be obtained analytically and more calculations on the response characteristic can be done (see Yoshizawa et al., 1982; Othmer and Watanabe, 1994; Xie and Othmer, 1996).

The PTC analysis is applicable to the finite ε (non-zero) case also. Usually, return maps or PTCs are useful for the analysis of such forced nonlinear dynamical systems (see Winfree, 1980; Guckenheimer and Holmes, 1983; Glass and Mackey, 1988). If a neuron model has a stable limit cycle (repetitive activity) and the external stimulus is applied to it for a short time, then the dynamics is reduced to the dynamics on the (one-dimensional) limit cycle and thus one-dimensional PTC analysis can be made. Even in the case of a (non-singular) non-oscillatory neuron, there exists an invariant (one-dimensional) curve to which the dynamics is reduced and the PTC analysis is applicable (Doi and Sato, 1995).

We have also considered the singular neuron model in the presence of both noise and sinusoidal stimuli. The return map or the PTC is naturally extended to the case of presence of noise and the iteration of the map to the iteration of a probability density function by an operator. The numerical computation of this operator is easy and can be done with any accuracy since this does not use numerical simulation of stochastic differential equations. So, we can analyze very delicate phenomena caused by noise. Using the method, stochastic phase lockings and stochastic bifurcations were analyzed, especially stochastic bifurcations were analyzed on the basis of spectral properties of the operator. In fact, stochastic bifurcations were clearly observed in the sense of our definition; the eigenvalues of the operator changed their values from complex to real abruptly (not smoothly) at the possible stochastic bifurcation point. More intensive discussion on the validity of this definition of a stochastic bifurcation is necessary in future research.

Noise, as expected, blurred the detailed bifurcation structure of the forced van der Pol oscillator (the special case of the simplified BVP neuron model). Noise, however, did not blur it uniformly; some main structure seemed to remain against noise (cf. the two-modes structure in Figure 16.14(b)). The dependency of the detailed phase-locking structures on the input amplitude A is rubbed out by noise and a simple dependency of the amplitude of the density functions on A are presented. In the presence of large noise, the neuron model transformed the amplitude of the input signal to the height of the probability density functions rather than the detailed structure, although this result is not so unexpected.

We have used the sinusoid as an external signal $v(t)$ and the signal was applied to the recovery variable y rather than the membrane voltage x. As seen in section 3, the application of $v(t)$ to y is equivalent to that of a 'filtered' signal $\int v(t)\,dt$ to x. Sinusoids are invariant under this filtering. If we introduce more realistic input signal, then the response characteristics of the BVP neuron to the input may depend on the choice of the variables to which the input is applied to, since generally $rv(t)$ is different from $\int v(t)\,dt$. Thus, we considered the sinusoidal input as the simplest case. Note that our analysis framework does not depend on this specific choice of the input signal, and is applicable to any periodic signal.

6. CONCLUSIONS AND FUTURE PERSPECTIVES

6.1 Time Scales and the Shape of External Stimuli

In the singular limit $\varepsilon = 0$, the fast dynamics, which may correspond to the fast activation process of the conductance of the Na^+ ion channel, is treated as 'infinitely fast'. So the dynamics of the Na^+ activation is supposed to be faster than any other dynamics of a neuron. If the external stimuli is pulsatile, i.e. the injection of external current is much faster than the Na^+ activation, then we have to consider another situation. In fact, the response characteristics of the (non-singular) BVP neuronal model depends greatly on the value of ε in the case of such an impulsive stimulation (Doi and Sato, 1995). Thus, in order to use a simplified neuron model, it is important to consider the time scales of the dynamics in a neuron.

The response characteristics may also depend on the specific choice (shape) of the external signal $v(t)$ and on the choice of variable (x or y in our model) to which $v(t)$ is applied. Note that an application of $v(t)$ to the variable x is equivalent to that of $dv(t)/dt$ to y if $v(t)$ is differentiable. So, the input of a rectangular pulse to x cannot be converted to the input of a signal to y; the time scales are important also in this sense.

The response characteristics can be more complicated if a non-periodic signal is considered. In fact, 'non-chaotic' strange attractors have been found in the van der Pol oscillator forced by a two-frequency quasi-periodic signal (Pokorný et al., 1996).

6.2 Asymmetry of Waveforms

We briefly discuss the importance of asymmetry of the waveform. A more asymmetrical oscillator seems to be more easily locked to an input. The asymmetry of a waveform can be considered as one aspect of a neuron (Kepler et al., 1990; Meunier, 1992; Nomura et al., 1994b). In the case of $a=0$, the BVP model is equal to the van der Pol oscillator and the waveform of 'membrane voltage' $x(t)$ is symmetric and does not resemble that of a usual neuron. We, however, consider that the van der Pol relaxation oscillator still preserves an aspect of neuronal realism since it possess slow-fast dynamics which is typical of the dynamics found in neurons. The more physiological (Hodgkin–Huxley-type) model of a Purkinje fiber of a heart has such characteristics as a relaxation oscillator (Noble, 1975; Cronin, 1987) and its waveform resembles that of the van der Pol oscillator.

We intensively studied the noise effects on the phase lockings and bifurcations of the forced BVP model in the symmetric case $a=0$, i.e. the van der Pol oscillator. What happens if the parameter is changed to an asymmetric oscillator ($0<|a|<1$), and to a non-oscillatory neuron ($|a|>1$)? More intensive investigations concerning the interplay between asymmetry and noise is also a future subject of relevance to neurobiology.

6.3 Neurons and Neuronal Oscillators

Another important character of a neuron is its excitability or firing behavior. As shown in section 3.2, there are no essential differences between the response characteristics of the non-oscillatory BVP neuron model and that of the BVP neuronal oscillator to sinusoidal input if the amplitude of the input is sufficiently large. Then what are the differences among the signal-transmission characteristics of non-oscillatory neurons, neuronal oscillators, and other (biological or non-biological) nonlinear oscillators? This question still remains to be answered.

6.4 Integrate-and-Fire Models

In the singular BVP model, orbits move on the x-nullcline switching between two branches of the nullcline (see Figure 16.5). If we consider only right branch; if orbits jump to the point A on the right branch rather than to C on the left branch when orbits reach the threshold point B, then the model is same as the so-called integrate-and-fire model (Glass and Mackey, 1988). Thus our framework is straightly applicable to the analysis of the noise effect on the forced integrate-and-fire model (Keener et al., 1981) which is revived in the context of stochastic resonances (Bulsara et al., 1996). Note that the singular BVP model is thus considered as a 'complex' of two integrate-and-fire neuron models.

6.5 Bursting Oscillators

If the period of the sinusoidal input is long, then the BVP neuron model produces a bursting oscillation (see Figure 16.6(d)), in which an active phase with successive spikes and a silent phase without spikes alternate. In the present work, the sinusoid is considered as an 'external' signal. If we regard the sinusoidal term an 'intrinsic' term, then the forced BVP model (3) is considered as a burst neuron model (see Baer et al., 1995 for a simple burster model). Thus our analysis framework can analyze the noise effects on bursting oscillations, which seems an interesting future subject.

6.6 Network of Neurons

We have considered a single neuron (uni-directionally) coupled to the external signal and the phase of sinusoidal input was observed whenever the forced neuron reached a fixed point (the point A of the x–y phase plane). In the case of a pair of two integrate-and-fire neurons coupled in both directions, the phase of one neuron can similarly be defined when another neuron reaches a certain state (e.g. threshold) (Pérez and Lomnitz-Adler, 1988; Mirollo and Strogatz, 1990). Thus our approach using the operator on the probability density function of the phase, can seemingly be extended to the two coupled neurons. In the case of a network of more neurons, we have to consider the 'vector' of phases (Mirollo and Strogatz, 1990). Thus, in the presence of noise, the operator on the multi-dimensional probability density function of the vector of phases should be treated. This seems to be a challenging future subject.

6.7 Noise in Neural Systems

Noise in neural systems cannot be ignored. Does noise have positive role in neural systems? Or, do the neural systems just work well in such a 'bad environment' of large noise? This seems hard to answer.

Some of such studies about noise effects are arbitrarily selected and summarized: The bursting oscillation of the pancreatic β-cells is responsible for insulin secretion from the islets of Langerhans. Single β-cells, however, do not burst and have a noisy irregularity. The coupling between the cells conquer this difficulty of bursting by noise; the coupled cluster of β-cells can burst (Chay and Kang, 1988; Sherman et al., 1988) (noise has a negative role in this example). Noise can suppress deterministic irregularity; noise can control the chaos (Doi, 1989; Rajasekar and Lakshmanan, 1993); in this case noise plays a positive role. In the neural networks or in the population of coupled neurons, all the neurons may present a synchronized behavior if the coupling is sufficiently strong. If the coupling strength is moderate, the population of neurons may divide into several subgroups (clusters), in each subgroup of which, neurons are synchronized. Noise can control the size of the clustering in the network of neuron models of the reticular thalamic nucleus (Golomb and Rinzel, 1994).

Recently the enhancement of the signal transduction induced by noise has been collecting much attention in the context of 'stochastic resonance' (Moss et al., 1993; Mannella and McClintock, 1995; Wiesenfeld and Moss, 1995). In fact, noise enhances the coherency of output spike trains of the crayfish mechanoreceptor cells with a weak input signal (Douglass et al., 1993) and that of output current through an array of ion channels with the weak transmembrane potential (Bezrukov and

Vodyanoy, 1995). These examples show that noise may have an important role in the detection of weak signals. Note, however, that the noise must carefully be adjusted in a suitable intensity to achieve such a high transmission performance; the existence of such a delicate tuning in biological systems is hard to imagine. In a large population of BVP (or FHN) model neurons, the stochastic resonance can occur without this delicate tuning of noise intensity (Longtin, 1993; Collins et al., 1995a,b).

PROBLEMS

16.1. If we ignore the dynamics on the left branch of the N-shaped x-nullcline in the x–y phase plane of Fig. 16.5 (i.e. the orbit jumps from the point B to A rather than to C), then the BVP model becomes an integrate-and-fire model. Use the method stated in section 4 (c.f. Tateno et al., 1995) to investigate the noise effect on the sinusoidally-forced integrate-and fire model.

16.2. The BVP model (3) is an (asymmetric) combination of two integrate-and-fire neuron models. Compare the response characteristics of the BVP neuronal oscillator to a sinusoidal input to that of the integrate-and-fire neuronal model. How does the parameter a which controls the symmetry of the BVP model affect the stimulus-response characteristics? Discuss the possible role of the asymmetry in neuronal signal transmissions.

16.3. If the external sinusoidal signal is considered as an intrinsic property of a cell and the period of the sinusoid is large, then the forced BVP model (3) is a simple model of a bursting neuronal oscillator. Study the effect of noise on the burster. The number of spikes per one cycle of the bursting oscillation may fluctuate by noise. How does the number fluctuate and what does the probability distribution of the number becomes like?

16.4. Extend the method of section 4 to analyze the noise effects on the sinusoidally-forced coupled two neurons and provide a discussion about the possibility of the extension to the case of three or more coupled neurons.

16.5. A sinusoidal signal was used as an external input $v(t)$. Consider the case of periodic pulsatile (rectangular) signal (Yoshizawa et al., 1982):

$$v(t)=\begin{cases} A & \text{if } nT\leqslant t\leqslant nT+w\,(n=0,1,2,\ldots) \\ 0 & \text{otherwise} \end{cases}$$

and the case of more realistic input signal of the train of the so-called alpha functions:

$$v(t)=\sum_{n=0}^{\infty} v_0(t-nT)$$

$$v_0(t)=\begin{cases} Ate^{-t/\alpha} & \text{if } t\geqslant 0, \\ 0 & \text{otherwise.} \end{cases}$$

How does the response characteristics depend on the shape of the input?

16.6. This chapter considered the space-clamped neuron model only. Pulse trains usually suffer modulation along a neuronal axon (Horikawa 1991, 1992, 1993).

Analyze the response characteristics of the forced (piecewise linearized) FHN axon:

$$\varepsilon \frac{\partial u(x,t)}{\partial t} = v - u + 5/6\{|u+1| - |u-1|\} + D\frac{\partial^2 u}{\partial x^2},$$

$$\frac{\partial v(x,t)}{\partial t} = -u + a + A\sin\{2\pi(t/T+\theta_0)\} + \sigma\frac{\mathrm{d}W(t)}{\mathrm{d}t}$$

in the absence ($\sigma = 0$) and the presence ($\sigma \neq 0$) of noise.

16.7 As a research project examine the differences and similarities between the response characteristics of 'simple' neuron models, such as the leaky integrator model (cf. Scharstein, 1979) and the BVP models discussed in this Chapter.

REFERENCES

Aihara, K., Matsumoto, G. and Ikegaya, Y. (1984) Periodic and non-periodic responses of a periodically forced Hodgkin–Huxley oscillator. *J. Theor. Biol.*, **109**, 249–269.

Alexander, J.C., Doedel, E.J. and Othmer, J.C. (1990) On the resonance structure in a forced excitable system. *SIAM J. Appl. Math.*, **50**, 1373–1418.

Arnold, L. (1995) Random Dynamical Systems. In *Dynamical Systems*, Lecture Notes in Math., edited by R. Johnson, Vol. 1609, pp. 1–43. Springer.

Baer, S.M., Rinzel, J. and Carrillo, H. (1995) Analysis of an autonomous phase model for neuronal parabolic bursting, *J. Math. Biol.*, **33**, 309–333.

Baer, S.M. and Tier, C. (1986) An analysis of a dendritic neuron model with an active membrane site, *J. Math. Biol.*, **23**, 137–161.

Barnes, B. and Grimshaw, R. (1997) Analytical and numerical studies of the Bonhoeffer Van der Pol system, *J. Austral. Math. Soc. Ser. B*, **38**, 427–453.

Bezrukov, S.M. and Vodyanoy, I. (1995) Noise-induced enhancement of signal transduction across voltage-dependent ion channels. *Nature*, **378**, 362–364.

Braaksma, B. and Grasman, J. (1993) Critical dynamics of the Bonhoeffer–van der Pol equation and its chaotic response to periodic stimulation. *Physica*, **D 68**, 265–280.

Bulsara, A.R., Elston, T.C., Doering, C.R., Lowen, S.B. and Lindenberg, K. (1996) Cooperative behavior in periodically driven noisy integrate–fire models of neuronal dynamics. *Phys. Rev. E.*, **53**, 3958–3969.

Chay, T.R. and Kang, H.S. (1988) Role of single-channel stochastic noise on bursting clusters of pancreatic β-cells. *Biophys. J.*, **54**, 427–435.

Clay, J.R. (1976) A stochastic analysis of the graded excitatory response of nerve membrane. *J. Theor. Biol.*, **59**, 141–158.

Collins, J.J., Chow, C.C. and Imhoff, T.T. (1995a) Stochastic resonance without tuning. *Nature*, **376**, 236–238.

Collins, J.J., Chow, C.C. and Imhoff, T.T. (1995b) A periodic stochastic resonance in excitable systems. *Phys. Rev.* **E 52**, R3321–R3324.

Cronin, J. (1987) *Mathematical Aspects of Hodgkin–Huxley Neural Theory*. Cambridge University Press.

DeFelice, L.J. (1981) *Introduction to Membrane Noise*. New York: Plenum Press.

Degn, H., Holden, A.V. and Olsen, L.F. (eds.) (1987) *Chaos in Biological Systems*. New York: Plenum Press.

Doi, S. (1989) A chaotic map with a flat segment can produce a noise-induced order. *J. Stat. Phys.*, **55**, 941–964.

Doi. S., Inoue, J. and Kumagai, S. (1998) Spectral analysis of stochastic phase lockings and stochastic bifurcations in the sinusoidally-forced van der Pol Oscillator with additive noise, *J. Stat. Phys.*, **90**, 1107–1127.

Doi, S. and Sato, S. (1995) The global bifurcation structure of the BVP neuronal model driven by periodic pulse trains. *Mathematical Biosciences*, **125**, 229–250.

Douglass, J.K., Wilkens, L., Pantazelou, E. and Moss, F. (1993) Noise enhancement of information transfer in crayfish mechanoreceptors by stochastic resonance. *Nature*, **365**, 337–340.

FitzHugh, R. (1961) Impulses and physiological states in theoretical models of nerve membrane. *Biophy. J.*, **1**, 445–466.

Fox, R.F. (1997) Stochastic versions of the Hodgkin–Huxley equations. *Biophys. J.* **72**, 2068–2074.

Gantmacher, F.R. (1959) *The Theory of Matrices*, Vol. II. Chelsea Pub.

Gardiner, C.W. (1983) *Handbook of Stochastic Methods for Physics, Chemistry and the Natural Sciences*. Springer.

Glass, L., Graves, C., Petrillo, G.A. and Mackey, M.C. (1980) Unstable dynamics of a periodically driven oscillator in the presence of noise. *J. Theor. Biol.*, **86**, 455–475.

Glass, L. and Mackey, M.C. (1988) *From Clocks to Chaos, The Rhythms of Life*. Princeton University Press.

Golomb, D. and Rinzel, J. (1994) Clustering in globally coupled inhibitory neurons. *Physica*, **D 72**, 259–282.

Grasman, J. and Jansen, M.J.W. (1979) Mutually synchronized relaxation oscillators as prototypes of oscillating systems in biology. *J. Math. Biol.*, **7**, 171–197.

Grasman, J. and Roerdink, J.B.T.M. (1989) Stochastic and chaotic relaxation oscillations. *J. Stat. Phys.*, **54**, 949–970.

Guckenheimer, J. and Holmes, P. (1983) *Nonlinear Oscillations, Dynamical Systems, and Bifurcation of Vector Fields*. Springer.

Guevara, M.R., Glass, L., Mackey, M.C. and Shrier, A. (1983) Chaos in neurobiology, *IEEE Trans.*, **SMC-13**, 790–798.

Hadeler, K.P., an der Heiden, U. and Schumacher, K. (1976) Generation of the nervous impulse and periodic oscillations. *Biol. Cybern.*, **23**, 211–218.

Hassard, B. (1978) Bifurcation of periodic solutions of the Hodgkin–Huxley model of the squid giant axon. *J. Theor. Biol.*, **71**, 401–420.

Hodgkin, A.L. and Huxley, A.F. (1952) A quantitative description of membrane current and its applications to conduction and excitation in nerve. *J. Physiol.*, **117**, 500–544.

Holden, A.V. (1976a) The response of excitable membrane models to a cyclic input. *Biol. Cybern.*, **21**, 1–7.

Holden, A.V. (1976b) *Models of the Stochastic Activity of Neurons*. Berlin: Springer.

Horikawa, Y. (1991) Noise effects on spike propagation in the stochastic Hodgkin–Huxley models. *Biol. Cybern.*, **66**, 19–25.

Horikawa, Y. (1992) Spike propagation during the refractory period in the stochastic Hodgkin–Huxley model. *Biol. Cybern.*, **67**, 253–258.

Horikawa, Y. (1993) Noise effects on spike propagation during the refractory period in the FitzHugh–Nagumo model. *J. Theor. Biol.*, **162**, 41–59.

Horsthemke, W. and Lefever, R. (1984) *Noise-Induced Transitions, Theory and Applications in Physics, Chemistry, and Biology*. Berlin; Springer.

Kaplan, D.T., Clay, J.R., Manning, T., Glass, L., Guevara, M.R. and Shrier, A. (1996) Subthreshold dynamics in periodically stimulated squid giant axons. *Phys. Rev. Lett.*, **76**, 4074–4077.

Keener, J.P., Hoppensteadt, F.C. and Rinzel, J. (1981) Integrate-and-fire models of nerve membrane response to oscillatory input. *SIAM J. Appl. Math.*, **41**, 503–517.

Kepler, T.B., Marder, E. and Abbott, L.F. (1990) The effect of electrical coupling on the frequency of model neuronal oscillators. *Science*, **248**, 83–85.

Lasota, A. and Mackey, M.C. (1994) *Chaos, Fractals, and Noise: Stochastic Aspects of Dynamics*. Springer.

Longtin, A. (1993) Stochastic resonance in neuron models. *J. Stat. Phys.* **70**, 309–327.

Mannella, R. and McClintock, P. (eds.) (1995) Proceedings of the International Workshop on Fluctuations in Physics and Biology: Stochastic Resonance, Signal Processing, and Related Phenomena, *Nuovo Cimento*, **17D**.

Matsumoto, G., Aihara, K., Ichikawa, M. and Tasaki, A. (1984) Periodic and nonperiodic responses of membrane potentials in squid giant axons during sinusoidal current stimulation. *J. Theor. Neurobiol.*, **3**, 1–14

Meunier, C. (1992) The electrical coupling of two simple oscillators: load and acceleration effects. *Biol. Cybern.*, **67**, 155–164.

Mirollo, R.E. and Strogatz, S.H. (1990) Synchronization of pulse-coupled biological oscillators. *SIAM J. Appl. Math.*, **50**, 1645–1662.

Moss, F., Bulsara, A. and Shlesinger, M. (eds.) (1993) Proceedings of the NATO Advanced Research Workshop on Stochastic Resonance in Physics and Biology. *J. Stat. Phys.*, **70**(1/2).

Moss, F. and McClintock, P.V.E. (1987) *Noise in Nonlinear Dynamical Systems*, Vol. 1–3. Cambridge University Press.

Nagumo, J., Arimoto, S. and Yoshizawa, S. (1962) An active pulse transmission line stimulating nerve axon, Proc. Inst. Radio Engineers, **50**, 2061–2070.

Noble, D. (1975) *The Initiation of the Heartbeat*. Oxford University Press.

Nomura, T., Sato, S., Doi, S., Segundo, J.P. and Stiber, M.D. (1993) A Bonhoeffer–van der Pol oscillator model of locked and non-locked behaviors of living pacemaker neurons. *Biol. Cybern.*, **69**, 429–437.

Nomura, T., Sato, S., Doi, S., Segundo, J.P. and Stiber, M.D. (1994a) Global bifurcation structure of a Bonhoeffer van der Pol oscillator driven by periodic pulse trains. Comparison with data from an inhibitory synapse. *Biol. Cybern.*, **72**, 55–67.

Nomura, T., Sato, S., Doi, S., Segundo, J.P. and Stiber, M.D. (1994b) A modified radial isochron clock with slow and fast dynamics as a model of pacemaker neurons. Global bifurcation structure when driven by periodic pulse trains. *Biol. Cybern.*, **72**, 93–101.

Okuda, M. (1981) A new method of nonlinear analysis for threshold and shaping actions in transient state. *Prog. Theor. Phys.*, **66**, 90–100.

Othmer, H.G. and Watanabe, M. (1994) Resonance in excitable systems under step-function forcing I. Harmonic solutions. *Adv. Math. Sci. Appl.*, **4**, 399–441.

Pérez Pascual, R. and Lomnitz-Adler, J. (1988) Coupled relaxation oscillators and circle maps. *Physica D*, **30**, 61–82.

Pokorný, P., Schreiber, I. and Marek, M. (1996) On the route to strangeness without chaos in the quasiperiodically forced van der Pol oscillator. *Chaos, Solitons & Fractals*, **7**, 409–424.

Rajasekar, S. (1996) Dynamical structure functions at critical bifurcations in a Bonhoeffer–van der Pol equation. *Chaos, Solitons & Fractals*, **7**, 1799–1805.

Rajasekar, S. and Lakshmanan, M. (1988a) Period doubling route to chaos for a BVP oscillator with periodic external force. *J. Theor. Biol.*, **133**, 473–477.

Rajasekar, S. and Lakshmanan, M. (1988b) Period-doubling bifurcations, chaos, phase-locking and devil's staircase in a Bonhoeffer–van der Pol oscillator. *Physica*, **D 32**, 146–152.

Rajasekar, S. and Lakshmanan, M. (1993) Algorithms for controlling chaotic motion: application for the BVP oscillator. *Physica*, **D 67**, 282–300.

Ricciardi, L.M. (1977) *Diffusion Processes and Related Topics in Biology*. Berlin; Springer.

Rinzel, J. (1978) On repetitive activity in nerve. *Fed. Proc.*, **37**, 2793–2802.

Sato, S. and Doi, S. (1992) Response characteristics of the BVP neuron model to periodic stimuli. *Math. Biosci.*, **112**, 243–259.

Scharstein, H. (1979) Input–Output relationship of the leaky-integrator neuron model. *J. Math. Biol.* **8**, 403–420.

Sherman, A., Rinzel, J. and Keizer, J. (1988) Emergence of organized bursting in clusters of pancreatic β-cells by channel sharing. *Biophys. J.*, **54**, 411–425.

Skaugen, E. and Walloe, L. (1979) Firing behaviour in a stochastic nerve membrane model based upon the Hodgkin–Huxley equations. *Acta Physiol. Scand.*, **107**, 343–363.

Takabe, T., Aihara, K. and Matsumoto, G. (1988) Response characteristics of the Hodgkin–Huxley equations to pulse-train stimulation. *Trans. of IEICE Japan*, **J71-A**, 744–750 (in Japanese).

Takahashi, N., Hanyu, Y., Musha, T., Kubo, R. and Matsumoto, G. (1990) Global bifurcation structure in periodically stimulated giant axons of squid. *Physica*, **D 43**, 318–334.

Tateno, T., Doi, S., Sato, S. and Ricciardi, L.M. (1995) Stochastic phase-lockings in a relaxation oscillator forced by a periodic input with additive noise—A first-passage-time approach. *J. Stat. Phys.*, **78**, 917–935.

Tuckwell, H.C. (1988) *Introduction to Theoretical Neurobiology*. Cambridge: Cambridge University Press.

Tuckwell, H.C. (1989) *Stochastic Processes in the Neurosciences*. Philadelphia: SIAM.

van der Pol, B. (1926) On "relaxation-oscillations". *Phil. Mag.*, **2**, 978–992.

Wang, W. (1989) Bifurcations and chaos of the Bonhoeffer–van der Pol model. *J. Phys. A: Math. Gen.*, **22**, L627–L632.

Wiesenfeld, K. and Moss, F. (1995) Stochastic resonance and the benefits of noise: from ice ages to crayfish and SQUIDs. *Nature*, **373**, 33–36.

Winfree, A.T. (1980) *The Geometry of Biological Time*. New York: Springer-Verlag.

Xie, M. and Othmer, H.G. (1996) Resonance in excitable systems under step-function forcing II subharmonic solutions and persistence. *Physica D*, **98**, 75–110.

Xu, J.-X. and Jiang, J. (1996) The global bifurcation characteristics of the forced van der Pol oscillator. *Chaos, Solitons & Fractals*, **7**, 3–19.

Yasin, S., Friedman, M., Goshen, S., Rabinovish, A. and Thieberger, R. (1993) Intermittency and phase locking of the Bonhoeffer–van der Pol model. *J. Theor. Biol.*, **160**, 179–184.

Yoshizawa, S., Osada, H. and Nagumo, J. (1982) Pulse sequences generated by a degenerate analog neuron model. *Biol. Cybern.*, **45**, 23–33.

Yu, X. and Lewis, E.R. (1989) Studies with spike initiators: linearization by noise allows continuous signal modulation in neural networks. *IEEE Trans. Biomed. Eng.*, **36**, 36–43.

CHAPTER 17
DYNAMICS OF ASSOCIATIVE MEMORY NETWORKS

Alan M.N. Fu and Hong Yan

1. INTRODUCTION

The detailed structure of the human central nervous system began to be unravelled only a century ago. In the nineteenth century there were two schools contending for scientific acceptance (Müller et al., 1995). The reticularists considered that the brain was a continuous, uninterrupted network of nerve fibres while the neuronists assumed that the brain was composed of many single, interconnected cellular units, the neurons. The contest between these two doctrines was determined with the advent of a new technique for the staining of nerve fibres by means of a bichromate silver reaction invented by Camillo Golgi around 1880. This technique was successfully applied by Santiago Ramon y Cajal in 1888 to disprove the doctrine of reticularism by discovering the tiny gaps between individual neurons (Müller et al., 1995).[1] The detailed study of the internal structure of neural cells, especially since the discovery of the electron microscope about half century ago, has revealed that all neurons are constructed from the same basic parts, independent of their size and shape. A generic neuron consists of four parts: cell body, axon, dendrites and synapses. The dendrites serve as receptors for signals from adjacent neurons, the cell body acts as the central operator of signals to generate the neural activity which is transmitted along axons to other nerve cells or muscle fibres via the synapses. This functionalism can be reproduced by an 'artificial' neuron shown in Figure 17.1.

If we think of the human brain as a dynamical system (Kosko, 1992), by focussing on the neuron one can investigate the behaviour and functions of this dynamical system by modeling it as a neural network. A neural network consists of a large number of mutually connected neurons with the ability of integrating information received by their dendrites from other neurons and make a nonlinear input–output relation to transmit their activities to other neurons along axons via synapses (interconnected weights of networks). Thus neural networks are cellular networks which have a natural propensity for storing experiential knowledge that is retained in the form of stable states, or cycles of states, of networks by adjusting synapse weights in a dynamical process. This natural propensity is referred to as the *learning ability* of neural networks. The learning ability of the human brain represents its capability to capture the pertinent information, i.e. a set of features of an object

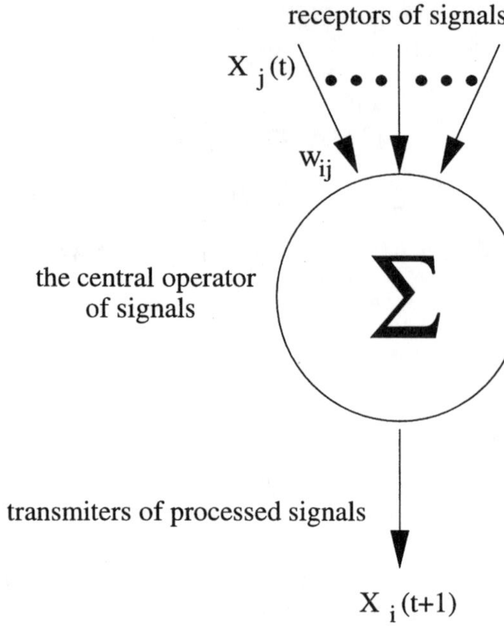

Figure 17.1 A neuron represents an elementary processing unit in which the dendrites of the neuron act as receptors of signals, the soma serves as the central operator and the axon terminals act as transmitters of processed signals, where the Σ represents the summation of all incoming signals converging at the soma. The activity $h_i(t)$ of the ith neuron at the tth step is the instantaneous linear sum of the presynaptic inputs $X_j(t)$, $j=1, 2\ldots, n$ multiplied by their corresponding synaptic weights w_{ij} (see (17.4) in text).

under consideration. This information is then remembered and retrieved in future when a similar object related to the original information appears. A function of neural networks is to retrieve these stable states or cycles in response to incomplete information, the stimuli to the networks.

The first mathematical model of neural networks was proposed by McCulloch and Pitts (1943) based on the idea that neurons operated in an all-or-none fashion. This model introduced the step threshold function currently used in many neural network models. Although this model created great excitement by showing that a neural network could perform rather sophisticated operations from simple computations, it lacked of the ability to learn. Rosenblatt (1958, 1962) generalized the McCulloch–Pitts model by adding the learning ability and it was called a *perceptron*. Rosenblatt showed that the perceptron was capable of learning random associations, i.e. in an environment of random stimuli the perceptron whose synaptic weights are randomly determined can learn to associate specific responses to specific stimuli. However, it has some serious computational limitations. For example, the two-layer perceptron was proven to be able to separate inputs into two classes only if the two classes were linearly separable[2] (Block, 1962). Minsky and Papert (1969) showed that the perceptron could not be used for complex logic functions, so it could not separate nonlinearly separable inputs into different classes.

Hopfield (1982) proposed an auto-associative memory model that is a recurrent dynamical system in which the interconnection (synaptic) weights among neurons are generated by the Hebbian learning rule (Hebb, 1949) and each neuron is represented by a binary element with values 1 and 0 (or -1) to indicate the neuron being in firing and not-firing states, respectively. This model is an asynchronous system, that is, in each time step only one neuron of the network updates its state. Grossberg (1988) showed that the Hopfield model is a subset of a more general class of earlier models (Cohen and Grossberg, 1983). The key point of Hopfield models is the use of an energy function, i.e. the Lyapunov function, to deal with equilibrium properties of networks. Peretto and Niez (1986) proposed a dynamical analysis of asynchronous associative memory networks by means of a statistical method rather than the energy function. An auto-associative memory network is also referred to as a dynamic associative memory (Hassoun, 1993).

The Hopfield model can solve complex problems (see e.g. Hopfield and Tank, 1985) because the model always approaches *a stable state* from a certain initial state (i.e. from an initial state to a stable state at which the energy of the system is a local or global minimum) and can operate as a *content addressable memory*.[4] The synaptic weights can be selected in such a way that the stable states of the network are in fact the memory states.

The Hopfield model can be discrete and continuous.[5] There is no evidence that real neurons have binary response from neuro-biological literature so the continuous model is more biologically plausible and generally accepted. Although the discrete model is not much like real neurons in the nervous system their analyses are easier than those of continuous ones; also, there is a connection between the stable states of the discrete and continuous models (Hopfield, 1984). Thus, investigation of discrete Hopfield models is of significance in both theoretical analyses and practical applications. The Hopfield model is not a model of real biological neurons, but nevertheless it can reproduce several properties of real neurons such as: the model has learning ability (i.e. the function of storing information) the stored information can be retrieved by a hint (i.e. by incomplete information), it therefore provides a good starting point in the mathematical modeling of real neuronal behaviour.

If the asynchronous updating rule in Hopfield models is replaced by the synchronous one the related networks are referred to as Hopfield-type networks (Little, 1974; Amari, 1972, 1974). Since the interconnection weights of both Hopfield and Hopfield-type networks are constructed based on the Hebbian learning rule (Hebb, 1949), they will be referred to simply as associative memory networks (AMN). The stable states of a Hopfield model are obtained by minimizing its energy function in a dynamical process (Hopfield and Tank, 1985; Shen and Ip, 1997; Lee et al., 1997), while those of a synchronous AMN (Hopfield-type neural network) are achieved directly from a dynamical process without the evaluation of an energy function (Fu and Yan, 1995). The asymptotic stationarity of the dynamics of a synchronous AMN with noise acting on the membrane potentials was studied by Wang and Sheng (1996). It was found that a noise source which affects the cellular potentials improves the performance of neural networks (Buhmann and Schulten, 1987). To enhance the recognizing capability of Hopfield models Braham and Hamblen (1988) introduced positive feedback (from a neuron to itself) to the original Hopfield model. This modified model is more sensitive to input variation and concurrently less dependent on the stored memories correlation. However, there is no apparent biological evidence to support the existence of 'positive feedback' and 'decision neurons' that were used in the Braham and Hamblen's model.

How information is stored in AMNs and mechanisms to increase the capacity of storing information in this kind of memory model are interesting issues and many researchers, such as Anderson (1968, 1972), Palm (1980), Cottrell (1988), Kohonen (1989), Treves and Rolls (1992), Li, 1996, Bal et al., 1994, Frolov et al., 1997, have made contributions in this area. AMNs have been used to mimick biological associative memory. For example, central pattern generators in biological systems were modelled as associative memory models (Kleinfeld and Sompolinsky, 1989) in which collective outputs consist of temporally coherent patterns of linear or cyclic sequences of state. The olfactory bulb served as an AMN (Hendin et al., 1997) in which the memories defined by Hebbian mitral assemblies are activated via the interactions with the inhibitory granule cells, and the CA3 pyramidal neurons in the hippocampus was modelled as a synchronous AMN (Bennett et al., 1994) in which the synaptic weights are not symmetric. Despite similarities between associative memory networks and central nervous functions, such as place learning in the hippocampus (O'Keefe, 1983; Bennett et al., 1994; Rolls, 1995; Rolls and Treves, 1998; Tsodyks and Sejnowski, 1995), olfaction (Gelperin et al., 1985; Haberly, 1985; Baird, 1986; Hendin et al., 1997) and visual processing (Koch et al., 1986), the models still need to be verified at the level of neurophysiology.

The dynamical behaviour and rigorous mathematical foundation of Hopfield-type neural networks with randomly generating interconnection weights was first proposed by Amari (1972, 1974) using a statistical neurodynamical approach rather than the Ising spin system of statistical mechanics (see e.g. Amit et al., 1985a,b). Amari and Maginu (1988) proposed a statistical neurodynamical analysis of Hopfield-type neural networks and showed that the statistical neurodynamical method is useful for analysing the behaviour of associative memory models. Since then many researchers (such as Amari, 1988; Zagrebnov and Chvyrov, 1989; Patrick and Zagrebnov, 1991a,b; Nishimori and Ozeki, 1993; Nishimori and Opris, 1993; Yanai and Sawada, 1990; Gibson and Robinson, 1992; Morita, 1993; Yoshizawa, et al., 1993; Komlós and Paturi, 1993; Okada, 1995, 1996; Fu, 1994; Fu and Yan, 1995; Thomas et al., 1996; Yanai and Amari, 1996; Ma, 1997) have worked on dynamical analyses of Hopfield-type neural networks based on the statistical neurodynamical method. This chapter is focuses on the long-term behaviour of synchronous AMNs. In section 2, the Hopfield neural network model are given, in section 3 we present an extension based on the Amari and Maginu theory. In section 4, the prediction of stable states of a synchronous AMN is given. The discussion is given in section 5 and conclusions and future perspectives are given in section 6.

2. MATHEMATICAL DESCRIPTION OF THE HOPFIELD NEURAL NETWORK MODEL

In the last four decades various kinds of neural networks have been developed to model the human brain, such as the perceptrons, recurrent networks, multi-layer networks, self-organizing networks, BSB (brain-state-in-a-box) model, Hopfield networks and Hopfield-type networks (see Cowan and Sharp (1988) for an overview). In this chapter, our work is focused on the Hopfield and Hopfield-type networks because it is felt that they provide the greatest realism in modeling the human brain. The Hopfield-type networks considered here have the same structures as Hopfield networks but with different updating rules (one is synchronous and other is asynchronous). Thus, the concepts and notations used in both networks are identical. In view of the basic structure of a real neuron (Figure 17.1) a neural network is characterized by the following three parts:

1. isopotential neurons (i.e. point-like structures),
2. interconnection (synaptic) weights,
3. updating rule.

For simplicity some necessary notations for describing the network are:

1. Let n denote the total number of neurons in the network.
2. Let $X_i(t)$ denote the state of the ith neuron at time t, which can be either continuous or discrete. A neuron having a continuous value state has a graded response while a neuron with a bivalent state is binary, i.e. excited or quiescent, in a response. In this chapter, the bivalent state is represented by $X_i(t)=1$ or $X_i(t)=-1$, depending on whether the ith neuron is in an excited or quiescent state at time t.
3. The state of the network at time t is represented by the n-dimensional vector $\mathbf{X}(t)=(X_1(t), \ldots, X_n(t))^{\mathrm{T}}$ where T denotes the transpose.
4. Let w_{ij} denote the synaptic weight between the ith and the jth neurons, which represents the effect on the ith neuron from the jth neuron, and $w_{ji}=w_{ij}$ and $w_{ii}=0$.

The dynamical behaviour of a Hopfield network is completely determined by its energy function and updating rule. The energy function of a discrete Hopfield model is defined as

$$E(t)=-\frac{1}{2}\sum_{i=1}^{n}\sum_{j=1}^{n} w_{ij}X_i(t)X_j(t), \tag{17.1}$$

where the synaptic weight w_{ij} is given by

$$w_{ij}=\begin{cases}(1/n)\sum_{\alpha=1}^{m} s_i^\alpha s_j^\alpha & \text{if } j\neq i,\\ 0 & \text{otherwise,}\end{cases} \tag{17.2}$$

where the n-dimensional vectors $\mathbf{S}^\alpha=(s_1^\alpha, \ldots, s_n^\alpha)^{\mathrm{T}}$ ($\alpha=1, \ldots, m$) is the αth state of the network and each component of $\mathbf{S}^\alpha$ is a random variable whose value is equal to either $+1$ or -1. The m patterns $\mathbf{S}^\alpha$, $\alpha=1, \ldots, m$ are referred to as memory patterns of the network and the mechanism by which they are stored in the network is given in (17.2).

The updating rule of the network is

$$X_i(t+1)=\mathrm{Sgn}[h_i(t)], \tag{17.3}$$

where Sgn[x] is the signum function taking value $+1$ or -1 depending on whether $x>0$ or otherwise, and $h_i(t)$ represents the total input to the ith neuron from the network and is referred to as the activity of the ith neuron in the network at time t defined by

$$h_i(t)=\sum_{j=1}^{n} w_{ij}X_j(t). \tag{17.4}$$

Based on the updating rule, the n-dimensional vector $\mathbf{X}(t)=(X_1(t),\ldots,X_n(t))^{\mathrm{T}}$ represents the output of the network at time t and the input of the ith neuron from the network at the time $t+1$ is the activity $h_i(t)$.

Figure 17.2 shows a schematic diagram of a Hopfield model or a Hopfield-type neural network composed of N neurons. It shows a full interconnection network in which the input to the ith neuron at time t is the synaptic weights sum of states of all neurons at time $t-1$, i.e. the activity $h_i(t-1)$ of the network given in (17.4), in the absence of external excitation. This structure implies that Hopfield models or Hopfield-type neural networks possess emergent collective computational abilities.

For a continuous Hopfield network, the energy function is defined as (Hopfield, 1984):

$$E(t)=-\frac{1}{2}\sum_{i=1}^{n}\sum_{j\neq 1}^{n} w_{ij}X_i(t)X_j(t)+\sum_{i=1}^{n}\frac{1}{R_i}\int_0^{X_i(t)} g_i^{-1}(v)\,\mathrm{d}v, \tag{17.5}$$

where $g_i(v)$ is the sigmoid function; R_i is the transmembrane resistance; $X_i(t)$ is the output of the ith neuron at time t; the activity $u_i(t)$, i.e. the input to the ith neuron at time $t+1$, is given by the differential equation

$$C_i\frac{\mathrm{d}u_i(t)}{\mathrm{d}t}=\sum_{j=1}^{n} w_{ij}X_j(t)-\frac{u_i(t)}{R_i}, \tag{17.6}$$

where C_i is the capacitance of the ith cell membrane. This means that the output of the ith neuron depends on the rate of change of the activity of the ith neuron. Hopfield (1984) showed that the relationship between the input and output of the ith neuron, i.e. the solution of (17.6) is

$$u_i(t)=g_i^{-1}(X_i(t)). \tag{17.7}$$

When t is discrete, the differential equation (6) is replaced by the following difference equation:

$$u_i(t+1)=\frac{1}{C_i}\sum_{j=1}^{n} w_{ij}X_j(t)+\left(1-\frac{1}{C_iR_i}\right)u_i(t), \tag{17.8}$$

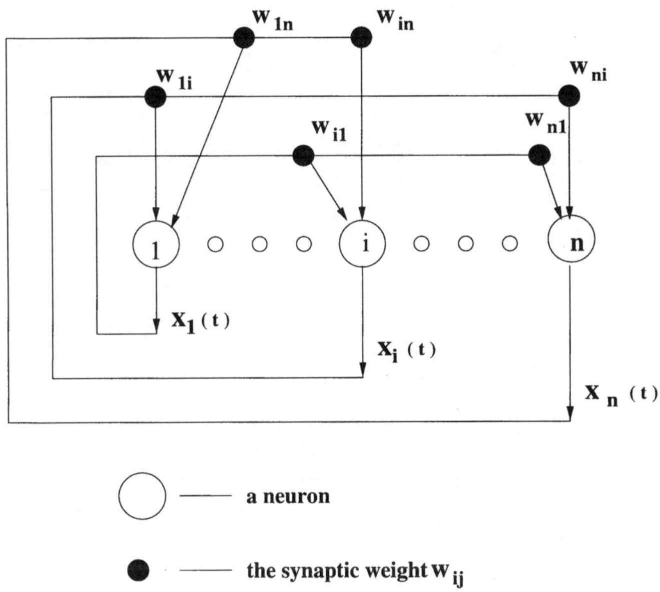

Figure 17.2 The structure of a Hopfield model or a Hopfield-type neural network with N neurons, where w_{ij} is the synaptic weight between the ith and jth neurons and each neuron serves as an operator of the sum of products of synaptic weights and states of the network.

thus (17.6) (or (17.8)) and (17.7) are the updating rules of states of the Hopfield network with a continuous model. If the Hopfield network arrives at a stable state at time $t=t_s$, from the (17.6) we have

$$u_i(t)=R_i \sum_{j=1}^{n} w_{ij}X_j(t) \equiv \text{constant}, \quad \text{as } t \geqslant t_s. \tag{17.9}$$

The general structure of an analog continuous Hopfield model is shown in Figure 17.3 in which the neurons are modelled as amplifiers in conjunction with feedback circuits (see (17.8)) comprised of wires, resistors and capacitors organized so as to model the axons, dendritic arborization and synapses connecting the different neurons (Hopfield and Tank, 1985). Note that since the current is continuous, the discrete model cannot be completely simulated by the analog circuit. That is one of the reasons that the Hopfield continuous model has more biological significance than the discrete one.

Hopfield (1982) showed that the energy function $E(t)$ defined by (17.1) decreases until it reaches a (local or global) minimum when the time t increases. In other words, stable states of the Hopfield network can be found by minimizing $E(t)$. The computational strategy of the Hopfield network is: (i) the network converges to stable states, which are related to stored information; (ii) the network finds stable states because of its dynamical characteristics; and (iii) the stable states are the

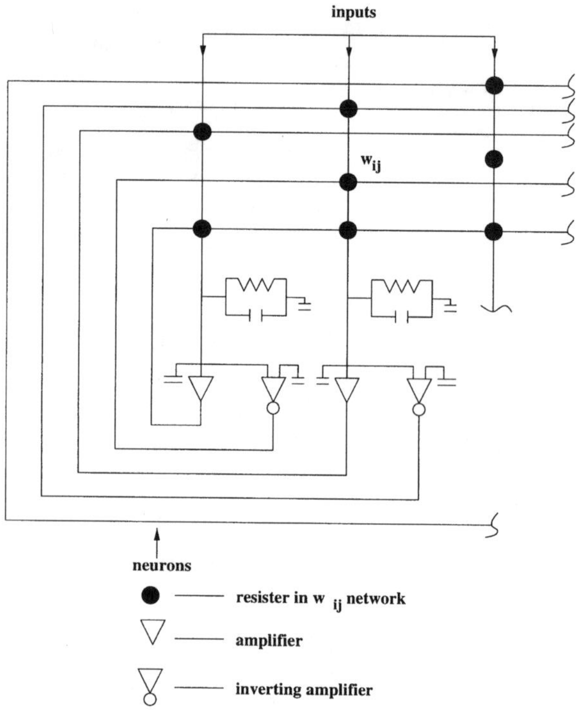

Figure 17.3 The structure of Hopfield networks with continuous models—the analog circuit, where black circles at intersections represent resistive connections (w_{ij}) between outputs and inputs. Connections between inverted outputs and inputs represent negative (inhibitory) connections. (Redrawn from Hopfield and Tank, 1985).

energy minima. To solve a practical problem with a Hopfield network model, the energy function should be generated from the constraints of the given problem. Hopfield and Tank (1985) provided a good example of how to define an energy function for the Traveling-Salesman Problem. The key point in Hopfield networks is how to define related energy functions and minimize them, while that in Hopfield-type networks is how directly to analyse the dynamic behaviour and stable states in a dynamical process. Our study will begin with the main result of the theory proven by Amari and Maginu (1988) using the statistical neurodynamics method, which is introduced briefly in the next section.

3. MATHEMATICAL DESCRIPTION OF THE HOPFIELD-TYPE NEURAL NETWORK

As mentioned in section 2, the synaptic weights matrix W of a Hopfield-type network is the same as that of a Hopfield network whose element w_{ij} is given by (17.2). Since s_i^{α} ($\alpha = 1, \ldots, m$, $i = 1, \ldots, n$) are independent random variables so also are

w_{ij} $(i,j=1,\ldots,n)$. Thus, Hopfield-type networks are stochastic recurrent dynamical systems and their dynamic behaviour must be considered using the statistical method as in the contributions made by Amari (1972, 1974) and Little (1974). In this section we follow the theoretical result from Amari and Maginu (1988) very closely.

To analyse the dynamic and stable states of a Hopfield-type network, we suppose only one of the memory patterns is the real memory stored in the network which we want to recall, without loss of generality, say $\mathbf{S}^1$, and others denote noise terms (spurious information), which are generated randomly with the probability

$$P\{s_i^\alpha=1\}=P\{s_i^\alpha=-1\}=\tfrac{1}{2}, \quad \alpha=1,\ldots,m, \; i=1,\ldots,n. \tag{17.10}$$

The dynamics of the neural network are determined by its update rule. In state space, the distance between the real memory $\mathbf{S}^1$ and current state $\mathbf{X}(t)$ is measured by the quantity $A(t)$ called a main overlap defined as

$$A(t)=\frac{1}{n}\sum_{j=1}^{n} s_j^1 X_j(t) \tag{17.11}$$

which is the measure of the similarity between the real memory $\mathbf{S}^1$ and the current state $\mathbf{X}(t)$ of the neural network at time t. Since the networks are stochastic dynamic recurrent systems, the states $X_j(t)$ $(j=1,\ldots,n)$ and main overlap $A(t)$ are random processes. Let $a(t)$ denote the expectation (mean) of $A(t)$, which is a deterministic function. Clearly, $-1\leqslant A(t)\leqslant 1$. The larger the value of $A(t)$ the more similar (or closer) are $\mathbf{S}^1$ and $\mathbf{X}(t)$. More precisely, if $A(t)=z$ then $50(1-z)\%$ of the components of $\mathbf{X}(t)$ are different from those of $\mathbf{S}^1$. If $A(t)=1$ means then the current state $\mathbf{X}(t)$ and the real memory $\mathbf{S}^1$ are identical. Let $d(\mathbf{X})$ denote the normalized Hamming distance between $\mathbf{S}^1$ and $\mathbf{X}(t)$, i.e.

$$d(\mathbf{X})=\frac{1}{2n}\sum_{i=1}^{n} |X_i(t)-s_i^1| \tag{17.12}$$

then we can easily verify the following relation:

$$A(t)=1-2d(\mathbf{X}). \tag{17.13}$$

Hence,

(i) $d(\mathbf{X})=0 \Leftrightarrow A(t)=1$ with the probability one, and
(ii) $d(\mathbf{X})=1 \Leftrightarrow A(t)=-1$ with the probability one,

where the symbol $\Leftrightarrow$ means the two equations on both sides are equivalent. We can use $A(t)$ or $d(\mathbf{X})$ to measure the similarity between $\mathbf{X}(t)$ and $\mathbf{S}^1$. Note that both $A(t)$ and $d(\mathbf{X})$ are random processes.

The activities of the ith neuron of the network at time t given in (17.4) can be rewritten as

$$h_i(t)=s_i^1 A(t)+N_i(t), \tag{17.14}$$

where the quantity $N_i(t)$ is defined as

$$N_i(t)=\frac{1}{n}\sum_{\alpha\neq 1}^{m}\sum_{j\neq 1}^{n} s_i^\alpha s_j^\alpha X_j(t) \tag{17.15}$$

which does not involve the components of the real memory $\mathbf{S}^1$ and denotes the effect of noise on the ith neuron (called noise terms of the network). The noise term $N_i(t)$ is a random process and its mean and standard deviation is denoted by $\mu_N(t)$ and $\sigma_N(t)$, respectively.

We can see that a Hopfield-type network involves $n+1$ random processes, one is the main overlap $A(t)$ and the remainder are the noise terms $N_i(t)$ $(i=1,\ldots,n)$. Under the following two assumptions:

1. the main overlap $A(t)$ satisfies $P\{A(t)=a(t)\}=1$, where $a(t)=E[A(t)]$, the expectation of $A(t)$; in other words, the main overlap is assumed to be a deterministic function rather than a random process;
2. the noise terms $N_i(t)$ $(i=1,\ldots,n)$ are independent random processes which obey the identical Gaussian distribution with mean $\mu_N(t)=0$ and standard deviation $\sigma_N(t)$;

Amari and Maginu (1988) showed that the dynamic behaviour of Hopfield-type neural networks are determined by the following simplified macroscopic state[6] equations with two macroscopic variables, i.e. the mean $a(t)$ of the main overlap $A(t)$ and the standard deviation $\sigma_N(t)$ of the noise terms $N_i(t)$:

$$a(t+1)=\frac{1}{\sqrt{2\pi}}\int_{-\infty}^{a(t)/\sigma_N(t)}\exp(-0.5v^2)\,dv-\frac{1}{\sqrt{2\pi}}\int_{-\infty}^{-a(t)/\sigma_N(t)}\exp(-0.5v^2)\,dv, \tag{17.16}$$

$$\sigma_N^2(t+1)=r+\frac{2}{\pi}\left[\exp\left(-0.5\frac{a^2(t)}{\sigma_N^2(t)}\right)\right]^2+\frac{2\sqrt{2}ra(t)a(t+1)}{\sqrt{\pi}\sigma_N(t)}\exp\left(-0.5\frac{a^2(t)}{\sigma_N^2(t)}\right), \tag{17.17}$$

where

$$r=m/n, \tag{17.18}$$

the ratio of the total number m of memory patterns in the neural network to the total number n of neurons, known as *the pattern ratio* of the neural network.

Other researchers such as Zagrebnov and Chvyrov (1989), Patrick and Zagrebnov (1991a,b), and Nishimori and Ozeki (1993), have also worked on the same topic. As mentioned in Fu (1994), all the above theoretical analyses of the dynamics of a network with a small pattern ratio ($r<0.15$) are in agreement with computer simulations when the initial main overlap is larger than a certain critical value. When working with a larger r system these analyses are only correct for only the first three steps. The storage capacity of the network obtained by Amari and Maginu (1988) or by Zagrebnov and Chvyrov (1989) is in agreement with those by Hopfield (1982) and Amit et al. (1985a). Okada (1995) proposed a theoretical analysis of a Hopfield-type network using the same two-variable dynamical system as that in the previous theoretical models but has taken account of the direct correlation between the noise terms at different time steps. This analysis is also in agreement with computer simulations in evaluating the storage capacity of the network. Using identical assumptions Yanai and Amari (1996) proposed a theoretical analysis of networks with nonmonotonic neurons and showed that the capacity of networks is enhanced remarkably. Note that in all the above theoretical models except Amit's, the main overlap $A(t)$ is assumed to be a deterministic function $a(t)$. However, this assumption only holds if r is very small and the initial main overlap is large enough (Fu and Yan, 1995). This is why the above theoretical analyses do not work in predicting the stable states of the system when r is not very small. The properties of the main overlap $A(t)$ have been studied by Amit (1989) and Fu and Yan (1995) theoretically and numerically. Although with $r < 0.13$ and initial main overlap large, the variance of $A(t)$ is not equal to zero, which means that $A(t)$ is not a deterministic function. Thus, the randomness of $A(t)$ should be taken into account in analysing the dynamics of the system by the statistical neurodynamics method. It is necessary to point out that in the Amari and Maginu theory, it is the main overlap $a(t)$ that converges to either 1 or 0 when t increases. In other words, the Amari and Maginu theory corresponds to a *simple system* (i.e. two values system) at stable states, as does the Okada theory (Okada, 1995). However, computer simulations (Fu and Yan, 1995; Fu, et al., 1995) show that $a(t)$ in a Hopfield-type neural network is distributed over the interval [0 1], i.e. the network corresponds to a multivalued system at stable states. Thus, the stable states of Hopfield-type neural networks resulting from the Amari and Maginu's and Okada's theories are described by characteristic functions of classical sets, while those obtained from computer simulations are described by fuzzy sets.

The theoretical work undertaken by Fu (1994) extends the Amari and Maginu theory by taking into account the randomness of the main overlap $A(t)$ and the mean of the noise terms $N_i(t)$. This work is composed of the following two assertions. It will be presented in the next section that the proposed theory is more effective in predicting stable states of the network than that of Amari and Maginu.

Assertion I. Suppose the following three conditions are satisfied:

(i) the components $X_i(t)$ $(i=1,2,\ldots,n)$ of the state of the network are exchangeable random variables;
(ii) the noise terms $N_i(t)$ $(i=1,2,\ldots,n)$ are independent exchangeable identical normal random variables with expectation $\mu_N(t)$ and standard deviation $\sigma_N(t)$; and
(iii) $P\{A(t)=a(t)\}=1$.

Then the governing equations for the macroscopic state of the neural network should include the noise terms by the additional equation as below:

$$\mu_N(t+1)=\frac{\sqrt{2}ru(t)}{\sqrt{\pi}\sigma_N(t)}\exp(-0.5(a(t)+\mu_N(t))^2\sigma_N^{-2}(t)). \tag{17.19}$$

Assertion II. If assumptions (i) and (ii) above together with the following two conditions are held:

(iv) the main overlap $A(t)$ is normally distributed with expectation $a(t)$ and standard deviation $\sigma_A(t)$;
(v) the main overlap $A(t)$ and noise terms $N_i(t)$ $(i=1,2,\dots,n)$ are independent random variables;

then the macroscopic state of the neural network is determined by (17.17) and (17.19) together with the following two recurrent equations:

$$a(t+1)=\frac{1}{2\pi\sigma_A(t)}\int_{-1}^{1}\mathrm{e}^{-0.5(x-a(t))^2\sigma_A^{-2}(t)}\int_{-(x+\mu_N(t))/\sigma_N(t)}^{(x+\mu_N(t))/\sigma_N(t)}\mathrm{e}^{-0.5v^2}\,\mathrm{d}v\,\mathrm{d}x, \tag{17.20}$$

$$\sigma_A^2(t+1)=\frac{1}{(2\pi)^{1.5}\sigma_A(t)}\int_{-1}^{1}\mathrm{e}^{-0.5(x-a(t))^2\sigma_A^{-2}(t)}\left(\int_{-(x+\mu_N(t))/\sigma_N(t)}^{(x+\mu_N(t))/\sigma_N(t)}\mathrm{e}^{-0.5v^2}\,\mathrm{d}v\right)^2\mathrm{d}x$$
$$-a^2(t+1). \tag{17.21}$$

Obviously, Assertion II reduces to Assertion I when the density function $p_A(x,t)$ is a δ-function. A full derivation of (17.19)–(17.21) is given in the Appendix 17.1.

Assertions I and II which are dynamical systems with three and four variables present two mathematical models of a Hopfield-type neural network for analysing its dynamic behaviours in macroscopic state, respectively. When $\mu_N(t)\equiv 0$ Assertion I reduces to (17.16)–(17.17) as shown in Amari and Maginu (1988).

It has been shown (Fu, 1994) that the theoretical work provides reliable dynamical analyses for a Hopfield-type network at the first few steps in a dynamic process and it also has an advantage over the Amari and Maginu theory when the pattern ratio r is large. Furthermore, it will be illustrated later that the theoretical work presents a multivalued system to describe the stable states of a Hopfield-type network, which is essentially different from the simple systems (Amari and Maginu, 1988; Okada, 1995). The explanation of why it is effective in predicting the stable states of a Hopfield-type network is given in section 4.

It was also shown (Fu, 1994) that the analyses based on the theory are in agreement with computer simulations, which not only effectively analyses the dynamics of Hopfield-type networks with a small r as in other theories, but also effectively provides a correct analysis of the dynamic behaviours for Hopfield-type networks with a large r (note that other theories are effective up to the first three steps only in this case). It will be shown in section 4 that the proposed theory corresponds to a multivalued system[7] at stable states, which is in agreement with computer simulations.

4. PREDICTION OF STABLE STATES OF SYNCHRONOUS AMNs

Before studying the statistical properties of synchronous AMNs by computer simulations, we recall that a synchronous AMN is a recurrent random dynamical system. Based on the previous assumption (see section 3), the main overlap $A(t)$ in

(17.11) and noise terms $N_i(t)$, $i=1,\ldots,n$ in (17.15) are the random processes defined in the probability space $(\Omega, \mathfrak{F}, P)$, where $\Omega=\{-1,1\}^n$ is an n-dimensional Bernoulli space, $\mathfrak{F}$ is the minimal σ-algebra generated by the basis of Ω, i.e. the set of subset $\{\Lambda\}$ whose element Λ is defined as

$$\Lambda=\{\omega\colon \omega=(\omega_1, \omega_2, \ldots, \omega_n)^{\mathrm{T}}\}, \tag{17.22}$$

ω_i taking $+1$ or -1, $i=1,\ldots,n$ and the probability of Λ is given by

$$P\{\Lambda\}=\beta^{\Sigma_{i=1}^{n}(1+\omega_i)/2}(1-\beta)^{\Sigma_{i=1}^{n}(1-\omega_i)/2}, \tag{17.23}$$

where $\beta=P\{\omega_i=1\}=\frac{1}{2}$. Notice that the memory patterns are $\mathbf{S}^k\in\{\Lambda\}$, $k=1,\ldots,m$ and that when they are fixed in a specified simulation one obtains the sample functions, i.e. the realization of $A(t)$ and $N_i(t)$ given by (17.11) and (17.15), respectively. Since $\mathbf{S}^k$ $(k=2,\ldots,m)$ is a random variable, so is $X_i(t)$ $(i=1,\ldots,n)$ and therefore $A(t)$ and $N_i(t)$ are random variables for fixed t. In view of the definition of a random process (Doob, 1953), the main overlap $A(t)$ and noise terms $N_i(t)$ are random processes defined in the probability space $(\Omega, \mathfrak{F}, P)$. What follows is a presentation of the statistical properties of $A(t)$ and $N_i(t)$ in terms of numerical calculations by the synchronous AMN.

To simulate the dynamics of the neural network, the initial external excitement, i.e. the initial state, $\mathbf{X}(0)$, of the neural network must first be given. To simplify the analysis of the dynamics, $\mathbf{X}(0)$ can be constructed deterministically. Without loss of generality, $\mathbf{X}(0)$ may be expressed in the following form:

$$\mathbf{X}(0)=(\underbrace{1,1,\ldots,1}_{n'}, \underbrace{-1,-1,\ldots,-1}_{n-n'})^{\mathrm{T}}, \tag{17.24}$$

where $0\leqslant n'\leqslant n$ is a positive integer, which is a distortion of the real memory

$$\mathbf{S}^1=(1,1,\ldots,1)^{\mathrm{T}}. \tag{17.25}$$

The updated states $X_i(1)$ $(i=1,2,\ldots,n)$ are obtained from (17.3). This process is repeated with $\mathbf{X}(1)$ as the new initial state and thus the progressive recall sequence $\mathbf{X}(0)$, $\mathbf{X}(1)$, $\mathbf{X}(2)$, …, is generated.

To analyse the dynamics of a neural network, a set of statistical quantities concerning the main overlap and noise terms is required. In order to obtain these statistical quantities it is necessary to collect a set of samples of the data for the main overlap and noise terms by repeated computer simulation.

Figure 17.4(a) demonstrates the theoretical analysis and experimental results of a system with a small pattern ratio ($r=0.08$). The crosses represent 50 independent computer simulations by the system with $n=400$ and $a(0)=0.7$. The solid line denotes the expectation of the main overlap obtained from the computer simulations. The other curves in the figure represent the theoretical results based on Assertions I and II, and Theorem 2 in Amari and Maginu (1988), respectively. The graph shows that all theoretical analyses are successful in predicting the stable states of the system when its pattern ratio r is small.

Figures 17.4(b) and 17.4(c) show the theoretical results and computer simulations of the systems with the same parameters as those in Figure 17.4(a) except that r has been changed to 0.3 and 0.5, respectively. The graphs show that the theoretical results based on Assertions I and II are in agreement with the computer simulation in evaluating the average stable state, but that results based on the theory of Amari and Maginu (1988) do not hold in these cases.

Figure 17.4(d) shows the theoretical analyses and experimental results by the system which has the same parameters as those in Figure 17.4(d) except that r is 0.16 instead of 0.08. The results indicate that no theoretical analyses are effective in predicting the average stable state of the system in this case (a system with an intermediate pattern ratio). Notice the strange 'jumps' of the curve obtained by Assertion II appear in this case, which is strong evidence that the main overlap does not obey a normal distribution.

Experiments have shown that the overlap $A(t)$ obeys approximately normal distribution for the network with a small or large r (see Figures 17.5(c) and 17.5(e)) when the initial state of the network is close to the real memory pattern, but not for the network with an intermediate pattern ratio $(1.5<r\leqslant 0.2)$ even though the initial state is quite close to the real memory pattern. Experiments have also shown that

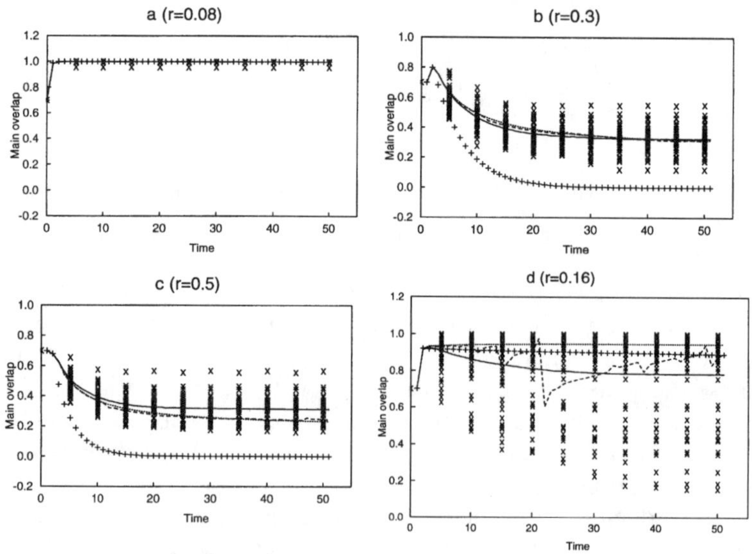

Figure 17.4 Main overlaps $A(t)$ as random processes in the networks with $r=0.08$ (a), $r=0.3$ (b), $r=0.5$ (c) and $r=0.16$ (d), respectively, and with the identical input having initial main overlap $a(0)=0.7$. The cross ($\times$) lines present the computer simulations; the solid lines present the expectation of $A(t)$; the plus (+) lines present the result obtained by the theory (Amari and Maginu, 1988); and the dotted (···) and dashed (---) lines present the results obtained by Assertions I and II, respectively.

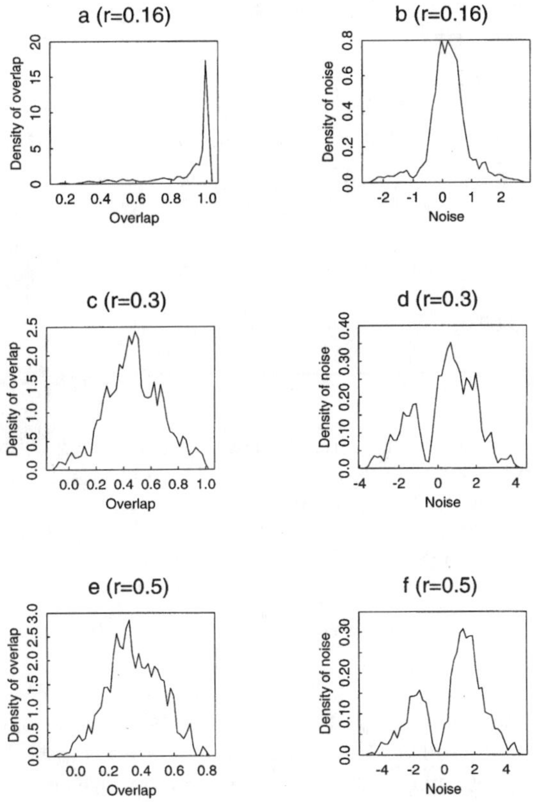

Figure 17.5 Densities of overlap and noise terms of networks with pattern rates $r=0.16$ (a and b), $r=0.3$ (c and d) and $r=0.5$ (e and f), respectively, and having an identical input with an initial main overlap $a(0)=0.7$.

the noise terms in a network have the following properties: they approximately obey a normal distribution when the network has a small or an intermediate r and have a mixed normal distribution if r is large (see Figures 17.5(b), 17.5(d) and 17.5(f)).

It follows that the hypothesis that noise terms are normal means that the actual density function (mixed normal) represented by the solid line is approximated by a normal distribution shown with the dashed line in Figure 17.6. Since AMNs are insensitive to small errors in an input pattern, the conditions of assumption (ii) in Assertions I and II are still acceptable in this case. Furthermore, relatively small correlation between the main overlap and noise terms and small deviation of the main overlap are found in a large r system (Fu and Yan, 1995), so the effect of the correlation between the main overlap and noise terms can be ignored and assumption (v) in Assertion II is approximately satisfied as well.

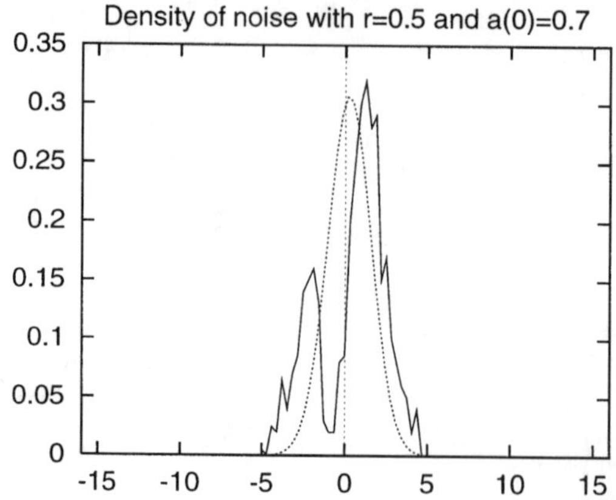

Figure 17.6 The approximation of the density of noise terms in the network with the pattern rate $r = 0.5$ and initial main overlap $a(0) = 0.7$ by a normal distribution. The solid line represents the density function of noise terms in the network obtained from the computer simulations and the dotted line represents the normal distribution used to approach the density function of noise terms.

5. DISCUSSION

We have studied the long-term behaviour (i.e. stable states) of synchronous AMNs. Although Hopfield models and synchronous AMNs are identical in their structure, they have important dynamical differences which we wish to highlight. We will also discuss the capacity problem of a AMN as well as the noise effect of neural networks and the main differences in dynamics between the proposed theoretical analyses and the theory of Amari and Maginu (1988).

(i) The Stable States

The most important property of synchronous AMNs is that the networks always converge to stable states within about 40 iterations. This property is quite in agreement with the fact that it takes only a few steps for a human to recall specific information. The recalling result may be one of the three cases: full or part of the information is called, or nothing is recalled. The process in any case is completed in finite time. However, it is not known in advance how long it will be taken for a Hopfield neural network to arrive at a stable state. In fact, the time taken for the network to arrive at a stable state depends on not only an initial input (stimulus) but also the order of the firing neurons. It may be a long time for the network to converge to a stable state. Thus, the synchronization behaviour of AMNs is the key property to ensure good dynamical behaviour of AMNs.

(ii) The Synchronization

The essential difference between synchronous AMNs and Hopfield neural networks is that the Hopfield model is an asynchronous system. It was found (Miles and

Wong, 1983; Miles et al., 1984) that burst firing of neurons within the CA3 region of a human brain is generated synchronously. The most obvious population behaviours in the hippocampus are those in which neuronal firing is synchronized (Traub and Miles, 1991). In fact, a neuron is much more sensitive to synchronous activation. This property has been studied by many researchers, such as Abeles (1982, 1991), Abeles et al. (1994), Bal et al. (1994) and Traub et al. (1994). It was pointed out by Peretto (1984) that synchronous AMNs could possibly perform serial processing (i.e. the ability for a network to produce a whole string of meaningful states given the first states of the string). Peretto (1984) also claimed that the functioning of real neural networks is neither purely synchronous as in Little (1974) model nor purely asynchronous as in the dynamics associated with the Hopfield model. Like most of the associative memory networks the synchronous AMNs are generated based on the Hebb (1949) rule. However, it was claimed by Alkon et al. (1993) that none of the associative memory networks with Hebbian synapses have demonstrated any pattern recognition capacity, memory capacity, or robustness of biological networks. Alkon et al. (1993, 1994) proposed associative memory networks by non-Hebbian synapses and retrieval in biological systems derived from neurophysiological experiments on both invertebrate and vertebrate animals. Meir and Domany (1987) presented a statistical analysis of the Hopfield-type networks with time-dependent synaptic weights and found that none of the stored memory patterns can be retrieved by the networks when the pattern ratio r of the network is larger than 0.269. The dynamics of Meir and Domany's model seem to be similar to those of the synchronous AMNs.

(iii) The Capacity of Synchronous AMNs

The capacity of synchronous AMNs can be described by their pattern ratios r. As seen in Section 4, the dynamics of synchronous AMNs is determined completely by the pattern ratio r and input (initial stimulus). The study of dynamics of synchronous AMNs is usually achieved in terms of the analyses of the main overlap $A(t)$. In most of the theoretical models proposed by Amari and Maginu (1988) and Okada (1995) for synchronous AMNs, the main overlap $A(t)$ is treated as a deterministic function $a(t)$, so the limiting operation, $\lim_{t\to\infty} a(t)$, means an ordinary convergence process. However, it is not reasonable to assume that the main overlap is a deterministic function when the pattern ratio r is not small, say $r>0.1$, because in this case $P\{A(t)=a(t)\}\neq 1$ although the initial main overlap $a(0)$ is large. This fact leads to the following two problems:

(a) For $r=0.13$, we have found that the probability $P\{A(t)=1\}$ does not converge to 1 as t increases although the input of the network has a very large main overlap. This means that there are some samples of $A(t)$ which do not converge to 1 as t increases.

(b) On the other hand, for $0.15<r<0.2$, we have found that there are still some samples of $A(t)$ which converge to 1. An example of this case is given in Figure 17.4(d).

This means that the memory patterns stored in synchronous AMNs with $r=0.13$ are not retrieved 100%. Although some information stored in synchronous AMNs with

$r=0.16$ can still be retrieved. Obviously, it is not adequate to characterize the storage capacity of synchronous AMNs by the variable r only.

Let $a_s[a(0), r]$ and $a_t[a(0), r]$ represent the expectations of the main overlap $A(t)$ at stable states resulting from computer simulations and the proposed theoretical analyses, respectively, i.e. $\lim_{t\to\infty} E[A(t)] = a_s[a(0), r]$ and $\lim_{t\to\infty} a(t) = a_t[a(0), r]$ where $a(t)$ is obtained theoretically. For a given $a(0)$, let

$$R_\alpha^{a(0)} = \{r: a_s[a(0), r] \geqslant \alpha, 0<r<1\},\ 0<\alpha<1, \tag{17.26}$$

obviously $R_\eta^{a(0)} \supset R_\alpha^{a(0)} \supset R_1^{a(0)}$, where $\eta<\alpha<1$. We define

$$r_\alpha(a(0)) = \begin{cases} \mathrm{Sup}_r\{r:\ r\in R_\alpha^{a(0)}\} & \text{if } R_\alpha^{(a(0))} \neq \phi, \\ 0 & \text{otherwise.} \end{cases} \tag{17.27}$$

When the initial $a(0)$ is large, $r_\alpha[a(0)]$ can be used to measure the storage capacity of the network. Note that the *large* $a(0)$ is a vague concept and needs to be defined in a further study. The measure $r_\alpha[a(0)]$ is associated with the reliability coefficient α, which is different from the traditional storage capacity measure of the network. Computer simulations show that $r_1[a(0)]<0.13$ and $r_\alpha[a(0)]>0.16$ when $a(0)=0.7$ and $\alpha\neq 1$.

It was shown by Amari (1989) and Frolov et al. (1997) that sparsely[8] encoded associative memory networks not only improve the capacity but also enhance the size of a basin of attraction around each memorized pattern when and only when an activity control mechanism is attached to it.

(iv) The Effection of a Noise Source in Neural Networks

The performance of neural networks can be improved by adding noise source to membrane potentials (Buhmann and Schulten, 1987). This property was confirmed by Little and Shaw (1975) in which a storage capacity was enhanced when noise had been added to their model network. We do not know what kind of noise contributes to brain function. The following three cases are possible (Buhmann and Schulten, 1987): (a) membrane noise originating from the stochastic closing and opening of ion channel; (b) the noise which originates from the release of neurotransmitter; (c) the noise connected with afferent synaptic signals. Thus, based on the effect of a noise source, we have an additional parameter, i.e. noise term, for adjusting to achieve a desirable neural network (Fu et al., 1995).

(v) The Main Differences in Dynamics between Our Theoretical Analyses and Amari and Maginu's

The proposed theoretical analyses and Amari and Maginu theory are shown in Table 17.1. We can see that:

- the long-term dynamic behaviour of synchronous AMNs obtained from the computer simulations are closer to those derived from the proposed theoretical analyses than those from the Amari and Maginu theory (see Table 17.1 and Figure 17.4);
- all the theoretical analyses (i.e. the proposed theory and Amari and Maginu theory) are in agreement with computer simulations when synchronous AMNs

Table 17.1 The comparison between computer simulations and theories

	Computer Simulations	The Proposed Theory	Amari and Maginu Theory
r is small and $a(0)$ is large enough	$\lim_{t\to\infty} A(t)=1$ with the probability one	It is the same as the computer simulations	It is the same as the computer simulations
$0.15<r\leqslant 0.2$ and $a(0)$ is large	$\lim_{t\to\infty} E[A(t)]=a_s(a(0),r)$ $0<a_s(a(0),r)\leqslant 1$	$\lim_{t\to\infty} a(t)=a_t(a(0),r)$ $0<a_t(a(0),r)\leqslant 1$ $a_s(a(0),r)\neq a_t(a(0),r)$	$\lim_{t\to\infty} a(t)=0$ $\lim_{t\to\infty} a(t)=0$
$0.2\leqslant r\leqslant 0.5$ and $a(0)$ is large	$\lim_{t\to\infty} E[A(t)]=a_s(a(0),r)$ $0<a_s(a(0),r)<1$	$\lim_{t\to\infty} a(t)=a_t(a(0),r)$ $0<a_t(a(0),r)<1$ $a_s(a(0),r)\doteq a_t(a(0),r)$	$\lim_{t\to\infty} a(t)=0$ $\lim_{t\to\infty} a(t)=0$

with r small ($r<0.15$) and $a(0)$ being large enough (see the second row in Table 17.1 and Figure 17.4(a));

- no theoretical analyses is appropriate when the synchronous AMNs have an intermediate r ($0.15<r\leqslant 0.2$) (see the third row in Table 17.1 and Figure 17.4(d));
- the proposed theoretical analyses effectively predict the stable states of synchronous AMNs when the pattern rate r is large ($0.2<r$), but in this case the Amari and Maginu theory does not (see the fourth row in Table 17.1 and Figures 17.4(b) and 4(c)).

6. CONCLUSIONS AND FUTURE PERSPECTIVES

The behaviour of stable states of synchronous AMNs has been studied using both theoretical analysis and computer simulation. The results show that the proposed theoretical model can effectively predict the stable states of the AMN with small and large pattern ratios r (see Figure 17.4). The proposed model of synchronous AMNs has the following distinct features:

1. The AMN can retrieve the entire information from incomplete knowledge (see Figure 17.4(a)). This is in agreement with the human being recalling something from a hint.
2. The AMN always arrives at a stable state within about 40 iterations (see Figure 17.4). This property is also in agreement with the human ability to recall something usually in finite time.
3. The stable state of the AMN is multivalued rather than binary. This property is also similar to human recalling behaviour, they may recall an entire information, a portion of an information or none at all, i.e. the result that a human recalls is not binary.

Since the stable state of a synchronous AMN reflects the information that is stored in the AMN, so the fuzzy set theory (Zadeh, 1965) seems to be a powerful tool for studying the behaviour of the human brain. It is claimed that 'Fuzzy systems start from highly formalized insights about the psychology of categorization and the structure of categories found in the real world (Kosko, 1992). Therefore, the 'theory of fuzziness' as developed is an abstract system that makes no further claims about biological or psychological plausibility'. Thus it is very interesting and also an important direction of neuroscience that fuzzy neural systems are used to investigate the cognitive abilities of man and his extensive memorizing capabilities.

PROBLEMS

17.1. Both Hopfield-type neural networks and Hopfield neural networks are feedback associative memory models. What are the similarities in their structures? What are the difference in their dynamics? Do they have different behavior in their stable states?

17.2. Based on the definition of the main overlap $A(t)$, i.e. (17.11), $A(t)$ is a random process and its expectation is denoted by $a(t)$. Suppose the components $X_i(t)$, $i=1,\ldots,n$ of the state of the network are exchangeable random variable, i.e. they have identical distributions,

(i) if the memory pattern $s_j^1 = 1$ where $j=1,\ldots,n$, then we have formulae (A17.14) and (A17.15), i.e.,

$$P\{X_i(t)=1\}=\frac{1+a(t)}{2},$$

$$P\{X_i(t)=-1\}=\frac{1-a(t)}{2},$$

prove the above two equations;

(ii) if the memory pattern $s_j^1 = -1$ where $j=1,\ldots,n$, then what are $P\{X_i(t)=-1\}$ and $P\{X_i(t)=-1\}$?

(iii) prove $E[X_i(t)]=a(t)$ in case (i) and $E[X_i(t)]=-a(t)$ in case (ii).

17.3. (i) AMNs can be considered being dynamic systems. Explain that the stable states of the systems are determined completely by their initial main overlap $a(0)$ and the pattern ratio r.

(ii) Explain that the pattern ratio r is a quantity to characterize the structure of AMNs while the initial main overlap $a(0)$ is a quantity to describe the strength of an external excitation of AMNs.

17.4. (i) For an arbitrary pair $(a(0), r)$, where $0 \leqslant a(0) \leqslant 1$, $0 < r \leqslant 1$, based on question 17.3, a specific *AMN* has been given. Can we say that the pairs $\{(a(0), r)$: $0 \leqslant a(0) \leqslant 1, 0 < r \leqslant 1\}$ are a fuzzy set defined on the universal set $0 \leqslant a(0) \leqslant 1$ and $0 < r \leqslant 1$ where $a_i(a(0), r)$ (see section 5) is considered being the membership function?

(ii) If the statement in (i) is true, can we classified *AMN*s into different classes using fuzzy set relations and their operations?

17.5 The Hopfield model is a simple model of associative memory. As a research project address the question of what happens if the formal neurons used in this Chapter (cf. Figure 17.1) are replaced by 'spiking' neurons. [Hint: See Gerstner and Van Hemmen, 1992].

NOTES

1. Golgi and Ramon y Cajal shared the 1906 Nobel prize in medicine for their discoveries.
2. Suppose an input is an n-dimensional vector which can be considered being a point in an n-dimensional space. A set of inputs is a set of points in n-dimensional space. If the points in the space can be separated into two classes by a hyperplane then the related scheme is known to be linearly separable and to be nonlinearly separable otherwise where the hyperplane is determined by an activity of a neural network and a given threshold, i.e. the hyperplane is given by the equation: *the activity equal to the threshold*.
3. A collective relaxation search is an iterative process of a system in which the state of the system converges to the stable state that reflects the stored memory of the system when the iterative number increases.
4. A content addressable memory is an n-dimensional vector, a stable state of neural networks, which reflects the information stored in the network distributed over the synaptic weights and which can be retrieved when an input closes the stable state.
5. A neural network whose state at time t is denoted by $X(t)$ is known as a discrete or continuous neural network if the components of $X(t)$ are discrete or continuous, respectively.
6. The Hopfield-type neural network is composed of a large number of neurons which are connected to each other. The state of each neuron in the network indicates microscopic behaviour in the network, while the main overlap $A(t)$ and noise terms $N_i(t)$ which are the integrated measures of the network obtained through considering the states of all neurons and synaptic weights of the network are the quantities used to describe the macroscopic state of the neural network.
7. A dynamical system whose main overlap $A(t)$ is not binary at stable states in general is referred to as a multivalued system.
8. If most of the components of encoded pattern vectors to be stored are 0 and only a small ratio of the components are 1, the encoding method is said to be sparse.

REFERENCES

Abeles, M. (1982) Role of cortical neuron: Integrator or coincidence detector? *Isr. J. Med. Sci.*, **18**, 83–92.

Abeles, M. (1991) *Corticonics: Neural Circuits of the Cerebral Cortex.* New York: Cambridge University Press.

Abeles, M., Prut, Y., Bergman, H. and Vaadia, E. (1994) Synchronization in neuronal transmission and its importance for information processing. In *Temporal Coding in the Brain,* edited by G. Buzsáki, R. Llinás, W. Singer, A. Berthoz and Y. Christen. Berlin: Springer-Verlag.

Alkon, D.L., Blackwell, K.T., Barbour, G.S., Werness, S.A. and Vogl, T.P. (1994) Biological plausibility of synaptic associative memory models. *Neural Networks*, **7**, 1005–1017.

Alkon, D.L., Blackwell, K.T., Vogl, T.P. and Werness, S.A. (1993) Biological plausibility of artificial neural networks: learning by non-Hebbian synapses. In: *Associative Neural Memories*, edited by Mohamad H. Hassoun. New York: Oxford University Press.

Amari, S. (1972) Characteristics of random nets of analog neuron-like elements. *IEEE Trans.*, **SMC-2**, 643–657.

Amari, S.-I. (1974) A method of statistical neurodynamics. *Kybernetik*, **14**, 201–215.

Amari, S.-I. (1988) Associative memory and its statistical neurodynamical analysis. In, Neural and Synergetic Computers, edited by H. Haken. Berlin: Springer-Verlag.

Amari, S.-I. (1989) Characteristics of sparsely encoded associative memory. *Neural Networks*, **2**, 451–457

Amari, S.-I. and Maginu, K. (1988) Statistical neurodynamics of associative memory. *Neural Networks*, **1**, 63–73.

Amit, D.J. (1989) *Modeling Brain Function*. Cambridge: Cambridge University Press.

Amit, D.J., Gutfreund, H. and Sompolinsky, H. (1985a) Storing infinite numbers of patterns in a spin-glass model of neural networks. *Phys. Rev. Lett.*, **55**, 1530–1533.

Amit, D.J., Gutfreund, H. and Sompolinsky, H. (1985b) Spin-glass models of neural networks. *Phys. Rev. A.*, **32**, 1007–1018.

Anderson, J.A. (1968) A memory storage model utilizing spatial correlation functions. *Kybernetik*, **5**, 113–119.

Anderson, J.A. (1972) A simple neural network generating an interactive memory. *Math. Biosci.*, **14**, 197–220.

Baird, B. (1986) Nonlinear dynamics of pattern formation and pattern recognition in the rabbit olfactory bulb. *Physica D*, **22**, 150–175.

Bal, T., von Krosigk, M. and McCormick, D.A. (1994) From cellular to network mechanisms of a thalamic synchronized oscillation. In *Temporal Coding in the Brain*, edited by G. Buzsáki, R. Llinás, W. Singer, A. Berthoz, and Y. Christen. Berlin: Springer-Verlag.

Bennett, M.R., Gibson, W.G. and Robinson, J. (1994) Dynamics of the CA3 pyramidal neuron autoassociative memory network in the hippocampus. *Phil. Trans. R. Soc. Lond.*, B **343**, 167–187.

Block, H.D. (1962) The Perceptron: a model for brain functioning. *I, Rev. Mod. Phys.*, **34**, 123–135.

Braham, R. and Hamblen, J. O. (1988) On the behavior of some associative neural networks. *Biol. Cybern.*, **60**, 145–151.

Buhmann, J. and Schulten, K. (1987) Influence of noise on the function of a "physiological" neural network. *Biol. Cybern.*, **56**, 313–327.

Cohen, M.A. and Grossberg, S. (1983) Absolute stability of global pattern formation and parallel memory storage by competitive neural networks. *IEEE Trans.*, **SMC-13**, 815–826.

Cottrell, M. (1988) Stability and attractivity in associative memory networks. *Biol. Cybern.*, **58**, 129–139.

Cowan, J.D. and Sharp, D.H. (1988) Neural nets. *Quart. Rev. Biophys.* **21**, 365–427.

Doob, J.L. (1953) *Stochastic Processes*. New York: John Wiley & Sons.

Frolov, A.A., Husek, D. and Muraviev, I.P. (1997) Informational capacity and recall quality in sparsely encoded Hopfield-like neural network: analytical approaches and computer simulation. *Neural Networks*, **10**(5), 845–855.

Fu, A.M.N. (1994) Statistical analysis of an autoassociative memory network. *Neural Comput.*, **6**, 837–841.

Fu, A.M.N. and Yan, H. (1995) Distributive properties of main overlap and noise terms in autoassociative memory network. *Neural Networks*, **8**, 405–410.

Fu, A.M.N., Yan, H. and Suters, M.G. (1995) Flexible pattern matching using a Hopfield–Amari neural network. *Opt. Eng.*, **34**, 2467–2474.

Gelperin, A., Hopfield, J.J. and Tank, D.W. (1985) The logic of limax learning. In *Model Neural Networks and Behavior*, edited by A.I. Selverston. New York: Plenum Press.

Gerstner, W. and van Hemmen, J.-L. (1992) Associative memory in a network of 'spiking' neurons. *Network*, **3**, 139–164.

Gibson, W.G. and Robinson, J. (1992) Statistical analysis of the dynamics of a sparse associative memory. *Neural Networks*, **5**, 645–661.

Grossberg, S. (1988) Nonlinear neural networks: Principles, mechanisms and architectures. *Neural Networks*, **1**, 17–61.

Haberly, L.B. (1985) Neuronal circuitry in olfactory cortex: Anatomy and functional implications. *Chem. Senses*, **10**, 219–238.

Hassoun, M.H. (1993) Dynamic associative neural memories. In *Associative Neural Memories Theory and Implementation*, edited by M.H. Hassoun. Oxford: Oxford University Press.

Hebb, D.O. (1949) *The Organization of Behavior*. New York: Wiley.

Hendin, O., Horn, D. and Tsodyks, M.V. (1997) The role of inhibition in an associative memory model of the olfactory bulb. *J. Comput. Neurosci.*, **4**, 173–182.

Hopfield, J.J. (1982) Neural networks and physical systems with emergent collection computational abilities. *Proc. Natl. Acad. Sci. USA*, **79**, 2554–2558.

Hopfield, J.J. (1984) Neurons with graded response have collective computational properties like those of two-state neurons. *Proc. Natl. Acad. Sci. USA*, **81**, 3088–3092.

Hopfield, J.J. and Tank, D. W. (1985) "Neural" computation of decisions in optimization problems. *Biol. Cybern.*, **52**, 141–152.

Kleinfeld, D. and Sompolinsky, H. (1989) Associative network models for central pattern generators. In *Methods in Neuronal Modeling*, edited by C. Koch and I. Segev. MIT Press.

Koch, C., Marroquin, J. and Yuille, A. (1986) Analog "neuronal" networks in early vision. *Proc. Natl. Acad. Sci. USA*, **83**, 4263–4267.

Kohonen, T. (1989) *Self-Organization and Associative Memory*. 3rd edition New York: Springer-Verlag.

Komlos, J. and Paturi, R. (1993) Convergence analysis of associative memories, in *Associative Neural Memories* M. H. Hassoun, ed., Oxford University Press, 184–196.

Kosko, B. (1992) *Neural Networks and Fuzzy Systems*. Englewood Cliffs, New Jersey: Prentice Hall.

Lee, J.S., Chen, C.H., Sun, Y.N. and Tseng, G.H. (1997) Occluded objects recognition using multiscale features and Hopfield neural network. *Patt. Recogn.*, **30**, 113–122.

Li, S.Z. (1996) Improving convergence and solution quality of Hopfield-type neural networks with augmented Lagrange Multipliers. *IEEE Trans. on Neural Networks*, **7**, 1507–1516.

Little, W.A. (1974) The existence of persistent states in the brain, *Math. Biosci.*, **19**, 101–120.

Little, W.A. and Shaw, G.L. (1975) A statistical theory of short and long term memory. *Behav. Biol.*, **14**, 115–133.

Ma, J. (1997) Simplex memory neural networks. *Neural Networks*, **10**, 25–29.

McCulloch, W.S. and Pitts, W. (1943) A logical calculus of the ideas immanent in nervous activity. *Bull. Math. Biophys.*, **5**, 115–133.

Meir, R. and Domany, E. (1987) Exact solution of a layered neural network model. *Phys. Rev. Lett.*, **59**(3), 359–362.

Miles, R. and Wong, R.K.S. (1983) Single neurones can initiate synchronized population discharge in the hippocampus. *Nature*, **306**, 371–373.

Miles, R., Wong, R.K.S. and Traub, R.D. (1984) Synchronized after-discharges in the hippocampus: contribution of local synaptic interaction. *Neuroscience*, **12**, 1179–1189.

Minsky, M.L. and Papert, S.A. (1969) *Perceptrons: An Introduction to Computational Geometry*. Cambridge, MA: MIT Press.

Morita, M. (1993) Associative memory with nonmonotone dynamics. *Neural Networks*, **6**, 115–126.

Müller, B., Reinhardt, J. and Strickland, M.T. (1995) *Neural Networks*. Berlin: Springer.

Nishimori, H. and Ozeki, T. (1993) Retrieval dynamics of associative memory of the Hopfield type. *J. Physics, A,* **26**, 859–871.

Nishimori, H. and Opris, I. (1993) Retrieval process of an associative memory with a general input–output function. *Neural Networks*, **6**, 1061–1067.

Okada, M. (1995) A hierarchy of macrodynamical equations for associative memory. *Neural Networks*, **8**, 833–838.

Okada, M. (1996) Notions of associative memory and sparse coding. *Neural Networks*, **9**, 1429–1458.

O'Keefe, J. (1983) Spatial memory within and without the hippocampal system, In *Neurobiology of the Hippocampus*, edited by W. Seifert. New York: Academic Press.

Palm, G. (1980) On associative memory. *Biol. Cybern.*, **36**, 19–31.

Patrick, A.E. and Zagrebnov, V.A. (1991a) On the parallel dynamics for the Little–Hopfield model. *J. Stat. Phys.*, **63**, 59–71.

Patrick, A.E. and Zagrebnov, V.A. (1991b) A probabilistic approach to parallel dynamics for the Little–Hopfield model. *J. Phys. A*, **24**, 3413–3426.

Peretto, P. (1984) Collective properties of neural networks: a statistical physics approach. *Biol. Cybern.*, **50**, 51–62.

Peretto, P. and Niez, J.J. (1986) Stochastic dynamics of neural networks. *IEEE Trans. Syst., Man, Cybern.*, **SMC-16**, 73–83.

Rolls, E.T. (1995) A model of the operation of the hippocampus and entorhinal cortex in memory. In *International Journal of Neural Systems—supplementary issue*, edited by D.J. Amit, P. del Giudice, B. Denby, E.T. Rolls and A. Treves. Singapore: World Scientific, 51–70.

Rolls, E.T. and Treves, A. (1998) Neural Networks and Brain Function. Oxford: Oxford University Press

Rosenblatt, F. (1958) The perceptron: a probabilistic model for information storage and organization in the brain. *Psychol. Rev.*, **65**, 386–408.

Rosenblatt, F. (1962) *Principles of Neurodynamics*. New York: Spartan.

Shen, D. and Ip, H.H.S. (1997) A Hopfield neural network for adaptive image segmentation: An active surface paradigm. *Patt. Recogn. Lett.*, **18**, 37–48.

Thomas, C.M., Gibson, W.G. and Robinson, J. (1996) Stability and bifurcations in an associative memory model. *Neural Networks*, **9**(1), 53–56.

Traub, R.D., Jefferys, J.G.R. and Miles, R. (1994) Common principles in three experimental epilepsies. In *Temporal Coding in the Brain*, edited by G. Buzsáki, R. Llinás, W. Singer, A. Berthoz, and Y. Christen. Berlin: Springer-Verlag.

Traub, R.D. and Miles, R. (1991) *Neuronal Networks of the Hippocampus*. Cambridge: Cambridge University Press.

Treves, A. and Rolls, E.T. (1992) Computational constraints suggest the need for two distinct input systems to the hippocampal CA3 network. *Hippocampus* **2**, 189–200.

Tsodyks, M. and Sejnowski, T. (1995) Associative memory and hippocampal cells. In *International Journal of Neural Systems—supplementary issue*, edited by D.J. Amit, P. del Giudice, B. Denby, E.T. Rolls and A. Treves. Singapore: World Scientific.

Wang, T. and Sheng, Z. (1996) Asymptotic stationarity of discrete-time stochastic neural networks. *Neural Networks*, **9**, 957–963.

Yanai, H.-F. and Sawada, Y. (1990) Associative memory network composed of neurons with hysteretic property. *Neural Networks*, **3**, 223–228.

Yanai, H.-F. and Amari, S.-I. (1996) Auto-associative memory with two-stage dynamics of nonmonotonic neurons. *IEEE Trans. Neural Networks*, **7**, 803–815.

Yoshizawa, S., Morita, M. and Amari, S.-I. (1993) Capacity of associative memory using a nonmonotonic neuron model. *Neural Networks*, **6**, 167–176.

Zadeh, L.A. (1965) Fuzzy sets. *Inform. Contr*., **8**, 338–353.

Zagrebnov, V.A. and Chvyrov, A.S. (1989) The Little–Hopfield model: Recurrence relations for retrieval-pattern errors. *Sov. Phys.-JETP*, **68**, 153–157.

APPENDIX 17.1: Derivation of (17.19)–(17.21)

1. Derivation of (17.19)

Based on the definition of the main overlap

$$A(t+1)=\frac{1}{n}\sum_{j=1}^{n} s_j^1 X_j(t+1). \tag{A17.1}$$

Since $s_j^1=1$, $j=1,2,\dots,n$, (A17.1) simplifies to

$$A(t+1)=\frac{1}{n}\sum_{j=1}^{n} X_j(t+1). \tag{A17.2}$$

Using (17.3) and (17.14) we have

$$X_j(t+1)=\mathrm{Sgn}(A(t)+N_j(t)). \tag{A17.3}$$

Based on the definition of $N_i(t)$, we have

$$N_i(t+1)=Q_{ij}(t+1)+\frac{1}{n}s_i^\alpha s_j^\alpha X_j(t+1), \tag{A17.4}$$

where

$$Q_{ij}(t+1)=\frac{1}{n}\sum_{\beta\neq 1}^{m}\sum_{k\neq i,j}^{n} s_i^\beta s_k^\beta X_k(t+1)+\frac{1}{n}\sum_{\beta\neq 1,\alpha}^{m} s_i^\beta s_j^\beta X_j(t+1). \tag{A17.5}$$

For n large, $Q_{ij}(t+1)\approx N_i(t+1)$, hence $Q_{ij}(t+1)$ and $N_i(t+1)$ are approximately identical random processes. In view of assumption (ii), we have

$$Q_{ij}(t+1)\sim N(\mu_N(t+1),\sigma_N(t+1)). \tag{A17.6}$$

Letting

$$Z_{ij\alpha}(t+1)=s_i^\alpha s_j^\alpha X_j(t+1), \tag{A17.7}$$

we obtain

$$N_i(t+1)=\frac{1}{n}\sum_{\alpha\neq 1}^{m}\sum_{j\neq i}^{n} Z_{ij\alpha}(t+1) \tag{A17.8}$$

and therefore

$$E[N_i(t+1)]=\frac{1}{n}(m-1)(n-1)E[Z_{ij\alpha}(t+1)]. \tag{A17.9}$$

Based on (A17.3) and (A17.4) we have

$$X_j(t+1)=\mathrm{Sgn}\left[A(t)+Q_{ji}(t)+\frac{1}{n}s_j^\alpha s_i^\alpha X_i(t)\right]. \tag{A17.10}$$

Using assumption (iii) in Assertion I yields,

$$X_j(t+1) = \mathrm{Sgn}\left[a(t) + Q_{ji}(t) + \frac{1}{n} s_j^\alpha s_i^\alpha X_i(t) \right] \tag{A17.11}$$

and

$$Z_{ij\alpha}(t+1) = s_i^\alpha s_j^\alpha \mathrm{Sgn}\left[a(t) + Q_{ji}(t) + \frac{1}{n} s_j^\alpha s_i^\alpha X_i(t) \right]. \tag{A17.12}$$

Taking the expectation of both sides of (A17.12) and using (17.10) we obtain

$$\begin{aligned} E[Z_{ij\alpha}(t+1)] &= E[E(Z_{ij\alpha}(t+1) \mid s_i^\alpha s_j^\alpha] \\ &= \frac{1}{2} E\left[\mathrm{Sgn}\left[a(t) + Q_{ji}(t) + \frac{1}{n} X_i(t) \right]\right] \\ &\quad - \frac{1}{2} E\left[\mathrm{Sgn}\left[a(t) + Q_{ji}(t) - \frac{1}{n} X_i(t) \right]\right]. \end{aligned} \tag{A17.13}$$

Note the following formulae:

$$P(X_i(t) = 1) = \frac{1 + a(t)}{2}, \tag{A17.14}$$

$$P(X_i(t) = -1) = \frac{1 - a(t)}{2}. \tag{A17.15}$$

In view of (A17.14) and (A17.15), the first term in (A17.13) now becomes

$$\begin{aligned} E\left[\mathrm{Sgn}\left[a(t) + Q_{ji}(t) + \frac{1}{n} X_i(t) \right]\right] &= E\left[E\left[\mathrm{Sgn}\left[a(t) + Q_{ji}(t) + \frac{1}{n} X_i(t) \right] \mid X_i(t) \right]\right] \\ &= \frac{1 + a(t)}{2}\left(2P\left(a(t) + Q_{ji}(t) + \frac{1}{n} > 0 \right) - 1 \right) \\ &\quad + \frac{1 - a(t)}{2}\left(2P\left(a(t) + Q_{ji}(t) - \frac{1}{n} > 0 \right) - 1 \right). \end{aligned} \tag{A17.16}$$

Using (A17.6) we now obtain

$$\begin{aligned} P\left(a(t) + Q_{ji}(t) + \frac{1}{n} > 0 \right) &= P\left(Q_{ji}(t) > -a(t) - \frac{1}{n} \right) \\ &= 1 - P\left(\frac{Q_{ji}(t) - \mu_N(t)}{\sigma_N(t)} < -\frac{a(t) + \mu_N(t) + (1/n)}{\sigma_N(t)} \right) \\ &= 1 - \Phi\left(-\eta(a(t), t) - \frac{1}{n\sigma_N(t)} \right), \end{aligned} \tag{A17.17}$$

where

$$\eta(a(t),t)=\frac{a(t)+\mu_N(t)}{\sigma_N(t)},$$

$$\Phi(y)=\frac{1}{\sqrt{2\pi}}\int_{-\infty}^{y} \mathrm{e}^{-(1/2)x^2}\,\mathrm{d}x. \tag{A17.18}$$

Applying a similar procedure yields

$$P\left(a(t)+Q_{ji}(t)-\frac{1}{n}>0\right)=1-\Phi\left(-\eta(a(t),t)+\frac{1}{n\sigma_N(t)}\right). \tag{A17.19}$$

Substituting (A17.17) and (A17.19) into (A17.16) we have

$$E\left[\mathrm{Sgn}\left[a(t)+Q_{ji}(t)+\frac{1}{n}X_i(t)\right]\right]=1-(1+a(t))\,\Phi\left(-\eta(a(t),t)-\frac{1}{n\sigma_N(t)}\right)$$
$$-(1-a(t))\,\Phi\left(-\eta(a(t),t)+\frac{1}{n\sigma_N(t)}\right). \tag{A17.20}$$

Adopting a similar procedure we find that the corresponding result for the second term in (A17.13) is

$$E\left[\mathrm{Sgn}\left[a(t)+Q_{ji}(t)-\frac{1}{n}X_i(t)\right]\right]=1-(1+a(t))\,\Phi\left(-\eta(a(t),t)+\frac{1}{n\sigma_N(t)}\right)$$
$$-(1-a(t))\,\Phi\left(-\eta(a(t),t)-\frac{1}{n\sigma_N(t)}\right). \tag{A17.21}$$

Substituting (A17.20) and (A17.21) into (A17.13) we obtain

$$E[Z_{ij\alpha}(t+1)]=-\frac{1+a(t)}{2}\left[\Phi\left(-\eta(a(t),t)-\frac{1}{n\sigma_N(t)}\right)-\Phi\left(-\eta(a(t),t)+\frac{1}{n\sigma_N(t)}\right)\right]$$
$$-\frac{1-a(t)}{2}\left[\Phi\left(-\eta(a(t),t)+\frac{1}{n\sigma_N(t)}\right)-\Phi\left(-\eta(a(t),t)-\frac{1}{n\sigma_N(t)}\right)\right]$$
$$\approx\frac{1}{n\sigma_N(t)}(1+a(t))\frac{\mathrm{d}\Phi(x)}{\mathrm{d}x}\Big|_{x=-\eta(a(t),t)}$$
$$-\frac{1}{n\sigma_N(t)}(1-a(t))\frac{\mathrm{d}\Phi(x)}{\mathrm{d}x}\Big|_{x=-\eta(a(t),t)}$$
$$=\frac{2a(t)}{n\sigma_N(t)}p(-\eta(a(t),t)), \tag{A17.22}$$

where

$$p(x) = \frac{1}{\sqrt{2\pi}} e^{-(1/2)x^2}. \tag{A17.23}$$

Finally, upon substituting (A17.22) into (A17.9) we obtain the mean of the noisy term

$$\mu_N(t+1) = \frac{\sqrt{2r a(t)}}{\sqrt{\pi \sigma_N(t)}} \exp(-0.5(a(t) + \mu_N(t))^2 \sigma_N^{-2}(t)). \tag{A17.24}$$

From the deductive process above we know that a necessary condition for (A17.24) to be valid is

$$n \gg \sigma_N^{-1}(t). \tag{A17.25}$$

Thus (17.19) is achieved.

2. Derivation of (17.20)

Substituting (A17.3) into (A17.2) and then taking the expectation of both sides of the resulting equation and using assumption (ii), we obtain

$$\begin{aligned} a(t+1) &= E\left[\frac{1}{n}\sum_{j=1}^{n} \mathrm{Sgn}(A(t) + N_j(t))\right] \\ &= \frac{1}{n}\sum_{j=1}^{n} E[\mathrm{Sgn}(A(t) + N_j(t))] \\ &= E[\mathrm{Sgn}(A(t) + N_j(t))] \\ &= E[E[\mathrm{Sgn}(A(t) + N_j(t)) | A(t)]]. \end{aligned} \tag{A17.26}$$

Based on assumption (iv), the main overlap $A(t)$ is normally distributed with mean $a(t)$ and variance $\sigma_A^2(t)$, hence

$$\begin{aligned} a(t+1) &= \int_{-1}^{1} p_A(x,t) E[\mathrm{Sgn}(A(t) + N_j(t)]\,\mathrm{d}x \\ &= \int_{-1}^{1} p_A(x,t)[P(x + N_j(t) > 0) - P(x + N_j(t) < 0)]\,\mathrm{d}x \\ &= \int_{-1}^{1} p_A(x,t)(\Phi(\eta(x,t)) - \Phi(-\eta(x,t)))\,\mathrm{d}x, \end{aligned} \tag{A17.27}$$

where

$$p_A(x,t) = \frac{1}{\sqrt{2\pi}\sigma_A(t)} e^{-(x-a(t))^2/2\sigma_A^2(t)},$$

$$\eta(x,t) = \frac{x + \mu_N(t)}{\sigma_N(t)}.$$

The derivation of (17.20) is achieved.

3. Derivation of (17.21)

Finally, we derive formula (17.21). Squaring both sides of (A17.2) and using (A17.3) we have

$$A^2(t+1)=\frac{1}{n^2}\sum_{j=1}^{n}[\mathrm{Sgn}(s_j^1 A(t)+N_j(t))]^2+\frac{1}{n^2}\sum_{j\neq l}^{n}\sum_{l=1}^{n}\mathrm{Sgn}[A(t)+N_j(t)]\,\mathrm{Sgn}[A(t)+N_l(t)]$$

$$=\frac{1}{n}+\frac{1}{n^2}\sum_{j\neq l}^{n}\sum_{l=1}^{n}\mathrm{Sgn}[A(t)+N_j(t)]\,\mathrm{Sgn}[A(t)+N_l(t)]. \quad \text{(A17.28)}$$

Based on assumptions (ii), (iv) and (v) in Assertion II (see section 3) and taking the expectation on both sides of (A17.28) yields

$$\begin{aligned}E[A^2(t+1)]&=\frac{1}{n}+\frac{n(n-1)}{n^2}E[\mathrm{Sgn}(A(t)+N_j(t))\,\mathrm{Sgn}(A(t)+N_l(t))]\\&\doteq E[\mathrm{Sgn}(A(t)+N_j(t))\,\mathrm{Sgn}(A(t)+N_l(t))]\\&=E[E[\mathrm{Sgn}(A(t)+N_j(t))\,\mathrm{Sgn}(A(t)+N_l(t))\,|\,A(t)]]\\&=\int_{-1}^{1}p_A(x,t)(E[\mathrm{Sgn}(x+N_j(t))])^2\,\mathrm{d}x. \quad \text{(A17.29)}\end{aligned}$$

Noting that

$$\begin{aligned}E[\mathrm{Sgn}(x+N_j(t))]&=P(x+N_j(t)>0)-P(x+N_j(t)<0)\\&=\Phi(\eta(x,t))-\Phi(-\eta(x,t)), \quad \text{(A17.30)}\end{aligned}$$

enables us to write (A17.29) in the following form:

$$E[A^2(t+1)]=K_2(a,\sigma_A,\mu_N,\sigma_N,t), \quad \text{(A17.31)}$$

where

$$K_2(a,\sigma_A,\mu_N,\sigma_N,t)=\int_{-1}^{1}p_A(x,t)(\Phi(\eta(x,t))-\Phi(-\eta(x,t)))^2\,\mathrm{d}x. \quad \text{(A17.32)}$$

Hence, the variance of the main overlap is

$$\sigma_A^2(t+1)=K_2(a,\sigma_A,\mu_N,\sigma_N,t)-a^2(t+1). \quad \text{(A17.33)}$$

This gives a proof of (17.21).

Note that the recurrent equation of $\sigma_N(t)$ is the same as (17.17) shown in (Amari and Maginu, 1988) hence the effects of both the randomness of $A(t)$ and the means of $N_i(t)$, $i=1,\dots,n$ on the variance $\sigma_N^2(t)$ are ignored in our theoretical analyses.

CHAPTER 18

ANALYSIS AND SYNTHESIS OF MORPHOLOGICALLY REALISTIC NEURAL NETWORKS

Luciano da F. Costa, Roberto M. Cesar Jr., Regina C. Coelho and Júlia S. Tanaka

... *the functional superiority of the human brain is intimately linked up with the prodigious abundance and unaccustomed wealth of forms of the so-called neurons with short axons.*

Santiago Ramón y Cajal, *Recollections of My Life, 1937*

1. INTRODUCTION

As indicated in the above quotation, the amazing morphological variety exhibited by neural cells has long been recognized as an important factor determining the mammals' brain emergent behavior. However, despite the isolated approaches reported in the literature, the systematic study and analysis of neural shape has been relatively overlooked in favor of electric-differential models such as those based on the transmission line or cable analogy. The main shortcoming of such a tendency is that the neural shape plays an important role in defining and constraining neural behavior, and should thus be carefully considered, jointly with electric-differential issues, in modeling and simulation approaches. However, all compartmental approaches to neural modeling, such as NEURON (Hines and Carnevale, 1997) and GENESIS (Bower and Beeman, 1994), have failed to fully describe the true shape of single neurons, in the sense that any set of neurons presenting the same branching pattern, with the respective dendritic segments having the same average diameters and lengths, but with different shapes, will yield virtually identical multicompartmental models. Several important morphological features, such as the branch angles between dendrites, are not taken into consideration in current existing models of 'realistic neurons', leaving a 1D caricature of the cells. A similar tendency has also been observed as far as developments in artificial neural networks are concerned, which have been mostly restricted to representing neurons as 'points' interconnected through simple and regular schemes (see Chapter 17). Amongst the few exceptions that have explicitly taken shape factors into account, we have the self-organizing models proposed by Kohonen (1990), which incorporate the spatial relationship between cells in cortical surfaces. The present Chapter will aim at incorporating morphological information into account, trying to circumvent as many as possible of the above caveats and allowing for 2D and 3D characterization of single neurons and networks.

The multiscale paradigm, see for instance (Cesar and Costa, 1997) provides a particularly interesting perspective from which this interplay between neural shape and function can be approached: at *small spatial scales*, the diameter and length of the dendritic segments, together with the intrinsic electrical properties of membrane (R_m, R_i, C_m), determine the electrotonic properties of the dendritic arborization; at *intermediate spatial scales*, the branching pattern of the dendritic and axonal arborizations define the patterns of interconnection between neural cells; at *large spatial scales*, the relative position of the cells, as well as the size of their arborizations, bear close relationship with macroscopic features of the cell, such as the shape and size of the associated respective fields, and even the temporal evolution of the overall neural activity. The principal problem in investigating such phenomena is that there is no definitive and comprehensive set of shape measures which could be applied to generic problems in neuroscience. While approaches such as density of branches (Caserta et al., 1995; Dann, 1988; Dacey, 1993; Troilo et al., 1996), fractal dimension (Coelho and Costa, 1996; Morigiwa et al., 1989; Montague and Friedlander, 1991; Panico and Sterling, 1995; Porter et al., 1991; Smith et al., 1989, 1996), and dendrograms (Sholl, 1953; Poznanski, 1992; Schutter and Bower, 1994; Turner et al., 1995; Velte and Miller, 1995) have been proposed and used in the literature, there have been few attempts at fully validating and characterizing the potential of such shape measures. In fact, the concept of dendrogram, by providing a particularly comprehensive representation of the neural cell, deserves some special attention. It seems reasonable to assume that a dendrogram consists of a data structure, namely a binary tree, which, in addition to hierarchically describing the branching structures, is enriched with additional information about each dendritic segments, such as its diameter, length, roughness, density of spines, etc, which may be visualized in some way.

The present paper describes a number of concepts and techniques aimed at quantifying the morphometric characterization of neurons in order to lead towards the modeling and simulation of morphologically more realistic neural networks. It starts by discussing how the critical points, namely the branch and endpoints, in neural arborizations can be detected (i.e. how a neural cell can be *segmented*) in a semi-automated fashion, and follows by presenting how such results can be combined with a computer theoretical approach in order to allow the semi-automated generation of dendrograms. In addition, the multiscale shape analysis techniques described in the next section can be applied in a number of alternative problems of neural shape characterization (Cesar and Costa, 1997). Next, the important issues of area of influence and spatial coverage by neural cells is addressed in terms of new shape measures, including Minkowski's sausages and influence histograms. Results regarding the classification of neural cells (e.g. Ramon-Moliner, 1962), more specifically cat retinal ganglion cells, by using the aforementioned shape features and two alternative clustering approaches are also presented, which is followed by the description of an approach to the synthesis, modeling and simulation of morphologically more realistic neural structures.

2. MORPHOMETRIC CHARACTERIZATION OF SINGLE NEURONS

At intermediate spatial scales, the more important shape features exhibited by neural cells include the position of the branch points and extremities exhibited by the dendritic and axonal arborizations. Such points are particularly important as subsidy for the automatization of dendrogram generation and as features for the classification of neural cells, to be addressed in the following sections.

Shape analysis of neural cells is here carried out in terms of its *outer contours* (Cesar and Costa, 1997). Firstly, the neural cell of interest in the digital image must be separated from the other cells and background. Next, beginning from a starting point, the contour of the cell has to be traced, while the Cartesian coordinates (x and y) of each point along this contour are recorded as a function of the position from the starting point. Figure 18.1(a) presents a neural cell while Figure 18.1(b) shows the parametric signals $x(t)$ and $y(t)$ describing its contour $C(t)=(x(t), y(t))$, where the parameter t usually stands for the arc length parametrization of C. A more efficient representation of the contour is obtained by expressing the contour C as the complex signal $u(t)$, i.e. $u(t)=x(t)+\mathrm{i}y(t)$, $\mathrm{i}=\sqrt{-1}$. The adopted methods for shape analysis are based on the multiscale curvature methods (derived from the *curvegram*) introduced in (Cesar and Costa, 1996, 1997) and on the *continuous wavelet transform* or W-representation (Antoine et al., 1997).

The multiscale curvature description of $C(t)$, namely the *curvegram* (Cesar and Costa, 1997), is given by (18.1). The curvegram $k(t,a)$ describes the curvature of $C(t)$ analyzed at scale a.

$$k(t,a)=\frac{-\mathrm{Im}\{\dot{u}(t,a)\,\ddot{u}^*(t,a)\}}{|\dot{u}(t,a)|^3} \tag{18.1}$$

where $\mathrm{Im}(z)$ and z^* denote the imaginary part and complex conjugate of $z\in\mathbb{C}$, while $\dot{u}(t,a)$ and $\ddot{u}(t,a)$ denote the first and the second derivatives (with respect to t) of the smoothed version $u(t,a)$ of $u(t)$, i.e. $u(t)$ analysed at scale $a\in R, a>0$ (Cesar and Costa, 1996, 1997).

The segmentation of the neural contour is carried out by choosing an appropriate scale a_0, followed by two threshold operations, one for the detection of the extremities of the neurons (defined by negative minima of curvature, which characterize convexities, assuming that the contour is traversed clockwise), and other for the detection of the branch points (defined by positive maxima of curvature characterizing concavities, under the above mentioned assumption). It is important to note that, if the contour is traversed counterclockwise, it suffices to define the extremities as positive maxima and the branch points as negative minima.

The second multiscale representation of contours that can be adopted is the so called W-representation, defined as the *continuous wavelet transform* of the signal $u(t)$:

$$U[\Psi,u](b,a)=U_\Psi(b,a)=\frac{1}{\sqrt{a}}\int_{-\infty}^{\infty}\Psi^*\left(\frac{t-b}{a}\right)u(t)\,\mathrm{d}t$$

where $\Psi(t)$ is the analyzing wavelet (Antoine et al., 1997), a is the *scale* and b is the *position along the curve*, $a,b\in R, a>0$. Figures 18.1(c) and (d) present, respectively,

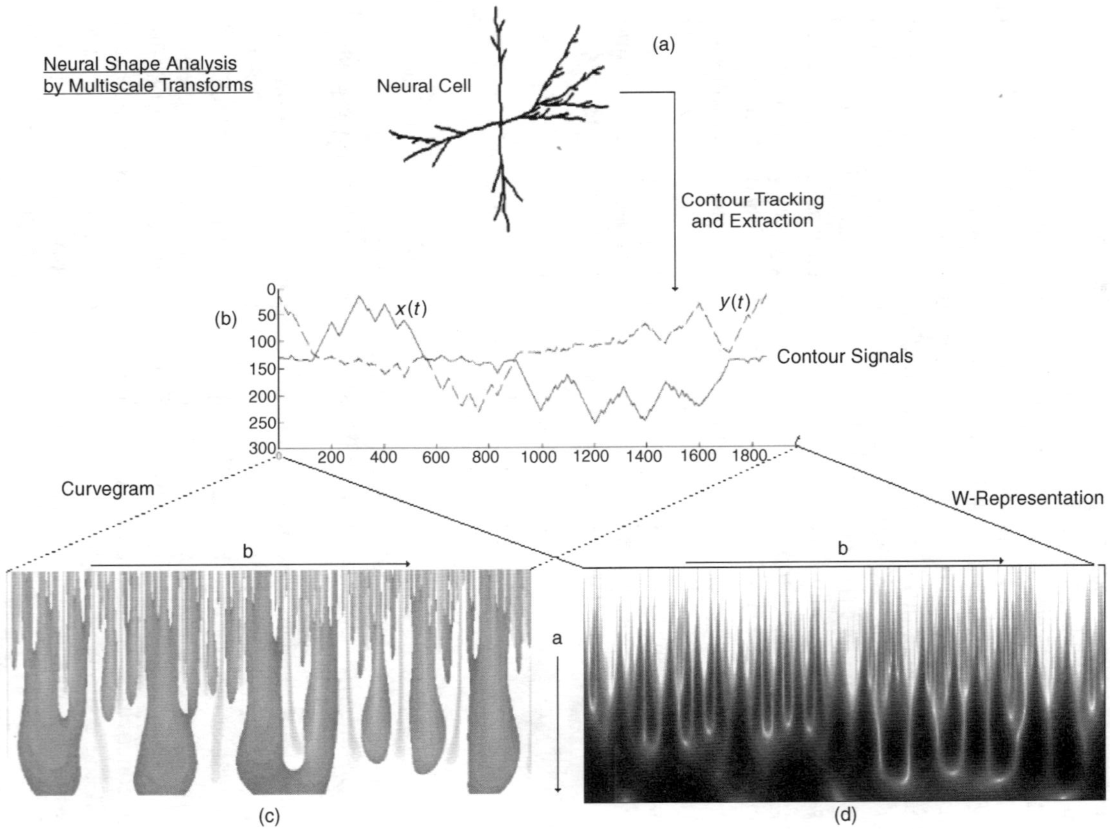

Figure 18.1 General scheme for multiscale analysis of neural shape: (a) A digital image of a neural cell; (b) The parametrized representation of the contour of the cell in (a) in terms of the signals $x(t)$ and $y(t)$ (the axes are measured in pixels; see text); (c) The respective curvegram obtained by applying (18.1). The respective axes are indicated in the figure; (d) The magnitude of the *W*-representation (see text for further details).

the curvegram and the modulus of the W-representation of the neuron in Figure 18.1(a). Both the curvegram and the W-representation of neural boundaries can be used to derive *multiscale energy measures* which are used as features for characterization of shape complexity. The *normalized multiscale bending energy* can be defined from the curvegram as (Cesar and Costa, 1997):

$$NMBE(a) = \frac{P^2}{N} \sum_{n=0}^{N-1} k(a,n)^2$$

where $k(n,a)$ is the sampled version of the curvegram (18.1) for a digital contour of N discrete points along the neural contour and P is the perimeter of such a contour. On the other hand, the *normalized multiscale wavelet energy* is defined as:

$$NMWE_\psi(a) = \frac{1}{K^2} \frac{1}{N} \sum_{n=1}^{N} |U_\psi(n,a)|^2$$

where $U_\psi(n,a)$ is also the sampled version of the W-representation and K is the length of the largest side of the minimum enclosing rectangle of the neural cell (Cesar and Costa, 1997). These energies have been used in the generation of features for automatic neurons classification based on their shape. Furthermore, psychophysical experiments have shown that these measures are in accord with shape complexity evaluation by humans.

The framework for semi-automated generation of dendrograms is organized as follows. Figure 18.2(a) shows a dendritic whole branch typically found in cat retinal ganglion cells (α type), from which the dendrogram is to be generated. The branch points and extremities along the whole dendritic branch outline, which form the core of the dendrogram generation algorithm, are indicated in the image by the symbols b_1, b_2, b_3, b_4 and b_5 (branch points), and e_1, e_2, e_3 and e_4 (extremities). As it can be noted, the points b_1 and b_5 indicate the place where the whole branch leaves the soma. On the other hand, b_2, b_3 and b_4 indicate the branch points where the whole dendritic branch bifurcates, leading to the extremities e_1, e_2, e_3 and e_4. Therefore, the whole branch outline is coded by the string $b_1e_1\ b_2e_2\ b_3e_3\ b_4e_4b_5$. Furthermore, the whole branch hierarchy indicates that the dendrogram, as a function of the bifurcation points, should be generated from the initial and final branch points (b_1 and b_5) to b_2; from b_2 to the extremity e_1 and to the branch point b_3; from b_3 to the extremity e_2 and to the branch point b_4; and finally from b_4 to the extremities e_3 and e_4.

The definition of a suitable formal grammar (Fu, 1982) has allowed the development of a parsing algorithm whose derived parsing tree reflects the aforesaid structure of the whole branch, as shown in Figure 18.2(b). In order to recognize this structure in the resulting ternary tree, observe that the first node of the tree has three children, two leaves corresponding to the symbols b_1 and b_5 of the initial coding string (the branch points near the soma), and a middle node which expands itself in other three nodes. This middle node corresponds to the first bifurcation of the whole dendritic branch and, in fact, each expanding node in the tree corresponds to a

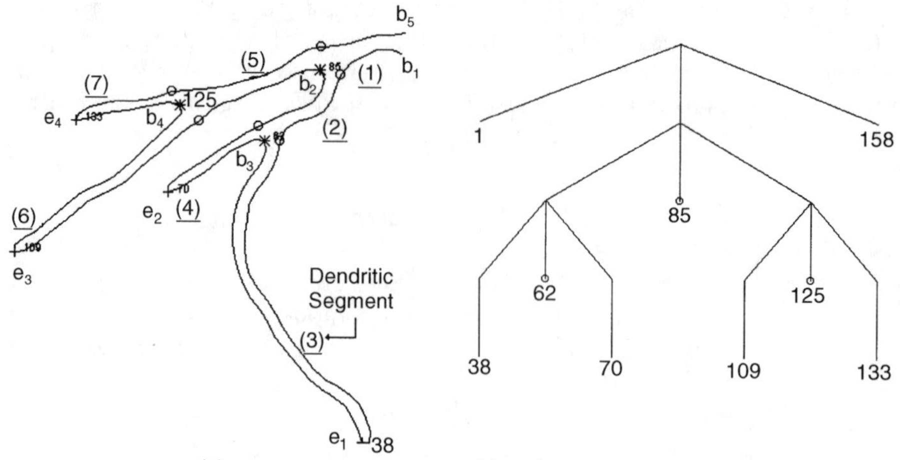

Figure 18.2 Semi-automated generation of dendrograms: (a) A typical dendritic whole branch in cat retinal ganglion cells. The syntactic shape analysis of the dendritic tree shown in (a) leads to a parsing tree, which is shown in (b). This parsing tree is a representation that completely describes the branching structure of the dendritic tree. Therefore, the dendrogram can be obtained in a straightforward manner by traversing the parsing tree and measuring the desired physical properties (such as length or average thickness) in the original contour.

branch point. In this case, the node also expands itself into other three nodes, a middle node that indicates the current branch point (i.e. b_2), a leftmost node, which corresponds to the branch point b_3 in the whole dendritic tree, and a rightmost node, which corresponds to the extremity e_1. The remaining of the parsing tree may be explained in an analogous way and, as it can be readily observed, it completely reflects the hierarchical structure of the dendritic whole branch. The dendrogram is constructed following the structure of the parsing tree. The central point in the construction of the dendrogram is the correct determination of the dendritic segments in the main tree, as well as how they are properly structured, which is achieved straightforwardly from the parsing tree. The first dendritic segment is that defined between the pair b_1 and b_5 (the branch points near the soma) and the branch point b_2. Therefore, the desired features (length, average thickness, bending energy and so on) can be measured from this first segment and represented in the first stroke of the dendrogram. At this point, the dendrogram bifurcates, and the desired features are measured from the next two dendritic segments indicated by the parsing tree: the segment between the b_2 and b_3, and the segment between b_2 and e_1. The algorithm performs the derivation of the dendrogram until the parsing tree has been completely explored. The complete dendrogram for the dendritic whole branch of Figure 18.2(a) can be seen in Figure 18.3. This dendrogram codes three different features of the dendritic whole branch: the segments length (coded by the length of

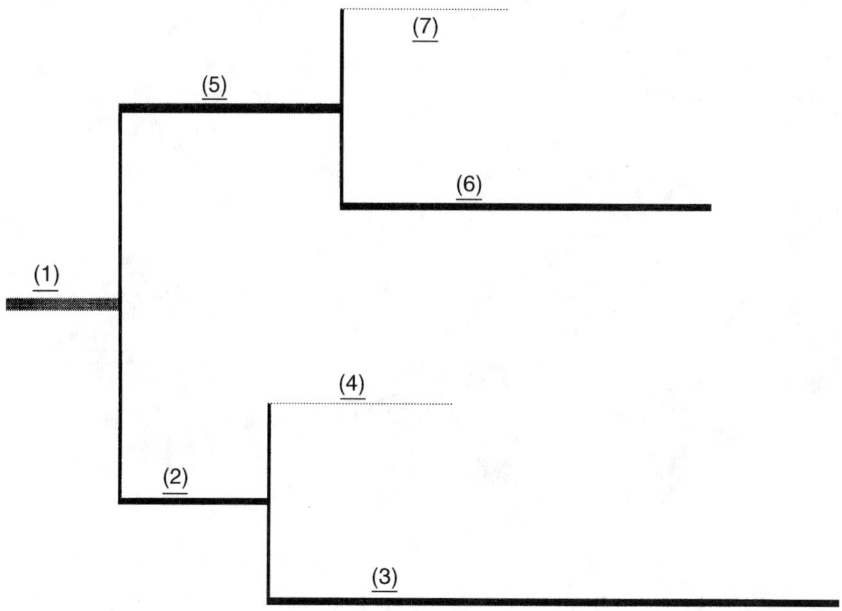

Figure 18.3 The dendrogram obtained for the whole branch in Figure 18.2(a), created from of the parsing tree shown in Figure 18.2(b). The dendrogram represents each segment *length* (expressed in terms of the respective horizontal line), average *thickness* (expressed in terms of the thickness of the respective horizontal line) and *mean bending energy* (expressed by the gray level intensity, with darker intensities representing higher energies). Refer to text for further details.

the horizontal lines of the dendrogram); the average thickness of each dendritic segment (coded by the thickness of each horizontal line of the dendrogram); and the mean bending energy (Cesar and Costa, 1997) of each dendritic segment (coded by the gray level intensity of each dendritic segment, where darker intensities code segments with high energy).

3. DEFINING AND MEASURING THE AREA OF INFLUENCE AND SPATIAL COVERAGE BY NEURAL CELLS

At intermediate spatial scales, one of the most important shape features exhibited by neural cells include their *area of influence* and *spatial (or territorial) coverage.* Although being intuitive, no clear definition of these concepts have been found in the literature. As far as the area of influence of a cell is concerned, it has been often defined in terms of the polygon which is obtained by joining the cell extremities (Vaney, 1994). As illustrated in Figure 18.4(a), (b), such an involving polygon is extremely sensitive to small changes in the cell shape, in such a way that an

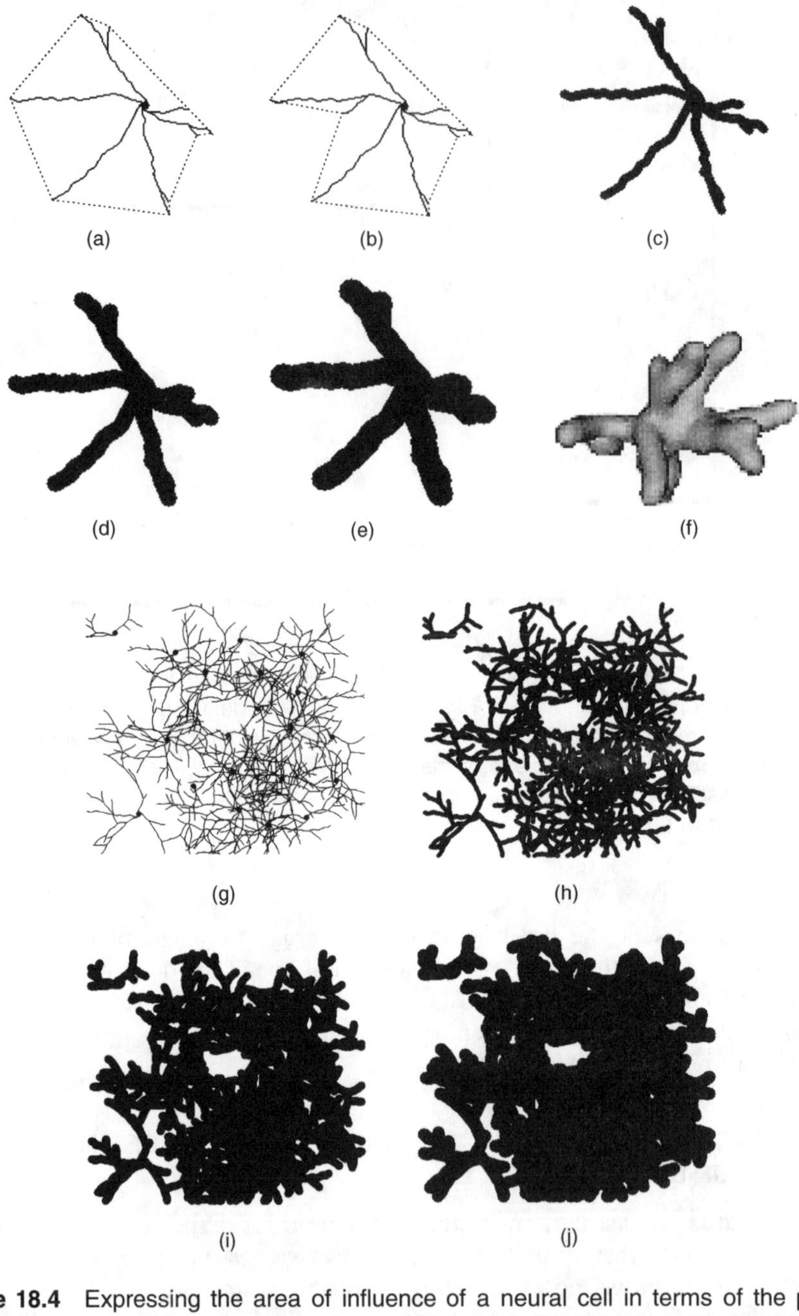

Figure 18.4 Expressing the area of influence of a neural cell in terms of the polygon defined by its extremities presents serious problems (a), (b) which are alleviated by using the Minkowski's sausage approach (c)–(e). In (f) is presented an example of 3-D Minkowski sausage. The Minkowski's sausages for the neural network in (g) are presented in (h)–(j).

alternative and more stable approach has been proposed by (Costa, 1995), which defines the area of influence in terms of the convex hull (i.e. the smallest convex contour containing the shape) of the cell. However, both these approaches fail to take into account the internal shape of the cell. Actually, the key concept underlying the area of influence is the fact that such an area depends on the type of biological influence, such as gradients of chemoattractors (Goodman, 1996; Lavigne, 1996) or electrical potential fields (Bedlack et al., 1992; Brown and Loew, 1994), one is interested to investigate. From this perspective, it becomes clear that a sound way to define the area of influence of a neural cell is as the set of points lying up to a maximum distance D from the cell (the minimal distance). Such inflated representations of the original cell are called Minkowski's sausages (Tricot, 1995). Figure 18.4(c)–(e) illustrate the areas of influence obtained for the neural cell in Figure 18.4(a) with respect to $D=3,6$ and 9. Figure 18.4(f) presents an example of 3D Minkowski sausage. The concept of influence area can also be extended to neural networks, as presented in Figure 18.4(g)–(j), with respect to the same values of D as above (i.e. $D=3,6$ and 9).

As plant roots, neural arborizations are meant for effectively sampling the surrounding space, hence the notion of spatial coverage. The above developed concept of area of influence in terms of Minkowski's sausages provide an interesting motivation for attacking the concept of spatial coverage. Since these sausages provide an indication of the area of influence of a neural cell with respect to a specific distance D, it is reasonable to investigate how this area grows as D is monotonically increased. While for a single point the area of influence grows steadily, it tends to saturate for dense shapes such as filled regions. Thus, if the log–log graph exhibiting the number of covered points in terms of D is plotted, curves with higher slopes are obtained for simpler shapes. In fact, the quantity $D_F=2-slope$ can be verified to correspond to the Bouligand–Minkowski fractal dimension (Peitgen and Saupe, 1988; Falconer, 1990) of the shape in question. In addition to such a nice conceptual interpretation, the Minkowski's sausage method for estimation of fractal dimension completely avoids the influence of the position and orientation of the grid implied by the traditionally used, e.g. (Montague and Friedlander, 1991; Morigiwa et al., 1989; Panico and Sterling, 1995; Smith et al., 1996), box-counting technique.

The above discussed approach to spatial coverage can be extended even further in order to express with increased accuracy the biological phenomenon of spatial coverage by neural cells, which will be done in a progressive fashion, as described in the following. Assuming a single point as the only source of influence, it is clear that its effect (e.g. the diffusion of a chemoattractor, the probability of having a synaptic contact during morphogenesis, or the electric field established around the cell) will extend radially and isotropically, according to a profile that is specific to the kind of effect under analysis (e.g. in the case of a 2D electric potential field, this profile will be a decaying exponential). Let us call this spatial profile the *point-spread-function* of the influence process. In other words, the point-spread-function describes how the neuron would influence the surrounding space in case it were a single point. In the case of more complex structures, such as neurons, standard results from systems theory (see Appendix 18.1) allows us to calculate the overall area of influence by convolving the point-spread-function defining the field of influence with the neural shape, in such a way that a cloud of points results around the neuron. Such clouds are illustrated in Figure 18.5(d)–(f) with respect to three neural cells with increasing complexity. Influence fields can also be defined for neural networks, as illustrated in Figure 18.5(h). The most interesting advantages of such an approach is its biological soundness and the explicit acknowledgment that the spatial coverage induced by a specific neural cell is *relative* to the phenomenon one is interested to model. Consequently, the spatial coverage defined by a specific cell will vary depending on if we are modeling the probability of synaptic contact during morphogenesis, expressing the distribution of electric field governing the development of growth cones, or predicting the receptive field exhibited by a specific cell. In this latter situation the receptive field of a specific cell, let us say a ganglion cell in the retina, can be foreseen (assuming topographical mapping between stimuli spaces) by using the receptive field of the afferent input (e.g. bipolar cells) as point-spread-function. It is also possible to consider inhomogeneous distributions of influence, such as in the case where the 'point-spread-function' depends on the electric charge distribution along the cell membrane. Although such situations can not be treated by convolving the shape with a unique point-spread-function, it can still be approached directly at the image space by linearly adding the influences defined by each of the neuron constituent elements, suitably weighted by the local charge intensity.

However nice, the full representation of the spatial coverage implies too many data, i.e. virtually all the points in the generated cloud, in such a way that a more global and compact representation would be desirable for finalities such as general characterization and neural cell classification, allowing a smaller number of shape features to be taken into account. While the variance and entropy of the influence magnitude at each point provide two interesting global measures, it is also possible to represent the distribution of such magnitudes in terms of histograms, henceforth called *influence histograms*, which allow a good compromise between the microscopic representation allowed by the 'clouds' and fully global measures such as entropy. Figure 18.5(g) illustrates the influence histograms obtained for the neural cells in Figure 18.5(a)–(c), where H is the number of points under similar influence and S is the influence magnitude. Basically, an influence histogram represents the number of lattice points (alternatively the relative proportion of such points) inside the influence area defined by a specific cell that are under similar influence by the cell. Thus, we have from the influence histograms in Figure 18.5(g) that, for the cell in Figure 18.5(c), 2500 points inside the influence are suffering similar influence of about 30 relative units. Interestingly, more complex cells tend to produce broader histogram profiles, indicating a more effective influence of the cell over its surrounding space.

The concept of spatial coverage has been extended to express vectorial influence (i.e. the orientation and direction of the influence at each point), such as that implied by electric fields. Figure 18.6 presents the electric field established around the neuron in Figure 18.4(a) assuming exponentially decaying magnitude and uniform charge distribution along the neural cell surface (i.e. each pixel would act as an electrical charge). There are two main nice features exhibited by such a kind

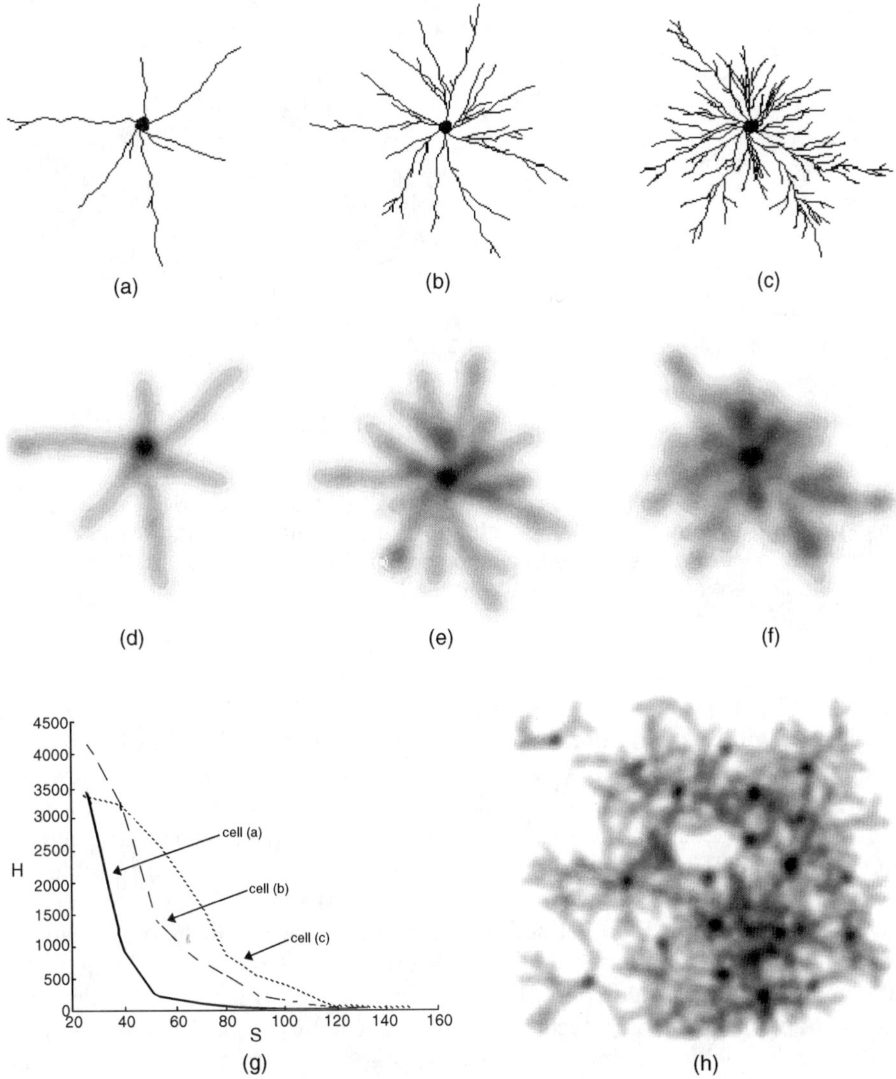

Figure 18.5 Three neural cells presenting increasing complexity (a)–(c), their respective influence fields (d)–(f), and the obtained influence histograms (g). The influence field for the neural network in Figure 18.4(g) is presented in (h). The overall area of influence, represented by the cloud of points around the figure, is obtained by convolving the point-spread-function, that defines the field of influence, with the neural shape. The resulting image provides a faithful estimation of the spatial coverage by the trophic field defined around the neuron or neural network.

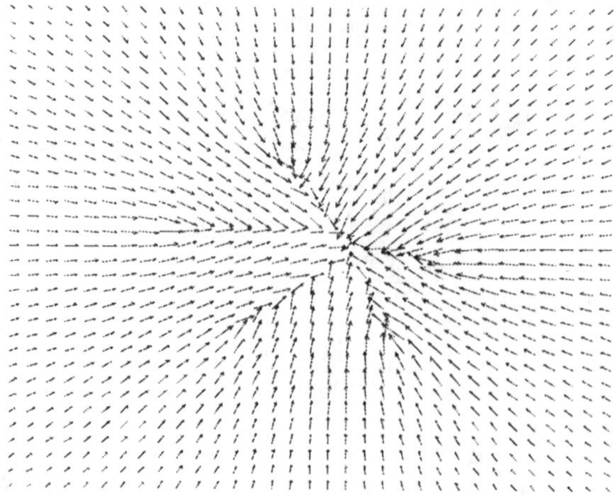

Figure 18.6 The vector field (2-D electric field) established around the cell in Figure 18.1(a), assuming uniform distribution of electric charge along the neural surface. The possible paths of growth-cones can be foreseen by the lines of force (i.e. the curves characterized by having the tangent at each point parallel to the field).

of representation: (a) it allows comprehensive static and dynamical simulation of the electrochemical interaction along the cell membrane (using a combination of equations from electromagnetism and diffusion theory), and of the electrotonic distribution and ionic densities in and outside the cell; and (b) it provides a biologically sound basis for simulating the development of growth cones and neurites (van Ooyen & van Pelt, 1996; van Oss & van Ooyen, 1997). It is also possible that such vector fields play an important role at the neural basis of visual attention, in the sense that the vector field defined by a specific shape would help to guide the processes underlying visual attention. In such case, the high curvature portions of a specific image contour would concentrate the lines of force and consequently act as a special attention attractor, which is in agreement with psychophysical evidences. Such a possibility, which is being currently investigated, presents direct relationship with the fields established at the ganglion cells in the retina.

Being an important feature of biological neurons, the concept of spatial coverage also provides an interesting means for investigating and characterizing neural networks explicitly incorporating neural shape. Two aspects are particularly relevant regarding such applications: (a) during morphogenesis, the spatial coverage exhibited by a cell can provide an indication of the density of synapses along the cell; and (b) the receptive field of each specific cell in topographically organized spaces can be predicted from the area of influence and spatial coverage.

In addition to providing valuable resources for the morphological characterization of neural cells and structures, influence histograms have also proven to be particularly useful for the automatic classification of neural cells, which is an important problem in neuroscience. We have investigated the use of influence histograms, as well as alternative measures such as fractal dimension, cell diameter (i.e. the largest distance between any two points belonging to the cell), and area of convex hull, as classification features with respect to a database of cat retinal ganglion cell images extracted from the literature (Boycott and Wässle, 1974; Dann et al., 1988; Fukuda et al., 1984; Kolb et al., 1981; Leventhal and Schall, 1983; Saito, 1983; Wässle et al., 1981a, b; Watanable et al., 1993). The results obtained by using two clustering techniques, namely *k*-means and Ward's hierarchical approach (Anderberg, 1973; Romesburg, 1990), are illustrated in Figure 18.7. The first overall impression is that the fractal dimension estimated by the box-counting procedure tended to present the worst results. The area of the convex hull allowed better classification, independently of the adopted clustering procedure. The best results were obtained for the diameter and influence histograms, though the respective performances depended substantially on the clustering approach, a phenomenon that is very likely caused by statistical fluctuations. While the best performance was achieved by considering influence histograms and Ward's method, it is remarkable to verify that the diameter, a simple and global measure, has also yielded good results when processed by *k*-means. It should also be observed that the relatively high obtained misclassification rates are very likely explained by the fact that the class assignment used as comparison standard was produced subjectively by diverse authors, often by using different criteria.

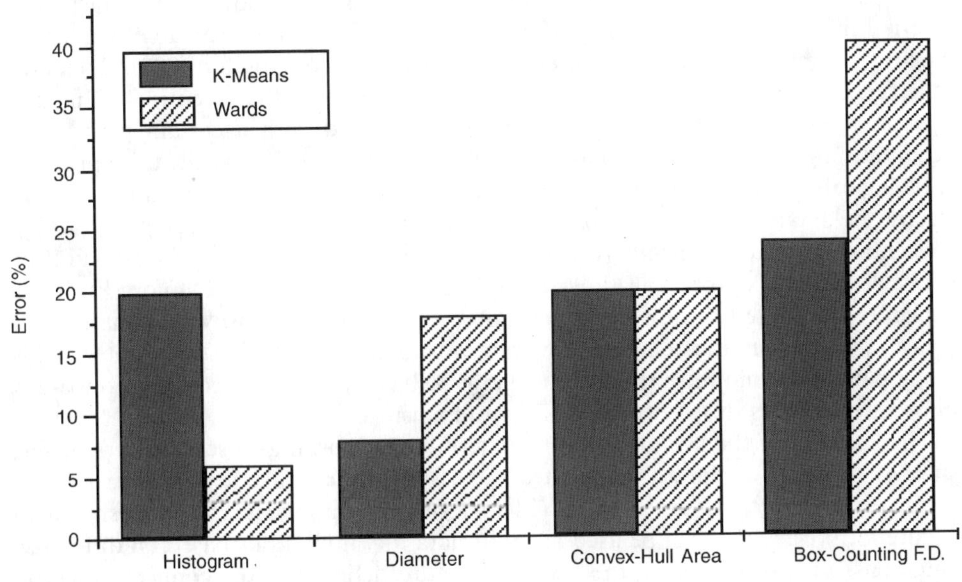

Figure 18.7 Misclassification rates (errors) obtained for the two clustering techniques (*k*-means and Ward's method) in terms of several shape features. We can observe that the diameter and influence histograms using the *k*-means and Ward's hierarchical methods respectively. The influence histograms presented the best results.

4. SYNTHESIZING MORPHOLOGICALLY REALISTIC NEURAL NETWORKS FOR MODELING AND SIMULATION

One of the keys to the problem of generating morphologically realistic neural cells consists in adopting a *statistical approach* (see Hamilton, 1993, for a different approach), in the sense that cells are produced by taking into account conditional distribution probability functions describing the principal shape features of interest. Of course, such statistical descriptions have to be extracted from a representative population of real cells, a process that can be particularly challenging and time consuming by itself, since it may imply the analysis of hundreds or even thousands of neural cells. In fact, several of the results described in the previous sections, particularly the semi-automated generation of dendrograms, provide effective means for obtaining such statistical descriptions. The first step in generating neural cells consists in selecting a good set of shape features, which has to be done in terms of the expected degree of accuracy and biological phenomena to be modeled. Then, after the statistical descriptions have been experimentally estimated, Monte-Carlo-based methods can be developed to produce the description of synthetic cells. Undoubtedly, one of the most adequate and comprehensive representations to be adopted in the generation of neural cells are dendrograms, which allows us to determine the statistical description of the adopted shape features conditionally to the branching order.

We have been working on the problem of neural shape generation by using stochastic-syntactic approaches (Fu, 1982). Figure 18.8 presents three 3-D cells obtained by using our developed framework, which also allows the inclusion of external influences such as trophic fields, as shown in Figure 18.9. Figure 18.10(a) illustrates a relatively complex centric neural structure composed of 200 2-D neural cells generated in order that their diameter decreases with the eccentricity. In such structures, synaptic connections have been established with a certain probability every time a growing axon touches a dendritic segment or soma. While this approach is biologically realistic, alternative mechanisms incorporating trophic gradients and electric fields are also being investigated. Trophic fields corresponding to varying concentration of ions, chemoattractors or electric fields emanating from neural or glial cells (Challacombe et al., 1996; Kolodkin, 1996; Stirling and Dunlop, 1995) are explicitly considered during the generation process, in such a way that the growth cones at the extremities of the developing axons are guided by the trophic fields along paths that minimize the overall energy, which can be defined in various ways in order to express different developing mechanisms.

The strength of the synapses can be determined according to several hypotheses, such as in terms of the arc length from the synaptic contact to the axon hillock. Such neural structures are represented in terms of graphs such as that shown in Figure 18.10(b), which can be used as input data for simulations based on difference equations. In this approach, the neural structure defined by the connection graph such as that in Figure 18.10(b) is represented in terms of a system of linear (or non-linear) differential equations in such a way that the neural activity at each cell is represented as a state variable, and the synaptic connections, established as above mentioned, define the interactions between such state variables. The evolution of the

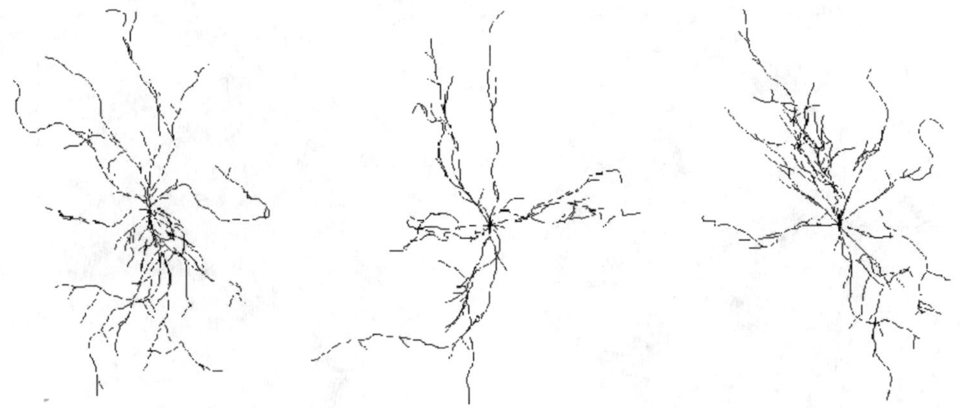

Figure 18.8 Three 3-D neural cells obtained by our stochastic-syntactic framework for synthesis of neural shapes. Various average numbers and positions of the branches, as well as distinct variation of the orientation evolution of the drawing vectors were considered.

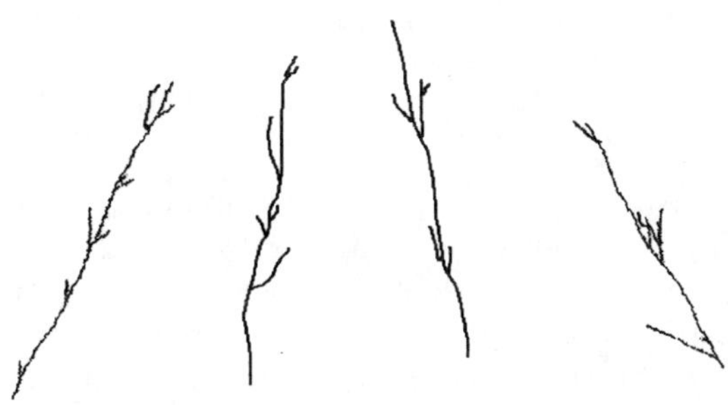

Figure 18.9 Dendritic development under influence of a trophic field (e.g. chemoattractor or electric field). Such growing mechanisms are particularly important for modeling the development of neural structures under the influence of chemical markers and NGF (neurotrophic growth factor).

neural activity throughout the domain is obtained by adding to each state variable the weighted contribution of the integrated afferent activity, which is done in time steps. This modeling approach is formally specified by (18.2), where y_i is the activity at cell i at time t, $w_{i,j}$ is the synaptic weight for the connection between cell i and j, and $g()$ is an optional non-linear function (see Appendix 18.2):

$$y_i^{t+1} = g\left(\sum_{\substack{j \mid j \text{ is} \\ \text{connected to } i}} w_{i,j} \cdot y_j^t \right) \tag{18.2}$$

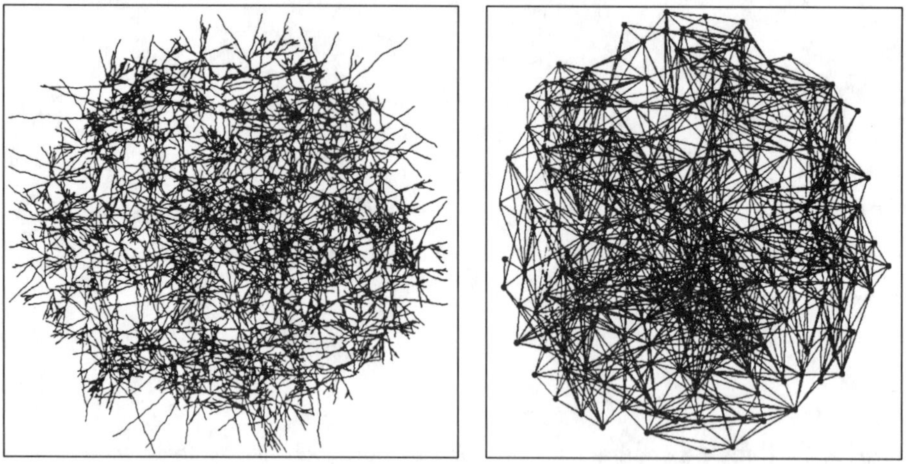

(a) (b)

Figure 18.10 Synthetic neural structure containing 200 2-D neurons: (a) In this structure, every time an axon growing finds a dendritic segment or soma in its way, a synaptic connection is established. The interconnection graph respective to the figure (a) is presented in (b). Such graphs are essential during simulations of the neural activity along the neural structures and can be used as input data for simulations based on difference equations. Refer to text for further details.

While the analytic investigation of regularly distributed and connected neural cells is feasible, the analysis of morphologically realistic cells, which are inherently irregular, becomes quite complex, implying experimental simulation approaches. Figure 18.11 presents the profile of neural activity obtained by the presentation of a straight stimulus to the neural structure in Figure 18.10, i.e. the straight line defines the initial condition ($t=0$) of the difference equations (see (18.2)). Amongst the additional mechanisms that can be incorporated into the above outline framework for synthesis and modeling of morphologically realistic neural networks, it would also be possible to implement Hebbian mechanisms, in such a way that the synaptic weight, even though constrained by the electrotonic properties defined by the distance to the axon hillock, would be allowed to evolve in terms of the respective history of neural activity. Other possibilities would include the consideration of asymmetric synapses and the dynamic evolution of the neural shape during simulation, including the establishment of new synapses and even apoptosis (Alberts et al., 1994). We are currently investigating how developmental mechanisms based on such effects could be defined in such a way as to produce controlled variation of neural shape, such as the gradient of cell diameters or preferred orientation along cortical surfaces.

Such modeling and simulation approaches have been verified to open completely new and exciting possibilities for investigating the interplay between neural shape

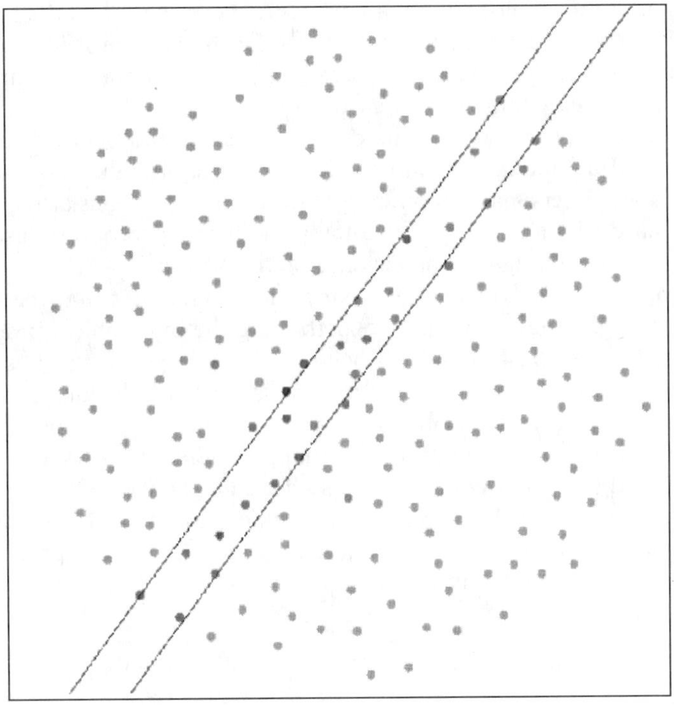

Figure 18.11 Profile of neural activity obtained for the presentation of a straight stimulus to the structure in Figure 18.10 (for the sake of better visualization, only the borders of the stimulus has been drawn). The peak of the neural activity is represented by darkest soma (near the middle of the portion of the stimulus).

and function. For instance, it has been verified that anisotropies (i.e. different properties at different orientations) in the size of the dendritic arborizations centric neural domains, such as that shown in Figure 18.10, as well as in the relative position of the cells, can yield interesting overall dynamic behavior. While structures presenting cells with decreasing diameters (as the eccentricity increases) tend to favor sharper profiles (asymptotic convergence) of neural activity, irregularly distributed cells tend to produce localized eigenmodes (i.e. convergent profiles), such as in irregular crystals. These are clear indications corroborating beyond any doubt the relationship between neural shape and function at macroscopic scale spaces. We have just started investigating non-linear versions of such morphologically realistic neural structures, which is expected to entail a number of interesting findings.

5. DISCUSSION

This chapter has addressed a series of important issues related to the shape of neural cells and networks, including the quantification of shape features and the modeling and simulation of morphologically realistic neural cells. As far as the former issue is concerned, it should be observed that the algorithmic approach adopted in the present work allows the characterization of a number of important shape features in neural cells, including complexity, area of influence, spatial coverage, and dendrograms. It should be observed that the described techniques demand relatively little computational effort, the most complex algorithms requiring at most some minutes in a well-configured personal computer system. Implementations of these algorithms in dedicated hardware and/or optical computing in order to obtain faster execution time is also possible, though at the expense of versatility.

It should be clear from the results reported in this chapter that the neural shape plays an exceedingly relevant role, by constraining the biochemical processes in and outside the cell, in establishing neural behavior. It is important to observe that the interplay between neural shape and function is particularly interesting when the afferent space and/or neural network layer present some inherent geometrical organization, such as in topographically mapped domains. In such situations, in addition to directly correlating with their associated receptive fields, the neural shape will also determine how the input space is sampled. For instance, elongated neural cells will be inherently suited to integrating the input stimuli space in a fashion similar to the Radon transform:

$$h_{\theta,\rho} = \iint f(x,y)\, \delta(x\cos\theta + y\sin\theta - \rho)\, \mathrm{d}x\, \mathrm{d}y$$

where $f(x,y)$ is the visual field, θ and ρ are respectively the normal angle and distance of a line in polar representation, and $\delta()$ is the Delta-Dirac function extending along a straight line. Moreover, the output of a specific cell will be proportional to the degree of matching between the stimuli and the overall neural shape.

While the interplay between neural shape and function can be investigated at several spatial scales, such as by considering the relative position and size of cells at more macroscopic spatial scales (as discussed in Section 5), by adopting morphologically reduced models such as in (Poznanski, 1992), or even by analysing how synaptic density is influenced by the shape of spines, it is only by integrating all these approaches that a more complete understanding of neural structures is likely to emerge. Thus, the extent in that neural shape must be considered in order to obtain biologically realistic neural cells is relative to the phenomena being considered—the broader the spatial scales considered in the investigation, the more comprehensive the accomplished advances will be. In fact, considering that the brain processes may present chaotic nature, unmeasurable perturbations at the quantum mechanics level may be amplified by chaos, which dooms at the outset our ambition of trying to fully predict long-term neural behavior. If that is true, the incorporation of very microscopic shape information, even at the molecular level, will enhance the predicting power of more morphologically realistic neural models. It should be also observed that, although the emerging neural behavior can be understood as the final result of redundant processes taking place among large assemblies of neural cells, the constraints imposed by the shape of neural cells at more microscopic scales are still of paramount importance because of at least the following two aspects: (i) the above mentioned amplification of small perturbations by non-linear and possibly chaotic real neurons; and (ii) the interconnection patterns characterizing such large neural assemblies are the product of local interactions defined at the spatial scale of neural cell processes.

6. CONCLUSIONS AND FUTURE PERSPECTIVES

The current article has presented an *integrated computational approach* to the comprehensive analysis, modeling and simulation of biologically more realistic neural networks, with special attention focused on neural morphology. It has become clear that the first step implied by such activities consists in defining a suitable set of shape features, as well as methods for their effective estimation. The curvature and wavelet-based approaches, as well as the biologically sound alternatives proposed for quantifying the area of influence and spatial coverage by neural cells discussed in the present paper represent important advances in that direction. Efforts aimed at the semi-automated generation of dendrograms, one of the most comprehensive approaches to the representation of neural cells, have also been reported, which allow more comprehensive and faster neural shape analysis, classification, and generation. The problem of neural cell classification has also been approached in terms of two clustering techniques applied while considering a number of alternative shape features, with particularly encouraging results obtained for influence histograms. Finally, the exciting possibilities allowed by the more comprehensive incorporation of morphological features in neural networks have been illustrated through the use of a stochastic-syntactic approach, which has led to a number of interesting simulation findings clearly indicating the strong interplay between neural shape and function. Additional tools traditionally adopted for biological shape analysis such as skeletons (Blum, 1973), statistical methods (Bookstein, 1978; Bookstein et al., 1985), and alternative approaches based on differential geometry (Todd, 1986) and on diffusion processes (Meinhardt & Klingler, 1987) are also under consideration for the analysis of neural cells. Further efforts are being made in all the discussed research lines, and new results should be futurely reported.

PROBLEMS

18.1. Propose new multiscale shape descritors that could be obtained from the multiscale representations (curvegram and W-representation) defined in Section 2.

18.2. Propose an algorithm for dendrogram extraction from the neural skeleton. Compare it with the algorithm based on the neural contour.

18.3. Compare the concept of spatial coverage by neural cells, roots and arborizations in trees, the bronchial tubes in vertebrates, the reticular structure underlying the wings of many insects, and hydrographic basins. What are the common and distinct aspects regarding these cases? How could the techniques for spatial coverage characterization proposed in the current paper be extended for application to the other above situations?

18.4. Apply the box-counting and Minkowski's sausages approaches to the determination of the fractal dimension of curves with known fractal dimension.

Compare the experimentally obtained results as far as accuracy and execution time are concerned. Suggest other measures which could be used in order to express the structural complexity and spatial coverage by neural cells.

18.5. Lines of field defined by electric field are known never to cross one another. Could the same be said regarding lines of fields defined by fields exhibiting other kinds of magnitude decay with distance? Suggest a technique allowing the determination of the density of lines of field reaching each point in the neural outline. Would such a measure bear any correlation with the expected rate of synaptic connections along the cell contour? How could such an approach be extended to 3-D cells?

18.6. Are the statistical descriptions of the branch angles and dendrite segments, organized conditionally to the branching order, enough to describe the shapes of all possible types of 2-D neurons? If not so, propose additional shape features that would make such a mapping complete (i.e. each neural shape would have a unique data description in terms of the proposed shape features, and vice-versa). Propose a set of features allowing the complete statistical description of 3-D neural cells. How could such artificially produced neural cells be validated with respect to their biological counterparts?

18.7. Propose an adaptive scheme of dendrite growth, acting during the neurogenesis stage, which would be capable of producing larger cells near the center of centric domains such as that described in this paper. How could such a process reflect the history of neural activity in each domain?

18.8. What sort of behavior could be expected (in terms of neural activity peaks, such as in the paper) if the neural cells presenting maximum diameter are near the circle centered at the domain origin and having (the circle) radius equal to the half of the radius of the centric domain (i.e. the larger neurons arranged as ring, with the smaller ones inside and outside this ring)? Could a general qualitative law relating the position of maximum size and the sharpness of neural activity be obtained?

18.9. Describe how genetic algorithms (see, for instance, Back (1996)) could be used in order to find optimal and sub-optimal sets of shape features allowing suitable classification of neural cells.

ACKNOWLEDGMENTS

Luciano da F. Costa is grateful to FAPESP (Procs. # 96/05497-3 and 94/04691-5) and CNPq (Proc. # 301422/92-3), for financial help, and to Dr Toby Velte, Prof. R. Miller and Prof. L.C. Ioriatti for the collaboration and kind supply of data on neural cells. Roberto M. Cesar Jr. is indebted to Roberto Lotufo, Jean-Pierre Antoine for partial computational support and to FAPESP (97/04186-7) for the financial support. Roberto M. Cesar Jr. and Júlia S. Tanaka are grateful to CAPES for financial support Regina C. Coelho thanks CNPq (Proc. 141483/93-8) for financial help. The authors are also grateful to Dr Rosane Minghim and Dr Maria Cristina F. de Oliveira for the collaboration on the visualization of the tridimensional neurons.

REFERENCES

Alberts, B., Bray, D., Lewis, J., Raff, M., Roberts, K., Watson, J.D. (1994) *Molecular Biology of the Cell*, 3rd Ed., Garland Publishing, New York.

Anderberg, M.R. (1973) *Cluster Analysis for Applications*, Academic Press, London.

Antoine, J.-P., Barache, D., Cesar Jr., R.M. and da F. Costa. L. (1997) Multiscale shape analysis using the continuous wavelet transform, *Signal Processing*, **62**, 265–290.

Back, T. (1996) Evolutionary Algorithms in Theory and Practice. Oxford: Oxford University Press.

Bedlack Jr., R.S., Wei, M.-D. and Loew, L.M. (1992) Localized membrane depolarizations and localized calcium influx during electric field-guided neurite growth, *Neuron*, **9**, 393–403.

Blum, H. (1973) Biological shape and visual sciences, *J. Theor. Biol.*, **38**, 205–287.

Bookstein, F.L. (1978) *The Measurement of Biological Shape and Shape Change*, Springer-Verlag, Lecture Notes in Biomathematics, Berlin.

Bookstein, F.L., Chernoff, B., Elder, R.L., Elder, J.L., Humphries Jr., J.M., Smith, G.R. and Strauss, R.E. (1985) *Morphometrics in Evolutionary Biology – The Geometry of Size and Shape Change, with Examples from Fishes*, The Academy of Natural Sciences of Philadelphia, Ann Arbor, Michigan.

Bower, J.M. and Beeman, D. (1994) *The Book of GENESIS*, Springer-Verlag, New York.

Boycott B. B. and Wässle, H. (1974) The morphological types of ganglion cells of the domestic cat's retina, *J. Physiol. (Lond.)*, **240**, 397–419.

Brown, M.J. and Loew, L.M. (1994) Electric field – directed fibroblast locomotion involves cell surface molecular reorganization and is calcium independent, *J. Cell Biol.*, **127**, 117–128.

Caserta, F., Stanley, H.E., Eldred, W.D., Daccord, G., Hausman, R.E., Nittmann, J. (1990) Physical mechanisms underlying neurite outgrowth: a quantitative analysis of neuronal shape, *Phys. Rev. Lett.*, **64**, 95–98.

Caserta, F. and Eldred, E.D. and Fernandez, E., Hausman, R.E. and Stanford, L.R., Bulderev, S.V., Schwarzer, S. and Stanley, H.E. (1995) Determination of fractal dimension of physiologically characterized neurons in two and three dimensions, *J. Neurosci. Meth.*, **56**, 133–144.

Cesar Jr., R.M. and da F. Costa, L. (1996) Towards effective planar shape representation with multiscale digital curvature analysis based on signal processing techniques, *Pattern Recognition*, **29**, 1559–1569.

Cesar Jr., R.M. and da F. Costa, L. (1997) Application and assessment of multiscale bending energy for morphometric characterization of neural cells. *Review of Sci. Instrum.*, **68**, 2177–2186.

Challacombe, J.F., Snow, D.M. and Letourneau, P.C. (1996) Hole of the cytoskeleton in growth cone motility and axonal elongation, *Sem. Neurosciences*, **8**, 67–80.

Coelho, R.C. and da F. Costa., L. (1996) On the application of the Bouligand-Minkowski fractal dimension for shape characterization, *Appl. Signal Process.*, **3**, 163–176.

Costa, L. da F. (1995) Computer vision based morphometric characterization of neural cells. *Rev. Sci. Instrum.*, **66**, 3770–3773.

Dacey, M.D. (1993) The mosaic of midget ganglion cells in the human retina, *J. Neurosci.*, **13**, 5334–5355.

Dann, J.F., Buhl, E.H. and Peichl, L. (1988) Postnatal dendritic maturation of alpha and beta ganglion cells in cat retina, *J. Neurosci.*, **8**, 1485–1499.

Falcolner, K. (1990) *Fractal Geometry – Mathematical Foundations and Applications*, John Wiley & Sons, New York.

Fu, K.S. (1982) *Syntactic Pattern Recognition and Applications*. Englewood Cliffs, NJ: Prentice-Hall.

Fukuda, Y., Hsiao, C.-F., Watanabe, M. and Ito, H. (1984). Morphological correlates of physiologically identified Y-, X-, and W-cells in cat retina, *J. Neurophysiol.*, **52**, 999–1013.

Goodman, C.S. (1996) Mechanisms and molecules that control growth cone guidance, *Annu. Rev. Neurosci.*, **19**, 341–377.

Hamilton, P. (1993) A Language to describe the growth of neurites, *Biol. Cybern.*, **68**, 559–565.

Hines, M.L. and Carnevale, N.T. (1997) The NEURON Simulation Environment. *Neural Comput.*, **9**, 1179–1209.

Kohonen, T. (1990) The self-organizing map *Proc. IEEE*, **78**, 1464–1480.

Kolb, H., Nelson, R. and Mariani, A. (1981). Amacrine cells, bipolar cells and ganglion cells of the cat retina: A Golgi study, *Vis. Res.*, **21**, 1081–1114.

Kolodkin, A.L. (1996) Growth cones and the cues that repel them, *Trends Neurosci.*, **19**, 507–513.
Lavigne, M.T. and Goodman, C.S. (1996) The molecular biology of axon guidance, Science, **274**, 1123–1133.
Leventhal, A.G. and Schall, J.D. (1983) Structural basis of orientation sensitivity of cat retinal ganglion cells, *J. Comp. Neurol.*, **220**, 465–475.
Meinhardt, H. and Klingler, M. (1987) Pattern formation by coupled oscillations: the pigmentation patterns on the shells of molluscs. In, E. Teramoto and M. Yamaguti (eds.), *Mathematical Topics in Population Biology, Morphogenesis and Neurosciences*, Springer-Verlag, Lecture Notes in Biomathematics, Berlin.
Montague, P.R. and Friedlander, M.J. (1991) Morphogenesis and territtorial coverage by isolated mammalian retinal ganglion cells. *J. Neurosci.*, **11**, 1440–1457.
Morigiwa, K., Tauchi, M. and Fukuda, Y. (1989) Fractal analysis of ganglion cell dendritic branching patterns of the rat and cat retinae, *Neurosci. Res.*, Suppl. 10, S131–S140.
Panico, J. and Sterling, P. (1995) Retinal neurons and vessels are not fractal but space-filling, *J. Comp. Neurol.*, **361**, 479–490.
Peitgen, H.-O. and Saupe, D. (1998) *The Science of Fractal Images*. Springer-Verlag, New York.
Porter, R., Ghosh, S., Lange, G.D. and Smith Jr., T.G. (1991) A fractal analysis of pyramidal neurons in mammalian motor cortex. *Neurosci. Lett.*, **130**, 112–116.
Poznanski, R.R. (1992) Modelling the electrotonic structure of starbust amacrine cells in the rabbit retina: functional interpretation of dendritic morphology. *Bull. Math. Biol.*, **54**, 905–928.
Ramon-Moliner, E. (1962) An attempt at classifying nerve cells on the basis of their dendritic patterns, *J. Comp. Neurol.*, **119**, 211–227.
Romesburg, H.C. (1990) *Cluster Analysis for Researchers*, Robert E. Krieger Publishing Company, Malabar, Florida.
Saito, H.A. (1983) Morphology of physiologically identified X-, Y-, and W-type retinal ganglion cells of the cat, *J. Comp. Neurol.*, **221**, 279–288.
Scholl, D.A. (1953) Dendritic organization in the neurons of the visual and motor cortices of the cat, *J. Anat.*, **87**, 387–406.
Schutter, E. De and Bower, J.M. (1994) An active membrane model of the cerebellar Purkinjie cell I. simulation of current clamps in slice. *J. Neurophysiol.*, **71**, 375–400.
Smith Jr., T.G., Lange, G.D. and Marks, W.B. (1996) Fractal methods and results in cellular morphology – dimensions, lacunarity and multifractals, *J. Neurosci. Meth.*, **69**, 123–136.
Smith Jr., T.G., Marks, W.B., Lange, G.D., Sheriff Jr, W.H. and Neale, E.A. (1989) A fractal analysis of cell images, *J. Neurosci. Meth.*, **27**, 173–180.
Stirling, R.V. and Dunlop, S.A. (1995) The dance of the growth cones – where to next?, *Trends Neurosci.*, **18**, 111–115.
Todd, P.H. (1986) *Intrinsic Geometry of Biological Surface Growth*, Springer-Verlag, Lecture Notes in Biomathematics, Berlin.
Tricot, C. 1995. *Curves and Fractal Dimension*, Springer-Verlag, Paris.
Troilo, D., Xiong, M., Crowley, J.C. and Finlay, B.L. (1996) Factors controlling the dendritic arborization of retinal ganglion cells. *Visual Neurosci.*, **13**, 721–733.
Turner, D.A., Li, X.-G., Pyapali, G.K., Ylinen, A. and Buzsaki, G. (1995) Morphometric and electrical properties of reconstructed hippocampal CA3 neurons recorded *In Vivo*. *J. Comp. Neurol.*, **356**, 556–580.
van Ooyen, A. and van Pelt, J. (1996) Complex periodic behaviour in a neural network model with activity-dependent neurite outgrowth, *J. Theor. Biol.*, **179**, 229–242.
van Oss, C. and van Ooyen, A. (1997) Effects of inhibition on neural network development through active-dependent neurite outgrowth, *J. Theor. Biol.*, **185**, 263–280.
Vaney, D.I. (1994) Territorial organization of direction-selective ganglion cells in rabbit retina. *J. Neurosci.*, **14**, 6301–6316.
Velte, T.J. and Miller, R.F. (1995) Dendritic integration in ganglion cells of the mudpuppy retina. *Visual Neurosci.*, **12**, 165–175.
Wässle, H., Boycott, B.B. and Illing, R.-B. (1981a) Morphology and mosaic of On- and Off-Beta cells in the cat retina and some functional considerations, *Proc. R. Soc. Lond. B*, **212**, 177–195.

Wässle, H., Peichl, L. and Boycott, B.B. (1981b) Morphology and topography of On- and Off-alpha cells in the cat retina, *Proc. R. Soc. Lond. B*, **212**, 157–175.

Watanabe, M., Sawai, H. and Fukuda, Y. (1993) Number, distribution, and morphology of retinal ganglion cells with axons regenerated into peripheral nerve graft in adult cats, *J. Neurosci.*, **13**, 2105–2117.

APPENDIX 18.1: Basic Concepts in Linear Systems

Let $S\{f(x,y)\}$, where $f(x,y)$ is a two-dimensional input (e.g. an image), be linear system, i.e. $S\{a.f(x,y)+b.g(x,y)\}=a.S\{f(x,y)\}+b.S\{g(x,y)\}$, defined by its impulse response $I(x,y)=S(\delta\{x,y\})$. Let S be also shift invariant, i.e. $S(f\{x,y\})=h(x,y)$ implies $S(f\{x-\alpha,y-\beta\})=h(x-\alpha,y-\beta)$.

Under these conditions, the system response to any possible input $f(x,y)$, i.e. $R(x,y)=S(f\{x,y\})$, can be given by the convolution between $f(x,y)$ and the system impulse response, i.e.:

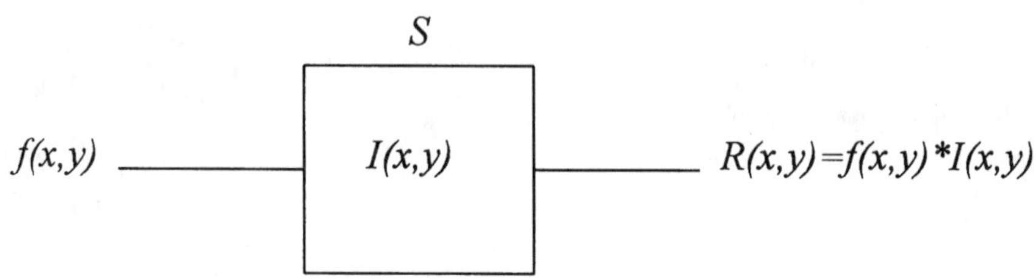

In the present case, the influence produced by a neural cell is understood as the effect of a linear, shift invariant (i.e. the influence is uniform throughout the image space), system having as impulse system (i.e. the point-spread-function) a specific profile of influence which would be produced if the image were a single pixel. Thus, the influence field can be obtained by convolving the image of the neural cell with the influence profile. This situation is similar to the effect of blurring produced by an imaging system. In case the point-spread-function is broad, it can be convenient to perform such convolutions and correlations in the Fourier domain. Since the influence profiles are usually circularly symmetric, the convolution equals correlation.

APPENDIX 18.2: Nonlinear Transfer Functions

Typical nonlinear function commonly used for $g()$ in (18.2) include the step function $h(x)$, the limiting function $t(x)$, and the sigmoid $s(x)$, which are exemplified by:

$$h(x)=\begin{cases}1 & if\ x\geqslant 0\\ 0 & otherwise\end{cases}$$

$$t(x)=\begin{cases}a & if\ x\geqslant a\\ x & if\ -a\leqslant x<a\\ -a & if\ x<-a\end{cases}$$

$$s(x)=\frac{1}{1+e^{-x}}$$

INDEX

C

D

H

I

K

L

Color Plate I Schematic illustration of the vertebrate retina. The principal cell types are distinguished by different colours. See the text for details. Inset: Cross section of the retina. (Reprinted with permission from Matsumoto, T., Kobayashi, H. and Yagi, T. (1993) *Proceedings of IEICE*, **76**(7), 783–791.) *see p.* 268

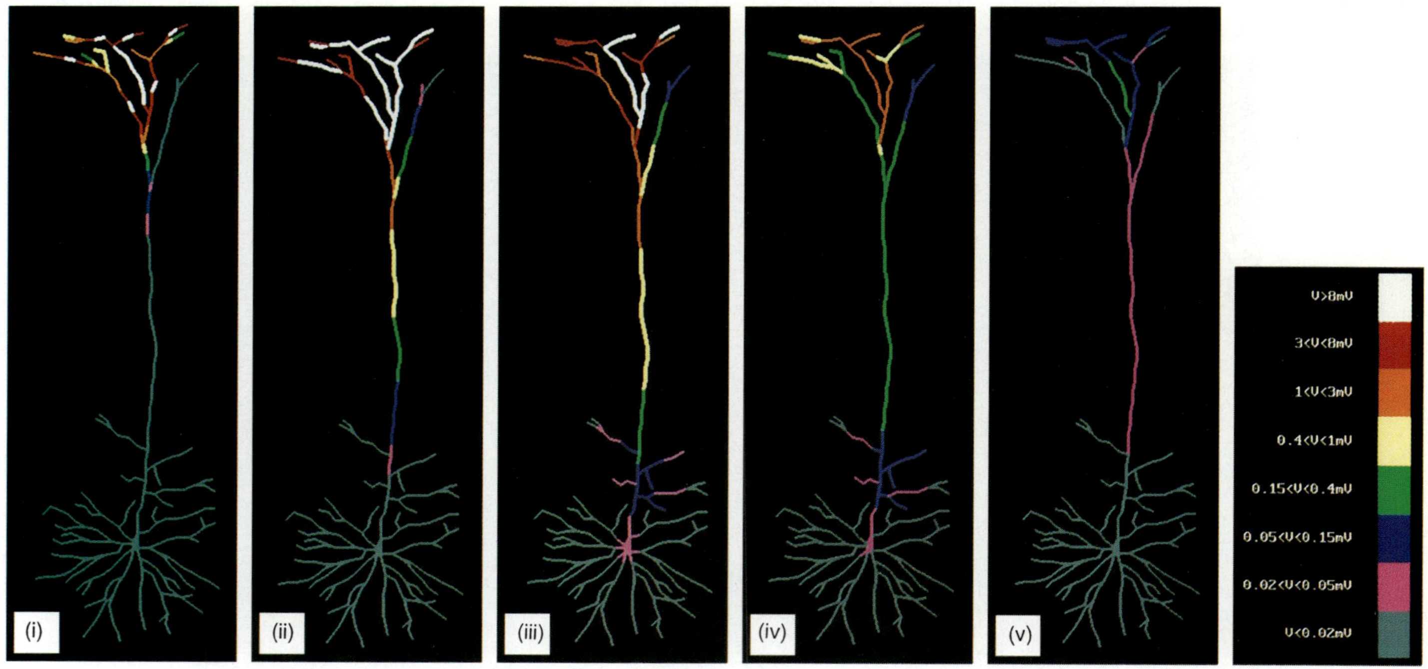

Color Plate II The distribution of depolarization over the pyramidal cell as a function of time. The colors show the distribution using the log scale illustrated in the calibration. (a) The distribution following eight inputs to the apical dendrites and only passive conduction: (i) 0.4 ms; (ii) 1.0 ms; (iii) 3.5 ms; (iv) 9.0 ms; (v) 18.0 ms. *see p.* 367

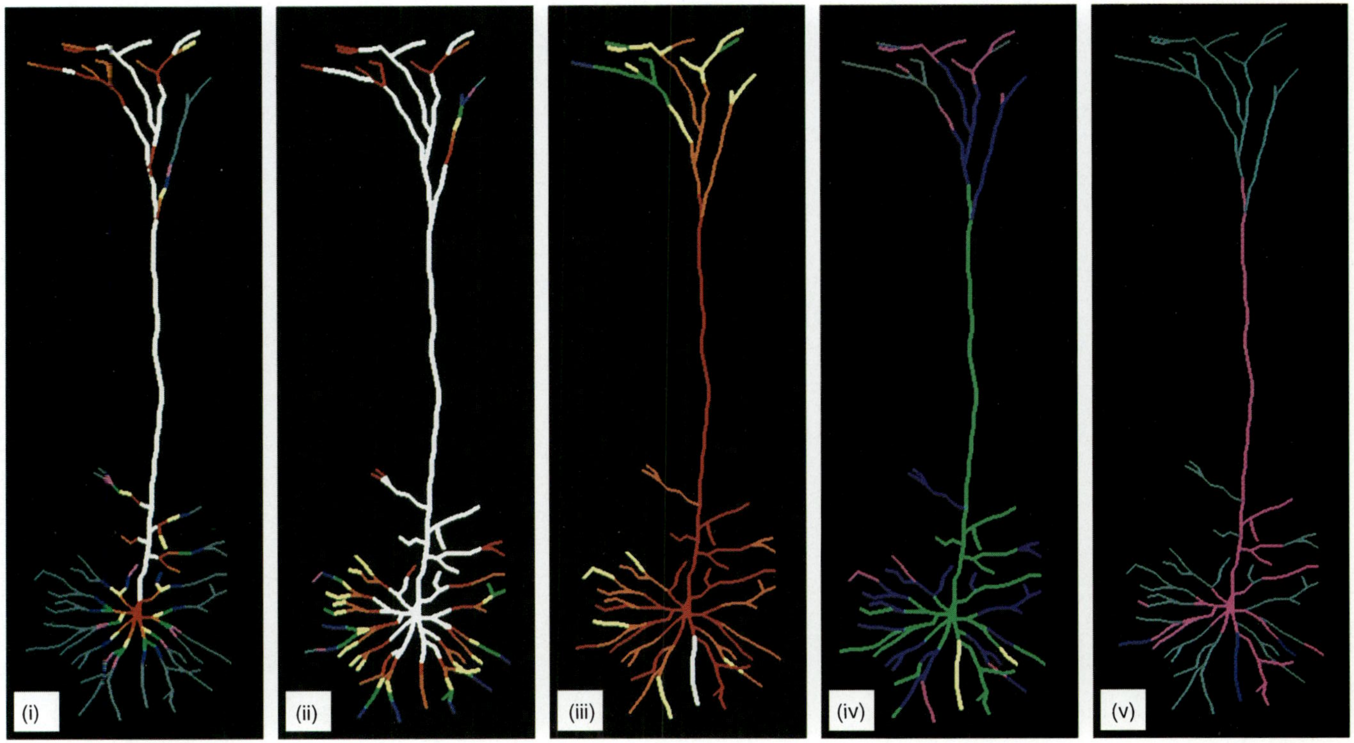

Color Plate III The distribution following eight inputs to the apical dendrites but with active conduction along the climbing apical dendrite: (i) 2.0 ms; (ii) 4.5 ms; (iii) 11.0 ms; (iv) 18.0 ms; (v) 22.0 ms. The threshold for initiation of active spikes at the apical nexus was set at only 5 mV to allow the small number of inputs to initiate firing. *see p.* 368

£65.00
5/99
24 0470332 1